LUTION *of the* LIBERTY MOTOR

rd Aviation Engine

"905"-No. 2

ediate forerunner of
ll-steel cylinders were
power at less weight,
plify production by
Comparison of this
rty motor is convinc-
rty's origin.

ril, 1917

The Packard-Built Liberty Motor

MODEL B

The world's finest aviation engine, built on the Twin Six principle and of Packard design—the Liberty-12 airplane motor, Model B. It is fitted with reduction gearing between the engine crankshaft and the propeller, similar to that used in Packard 905 and Packard 905-No. 2.

Built January, 1918

The Packard-Built Liberty Motor

MODEL A

Packard principles of design are apparent in the structure of Liberty-12, Model A. Many of the tools used in building the Packard 905 and 905-No. 2 were used in its construction, and many of its parts were interchangeable with corresponding parts of its Packard prototypes.

Built June, 1917

This L-8 was the first Liberty built. It was delivered to the National Bureau of Standards on July 3, 1917. It is now in the collection of the Smithsonian Institution. (Courtesy of the Smithsonian)

A Technical & Operational History of the Liberty Engine

Tanks, Ships and Aircraft 1917 - 1960

Robert J. Neal

Author of *Smith & Wesson 1857-1945*, *Packards At Speed* and *Master Motor Builders*

Specialty Press
PUBLISHERS AND WHOLESALERS

Specialty Press
39966 Grand Avenue
North Branch, MN 55056
Phone: 651-277-1400 or 800-895-4585
Fax: 651-277-1203
www.specialtypress.com

Edited by Joe Hopkins
Design by Robert J. Neal

ISBN-13 978-1-58007-149-9
Item No. SP149

Library of Congress Cataloging-in-Publication Data

Neal, Robert J.
A technical & operational history of the Liberty engine : tanks, ships and aircraft 1917-1960 / Robert J. Neal.
p. cm.
Includes bibliographical references and index.
ISBN 978-1-58007-149-9
1. Liberty engines--History. 2. Airplanes--Motors--History. 3. Tanks (Military science)--Motors--History. 4. Marine engines--History. I. Title. II. Title: Technical and operational history of the Liberty engine.
TL703.L5N43 2008
621.43--dc22

2008031277

Printed in China
10 9 8 7 6 5 4 3 2 1

Dedicated to my ever patient and ever supportive wife, Donna C. Neal. This is the third major historical undertaking she has endured and encouraged. Few men are fortunate enough to have such a mate.

Front Cover Photo: Minor engine maintenance, 1925 style, on a Sikorsky S-29A Liberty. Just below the left # 1 cylinder can be seen the Johnson Motor Products rebuild data plate. Courtesy of Peter Bowers.

Table of Contents

Preface

In 1990 I began a serious study of the non-automotive engines of the Packard Motor Car Company. This resulted in the publication of two volumes devoted to that subject, *Packards At Speed* in 1995 and *Master Motor Builders* in 2000. That study is also the reason this volume devoted to the history of the Liberty engine is in print.

As the reader will see, Packard was intimately associated with the design of the U. S.-built Liberty and thus also with later Liberties produced in Russia and England. Thus my study and writing about Packard's aero engines led directly to the Liberty as it represented the third aero engine design built by Packard and, although Packard and its engineers could not claim total responsibility for the design of the Liberty, the company could and did claim responsibility for a major portion of it as well as being its largest producer during World War I.

The Liberty was born of necessity when the United States was drawn into a conflict that had been raging in Europe since 1914. It was rapidly becoming apparent that aircraft would play an important part in the winning of the war and all participants were building up their air arm as rapidly as possible. Initially aircraft were found most useful for observation purposes and were in the form of lighter-than-air craft, some powered and some simply anchored in place with a cable to the ground.

The heavier-than-air craft was also fairly well developed by then and could also be used for observation of what the enemy was up to behind the fighting lines. It could not stay in the air very long and was not as useful in the observation capacity as was the lighter-than-air craft which could remain aloft for long periods of time.

It was maneuverable, however, and it was soon discovered that if the airplane's pilot or observer were armed he could often shoot down the enemy's observation balloons. One thing leads to another, as the old saying goes, and it was soon learned that if you sent up an airplane to shoot down the enemy's observation balloons, the enemy would send up an airplane of their own to do the same to you and also try to shoot down your airplane.

In addition it was soon found that while flying over enemy territory for observations purposes one might have the opportunity to inflict some damage, either physical or moral, by dropping upon him some kind of explosive. Thus was born the bomber.

The more the airplane was used, the more uses were found for it. This resulted in a rapid proliferation of different designs for different purposes or different designs by different aero engineers for the same purpose. Thus by the time the United States joined the war effort on the side of the allies there were a great number of airplanes and airplane designs in use. Lack of standardization in types of airplanes and thus also types of engines to power them was a major obstacle to volume production and to maintenance of those aircraft. Maintaining stocks of repair parts and trained personnel for such a large number of different designs was an overwhelming task.

The Liberty was an attempt to standardize aircraft engine design. The theory was to design an engine which could be built in several sizes and thus power airplanes for any purpose from training to bombing. The differences in sizes would be obtained by using different numbers of cylinders of the same design. A large number of other parts would also be used in common by all resulting sizes of the engine series.

The overall result was to be several sizes of available engines with many parts interchangeable between models as well as within models (i.e., multiples of the same individual cylinder-piston-rod-valves). These parts were to be built to tolerances that would allow a part built by any one of several manufacturers to be used in any engine built by any manufacturer.

The result was to be high volume, low cost, manufacture and engines of several sizes which could be maintained by the same mechanic with many of the same parts.

Although the Liberty was never put into volume production in any but its twelve cylinder form, the purpose was accomplished.

The engine was "state-of-the-art" for its day. Its use far outlived any such claim and the reader will see that in its original form it saw active military use into the 1930s and that it was put back into production by two other countries and again saw active military use through World War II.

In addition it provided noble service in a multitude of civilian uses and still does even today some ninety years after the first engine ran.

I will attempt to document the inception, design, production and every major use of this engine. This was indeed an engine that was used on land, sea and in the air and by a great many nations. It was put into mass production by both Russia and Great Britain and was used to power a great variety of vehicles.

I can think of no other piston engine design which gave as long an active military service as did the Liberty.

I am a lover of statistics. I gathered a monumental amount of statistical information in the researching of this volume, as I did while researching my prior two books, *Packards At Speed* and *Master Motor Builders*. Once again I have tried to save for future historians as much of this information as I could in the form of appendices. Some will care little about it but others will bless me for preserving it as I did others for such information preserved in their writings.

This picture was taken in Detroit, Michigan, at the Board of Commerce at the unveiling of the 10,000th Liberty engine on Saturday, October 12, 1918. From left to right those witnessing the event were: Col. E. A. Deeds (former Chief of the Equipment Division), Lt. Col. J. G. Vincent (Chief of Engineering Airplane Engineering Division), Lt. Col. L. S. Horner (Chief Executive Staff BAP), Mr. Howard E. Coffin (Chairman of APB), Major James G. Heaslet (Chief Detroit District), Lt. Harold H. Emmons, U. S. N. (Chief of BAP Engine Production), Mr. Alvan Macauley (Packard), Mr. R. H. Collins (Cadillac), Mr. F. E. Moscovies (Nordyke & Marmon), Mr. O. E. Hunt (Associate Chief of Detroit District), Mr. W. C. Leland (Lincoln), Mr. T. P. Myers, and Mr. A. A. Landon (Chief of Production Division). (Author's collection)

Acknowledgments

Over a period of 35 years I have written four major historical reference books. All have required large amounts of research of both public and private documents dating back into the middle eighteen hundreds in some cases and to the early nineteen hundreds in the case of this book.

Such a task invariably requires the aid of many other historians as well as libraries, museums and government archives and those who staff them.

As I have stated in prior works, writing historical reference books of the type I take on is a great deal like trying to assemble a very large puzzle from parts scattered worldwide. It takes a lot of work and a great deal of time to find enough pieces to complete the puzzle sufficiently well to get a fairly good picture. Given enough perseverance and a certain amount of luck, one can build a picture that is so close to complete that he can fill in the blanks left with very good accuracy.

If I were asked to compare this Liberty history project to such a puzzle I would say it is perhaps comparable to a 5,000 piece puzzle, with five random missing pieces.

One must understand, however, that I say this in reference to what I intended this book to document regarding Liberty history and not in reference to everything there is to know about Liberty history. The Liberty was a very unique engine. This was not so much because of its mechanical design as it was because of its unique place in history which resulted in its very long useful life. Even though this study has resulted in a major size book, it certainly does not say everything there is to say about the Liberty engine and its uses.

I will first express my thanks to those individuals who provided major amounts of help with this project. Sometimes it was solely in the form of large amounts of reference material and sometimes in the form of both reference material and knowledge of the inner workings of the Air Service gathered over years of research thus saving me a great amount of time and effort "re-inventing the wheel" so to speak.

In alphabetical order rather than in order of importance, they follow.

Peter M. Bowers was well-known to aviation historians as a walking "aviation history book." Pete has once again provided much help to me in the form of access to his massive collection of aviation images and early Air Service technical publications. In addition, he was of great aid in compiling a comprehensive list of every type of airplane that was ever powered by a Liberty engine. We all lost a premier aviation historian with the passing of Peter Bowers on May 27, 2003.

Hugo Byttebier, a Belgian historian of some note, provided me with his notes on the Liberty accumulated over a period of many years as well as being kind enough to read over the manuscript of Chapter 1 and offer criticism and correction. Hugo has been a serious student of early aero engines for many years and has written extensively on the subject.

Robert B. Casari has also once again come to my aid. Bob is without doubt the foremost authority on Army aircraft and engine contracts of the period 1908 to 1923. This includes not just the contracts themselves but actual delivery and acceptance of the items. It was only with his help that I could put together a complete story of the myriad of contracts for the Liberty in all of its variations.

Jamie Fairchild, a British historian and engine enthusiast, first asked me some questions regarding the Liberty engine because he was thinking of acquiring one, then he got caught up in my quest for information about the Nuffield-built Liberty. He did much research about that engine and passed it along to me. In doing so, he added much to my knowledge of it.

Knut Erik Hagen of Norway was kind enough to furnish detailed information regarding that country's use of the Liberty-powered Douglas DT-2 aircraft.

Jon Hornbostel's interest and expertise is with early United States tanks. Although there is considerable interest in United States armor of the World War II period and beyond, there are few experts on our early efforts in that field of endeavor and little scholarly work devoted to the subject. Armor, both in this country and abroad, was one of the areas in which Liberty engine power was used. Jon provided a great amount of information for my study in this area.

John M. Leonard is an Allison enthusiast primarily, but has a wide interest in early aero engines. He is the newsletter editor of the Allison Branch of the Rolls-Royce Heritage Trust. John provided much valuable information regarding Allison's various improvements to the Liberty as well their involvement in the air-cooled and inverted water-cooled versions. It was only with his help that I could put together a complete history of Allison's involvement in the Liberty engine's evolution.

William Lewis is an aeronautical historian who has been collecting material related to 1950 and earlier aircraft and their engines for a great many years. This material relates to early aviation in about any way one could imagine and ranges from early books on the subject to reports of factories producing airplanes or engines to extensive government reports on the subject. Access to Bill's material helped answer many questions about early Air Service development work with the Liberty as well as some about the producers.

Dave and Roberta Vincent Mocabee provided the author with complete access to their extensive collection of early Packard material. This material came from the personal archives of Roberta's father, Charles H. Vincent, and uncle, Col. Jesse G. Vincent. Charles Vincent was an engineer who worked on the Liberty both at Packard and the Bureau of Aircraft Production and Col. Vincent was its co-designer and headed the Aircraft Engineering Division during World War I.

Kimble McCutcheon, author, hostorian and president of the Aircraft Engine Historical Society, has recently been a moving force in an effort to catalogue the USAF Engineering Division records at the U.S. National Archives II. While doing so he has also gathered an extensive amount of research data on aero development between 1917 and the end of the Liberty engine era, which he made available to the author. That material added significantly to the content of this volume.

Peter Seymour, British automotive author and historian, was quite helpful in supplying information regarding Lord Nuffield (W. R. Morris), Morris Motors and Nuffield Mechanizations. The Nuffield organization were the British producers of the Liberty during the WW II era.

Alan D. Toelle provided much help in the form of documented Liberty engines and DH 4 aircraft that went to the Romorantin Air Service Depot in France during WW I and their disposition as well as information on the use of the Liberty in French aircraft.

Graham White, noted aero historian and author, has been studying the design evolution of the turbosupercharger for some years and was kind enough to provide me with considerable information relative to those types used on the Liberty.

In addition to the above mentioned historians, I also received much useful information or leads thereto from the following individuals:

John A. Bluth
Jim Dillon
William Dykstra
David Fletcher
Dan Hagedorn
Wesley B. Henry
Jack Hoffman
George F. Hofmann, Ph.D.
Leo Opdycke
Dave Ostrowski
Al Shinnerer
Daniel D. Whitney
John Wills
Joan Zigmunt (late)

I also did research at, or otherwise obtained information from, a number of libraries, museums and research archives across the country. I found their staff in all cases to be friendly and helpful. The amount of information I acquired from each location varied from minimal to voluminous. They are listed in alphabetical order following:

The Air Force Museum - Wright-Paterson A.F.B. - Dayton, OH
British Motor Industry Heritage Trust - United Kingdom
Museum of Flight - Seattle, WA
National Air and Space Museum - Washington, DC
The National Archives - Washington, DC
The National Archives - United Kingdom
The National Automotive History Collection - Detroit Public Library - Detroit, MI
The Navy Museum - Washington, DC
Rolls-Royce-Allison - Indianapolis, IN
The Seattle Public Library - Seattle, WA
The Studebaker National Museum - South Bend, IN
The Tank Museum - Bovington Dorset, England

Illustrations for the book were gathered from a great number of sources and are acknowledged in the captions of each as they appear.

Left: Cover of the first manual for the Liberty 12, issued in 1917. Right: Cover of the August 1919 version of the Liberty 12 manual. (Both from the authors collection)

Chapter 1

Before the Liberty

Successful powered flight can be traced from the Wright's December 17, 1903 Kitty Hawk, North Carolina, effort. From that time on, the development of heavier-than-air flight made steady progress. That progress required new-and-improved designs of both the airframe and its power plant.

These improvements, for the most part, progressed more rapidly in Europe than here in America. I will make no attempt here to explain why that occurred, but it did. As a result, both civilian and military aircraft development in America was well behind that of Europe when war broke out there in 1914.

The war rapidly proved that the airplane could be a useful, and even decisive, tool to the military. The governments of the principal belligerents began to pour money into the development of aircraft and their engines in what seemed a never-ending stream. Since engine development is a very expensive process, its speed had always been controlled to some considerable extent by the private resources available to fund it. With the much larger resources of the various governments now available, progress accelerated at a rapid pace.

I will attempt to give the reader a brief overview of the progress of aeronautical power plant design and production from 1903 to early 1917 with emphasis on the WW I era from 1914 to 1917. Hopefully this will allow you to see the Liberty of 1917 in proper prospective regarding the state of the art at the time.

Wright's 1903 four-cylinder engine of 201.1 cu. in. This engine developed 12 HP @ 1,090 RPM and weighed 179 lb. (Courtesy of the Smithsonian)

The Wright engine of 1903 was fairly well advanced for its time and was certainly reliable enough for the job it had to do, power the Wright Flyer for periods measured in minutes.

Its most important design consideration was its power-to-weight ratio. It had to produce enough power to lift the aircraft, its pilot and itself off the ground. The state of the art in gasoline internal combustion engines at the time was such that the Wrights found it necessary to design their own as none suitable were available elsewhere.

To give the reader an idea of the current state of the art in automotive engines, which would have been the easiest and most logical source of a power plant, I will quote from an article in the May 1928 issue of the journal *Slipstream* by Charles E. Taylor, the Wright's mechanic who helped design and build the first engine.

"The automobile manufacturers were out of the picture then and the Oldsmobile firm was the only one I was familiar with at the time. We tried to get a motor built there but they couldn't make one near the low weight we wanted. The old one-lunger auto engines of that day really weighed more than our entire flying machine with the first motor installed."

Thus the Wrights found it necessary to design and build their own. The result was a water-cooled engine of four cylinders mounted horizontally in-line. The bore and stroke were 4" x 4" giving a displacement of 201.1 cu. in. Ignition was by low-tension magneto with make-and-break spark. It used automatic inlet valves and exhaust valves driven by rockers operated by the camshaft which was in turn chain-driven from the end of the crankshaft.

The engine had no carburetor and gasoline was fed to the motor by gravity and vaporized by running over the heated surface of the cylinder water jackets. Inlet and exhaust valves were enclosed by boxes which had no cooling facilities. As a result the box, valves and intake chamber ran red-hot after a short period of operation. The result was that the initial power capability of 16 HP was shortly reduced to 12 HP because of heating and expansion of the intake mixture. What little engine control that was available was gained by changing ignition timing by use of a small lever which altered the ignition system cam position. This control was not available during flight and the engine ran only at full speed during that time.

The engine weighed 179 pounds without water and thus produced a power-to-weight ratio of about 14.92 lb./HP. Although it was quite light for the time, the Wrights and others were shortly able to produce engines with far greater power, better power-to-weight ratios, better control and greater reliability.

Between 1903 and 1905 the Wrights improved their flying machine and its engine sufficiently to be able to stay in the air for more than 30 minutes and travel a distance of almost 25 miles. In the mean time no other man and airplane managed more than very short flights in machines that were barely controllable. On November 9, 1907 Henri Farman flew a Voisin biplane 3,368 ft. in 1 minute 14 seconds. The engine he used was a Antoinette of 35/40 HP.

The Antoinette engines were built in Paris by Levavasseur in a number of configurations beginning in 1905 and were used by many of the successful European flyers from about 1905 to 1910. The 35/40 HP model used by Farman developed 1 HP for each 3.9 lb. of weight and was a V-8 of 4.13" bore and 4.13" stroke, giving a displacement of 442.62 cu. in. These early designs had cast-iron cylinders and aluminum heads and were built for extreme lightness. Reliability suffered as a result.

Starting in 1909 Levavasseur took steps to remedy the weaknesses of his early designs by introducing an all-steel cylinder and making other changes. The bore was increased to 4.33" which raised the displacement to 486.56 cu. in., weight to 250 lb., power to 65 HP and resulted in a power-to-weight ratio of 3.85 lb./HP. Power was up and so was reliability. Other models were built ranging in power from 24 to as high as 134 HP. All were water-cooled V types of 8 and 16 cylinders.

1908 brought rapid development of both the airplane and its engine. Renault built an air-cooled V-8 of 45 HP which had a 3.54" bore and a 4.72" stroke, giving it a displacement of 371.68 cu. in. It weighed 320 lb. resulting in a power-to-weight ratio of 7.1 lb./HP.

In addition, Curtiss finished his A-8 model which was also an air-cooled V-8. This engine displaced 215.68 cu. in., weighed 140 lb. and was rated at 30 HP @ 1,800 RPM, giving a power-to-weight ratio of 4.66 lb./HP. It was this engine that powered the Curtiss "June Bug" in a flight of 1 minute 43 seconds on July 4, 1908 at Hammondsport, NY.

Glenn Curtiss, of course, went on to become the next most famous American aircraft designer and flyer to the Wrights. He did not continue for long with the V-8 air-cooled design, but in late 1908 switched to water-cooled designs, with a four-cylinder being first, the model D-4, and the model D-8 90° V-8 following shortly afterward. Both of these engines had cast iron cylinders with monel-metal water jackets. The D-4 model had a bore and stroke of 3 5/8" x 4". The D-8 bore was slightly larger at 3 3/4". Displacements were 143.1 and 353.4 cu. in. respectively and rated powers were 25 and 50 HP @ 1,800 and 1,600 RPM. Weights of the two engines resulted in power-to-weight ratios of 5.2 and 3.3 lb./HP.

Curtiss built a number of other engines of 4, 6 and 8 cylinders from 1908 to 1914, more notably the 90° V-8 model L with a bore and stroke of 4" x 5", which evolved into the model O and the model OX of the same dimensions. The model OX became available in 1914 and was

This 93 lb. 8 cylinder Levavasseur Antoinette engine developed 24 HP @ 1,400 RPM and was built from 1905 to 1907. It had an aluminum crankcase, water-cooled steel cylinders with brass water jackets and aluminum heads. (Courtesy of the Smithsonian)

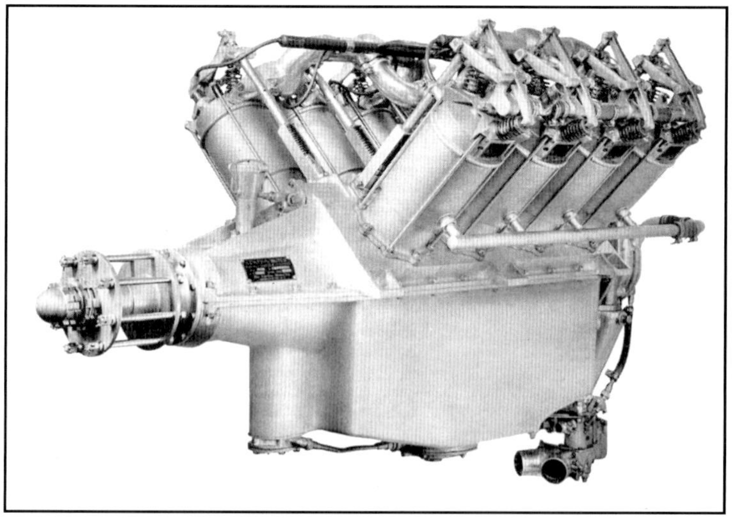

The Curtiss OX-5. This version of the OX series of engines was used extensively by the U. S. Services beginning in 1917. The 90 HP engine developed 1 HP per 4.33 lb. of weight. Because of its extensive use, this engine obtained fame almost equal to that of the Liberty. (Author's collection)

rated at 90 HP @ 1,400 RPM. It produced 1 HP for each 4.33 lb. of weight. It was a reliable engine for its day and saw extensive use by the services as training plane power, most notably in the Curtiss JN-4 "Jenny". In addition to those OX engines supplied to the U. S. Services, 2,209 were sold to the British government between 1915 and 1918. As post WW I surplus, this engine/plane combination became the backbone of the new wave of private flyers and barnstormers.

Although the OX series of Curtiss engines were certainly the best known of this company's line up to the end of WW I, they built others of note during the period of 1914 to 1917. Among them was the model V which came out in 1915. This was also a 90° V-8 but had a bore and stroke of 5" x 7" giving a displacement of 1099.56 cu. in. It was rated at 160 HP @ 1,100 RPM. This model was improved upon and became the model VX in 1916 rated at 180 HP @ 1,400 RPM. The VX was improved upon and was replaced in late 1916 by the V2 which was rated at 200 HP @ 1,400 RPM. All had the same bore and stroke.

In 1917 Curtiss also built a 12 cylinder version of the 5" x 7" V2 called the V4 in experimental quantities. This engine was rated at 350 HP @ 1,600 RPM and weighed 1,086 lb. giving a weight of 3.1 lb. for each horsepower.

The Hall-Scott Motor Car Co. of San Francisco was another American aircraft engine builder that achieved success during the period 1910 to 1917, starting with their model A-1, which was a four-cylinder engine of 4" bore and 5" stroke. It was a water-cooled engine as were all Hall-Scott aero engines to follow. The engine's 251.32 cu. in. displacement produced 40 HP @ 1,500 RPM and weighed 165 lb. Power-to-weight ratio was therefore 4.1 lb./HP.

This engine was followed in 1911 by a 4" x 4" 90° V-8 of 402.08 cu. in. displacement called the A-2 model. This engine developed 60 HP @ 1,400 RPM and weighed 260 lb. It thus produced 1 HP for each 4.33 lb. of weight. This engine was followed progressively by the A-3 and A-4 models. Both were also 90° V-8 designs. The A-3 was very much like the A-2 but with a 4" bore and a 5" stroke. The added stroke increased the displacement to 502.64 cu. in. and the output to 80 HP @ 1,400 RPM and only increased the weight to 290 lb. This gave an exceptionally good power-to-weight ratio of 3.63 lb./HP. The A-4 model came out in 1914 and had a bore of 5" and a stroke of 5" for a displacement of 785.4 cu. in. Output was 120 HP @ 1,500 RPM. Power-to-weight ratio was 4.46 lb./HP.

The first Hall-Scott engine to see extensive use was the 1915 A-5 model. This was an in-line 6 cylinder engine of 5" x 7" bore and stroke giving a displacement of 824.67 cu. in. Introduced in 1915, this engine was sold to friendly nations overseas in quantity and in small lots to the U.S. Services. It was rated at 125 HP @ 1,250 RPM and weighed 525 lb., giving a power-to-weight ratio of 4.2 lb./HP. This was Hall-Scott's first 5" x 7" engine.

Developed shortly afterward was the A-7 model (there was no production A-6). This was a four-cylinder version of the A-5 which weighed 410 lb. and developed 100 HP @ 1,400 RPM. Being rated at a higher RPM than the A-5 resulted in a power-to-weight ratio of 4.1 lb./HP in spite of the high weight of 410 lb. This engine also saw much service by early 1917.

By mid-1916 Hall-Scott had increased the power of both the A-5 and A-7 by increasing the bore by 1/4" to 5 1/4". Other minor changes were also made and the resulting models were called the A-5a and A-7a. Outputs

The Curtiss V2. A modest number of these engines were purchased by the U. S. Services in 1916-17. They used the same 5" x 7" bore and stroke as did the later Liberty series of engines. (Author's collection)

The Hall-Scott model A-2 engine of 1911. This 60 HP V-8 had exactly the same power-to-weight ratio as did the Curtiss OX, 4.33 lb./HP. (Author's collection)

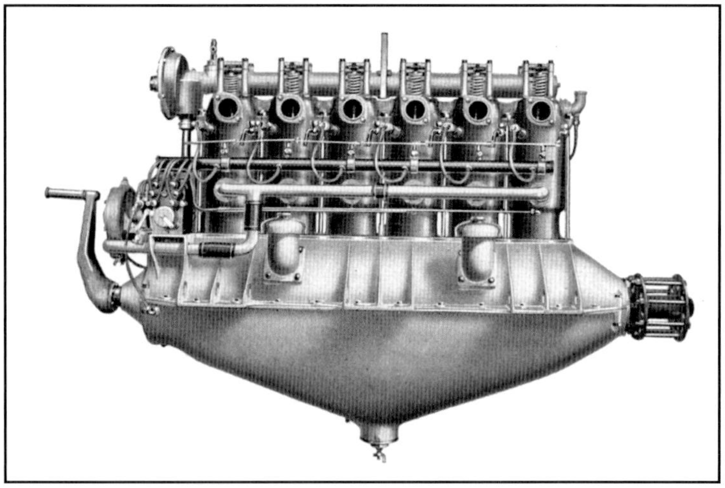

Hall-Scott A-5 of 1915. Some 300 of this model were sold to Russia in addition to smaller quantities purchased by other European nations and the U. S. Services. This was Hall-Scott's first 5" x 7" bore and stroke engine. (Author's collection)

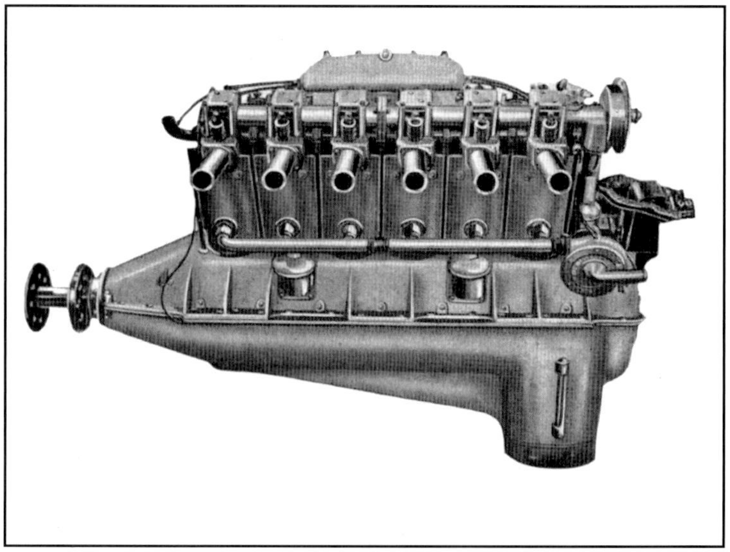

The Hall-Scott A-8 experimental engine of April 1917. This engine had a bore and stroke of 5" x 7" which gave it the same displacement as the Liberty 12. (Author's collection)

were 150 HP @ 1,250 RPM and 165 HP @ 1,475 RPM for the A-5a, and 100 HP @ 1,400 RPM for the A-7a. Both of these models were ordered in quantity by the U.S. Services.

Hall-Scott's last pre-WW I engine was the A-8 model. This was their first V-12 engine and also their first attempt at designing a high-powered aero engine. They chose the 5" x 7" bore and stroke they had first used on their A-5 engine in 1915. Displacement was 1,649.34 cu. in. and weight was claimed to be about 1,000 lb.

E. J. Hall, who would become co-designer of the Liberty, was its principal designer and the engine had just been completed and been given initial testing in late April of 1917. Below is an ad placed by Hall-Scott in the November 15, 1917 issue of *Aviation*. In it they claimed 300 HP for the engine. When performance information was released by Hall-Scott in 1919, it was, in the author's opinion, somewhat exaggerated at 450-500 HP in an attempt to make Hall-Scott look good and Packard look otherwise. I will discuss this topic in more detail in Chapter 2. It is mentioned here primarily as a point of reference in Hall-Scott's series of aero engines since it was strictly experimental at this point.

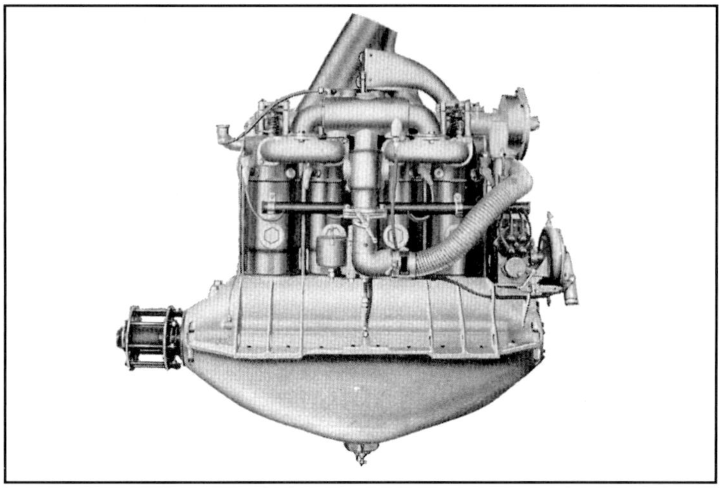

Hall-Scott A-7a. This engine was a popular one for training planes of the period of 1917-1918. (Author's collection)

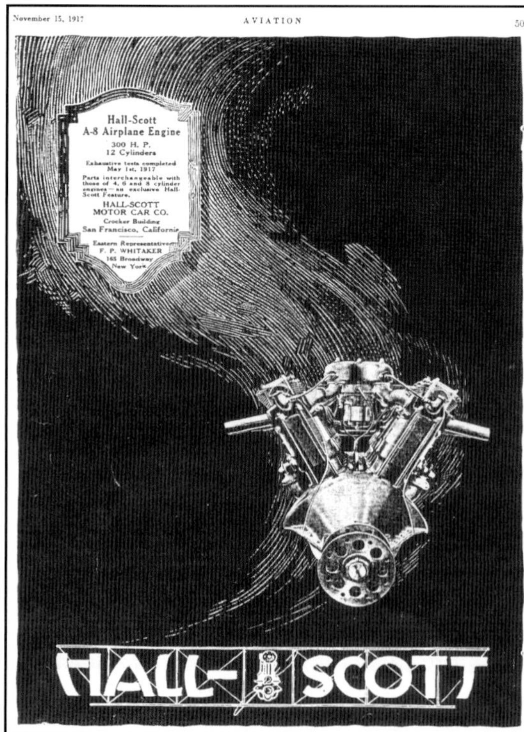

November 15, 1917 Aviation magazine ad in which Hall-Scott claimed 300 HP for the A-8 engine. In later publications, 450-500 HP was claimed. (Courtesy of Ric Dias)

Although there were numerous manufacturers of water-cooled aero engines in this country prior to 1917, not many met with much success. The Air Service purchased several Sturtevant airplanes in the 1916 - 1918 period powered by Sturtevant 5 and 5-A engines. Both seemed to have given reasonably good service and are cited here as examples of the state of the art in the United States at this time.

Both model 5 engines were 90° V-8 designs of 4" bore and 5 1/2" stroke. The resulting displacement of 552.88 cu. in. gave an output of 140 HP @ 2,000 RPM. These engines had reduction gears with a 1.68:1 ratio thus giving a propeller speed of 1,200 RPM at rated engine speed.

The model 5 had semi-steel L-head cylinders cast in pairs with valves operated by push-rods from the central camshaft located in the Vee. The crankshaft rode on three main bearings. The engine weighed 580 lb. for a power-to-weight ratio of 4.14 lb./HP.

Sturtevant 5 V-8 engine of 140 HP built by the B. F. Sturtevant Company of Hyde Park, Boston. Notice the similarity to the Model 5-A pictured below. (Author's collection)

Sturtevant 5-A V-8 engine of 140 HP and power-to-weight ratio of 3.67 lb./HP. This 1916-1917 era engine was used in modest quantities by the Air Service in Sturtevant airplanes. (Author's collection)

The model 5-A was of similar design except that the cylinder pairs were cast of aluminum and fitted with steel liners. The valves were changed to overhead, operated by rocker arms. The resulting weight was 514 lb. and the power-to-weight ratio was 3.67 lb./HP.

The above mentioned engines constituted the most successful and highly developed of the American water-cooled aero engines up to about 1917. Although a few were built with outputs above 200 HP, they were experimental rather than proven production engines and thus we will not be concerned with them here.

Packard was one of the producers of an engine of above 200 HP, but it was still in the experimental stages. I will discuss that engine, the 1A-905, in some detail in Chapter 2 because it was designed by Jesse G. Vincent, one of the designers of the Liberty. Actually the 1A-905 was the third model of a series of 905 cu. in. V-12 engines designed by Vincent and built by Packard. The first model had been completed in late 1916 and submitted to numerous tests.

As a result of this testing a second model of the same size was designed and it was completed in April of 1917. It was subjected to testing which included its extensive use in a race car.

Packard 1A-905 engine of 1917. This 285 HP V-12 was introduced in April of 1917. (Author's collection)

In the mean time the third model, the 1A-905, was designed and the actual engine completed in May of 1917. All engines of the series were 40° V-12 designs. The 1A-905 weighed 990 lb. and developed 285 HP @ 2,000 RPM. It utilized a 1.68:1 reduction gear. Bore and stroke were 4" x 6" giving a displacement of 904.8 cu. in. Power-to-weight ratio was 3.47 lb./HP.

Air-cooled aero engines had followed two lines of development, stationary and rotary. The stationary types consisted of in-line and fan or radial. The rotaries were, of necessity, all radials.

The stationary designs rapidly reached their practical power limits of about 150 HP because of the cylinder cooling limits imposed by the iron or steel composition of their cylinders. These metals are rather limited in their ability to conduct heat from combustion inside the cylinder to the air outside the cylinder.

In the in-line category the 80 HP Renault, 90 HP R. A. F. 1A, and the 150 HP R. A. F. 4A all saw extensive use from 1915 until the end of the war.

The 80 HP Renault was a 90° V-8 with a displacement of 548.72 cu. in. and weighed 463 lb. giving a power-to-weight ratio of 5.78 lb./HP.

The 90 HP R. A. F. 1A was a 90° V-8 engine of 537.4 cu. in. which would develop 110 HP @ 1,800 RPM and weighed 440 lb. Its power-to-weight ratio was thus 4 lb. /HP.

The 150 HP R. A. F. 4A was a 60° V-12 engine of 806.15 cu. in. displacement which would develop 150 HP @ 1,800 RPM and weighed 680 lb. Power-to-weight ratio was 4.53 lb./HP.

It was not until the finned aluminum cylinder was perfected in the 1920s that the stationary radial air-cooled engine could be made into a high power one. Stationary air-cooled radial engines would not see significant use until after the end of the war so we will not be concerned with them here.

The power limit of the air-cooled rotary was higher than that of the stationary radial engine because of its basic design. The rotary is an air-cooled engine with rotating cylinders. A central single throw crankshaft carries a master rod plus several slave rods attached thereto. Each rod carrys a piston which is fitted in a cylinder. The cylinders are mounted radially around a crankcase. Generally these engines had either 7 or 9 cylinders. These cylinders were machined steel with integral cooling fins.

R. A. F. 1A air-cooled V-8 of 90 HP. (Author's collection)

R. A. F. 4A V-12 air-cooled engine of 150 HP. (Author's collection)

130 HP Clerget shown without the carburetor mounted on the right end of the crankshaft. (Author's collection)

The unique feature of the rotary is that its crankcase and cylinders rotate around the crankshaft. We are used to having the crankcase of an engine mounted solidly to whatever vehicle it is mounted in and the crankshaft rotating inside it carrying the rods upon their journals with these rods carrying their individual pistons up and down in the cylinders, which are in turn fixed to the crankcase. In this case the power output is taken from the end of the crankshaft. In the case of an airplane a propeller is mounted either on the end of the crankshaft or on the output shaft of a reduction gear which is powered from the crankshaft.

Since on the rotary the crankcase and cylinders rotate and the crankshaft stays still, we must mount the crankshaft solidly to the vehicle. To enable this, the crankshaft extends for some distance out of the rear of the engine and flanges are provided on this end for that purpose.

Thus mounted, the shaft remains attached solidly to the vehicle and the crankcase and cylinders rotate around it. The result is that most of the engine rotates while it is running. The finned cylinders are not only progressing forward through the air when the airplane is flying, but rotating like a fan. Cylinder movement relative to the air surrounding it is thus great and cooling is far better than is possible if the cylinders must be cooled only by air movement caused by the forward movement of the airplane and air blast from the propeller.

This type of construction, however, has numerous drawbacks and reaches practical limits not much beyond the 150 HP size. We will not attempt to go into the purely technical problems that had to be dealt with in this design, but one that was perhaps the most limiting we will point out. Engines of this type will weigh on the order of 2.5 to 3 lb./HP. That means a well-designed 150 HP engine will weigh perhaps 375 lb.

A large part of this mass is contained in the rotating portion of the engine. That rotating portion becomes a huge gyroscope. The heavier it is, the more gyroscopic action it exerts upon the airplane. This action can severely effect attempts to change the direction of the airplane in flight.

In spite of the drawbacks of the rotary design, it saw extensive use during WWI. Almost all the successful rotary designs were European. With the end of the war, however, it was realized that the development of the rotary had reached its zenith and little further effort was expended upon this design. The stationary radial air-cooled and in-line water-cooled types would take over as the designs with further promise.

Although, as we have seen, well proven production water-cooled aero engines in the United States were limited to the under 200 HP class, European designs had progressed toward more power. The impetus of war since 1914 had produced a good deal more effort on their part than on ours.

150 HP Hispano-Suiza Model A built by the Wright-Martin Aircraft Corp. in the United States. These engines were built in France, England and the United States during and after the war. Pictures from the Wright-Martin instruction book of November 1918. (Author's collection)

Next let us consider the monobloc-type water-cooled engine. The most successful of the engines of this type was the Hispano-Suiza. The first model of this series of 90° V-8 engines was introduced in 1915 and rated at 150 HP @ 1,450 RPM and had a bore of 4.72" and a stroke of 5.12" resulting in a displacement of 718.88 cu. in. Dry weight was 445 lb. and the resulting power-to-weight ratio was 2.96 lb./HP. This was an excellent achievement for the time.

In monobloc construction, the cylinder banks and their cooling jackets are cast in a single aluminum block. These blocks, two in the case of the V-8 types, are bolted to the aluminum two-piece crankcase. In the case of the Hispano-Suiza, the valves (two per cylinder) are operated by a single overhead cam on each cylinder bank. Steel cylinder barrels are screwed into the aluminum block to carry the cast aluminum pistons.

This engine had its power increased during the course of its wartime use in two ways: first, its compression ratio was increased; second, operating RPM was increased and reduction gears fitted to keep propeller speed within an efficient range; third, a combination of both. The resulting power outputs were 180 HP, 200 HP and 220 HP from the same original displacement.

In all three cases the increased stress and heat caused by the increase in power resulted in overheating and stress failures not encountered in the original 150 HP model. It took most of the remainder of the war years to get these problems under control.

When this was accomplished the Hispano-Suiza series of engines became one of the best and most reliable of the water-cooled types. Various models were built with displacements up to 1,127.36 cu. in. and rated output of 325 HP @ 1,800 RPM.

Now let us switch to a bit of history regarding British wartime procurement of aero engines. First we must realize that the aero industry was in its infancy when the war started in 1914. As a result, numerous designs were proposed and accepted for use by the military services. This resulted not only in many different designs of airplanes but almost as many different engine designs to power them. By early 1917 there were almost 40 different designs in use. This not only created a logistical nightmare trying to provide maintenance and repair parts and expertise, but also made mass production difficult. It was decided that to overcome this problem there would have to be some standardization of engine types.

In addition, in late December of 1916 the British Army Board had sanctioned a doubling of the size of the Royal Flying Corps which in turn would require a huge increase in engine and aircraft production. It was calculated that to meet this demand engine production would have to be raised from its current 600 per month to 2,000 per month.

As part of the standardization scheme, an attempt was made to take each category of engine design, settle upon a small group of the best designs and then put them in production or keep them in production, as the case may be.

Sunbeam Arab I. This is the geared version of the ill-fated British monobloc engine. Although meant to be an improvement of the Hispano-Suiza, it proved ultimately to be a failure and a significant setback to the British aeronautical program during WW I. This 220 HP engine weighed 530 lb. or 2.4 lb./HP, rather remarkable for the time. (Author's collection)

One of the categories was the 150/200 HP water-cooled engine. In this group were four engines thought to be of comparable qualities. They were a Sunbeam V-8 (eventually to be called the "Arab"), a Sunbeam of six-cylinder design, the six-cylinder B. H. P. (Beardmore-Halford-Pullinger) and the Hispano-Suiza. In addition there were Rolls-Royce designs already in production which will be discussed later.

The decision of which to put in production was put before the Advisory Committee for Aeronautics. At the end of January of 1917 they recommended the V-8 Sunbeam as the best, in spite of some objections raised because of as yet unproved new features and production requirements of this new design. The B. H. P. was also recommended. Both of these engines were of aluminum monobloc design.

The 1915 Hispano-Suiza engine inspired a number of designers to bring out new monobloc engines. They recognized it as a design with great promise and they were ultimately correct. The aluminum monobloc design would become the dominate high-power water-cooled aero engine until the piston aero engine was replaced by the turbojet after WW II.

Great Britain was one of the first to attempt a new monobloc design. The Sunbeam company proposed to the British Air Board a 90° V-8 of the same dimensions as the Model A Hispano-Suiza and promised an output of 200 HP as compared to the 150 HP output of the Model A. The design of the Sunbeam was an advance from the original Hispano-Suiza but many unexpected problems were encountered and by January of 1917 the engine was still unable to produce the promised 200 HP.

The design had seemed so promising, however, that the Air Board, as mentioned above, had selected it as the preferable 200 HP-class engine for production.

The engine was ordered into mass production before it was actually a proven design. Thus they were committed to it at the expense of others. Unfortunately its problems proved formidable, the principal one of which turned out to be a resonant torsional vibration inherent to the design of the engine. Vibration could become so intense that it would literally shake the engine apart and engines were know to actually break apart in the air.

In addition, problems with production of acceptable aluminum castings for the blocks had also caused significant production difficulties.

Although the severity of the vibration problem was eventually reduced, it was never licked. Only a complete redesign of the engine could have accomplished it. In the mean time the program was based on the availability of 1,800 engines by the end of 1917. Only 81 had been delivered by that time and an additional 1,114 by the end of 1918. 4,400 had been ordered.

The B. H. P. was also ordered into production and a contract for 2,000 was given to the Siddeley Deasy Company. Production of this engine also proved difficult. The most formidable problem was obtaining acceptable aluminum cylinder blocks. Initially more than 90% of them were rejected for defects of one type or another.

In addition, some delays were encountered because of the necessity of redesign to better facilitate mass production.

Once the engine did get into heavy production and began to see extensive use, it had trouble with burnt exhaust valves. Curing this problem took some time and it was spring of 1918 before the engine got into full production. It too was a setback to the British production program, although not as serious a one as was the Sunbeam Arab. This engine was to be named the "Puma."

The third monobloc engine to be produced was, of course, the Hispano-Suiza. This was to be accomplished in two ways. At the insistence of the Admiralty an order for 8,000 of these engines was given to the Mayen Company in France and funds advanced by the British Government (almost £2 million) to build a factory to produce them.

This had occurred in spite of the objections of Lord Curzon who expressed great fears of the country being left at the end of the war with vast surpluses of engines for which it would have no use. The Air Board over-ruled his objection with a "better safe than sorry" attitude and, as it turned out, the engines thus obtained late in 1917 and beyond helped relieve the tremendous pressures caused by the failures in production at home.

In addition to the French Hispano-Suiza engines, a contract was given to the Wolseley Company to produce 400 of the 150 HP type. That order was later increased to 1,100. Most of these were actually delivered as 180 HP engines known as the Wolseley "Viper." Wolseley was also contracted to produce the 200 HP version of this engine known as the Wolseley "Adder." Deliveries of these two engines did not commence until the first quarter of 1917. The quantity delivered was much below quotas and remained so until the first quarter of 1918, thus compounding problems with obtaining engines in the 200 HP class.

Let us digress a bit and discuss the type of design that was eventually chosen as the basis for the Liberty series of engines, water-cooled engines with individual steel cylinders.

In all fairness, this type of engine must be credited to the Daimler Company of Germany who first used the design with success in 1911/12. These Mercedes engines, as they were called, utilized machined forged steel cylinders with valve ports welded into the head at an angle in such a way that when the head was then welded onto the top of the cylinder it formed a hemispherical combustion chamber. Sheet steel water jackets were then welded to the outside of the cylinders.

The overhead valves were driven by rocker arms which were activated by a camshaft which was mounted in a housing bolted across the tops of the in-line cylinders. These cylinders were bolted to an aluminum crankcase. The most common and most successful configuration used by the company was the in-line six.

The six-cylinder 160 HP Mercedes D-IIIa became a standard in 1915 and continued to be so until the end of the war. Its 5.51" bore and 6.3" stroke produced a displacement of 901.32 cu. in. Weight was 618 lb. and power-to-weight ratio was 3.86 lb./HP. Although not outstanding in this category, the engine was extremely reliable and was used in several types of aircraft.

The B. H. P. Siddeley Puma. A 230 HP engine weighing 648 lb. or 3.1 lb./HP. (Author's collection)

260 HP Mercedes D-IVa of 1916. This engine and its predecessor, the 1915 D-IIIa, were the basis from which sprang a number of individual steel-cylinder designs. (Author's collection)

300 HP Renault model 12 Fe of 1916. This 50° V-12 utilized a cylinder construction similar to that of the 1A-905 Packard. (Author's collection)

In 1916 Daimler introduced the D IVa with a bore and stroke of 6.3" x 7.09" giving a displacement of 1,326.06 cu. in. This engine developed 260 HP and weighed 936 lb. The resulting power-to-weight ratio was 3.9 lb./HP.

The French did not let the Mercedes-type engine go unnoticed and Renault, who had formally concentrated primarily on air-cooled engines, took up the design. They first built a six-cylinder engine but in 1916 advanced to a V-12. This engine had its cylinders built up in pairs, much like the 1A-905 Packard. It was a 50° V whereas the Packard was a 40° V.

This model 12 Fe engine had a bore and stroke of 4.92" x 5.91" for a displacement of 1,348.3 cu. in. and weighed 794 lb. Its output of 300 HP gave it a quite creditable power-to-weight ratio of 2.65 lb./HP.

Another French builder to follow the Mercedes lead was Lorraine-Dietrich. They started in 1914 by building a six-cylinder engine very similar to the Mercedes D-I. Their chief designer, Marius Barbarou, had been a Mercedes engineer from 1902 to 1904.

In 1915 they developed a 90° V-8 with a bore and stroke of 4.72" x 5.51" giving a displacement of 771.28 cu. in. and an output of 160 HP. This engine was submitted to considerable refinement, including an increase in stroke to 6.69". The resulting engine displaced 936.48 cu. in. and was rated at 220 HP @ 1,350 RPM. However, it could develop 270 HP @ 1,700 RPM. It weighed 545 lb. which gave a power-to-weight ratio of 2.2 lb./HP, which was about the best so far obtained in a water-cooled engine which had, supposedly, passed its test stage and was ready for production. Great things were expected of it.

As was the case with the Renault, this engine, and a V-12 of the same design which shortly followed it, had its cylinders built-up in pairs. The V-12 displaced 1,404.72 cu. in. and developed 370 HP. Although some 30 examples of the V-8 had been built and placed on test by early 1917 and the V-12 reached the same point a few months later, the V-8 was just beginning to become available for aircraft use when the war ended in late 1918 and the V-12 model was still in the testing stage.

270 HP V-8 Lorraine-Dietrich which was highly regarded as a promising new design of early 1917. (Author's collection)

370 HP V-12 Lorraine-Dietrich which was an enlarged version of the 270 HP V-8. Although not in production in time for use during the war, this engine would eventually see much reliable service in France. (Author's collection)

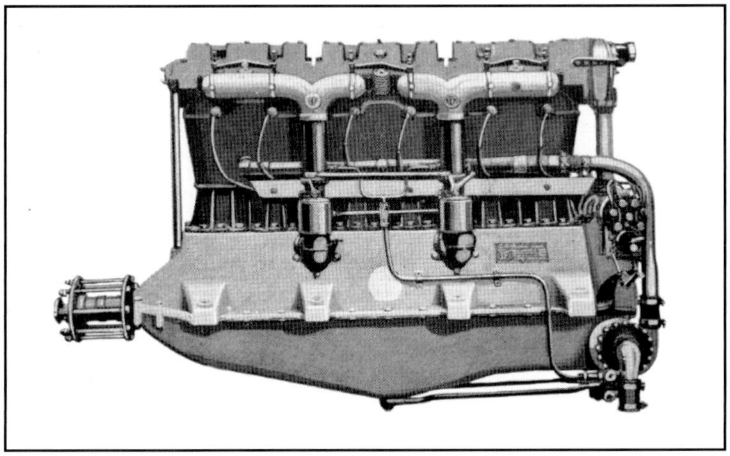

Isotta-Fraschini V-4-B. This 170 HP cast iron cylinder in-line six was heavy but reliable. (Author's collection)

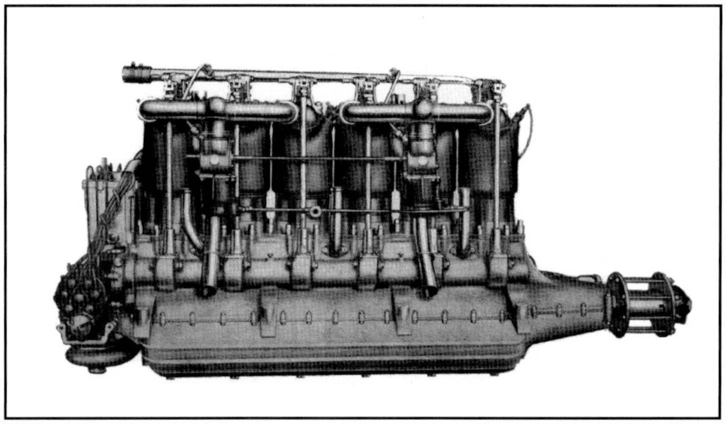

The 160 HP Beardmore was an enlargement and minor improvement of the 120 HP Astro-Daimler. (Author's collection)

The Italian Fiat A-12 in-line six-cylinder engine came into use in 1916. This was a 6.3" x 7.1" bore and stroke engine of 1,326 cu. in. which delivered 265 HP @ 1,500 RPM. This same engine was updated with minor refinements in 1917 and became the A-12 bis rated at 300 HP @ 1,600 RPM. The individual cylinders of the A-12 models were made up of steel forgings which had the valve ports and water jackets welded in place. The A-12 bis weighed 920 lb. and thus had a power-to-weight ratio of 3.1 lb./HP.

The engine saw service by both the Italians and British but did not gain a particularly good reputation for reliability. The British ordered 2,000 of these models but only 610 had been delivered by the end of 1918.

Before we leave the Italian engines we should mention the Isotta-Fraschini V-4-B. This was another in-line six-cylinder engine with cast iron cylinders built up in pairs with welded on steel water jackets. This engine had a bore and stroke of 5.12" x 7.09" for a displacement of 875.22 cu. in. and was rated at 170 HP @ 1,450 RPM. It weighed 605 lb. to give a power-to-weight ratio of 3.6 lb. /HP. This was rather heavy for the time but was perhaps compensated for by its reputation for reliability.

Fiat A-12 bis. This 300 HP in-line six-cylinder engine and its predecessor, the A-12, saw use by both the Italian and British services. (Author's collection)

Lastly we will discuss briefly the British accomplishments up to 1917 in the area of individual cylinder designs.

The William Beardmore Co. of Scotland had secured a license in 1914 to build the 120 HP Astro-Daimler engine of Germany. Beginning in 1915 they built this engine in large quantities and supplied the Royal Air Force with them for use on the R. A. F.-designed F. E. 2b pusher. (The British generally referred to it as the 120 HP Beardmore.)

They very soon set about to enlarge and improve upon the 120 HP engine. The principal change was to enlarge the bore from 5.12" to 5.59". In so doing the displacement was increased from 850.5 cu. in. to 1,014.54 cu. in. and the power increased to 160 HP @ 1,250 RPM. The engine was capable of 180 HP @ 1,400 RPM. With a weight of 600 lb. this engine gave a power-to-weight ratio of 3.33 lb./HP.

This individual cylinder engine had cast iron cylinders with welded-on water jackets. The overhead valves were operated by push rod-driven rocker arms from a side-mounted camshaft. This engine was used with success and 2,556 had been delivered to the services by the end of 1918.

The last of the British designs we will look at are those of Rolls-Royce. This company expanded upon the Mercedes-type single-cylinder construction, and did so with great success. The principal shortcoming of the Rolls-Royce engines was that although they were precision high quality products, they were almost hand-built by master mechanics. The result was an engine that performed very well indeed, but one that could not be produced with speed. Thus output of the factories was always behind demand. In addition, when repairs or rebuild was required, engines had to be sent back to the factory because proper equipment and the master mechanics required for their use were only available at the factory. Individual replacement parts in a great many cases had to be hand-fitted to each individual engine.

Rolls-Royce was a luxury car builder who had never designed or built an aero engine before the war started. The British war office approached the company with a request that they build the Royal Aircraft Factory-(R. A. F.) designed air-cooled V-8. F. H. Rolls did not like the idea of building a cast-iron cylinder air-cooled engine and felt that the water-cooled steel cylinder design had much more promise.

The government saw to it that they received one of the 1914 Mercedes Grand Prix cars which was currently in England after having been on display at various automotive showrooms. Its engine was of the same design as the then current Mercedes aero engines. Rolls engineers tore the engine down for study and then proceeded with the design of a six-cylinder engine of their own which served primarily as a test engine.

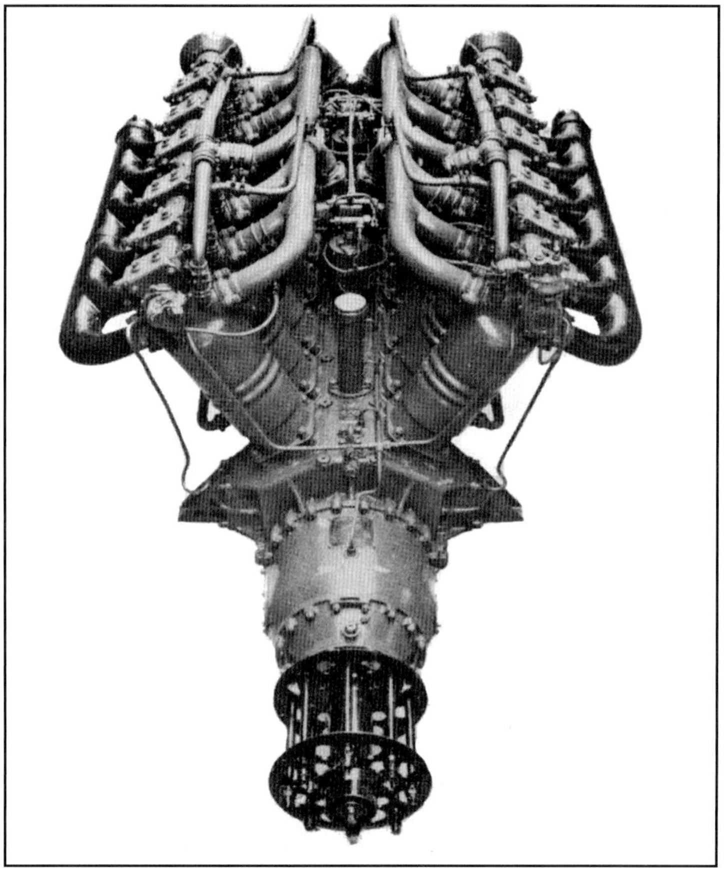

Rolls-Royce Eagle, produced from 1915 - 1922 in models I through VIII. One of the best individual cylinder water-cooled engines developed up to the end of WW I. (Author's collection)

The British Admiralty expressed a desire for more power than was available from six cylinders and asked that they design an engine of at least 200 HP. They did so and the result was the 240 HP Eagle. This was a 60° V-12 utilizing steel cylinders with welded-on sheet steel water jackets. Bore and stroke were 4.5" x 6.5" giving a displacement of 1,240.56 cu. in. Valves were overhead, two per cylinder, driven by rocker arms from overhead camshafts. Weight was 900 lb. with epicyclic reduction gear. Power-to-weight ratio was 3.75 lb./HP.

The first model of this design, the Eagle I, was first run on test in March 1915 and the first production engines delivered in October of that year.

During the course of the war this engine was constantly improved upon and went through a series of models. The Eagle I gave 240 HP @ 1,600 RPM utilizing two carburetors, one to feed each bank of six cylinders. 104 were built in 1915 - 1916.

The Eagle II produced 260 HP @ 1,600 RPM and used four carburetors feeding three cylinders each. 36 were built in 1916.

The Eagle III was rated at 270 HP @ 1,600 RPM and had minor carburetion changes from the Eagle II design. 110 were built in 1916 - 1917.

The Eagle IV was also rated at 270 HP @ 1,600 RPM but went back to the two carburetor setup. 150 were built in 1916 - 1917.

The Eagle V went back to four carburetors with output increased to 285 HP @ 1,600 RPM. 100 were built in 1916 - 1917.

The Eagle VI, with increased RPM and ignition and other changes, raised output to 308 HP @ 1,650 RPM. 300 were built in 1917.

The Eagle VII had refinements but remained at the same power rating. 200 were built in 1917 - 1918.

The Eagle VIII had increased compression and carburetion refinements and gave an output of 360 HP @ 1,800 RPM. Weight of this model was 960 lb., thus the power-to-weight ratio at this rating was 2.7 lb./HP. Weights increased with each model change and the Eagle VIII had the best power-to-weight ratio of the series. It took until well into 1917 before this point of refinement was reached. 3,302 were built from 1917 - 1922. Of these approximately 3,080 were built before the end of the war.

Not long after the Eagle was put on tests a second 60° V-12 of smaller size but very similar design was drawn up. This 4" x 5.75" bore and stroke engine displaced 866.5 cu. in. and in its original series I form, called the 190 HP model and later the Falcon, developed 230 HP @ 2,000 RPM. It weighed 710 lb. thus giving a power-to-weight

Rolls-Royce Falcon III, the last of a series that was introduced in 1916 and continued in production until 1927. (Courtesy of Rolls-Royce)

ratio of 3.1 lb./HP. Production deliveries began in the second quarter of 1916 and 250 were built.

The Falcon II with somewhat revised carburetion was introduced in 1917 and developed 253 HP @ 2,000 RPM. 250 were built.

The Falcon III replaced the II in mid 1917 and stayed in production until 1927. This engine had revised carburetion and ignition and a higher compression ratio. Output was thus raised to 285 HP @ 2,200 RPM and with a weight of 715 lb. the power-to-weight ratio was 2.5 lb. /HP. A total of 1,685 were built of which 1,132 were delivered before the end of the war.

So as not to abruptly cut off the general history of British aero engines as related to this period I will carry on a bit farther. The reader will soon see that it relates directly to decisions to adopt the soon to be designed Liberty to British use.

The June 1917 daylight German air raid of London brought a rapid government decision to again instigate a doubling of the British Air Services, both Army and Navy. It had only been some six months since they had planned the first doubling. The 2,000 engines per month goal now became one which it was estimated might eventually reach 4,500 engines per month. Of these, 3,500 were to come from British sources, 500 from America (the United States and Canada), and 500 from France.

One of the principal goals was to carry the air war to Germany in the form of heavy bombing of their war industry. To that end the British needed bomber-type aircraft. These were classed as day bombers and night bombers. The majority of aircraft anticipated for use for these purposes for mid 1917 and beyond were to be powered by engines of 200 HP and above. As we shall see, the Liberty engine had been designed and heavy production anticipated by the time plans were being laid for this second large expansion of British air power and its use would eventually figure significantly in Britain's planned armada of bombing planes.

In summary we can see from the prior narrative and from the chart on page 14 that the state of the art in aero engine design in early 1917 certainly indicated that there were no air-cooled designs suited for development into a series of standardized engines of from 100 to 400 HP, which was to be the goal set for the design of what was to become the Liberty series of engines.

Thus the water-cooled engine had been chosen as the basic type. In this category were the aluminum monobloc and the individual and dual cylinder types.

The Hispano Suiza monobloc had been successfully built in the 150 HP size by the French but other efforts to use the design were proving difficult. Casting large aluminum monoblocs was a new science and was one it would likely be wise to avoid tackling if one wished to bring a new engine design rapidly into production.

That then leaves the individual or dual-cylinder water-cooled engine as the most logical choice for a new engine using only proven design features.

What was the goal to shoot at? What can be considered as successful engines of recent design had an average power-to-weight ratio of 3.34 lb./HP. Only four had ratios below 3 lb. The best was the Lorraine-Dietrich V-8 of 2.2 lb. and, as it turned out, this was not the proven engine that was first thought. The Liberty was to beat it into production. The next best were the Renault 12-fe at 2.7 lb. and the Rolls-Royce Eagle V at 2.8 lb.

There was far more to consider than the basic cylinder and cooling design, of course, and we shall see what features were chosen and what kind of engine they produced in the following chapters.

General state of aero engine development as of the first quarter of 1917

(The general state of the art as indicated by a representative group of proven designs)

Air-cooled rotary

Name	Bore & Stroke			HP per Cylinder	Pounds per HP	Engine HP
130 HP Clerget	4.72	x	6.3	14.45	2.5	153 max.
120 HP Le Rhone	4.41	x	6.7	13.33	2.7	120 max.

Note: This type of engine was not suitable for development into a power plant of more than about 200 HP.

Air-cooled in-line

R. A. F. 4a	3.94	x	5.51	12.5	4.53	150 max	60° V-12

Water-cooled aluminum monobloc

Hispano-Suiza	4.72	x	5.12	25.00	2.35	200 rated	90° V-8
Siddeley Puma	5.71	x	7.48	40.00	2.7	240 max	IL-six

Note: Other attempts at monobloc engine construction were made during the war but without success until the end of 1918. This engine design eventually became the most successful of the water-cooled types.

Water-cooled individual or dual cylinder

Curtiss V2	5	x	7	29.38	2.9	235 max	90° V-8
Curtiss V4 *	5	x	7	29.16	3.1	350 max	90° V-12
Hall-Scott A-7	5	x	7	25.00	4.0	100 max	IL-four
Hall-Scott A-7a	5 1/4	x	7	27.50	3.7	110 max	IL-four
Hall-Scott A-5	5	x	7	25.00	3.8	150 max	IL-six
Hall-Scott A-5a	5 1/4	x	7	27.50	3.4	165 max	IL-six
Sturtevant 5-A	4	x	5.5	17.50 **	3.7	140 rated	90° V-8
Packard 1A-905 *	4	x	6	23.75 **	3.5	285 max	40° V-12
Mercedes D IIIa	5.5	x	6.3	26.66	3.9	160 rated	IL-six
Mercedes D IVa	6.3	x	7.1	43.33	3.9	260 rated	IL-six
Renault 12-Fe	4.9	x	5.9	27.25 **	2.7	327 max	50° V-12
Lorraine-Dietrich	4.7	x	6.7	33.75 **	2.2	270 max	90° V-8
Lorraine-Dietrich *	4.7	x	6.7	33.33 **	2.0	400 max	60° V-12
Fiat A-12 bis	6.3	x	7.1	50.00	3.1	300 max	IL-six
Isotta-Fraschini V-4-B	5.1	x	7.1	28.33 **	3.6	170 rated	IL-six
Beardmore 160 HP	5.6	x	6.9	30.00	3.3	180 max	IL-six
Rolls-Royce Falcon I	4	x	5 3/4	19.17	3.1	230 max	60° V-12
Rolls-Royce Eagle V	4 1/2	x	6 1/2	26.83	2.8	322 max	60° V-12

Note: engines with * were experimental in early 1917 but are noted for reference. The Curtiss V4 was never developed farther and had an odd 90° included angle for a V-12. The Packard 1A-905 was also not developed farther. The Lorraine-Dietrich V-8, although it was thought a proven engine in early 1917, took until late 1918 to actually reach the front in an airplane. The Lorraine-Dietrich V-12 was developed into a good engine, but it took until after the end of the war to do so.
Engines with ** had dual cylinders.

Chapter 2

The Birth of the Liberty

At the beginning of 1917 the United States stood fourteenth among the world's nations in terms of aviation. In 1903 the Wright brothers had propelled us to first place insofar as controlled, powered flight of heavier-than-air craft was concerned. Europe seemed far more affected by this accomplishment than did the United States.

In this country we treated flight as a novelty to be used, for the most part, by the affluent for excitement and pleasure. Since there was little market for airplanes, development was slow. The military took scant interest and received minimal support in even that by the controller of the purse, Congress.

EARLY U.S. MILITARY AVIATION DEVELOPMENT

Probably the earliest individual to gain significant success in the design of heavier-than-air craft in this country was Samuel P. Langley whose models of his "Aerodrome" design were flying as early as 1896. Under a U.S. government contract in the amount of $50,000, he attempted to construct a full-scale, man-carrying aircraft of the "Aerodrome" design. Langley was at that time Secretary of the Smithsonian Institution.

This effort was culminated in two failed attempts to take off from a houseboat on the Potomac River on October 7 and December 8, 1903. With better preparation the effort may have succeeded but instead it resulted in harsh criticism from the public and Congress. The airplane was not accepted by the Army, to whom it was to have gone, but it represented the first aircraft development paid for by public funds.

Langley had hardly recovered composure from his failure when the Wright brothers succeeded with their better prepared effort at Kitty Hawk on December 17, 1903.

The failure of Langley left its effects upon both the Army and Congress and no funds were forthcoming for military aircraft until 1908. On February 10, 1908 the Wright brothers proposed to build a military aircraft capable of carrying two persons aloft at 40 MPH for the sum of $25,000. The result was the Wright Model A which completed its required tests in September of 1908 and was accepted on August 2, 1909. Total aeronautical expenditures for 1907-1908 was $30,000.

This Wright Model A was the Army's first heavier-than-air craft and was their sole example for two years. It served to train the Army's aviators during that period.

On August 1, 1907, an Aeronautical Division had been established in the Office of the Chief Signal Officer of the Army Signal Corps. Captain Charles de Forrest Chandler was its first chief. He ran the Division with the aid of two enlisted men.

Annual expenditures on Army aeronautics in 1909, 1910 and 1911 were minimal, at best. It was fiscal year 1912 before a specific aeronautical appropriation ($125,000) was made. Appropriation for 1913 was $100,000; for 1914 was $125,000; for 1915 was $250,000.

Army aviation was given official recognition on July 18, 1914 when an act of Congress established it as the Aviation Section of the Signal Corps. At that time its strength was 60 officers and 260 enlisted men. By this time about 29 airplanes had been delivered to the Army. It was also very close to the outbreak of WW I in Europe (the assassination of Austrian Archduke Ferdinand on June 28, 1914).

The Navy had also shown an interest in Langley's aircraft and had followed its development and failure. An inter-service board had been appointed to investigate the military possibilities of this aircraft. Official Navy interest surfaced again when they assigned two officers, Lt. George C. Sweet and Naval Constructor William McEntee to observe the September, 1908 test of the Wright Model A at Fort Meyer.

As a result of the report of these two officers, on December 2, 1908, Rear Admiral William S. Cowles, Chief of the Bureau of Steam Engineering, recommended to the Secretary of the Navy that "a number of aeroplanes be purchased to operate from a ship's deck —." The recommendation resulted in no action on the Navy's part.

On August 16, 1909, just after acceptance of the Wright Model A by the Army, the Navy's Bureau of

Wright Model A at Fort Meyer, Virginia in September, 1908. This was the U.S. Army's first heavier-than-air airplane.
(Courtesy of the Air Force Museum)

Equipment requested authority to advertise for the construction of two airplanes for the Navy. This request was also disapproved.

As both civilian and Army use of airplanes continued to increase, so did interest by Naval personnel. In response, on September 26, 1910, the Secretary of the Navy appointed Captain W. I. Chambers as the officer to whom all questions and correspondence related to aviation should be addressed. Chambers became, in effect, officer in charge of Naval aviation and remained so for the next three years.

Although not himself a flyer, Captain Chambers was deeply interested in the possibilities of Naval aviation. He soon tried to interest the Wrights in attempting a flight from the deck of a United States warship but Wilbur Wright declined to make such an attempt. The Curtiss Airplane and Motor Company was approached with the idea and they accepted.

Captain Chambers had a platform constructed on the forward deck of the cruiser *U.S.S. Birmingham* anchored at Hampton Roads, Virginia and on November 14, 1910, Curtiss test pilot Eugene B. Ely successfully flew a 50-horsepower Curtiss landplane from it.

As a result of these events, Glenn Curtiss sent a letter to Captain Chambers offering to instruct a Navy officer as a pilot free of charge. The offer was accepted and Lt. Theodore G. Ellyson was sent to the Curtiss flying school in California. Ellyson completed his training successfully and went on the become Naval Aviator No. 1.

The Navy's first aviation appropriation was $25,000 for 1911-1912. With it the Navy purchased its first airplanes, two Curtiss Model E pusher planes and one Wright. The two Curtiss planes were delivered on July 1 and 13, 1911, and the Wright, a twin propeller pusher, some months later.

The next two years saw modest but steady progress in Naval aviation. Appropriations for aviation use were under $20,000 per year and by March 1914 the Navy had taken delivery of 14 airplanes, of which only 10 remained in service.

On July 1, 1914, Naval Aviation was formally recognized with the establishment of an Office of Naval Aeronautics in the Division of Operations under the Secretary of the Navy. This was within two weeks of the official establishment of the Aviation Section of the Signal Corps. Both coincided with the outbreak of war in Europe. The total number of airplanes delivered to both services up to this time was about 43.

The Navy's Curtiss pusher AH-13 of early 1915. Note the dual controls for training and the 80 HP V-8 Curtiss which powered it. (Courtesy of the U.S. Navy)

AERONAUTICAL DEVELOPMENT IN EUROPE

The aero industry was in its infancy when war broke out in Europe in 1914, but the major nations there had taken more interest in flying than had the people of the country in which powered flight was born. As a result, a good deal of progress had been made in Europe and the onset of the war spurred it onward at a rapid pace.

What was now required was new airframe designs and engines of higher power, lighter weight and more reliability to power them.

The design of new airframes was considerably easier and the building of a new design for testing was a far simpler matter than was the case for the engines that powered them. Airframe designs have been known to have been laid out and a sample built in as little as a few days and many were done in a period of a few weeks.

Following the design and testing of the airframe it could be put into production without undue delay. That is not to say this was a simple matter, but it was not a complex operation, did not require complex or precision machinery and could be accomplished by workmen needing only a modest amount of training.

The design of an aircraft engine was quite another matter. It required a far greater amount of time and likely was done by an engineer with a lot more formal education than was the case with the airframe. Building the first test engine or engines was also much more difficult. Precision machinery and skilled operators were required and months might elapse before the first engine was ready to be tested.

Even when that time came, it only marked the beginning of an arduous task which could take one, two, or even three years. This was the phase in which the design was proved or disproved. The weak points were found and strengthened and the design refined sufficiently to allow the engine to run reliably and produce the power for which it was designed.

This task was always a long one. Sometimes the design would prove to have flaws so serious that it could never be successfully put into production, but even if that were not the case, the job of putting a complex, precision aircraft engine into production was a major undertaking requiring large amounts of high precision heavy machinery which had to be operated by highly skilled labor. Both items were usually in short supply during wartime.

In the meanwhile the airframe was probably powered by obsolete but available engines which were already in production. The result was that airframe design invariably outran engine design, at least insofar as availability was concerned.

Since the demand in numbers and new types of aircraft continually escalated, the supply of engines never seemed to catch up. As a result, old engine designs were kept in use and, in many cases, in production even after new designs were available.

The end result was a proliferation of engine types and great difficulty in field maintenance because of the fact that each engine was different and was made of entirely different parts. The stocking of repair parts for so many different engines in numerous locations as well as having trained expert mechanics on each type at these locations was a logistical nightmare.

Between the outbreak of war in 1914 and the entry of the United States in April of 1917 (about three years), both England and France found themselves in the condition described above.

According to the official report *Aircraft Production Facts*, written by Col. G. W. Mixter and Lt. H. H. Emmons and published in January of 1919, "The best information available indicates that Germany entered the war with nearly 1,000 airplanes, France with about 300, and England with about 250."

The manufacturer of airplanes and their engines in France began in earnest in August of 1914, during which month 50 airplanes and 40 engines were produced. This was more airplanes than the combined services of the United States currently had in inventory. By the end of 1914 (five months of production) the total figure was 541 airplanes and 860 engines.

By the first of April 1917 the French had produced 15,482 airplanes and 28,666 engines. Although we do not have the statistics for this specific period, we know that for the entire period of the war the French produced 264 different prototypes of airplanes resulting in 38 different types put in service. W. H. Bradley, European technical correspondent for *Automobile* magazine, noted in an article published in the May 24, 1917 issue of the magazine that forty-six different kinds of aircraft engines were being manufactured in France.

The English, who started aeronautical production about a month later than did the French, had produced by the first of April, 1917, 2,760 airplanes, acquired 650 from France (included in the French production figure above) and 377 from the United States, for a total of 3,787. Their engine production over the same period totaled 12,125. They acquired an additional 3,601 from France (included in French production quoted above) and 542 from the United States for a total of 16,268 engines.

U.S. MILITARY AVIATION 1914-1917

During this same period of time the armed services of the United States had made minimal progress in the air, considering what was going on in Europe and the prospects that we might become involved.

By the declaration of war on Germany on April 6, 1917, the Army had acquired a total of 224 airplanes of all descriptions, none of which were suitable for fighting and probably no more than 200 were still in service. During the same period about 331 engines were purchased, including those in the airplanes and those used as spares. As noted above, builders within the United States had provided the British with 377 airplanes (primarily from Curtiss) and 542 engines (primarily from Curtiss).

The Aviation Section of the Signal Corps had a complement of 52 officers, 1,100 enlisted men, and about 200 civilian employees. By this time the Army had trained 139 men to fly, not all of which were still in service. Of this number only about 26 were considered really qualified pilots and, of course, none had flown anything but training-type airplanes.

Since July of 1914 Navy aeronautic appropriations had increased considerably. For the next two fiscal years they were $1,000,000 and $3,500,000 respectively. With this increased interest in the Naval air service, considerable progress was made toward establishing a useful arm of the Navy, even if it was soon to be found woefully inadequate in times of war. The $3,500,000 fiscal 1917 appropriation bill was signed into law by President Wilson on August 29, 1916 and also included authorization of the establishment of a separate Naval Flying Corps, consisting of 150 officers and 350 enlisted men. Because of opposition from the Office of the Chief of Naval Operations and the eventual declaration of war, the separate Corps was not established.

Those in charge of aviation matters in both the Army and Navy were deeply interested in increasing the aviation capabilities of their respective services and did whatever they could to do so. With little forthcoming in the way of appropriations from Congress, not much could be accomplished beyond planning. An effort in this direction was made by the Army's Chief Signal Officer when in October of 1916 Lt. Col. George O. Squier wrote to *Aviation and Aeronautical Engineering* magazine requesting they publish the following: "There is enclosed, herewith, a list of names of persons who it is understood have flown heavier-than-air machines in the United States. The addresses of few of these persons are known. Can you, through your journal or otherwise, make this list public, and at the same time ask those enumerated to communicate with the Officer-in-Charge of the Aviation Section, Signal Corps, War Department, Washington, D.C.? Their letter should state:

A. Nationality.
B. Age.
C. Type of aeroplane and control they have flown.
D. Whether or not they desire service with the United States Army.

"It is of the utmost importance that the War Department should be able to secure a complete list of American aviators, with their addresses, for its use in case of war or threatened war."

The list included all known flyers, even those whose addresses were already known to the War Department, and was published in the October 1, 1916 issue of the magazine. It included the names of 817 civilian aviators with their Aero Club of America license numbers (if known). The highest Aero Club license number listed was 349. (Glenn Curtiss held No. 1 and Orville Wright No. 4.)

It is interesting that there were 817 persons listed who had flown an airplane when so few airplanes had been produced in this country up to that time. Even though the list was built using any and all names from any source indicating a person who may have flown an airplane, it is still a large one.

By the time war was declared, the Navy had one air base (Pensacola, Florida), 54 training-type planes, three balloons and one blimp. Navy and Marine personnel assigned to aviation duty consisted of 48 officers who were Naval Aviators, 239 aviation-rated enlisted men and a small contingent of civilian personnel.

Thus we see that the total air services of the United States consisted, on April 6, 1917, of 100 officers, 1,339 enlisted men and about 254 training planes.

From the delivery of the first airplane to the Army in 1908 until the declaration of war on April 6, 1917, the total number of American airplanes delivered to the armed services of this country and those abroad had been 661. We can see from the above discussion that of these, 224 (about 200 still in service) had gone to the Army, 60 to the Navy (54 active and about 6 written-off) and 377 to England. The number of civilian-purchased airplanes built in America during this period is not accurately known, but it would be safe to say that a total production of 1,000 airplanes in this country prior to April of 1917 would be a liberal one. Obviously our aeronautical industry at this point in time was minimal.

PACKARD AERO ENGINES

Packard had been working on aero engine designs since as early as September of 1915. In the company board minutes of September 15, 1915, we find the first entry relative to these endeavors: "On motion duly made and seconded, the Management was authorized to continue their experiments with aeroplane motors, and to incur an expense in this connection, including cost of machinery, up to the sum of $20,000." From this point on, Packard continued working on aero engine designs and spending ever increasing sums in that effort. By May of 1917 they had expended some $400,000 in aero engine research.

Their first aero engine was ready for block test in February of 1916. It was a 60° V-12 of 299 cu. in. displacement rated at 100 HP @ 2,300 RPM and 130 HP @ 3,300 RPM. This engine was built to test the design rather than as an engine for flight. It received its principal testing as power for a race car and eventually two engines of this model were completed. Between 1916 and 1919 it was raced very successfully by Ralph DePalma and others (including Jesse G. Vincent). They referred to this engine as the "299" model.

Packard's second aero engine design was much larger and intended to eventually power an airplane. Known initially at Packard as the "905" because of its 905 cu. in. displacement, this too was a V-12 configuration engine, but this time with an included angle of 40°. The bore and stroke were 4" x 6" and the first engine of this design and displacement was tested at the factory as early as December 17, 1916, and publicly on a dynamometer on December 28, 1916. Witnessing the event were a number of Packard executives as well as A. Ludlow Clayton, owner of *Motor Age* Magazine, Captain James V. Martin, aviator and builder of aircraft, and Sidney D. Waldon. Waldon was acting as an advisor to Howard E. Coffin on the Council of National Defense as well as having by then enlisted in the Army Air Service. He was awaiting the granting of a commission as Captain.

Packard's first aero engine, the "299." Notice the use of two blocks of three cylinders on each bank of this V-12 engine. (Courtesy of Jim Dillon)

(Captain Martin had an interesting history worth mentioning here. Martin was born in 1883 and educated at the University of Virginia and Harvard. His initial interests were nautical. He worked for the U.S. Coast and Geodetic Survey in 1903 and by 1912 held a Navigation and Master of Ocean Vessels rating and thus the title of Captain. In addition to those considerable abilities and achievements, he had a passion for flying. He started flying in 1909 and was an instructor by 1911. He built his first airplane in 1911 and continued to design and build airplanes for some years after the end of WW I.)

This engine was capable of producing 248 HP @ 2,000 RPM and weighed about 900 lb. Cylinder banks were made up with two sets of three cylinders. Each cylinder was separately forged and then machined. Pressed steel water jackets were then welded to each set of three cylinders. Valves were overhead, two for each cylinder. They were driven by rockers which were driven by an overhead cam assembly rigidly mounted across the top of each cylinder bank. The camshafts were driven by a series of spur gears and the propeller was driven by a spur type reduction gear with a 1.82 to 1 ratio. The crankshaft was carried by three main bearings of 2 3/4" diameter. Widths were: front, 4 1/2"; center, 3 1/2"; rear, 2 3/4". Force-feed lubrication was used on all main, crankpin and camshaft bearings. Dual ignition was by magneto and both generator and starter were provided.

This engine was subjected to the usual block tests on the dynamometer, but, in addition, a rather unusual rig was set up on a Packard Model E truck in order to be able to mount a propeller on it and illustrate the engine's thrust abilities by propelling the truck. This was winter in

Dynamometer testing of the first Packard "905" engine at the Packard factory on December 28, 1916. From left to right the men are: W. T. Hunt, head of the testing area; William B. Stout, aircraft engineer and at that time heading Packard's aircraft engineering department; Capt. J. V. Martin, aviator and builder of aircraft; C. R. Norton, Packard sales manager; A. Ludlow Clayton, owner of* Motor Age *magazine; G. R. Bury, Packard assistant sales manager; H. H. Hills, Packard assistant general manager; Sidney D. Waldon, engineer and advisor to Howard E. Coffin of the Council of National Defense; Alvan Macauley, Packard president; Jesse G. Vincent, Packard vice president of engineering and the engine's inventor. (Courtesy of NAHC)

The first Packard "905" engine under test on a 5-ton Packard truck in January of 1917. Jesse G. Vincent stands at left, H. D. Church, chief truck engineer, sits in the cab and Capt. L. E. Goodier, Jr. stands at right. Wheels were marked so it could be seen that the truck was moving under power of the propeller while the wheels were locked still. (Courtesy of NAHC)

Detroit and the ground was covered with snow. While the engine was being used to move the truck on the Packard grounds, it was discovered that the engine would continue to move the truck at a rapid pace even with the brakes locked and the rear wheels stopped. Present as observer at these tests was Captain Lewis E. Goodier, Jr. of the Inspection Section of the Army Air Service and holder of Expert Pilot rating and ACOA license No. 200.

A Packard truck (probably this same vehicle) was later used numerous times as a test platform for the Liberty engine in its early development, in both V-8 and V-12 configurations.

At about the time the first model of the Packard "905" engine was ready for tests, Packard hired a chief engineer for their new aeroplane department. William B. Stout was employed for this position beginning December 1, 1916 at $600 per month. Stout had been chief engineer of both the Schumeier Motor Truck Co. and the Scripps-Booth Motor Car Co., had a long interest in aviation and had founded and been the first editor of *Aerial Age* magazine. Stout did not remain with Packard long, however. He resigned in August of 1917 to become a technical advisor to the Aircraft Production Board under Howard E. Coffin. Stout would become well known in the aviation field. He designed and built the first commercial all-metal airplane and founded the Stout All-Metal Airplane Co. in 1923. Its first plane was the Curtiss OX-5-powered Air Sedan and its second the Liberty 12-powered Air Pullman of 1923.

The Stout Metal Airplane Co. was purchased by Henry Ford on July 31, 1925 and became a division of the Ford Motor Co. It built a number of aircraft over the following years, the most famous being the "Ford Tri-motors."

Left to right: The Packard "905-1", "905-2" and "905-3" (1A-905). These three engines were built by Packard from late 1916 to early 1917. (Author's collection)

Stout remained active in aircraft and engineering for the remainder of his life. He was elected president of the SAE in 1935.

The first "905" engine was subjected to much testing, a great deal of which took place under the observation of various government officials. By January of 1917 Packard was already confident enough that this would be developed into a production engine that they displayed it publicly at the January Detroit Auto Show and at the February 8, 1917 Pan-American Aeronautic Exposition in New York. They had an eight-page brochure printed for distribution at those displays. In it they described in some detail the design of the engine but did not picture it. They stated that the engine was still under development and, although no definite production schedule had yet been established, one should be well under way within a short time.

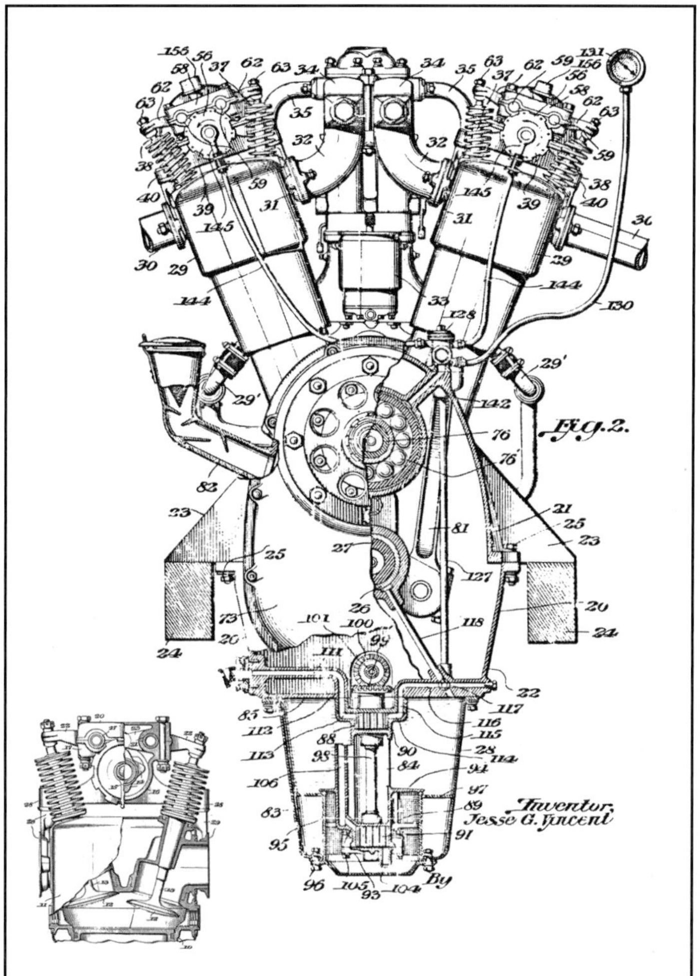

Sectional view of the Packard "905-3" (1A-905) engine of 1917. (Author's collection)

Although I am sure that Packard would liked to have sold the engine on the civilian market, its primary application would be for military use.

The December demonstrations and other observations by government authorities of the engine must have impressed them because it resulted in an order (No. 6202) placed on December 30, 1916, for two engines, one twelve and one six cylinder. The twelve would probably have been an improved version of the first "905" engine and the six a variation of it.

Before these engines could be completed and delivered, war was declared and the complexion of military requirements had changed considerably. The order was then canceled.

Just what the engines would have looked like, we do not know, other than they would have obviously been very much like the first "905" engine. However, Packard was working on an improved version of this engine and built a second example which was slightly different from the first. It was completed in April of 1917. The chances are very good that this engine would have been eventually delivered as the engine ordered on December 30 of 1916 had the order not been canceled.

This engine is referred to by Packard aero historians as the "905-2" engine, the first engine therefore became the "905-1."

While the "905-2" was in the process of build-and-test, a third version was on the design table. Vincent later stated "The third model of this '905' engine was started in April 1917, immediately after the War was declared (April 6, 1917). We were aiming particularly at a reduction of weight and simplification of manufacture, and intended to include improvements as a result of previous experience."

Quoting from the book *Master Motor Builders*, "The '905-3' was completed in May and incorporated a number of major changes from the prior two models. Cylinder blocks were made up as before but contained two instead of three cylinders. The principal advantage gained was to allow the use of four main bearings rather than the former three. The close spacing of the cylinders within a cylinder block leaves no space between rod journals for a main bearing. Sufficient space is only available between two cylinder blocks. Thus the first two models could only have three main bearings, one at each end and one between the cylinder blocks, as was also the case with the '299.' The intake manifold was redesigned and now carried two duplex updraft carburetors instead of one. The gear faces of the reduction gears were increased in width from 2" to 2 1/2" to reduce losses. The camshaft housing support was strengthened by having more supports and having the end supports placed closer to the end of the housing. The oil system was changed from wet to dry sump with external oil cooler and storage tank. The water pump was mounted on the rear of the engine and driven by the first gear in the train which drove the magnetos and camshafts. No generator was provided. The starter was mounted on top of the reduction gear housing and drove a ring gear mounted on top of the propeller hub. Weight was 990 pounds and 285 HP was developed at about 2,000 RPM."

HALL-SCOTT AERO ENGINES

Hall-Scott had been actively in the aircraft engine business since 1910 but the U.S. Military purchased no airplanes powered by their engines until 1915 when the Army purchased four Martin S Seaplanes powered by Hall-Scott A-5 engines and the Navy purchased one.

From that time until the declaration of war both the Army and Navy purchased small lots of training-type airplanes from several different aircraft designer/manufacturers which were powered by Hall-Scott A-5, A-5a, A-7 and A-7a models. The Navy's orders amounted to less than a dozen but those of the Army were somewhat more substantial, amounting to some 150 engines. Of the 150, only about 50 had been delivered by the declaration of war. Along with these orders, Hall-Scott also completed the delivery of a Russian order for 300 A-5 engines during 1916.

All of the Hall-Scott production designs were individual cylinder types using cast iron cylinders mounted on a cast aluminum crankcase split along the center line of the crankshaft main bearings. Cylinders were held down by long studs between lugs on the cylinder head and the upper half of the crankcase. Valves were overhead, driven by rocker arms in turn driven by a cam mounted in a housing which was bolted across the top of the bank of cylinders. The cam was driven by bevel gears on the top of a verticle shaft which was driven from the end of the crankshaft by another set of bevel gears. Lubrication was a wet sump type using a combination of pressure feed for the cam and main bearings and scupper lubrication of the rod journals. Oil pressure was maintained at about 20 lbs.

All of these engine models were rated between 1,250 and 1,450 RPM. Whether this was because of breathing and exhaust restrictions inherent to the design of the engines, or because of the inability of the lubrication system design (low pressures and scupper lubrication of rod journals) to cope with higher speeds and powers, we cannot positively say. We can state, however, that the best specific output obtained by the A-5/A-7 series of engines was .182 HP/cu. in.

This brings us to the last engine designed by Hall-Scott prior to the declaration of war. This engine was, according to a Hall-Scott publication issued after the war, "completed prior to May 1, 1917. Ready for final tests —" and called their A-8 model. It was a 60° V-12 utilizing 12 cylinders of the A-5/A-7 series engine design. Bore and stroke were 5" x 7" giving a displacement of 1649.34 cu. in. It weighed about 1,000 lbs. As can be seen from the illustrations below, the cylinder and valve train design was the same as the earlier engines from which this one was

Hall-Scott A-8 front view. (Courtesy of William Lewis)

Hall-Scott A-8 rear view. (Courtesy of William Lewis)

derived. Carburetion may have been improved and lubrication may have, as well. We can see, however, from the sectional drawing, that they were still using scupper lubrication of the rod journals. This practice was to be tried as well in the Liberty, found to be inadequate and replaced with pressure lubricated rod journals.

As initially announced in 1917 (see page 4) they gave this engine a 300 HP rating, probably at the same RPM as their prior engines, about 1,400.

In their 1919 publication *Pertinent Facts About the Liberty Motor*, Hall-Scott called this a 450-500 HP motor. By this time they would no doubt rate it at the speed at which the Liberty was currently rated or perhaps even at its maximum output. The Liberty at this time was rated at 400 HP @ 1,700 RPM and actually produced 421 HP at that speed and a maximum of 449 HP @ 1,940 RPM. If the A-8 produced 300 HP @ 1,400 RPM and had an average power curve one could expect it to have an output of about 380 HP @ 1,700 and 425 HP @ 1,940 RPM. Perhaps it might have produced the claimed 450-500 HP after a long period of development, but considering its great similarity in design to its predecessors, it does not seem likely they could have done so by 1919. In order to achieve the claimed 450-500 HP one would expect it would be necessary to apply a set of alterations similar to those tried on the Liberty 12 in late 1918 which obtained an output of 526 HP (see page 182). Even if the A-8 had been experimentally modified to obtain an output in this range, that would hardly justify calling it a 450-500 HP engine. Since no test reports or data other than that cited here has ever been uncovered, we must draw our own conclusions as to the actual statistics of the A-8.

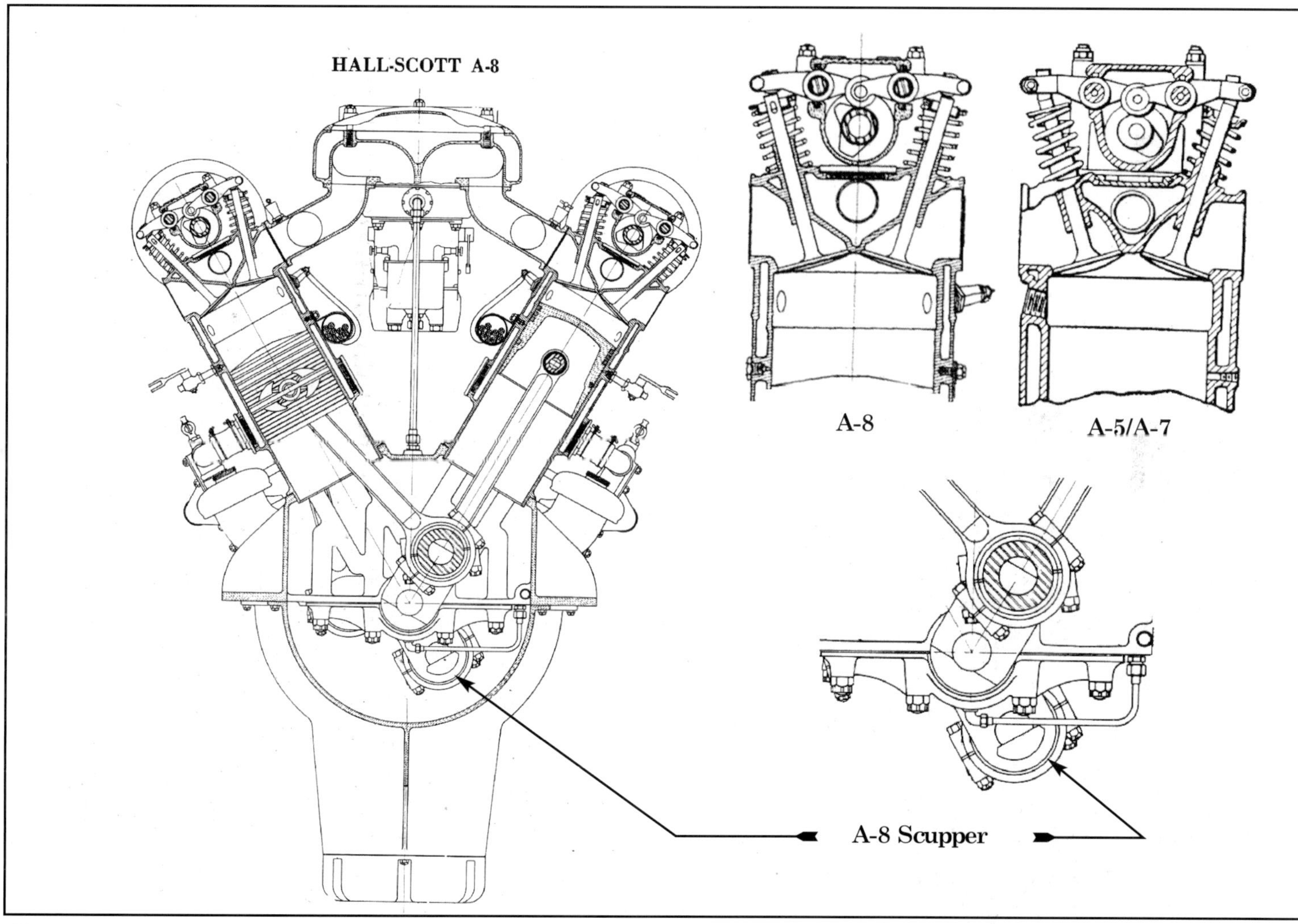

Above left is a sectional drawing of the 1917 Hall-Scott A-8. Above right are sectional drawings of the A-8 combustion chamber and valve mechanism beside the same section of a A-5/A-7 engine. It can easily be seen that the two are of the same design. Below right is an enlarged section of the crankshaft showing the rod journal of the A-8 with its scupper lubrication system. (Author's collection)

WAR AND THE AIRCRAFT PROGRAM

Some preparations had been made for the possibility that we would need a military air capability, but it had been limited to some government organizational changes and a survey of the capacity resulting from overseas orders and small military orders here.

The National Advisory Committee for Aeronautics (NACA) had been created by a rider to the Naval Appropriations bill for fiscal 1916 which passed on March 3, 1915. Army, Navy and civilian aeronautical personnel were assigned to sit on the committee and its objective was to advise government officials regarding all matters related to the development of aeronautics. This committee has endured in some form to this day, having now evolved into the National Air and Space Administration (NASA).

By the middle of 1916 an organization known as the **Council of National Defense** was put together to ascertain the capacity of this country's manufacturing facilities in case of war. The head of this council was Howard E. Coffin, vice president of the Hudson Motor Car Co. One of his assistants was Sidney D. Waldon, formerly vice president of the Packard Motor Car Co. and more recently (1915-1916) employed by the Cadillac Motor Car Co.

As chairman, Coffin directed the work of compiling an industrial index of nearly 30,000 United States manufacturers showing the general types and capacity of production of which each was capable. It was this work which laid the foundation for the rapid expansion of war production that was required following the declaration of war on April 6, 1917.

Coffin and Waldon had an early interest in aeronautics. Waldon was a charter member of the Aero Club of Michigan and later its president. He first flew as a passenger in a Wright plane in 1910 and helped purchase a plane and train the Michigan National Guard in its use in 1916.

The Aircraft Production Board as formed on May 16, 1917. Left to right: A. C. Cable, Sec., R. L. Montgomery, Sidney D. Waldon, Edward A. Deeds, Rear Admiral David W. Taylor, Brigadier General George O. Squier, Howard E. Coffin, chairman. (Courtesy of the Air Force Museum)

The United States officially entered the war on April 6, 1917 and thereafter things began to happen rapidly in the field of aeronautics.

On May 5, 1917 the Secretary of the Navy and Secretary of War created a Joint Technical Board on Aircraft whose purpose it was to determine the specifications and methods of inspection of all aircraft for the two services. The original members of that board were: Major (later Brig. General) Benjamin D. Foulois, Chairman, Captain V. E. Clark, and Captain Edgar S. Gorrell of the Army; and Lieutenant A. K. Atkins, Lieutenant J. H. Towers, and Lieutenant J. C. Hunsaker of the Navy. The official name of the board became the **Army and Navy Joint Technical Board on Aircraft.**

The **Aircraft Production Board** was created on May 16, 1917 by the Council of National Defense on the recommendation of the NACA. As originally formed, this board acted purely in an advisory capacity to recommend to the Army and Navy, in co-operation with the Joint Technical Board on Aircraft, requirements of the armed services for the war effort and how to accomplish the necessary production program. The first members were Howard E. Coffin, chairman; Brig. General George O. Squier, Chief Signal Officer, U.S.A.; Rear Admiral David W. Taylor, Chief Constructor, U.S.N.; Sidney D. Waldon, Edward A. Deeds, and R. L. Montgomery.

These men were chosen because of their positions and qualifications. General Squier was the Chief Signal Officer of the Army and as such was in charge of Army aviation. In addition, he had been in charge of Army aviation as a Lt. Colonel and later Colonel under the former Chief Signal Officer and had been in Europe prior to the war observing military aviation there. Rear Admiral Taylor was Chief Constructor of the Navy and thus responsible for all naval equipment. R. L. Montgomery was senior member of the Philadelphia law firm of Montgomery, Clothier and Taylor and was to serve as financial and business advisor of the board. Sidney D. Waldon was an engineer and long time aviation enthusiast and was already commissioned as Captain in the Aviation Section of the Signal Corps. He had been an official of the Packard Motor Car Company from 1902 -1915, serving as vice president from 1913 to 1915. In 1915 and 1916 he was a Cadillac Motor Car Co. executive. Edward A. Deeds was general manager of the National Cash

Benjamin D. Foulois

1879-1967

Enlisted volunteer engineer 1898-1901. Private to 1st Sergeant.
Second Lieutenant to Captain 1901-1916.
Learned to fly in 1910.
Commanded the First Aero Squadron in Mexico in 1916.
Major Aviation Section Signal Corps 1917.
Chief of Air Service A. E. F. 1917-1918 Brig. General.
Assistant Chief of Air Service 1918-1919.
Reverted to rank of Major and served as Air Attache to Germany 1919-1923. Lt. Col. 1923. Commander of Mitchel Field 1925-27.
From 1928 he served progressively as asst. to the Chief of Air Corps, Chief of Material Division, Commander of Air Corps exercises to Chief of the Air Corps 1931-1935.
Brig. General to Major General.
He retired in December of 1935.

George Owen Squier

1865-1934

Graduated from West Point in 1887 as Lieutenant. Earned his Ph. D. at Johns Hopkins University in 1893. Initially an artillery officer, he switched to the Signal Corps in 1898. 1st Lieutenant 1899, Captain 1901, Major 1903, Lt. Colonel and U.S. Military Attache at London 1912. Colonel and Officer in Charge of the Aviation Section of the Signal Corps 1916, Brigadier General and Chief Signal Officer February 1917. Major General October 1917. He was in charge of the Army Air Service from June 1916 until May 1918.
He remained Chief Signal Officer until his retirement as a Major General in 1924.

John Henry Towers

1885-1955

Graduated from the U.S. Naval Academy 1906.
Naval Aviator No. 3 - 1913. Ensign.
Set up and was executive officer of the first Naval Air Station at Pensacola, Florida - 1914. Lieutenant (j.g.).
Assistant Naval Attache in London - 1914-1916. Lieutenant.
Assistant Director of Naval aviation - 1917-1919. Lieut. Commander - Commander.
Commanded the transatlantic flight of the NC boats and was the flight commander of NC-3. 1919
Various sea duties and commands. 1920-1939 raising to the rank of Captain and becoming asst. Chief of the Bureau of Aeronautics.
Chief of the Bureau of Aeronautics - 1939. Rear Admiral.
Commander, Air Force, Pacific Fleet - 1942. Vice Admiral.
Various Pacific Fleet Task Force Commands - 1943-1945. Commander 5th Fleet - Commander in Chief Pacific Fleet - 1946-1947. Admiral.
Retired December 1947.

Register Co. and President of the Dayton Engineering Laboratories Company (DELCO), was very much interested in aviation, had known the Wrights, who were also from Dayton, and had organized the Dayton Wright Airplane Co. with C. F. Kettering, H. E. Talbott and H. E. Talbott, Jr. Deeds was a trained engineer and an industrialist and well qualified for the job he would soon undertake.

The French sent a military mission to this country which arrived in Washington, D. C. on April 24, 1917. The military section was headed by Marshal Joffre and the diplomatic by former Prime Minister Rene Viviani. The mission consisted of 400 officers and technicians. The aviation section of the mission was made up of 25 persons headed by Major Joseph Tulsane.

The British mission, headed by Lord Northcliffe, arrived in Washington about a month later. The mission's technical head, Wing Commander W. Briggs, chief engineer for the Royal Naval Air Service, was later attached to the Aircraft Board as technical advisor to the Engineering Division, Bureau of Aircraft Production.

On April 3, 1917 (three days prior to the declaration of war) the NACA sent a telegram to fourteen engine manufacturers to determine what manufacturing facilities each had available and asking for estimates of their production capabilities should they receive production contracts. The Companies were:

Ashmuson Manufacturing Co. of Providence, RI
Atwood Aeronautic Co. of Williamsport, PA
Dayton Aero Motors Co. of Dayton, OH
Duesenberg Motor Co. of St. Paul, MN
General Vehicle Co. of Island City, NY
Gyro Motor Co. of Washington, DC
Hall-Scott Motor Car Co. of San Francisco, CA
Packard Motor Car Co. of Detroit, MI
The Roberts Motor Mfg. Co. of Sandusky, OH
Springfield Motor Co. of Springfield, MA
The Sterling Engine Co. of Buffalo, NY
B. F. Sturtevant Co. of Hyde Park, MA
Van Blerck Motor Co. of Monroe, MI
Wisconsin Motor Co. of Milwaukee, WI

Interestingly enough, two of the major builders of aircraft engines in this country, the Curtiss Airplane and Motor Co. and the Wright-Martin Co. were not among those cabled. Probably because they were already so deeply involved in government aero business that their capabilities were well-known.

Packard responded not only with potential production information but also that they had "perfected a 225 HP motor" and "would have a standardized motor ready to deliver May 15. Want definite order so as to start

Chronology 1914-1917

June 1914 - War starts in Europe.

July 1914 - U.S. Army and Navy aviation organized. Total aircraft = 43.

December 1914 - French and English together have about 797 airplanes.

February 1915 - Waldon leaves Packard.

February 1915 - Vincent moves from chief engineer to vice president of engineering of Packard.

March 1915 - NACA created.

February 1916 - Packard completes their first aero engine, the "299."

June 1916 - Council of National Defense created. Waldon a member.

December 1916 - Packard completes the "905-1."

December 30, 1916 - Army contracts for two Packard aero engines.

February 19, 1917 - Waldon commissioned Captain.

April 1917 - Packard completes the "905-2."

April 6, 1917 - U.S. enters WW I. The Army and Navy have 254 airplanes between them.

April 6, 1917 - French and English together have built or acquired about 18,784 airplanes since the beginning of the war.

April 7, 1917 - U.S. mission to Europe is considered with Vincent as member. It is canceled when word of the pending French and English missions to this country is received.

April 11, 1917 - Army contracts for two V-12 Packard aero engines for the J. V. Martin "Cruising Tractor."

April 24, 1917 - French military mission arrives.

April 26, 1917 - Vincent tells Waldon of new design features to be in "905-3."

May 1917 - Hall-Scott completes the A-8.

May 1917 - Packard completes the "905-3."

May 5, 1917 - Army - Navy Joint Technical Board created.

May 16, 1917 - Aircraft Production Board created. Capt. Waldon a member.

May 16, 1917 - Vincent wires Waldon of "905-3" (1A-905) test results.

May 23, 1917 - E. J. Hall met with Waldon, Montgomery and Deeds regarding order for 1,000 A-7a engines.

May 24, 1917 - Cablegram received from French Premier Ribot with French aeronautical program request.

May 27, 1917 - Huge new aircraft production program proposed by the Joint Army-Navy Technical Board and approved by the Secretaries of War and Navy.

May 28, 1917 - Conference with Vincent, Coffin, Deeds, Waldon and Chapin regarding engines.

May 29, 1917 - Hall joins conference on engines and Liberty plans are started.

June 4, 1917 - Finished Liberty drawings shown to joint boards.

June 6, 1917 - Packard loans Vincent to government for three months.

June 17, 1917 - Bolling Commission sails for Europe with Major Bolling as its head.

July 3, 1917 - First Liberty 8 arrives at Bureau of Standards.

August 15, 1917 - Vincent commissioned Major.

August 24, 1917 - Deeds commissioned Colonel.

August 26, 1917 - Waldon commissioned Colonel.

October 1917 - Hall commissioned Major.

continuous production." Packard had designed this engine with the idea that it would serve the United States as a military aircraft engine and obviously hoped to be able to build it in quantity if the need arose. By April 3 it was becoming painfully obvious that the need was imminent.

During this period Sidney Waldon was, under the direction of Howard Coffin, handling most matters concerned with preparations for the production of military aeronautical items in this country. He had received a commission as Captain in the Signal Corps Air Service on February 19, 1917.

Capt. Waldon received a letter dated April 4 from W. H. Bradley, foreign technical correspondent for *Automobile* Magazine, who was at that time in France, but who had been in England as well, covering the war. Bradley was a "motor expert" and was attempting to give advice to Waldon, as current head of the American effort to mobilize an air arm, as to the current needs for aircraft motors in Europe as he saw it. Most of it dealt with the prospects of American manufacturers building French-designed engines. Waldon answered the letter on May 2.

Bradley replied to Waldon in a letter dated May 19. In it he went to some lengths to explain just what powers had been considered satisfactory and were available from the start of the war in 1914 up to the present and what would seem to be required in the future. He went on to describe just what types of engines were in production there and which seemed to have the most promise. He finished by giving his recommendation as to what he felt were promising new French designs which might be taken up by factories here as a military engine for series production.

Regarding power, he stated that 80 HP had been average in 1914 but that had risen currently to 130 HP. The largest engine in series production was rated at about 250 HP and within a year 400 HP engines would be required.

He described the rotary type as useful where powers beyond 100 HP are not required but noted that this type was not suitable for development to higher powers and that water-cooled designs were starting to replace them for use in the smaller aircraft in which they predominated.

From this point he described at length the general types of water-cooled engines which were successful and showed any promise as the foundation of the future higher powered engines required. First he described the "Mercedes" type. This is the individual (or dual) cylinder type with welded-on sheet metal water jackets and overhead valves operated by a camshaft mounted above the cylinder banks. He noted the type as being produced by Mercedes, Dietrich, Fiat and others. He accessed it as being somewhat heavy as well as being noted for problems with rocker arm design and water jacket leaks.

The second design discussed was called the "Peugeot" type. This design used cast iron monobloc cylinders and was now considered too heavy as well as having several other noted problems. He stated that experiments had been carried out to cast the cylinder blocks of aluminum and use direct operated valves, but with little success.

The third type described was the "Hispano-Suiza" design. This, he said, was the preferred type and that of which the largest numbers would be built in the next year. He noted that the 150 HP model was very successful but the 200 HP one "has proved very defective." (This engine was discussed in some detail on pages 7 and 8.)

This was the general design Bradley felt held the most promise but his suggestion was not to produce an enlarged Hispano-Suiza but one of two new monobloc designs which would eliminate the difficulties with it. The first engine he proposed as his "Proposition A" and was a monobloc Y-12 four valve French design with reduction gears which was to produce the required 400 HP. It is believed that this engine was the French Abadal Y-12 which was produced experimentally during the war but never put into production, although Bradley claimed the French government had great interest in it and was urging its designer to proceed with its development with all available speed. It had a bore and stroke of 4.72" x 5.51" resulting in a displacement of 1,156.92 cu. in. As eventually built it was rated at 350 HP @ 2,500 RPM and 400 HP

Bradley's "Proposition A", the experimental 1,156.92 cu. in. 350 HP Y-12 Abadal of French design. (Author's collection)

was claimed (but doubted by prominent engineers of the period). Its weight of 705 lbs. resulted in a very creditable ratio of 2.01 lb./HP.

The second proposed design was "Proposition G," which was a monobloc V-8 of the same general design as the Hispano-Suiza but with that engine's faults supposedly eliminated. This Maurice Sizaire design was also supposed to be under serious consideration by the French government. It was similar in many respects to the Hispano-Suiza but with different dimensions plus a rather unusual valve arrangement. It was a 90° V-8 with a bore and stroke of 5.12" x 6.73" for a displacement of 1,107.63 cu. in. The valve design incorporated a single very large intake valve driven directly by its cam lobe. The exhaust valve was a reciprocating sleeve whose motion was imparted by two small cranks on the overhead camshaft. It also had auxiliary exhaust ports formed near the bottom of the cylinder similar to a two-stroke engine. The engine was rated at 350 HP @ 2,000 RPM, which was 50 HP short of what was to have been Bradley's goal. A reduction gear was provided to allow a normal propeller speed of 1,230 RPM at rated engine speed. The engine was designed to allow the mounting of a 37mm gun to be fired through the hollow propeller shaft. Weight was claimed to be 550 lbs. which, if true, gave it a rather astounding power-to-weight ratio of 1.57 lb./HP.

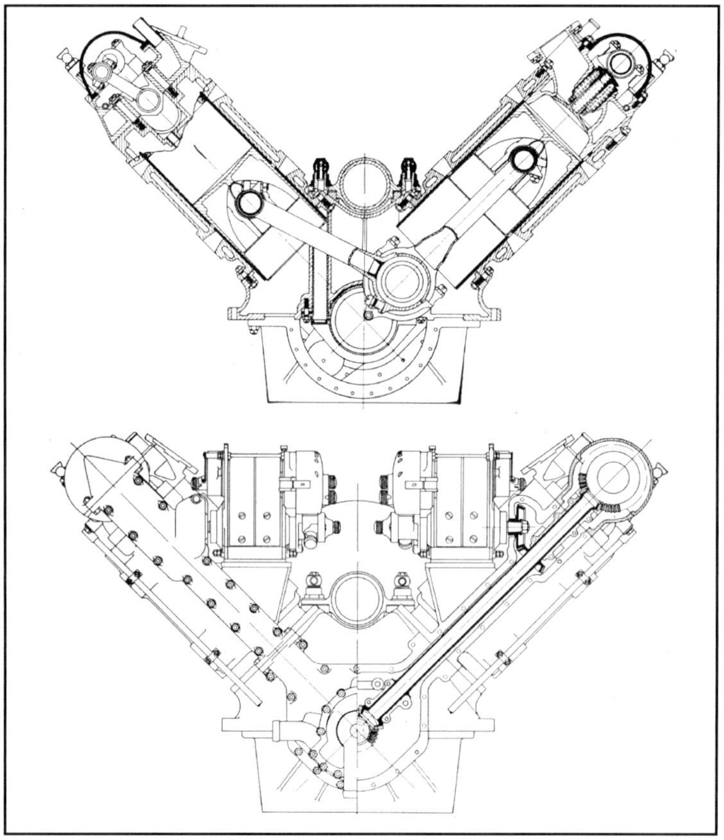

Bradley's "Proposition G", the experimental French Sizaire 90° V-8 of 1,107.63 cu. in. and 350 HP. (Author's collection)

Neither of these suggested designs were pursued by Waldon and evidently the French government's interest waned as well, as neither was ever put into production.

On April 7 Waldon suggests that the government send a mission to Europe to study the motor and plane situation. He suggested sending Mr. Zimmershied, head of the Washington branch of the SAE, Vincent, Mr. Dunham and Capt. Edgar S. Gorrell, an Air Service officer who had graduated from the Military Academy and gone on to earn a master of aeronautical engineering degree at M. I. T. Packard granted Vincent a three-month leave of absence for that purpose on April 12.

Vincent kept Waldon constantly updated as to the progress being made on Packard's "905" engines. On April 10 he wired that the engine had developed 242 HP on tests and 250 HP would be easy to accomplish. On April 26 he wired that he wanted to meet with Waldon, Major Henry Souther, officer in charge of the aircraft engineering and soon-to-be OIC of the Aircraft Engineering Division, Capt. Virginius E. Clark, pilot and chief aeronautical engineer for the Army and Charles B. King, who was in charge of engine design. In that wire he said that he had decided to use dry sump and a four-bearing crankshaft in the "905" in order to save weight, improve lubrication, lengthen life and make cylinder blocks interchangeable in 8 and 12 cylinder designs. Although he did not mention it, and may not have been aware of it at the time, this same dual-cylinder four-bearing layout was being developed in France by Lorraine-Dietrich.

On May 2 Capt. Waldon wrote General Squier that a month ago he was considering sending a mission to Europe but the matter was dropped on account of the coming of the French and English missions. Since that time they had had many conferences with mission representatives but all their information is second-hand. He suggests that Capt. Clark and Capt. Gorrell go abroad to collect information. This resulted about a month later in the sending of the Bolling Commission to Europe to study the aircraft situation.

On May 5 the **Joint Army-Navy Technical Board** was created with Capt. Clark and Capt. Gorrell and others as members. (See page 25.)

On May 7 Waldon wrote General Squier that it may be necessary to commandeer drawings of French engines, place them in factories where we want them built and refer the French companies owning the rights to the Court of Claims for compensation, because of the absurdly high prices being asked for production rights.

A few examples of the royalties asked per 1,000 engines built by owners of the rights to various engine design's are:

Clerget engines	$700,000
Gnome engines	$1,215,000
Sunbeam (including cost of engine and royalty)	
200 HP-type	$7,000,000
300 HP-type	$8,000,000
Le Rhone	$1,375,000

A little arithmetic tells us that this ranges from $700 to $1,375 per engine in royalties on those engines where royalties alone were quoted. In the case of the Sunbeam engines we cannot separate the royalty from the cost of the engine since only quotes for complete engines were given. In that case the 200 HP engine would cost $7,000 each, including royalty, and the 300 HP one would be $8,000 with royalty. However, we can get some idea of the charges by the actual costs of the Liberty engine produced a few months later in this country.

Engine costs were, for the most part, based upon horsepower rating. The higher the power, the higher the cost. The 400 HP Liberty in series production was initially priced at $5,625 and rapidly fell to $4,000 with, of course, no royalty paid.

At this point there had been no decision as to just what aircraft or engines we should build in this country or acquire from our allies in Europe but, as Col. Waldon would later state, "Agents for the various European manufacturers swarmed through the offices of the Aircraft Board. Each of them could produce unanswerable demonstrations of the superior excellence of his article, but it soon developed that practically all of them called for the payment of very considerable royalties."

For about six months there had existed a pseudo Aircraft Production Board and Capt. Waldon had pretty much been "it."

Now, on May 16, some five weeks after the declaration of war, a formal Aircraft Production Board was formed. (See page 25.) Its members consisted of chairman Coffin, who was also chairman of the Council of National Defense, vice chairman Waldon, who had handled the majority of matters related to the subject up to this time and was thus currently the best informed on the situation, Rear Admiral David W. Taylor, chief constructor of the Navy and responsible for all Navy ships and aircraft, Brig. General George O. Squier, chief signal officer and responsible for the Army Air Service, Edward A. Deeds, an engineer and industrialist with all the qualifications to deal with the overall control and coordination of production and distribution of aircraft materiel for the entire

General Aeronautical Organization as of April 6, 1917

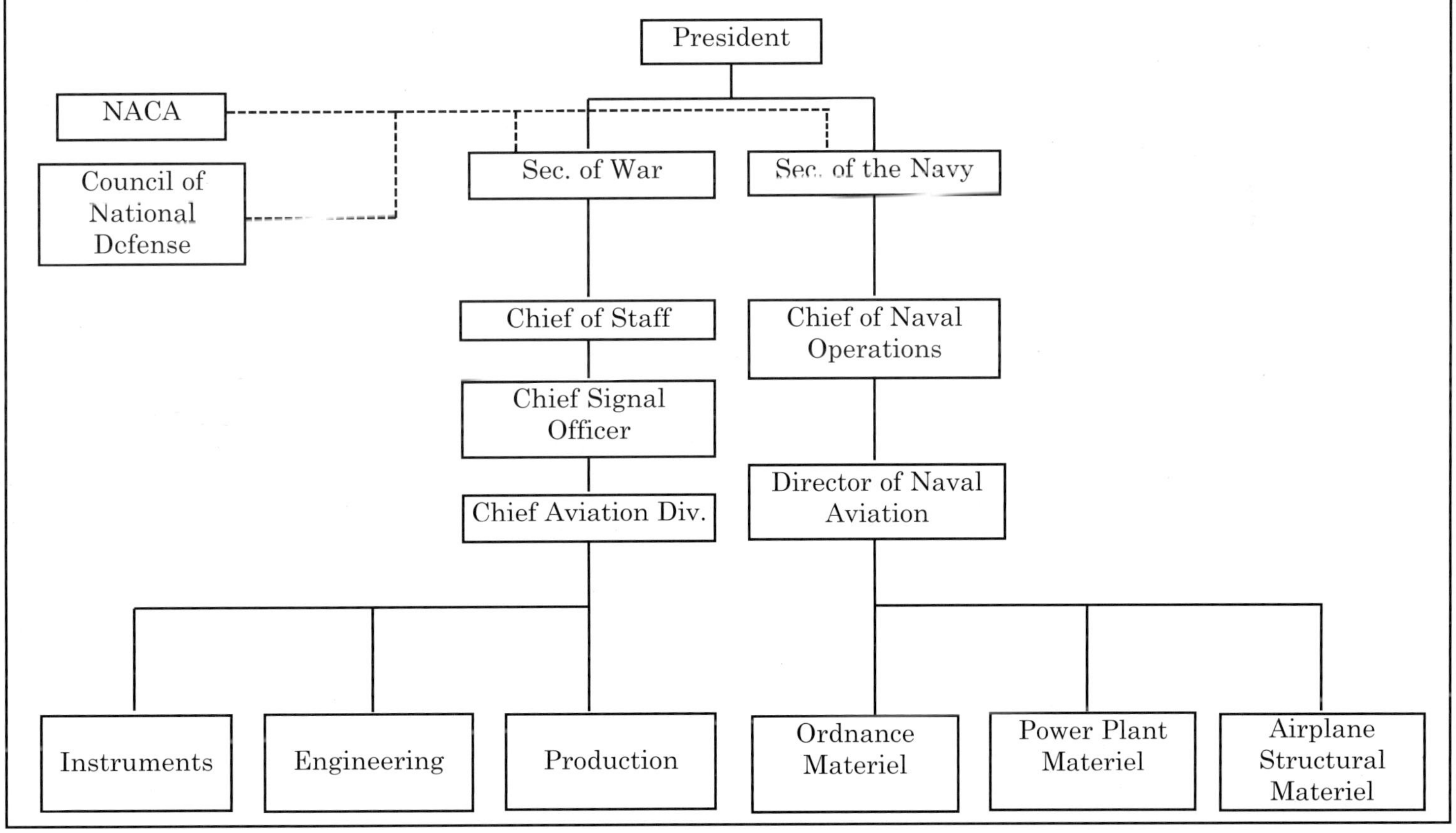

American war effort and R. L. Montgomery, a prominent and experienced lawyer and banker who could handle the legal and financial complications likely to come up in the execution of such large and complex contracts as were soon to be let.

At the first meeting of this board, held in the office of General Squier, it was resolved that Mr. Coffin should organize a commission to proceed at once to England and France to study the aircraft and engine situation. In addition it was resolved to place an order for 1,000 Hispano-Suiza engines with Wright-Martin and 1,000 A-7a engines with Hall-Scott with prices subject to negotiation.

On May 18 Waldon received from Vincent more information on the tests of the Packard "905-3" engine along with a data book on the engine, which Packard had now named the 1A-905 model. This was a system Packard would continue to use for model designations of non-automotive engines for the remainder of its existence. (1=first design of a series for a particular engine; A=aero, an M was used for marine engines; 905=displacement in cubic inches.) Also on this date the Aircraft Production Board (APB) resolved that specifications and plans be obtained for the Clerget 130 HP motor so that it may be produced in this country and that the Curtiss OX-5 and Hall-Scott A-7a be put into production as training engines.

Ordering of the Hispano-Suiza, OX-5 and A-7a engines was the topic of discussion numerous times following the above mentioned meetings. Resulting action involved, of course, much planning and, in some cases, advances of funds to build or equip factories for production. In the case of these three engines, production contracts were signed with Nordyke & Marmon for the production of 1,000 Hall-Scott A-7a engines on June 21, 1917 (order No. 8440); with Curtiss Airplane & Engine Co. for 600 Curtiss OX-5 engines on June 30, 1917 (order No. 8435); with Wright-Martin for 500 Hispano-Suiza Model A engines on July 14, 1917 (order No. 8807); and with Hall-Scott for 750 Hall-Scott A-7a engines on July 26, 1917 (order No. 8981) and an additional 1,000 on December 27, 1917 (order No. 30127).

On May 23 a conference was held with E. J. Hall, R. L. Montgomery, S. D. Waldon and E. A. Deeds in attendance to discuss the potential order for 1,000 A-7a engines. While in Washington, Hall attended other meetings regarding various Naval orders for aircraft with Hall-Scott engines, and on the morning of May 28 boarded a train en route to his home and factory in California.

Also on May 23 the initial production program to equip the Navy with the aircraft necessary for war was recommended by the Joint Army-Navy Technical board. It was to consist of 300 school machines, 200 service seaplanes, 100 speed scouts and 100 large seaplanes. The N-9 and R-6 were noted as most satisfactory for school and service seaplanes but no designs were developed sufficiently to allow recommendations for any other types.

On May 24 a cablegram was received in Washington from French Premier Ribot in which he recommended that America should plan to send to the French front by the spring of 1918 4,500 airplanes, 5,000 pilots and 50,000 mechanics, with the necessary accompaniment of support personnel and materiel. It further stated that to attain the goals they felt necessary, this country would have to plan to build 2,000 airplanes and 4,000 engines per month. By the end of the first six months of 1918, 16,500 service airplanes and 30,000 engines would be needed. The cablegram was referred to the Joint Army-Navy Technical Board for consideration.

Also on May 24 the Aircraft Engineering Division of the Signal Corps was created. Major Henry Souther, who had recently been ordered to active duty from the Signal Officer's Reserve Corps, was designated as Officer in Charge of the division. Capt. V. E. Clark was in charge of airplane design and Charles B. King was in charge of engine design.

On May 26 the French and English aeronautical missions visited Detroit and spent the day going through the various automobile factories. Among them was the Packard factory. When they arrived there, Vincent showed them through the Aircraft Engine Department and explained the work Packard had been doing in the past two years. He also showed them their latest engine, the 1A-905, and described its specifications. Their response

Jesse G. Vincent
1880-1962

Self-educated mechanical engineer.
Completed I. C. S. course in mechanical engineering in 1902.
Superintendent of Inventions at Burroughs 1903-1910.
Chief Engineer of Hudson Motor Car Co. 1910-1912.
Chief Engineer of Packard Motor Car Co. 1912-1913.
Vice President of Engineering PMCC 1913-1917.
Major and Lt. Col. U.S. Army 1917-1918.
Promoted to Colonel Officers Reserve Corps 1919.
Vice President of Engineering PMCC 1919-1948.
Executive Vice President PMCC 1949-1950.
Retired in 1951.
Served on the PMCC Board of Directors 1946-1954.
Learned to fly in the Army in 1918 and was an active pilot until 1945, owning at least a half dozen airplanes.
A very prolific engineer, he held 42 patents by 1917 and 202 by the time he retired.
He was also a boat racer of considerable note, having won the Gold Cup in 1922 and 1923 and the Detroit Sweepstakes races of 1925.

was that although it was a fine engine that appeared to be well-designed and made, its power of 285 HP and power-to-weight ratio of just under 3 1/2 lb./HP would prevent it from being of value in a fighting airplane. A study of the chart on page 14 showing the characteristics of those European engines of the same general power range verifies this in that it would at best be competitive with like power engines then in production and use. It would not make sense to bring a new engine into production that would add yet another design to develop that would not make a significant advance in either power or lightness.

That evening a number of Detroit business men hosted a dinner at the Detroit Club for the members of the commissions and Vincent was invited to be present. After dinner Vincent had a long discussion with members of the commission in which they gave him many valuable points with respect to the latest developments in Europe.

Vincent later noted, "I did not sleep very much that night because I realized that there was not a single concern in the United States working on aircraft engines that would be of any real value on the fighting line." In making such an observation, he was including Packard.

The next morning, May 27, Vincent picked up the May 24 issue of the *Automobile* and while thumbing through it, came upon an article on the state of military aeronautics in Europe. This article was written by the same Mr. Bradley with whom Waldon had been corresponding regarding which engine it would be wise for the United States to produce for the war effort.

Elbert John Hall
1882-1955
Self-educated mechanical engineer.
Completed I. C. S. course in steam engineering in about 1900.
Gained training and experience in engine design with the Henshaw-Buckley Co. and the Burton Co. of San Francisco 1900-1905.
Designed autos for the Heine-Velox Co. of San Francisco 1905-1906.
Designer-manager of the Comet Automobile Co. 1907.
With Bert C. Scott, formed the Hall-Scott Motor Car Co. in 1910 to build industrial locomotives, interurban car bodies, automobile engines and aviation engines.
Became vice president and chief engineer.
Major and Lt. Col. U.S. Army 1917-1918.
Returned to Hall-Scott in former positions 1919.
Sold Hall-Scott MCC to American Car & Foundry in 1926 and remained as VP until 1930.
Joined with Norman de Vaux to build De Vaux autos 1930-31. He worked as a design and advisory engineer from 1932 until 1940.

In the article Bradley pointed out the fact that there were about thirty-seven different kinds of engines being manufactured in England and forty-six in France. He further stated that this multitude of designs not only resulted in high price and low production but made field maintenance extremely difficult due to the requirement of spare parts for so many engine types.

Vincent then laid down the magazine, got into his car and drove to the home of Packard's president, Alvan Macauley. He had come to the conclusion that the way to make sure that this country did not put itself into the same situation was to design and build a standardized line of aircraft engines. When he recorded his thoughts on the subject in his *History of the Development of the U.S.A. Standardized Aircraft Engines* he did not go into any detail as to just how he visualized these engines. About all we know of his prior thoughts on the subject was his mention in his letter to Waldon of April 10 in which he indicated he was designing the latest "905" engine so that its cylinder blocks could be used on both V-8 and V-12 models. Although not mentioned, they could, of course, also have been used for four and six-cylinder engines.

When he reached Macauley's home he discussed the matter with him. After a while Macauley saw the importance of the argument for a standardized line of engines and suggested they send a long telegram to Mr. Coffin in Washington telling him what they felt should be done. Vincent suggested that it would be very difficult to get such a point across via that method and that it would be better if he went to Washington and explained it in person. He then pointed out to Macauley that if he could get to the train station within twenty minutes he could catch the noon train from Detroit to Washington. Macauley got him to the station on time and he left on Sunday, May 27 at noon.

Also on May 27 the Joint Army-Navy Technical Board approved the suggestion of the French premier as put forth in his cable. The board stated that in order to accomplished the program this country would have to construct approximately 12,000 service airplanes and 24,000 service engines plus approximately 10,000 training airplanes and approximately 21,000 training engines; the total was exactly 22,625 airplanes and 45,250 engines. No particular date was set for the attainment of these goals. The report was approved by the Secretaries of War and Navy on May 27, 1917.

Vincent found Mr. Roy D. Chapin, president and one of the founders (with Howard E. Coffin) of the Hudson Motor Car Co., on the train. The two of them discussed the general situation in Washington and the aircraft situation in particular.

Vincent and Chapin arrived in Washington during the morning of Monday, May 28 and immediately went to the

HISTORY OF THE DEVELOPMENT
OF THE
U.S.A. STANDARDIZED AIRCRAFT
ENGINES.
BY
J. G. VINCENT.

Title page of Vincent's 1917 history of the Liberty. (Courtesy of the Air Force Museum)

Lafayette Hotel and had a conference with Coffin, Deeds and Waldon. Vincent explained his theory regarding the design and production of a standardized series of aircraft engines. He pointed out that the Packard 1A-905 was too heavy as designed to meet the current military needs since it had been designed some two years prior, when power requirements were not as high, power-to-weight ratios of under 4 lb./HP were considered adequate, and the higher reliability gained by higher weight was considered more important. He noted that the weight of the engine could be easily reduced if a better power-to-weight ratio was now considered more important than a greater safety factor.

It seems evident, upon looking back at the incident with current knowledge, that it did not take long for Vincent to realize that to redesign the 1A-905 engine to produce what was now considered the desired output power would require such extensive changes that they might seriously consider starting from scratch. Not only were the power requirements much higher than those Packard had been working toward, but they were considering a much wider range of engine sizes emerging from a single basic design.

After some considerable discussion between all present, Deeds suggested that Vincent and Hall should get together and consider the aircraft engine requirements of the country and make a joint report. Hall had left by train that morning for a return trip to California. They wired Hall in Cleveland and had him catch the next train back to Washington. Although it is not recorded just how they caught up with Hall in Cleveland, he was probably scheduled to stop at the Park Drop Forge Co. of that city and discuss business, as they were the producers of Hall-Scott engine crankshafts and shortly would make those used in the prototype Liberty engines.

We can see by the sequence of events that with the aircraft production program proposal set by the Joint Army-Navy Technical Board and approved by the Secretaries of War and Navy, there was suddenly a tremendous multiplication of pressure on the Aircraft Production Board.

According to the *History of the Liberty Engine* submitted on June 6, 1918, by Lieutenants Porter, Benet and Kent, at a meeting in the early morning of May 29 in the office of the Chief Signal Officer, General George Squier, it was agreed to concentrate on 4, 6, 8, and 12 cylinder engines with standard cylinders. Hall arrived back in Washington later in the morning of that day and he and Vincent got together in Deeds' suite of rooms at the New Willard Hotel in the early afternoon to begin plans for this new series of engines.

There has always been some controversy among aero historians as to just where the idea of producing a standardized engine or a series of standardized engines originated. We will never know for sure, but certain facts can be presented and conclusions drawn. We have four sources of information that make any reference to where the idea of a line of standardized engines may have originated.

The **first source** is the writings of Jesse G. Vincent. Vincent was a prolific writer and he wrote much during and about the First World War, most of which was regarding the Liberty engine, the Air Service and his involvement in both. His earliest writings are found in a document he titled *History of the Development of the U.S.A. Standardized Aircraft Engines.* He probably started keeping records about his involvement in the Liberty engine design from the first (May 26, 1917). He prefaces this document as follows: "As a matter of general interest, I am going to attempt to clearly record the history of the development of the U.S.A. standardized aircraft engines from time to time, as this very interesting and important development is carried forward. I will attempt to cover only the important steps in the development and will make every possible effort to be absolutely accurate in my statements." It appears that he must have made the decision to do this sometime in early June of 1917. He notes that the first chapter was dictated on July 7, 1917. From that point on it appears that new chapters or portions thereof are dictated every two to four weeks. This manuscript contains 103 double-spaced typewritten pages and covers the period from May 26, 1917 to about February 6, 1918. This is followed by a 23-page single-spaced letter from Vincent to Lt. H. H. Emmons dated December 4, 1918 outlining (at Emmons request) "why the Liberty Motor was developed and what has been accomplished by its development." The document survives in the collection of the Air Force Museum at Wright-Patterson Field.

In this manuscript he states that he presented Packard's president, Alvan Macauley, his argument for "a standardized line of aircraft engines" for use by the armed services of this country. Later he told Chapin, Coffin, Waldon and Deeds that if the automotive companies of this

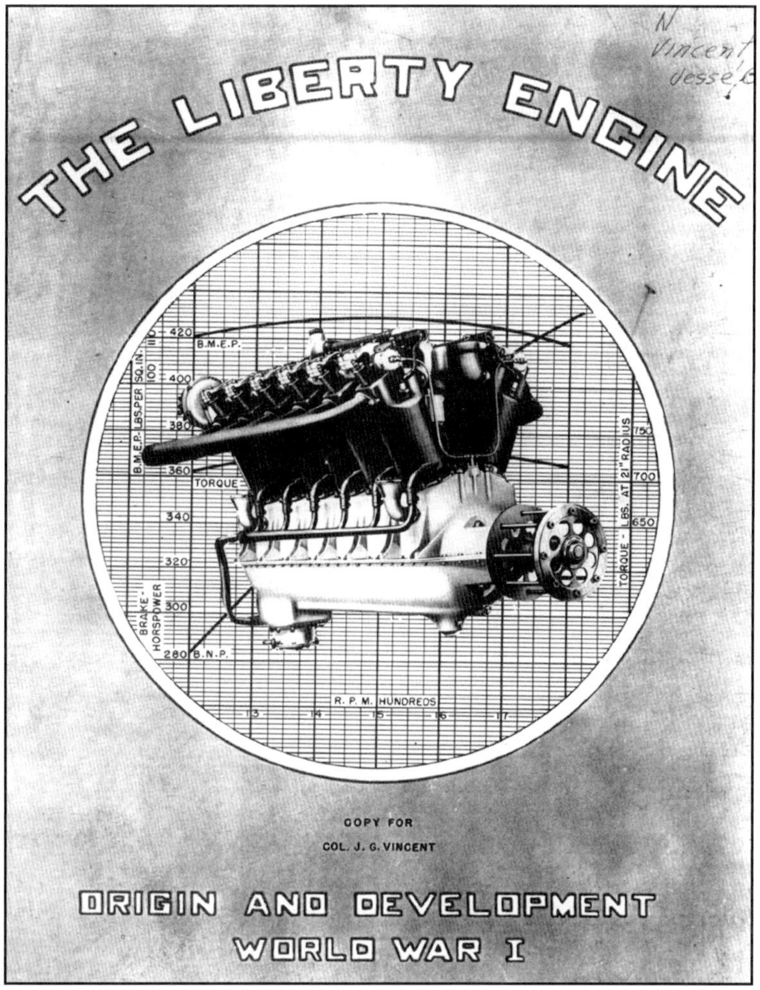

Title page of Vincent's 1945 history of the Liberty. (Courtesy of the Air Force Museum)

country were to produce engines in large numbers, they must be furnished with "proper standardized designs." After this point E. J. Hall is contacted to return to Washington and the two men start on a series of designs.

Vincent kept diaries during most of his active career. The earliest known (although he very likely kept them earlier) start on January 29, 1918. This series continues to December 4, 1918 and thus picks up at about the time the above-mentioned document ends. This diary series is also in the Air Force Museum collection.

Vincent wrote another and more complete coverage of his involvement with the Liberty engine which was published in a very small quantity by the Engineering Division of the Packard Motor Car Co. in November of 1945.

The title of this document was *The Liberty Engine Origin and Development World War I.* It is much more detailed than is the *History of the Development of the U.S.A. Standardized Aircraft Engines* but the early section of it is closely based upon the first document. This second document starts at the same date as did the first history, but continues until Vincent's discharge from the Army in December of 1918. His statements regarding standardized engines are almost exactly the same as those made in the first document. It went on to cover his career and the Liberty in some detail up to the end of the war. It contains 195 pages.

In a presentation at the annual meeting of the SAE on February 4-6, 1919, titled *The Liberty Aircraft Engine*, Vincent made exactly the same statements regarding "standardized engines" as are quoted above.

The **second source** is a document titled *History of the Liberty Engine* written by 1st Lt. Harold E. Porter, 2nd Lt. William R. Benet and 2nd Lt. Warner W. Kent of the Army Signal Corps Personnel Department at the direction of the Bureau of Aircraft Production. This request was probably made in early May of 1918 and was completed on June 6, 1918. The opening statement of the document was "The within History of the Liberty Motor has been compiled from original sources and is believed to be as comprehensive and accurate as the circumstances of its creation permitted."

This history contained 157 pages and documented many facets of the production of the Liberty. Probably the most important single feature was 32 pages which contained a section titled "Liberty Motor Chronology." It contained dated extracts from various documents ranging from letters to minutes of meetings, from a sentence to a paragraph, of what the writers considered important decisions regarding the engine. The first was dated April 3, 1917 and the last May 18, 1918. The last page was a tabulation of engine deliveries of Liberties and all other types during April 1918, May 1918 and total deliveries up to May 31, 1918.

Between April 3 and May 30, 1917 there are seventeen entries. The first entry with any indication of standardizing engine designs is one of April 26 from Vincent to Waldon where Vincent states in a wire that he has decided on some design improvements in the Packard "905" engine. He "decided on 4 bearing crankshaft in 12 cylinder, in order to save weight, improve lubrication, lengthen life and make cylinder blocks interchangeable - 8 & 12."

On May 28 "Conference Coffin, Deeds, Waldon, Vincent & Chapin" is noted.

On May 29 "Meeting O.C.S.O. on standardizing motor. Agreed concentrate on a few types of engines with standard cylinders, 4, 6, 8, 12. Hall joins conference. Plans begin." O.C.S.O. means Office of Chief Signal Officer - General Squier. This is the first time there is a mention of "standardization." This was after Vincent had presented his plan to all the persons noted above as being at the May 28 conference. All seemed to have agreed on the plan on May 29.

The **third source** is the eight-volume *History of the Bureau of Aircraft Production* written sometime after the end of the war and compiled from innumerable sources, and includes written commentaries from civilians and officers who were involved during the war. On pages 1891-1892 of volume 7 it indicates that Deeds had made the decision and discussed it with Waldon before Vincent came to Washington. Just what was the source of the information gathered some two years after the fact, we cannot say.

It should also be noted that the document previously noted as the first source was reproduced in its entirety in the *History of the Bureau of Aircraft Production* beginning on page 2040. Thus this document contains both views.

The **fourth source** is the biography of Edward A. Deeds titled *Colonel Deeds Industrial Builder* written by Isaac F. Marcosson and published in 1947. On pages 234 through 238 Marcosson goes through several pages of eloquent dissertation leading up to why Deeds thought up the idea of having a standardized engine designed and called Vincent and Hall to Washington to do the job. It is certainly possible that Deeds could have thought of the idea as he had enough information at hand to have come to the same conclusion as had Vincent. His engineering expertise was not in the line of engine design but he was certainly capable of coming to such a conclusion. In the case of Marcosson's dissertation, however, there are many holes in his story. He indicates Deeds called Vincent to Washington to design the engine for which he had the idea and does not mention the May 28 meeting at all. This is definitely not the case as is well documented. Vincent came to Washington to present the idea of a standardized engine to Coffin and did so on May 28 whereas Marcosson indicates Vincent's first meeting with him was on May 29 and that he came to Washington at the request of Deeds. It is also well documented that Vincent went to Washington on his own volition, not at the request of Deeds. Marcosson also mentions a number of other things that supposedly happened earlier in May, some of which one would think would have been mentioned in the first Bureau of Aircraft Production *History of the Liberty Engine*, but none were. All indications in that document were that the only thing being considered up to May 28 was whose already-existing engine to put into production and how to do it. There was no indication that anyone within the organization was suggesting having a new engine designed.

It would certainly appear that the decision was finally made at the meetings of May 28, with Coffin, Deeds, Waldon, Vincent and Chapin present. At that point Hall was asked to return. On May 29, with probably the same group of men, plus General Squier, present, the decision was made to make any proposed design include 4, 6, 8 and 12-cylinder engines using the same cylinders.

Leaving the reader with this information to consider, I will go on to the story of the actual designing of the U.S.A. Standardized Aircraft Engines, soon to be known as the Liberty engines.

THE U.S.A. STANDARDIZED AIRCRAFT ENGINE IS DESIGNED

E. J. Hall arrived in Washington from Cleveland in the morning of May 29, 1917, and he and Vincent got together just before noon. By this time, as noted above, the decision had been made that efforts would be directed toward a line of engines with standard cylinders which could be arranged with 4, 6, 8 or 12 cylinders.

The two men had never met but knew each other by reputation. Perhaps this is a good point to take time out to make the reader aware that most of the automotive executives and engineers of that time did know each other. There would seem to be no doubt that this fact would occasionally have some effect upon decisions that would be made in the production of aircraft and their equipment during the war. I am not trying to intimate that it resulted in anything illegal regarding contracts or production to follow, but it certainly must have allowed some quicker or better decisions to be made regarding just which engineer would be best suited for some particular job or which company might best produce a certain item.

One example I can show, because I have the membership roster of 1915 for that club, is the fact that a large number of the corporate executives and high ranking engineers of the Detroit area (and even outside the area) belonged to the Detroit Athletic Club. As a matter of fact, this club's roster read like the "Who's Who of Detroit." I will list those I recognize as possibly associated with WW I aircraft production or design, either as an official of a company that produced aircraft materials or as a government official who had some control or connection with aircraft materials. (Club membership in 1917 was approximately 2,000.)

Name	Association
Alger, F. M.	Packard Director
Alger, Russell A.	Packard Vice Pres.
Beall, F. F.	Packard Vice Pres.
Bodman, Harry E.	Packard Director
Briggs, Walter O.	Briggs Body Co.
Chalmers, Hugh	Chalmers/Saxon Auto
Chapin, Roy D.	Hudson founder
Coffin, Howard E.	Hudson founder
Dodge, Horace E.	Dodge founder
Dodge, John F.	Dodge founder
Durant, W. C.	Gen. Motors

Eastman, Frank G.	Packard
Emmons, Harold H.	Studebaker/Attorney
Fisher, Laurence P.	Fisher Body
Ford, Edsel B.	Ford Motor Co.
Ford, Henry	Ford Motor Co.
Heaslet, James G.	Studebaker Gen. Mgr.
Hills, H. H.	Packard Sales Mgr.
Joy, Henry B.	Packard President
Joy, Richard P.	Packard Director
Kettering, Charles F.	DELCO/Dayton-Wright
Leland, Henry M.	Cadillac/Lincoln
Leland, Wilfred C.	Cadillac/Lincoln
Macauley, Alvan	Packard President
Moore, Charles J.	Packard Operating Mgr.
Nash, C. W.	Nash/General Motors
Newberry, Truman H.	Packard Director
Roberts, E. F.	Packard Manager
Robinson, F. R.	Packard Comptroller
Vincent, Jesse G.	Hudson/Packard
Waldon, Sidney D.	Packard/Cadillac
Wills, C. Harold	Ford Engineer

After Hall's return to Washington, Deeds asked that he and Vincent go to his suite of rooms in the New Willard Hotel (suite 201) and discuss in detail the airplane engine needs of the military services in preparation to making a joint report to the Aircraft Production Board on the subject. By middle afternoon the two men had made enough decisions between them that they had started "laying down two views of a proposed 8-cylinder aircraft engine."

The Detroit Athletic Club as it appeared in April of 1915. The club still occupies this building and it has changed little on the outside.
(Courtesy of John A. Bluth and the Detroit Athletic Club)

In order to accomplish more than rough sketching, they called Mr. Zimmerschied, chief of the Washington branch of the SAE, and asked if he could provide a draftsman and drafting materials. He was unable to locate an available draftsman but did send over his own drafting instruments. Hall and Vincent then went out and purchased two drafting tables and the necessary paper and returned to start doing their own drafting. Both men were well versed in the current state of the art of aero engines built in this country, and to some degree that of foreign engines.

In order that they would have the absolute latest information regarding European engines, they asked that Major Souther of the Signal Corps might obtain it for them. Major Souther then put Charles King to the task.

That evening Deeds called a conference in the suite at which a number of members of the French aviation mission were present. By that time Hall and Vincent had a number of questions about the latest French and English designs which were all answered.

Hall had taken on the task of laying out a longitudinal section of the proposed design and Vincent the transverse section. The next morning Mr. Zimmerschied turned over to the two men one of his volunteers to help with the drafting. This was a Mr. Schoonmaker of Pittsburgh. Schoonmaker had done no drafting for a number of years but took up the task without hesitation. Some years later, Schoonmaker would become chief engineer of the Dayton-Wright Company.

Schoonmaker continued with the formal drawings and as the day progressed, Vincent began drafting the report they were to present to the Aircraft Production Board. By the middle of the day of Thursday, May 31 the report had been typed and the longitudinal and transverse sectional drawings were well established. It had been about 48 hours since the two men had set out on the task.

At about three o'clock in the afternoon Sidney Waldon asked that the two men bring the report and their drawings and go with him to a meeting in General Squier's office in the Army-Navy Building. This was to be a joint meeting of the Aircraft Production Board and Joint Army-Navy Technical Board. Howard Coffin served as chairman. Waldon read the report and Hall and Vincent explained their idea for the series of engines using the drawings that had just been completed and noted that if it met with the board's approval, they could immediately start the process required to proceed with further layout drawings and determine construction features.

Washington, D. C., May 31, 1917

Aircraft Production Board,
Washington, D. C.

Gentlemen:

At your request we have made a careful study of the aircraft motor situation and hasten to submit our report as follows:

In order to get this report in your hands promptly we have condensed it as much as possible and have covered the essentials only.

In view of the fact that there are a number of good motors for training machines available, we have disregarded this type of motor and we have confined our attention strictly to the high efficiency, low weight per horsepower type, such as is necessary at the front.

In order that any motors that are built by this country may be of any value when received at the front, it is, of course, absolutely necessary that their efficiency be brought up to or a little beyond the best now available in Europe. The French and English Commission has enabled us to obtain this information by answering our questions very clearly and completely.

From information obtained from these gentlemen and from other sources we believe that the Lorraine-Dietrich is the coming motor in Europe. This motor has not been built in large quantities as yet, but some thirty had been constructed and carefully tested out at sea-level, and also at about 6,000 feet elevation. The important facts about this motor are as follows:

Eight cylinders: 120m/m bore by 170 m/m stroke. Cylinders made of steel with water jackets welded on. Motor is direct-drive and develops 250 H.P. at 1500 R.P.M. and 270 H.P. at 1700 R.P.M. The weight of the bare motor is 240 kilos, or approximately 528 lbs., while the weight of the motor complete, with radiator and water is 305 kilos or 671 lbs. There seems to be some reasonable doubt regarding the exact weight of the bare motor, as while the French Commission gave us the figure of 528 lbs., information from other sources indicates a weight of 552 lbs., probably some intermediate figure is more nearly correct, but in any event the motor gives a H.P. for approximately 2 lbs. of weight when figured at its maximum output of 270 H. P.

After obtaining this information and considering the matter very carefully, we next investigated the manner of testing such a motor as we knew that a motor of this type, could not be run at full power for long periods of time without developing serious trouble. Here again the French Commission gave us valuable information. They stated that in using a motor of this type, it is only run at full power for short periods of time while climbing or fighting and that all other times, it is run at speeds 200 to 300 R.P.M. slower. In view of the fact that the motor is built to run under these conditions, it is, of course, necessary to test it under similar conditions and they stated when trying out a new model of motor, it is their practice to mount a propeller which will just hold the motor down to maximum speed under full throttle. The motor is then run for 50 hours in periods of 6 to 8 hours each, but the motor is not run up to full speed for more than a total of 10 hours during this entire period, nor is it run more than 30 minutes at any single time under this condition. The other 40 hours running is under throttled condition, turning the same propeller 200 to 300 revolutions per minute less than maximum speed.

This information is of the utmost importance as it enables us to reduce all factors of safety and makes possible the light weight per H.P. now being obtained in Europe.

1

After obtaining this information, we immediately laid down a proposed motor which we believe can be produced promptly in large quantity in this country. Built carefully out of proper materials, this motor will have approximately the following characteristics and be as good, or a little better, than the Lorraine-Dietrich, which is not as yet really available abroad.

In laying down this motor, we have without reserve selected the best possible practice from both Europe and America. Practically all features of this motor have been absolutely proven out in America by experimental work and manufacturing experience in the Hall-Scott and Packard plants, and we are, therefore, willing to unhesitatingly stake our reputations on this design providing we are allowed to see that our design and specifications are absolutely followed.

The motor is to be of the eight-cylinder type with cylinders set at an inclined angle of 45°. The cylinders are of the individual type made of steel forgings with Jackets welded on. The bore is 5 inches and the stroke is 7 inches, giving a piston displacement of 1100 cubic inches. The crankshaft is of five-bearing type with all main bearings 2-3/8 inches in diameter, and all crank pin bearings 2-1/4 inches in diameter. The connecting rods are of the I-beam straddle type. The motor is of the direct-driven type with maximum speed of 1700 R.P.M. This motor will give a maximum output of 275 H.P. at 1700 R.P.M. It will weigh 525 to 550 lbs. but we feel very sure of the lower figure. It will have gasoline economy of .50 lbs. of fuel per H.P. hour or better; it will have an oil economy of .04 lbs. of oil per H.P. hour or better. Complete with water and radiator, this motor will not weigh more than 675 lbs. if a properly constructed radiator is used and placed high above the motor.

To obtain the above mentioned weights, it will be necessary to use the fixed type of propeller hub which has been thoroughly proved out by Hall-Scott practice. In order to obtain the above mentioned weights, it will also be necessary, as mentioned above, to use the very best material, workmanship and heat-treatment.

Complete detail and assembly drawings as well as parts lists and material specifications can be completed at the Packard factory under our direction in less than four weeks. We believe that a sample motor can also be completed in approximately six weeks if money is used without stint. As soon as the drawings, specifications and sample motor have been finished, complete information would, of course, be available, so that any high grade manufacturer could either make parts for this motor or manufacture it complete.

In laying down this design, we have had in mind the extreme importance of interchangeability as a well laid comprehensive program which has for its base interchangeability of important parts, such as cylinders, will speed up output and reduce ultimate cost to an astonishing extent. Europe is suffering right now from lack of uniformity of design, but it is too late for them to change their plan. We, however, can take a leaf out of their book and start right.

In the design which we have laid down, the cylinder, for instance, can be used to make 4, 6, 8 and 12-cylinder motors. As this is the most intricate part to make, immense facilities could be provided to produce them in large quantities for the use of many concerns who could manufacture the balance of the motor. Nearly all small parts and numerous large and important ones would also be interchangeable. This would not only speed up production, but would be of the utmost importance in connection with repairs and replacements. A full line of motors made according to this plan would line up about as follows:

TYPE	RATED H.P.	MAX. H.P.	WEIGHT	WEIGHT PER H.P.
4	110	135	375	2.7
6	165	205	490	2.3
8	225	275	535	1.9
12	335	410	710	1.7

Respectfully submitted,

2

Aeronautical Production Organization as of May 1917

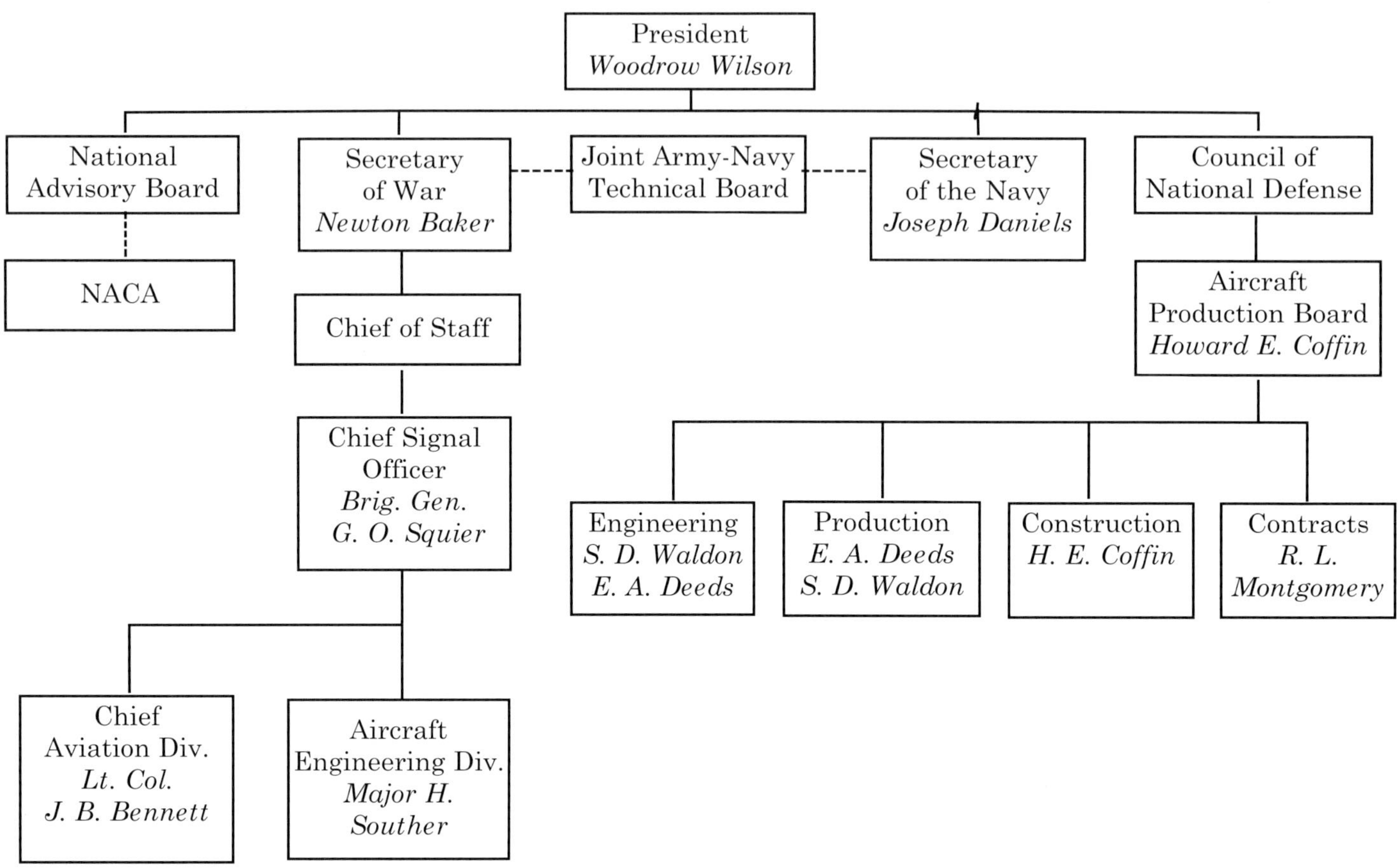

Authorization was given and after Vincent and Hall left, Vincent wired the Packard factory and had two layout men leave immediately for Washington. They arrived the next day, Friday, June 1. Hall, Vincent, Schoonmaker and the two layout men worked steadily over the weekend and by the afternoon of Monday, June 4, 1917, they were able to complete accurate layouts showing the longitudinal section, transverse section, rear elevation, and camshaft assembly views of the 8-cylinder version. These were then presented before a joint meeting of the Aircraft Production Board and the Joint Army-Navy Technical Board at about midnight that same day. At the meeting they went into the design in much more detail than was possible at the May 31 meeting. After providing the drawings and detailed information, Hall and Vincent left the meeting. A little later Deeds met the two and told them that it had been decided that they should build ten sample engines, five 8-cylinder and five 12-cylinder. These would be built by Packard and it was requested that the first be an 8-cylinder and completed as fast as possible. Deeds also informed Vincent that he had promised the board that the first engine would be completed in seven weeks from that date. Hall and Vincent had committed to approximately six weeks in their report. Vincent then sent his layout men back to Detroit with instructions to immediately start work on detail drawings of long-time parts, such as cylinders and crankcases.

Also during the continued session of the May 31 meeting, after Hall and Vincent left, it was resolved that the Aircraft Production Board "work out an agreement with certain motor manufacturers regarding a standardized motor." In addition, a "chart of proposed production" was asked for.

The input for this chart had to have been the Joint Army-Navy Technical Board report approved by the Secretaries of War and Navy on May 27 (see page 31). The minutes of the Aircraft Production Board noted on June 5 the resulting chart and its impact upon its members. Quoting the extract as printed in the early Liberty history: "Chart mentioned 5/31/17 shows need of 46,650 engines by June 30, 1918. Aircraft Production Board states has not sufficient powers and want to be empowered to make contracts. Situation grave."

One can certainly understand why the board thought the situation was "grave." They were suddenly in the position of trying to have 46,650 aircraft engines built within

This photograph of the full-size wooden model of the U.S.A. 8 engine, as it was then known, was taken at the Packard plant on June 17, 1917. Twenty-three individuals poised for this famous photograph. So far as we know, all are Packard engineers or executives. Unfortunately only eight can be identified. From left to right they are: Harry Gongwer, George Reifel and George Brodie. The next two are unknown. Under the left side of the picture is O. E. Hunt. The next five are unknown. Behind the engine in the center is Alvan Macauley followed by Charles H. Vincent and Jesse G. Vincent. The next six are unknown and seventh is J. R. Ferguson. The next man is unknown and the man on the extreme right is William B. Stout. (Courtesy of NAHC)

the next year by an industrial complex that had only built about 1,000 aircraft engines in the past eight years. On top of that, they could only recommend to those in power in the Army and Navy which companies should be contracted to build what items. After that, the usual bureaucratic process of accepting the recommendation or not and contract awarding would have to take place. They could foresee a nightmare of time-consuming red tape and frustration while trying to accomplish their task in the required time frame.

Vincent had kept Alvan Macauley, Packard president, informed as to the progress of the situation and they agreed to meet in New York, where Macauley was scheduled to be on business, on Thursday June 5. They discussed what had transpired and Vincent asked Macauley to return with him to Washington to talk with Deeds about it. They arrived in Washington in the morning of June 6.

On June 5 Hall went to Indianapolis to discuss engineering and production matters regarding the pending production contract with Nordyke and Marmon for 1,000 Hall-Scott A-7a engines (see page 30). He returned to Washington the next day.

Upon their arrival in Washington on June 6, Vincent and Macauley had a conference with Deeds and Waldon. Deeds went over the situation and convinced Macauley that the best course of action would be the production of a standardized line of aircraft engines based on the design put forth by Vincent and Hall. In spite of the discussion Macauley and Vincent had had the day before, evidently Macauley still held out some hope that production of a line of engines based upon Vincent's "905" design might be the solution to the problem.

As documented by Vincent in his history, "Mr. Deeds explained to Mr. Macauley the great advantage of standardizing a line of U.S.A. engines and after a little thought, Mr. Macauley agreed to this plan, although, of course, it meant passing up the long cherished plan of making Packard Aircraft engines as famous in the United States as Mercedes is in Germany, Rolls-Royce in England, etc."

Macauley also agreed to loan Vincent to the government for three months so that he could follow up on the engineering work required to build the 10 sample engines.

Deeds had already asked General Squier for $250,000 to finance the building of the 10 engines and received the promise of the appropriation. Actual receipt of it would take some time because of the usual government delays so Deeds asked Macauley if Packard would agree to finance the job in the mean time, that is that they would purchase all required materials and build the engines with their own funds until such time as government funds became available. Macauley agreed to do so and committed all engineering and production facilities needed for the job. In addition, Vincent was given authority to obtain all the necessary work and materials through commercial channels. Packard would submit an accounting for reimbursement when funds were available.

Without this commitment the proposed seven-week schedule for the first sample engine would never be met.

Hall and Vincent left Washington at noon on Thursday June 7 bound for Detroit and arrived at eight the next morning. They went to the Packard factory and checked on the work progress on the detailed drawings. O. E. Hunt, Packard's chief engineer, had already secured steel billets from Cleveland with which to make the cylinders for the first engine. Hall spent the day at the factory and left at 9:30 PM on the night boat for Cleveland to get crankshaft forgings made up by Hall-Scott's vender. In order to save time, the die blocks for the four-cylinder Hall-Scott A-7a crankshaft were used for the Liberty 8 and the six-cylinder A-5a die blocks for the Liberty 12. Doing so ruined the blocks for Hall-Scott use.

In order to complete the required detailed drawings as quickly as possible, Vincent called for volunteers from the various drafting departments to work over the weekend. Every man volunteered and as a result they had a very large force at work by noon Saturday. By the end of the day Sunday June 10, 86% of the detailed drawings were complete and thus on Monday morning he was able to

Right side view of the first Liberty built and delivered to the Bureau of Standards on July 3, 1917. The reader will be able to note a number of differences between this L-8 and the later production version. In this view the separate nose plate can be seen. (Courtesy of the Smithsonian)

have blueprints made and begin calling various automotive and accessory manufacturers to get the parts for the first eight-cylinder engine made.

The drawings for the upper and lower halves of the crankcase were far more difficult and were not finished until Friday June 15. They were near enough complete on the 13th to allow preliminary prints to be made so that the engineering department could start on pattern work and a wood model of the engine.

While this work was in progress, the Aircraft Production Board was having engineering offices set up at the Bureau of Standards in Washington under the direction of Dr. Stratton, the director. Vincent had put together an engineering organization drawn from a number of automobile factories in Detroit, including Cadillac, Dodge and Packard. He had arranged also for other engineers from the Pierce Company to meet them in Washington.

The wood model was finished and shipped to Dr. Stratton, director of the Bureau of Standards, on June 17th. Also on that day Vincent left for Washington with the Detroit-based staff he had assembled.

They arrived on Monday June 18th and occupied their new quarters. By the end of the week all equipment ordered had arrived and the staff went to work checking over the detailed drawings, setting up manufacturing limits, writing bills of material, etc.

In the meantime George Brodie of Packard had, at Vincent's direction, conferred with the engineering staff of the various factories in the Detroit area regarding any suggested changes or corrections in the drawings that might seem advisable for ease or speed of manufacture. He arrived in Washington with this material on Saturday June 30th.

Vincent's staff went to work with all possible speed on Monday morning July 2nd making any changes thought

Left side view of the first Liberty. Note the lack of crankcase vent by the left camshaft drive shaft, use of single instead of double strengthening webs and lack of data plate on rear surface of the upper crankcase. (Courtesy of the Smithsonian)

necessary. All corrections had been made by noon the next day and work was begun writing a complete bill of material and getting a complete set of up-to-date prints ready.

THE FIRST LIBERTY IS BUILT

At noon on Tuesday July 3rd Vincent was notified by telegram that the complete set of parts for the first sample engine had left Detroit at 5 PM the day before, accompanied by a man from the Packard Traffic Department to expedite its shipment. At 2 PM he received a phone message from Washington's Union Station that the parts had arrived.

They were trucked to the Bureau of Standards building and arrived at about 4 PM on July 3, 1917. They were examined and all were pleased with their appearance. They were accurately made and well-finished and showed no evidence of hurried workmanship in spite of the speed with which they were produced. Vincent, Mr. Hicks and Mr. Gongwer of the Packard Engineering Department spent that evening going over the parts to come up with any final suggestions regarding assembly or manufacture. The engine was probably assembled late that evening or early the next day.

Vincent asked his engineering staff if they would work on July 4th to incorporate all final changes in drawings and bills of material. They were completed and by 10 AM on the 5th of July about 3,600 prints had been made, trimmed and partially sorted, the entire bill of material written and blue-printed and all had been distributed to the various factories that were making the parts for the remaining eight and twelve-cylinder engines on order.

At this point I will let the reader know that the original order to build five sample eight-cylinder and five sample twelve-cylinder engines had been altered somewhat. The order now was to build six eight-cylinder and five twelve-cylinder engines and a wood model of each. The first eight-cylinder engine was to be used only for a manufacturing standard and to learn whatever possible to facilitate the process of manufacture and assembly. The remaining ten engines would be used for testing and design refinements.

Parts for this initial batch of engines were made principally by the following companies:

The General Aluminum & Brass Mfg. Co. of Detroit made the bronze-backed, babbitt-lined bearings.

The Cadillac Motor Car Co. of Detroit made the connecting rod and rocker arm assemblies.

The L. O. Gordon Mfg. Co. of Muskegon made the camshafts.

The Park Drop Forge Co. of Cleveland made the crankshaft forgings using Hall-Scott dies.

The first Liberty during assembly at the Bureau of Standards. (Courtesy of the Smithsonian)

The crankshafts were machined by Packard of Detroit.

All bevel gears were made by the Hall-Scott Motor Car Co. of San Francisco.

The Hess-Bright Mfg. Co. of Philadelphia made the ball bearings.

The Burd High Compression Ring Co. of Rockford, IL made the piston rings.

The Aluminum Casting Company of Cleveland made the die cast aluminum alloy pistons and machined them up to grinding.

The Rich Tool Co. of Detroit made the valves.

The Gibson Co. of Muskegon made the springs.

The Packard Motor Car Co. of Detroit made all patterns for aluminum castings which were then made by the General Aluminum & Brass Mfg. Co.

The Packard Motor Car Co. made most of the drop forge dies.

The Packard Motor Car Co. assembled and tested the engines.

During the period of about thirty days during which the first sample eight-cylinder engine was being produced, Vincent and Deeds submitted the plans to a great many eminent engineers for their opinion as to the soundness of the design and the feasibility of manufacturing it in quantity. Some made suggestions which were incorporated to aid in manufacture or tooling, but no basic design changes were suggested.

Among those men were:

Messrs. Manly and Taylor of the Curtiss Airplane and Motor Co.
Mr. H. M. Crane of the Simplex Co.
Mr. Fekete of the Hudson Motor Car Co.
Mr. Chryst of the Dayton Eng. Laboratories Co.
Mr. Riker of the Locomobile Co.
Mr. David Fergusson of the Pierce Co.
Mr. H. M. Leland of the Cadillac Co.
Mr. Burd of the Burd High Compression Ring Co.
Prof. Jeffries of the Aluminum Casting Co.
Messrs. F. F. Beall & E. F. Roberts of the Packard Motor Car Co.
Mr. C. Harold Wills of the Ford Motor Co.
Mr. Walter P. Chrysler of the Buick Motor Co.

U.S.A. Eight No. 2, the first engine to run, on Packard's outdoor test stand in Detroit. The date is July 25 or 26, 1917. The engine first ran on the 25th. Notice there is no data plate in its usual location on the upper rear of the crankcase. Just to the left of the post stands Henry M. Leland, soon to set up the Lincoln Company to produce Liberty engines. To his left, pointing to the control panel, is Alvan Macauley. To his left are F. F. Beall, V. P. of manufacturing, and E. F. Roberts, plant manager. The man at the left end of the engine is believed to be Frank H. Trego. All others are unidentified. (Author's collection)

On July 5th a final group of about 220 drawings was divided and sent to Cadillac, Dodge, Pierce and Packard to be traced on cloth. An additional 55 were retained by Vincent at his Bureau of Standards office to be done by his staff. His plan was to have a finalized set of drawings ready for manufacturers by July 18.

Parts for the second eight-cylinder and first twelve-cylinder engines were well under way and the second eight-cylinder was scheduled to be assembled and ready for test on July 25.

Authorization to build 10 sample engines (later changed to 11 and 2 wood models) was given on June 4 with a request the first engine be completed within 7 weeks (by July 22). The first engine had been delivered in 30 days.

As one might expect, work continued on finishing all changes required in order to have an absolute complete set of standardized drawings and specifications of the eight and twelve-cylinder models of the new engine. By July 21 the eight was pretty well cleaned up in that respect and the twelve was well along.

Vincent received a phone call from O. E. Hunt on Saturday the 21st of July that the second U.S.A. eight, referred to as the "first standardized eight-cylinder engine," would be completed and ready to run on Monday July 23rd. On the morning of the 23rd, Vincent called to check on the status and Hunt told him the engine was being worked-in on the dynamometer (being turned over by an electric motor in order to smooth up bearings, etc.) and that Hall had just arrived in town. Vincent directed that the motor be worked-in all day and that Hunt and Hall start it under its own power the next morning. Vincent planned to arrive at the Packard factory in the morning of Wednesday July 25th.

Vincent arrived at about 10 AM on the scheduled day and found the motor running in Packard's open-air laboratory. He joined in the testing and later reported that the engine ran smoothly and appeared to have no period of vibration up to 2,000 RPM. (Keep in mind that this testing was carried out on a solid mounting which could cover a modest amount of vibration. I say this because, as we shall see, some was reported in later tests.)

All comments regarding the initial tests run on the dynamometer and, shortly thereafter, mounted on a Packard truck set up for the purpose, indicated no cooling, lubrication, ignition, exhaust valve cooling, vibration or carburetion problems. By this time it is estimated the engine had been run some 20 hours or more, three quarters of it at 9/10 to full power. These early tests had all been made with low compression pistons. Unfortunately only one reference was documented by Vincent of the actual power output. He stated that "when the motor was new and not well worked in, it had pulled 214 HP at 1,350 RPM, and that this was extremely good with the low compression pistons which were used." And it would prove to be fairly close to the 206 HP that U.S.A. eight-cylinder No. 6 engine with low compression pistons developed at that same speed in a test on September 17, 1917. Although this was by no means proof that further refinements would not be required, it would certainly seem to indicate the basic design was sound.

Vincent then directed that O. E. Hunt have a test stand for the engine mounted on a Packard truck in much the same manner as had been done to test Packard's "905" engine. This way the engine could be tested with a propeller mounted.

U.S.A. Eight No. 2 mounted on a torque stand on a Packard truck at the Packard factory on July 27, 1917. From this view you can see that the nose plate used on engine No. 1 has been replaced by the pairs of long and short vertical through-bolts found on all the following eight-cylinder engines. Otherwise the engine was much the same as the first one. (Author's collection)

It was about 3 PM on Friday July 27 before the truck had been made ready and the engine mounted. A propeller was attached but the only one available was the one which had been designed for use on the Packard "905" engines. This propeller required about 200 HP to turn it at 1,200 RPM and was a little too large for the engine, but worked well nevertheless.

Vincent was anxious to see whether the engine would be easy to start by turning it over with a propeller with the battery ignition since one of the advantages of battery ignition over magneto is easy starting at slow cranking speeds. Hall did the honors and the engine started on the first pull.

A number of men had gathered to witness the engine actually turning a propeller. Among them were Henry M. Leland, Wilfred C. Leland, Charles King, Glenn Martin, Henry B. Joy and Alvan Macauley.

A number of tests were run during the remainder of the day, including one of an hour and a half duration. All went smoothly and trouble free. Vincent and King then sent separate telegrams to Washington outlining the results obtained.

U.S.A. Eight No. 2 on its test truck at the Packard factory on July 27, 1917. Left to right are F. F. Beall, E. F. Roberts, Henry B. Joy, Henry M. Leland and an unidentified man. Note the similarity to the test truck pictured on page 20. In this case two Packard truck radiators were used for cooling instead of one. (Author's collection)

Later that evening Vincent called Deeds and asked permission to have the complete outfit shipped to Washington for demonstration before the various government officials who were interested in military aviation matters. Permission was granted and on Saturday the 28th it was prepared for shipment. It left in a railway express car on the morning of Sunday the 29th.

Hall had left for Indianapolis on business on Friday night the 27th and Vincent left for Washington on Sunday the 29th. Vincent was delayed slightly in his trip when the engine of his train derailed 30 miles north of Baltimore.

The truck and engine arrived on the 30th and was unloaded on Tuesday the 31st and sent to the Bureau of Standards, where it arrived about noon. By this time Hall had arrived in Washington and so had Mr. Frank Trego, an engineer from Connecticut.

FRANK H. TREGO

The above is the first time Vincent mentions Trego with reference to the Liberty but this is not the first time the two have been associated. I will pursue a little of Trego's history here as I have never figured out just why this man was drawn so early into an interest in the Liberty engine. He had to have had "connections" somewhere in order to have been kept so well informed so early in the engine's development and to have eventually become a producer. As things turned out, he was certainly the least qualified of those eventually contracted to produce the engine. By he, I mean the factory and organization he set up for the purpose. Trego himself was probably a good engineer but, as you will see, his factory had difficulties with Liberty engine production from the very start.

Let us start with the first mention of Frank Trego in reference to the Liberty engine. On June 18, 1917, just 13 days after authorization was given to build ten sample engines, "Deeds and Montgomery were directed by the Aircraft Production Board to negotiate with Trego" regarding U.S.A. standardized engine production. There is no indication that at this time Trego had any facilities in place to manufacture the engines. Trego was the first potential producer mentioned in the chronological notes of the June 6, 1918 Liberty history and was awarded a contract on September 6, 1917.

As to why these events happened as they did, we can only speculate. Who was Frank Trego? He was a fairly well known automotive engineer and early auto racing enthusiast. Automotive historian D. J. Kava has found him mentioned in *Automobile* as early as 1908 as operating the timing equipment at the Algonquin Hill Climb and later at the Savannah races. He was almost killed in the former case when a Jackson automobile lost control and crashed into the timing tent.

In 1910 he went to work for the Hudson Motor Car Co. as manager of the Technical and Inspection Departments. The corporate notes contain numerous mentions of him until he leaves the company around 1912. Vincent was Chief Engineer of Hudson from 1910 to 1912. Vincent went to work for Packard as Chief Engineer in 1912. Frank Trego went to work for Packard shortly thereafter and shows up on Packard's corporate chart as a Research Engineer under Vincent in 1913. By 1915 he seems to have left Packard and struck out on other ventures. His employment by Packard resulted in three patents issued to him and assigned to Packard.

He also published two SAE papers in 1915 and one in 1920 as well as submitting published comments on Vincent's 1919 SAE paper on the Liberty.

Hudson had been started in 1909 by Roy D. Chapin and Howard E. Coffin, both veterans of the Olds Motor Works. Coffin would head the Council of National Defense under whose control was the Aircraft Production Board in 1917. He was also Chairman of the Aircraft Production Board. Chapin would also become a member of the Council of National Defense in late 1917 and serve as Chairman of the National Transport Committee.

Did all this have anything to do with Trego's early consideration for an engine production contract? We will probably never know, but it certainly looks likely. Coffin would have been the most likely influence since he had direct control of the Aircraft Production Board. Vincent knew Trego, of course, but at that point had no influence other than a like connection with his old Hudson bosses. If Vincent had wanted to peddle any influence, it would seem most likely he would have pushed for production by Packard. As a matter of fact, Packard's President Macauley would eventually begin to sweat whether or not the government would reimburse their considerable investment should a contract not be forthcoming. As it turned out, all initial contracts were offered to the various companies who would be the first producers within a period of one week.

In all fairness, however, I must add that in the aftermath of a considerable investigation that would eventually take place regarding possible fraud or malfeasance regarding production of aircraft and their engines, Trego Motors did not seem to come under any condemnation. If Trego got his contract through influence, no one found a problem with it.

Major Benjamin D. Foulois, Chairman of the Joint Army-Navy Technical Board and eventually Major General and Chief of the Air Corps. (Author's collection)

PRELIMINARY PREPARATIONS FOR PRODUCTION AND CHANGE OF DIRECTION

The Aircraft Production Board (APB) was letting no grass grow under their feet at this point, however. There are a number of important notes in the early Liberty history to indicate the board was, as soon as prototype production was authorized, taking steps to facilitate possible mass production. As just noted, negotiation with Trego for production was directed on June 18. On June 22 the Committee on Public Relations released information regarding the authorized building of the 10 prototype engines as well as the reasons for the plan of standardization of U.S. motors.

Up to this point in time everyone had agreed that the most needed engine size for the immediate future was going to be the 225 HP-class U.S.A. Eight and the production should be concentrated on that engine. Heavy need for the 335 HP class U.S.A. Twelve was not anticipated until 1919 and later.

They noted on July 3 that the Ford Company was experimenting with a method to make cylinders of steel tubing to speed production. Ford engineers had been among those who had studied the initial drawings of the proposed engines. As we shall see, Ford's idea bore much fruit as they did indeed perfect this method of making cylinders with great speed and at relatively low cost.

On July 7 General Squier asked the Lelands to get ready to manufacture the U.S.A. Eight engine and told them they could expect a "very substantial order as soon as funds available." On that same day he wrote a letter to Macauley of Packard noting that they should "advance work on U.S.A. 8A - after the motor is approved by Joint Army-Navy Technical Board a good sized order will be given your company."

General John J. Pershing, Commander in Chief of the AEF in Europe. (Author's collection)

On July 13 the APB received a telegram from General Pershing indicating that Allied requirements would seem to be for twice as many Liberty 12s as 8s. In response, Waldon wrote to Deeds that it now appeared that there was limited need of the 8 and that the 12 should receive the largest production.

Both Waldon and General Squier received letters from Macauley of Packard dated July 17. To Squier he wrote that Packard was prparing for production and to Waldon that they would prefer to manufacture a quantity of 8 cylinder models before taking on the 12, but in any case their facilities would be ready by October 1.

By late July Vincent was finishing up the production prints of both models and at the direction of the APB was sending sets to Packard, Trego, Henry M. Leland and Pierce-Arrow.

On July 28 Waldon wired Deeds and asked his opinion of placing contracts based on cost plus profit basis. It was also noted on that date that Major Foulois, Chairman of the Joint Army-Navy Technical Board, recommended bringing the eight to Washington for tests before the joint board and later to test it at altitude and to use the third eight for the first flight test.

PROTOTYPE PROGRESS

Now to get back to the production progress of the eleven prototype U.S.A. Standardized Aircraft Engines. The truck with U.S.A. Eight No. 2 mounted on it was set up and ready to run by early evening of Tuesday July 31 and Hall, Vincent and Trego proceeded to run a couple of short tests to make sure the engine was running well.

Between August 1 and 4 the engine was run on a number of occasions for various committees and individuals. These included a run of about an hour and a half for the Joint Army-Navy Technical Board on August 1, an hour run on the second for Army inspectors and other interested individuals, several short runs on Friday the third for members of the French Commission and Mr. Riker of Locomobile, and a final exhibition run on Friday August 4 for members of the Committee on Military Affairs of the House and Senate. Up to this time the engine had run more than 15 hours at wide open throttle and had always run without problems and without adjustments of any kind. The designers and all committees and boards who had witnessed its operation were well pleased.

The engine was dismantled on Monday August 6 and examined for condition, which was found to be good. The engine had been equipped with low-compression pistons up to this point but new high-compression ones had been ordered from the Aluminum Casting Co. of Cleveland for installation before tests were conducted at altitude on Pikes Peak. Colorado Springs, Colorado was to be the base for the Pikes Peak, Colorado altitude tests. The suggestion had been made by Major Foulois on July 28. Unfortunately the new pistons were found to have sand imbedded from the sand molds and therefore were unusable. The engine was therefore boxed and shipped to Packard to have new pistons and otherwise be checked over and then shipped to Colorado Springs. The test truck was shipped directly to Colorado Springs.

The first flight test was scheduled to be performed in a L. W. F. (Lowe, Willard and Fowler) Model F airplane at Mineola Field, New York. Vincent left Washington on Sunday August 12 for a trip to the L.W.F. plant at College Point, New York to look over the plane. He found it ready to accept the new engine. L.W.F. had been supplied with the original wood model for fitting so that the locations of all parts would be correct and ready to accept the engine when it arrived.

THE FIRST TWELVE IS RUN

Vincent left that evening on the train bound for Detroit where the first twelve was scheduled to be run the next day. He arrived on Monday morning August 13 to find the motor running with low-compression pistons under light load on the dynamometer to work it in. It was run in this manner for the remainder of the day. Hall arrived shortly thereafter.

Starting on Tuesday the 14th and continuing through Friday, Vincent made a number of tests followed by various changes in carburetor settings and even changes to the intake manifolds in order to get better fuel distribution and power. Initial tests had been with low-compression pistons and had produced about 230 HP @ 1,600 RPM. Obviously something was not working as it should since this was very close to the power output of the second eight-cylinder engine at the same speed.

In addition to the various carburetor and manifold changes, the pistons were replaced with high-compression ones. By Friday the engine would produce 346 HP @ 1,700 RPM and the same power at 1,800 RPM on the dynamometer.

It was then moved to the cradle stand in a newly-built outdoor test shed. It was here that the engine would be subjected to a 50-hour endurance run to determine its acceptance as a proven design. A number of tests were

U.S.A. Twelve No. 1 on an outdoor test stand at the Packard factory in August of 1917. (Courtesy of Roberta Vincent Mocabee)

tried totaling some seven hours. Vincent and Hall were accompanied by Lynn Reynolds, inspector for the Equipment Division.

At this point they decided the engine was ready for the 50-hour test. The problem now before them was how to approximate the test conditions recommended by the French (see page 36). They planned to run the engine in 5-hour increments, 4 1/2 hours at .9 throttle and 1/2 hour at full throttle. The difficulty was figuring out how to determine just what constituted .9 throttle. They felt the human element would enter to too great an extent. Thus they decided that a fairer and more consistent test would be to equip the engine with low-compression pistons and run it at full throttle for the entire 50 hours. This should prove the engine's abilities with more severe conditions than it would likely ever encounter during flight while certainly exceeding the requirements recommended by the French.

THE U.S.A. STANDARDIZED AIRCRAFT ENGINE BECOMES THE LIBERTY ENGINE

Although the precise date is not recorded, Rear Admiral David W. Taylor, a member of the Aircraft Production Board, presented the idea of calling this new series of engines *Liberty* engines. It seems evident that this must have occurred about the time the first twelve-cylinder engine was finished but after the third eight-cylinder was tested in the L.W.F. aircraft.

Vincent's running commentary refers to the engines by their early name through the testing of the third eight-cylinder engine and the initial testing of the first twelve. Vincent's reproduction of Lynn Reynolds' report of the 50-hour test indicates on the title page "Subject: Report - U.S.A. Twelve Engine, August 27, 1917."

Vincent's next chapter records his activities from September 3, 1917 to October 17, 1917 and from this point on, the engines are referred to as the *Liberty* engines.

In addition, the earliest photographs of the first Liberty 12 which show its data plate (page 48) show it with the name "LIBERTY Model A," as it appears on all following engines. We must conclude, therefore, that the change must have been made sometime in late August of 1917.

In Vincent's SAE paper *The Liberty Aircraft Engine*, which was presented at the annual meeting on February 4-6, 1919, he made the statement "As a matter of interest, I will state that Mr. Hall and I designated this line of engines as the U.S.A. Standardized line, but a little later I believe Admiral Taylor "dubbed" it the "Liberty" engine in one of the Aircraft Board meetings, and this name took so well that we were forced to adopt it, and change all our titles on the drawings."

Ser. No. 114,426. (CLASS 23. CUTLERY, MACHINERY, AND TOOLS, AND PARTS THEREOF.) UNITED STATES OF AMERICA, Washington, D. C. Filed Nov. 29, 1918.

LIBERTY

Particular description of goods.—Internal-Combustion Engines and Parts Thereof.

Claims use since on or about July 4, 1917.

Liberty trademark registration application as published in the February, 1919 United States Patent Gazette. It was issued as Number 125853 on February 11, 1919. (Author's collection)

In the discussion following that paper, Charles B. King made the following comment: "A few days after the design work had started at Washington, at a late supper at the Willard Hotel, I suggested to those present that the name Liberty engine be adopted. It is quite possible that Admiral Taylor may have made the title official at one of the Aircraft Board meetings." This being the case, it would seem that King must be credited with originating the name and Admiral Taylor with making it official. Although it has always been thought that the name originated as a result of the July 4 date on which the first Liberty 8 was displayed at the National Bureau of Standards, this must not be true since King's suggestion was made during the first week of June.

It would seem that there must have been government officials in high places who were impressed with the importance of the Liberty engine because on November 29, 1918 an application for registration of the trademark "Liberty" on internal combustion engines and parts thereof was filed on behalf of the United States of America. The trademark was issued on February 11, 1919, number 125853, and was the first instance of the U.S. government registering a trademark.

LIBERTY TWELVE NO. 1 PASSES 50-HOUR TEST

The engine was therefore removed from the test stand and sent to the experimental shop to have the pistons replaced and measure all working parts so as to have an accurate record of their condition before the start of the test.

After accomplishing the above and returning the engine to the test stand, several tests were run and some carburetor changes were found necessary because of the change to low-compression pistons. The first official 5-hour run was started at 5:20 PM on August 22, 1917. It was completed at 10:20 PM. During the latter part of the run the engine had backfired several times and after shutdown the problem was discovered to be caused by air in the fuel lines caused by a faulty water trap.

The second 5-hour run started at 12:55 AM August 23 and finished at 5:55 AM with no problems. The engine was inspected and 10 minutes later, at 6:06 AM, was started on the third 5-hour run which finished at 11:05 AM.

At the end of this run the engine was shut down and the old oil replaced by fresh. At 11:50 AM the fourth 5-hour run was started and this time the engine was not halted at the end of 5 hours but continued through the next run for a total of 10 hours, halting at 9:50 PM.

After an examination revealed no problems, the engine was started again at 11:00 PM on the 23rd and continued for another 10-hour run, halting at 9:00 AM on August 24. At this point 35 hours of the 50-hour test had been completed.

L.W.F.-F AS 2505 landing after the first flight of a U.S.A. Eight engine on August 29, 1917. (Courtesy of Peter Bowers)

The engine was examined very carefully and nothing was found wrong so it was started for a final run on the remaining 15 hours at 10:30 AM August 24 and was run until 1:30 AM August 25. Thus it completed the 50 hours with no forced stops and only five of the scheduled nine stops. During the run no adjustments were made, not even a spark plug having been changed.

The average RPM for the run was 1,584 and the average HP was 322. The engine was then pulled from the test stand and placed on a dynamometer and produced 337 HP @ 1,700 RPM and 346 HP @ 1,800 RPM.

L.W.F.-F AS 2505 at the L.W.F. field at Mineola, New York. Its engine was U.S.A. Eight No. 3 and this airplane was the first to fly with Liberty power on August 29, 1917. (Courtesy of Peter Bowers)

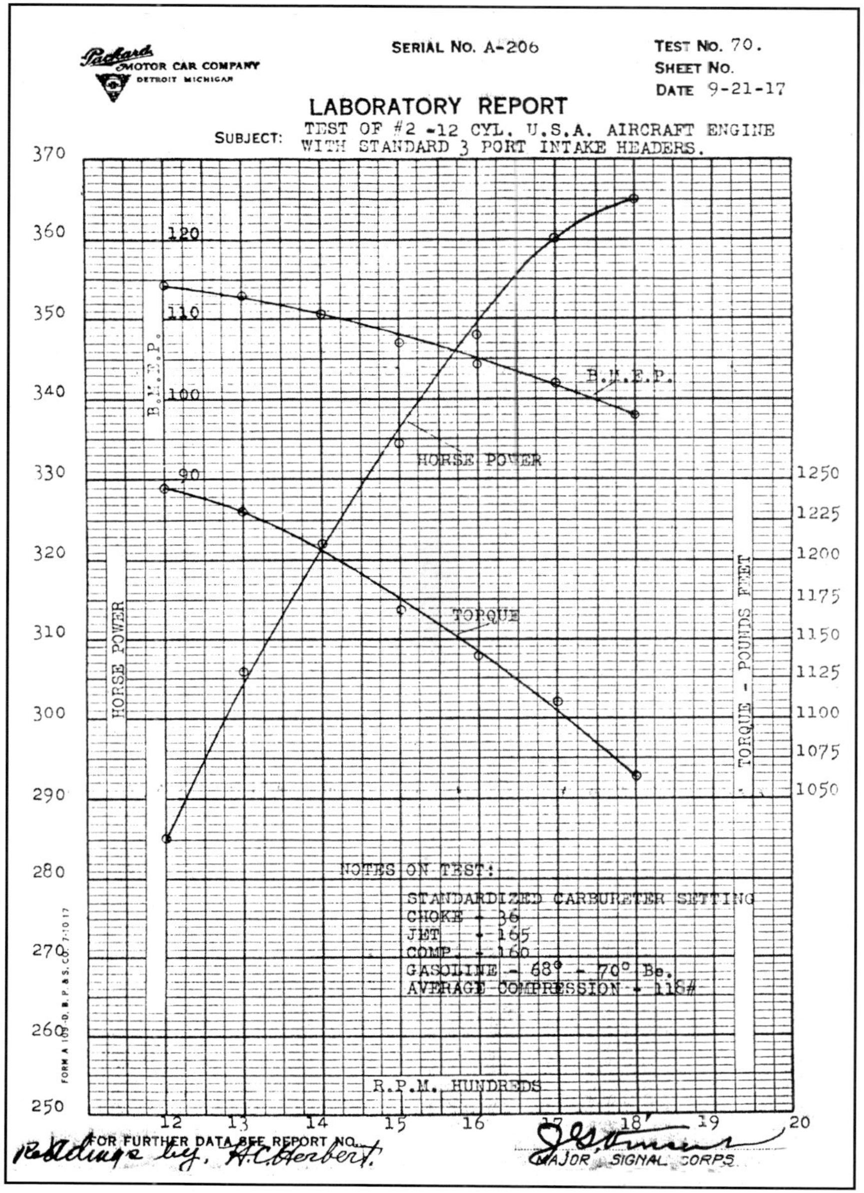

Power curve from the September 21, 1917 test of L-12A Exp. No. 2. The engine had low-compression pistons and developed 360 HP @ its rated speed of 1,700 RPM and 365 HP @ 1,800 RPM, at which point the engine was still not at its peak power. (Author's collection)

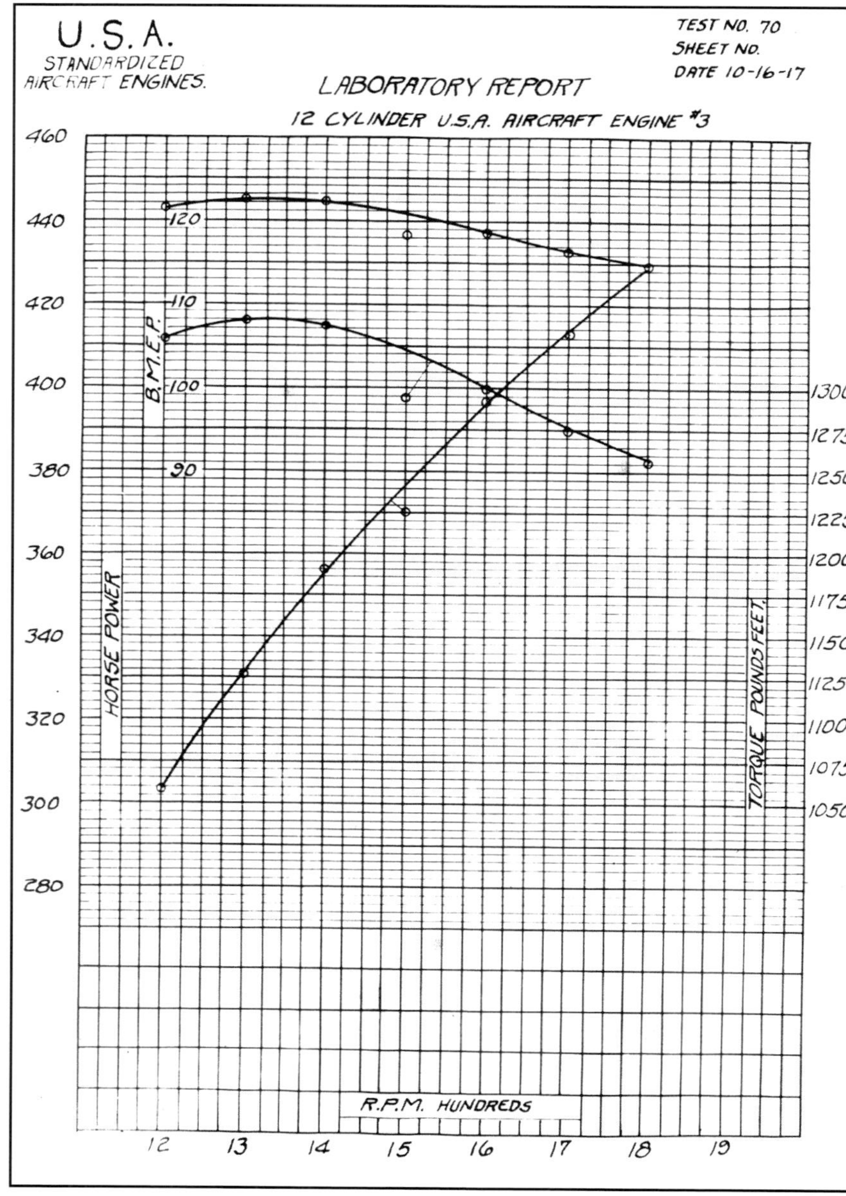

This power curve was sent to Lt. Col. Clark at Washington on October 25, 1917 by Capt. Howard C. Marmon, who was at that time in Dayton, Ohio. He noted that this was the engine that was now in the flying boat in Buffalo under Commander Westervelt's supervision. This was L-12A Exp. No. 3 with high-compression pistons. The test was run on October 16, 1917 and the engine developed about 412 HP @ its rated speed of 1,700 RPM and 430 HP @ 1,800 RPM at which point it had not peaked. (Author's collection)

Major Jesse G. Vincent, co-designer of the Liberty. Vincent was later commissioned Lt. Colonel and, in 1919, commissioned Colonel. (Author's collection)

It had been decided to go ahead with this run in spite of the fact that carburetion design was, at this early stage, far from finalized. Not only had the final choke size not been settled on but neither had the proper combination of jet and compensator sizes for the carburetors. In addition, the intake manifolds were not giving even distribution to all cylinders and the mixture had been run too rich on some cylinders to avoid running any too lean. In short, the engine was far from being in a final state of tune.

The reader should keep these early power accomplishments in mind when I later discuss claims of increased power accomplished by so called "changes in design" as more and more power was desired from the engine.

After the completion of the test, the engine was disassembled and measured for wear. Everything was found satisfactory and the condition of the engine was noted in detail by Lt. Reynolds for the test report.

The engine was then assembled and on Monday August 27 a 5-hour run was made for Navy Lt. Scofield which started at 7 PM and ended at 12 midnight. Upon completion of the test the engine was partially disassembled for inspection and Lt. Scofield wrote a report which was sent to Lt. Atkins in Washington.

The engine was then reassembled and packed and shipped to the Bureau of Standards in Washington on Tuesday August 28 so Vincent could carry out exhaustive intake header experiments in order to improve gas distribution.

Although not stated directly, the report of Hall and Vincent of the expectations of the U.S.A.-series of engines infers that "Rated H.P." would be output at probably 300 RPM under "Max H.P." "Max H.P." would be that achieved at 1700 RPM. (See pages 36-37.) Rated output of the high-compression twelve would thus be 335 HP @ 1400 RPM. Maximum power would be 410 HP @ 1700 RPM. The reader is cautioned here to realize that in almost all cases, and this is one of them, that when maximum power is referred to it meant that this was the maximum power that was to be required of the engine, not the maximum it was capable of delivering at full throttle.

Between those improvements devised by Vincent in Washington and others made by O. E. Hunt and E. J. Hall at the Packard factory in Detroit, power output of the U.S.A. Twelve engine had improved drastically within 30

Major Elbert J. Hall, co-designer of the Liberty. Hall was later commissioned Lt. Colonel. (Author's collection)

The test crew and equipment are posed for an official photograph on Pikes Peak. The group are, left to right, E. J. Hall, of Hall-Scott; unidentified driver; Lynn Reynolds, U.S. Inspector; V. I. Shobe, of the Zenith Carburetor Co.; Ferdinand Jehle, of the Aluminum Casting Co.; and E. L. Graham, U.S. Inspector. (Author's collection)

days to six weeks. (See power curves of Exp. No. 2 and Exp. No. 3 on page 51.)

As can be seen from observing the power curve of Exp. No. 3 generated on October 16, 1917, output at that time was 355 HP @ 1400 RPM and 414 HP @ 1700 RPM, which easily exceeded expectations. (See page 51.)

U.S.A. EIGHT NO. 3 IS THE FIRST LIBERTY TO FLY

Vincent left Detroit on the night of the 28th bound for New York to witness the flight in the L.W.F.- F of U.S.A. Eight No. 3 which he had had shipped from the Packard factory a few days before. He arrived on the 29th and found that the engine had been installed at the L.W.F. facilities at their flying field at Mineola, New York during the prior night. After checking things over, the plane was pulled out of the hanger at 3:15 PM and the engine started. It was allowed to run for a few minutes to make sure there were no leaks and then they prepared for its first flight.

In the words of Vincent, now Major Vincent, as he had been commissioned on August 15, although he had not learned of it until August 24 and would not be officially sworn in until September 4, "Promptly at 3:30, Aviator Blakely pulled the throttle wide open and left the ground in exactly three seconds. He made a beautiful 20-minute flight by himself and then came down to give us a chance to look it over. He reported that everything worked first-class and that the machine was fairly well-balanced, being slightly tail heavy with the engine wide open on account of the low center of thrust."

The flight was witnessed by Mr. Flint of L.W.F. and Mr. Crane of Crane-Simplex in addition to Vincent.

Over the next several days a number of flights were made and by Saturday September 1 they had accumulated a total of seven hours flying time. During that time they had had good success with performance of the airplane and engine. The only exceptions had been a worn rocker lever (it was found to have been made too soft) and a stopped-up fuel line.

Saturday evening Colonel Waldon and Major Horner arrived from Washington. Waldon, formally a Captain, had been commissioned Colonel on August 26. Several more flights were made and by the end of the day of Sunday September 2, 1917, a total of ten and a half hours had been flown. During the afternoon Major Kilner made two short flights in the airplane and became the first Army pilot to fly behind a Liberty engine.

With U.S.A. Eight No. 2 running, three of the test crew stand by while the rig is photographed in action. (Courtesy of Roberta Vincent Mocabee)

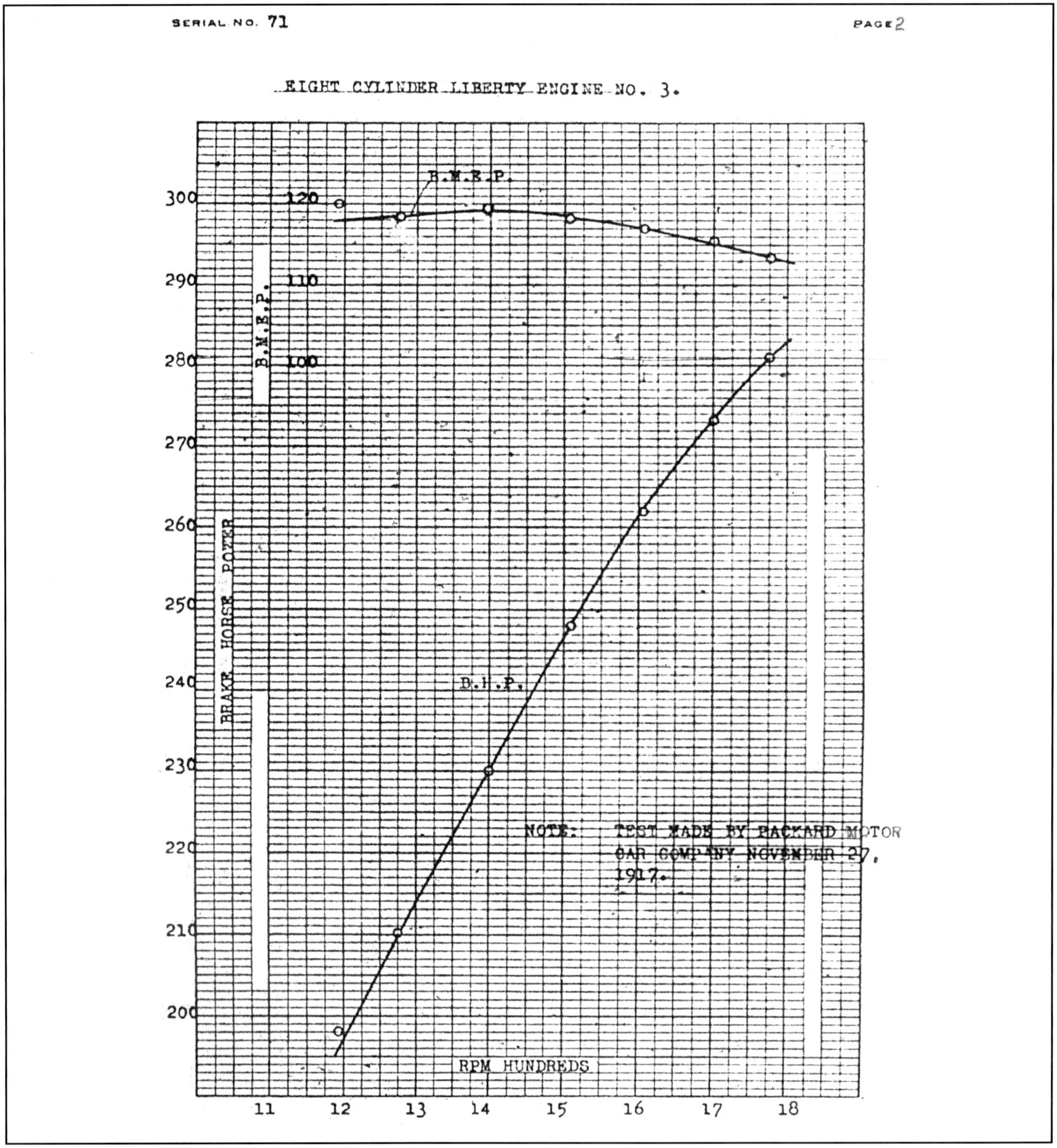

Power graph from a test run at the Packard plant on Liberty 8 Exp. No. 3 on November 27, 1917. The graph was copied by Capt. Lionel M. Woolson, BAP motor test engineer, and distributed to all interested engineers for informational purposes on July 11, 1918. (Author's collection)

The DH-4 prototype which on October 29, 1917, powered by L-12A Exp. No. 4, was the second L-12A-powered airplane to fly. Both the airplane and its engine survive in the Smithsonian. (Courtesy of the Smithsonian)

could reliably produce the continually increasing power demanded of it.

The first flight test of a Liberty 12, by now designated a Model A and generally referred to from this time as the Liberty 12A or L-12A, was made in a Navy Curtiss HS-1 flying boat. The engine was L-12A Exp. No. 3 and the flight was made from Curtiss' Buffalo, New York plant on October 21, 1917. The flight was successful and the Navy was very pleased with the performance. This flight and the tests that followed it led to the adoption of both the HS-1 and the Liberty 12 as a Navy standard.

A few days later, on October 29, 1917 the first Army flight test was carried out in the Dayton-Wright prototype DH-4 No. 1 flown from South Field at Dayton, Ohio. Actually it was reported that Dayton-Wright employee Howard Rinehart flew the plane and included in his testing numerous aerobatic maneuvers, including rolls, loops and inverted flight. During the latter, his passenger in the gunner's compartment, Archie Freeman, reportedly almost left the aircraft when the screws holding his seat in place stripped out of the frame to which it was attached.

This airplane was retained by Dayton-Wright and was used extensively for both testing and regular transportation duties throughout the war. During that time it accumulated some 1,078 hours during more than 4,000 flights. It traveled more than 111,000 miles and made 28

PIKES PEAK TEST OF U.S.A. EIGHT NO. 2

U.S.A. Eight No. 2 had its high-compression pistons installed at the Packard factory and was then shipped to Colorado Springs, Colorado to join the test truck setup which was awaiting it there. On September 2, 1917 the rig was driven up the mountain from its Colorado Springs base and set up for testing. Testing continued until September 26.

The engine was mounted on a torque stand and equipped with various sizes of propeller so that it could be tested at different speeds and powers.

Those present for the testing were E. J. Hall, Lynn Reynolds, U.S. Inspector; V. I. Shobe, of the Zenith Carburetor Co.; Ferdinand Jehle, of the Aluminum Casting Co., and E. L. Graham, U.S. Inspector. Although not mentioned in official accounts, there was obviously also a photographer on the trip.

In October of 1917, shortly after returning from this test trip, E. J. Hall was commissioned a Major in the Signal Corps. (He would serve in a number of capacities until the end of the war. Details of all of them were not well preserved, but I will document those which were in some way connected with the Liberty engine.)

As soon as Hall was commissioned, he returned to Detroit and was designated Associate Engineer of the Detroit District Office of the Aircraft Production Board, responsible for production at the Packard plant. In this position, it would be he and O. E. Hunt, Packard's chief engineer, who would be responsible for the early modifications that would be made to facilitate ease-of-manufacture and add strength where necessary in order that the engine

Curtiss HS-1. This is the type of aircraft used in the Navy's first test of the Liberty on October 21, 1917. This picture, however, is not of the actual airplane used for that test. The engine used was L-12A Exp. No. 3. (Courtesy of Peter Bowers)

trips out of the Dayton area to such locations as Philadelphia, Washington and New York. Both Orville Wright and Glenn L. Martin made flights in the plane during that time.

At the end of the war, on December 9, 1918, Colonel Waldon wrote to the Secretary of the Smithsonian, Dr. Charles D. Walcott, offering this airplane, "the Original No. 1 deHavilland Plane," to the Smithsonian for permanent display. The offer was accepted. The airplane was refurbished by Dayton-Wright before being shipped to the Smithsonian to more closely resemble the late production version of the DH-4 and thus some minor changes from the original configuration were made, including changing the original cream color.

The plane was fitted with a late Liberty engine but the original engine and propeller were shipped along with it. The plane has now been restored to its original configuration and color and its original engine and propeller installed. This airplane thus has two major historical attributes. It is the oldest surviving Liberty powered airplane; it is the oldest surviving American-built deHavilland DH-4; and it has the earliest surviving Liberty Twelve engine, L-12A Exp. No. 4. Thus the Smithsonian has in its possession L-8A Exp. No. 1 and L-12A Exp. No. 4.

Records of the completion of the last of the eleven prototype Liberty engines were not particularly well kept. Once the first Eight was tested, the first Twelve had passed its 50-hour test, and the initial flight tests had proved successful, it was not so much an historical event when another of the experimental engines was completed. In spite of this, we can date the completion of the last Eight and Twelve fairly closely. L-8A Exp. No. 6 was tested at the Packard plant by O. E. Hunt on September 10, 1917, and L-12A Exp. No. 5 on November 17, 1917. As we shall see in the next chapter, by the time the last experimental Liberty was completed, the engine had already been ordered into production. Packard, the first to complete a production engine, completed production engine No. 1 on November 19 and No. 2 during the last few days of November 1917. No. 1 passed inspection and was shipped on December 4.

Thus we can certainly say that with the completion of the last of the eleven prototypes and the early flight tests of them, ***THE LIBERTY ENGINE HAD INDEED BEEN BORN.***

Chronology July - December, 1917

July 31, 1917	L-8 Exp. No. 2 arrives at the Bureau of Standards on a Packard truck for demonstrations.
August 13, 1917	L-12 Exp. No. 1 first ran.
August 15, 1917	Vincent commissioned Major.
August 24, 1917	Deeds commissioned Colonel.
August 26, 1917	Waldon raised in grade from Capt. to Col.
August 22-25, 1917	L-12 Exp. No. 1 passes 50-hour test.
August 29, 1917	L-8 Exp. No. 3 flies in LWF airplane. This is the first Liberty engine to fly an airplane.
September 2, 1917	Major Kilner flies the L-8-powered LWF and becomes the first military officer to fly a Liberty-powered airplane.
September 3, 1917	Lincoln signs a contract (No. 1647) to produce 6,000 Liberty 12 engines.
September 4, 1917	Packard signs a contract (No. 1646) to produce 6,000 Liberty 12 engines.
September 6, 1917	Trego Motors signs a contract (No. 1648) to produce 500 Liberty 12 engines.
September 7, 1917	Nordyke & Marmon signs a contract (No. 1704) to produce 3,000 Liberty 12 engines.
September 10, 1917	L-8 Exp. No. 6 is tested.
October 16, 1917	L-12 Exp. No. 3 is tested.
October 21, 1917	L-12 Exp. No. 3 is the first Liberty 12 to fly an airplane when it powers a Navy Curtiss HS-1.
October 29, 1917	The prototype American DH-4 powered by L-12 Exp. No. 4 becomes the first Army airplane to fly with Liberty 12 power.
October 31, 1917	Ford signs a contract (No. 2129) to produce 5,000 Liberty 12 engines.
November 17, 1917	L-12 Exp. No. 5 is tested.
November 19, 1917	Packard completes production L-12 No. 1.
November 20, 1917	Duesenberg signs a contract (No. 2318) to produce 500 Liberty 12 engines.
December 11, 1917	General Motors signs a contract (No. 2266) to produce 1,000 Liberty 12 engines.

Aeronautical Production Organization as of October 1917

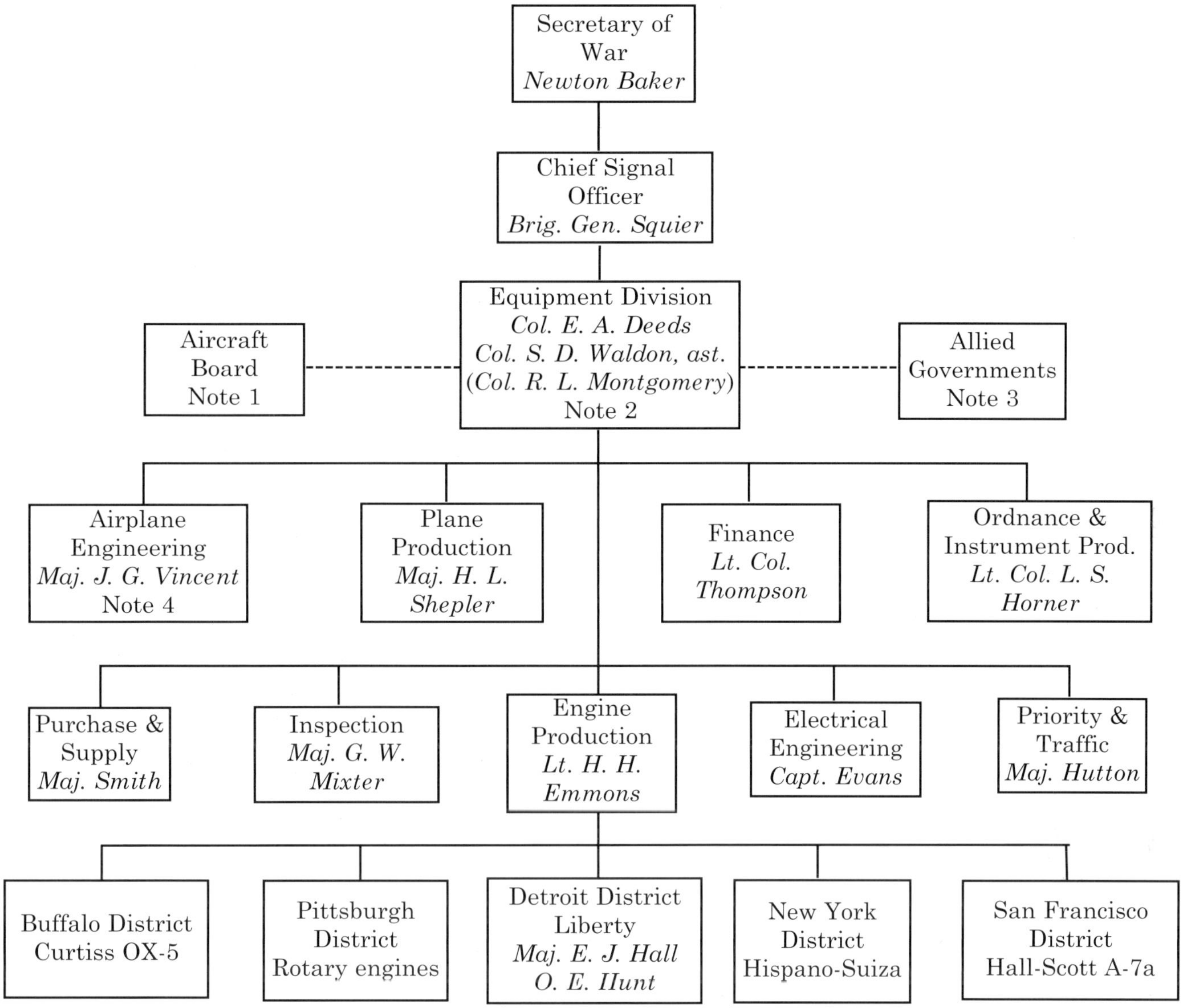

Notes*

1- In August of 1917 civilian personnel on the Aircraft Production Board were brought into the Signal Corps in officer grades and placed in charge of production. The Aircraft Production Board, which before this time was to be in charge of production, assumed an advisory roll and was renamed the Aircraft Board shortly thereafter.

2- The Equipment Division was formed as a result of the commissioning of these men and it then took over control of production of aircraft materials. In early November, Col. Waldon left for Europe on a fact finding mission and his position was filled by Col. R. L. Montgomery. Col. Waldon did not return until February 8, 1918.

3- Allied Government agencies acted in an advisory capacity regarding the types and design of needed aircraft.

4- The Airplane Engineering section succeeded and took over the duties of what were the separate plane design and engine design sections. Vincent had headed the Engine Design section, first as a civilian and then as a Signal Corps Major, and then became head of Airplane Engineering.

The above chart is not meant to be absolutely complete. It only covers those areas most directly associated with the design and production of airplanes and their engines. Organization had evolved through several changes between the May 1917 chart on page 38 and this one and it would continue to do so.

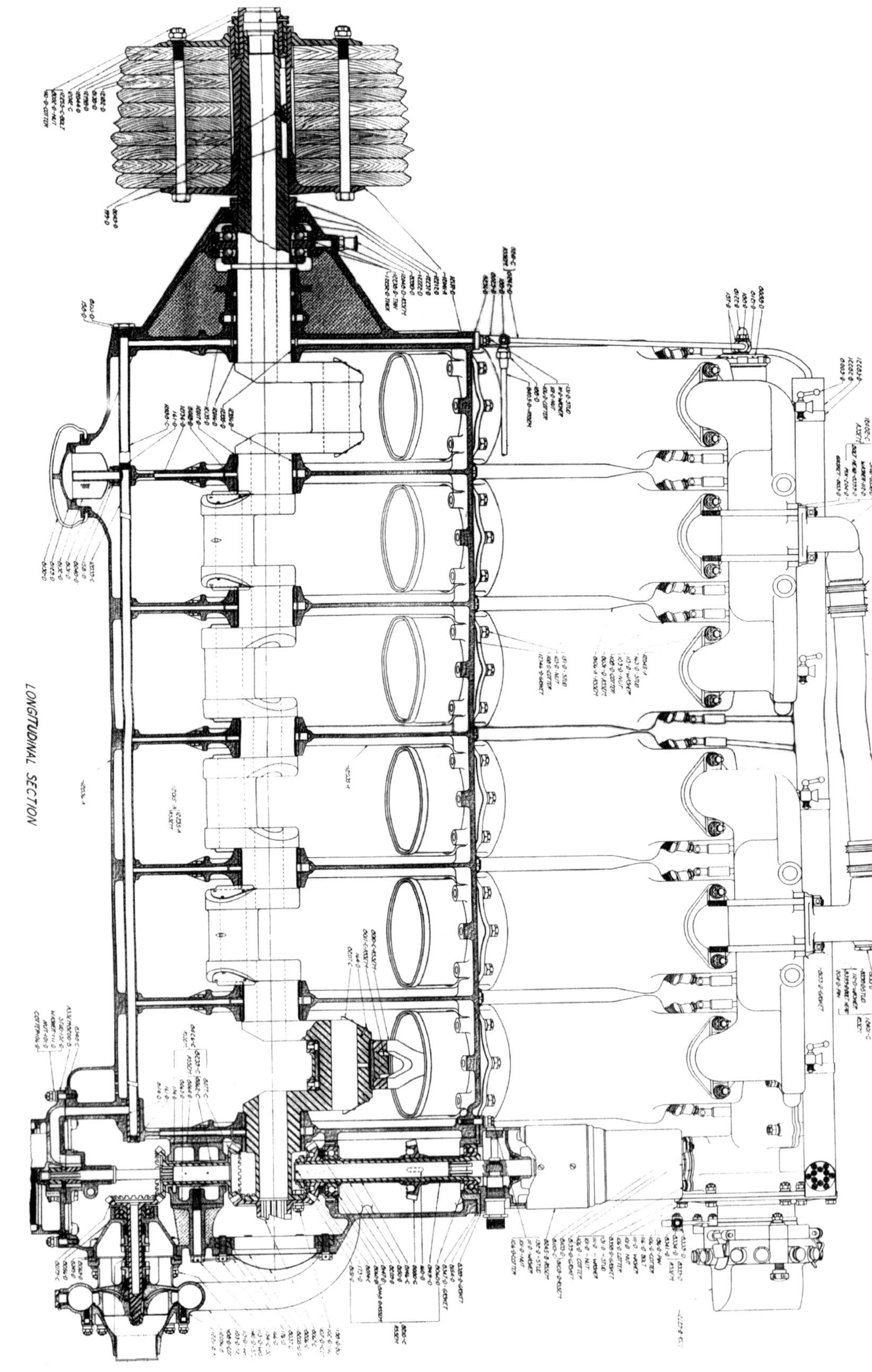

Longitudinal Section of the Liberty Twelve with part numbers shown. This drawing shows a very early design of the engine (about 11/1/17) with the first light crankshaft (Part Number 12015) with crank pins supplied oil by scuppers applied to the crank cheeks. The scuppers (Part Number 8207 C) can be seen applied to both sides of each crank pin cheek. The rear crank pin shows a cutaway of this oiling system and notes the Part Number of the scupper. It also shows the original design light connecting rods. (Courtesy of the Air Force Museum)

Chapter 3

Early Contracts and the Aircraft Investigations

On May 2, 1917, Capt. Waldon wrote General Squier that a month prior he had considered sending a mission to Europe to gather information regarding aeronautical requirements but had dropped the idea. (See page 28.) Circumstances now seemed to indicate that this proposal was once again valid and he suggested putting together such a mission and that it include Capt. Gorrell, one of his originally suggested members, and Capt. Clark. Capt. Clark was the chief aeronautical engineer of the Air Service and graduated from M. I. T. with a degree in that subject.

THE BOLLING MISSION

This time the idea was carried through. General Squier and Capt. Waldon, and others, put together a group of experts to go to Europe. On May 16, the Aircraft Production Board was officially organized. Its members, which included Capt. Waldon and General Squier, helped select the members of the commission. Its mission, as stated in the *History of the BAP*, was "that some well organized mission must be sent overseas to survey the aircraft situation there, determine what were Europe's most effective types of planes and motors, study out the program best suited to America's unlimited resources, and then handle the technical, legal, and diplomatic phases of the situation so as to promote the utmost harmony and cooperation between all the Allied air forces, and speed up the work of "winning the war in the air" by developing a concerted construction program among the Allies."

This became known as the Bolling Mission, named for its leader, Raynal C. Bolling. Bolling graduated from Harvard Law School in 1902 and was currently general counsel for United States Steel Corporation. He had been in the New York National Guard since 1907 and was first a Lieutenant and then Captain in charge of the first National Guard Aero Squadron since 1915. On June 9 he was commissioned a Major in the Air Service, designated to head the mission and made a member of the Bureau of Aircraft Production.

The six additional members were: Capt. Virginius E. Clark and Capt. Edgar S. Gorrell representing the Army; Commander G. C. Westervelt, the Navy's leading aeronautical engineer and naval designer, and Lt. W. G. Childs, recognized as the Navy's engine expert, representing the Navy; civilians Howard C. Marmon, one of the leading motor experts of the automotive industry and part owner of the Nordyke & Marmon Company, and Herbert Hughs, a Packard automotive engineer who had been recently in charge of that company's Paris Branch Office, whose intimate knowledge of French and Europe's best engineers, made him very useful.

In addition, five others were sent to give technical aid. They were: Robert Glendinning, a Philadelphia banker, now commissioned a Major in the Air Service, whose duty was to assist Major Bolling in completing financial negotiations and contracts and later to superintend the organization of a training school at Issoudun, France; C. H. Heilman, chief metallurgist of the Northway Motors Corp.; W. D. Hurley, steel expert and metallurgist of the Midvale Steel Co.; R. A. Vale, production engineer for the Dodge Motor Car Co.; and Mr. Stay, a casting expert for the Aluminum Casting Co.

The mission sailed from New York on the *SS Adriatic* on June 17, 1917, and arrived at Liverpool, England on June 26. They remained in England until July 2. They then went to France, staying there until July 15, at which time they left for Italy, and remained there until July 27, 1917, on which date they returned to Paris.

From not long after the mission's arrival in England until it officially ended its mission, information was sent back and forth across the Atlantic by cable and letter. On July 11 a conference was held in Paris with British, French, Italian and Bolling Mission representatives present. This meeting resulted in a manufacturing program for America. This was submitted to General Pershing's staff, who agreed, and so the recommendations returned to America represented the unanimous judgement of all concerned. A preliminary report was sent by cable on July 30.

The original membership of the mission broke up in mid-August of 1917 and each member submitted his report. The overall report submitted by Bolling, by now commissioned Lt. Col. (August 8, 1917), on August 13, 1917, became the general guide for the beginnings of American aircraft development and production.

MISSION IN ENGLAND

Before leaving England, Bolling requested the British to ship a number of items to the United States which he felt we would have a high interest in and some of which we might manufacture. Along with the request was the promise that the packing cases containing these items would not be opened here until word was given by

England that the patent rights issue had been settled. This procedure could provide a savings of a month in time by having the items in route while waiting for clearance of the patent issues. On July 1 Bolling cabled that some 14 items so requested were packed for shipment. Among them the following airplanes: de Havilland 4, SE-5, Sopwith Camel, Bristol Scout and a Bristol Fighter. The de Havilland 4, SE-5 and Sopwith had been recommended for study and possible production by Capt. Clark on May 2, 1917, more than a month prior to the formation of the Bolling Mission. This on-site study served to verify that prior recommendation.

MISSION IN FRANCE

After the mission's arrival in France on July 2, things went very slowly and unsatisfactorily. Major Bolling asked the French to ship a number of items to America for study. Among them were several planes. The French promised to ship all these items the following day and the mission cabled the United States on July 6 that they would be shipped immediately. In reality, the shipment did not leave France until well into August and one plane, the Spad, did not arrive in the United States until September 18th.

The mission was urged by the French to contract for some 20,000 airplanes and 30,000 engines. Later, contracts were signed for only about 25% of the French claimed capacity, 5,000 airplanes and 8,500 engines. Costs and delivery dates were agreed upon and contracts signed on August 30, 1917. The French guaranteed delivery of the material by the dates agreed upon.

The schedule called for the delivery of about 3,000 of the airplanes by May 1, 1918, and the remaining 2,000 by the end of June. Delivery was so poor that on May 3, 1918, an agreement was drafted between the French representatives and General Pershing arranging for cancellation of the August 30, 1917 contract because of French failure to deliver an appreciable quantity of the material ordered.

The mission recognized the great lack of facilities available in France for the off loading of material shipped there and also for assembly and test of any airplanes eventually shipped there for use by the American Expeditionary Forces (A. E. F.). Among other things, it was recommended that an assembly and test center be built at Romorantin.

As a follow-up to those recommendations, Col. Waldon went to France in November of 1917, arriving there November 12. He found conditions even worse than the Bolling Mission had been able to indicate. The following quote from his report indicated what conditions he found.

"Everything that we have in training, in equipment, in planes, in ordnance, should be carried to the very last possible operation before we transport a man or material to France. There are no facilities there. You have no idea of the conditions under which things have to be done in France. We have the time here, we have the men, we have the money, the food, the clothes, the housing, everything enabling us to carry every operation to the last degree. Over there, it is just like going into a desert. We have to provide everything. France is bare; it has nothing. A quartermaster at the line of communication told me a week before I left that five men had been in, saying that they had a keg of nails to sell of a certain size. He took them all up, but only one keg appeared — all had an option on the same keg."

As a result of this study, Colonel Waldon formulated plans for the huge industrial plant of the American Expeditionary Forces at Romorantin, where Air Production Center No. 2 was built.

MISSION IN ITALY

When the mission arrived in Italy they were extremely well received and co-operation could not have been better. The Italians agreed to send considerable material and a number of aircraft to the United States for study and to assure safe arrival, sent duplicates of most items on two different boats. They landed in the United States on August 15 and 16, 1917.

MISSION CONCLUSIONS AND REPORTS

As a matter of interest, by September, 1917, Capt. Gorrell had been commissioned Major, Howard Marmon had gone from civilian status to the rank of Captain in the Air Service and Capt. Clark had been commissioned Major and then Lt. Col. Clark and Marmon both returned to the United States early in October. Major Gorrell, Col. Bolling and a number of other members of the mission remained in Europe and took up other assignments. Major Gorrell was shortly commissioned Lt. Colonel.

As the Mission came in contact with more and more British and French aeronautical officers and engineers it became evident that the size engine of the Liberty series which would be most useful to the Allies would be the twelve. That fact was conveyed in a number of communications to Washington and its effect is seen in production decisions made by the Aircraft Production Board during this period.

Bolling commented in a letter of August 31, 1917, that "Your cable regarding the United States engine success delights us, as such success will do more than anything else to convince the Allies of our American ability to accomplish what is apparently impossible — Europe has not yet developed any engine of this size that is reliable or light-weight enough per horsepower to meet the pressing need. If the US-12 is successful, there will be an immediate and very great demand for it in the programs of our allies." This was in reference to the passing of its 50-hour test by that engine on August 25.

In his September 23, 1917, letter Bolling comments: "Maj. Gorrell is deeply impressed with the possibilities of the Bugatti engine, which the French and British have so far been slow to recognize." (We eventually attempted to put this engine into production as an alternate power plant to the Liberty, but with conspicuously bad results.) "You must forgive us overseas if our faith in the United States motor has been a bit faint, because we have been for four months among people who have not succeeded in doing anything of that sort, even in the pressure of three years war. Every successful engine here has taken a year, or even more, to develop. Many are not yet fully successful and capable of large quantity production." Here he was referring to reports of potentially having the Liberty engine in production before the end of 1917. This would have meant having an engine go from drawings to production in less than six months, twice the speed the Europeans had ever managed. This was in fact accomplished.

As noted, recommendations of the Bolling Mission resulted in the general direction taken by the aircraft production program of the United States regarding the producing of any foreign airplane or engine as well as having a significant effect on decisions regarding most aspects of the arming and training of the Air Service.

LT. COL. BOLLING

Upon the conclusion of his mission in Europe, Lt. Col. Bolling was appointed assistant to the Chief of Air Service, having charge of everything in the Zone of the Line of Communications. On December 12, 1917, Brig. Gen. B. D. Foulois relieved him of all other duties and put him in full charge, under the direction of the Chief of the Air Service, of all matters pertaining to the industrial relations between the A. E. F. Air Service, the Aircraft Production Board in Washington, and the English, French and Italian Air Boards.

Lt. Col. Bolling was killed in action on March 26, 1918 near Esterres, France. He was being driven in a staff car in route to an aerodrome when he and his driver were fired upon by German machine guns from both sides of the road. The car was disabled and he and the driver jumped for cover. After several minutes of being fired upon, two German officers advanced upon them and fired at the driver from close range. Bolling returned fire, killing one officer. The other fired at Bolling, killing him. Bolling was the first officer with the rank of Lieutenant Colonel to be killed in France. On June 28, 1918, the flying field at Anacostia, D. C. was named Bolling Field in honor of this officer.

PROGRESS TOWARD PRODUCTION

To keep from loosing continuity with our current subject, I will go back to the first mention of production plans for the engine now called the Liberty. Plans were started for the Liberty series of engines on May 29, 1917. On May 31 a report containing the proposal for the Liberty series of engines was presented before a joint meeting of the Aircraft Production Board and the Joint Army-Navy Technical Board. During that meeting, Navy members of the Technical Board stated that "the proposed engine would be merely for military battle and not adapted to Navy needs." This was an interesting statement when one considers the fact that some three months later, on October 21, 1917, the first L-12 to fly did so in a Navy

Colonel Sidney Waldon and Colonel Edward Deeds of the Aircraft Production Board. (Signal Corps photo)

Curtiss HS-1 airplane and that almost immediately the Navy adopted it as their standard engine for other than training purposes. This was quite a rapid turn around.

At that May 31 meeting it was also agreed that ten prototype Liberties would be built, five eights and five twelves. (Ultimately the order for eights would be changed to six.) Also at that meeting it was resolved that the Aircraft Production Board would "work out an agreement with certain motor manufacturers regarding a standardized motor." The names of the manufacturers were not stated.

The initial drawings were completed for the eight and twelve-cylinder models on June 4 and by June 6 work was begun on building the first prototype. Work began on production drawings at that same time. By July 5 the initial production drawings were complete. Early versions had been distributed among a number of potential manufacturers in the Detroit area to obtain advice on any corrections or changes they might deem advisable to better facilitate manufacture. One of these manufacturers was the Ford Motor Company and on July 3rd they reported that they were working on a method to make cylinders out of steel tubing rather than having to make them by boring and milling solid bar stock. They had successfully perfected this procedure by the time the Liberty went into production and it saved millions of dollars in cost of materials and production time and sped production tremendously.

On July 3rd the first prototype eight arrived in Washington at the Bureau of Standards. The remaining ten prototypes were completed one-by-one and submitted to testing in order to determine any changes deemed necessary before actual production began. The last prototype was completed in early November. But even before that happened, facilities for series production were being set up.

The first mention of actual attempts to find producers appears in the APB minutes of June 18, 1917, when Deeds and Montgomery are directed by the board to negotiate with Trego regarding producing the Liberty. The reason Trego was approached so early is a mystery. (The reader is directed back to page 46 for my thoughts on the subject.)

Obviously Packard was expecting to become a producer and I am sure the APB also assumed they would be and led them to so believe. As we shall see, however, Packard began to become anxious about it as time went by and no actual contract was proposed until September.

THE LELANDS PREPARE

On July 7 General Squier confirmed to the board that he had encouraged the Lelands to get ready to manufacture the Liberty 8, that a very substantial order would be given as soon as funds were available.

The Lelands had been kept informed as to the progress of the Liberty project. As expert motor engineers, they were approached for advice regarding the design of the engine and Cadillac, which they headed at the time, had built some of the parts for the prototype engines.

On April 7, the day after the declaration of war, the Lelands, Henry M. and son Wilfred C., had gone to see William Durant, the head of General Motors, of which the Cadillac Company was a part, to see if a building just completed for Cadillac and intended for constructing bodies could instead be devoted to the production of airplane engines for the government. The Lelands were highly patriotic and felt it their duty to use their best talents and the facilities of their company to aid in the winning of the war.

Durant refused and would not be persuaded otherwise. He told them he thought this was not our war and that General Motors would not do work for the government. (This attitude would, however, soon be found to be hastily arrived at and within a few months General Motors companies would be deeply involved in war work. Both Cadillac and Buick would soon be building Liberty engines.)

Henry M. Leland. (Courtesy of NAHC)

Alvan Macauley. (Author's collection)

The Lelands were very serious in their patriotism. They resigned from Cadillac on June 18, 1917, and by July 3 had completed all that was required to turn over their control to others. In the meanwhile they had purchased a group of small residences and small factory buildings on Holden Avenue in Detroit for the sum of $300,000. With an additional investment of some $900,000 they partially equipped a facility of 109,500 square feet. With that small factory and what final machine tools that would be found necessary after complete plans were available, they felt they would be able to produce about 20 Liberty 8 engines or 14 Liberty 12s per day.

PACKARD PREPARES

On the same day that Squier noted that he had told the Lelands to get ready to produce the Liberty 8, July 7, 1917, he also wrote Macauley of Packard to "advance work on the U.S. 8-A, after the motor is approved by Joint Army - Navy Technical Board a good sized order will be given your company." At this point in time, the first prototype 8 had just been delivered to Washington.

In the July 11 APB minutes is found a resolution saying that "30 days delay in tooling up in Cadillac (meaning, at that time, the production plans of the Lelands) and Packard will make a difference of 2,500 8's. Orders to be given Leland and Packard as soon as the motor is approved." This sense of urgency was passed along to both of these manufacturers who in turn pressed on in preparations using their own funds upon being assured that contracts and government funds would soon follow.

On July 9 Capt. Waldon contacted the Root and Vanderwort Co. and asked if they would be interested in producing a quantity of Liberty engines, that the board was anxious to get sources of supply lined up. There is no record surviving of this company ever being considered as a producer beyond this mention.

Under the date July 13 are several notes of interest: "Cablegram from Pershing: make twice as many 12's as 8's" "Vincent to Waldon: We have not mechanics to put over hand-made jobs on engines. Must manufacture." In other words, he was indicating that Packard must not be expected to produce large batches of hand-made engines. A contract that allowed installation of production equipment would be needed if quantity production was required. "Waldon to Deeds: Not sure 8 is right size or suited to any particular class of work here - 12 cylinder right size and should be largest production. 12 suited to DH and Breguet." Here Waldon is responding to Pershing's cablegram and other investigations of circumstances as they are developing in this country, and at this point things were moving rapidly.

We must assume that Washington had continued to convey the urgency of preparing for production to those in charge at Packard, but that they had also noted the swing toward thinking the 12 would be the more important size. On July 17 Macauley wrote letters to both General Squier and Capt. Waldon. To the General he wrote that Packard was going ahead "full tilt" and expected to be the first to produce in quantities, but that a "definite order was desired because there is considerable expenditure for materials." To Waldon he wrote "we prefer to manufacture 2,000 8's before tackling the 12. Problems will be easier to deal with in the 8-cylinder. All facilities will be available for this work by October 1."

On July 14 a contract was awarded to Wright-Martin for 500 150 HP Hispano-Suiza V-8 engines (order No. 8807). Wright-Martin was already well under way on an order for 450 of these engines for the French. They had received a sample engine in 1916 and also the aid of on-site Hispano-Suiza engineers in order to get production facilities set up. The engine did not reach production until about the same time the Liberty did, even though started a year earlier from an already designed and proven engine.

I will also note at this point that the initial order for 1,000 Hall-Scott A-7a 110 HP engines referred to on pages 30 and 39 was consummated with Nordyke and Marmon on June 21 (order No. 8440) and that an order for 600 90 HP Curtiss OX-5 engines was finalized on June 30 (contract No. 1448) with Curtiss. Curtiss was already in production of this model and thus deliveries began the next month. Hall-Scott's production facilities were already operating at capacity on foreign and Navy orders and thus the placement of the A-7a order with Nordyke and Marmon. That company was just getting production facilities set up and it was November of 1917 before deliveries started.

More orders for both of these engines would be placed in the following months in addition to further orders for various models of the Hispano-Suiza engine. Also an order for 2,500 Le Rhone 80 HP 9C models was placed with Union Switch & Signal (contract No. 2136) on September 25, 1917. Thus the decision was made early to produce large quantities of proven, and in many cases already in production, old designs covering the power ranges of the 4 and 6-cylinder versions of the newly designed U. S. A. engines.

In the board minutes of July 17 was a note that Major Bolling had urged increased production of the Hispano-Suiza engine but that inasmuch as the desired number could not be produced in time to meet requirements, the U. S. 8 had to be looked to for this need. Either these requirements eased, or Hispano-Suiza production improved more than expected, because no early contracts for the eight-cylinder Liberty materialized. At this point in time, Bolling had completed his tour of England and France and was in Italy.

On July 18 Deeds and Montgomery were directed to open negotiations with Trego for production of 12-cylinder models. The first mention of Trego had been a month earlier, on June 18, and no doubt the engine in mind at that time was the eight.

On July 19 the Aircraft Production Board directed Vincent to send Trego drawings of the eight and follow them with drawings of the twelve when they were completed. Vincent responded on July 20 that he was sending drawings of the eight to Trego, Henry M. Leland and Packard and the twelve drawings to Pierce-Arrow. The board replied telling him to also send the twelve drawings to Packard.

On July 24 the Aircraft Production Board wrote to Macauley of Packard telling him they were sending blueprints of both the eight and twelve-cylinder models. On the same day they received a letter from Macauley: "Asks for definite order because of great expense involved. Afraid government may change plans. In such a case, asks if they would reimburse for loss incurred."

On July 25 the Aircraft Production Board sent letters to Henry M. Leland, Frank Trego and Alvan Macauley telling them that complete sets of blueprints of both the eight and twelve-cylinder models are being forwarded and asks for suggestions and criticisms from the factories.

Ford engineers had been shown designs of the Liberty shortly after they were drawn up and had decided that they could devise a method to produce cylinders faster and cheaper than milling them from solid billets of steel. On July 3 the board minutes noted they were experimenting with making them from steel tubing. By August 7 Ford had made great strides with the process and Deeds wrote to Ford saying he understood the process was successful and that the company had been allocated production of 200,000 cylinders with a schedule of deliveries running from September of 1917 to July of 1918 of about 20,000 per month if Ford could handle that amount of production. Price was to be agreed upon later. They replied on August 13 that they expected that they could be producing 1,000 cylinders a day within three months. On August 14 the board noted that they were negotiating with Ford, Westinghouse and Drill for the production of all the cylinders that would be needed for the United States motor.

CONTRACT PLANNING BEGINS

On August 2, 1917, the Equipment Division of the Signal Corps was created and Deeds was placed in charge with Waldon as his assistant. By the end of the month both of these men had been commissioned Colonels in the Signal Corps and most other civilians on the Aircraft Production Board were commissioned as well.

Among those personnel brought on board at that time was Harold H. Emmons. Emmons was named Chief, Engine Production Section. He was a Detroit lawyer who had much experience in industrial organization and operation and was widely recognized as a top industrial executive. Emmons was a Lieutenant in the Michigan Naval Reserve and was activated in that rank for the remainder of the war. He did a superlative job of organizing a total of 23 engine production plants and 79 parts factories which handled all government aeronautical engine production during that period.

Lt. Emmons figured prominently in the history of all government controlled aeronautical engine production during the war. (I find it quite interesting that although that contribution was recognized by the presentation of the Distinguished Service Medal by the War Department, the Navy never saw fit to raise him in rank. Considering the responsibilities of his position and his performance in it, he should have been promoted to Commander.)

Following the war he left active service but remained in the Navy Reserves. He retired in the grade of Lt. Commander in 1924.

On July 28 Waldon wired Deeds and asked his opinion of placing contracts based on cost plus a profit of $500 on the 8 and $750 on the 12 with a bonus of 20% of savings under the average estimated cost and a similar discount from profit where cost exceeds estimate.

On August 22 Leland wrote the APB protesting the fact that, following the insistence of General Squier and the board, they had buildings, machinery, etc. ready but still no contract.

The form of contract noted above was approved in principle by the board and on August 24 a resolution was adopted whereby, at the suggestion of the Secretary of War and the Chairman of the War Industries Board, a cost estimate was to be secured from expert engineers. Mr. Rollin White of Cleveland was specified as one and the second was to be an engineer from Pierce-Arrow of Buffalo to be picked by the APB. The APB picked Henry May of Pierce-Arrow as the second man. Vincent and Emmons were to assist.

Henry May reported a cost estimate of $6,087 each for the 12-cylinder version of the engine on August 30 but as yet there was no report from Rollin White. The APB voted to go ahead with contracts with Lincoln and Packard for 6,000 Liberty 12 models each; 3,000 with Nordyke & Marmon and 500 with Trego Motors. The contracts offered set an estimated cost of each engine at $6,087 and a fixed profit of $913.05 plus 25% of any amount the estimated cost shall exceed the actual cost.

The actual resolutions that such Liberty 12 production contracts be offered to these companies were made to the Chief Signal Officer. (The Chief Signal Officer was the individual with power to execute these contracts. The APB could only recommend, but records seem to indicate that virtually all formal resolutions of the board were acted upon positively.) The resolution was as follows: 6,000 to Lincoln on August 31; 6,000 to Packard on August 31; 500 to Trego Motors on September 5; 3,000 to Nordyke & Marmon on September 7.

INITIAL CONTRACTS

The contracts were actually executed (signed) as follows:

No. 1647 - Lincoln 6,000: September 3, 1917
No. 1646 - Packard 6,000: September 4, 1917
No. 1648 - Trego Motors 500: September 6, 1917
No. 1704 - Nordyke & Marmon 3,000: September 7, 1917
Total-15,500

HOW *the* DECKS WERE CLEARED *for the* LIBERTY MOTOR

This photograph was taken on November 15, 1917. At that time the factory had machined, and for the most part, assembled six thousand engines, clutches and transmission cases for the entire season's run of Twin-Six passenger cars. This picture shows part of the accumulation of powerplants as they looked when stored in the main stock building. This caption and photograph are from the August 1918 issue of the "Packard house organ." (Author's collection)

PACKARD MOTOR CAR CO. SETS UP FOR PRODUCTION

Of the above manufacturers, Packard was the most advanced so far as ability to proceed with manufacturing was concerned. Packard's facilities, however, were relatively limited so far as any quantity production was concerned. They had been in the process of preparing to swing from automotive to Liberty engine production for some time, while simultaneously hand-building the experimental engines, the last of which would not be completed until about November 17. This involved not only converting what machinery they could to potential Liberty parts production, but shutting down automotive production in such a manner as to make storage of remaining stocks of parts easier and reduce potential losses involved in such a move.

Packard's current production automobile was the Third Series Twin Six. Production of it was curtailed and some 6,000 engines were assembled from parts on hand and stored in order to make space for proposed Liberty production and to make available what part of the automotive engine production machinery that could be used for that purpose. To quote from a letter from Alvan Macauley to the Honorable William L. Frierson, Assistant Attorney General, dated May 31, 1918, and written in response to an inquiry relative to engine deliveries not being up to contract schedules: "To do this (referring to those actions necessary to make Liberty production facilities available) meant the premature manufacture and accumulation of several million dollars worth of carriage motors and parts with a very severe strain upon our financial resources." It also was accompanied by much reduced automotive sales and thus reduced income.

Although they had considerable facilities available for limited production of these engines, they as yet had little of the large machinery necessary for full-scale production. Acquisition of such machinery would require both time and money and could not be ordered until an actual contract was in hand. They had, however, ordered some $400,000-worth of new machinery at their own risk prior to actually being offered a contract.

Their contract was signed on September 4, 1917 and called for the production schedule shown to the right. As will be seen, this schedule was overly optimistic for a number of reasons. Packard signed it on faith, even though the drawings and specifications Article I referred to as attached to the contract were not, and indeed final specifications were as yet nonexistent. It was understood by both parties that those conditions existed and, as Packard later stated, "It was the spirit of our undertaking that we were to continue to co-operate with the government in the further development of the design, and to this end we should attempt to produce motors before complete data were available, in order that, through the experience thereby gained, the specifications and details might be modified wherever necessary."

Initial Schedule for Packard Contract 1646
September 4, 1917

Month	Contract Schedule Quantity	Total to date
1917		
November	50	50
December	200	250
1918		
January	500	750
February	800	1550
March	1000	2550
April	1200	3750
May	1200	4950
June	1050	6000

Besides, they had the protection, they thought, of Article III of the contract which said, in part, "In view of the present uncertain factors inherent to the articles, and in the establishment of the industry of producing them, it is understood that this schedule may require revision...." Typical of government, both then and now, not much credence was attached to that clause when Packard was severely criticized sometime later for falling behind on the original schedule.

To put this particular point in time in perspective, we must realize that the second experimental L-12 had not been completed at the time this contract was signed and certainly no final production specifications and drawings were available. In addition, Packard had already made it clear that their production facilities would not be totally available for Liberty production until October 1.

To help offset the large expenditure needed for purchase and installation of heavy production machinery, on November 3, 1917, the government provided Packard an advance of $2,500,000 on their contract secured by a note in that amount at 5% interest.

A number of large production machines needed for connecting rod and crankshaft finishing which were ordered in August and September with delivery promised in October and November were not delivered until March of 1918. This is but the tip of Packard's iceberg of production problems and we will return to more of them later.

LINCOLN MOTOR CO. SETS UP FOR PRODUCTION

Although the APB had told the Lelands that they should prepare for production at the rate of about 14

Liberty 12 engines per day back in July, and they had done so, two months later the requirement had jumped to 70 engines per day. To be able to accomplish this would require additional plant and equipment expenditures of some $8,500,000. A complete new plant would be required.

The Lelands purchased a 50-acre plot of ground on Warren Avenue and made plans for the construction of an eight-building complex totaling 615,959 square feet. In the mean time, their initial plant on Holden Avenue was remodeled and equipped to begin operations on a small scale.

Construction on the new plant was begun in earnest on October 22, the installation of machinery was begun on January 1 and power was turned on in it on February 12. This construction had begun using funds borrowed and backed by notes signed by the Lelands and Joseph Boyer of the Burroughs Adding Machine Co.

By late November, Col. Robert L. Montgomery, who then controlled contracts and any financing required for them, had come to the conclusion that financing the required new factories and their facilities would soon overwhelm the private banking industry and that the government should itself provide the necessary loans to prevent that eventuality. (The reader is referred to the following discussion of the Nordyke & Marmon contract for some insight into the circumstances that led up to this decision and the contract renegotiations that followed.) As a result of this decision, the government loaned Lincoln $5,000,000 at 5% interest on December 10. It was to be used to provide what further finances as would be necessary to complete the new plant facilities, pay current obligations for machinery and materials and pay off their current $1,500,000 bank loan.

Lincoln faced somewhat the same situation as did Packard at this point in time. They had put considerable effort into preparations for production and had their small plant on Holden Avenue well along when their contract was signed, though it was by no means ready to go into production. Their contract had the same Article III with its apparent schedule protection as did Packard's. They also faced the same lack of specifications specified as existing in Article I as did Packard. They would soon also face the same condemnation by the government for failing to live up to their initial production schedule.

As can be seen from the prior two production schedules, Packard was expected to reach a capability of 1,200 engines per month and Lincoln at least 1,900. The 70 per day mentioned earlier as a Lincoln requirement would result in a capacity of 2,100 per month.

NORDYKE & MARMON SET UP FOR PRODUCTION

The Nordyke & Marmon Company of Indianapolis, Indiana, had been in the business of manufacturing flower milling machinery since the mid-eighteen hundreds and by 1904 could claim to be the world's largest producer. In 1902, under the direction of Howard C. Marmon, a young family member of the firm who had recently graduated from the University of California with a degree in mechanical engineering, the company added automobile manufacture to their line of work. Howard became chief engineer and principal designer and by 1917 the automotive section of the company would build some 4,500 cars.

Upon the declaration of war on April 6, 1917, Mr. Walter C. Marmon, president of the company, told the government that they would offer their services to the war effort in any way they could. After some study it was decided that their best services could be rendered in the production of aircraft engines.

Initial Schedule for Lincoln Contract 1647
September 3, 1917

Month	Contract Schedule Quantity	Total to date
1917		
November	5	5
December	80	85
1918		
January	160	245
February	275	520
March	700	1220
April	1400	2620
May	1900	4520
June	1480	6000

Initial Schedule for Nordyke & Marmon Contract 1647
September 7, 1917

Month	Contract Schedule Quantity	Total to date
1918		
January	25	25
February	125	150
March	550	700
April	700	1400
May	800	2200
June	800	3000

Nordyke & Marmon corporate officers. H. C. Marmon (upper right) would enter government service in 1917 as a member of the Bolling Mission, then be commissioned Captain in the Signal Corps. By the end of the war, he reached the rank of Lt. Col. (Author's collection)

At the time, Mr. F. E. Moskovics, vice president of the company, was in Washington working as an official of the Society of Automotive Engineers and was thus in close contact with the APB in an advisory capacity. On May 28 Mr. Moskovics was requested to meet with Major Souther, then Chief of Aircraft Engineering for the Signal Corps. Major Souther informed him that the Hall-Scott A-7a engine had been selected for production as a training plane engine and asked Nordyke & Marmon to take a contract to produce 1,000 of them. Mr. Moskovics later said that up until that time he had never heard of the Hall-Scott A-7a engine.

E. J. Hall shortly visited the Nordyke & Marmon plant and a suitable study was made by the plant engineering staff and the contract was signed on June 21, 1917. Production facilities were rapidly set up but the usual delays had to be dealt with, such as the lack of detail drawings and difficulty in obtaining production machinery on time. The company continued automotive and milling machine production in most of their current facilities and started construction of an entirely new single story factory to be used for aircraft engine production. The company received a $500,000 advance on the A-7a contract (No. 1516).

The A-7a contract was scheduled to be completed in October of 1917 and thus free production machinery and space for the production of the Liberty engines contracted for on September 7. In actuality, deliveries of these engines did not begin until November and the last of the lot was accepted by the Signal Corps in May of 1918. It was not until that time that the plant could be entirely switched over to Liberty production. In addition, they had underestimated the requirements for production of 3,000 Liberty 12 engines and additional production facilities had to be built. On December 10, 1917, the company received a government loan of $2,000,000 at 5% to pay off an existing loan of $1,500,000 and provide necessary working funds.

Nordyke & Marmon did not assemble their first Liberty engine until June 15, 1918. This was six months behind the initial schedule and brought them under the scrutiny of later investigations regarding delays. The delay did have its good point so far as Nordyke & Marmon was concerned — by the time they were ready to begin production, availability of detail drawings was no longer a problem and most required design changes had been made and the engine design was fairly stable.

TREGO MOTORS CORP. SETS UP FOR PRODUCTION

Trego Motors Corporation of New Haven, Connecticut appears to have been set up in early 1917 expressly for the purpose of building Liberty aircraft engines. City directories list it from 1918 through 1920. Frank H. Trego was listed as Vice President and Mr. A. Stroud Haxton of New York as President.

Initial Schedule for Trego Motor Co.
Contract 1648
September 6, 1917

	Contract Schedule	
Month	**Quantity**	**Total to date**
1918		
January	40	40
February	50	90
March	60	150
April	80	230
May	100	330
June	170	500

Frank H. Trego is believed to be the gentleman in the center observing the testing of L-8 No. 2. (Author's collection)

On July 25, 1917, APB representatives reported very favorably of Trego's plant facilities as follows: "Their plant was fairly well advanced though still in the development stage. The shop was not yet ready for producing motors but could at once begin on tools and fixtures. The engineering and designing rooms and tool rooms were found to be organized, equipped and running. The only doubt expressed was as to whether they would be able to produce engines in the quantities desired."

The company received two advances, one for $125,000 at 5% on January 9, 1918, and one for $100,000 at 6% on March 15, 1918. In addition, they sold capital stock in the amount of $1,025,000.

As with the other contractors, they were expected to complete deliveries on all engines by the end of June 1918. The reason June 30, 1918 was picked seemed to be based more on the requirements stated on the production chart presented to the APB on June 5, 1917, (see pages 31 and 38) than on the capabilities of the companies contracted.

FORD RECEIVES A PRODUCTION CONTRACT

Contracts for the first 15,500 Liberty 12 engines had been signed by September 7, 1917. At this point, it was still anticipated that a quantity of Liberty 8 engines would also be required as soon as possible.

On September 6 it was noted in the board minutes that no contracts should be placed for 8's until Ford had been contacted, as Major Hall had been led to believe Ford would build them at-cost. On September 9 Lt. Emmons wrote to Mayo of Ford: "Schedule of requirements of 8-cylinder engines changed to 5,000. Ford Company is desired to prepare for production as soon as possible. The entire 8-cylinder business is to be placed with the Ford Company." And on the following day is a note from Emmons to Mayo: "Refer to changes in contract as Ford waves any profit on 8-cylinder." On September 11 "Col. Deeds reported informally of successful negotiations with Mr. Ford for the building of the necessary number of US-8 engines."

Initial Schedule for Ford Motor Co.
Contract 2129
October 31, 1917

	Contract Schedule	
Month	**Quantity**	**Total to date**
1918		
April	200	200
May	800	1000
June	1000	2000
July	1000	3000
August	1000	4000
September	1000	5000

Obviously, the intentions that Lincoln and Packard would be the producers of the Liberty 8 mentioned on pages 62-63 had gone by the wayside. Perhaps it was because Ford had volunteered to build them at-cost.

Whatever may have been going on behind the scenes, it seems that none of it worked out. There is no further mention of Liberty 8 contracts or production during this period and Ford's name does not come up again as a producer until October 25. On that date a resolution was passed recommending a contract with Ford for 5,000 Liberty 12 engines. The contract was signed on October 31 (No. 2129). This contract was, however, placed at the standard conditions of all prior ones, not at-cost. Evidently those in charge soon realized what legal and moral complications could result if indeed Ford was allowed to take on a large government contract at-cost while other producers were paid a profit. It is also possible that Ford had second thoughts about the matter and changed his mind.

Delivery schedules were set up much more liberally than they had been for prior contractors. First deliveries were scheduled five months after the contract date whereas they were two months for Packard, two and a half for Lincoln, four for Trego and four and a half for Nordyke & Marmon.

CONTRACT RENEGOTIATION

Up to this point, contracts had been signed for the production of 20,500 Liberty 12 engines and, other than the original prototypes, no Liberty 8's.

Nordyke & Marmon had found it necessary to obtain additional production machinery which would cost in excess of $850,000 and was having difficulty in financing the purchase. They therefore requested that the Signal Corps purchase it for them. This resulted, on November 13, in a BAP resolution that this purchase be made and that in return, the contract be amended to reduce the standard 15% profit to 12 1/2%.

Before any of this was actually accomplished, it was suggested at the November 19 meeting that "as quantity production increased, the base price estimate might need to be reduced." This was a logical enough conclusion, except one must realize that at this time Packard was the only producer to have completed an engine. And that company had assembled its first production engine on that very day. Therefore, one would hardly think that the time had yet come to negotiate price reductions.

It is not recorded just how and why this subject built such momentum, but on November 27 Col. Montgomery brought up the question of a possible change in base price of the Liberty 12 and asked the opinion of the board as to whether it would be proper to renegotiate the price. After some considerable discussion, the board decided that in the case of any manufacturer who had asked the government for financial assistance, the government should negotiate a change in price from a base price of $6,087 plus 15% profit to a base of $5,000 plus 12 1/2% profit.

On December 10, 1917, Col. Montgomery reported that he had renegotiated the contracts of Packard, Lincoln and Nordyke & Marmon to those terms. The board followed with a resolution to modify the contracts of not only Packard, Lincoln and Nordyke & Marmon, but also Trego and a contract just negotiated with Ford for the production of 5,000 Liberty 12 engines. With the change in base price and percentage, the profit per engine was reduced from $913.05 to $625 with the incentive profit shares remaining as before. All of these companies either had been or would be advanced funds, except Ford.

The first contracts were signed in the first week in September. By December 10 only Packard had completed any engines. They had built eleven prototypes by hand and about the same number of production engines, although even these were almost hand-built engines at this point since Packard was far from having final machinery in place and even the final design specifications were not firm at this point. Thus, before production had even begun, profits had been reduced by about 32%.

Henry Ford. (Author's collection)

Test reports from Packard continued to be encouraging for both the eight and twelve. These were in many cases forwarded overseas and thus the confidence of both the AEF and the Allies that a successful combat engine was about to arrive was building. As a result, the Allies were beginning to cable requests for engine shipments. Notes in the APB minutes relative to requests for both test engines and production orders are frequent, as well as often being confusing.

As an example, on November 13 Italy expressed a desire to place an order for 2,000 engines with deliveries to begin in December. The APB replied that it would not be possible to meet such early deliveries but that Col. Deeds estimated that 3,000 engines could be supplied with deliveries beginning not later than June of 1918. (We will follow the various Italian requests and their results as we progress through this history. For now, we note this request because it was one of many and brought considerable pressure to bare on production from current producers as well as inciting further contracts.)

The APB minutes begin noting intentions to ship small numbers of sample engines to England, France and Italy as early as February. However, these shipments were continually delayed. Three were shipped to the AEF on January 26 and four to London on March 6. The primary reason for these delays was the desire to incorporate force feed oiling in place of scupper before shipping engines to the allies for testing. It had already been determined that the original scupper system was not going to be satisfactory. Ultimately, Italian shipments would be held up for lack of a signed contract.

DUESENBERG AND GENERAL MOTORS RECEIVE CONTRACTS

On November 15 the APB recommended a contract be placed with Duesenberg Motors for 500 Liberty 12 engines and that 1,000 be ordered from General Motors, to be built at the Buick plant. The Duesenberg order was executed on November 20 as contract No. 2318 and the General Motors order on December 11 as order No. 30071. On December 11, 1917, the APB recommended the Duesenberg order be increased to 1,000 and order No. 30126 was executed on about December 15 for an additional 500 engines. Total orders now stood at 22,500 Liberty 12 engines. General Motors was expected to begin deliveries in May of 1918. We do not know what was expected of Duesenberg. On January 4, 1918, before they neared production status, their contracts were replaced with one (contract 2318-B) to build 2,000 Bugatti engines. At this time, General Motors was given a contract (order No. 30206 dated January 21, 1918) for 1,000 Liberty 12 engines to replace the Duesenberg order. Although the contracts were with General Motors, it was noted in the board minutes on February 2 that Buick and Cadillac were expected to come into production in May, thus both had been setting up for production from the start. Total contracted thus remained at 22,500 Liberty 12 engines and, thus far, no Liberty 8 engines.

By February 19, 1918, the Italians had increased their request to 3,000 engines and the APB recommended contracts on behalf of the Italian Government for 3,000 Liberty 12 engines with deliveries scheduled as shown below.

In late January the British made inquiries as to whether they could expect deliveries of 300 engines during March, April and May and 3,000 between May and October if they ordered them. Colonel Deeds replied in the affirmative. That decision was shortly modified somewhat in that it was recognized that the delivery of the full 300 engines during the first three months was not going to be possible.

BRITISH CONTRACT FOR 3,000

In Late February the British contracted for 3,000 engines with the delivery schedule shown below.

The British would, before the end of the war, contract for a total of 8,000 engines and eventually take delivery of a little over one quarter of them.

In late March the French government had been shipped six engines for test purposes and held up placing a large order pending both testing and assurances of delivery if large quantities were ordered.

FRENCH GOVERNMENT ORDERS 250

On July 17 the APB adopted a resolution that a contract be entered between the United States Government and the French Government for the purchase of 250 Liberty 12 engines. Delivery schedules were not set at the time but instead "delivery to be made at such times as should be consistent with the requirements of our own government and commitments to allied governments." The price of the engines was to be the same as was quoted to the British Government: "The cost to be not less than the actual cost to the United States including 2 per cent of the amount paid to the manufacturer to cover the cost of inspection, testing, and supervision of manufacture and, in addition, the cost of any transportation expense."

Before the end of the war, the French would sign contracts for an additional 2,650 engines. The total contracted for was thus 2,900 plus the 6 test engines mentioned above. They also eventually placed verbal orders for an additional 6,000.

CURRENT STATUS OF ORDERS AND PRODUCTION

In order to be able to understand the reasoning behind some of the decisions that took place in early 1918, we must take a composite look at just what had occurred

Italian Potential Delivery Schedule
(Never executed)

1918			
July	150	October	600
August	300	November	750
September	450	December	750
	TOTAL 3,000		

British Delivery Schedule

1918			
March	10	July	500
April	20	August	600
May	150	September	700
June	300	October	720
	TOTAL 3,000		

relative to the placing of contracts and actual production since April of 1917.

In the chart below, we can see the history of the design, contracting, actual production and scheduled production from the declaration of war in April of 1917 up to June 30, 1918.

I picked June 30, 1918 because this was the target date to have produced the number of combat engines the French had said would be necessary by that date. On May 24, 1917, a cablegram had been received from French Premier Alexander F. Ribot indicating the United States should make all preparations necessary to be able to supply 16,500 service airplanes and 30,000 engines for the war effort by June 30, 1918. On May 27 the Joint Army-Navy Technical Board responded by recommending we build 9,900 training airplanes and 19,800 engines for them and 12,000 service airplanes and 24,000 engines for them, between January 1, 1918 and June 30, 1918. This was a direct response to the French request. This became known as Production Program II. It gave few specifics and mentioned service-type planes only by their proposed use. Interestingly, it was much more specific regarding training planes and 2,500 DH-4 aircraft were among the recommended training airplanes with the assumption that they would be powered by Rolls-Royce engines. Both the planes and engines were to be built here.

At this point in time, of course, the Liberty engine had not yet been designed, though it soon would be, as the initial drawings were completed by June 5.

By July 3, 1917 proposed production charts covering the period of July 1917 through June 30, 1918 had been laid out for those aircraft and engines they thought would be put into production. At this point, a run of 15,000 Liberty 8 engines was to be built by Packard, White and National. The first 100 were to be delivered in October of 1917. 6,000 Liberty 12 engines were scheduled to be built by Pierce and Ford, the first 200 also to be delivered in October. Later modification of this program called for the building of 5,000 Liberty six engines but the proposed airplane in which it was to be used was not stated. This is the only instance I have found of any proposal of a large order for the Liberty six engine. It was never acted upon.

No selection had been made as to airplane types to be built for use as observation and bombing, but the proposed

Initial Contracts and Production Schedule Versus Actual Production

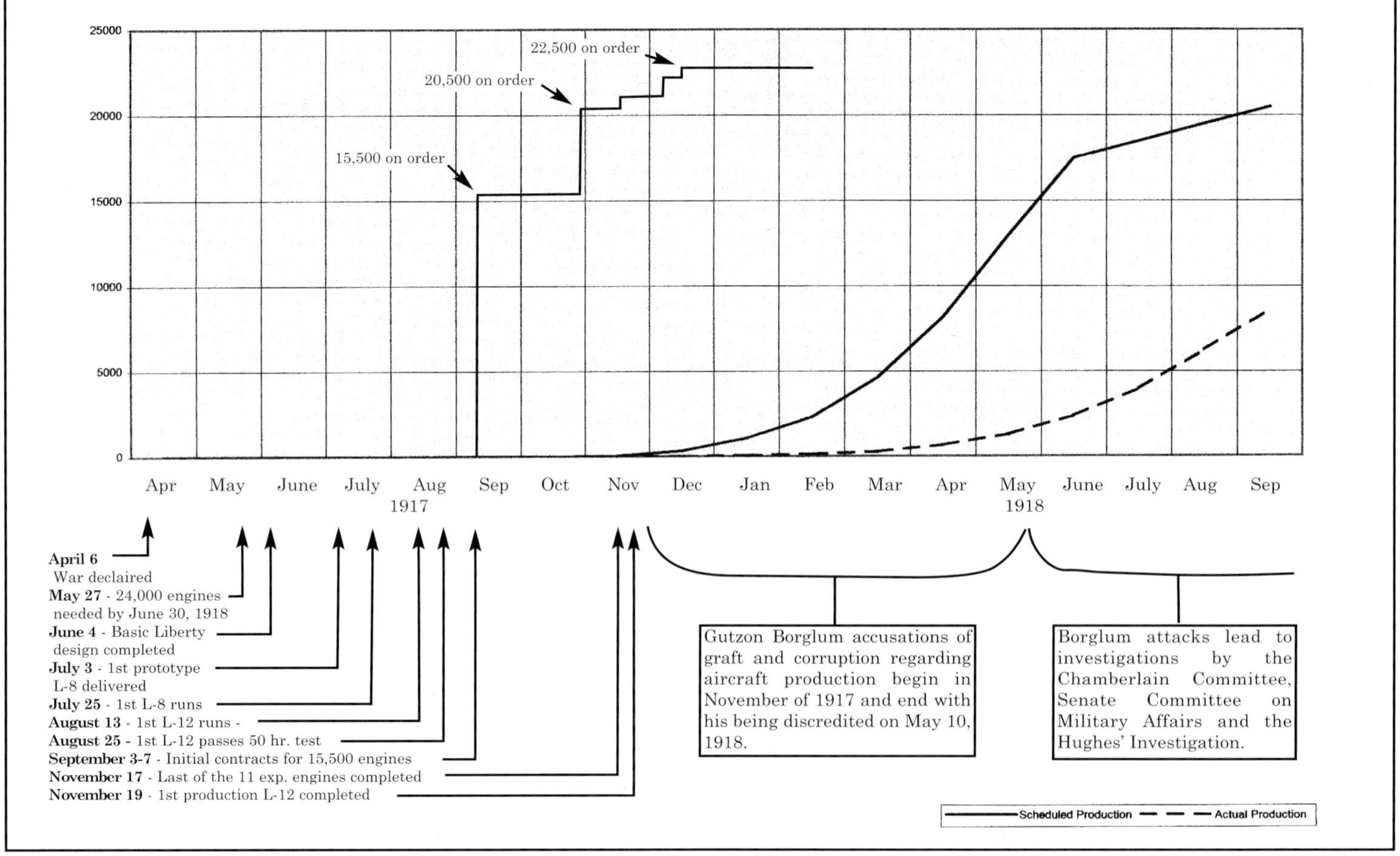

2,500 DH-4 airplane production had been charted with the first 100 to be delivered during November of 1917.

Production Program III was put in place on August 9, 1917 and it was upon this program that the initial Liberty engine order was based. Also, this program transferred the DH-4 from training status to combat service classification and increased the number to be produced by June 30, 1918 from 2,500 to 8,000. Actual airplane types were now specified and 1,000 L-12-powered Bristol Fighters, 1,000 L-12-powered Caproni bombers and 4,000 L-8-powered U. S.-built Spad Fighters were added to the list of airplanes to be completed by June 30. This totaled 14,000 Liberty powered planes requiring 16,000 engines, among them 4,000 Liberty eight types. The reader will recall that the eight cylinder model was considered during the period of engine contracting, but never finalized.

Of the 8,000 DH-4 airplanes called for in Production Program III, early orders were placed for 7,500 after much delay because of the eventual debate over whether to produce the DH-4 or the DH-9, its improved version. First deliveries were in February of 1918 and by the end of April only 28 had been delivered. A total of 517 had been delivered by the end of June 1918, whereas Production Program III called for 8,000 by that date. Liberty engine deliveries had reached 1,102 by that same date whereas initial contracts were for 15,500.

Production Programs were in a constant state of flux and at least nine officially designated ones had been generated before the end of the war. In addition, each program underwent constant modification to try to keep it up- to-date until another major new program was written. This was found necessary because of the constant changes in aircraft needs and designs as well as those dictated by discovering the realities of the ability or inability to produce various airplanes and their engines and accessories.

As can be seen from the Liberty contract and production chart on page 72 and the Aircraft Production Programs chart on page 74, great things were expected of the aircraft industry. At the advent of war in April of 1917 it had hardly even existed. By the time the initial Liberty contracts were signed and Production Program III was in place (August-September, 1917), it was expected that 10 months later (June of 1918) some 23,000 fully equipped American-built airplanes would be swarming the skies of the United States and Europe.

As to just why the military establishment and elected officials, even though advised by a multitude of industrial experts, would feel this country could supply the proposed amount of aeronautical equipment in the time-frame proposed is difficult to imagine. One might get a clue, however, from a statement found in the early minutes of the Aircraft Production Board: "If the automotive industry of this country can produce 100,000 automobile engines per month, as it now does, we can see no logical reason it cannot produce 3,000 aircraft engines per month." The same logic was applied to the production of airplanes as to their engines.

Actually, the overall logic proved to be acceptably accurate. The problem lay with the speed with which this production capacity could be put into motion. Production Program III (put in place in August of 1917) would require such production capacity by January of 1918 (six months). Airplane production was at about 2,000 per month when the war ended but would probably have reached 3,000 per month by mid-1919. Engine production fared much better. The last month of full-speed production of the Liberty was October 1918, during which 3,818 were delivered. An additional 1,479 aeronautical engines of other types were also delivered during that month for a total of 5,297.

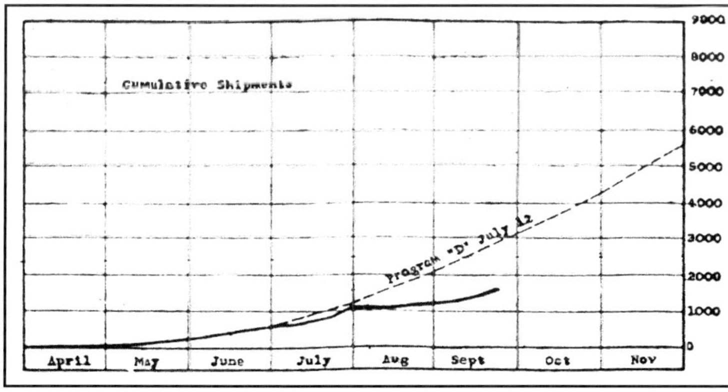

De Havilland 4 Planes

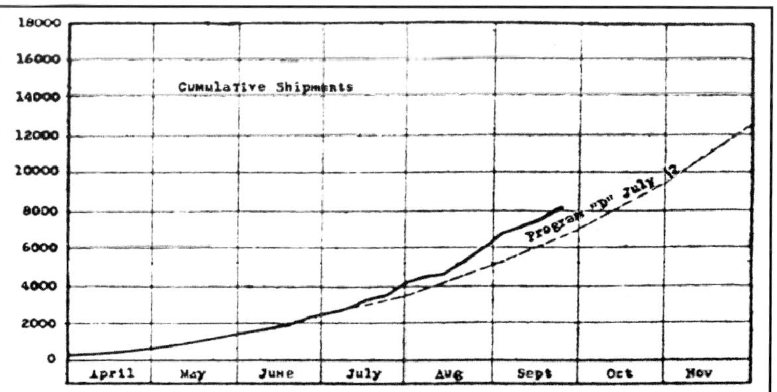

Liberty 12 Engines

The above charts are taken directly from the BAP weekly report of September 27, 1918 and show Program "D" (VIII) of July 12, 1918 verses actual production. Note that DH-4 production continues to fall behind plan whereas Liberty 12 production has finally managed to exceed plan. No American airplane production was ever held up for lack of a Liberty engine. (Author's collection)

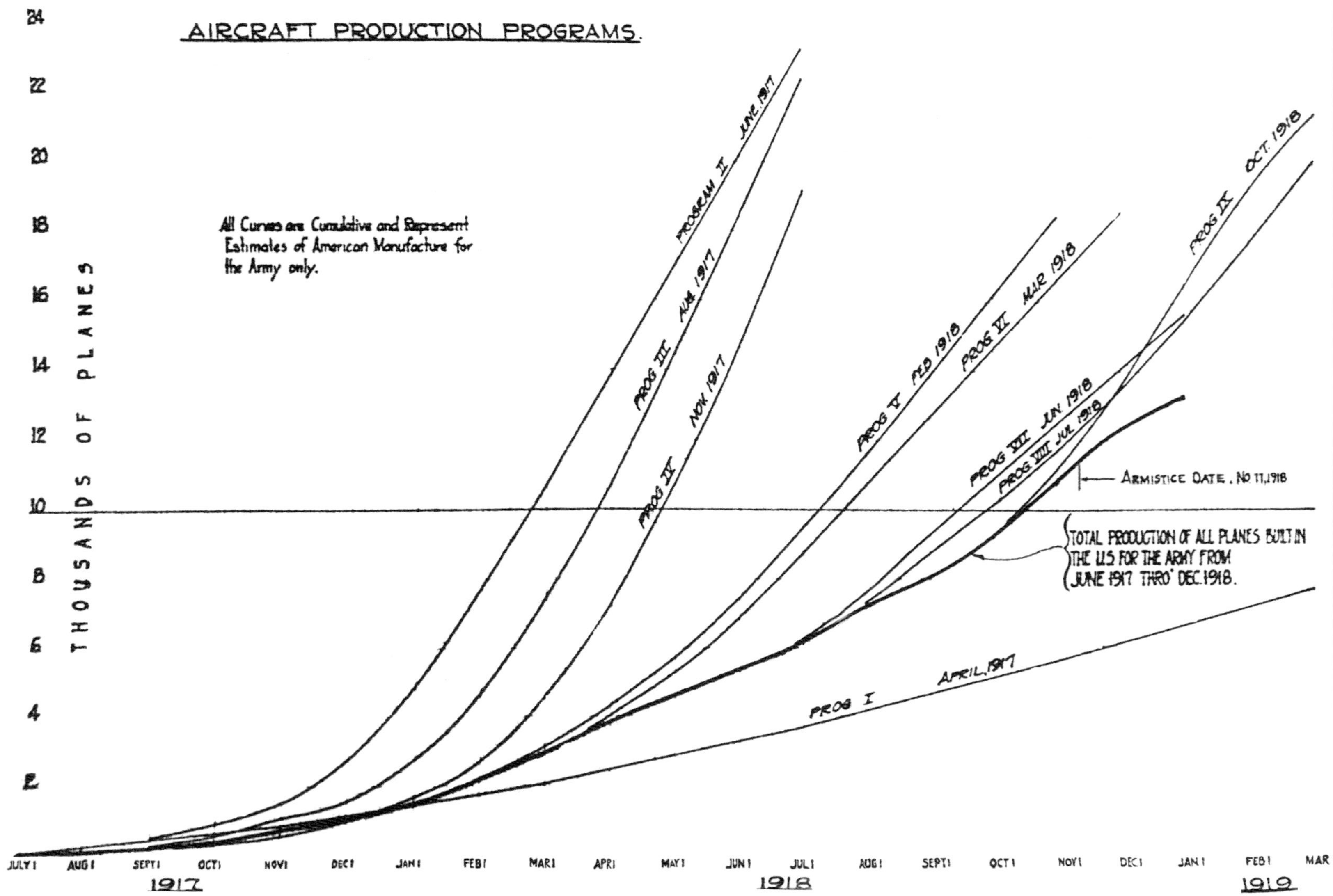

The above chart for the planned production of airplanes was extracted directly from the official BAP history compiled following the war by R. W. McFarland. The curve for each program initiates on the actual production line at the point in time it was put in place. Thus each plan's line represents contemplated production from its initiation onward. As can be plainly seen, each plan turned out to be overly optimistic. In no case did airplane production live up to the proposed plan. This chart shows the overall airplane production plan. Each individual airplane type had its own plan as did each type of airplane engine. The primary U. S. airplane using the Liberty engine was the DH-4. Its production remained behind schedule until the end of the war whereas Liberty 12 production exceeded plan from early July of 1918 onward. See the charts on page 73. Author's collection. **Note:** Program VIII on this chart was also referred to as Program D.

Aeronautical Production Organization as of February 1918

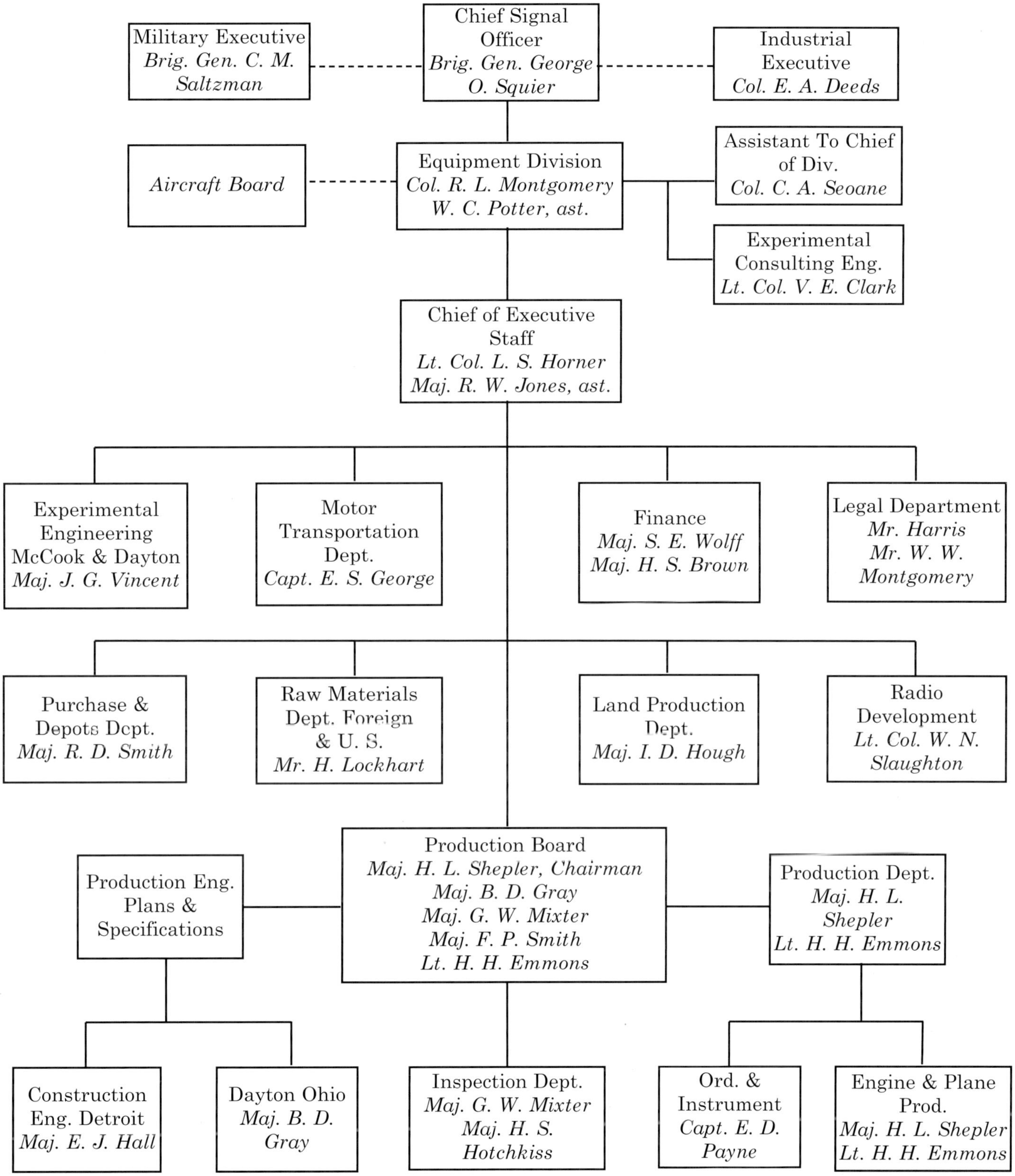

THE VARIOUS INVESTIGATIONS

The various aircraft investigations of World War I provided a major sensation but, in the end, minor revelations. What started as a personally driven attack on Col. E. A. Deeds by artist Gutzon Borglum, ended with investigations by the Chamberlain Committee, the Senate Committee on Military Affairs and, ultimately, the Hon. Charles E. Hughes' Justice Department Investigation.

I will cover these proceedings only briefly here but they were certainly a significant part of the history of the Liberty engine and thus cannot be entirely ignored. It is most likely that the accusations and the resulting investigations served only to cost the taxpayers money and impede the progress of aircraft production by directing manpower toward complying with investigative information requirements.

GUTZON BORGLUM

Gutzon Borglum was a sculptor of some note who had several reasons to dislike Col. Deeds and thus attempt to discredit him and the accomplishments, or lack of them, of the board which he headed. His earliest contact with Deeds seems to have been in 1912. Deeds headed a group of prominent Dayton citizens who were planning a memorial to the Wright brothers which was to be erected on Huffman prairie where they made their first flying experiments. Borglum was invited to submit a plan for a fitting bronze memorial and was paid a preliminary fee of $4,000. Before any final plan was adopted, the flood of 1913 devastated Dayton and the plan for the Wright memorial was put on permanent hold. Eventually it would become apparent that Borglum held Deeds responsible for the loss of this lucrative commission.

Several years later, in early 1917, Deeds once again contacted Borglum with a commission to sculpt a bronze bust of his recently deceased son Edward A. Deeds, Jr. After this was completed, Borglum suggested that Deeds build a $50,000 family mausoleum and proceeded to submit a design. In the mean time, Deeds became increasingly involved in the job of directing aircraft production efforts and Borglum's unsolicited plan was never acted upon.

Borglum was indeed a sculptor of considerable merit. He went on to win much acclaim as sculptor of the great Mount Rushmore. However, at this point in his life he envisioned himself also as an aeronautical engineer. He invented what he felt was a superior airplane of a so-called "Fish" type. It was brought before the aircraft board of which Deeds was chairman and rejected as not practical. Borglum was very much distressed over the rejection.

Borglum was, of course, not alone in his rejected aircraft proposal. A great number of others had offered various proposals to the aircraft board only to receive rejections. These proposals ranged from genuine attempts at better designs which were just not good enough, to outright attempts to profit from the seemingly endless supply of funds being spent on aeronautical readiness by selling anything the government might buy. Borglum was disgruntled, had a lot of company and some joined him in an effort to discredit those who had rejected them.

In November of 1917 Borglum wrote to Joseph P. Trumulty, Secretary to President Wilson alleging there was waste and mismanagement in the aircraft program. Trumulty showed the letter to the President who replied on December 5 that he was very much disturbed by the allegations and asked Borglum to please reply with the chief weaknesses he had observed in the program.

The sculptor did so with enough conviction that on January 2, 1918, President Wilson, with the concurrence of Secretary of War Baker, authorized Borglum to proceed on his own volition to investigate. A member of Baker's staff, Stanley King, was assigned to assist in gathering information. Although Borglum would later consider himself as authorized to conduct "an official" investigation on behalf of the President, such was not really the case. He had only been authorized to do so on his own behalf and report his findings should he desire to do so upon completion.

When Borglum's first accusations were made in November of 1917, the United States had been at war for about seven months. Production Program II (June 1917) which envisioned some 23,325 airplanes completed by June 1918 was five-months old. August 1917 had seen Program III come into being. In it were 29,675 airplanes to be finished by that same date. Among them were 125 DH-4 types scheduled to be finished by the end of November. No DH-4 airplanes had been delivered. The Liberty program was also moving slowly, but better than the DH-4 program.

By the time the President authorized Borglum to begin his investigation in January, some 335 Liberty engines were scheduled to have been delivered and about 22 had actually been received. About 550 DH-4 airplanes had been envisioned (when Programs II and III were put in place it was proposed to power the DH-4 with Rolls-Royce engines) and still none had been delivered.

Some public sentiments regarding the lack of results from the aircraft program were already surfacing when Borglum sent to the President and released to the *New York World* on March 21, 1918, a report which was carried in a two-part series with comments by reporter Herbert B. Swope. Among the comments made by Swope were "It is fair to say that the report, while undoubtedly revealing certain errors in calculation, shows a tendency to depend on individual opinions and intuitions rather than upon actual facts. It contains much hearsay." Swope's opinion was certainly well founded as may be seen by the following excerpts from that report.

> *"Our aeronautical Board of Production has been frankly and flatly a promoter's enterprise and it stands today an undeveloped organization. With all the Nation's resources at its command it has not produced a single service or battle plane that could be ordered across the lines.*
>
> *"The Production Board was provided the best materials, the Nation's purse, and unlimited power. Europe constantly, daily, informed them of her latest designs only to have these ignored until too late to serve. They have used an almost negligible portion of the available manufacturing facilities in the country in wood, metal products, and engines, and have antagonized the remaining large production by unfulfilled promises for work while over commissioning three or four favored organizations.*
>
> *"Nor have they hesitated to promise and advance money to interests where member or members were connected with them, and actually aided in building and overcommissioning companies where no facilities existed, awarding millions to companies hastily formed by business partners upon properties owned by members of the Board."*

The above three paragraphs were designed to paint conditions which had actually been a result of circumstances of the time to be as the result of shady operations and bad judgement. As to the accusations, aircraft and their engines were being produced as quickly as was possible under the existing circumstances. When war began for this country there were virtually no facilities in place with which to build neither airplanes nor their engines. No matter how much money one suddenly made available, factories would not appear ready for production by magic. Most of the required facilities had to be built from scratch and the money to do so was not always available to the companies chosen for the job, as has already been discussed. Had not advances been made, with the government's interest safely guarded, many of the required factories could not have been built in a timely manner, if at all.

In addition, it was only logical that many of the companies contracted for aeronautical production would have some connection with members of the Board or close associates. After all, the Board was purposely made up of industrialists and engineers who were familiar with the requirements of such production and would therefore likely be connected to many different companies suited to such production.

Borglum also complained that we did not adopt the best of the European engines for production instead of designing our own. A great deal of consideration had been given to doing just that but the idea was discounted in favor of designing the Liberty series for service planes and using, for the most part, already designed and in-production American engines for training planes (although the small versions of the Liberty could be put into production for that purpose should that be deemed necessary). European engines were built more or less by hand and did not lend themselves to American production methods. To adapt them to our methods would require considerable redesign and the drafting of complete new engineering drawings. This in itself would have required more time than starting with a design of our own which was done properly from the start.

Our choice was ultimately the correct one as was evidenced by Liberty production success as compared to the average time it took Europeans to get an engine from design to series production plus the conspicuous lack of success where we did attempt to produce European engine designs.

BAP research had revealed that it took from 12 to 18 months for a European aircraft engine to advance from design to series production. The Liberty took 6 months. When writing up its production history following the war, Lincoln made the following pertinent statement: "After one years development with 6,000 employees, the Lincoln Motor Company was producing at the rate of 50 motors a day. In contrast with such American methods, the leading English manufacture, with three years of aircraft engine experience and 10,000 employees, was producing at the rate of 5 engines per week." The manufacturer referred to was no doubt Rolls-Royce but other evidence indicates that Lincoln probably made a "slight error" in their calculation of the English producer's capacity. Col. Waldon and Lt. Emmons also commented on Rolls-Royce production capacity in their post-war report with the following statistics: "The engine which approaches nearest the Liberty in power and efficiency is the English Rolls-Royce. However, this weighs approximately 100 pounds more and delivers approximately 100 horsepower less than the Liberty, and its maximum production in England has never exceeded 70 per week, while its average has ranged from 40 to 45." Its average of 40 to 45 per week would approximate 6 per day. Perhaps Lincoln actually wanted to say "5 engines per day" rather than "5 engines per week."

In any case, the point is still well-made if approximately equal facilities and 40% less employees can produce a comparable engine at more than 8 times the rate.

We very seriously considered building Rolls-Royce engines in this country and spent about a year in various negotiations toward that end before making the final decision not to do so in July of 1918. (I will cover that subject in more detail later since this engine was considered first in place of the Liberty and then as a possible supplement.)

Borglum's accusations were aimed as much at Col. Deeds personally as at the production situation overall. Among the many accusations was the following:

> *"Colonel Deeds is listed as an engineer but investigation proves that he is trained in no profession. He is a successful salesman and promoter. — Deeds' powerful Teutonic personality seems to have completely hypnotized the trusting, scientific mind of General Squier. Deeds, not Squier, is controlling his statements if not his actions."*

Few believed there was any substance to Borglum's accusations, but they were very public, having been published by the newspapers, and very inflammatory. As a result, the President sent Borglum's report to Secretary of War Baker for analysis and comment. In part, his report to the President said:

> *"My general impression of Mr. Borglum's report is that it contains no facts beyond those frankly admitted by Government departments with reference to delays; that it makes no attempt to analyze the causes of delay other than to say that they are the result of corruption and disloyalty. The report contains nothing concerning the prospects for the immediate future. The charges against Col. Deeds are unsupported by evidence. Nevertheless I consider it would be well, in view of the charges, to have a full examination of Col. Deeds' record made as a matter of justice to him and the Department.*
>
> *"Admitting Mr. Borglum's statement of failure with respect to time, no analysis was submitted of causes which could be made the basis from which to draw his conclusions or make his recommendations.*
>
> *"Without denying Mr. Borglum's assertions that many things could have been done, and should have been done, more efficiently, Mr. Borglum does not seem to realize that the development of a new art on an enormous scale in a short time would inevitably be attended with some looseness of method and wrong personnel."*

Secretary Baker's report said, in effect, that Borglum's accusations were unsupported by fact and thus unfounded, topped off with the the invariably true statement that hindsight is generally better than foresight.

Borglum's public accusations continued in full force and General Squier and Col. Deeds demanded a full inquiry into the charges and Howard Coffin telegraphed the President requesting an investigation be made in order that the reputations of innocent men might not be ruined.

Borglum continued to operate as if he had been given full government authority to conduct an investigation on its behalf and began to complain to the President that he was not being given the co-operation from various departments, especially the Signal Corps, that his investigations required. The President's reply finally set straight what authority Borglum had and did not have. Unfortunately the damage had already been done. Among other things, his reply to Borglum said: "I never at any time constituted you an official investigator. I merely gave you the right to look into the matter of your own motion — ." He then asked that Borglum "put at the disposal of those whom I have constituted official investigators all the evidence that may be in your possession."

M. Ding cartoon from the* Kansas City Star *satirizing the Borglum charges. (Courtesy of the *Kansas City Star*)

BORGLUM DECEIT DISCLOSED

On May 10, 1918, Senator Thomas of Colorado submitted a large body of material, including a number of sworn statements by persons involved or with intimate knowledge, which proved Boglum's accusations to be not only without merit but motivated by revenge and self interest. This material was read into the Senate Record.

This material had already been routed through Howard Coffin, General Squier, Col. Deeds, Col. Montgomery, Secretary Baker and Vice President Marshall.

It revealed that Borglum intended to use the influence he believed he had with the President to get aeronautical contracts for a newly formed company in which he would be a silent partner. This was to take place after many companies who now had them were discredited by his accusations.

Even with these revelations made public, Borglum just "stonewalled" it all, claimed his accusers were liars and continued his attacks virtually unabated. They soon lost their effect since the news media no longer continued to be his tool of onslaught.

THE AIR SERVICE AND CONTROL OF AIRCRAFT PRODUCTION IS REORGANIZED

A number of reorganizations were begun during this period of time. The accusations which had been leveled at some individuals who were on the Aircraft Board and also executives of the Equipment Division had their effects during the rather extensive changes made.

The reorganization was not initiated by the air production scandals but by the realization that aircraft-related activities required of the Signal Corps during this very extensive wartime expansion was overwhelming the Corps itself and had now exceeded the overall requirements and workload of the Army organization from which it had originated.

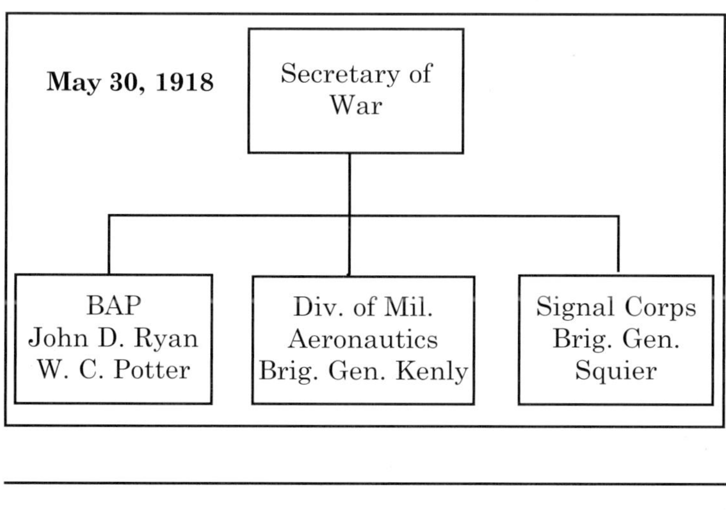

This problem, along with similar situations in other branches of the government, led to the passage in Congress on May 20, 1918, of the Overman Act. This act authorized the President, as Commander-in-Chief of all land and naval forces, to coordinate, consolidate or reorganize the executive bureaus, agencies, and offices under his command for the purpose of expediting the prosecution of the war.

The Equipment Division was extensively reorganized in late April and Mr. W. C. Potter took over as Chief. During the midst of this reorganization, Senate investigations and a Justice Department investigation were set in motion. As a result of this, Secretary Baker relieved Col. Deeds, Col. Montgomery and Col. Waldon of their current duties and assigned them to assist in any way they could in providing information and assistance to the various government investigators. It was made very plain that none of the three officers were in any way considered as being guilty of any illegal act nor suspended under charges but were detailed to aid in the inquiry. Although not stated, it also gave the administration safety from potential accusations of allowing these men to stay in command of aeronautical production should it later be found that they had indeed acted improperly.

Very shortly thereafter, aeronautics was separated from the Signal Corps and Brig. General William L. Kenly put in charge. At the same time the Equipment Division was dissolved and the newly formed Bureau of Aircraft Production (BAP) replaced it. Mr. W. C. Potter became its chief.

On May 4 it was announced that John D. Ryan would take over the position of Director of Aircraft production and head the BAP. He also became chairman of the Production Board. W. C. Potter would become Assistant Director.

On May 30 the Division of Military Aeronautics and the Bureau of Aircraft Production were officially separated from the Signal Corps.

THE CHAMBERLAIN COMMITTEE

On about May 15, 1918, Senator Chamberlain offered a resolution providing for a sweeping investigation of the war effort. After an amendment to limit the inquiry to aircraft, ordnance and the Quartermaster General's Department, it was passed and set in motion.

This became known as the Chamberlain Committee investigation. Although it carried on its work for some time, it seems to have made no significant discoveries nor reports.

SENATE COMMITTEE ON MILITARY AFFAIRS

Not long after the formation of the Chamberlain Committee, the Senate Committee on Military Affairs came into being. It also took up the subject of aviation production and released its report on August 22, 1918.

The principal conclusion of its investigation was that the aircraft production program had given "disappointing results." Although the report filled more than seven columns of the *New York Times* on the above date, it boiled down to three principal conclusions:

"1- That the airplane program was largely placed in the control of the great automobile and other manufacturers who were ignorant of aeronautical problems.

2- These manufacturers undertook the impossible task of creating a motor which could be adapted to all classes of flying craft. It is not too much to say that our airplane program has been largely subordinated to the Liberty motor.

3- We failed at the beginning of the war to adopt the common sense course of reproducing the most approved types of European machines in as great numbers as possible."

Ultimately all three conclusions were found to be erroneous. The first conclusion was correct as a statement of fact. However, we had virtually no aircraft industry to take on the task and thus it was given to the industry best suited to take it on. The second statement had a certain amount of truth in it but did not result in a great detriment to aircraft engine production. The Liberty was put into series production far faster than any European engine and, had the war continued, would have been one of, if not the most significant aero engine of the Allies.

As to item three, we did try to put some European engines into production here at the same time we were developing the Liberty. Our success ran from poor to very poor. The Wright-Martin-built Hispano-Suiza might serve as an example. This was a proven engine already in production in France. A sample engine, with French engineers to help set up for production, was brought here in 1916. It took thirteen months to get that engine into production. Another example was the Bugatti. Again, a sample engine accompanied by engineers was brought here. In this case it was done in December of 1917. At the end of one year only about 10 engines had been built. In the case of airplanes, we were unsucessful in the case of several models attempted, but we did put the British DH-4 into production with modest speed at first, but with notable success in the end.

THE HUGHES INVESTIGATION

Late in May the President decided to have a full legal inquiry into aircraft production and asked Charles E. Hughes to head it. Hughes had been Governor of New York and Associate Justice of the Supreme Court. He had run as the Republican presidential candidate against Wilson in the prior election. He had also been involved as a lawyer in several earlier government investigations.

This became the major investigation of the war and lasted about five months. 57,000 pages of testimony was taken and 280 witnesses examined. A final report was made on October 25, 1918.

The report alleged waste, extravagance, and costly experimentation with types of planes which later had to be abandoned but, contrary to some prior investigations, cleared almost everyone connected with aircraft production of graft or corruption and even concluded that the profits allowed on such were not excessive. Again, the accusations of waste, experimentation, etc. were made with the excellence of hindsight.

The one exception to the accusations of inefficiency seems to have been in the case of the Liberty engine, about which it said "It now appears to be conclusively established that the Liberty engine is a great success for observation and bombing planes, and for this purpose it has found high favor with the Allies."

Three individuals were singled out as candidates for prosecution by court-martial. They were Col. Deeds, Lt. Col. Vincent and Lt. Col. Mixter.

COLONEL E. A. DEEDS

In the case of Col. Deeds, the following charge was made:

"The evidence with respect to Colonel Edward A. Deeds should be presented to the Secretary of War to the end that Colonel Deeds may be tried by court-martial under articles 95 and 96 of the Articles of War for his conduct (1) in acting as confidential adviser of his former business associate H. E. Talbott of the Dayton-Wright Airplane Company, and in conveying information to Mr. Talbott in an improper manner with respect to the transaction of business between that company and the division of the Signal Corps of which Colonel Deeds was the head; and (2) in giving to the representatives of the Committee of Public Information a false and misleading statement with respect to the progress of aircraft production for the purpose of publication with the authority of the Secretary of War."

Secretary of War Newton D. Baker submitted all charges against Col. Deeds to the Judge Advocate General of the Army for complete investigation and recommendation as to what action should be taken in the matter. A Board of Review was appointed to examine all the evidence and obtain all possible additional facts relative to the charges and then render an opinion.

In a January 16, 1919 letter to the Chairman of the Committee on Military Affairs of the United States Senate, and a similar one to the Attorney General, he stated the results of this investigation along with his determination as to what action would be taken.

These two items extracted from that letter follow:

"The unanimous report of this Board of Review, approved by the Acting Judge Advocate General, recommends that Colonel Deeds be not tried by court-martial on any of the grounds suggested, and this recommendation has been approved by me.

"My duty as Secretary of War, with regard to any public servant under my jurisdiction, is clearly to bring about proper punishment for wrongdoing and equally clearly to protect those public servants whose conduct is faithful and upright against embarrassment, humiliation, or loss.

"Inasmuch as the purpose of Judge Hughes' suggestion has been accomplished, I have directed that all the records in this case be filed in the War Department and that this matter be considered as closed."

Lavish praise was heaped on Col. Deeds for his contribution to setting in motion the aircraft production program by all who were associated with him. Included were Secretary of War Baker, John D. Ryan, who replaced Deeds in early 1918 and eventually became Second Assistant Secretary of War in charge of all Army aviation personnel and production, Major General Squier, Major General Menoher, Chief of the Air Service, Rear Admiral D. W. Taylor and virtually every civilian and military officer with whom he had worked during the war.

LT. COLONEL VINCENT

The Hughes report had charged Col. Vincent with letting a contract with the Packard Motor Car Co. for the production of eleven prototype Liberty engines while he had an interest in the company. He recommended that Vincent be tried by court-martial for the offense. That interest was in the form of 429 shares of Packard stock of which he had failed to divest himself upon entering the service. This was without doubt a technical violation of regulations designed to prevent persons in position of authority from monetarily benefiting from their office. It was certainly a violation caused by oversight rather than any attempt to profit. The contract in question was for $249,159.10 and the end profit to Packard resulting from it was $13,586.81. The company had 118,159 shares of outstanding stock at the time. Vincent's 429 shares represented $55.14 equity in the company's profit.

On December 3, 1918, Attorney General T. W. Gregory wrote a letter to the President indicating there was no evidence to indicate the government was defrauded nor that there was any attempt to do so. He continued with the statement: "While there was a technical violation of the statute, I do not think the ends of justice require that Col. Vincent be prosecuted." "In view of all the circumstances, I recommend that a full and complete pardon be granted for the offense, as I believe it was unwittingly committed."

The President released information officially granting the pardon the next day.

LT. COLONEL GEORGE W. MIXTER

The Hughes report charged that Col. Mixter had improperly acted for the government in supervising inspection and production of aircraft and engines at the Curtiss Aeroplane and Motor Corporation while he owned 25 shares of preferred stock in the company valued at $2,500. Col. Mixter had owned a small amount of both common and preferred stock in that company and had sold his common stock upon entering Army service but had overlooked the ownership of the preferred. This stock payed fixed dividends.

Again the conclusion of the Hughes investigation was a recommendation that Col. Mixter be tried by court-martial for violation of section 41 of the Criminal Code.

Also on December 3, 1918, Attorney General Gregory wrote the President recommending a pardon for Col. Mixter for an offense unwittingly committed and from which no gain was made. It was also granted the next day.

These actions were covered in all the major newspapers of the country and virtually all heaped great praise on these two officers for the contributions they made to their country in time of war while thoroughly condemning the conclusions of Judge Hughes and the actions he recommended in these cases.

In the December 5, 1918, issue of the *New York Tribune* was a typical final comment regarding the matter. "Mr. Hughes' report expressly stated that "no ruling in the Federal Courts with respect to the applicability of the statute to such an officer or agent of the government has been found." It would seem as though the issue of a "pardon," for no crime committed, was an unfortunate method of clearing up a case where evidently complete exoneration was the intent."

Such would certainly seem to be the case. As a matter of fact, this remaining cloud upon Col. Vincent's record seems to have been the reason he did not receive the Distinguished Service Medal for his contributions to the war effort after having been recommended for it immediately following the war. A number of others were awarded this medal for lesser contributions.

The Borglum accusations started before any appreciable production of combat engines or airplanes had started. They ended in May of 1918. At that point the various government investigations began. The final one, the Hughes investigation, made its final report on October 25, 1918. The war ended seventeen days later on November 11, 1918. In the next chapter we will see what was accomplished during that period and to the end of production.

Aeronautical Production Organization as of November 1918

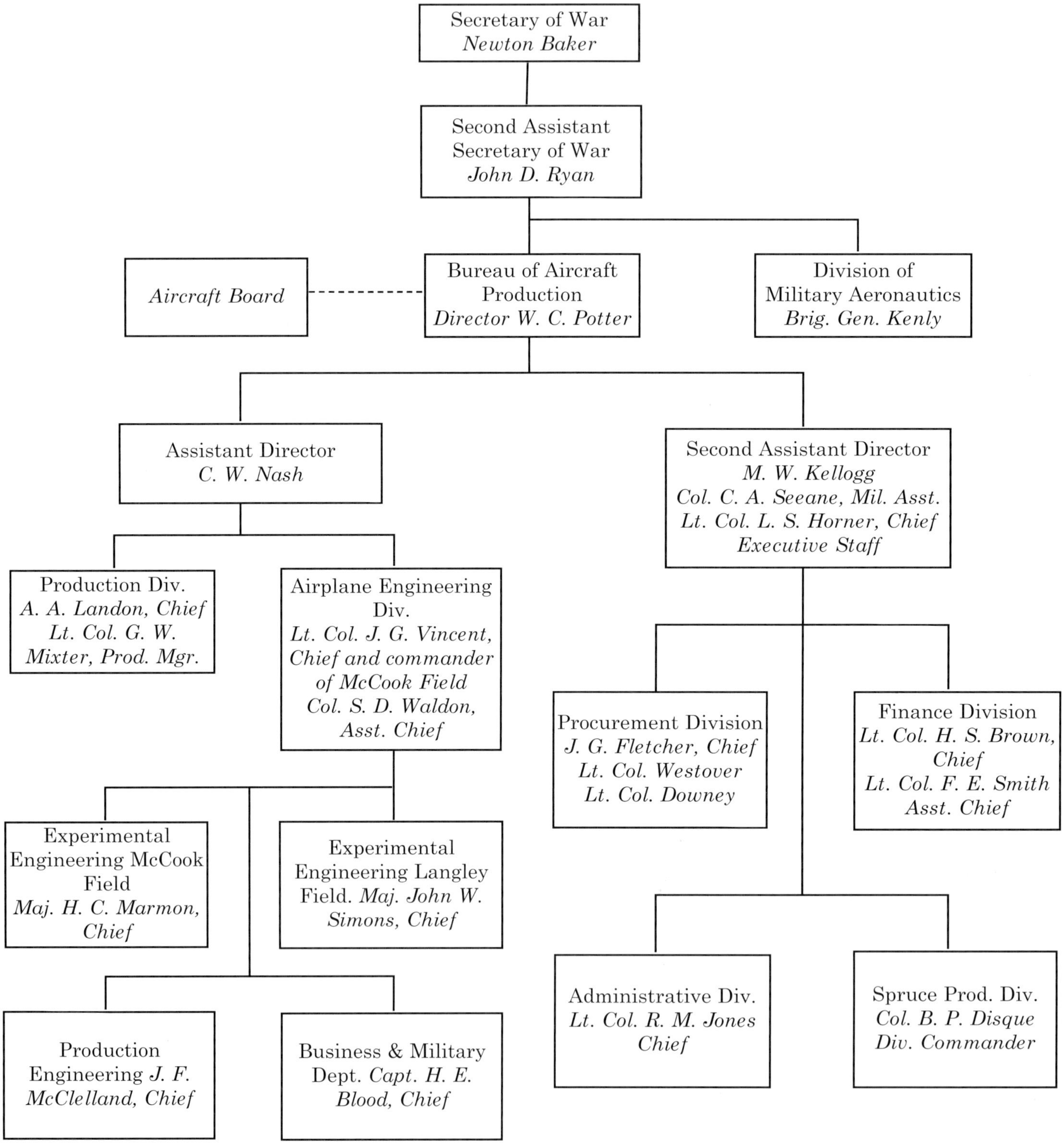

Chapter 4

Liberty Production and Late Contracts

Liberty production started with the contract with the Packard Motor Car Company for six L-8 and five L-12 prototype engines. A complete set of parts for the first L-8 prototype left the Packard factory bound for the National Bureau of Standards in Washington on July 2, 1917. They arrived there on July 3rd and were assembled into a completed engine late on July 3rd or early in the day of July 4, 1917. This first engine served as a sample and was not run. L-8 Exp. No. 2 was the first running engine and it was first tested on July 25.

L-12 Exp. No. 1 was completed on August 12 and first ran on August 13, 1917 and passed its 50-hour type-test on August 25, 1917.

L-8 Exp. No. 3 was probably completed by August 20, 1917 or earlier. It had been shipped from the Packard factory to the L. W. F. Field in Mineola, New York, had been installed in an airplane and was ready to fly by August 29. There seem to be no remaining records as to the dates L-8 Exp. No. 4 and 5 were completed and the earliest report of Exp. No. 6 is a test conducted by O. E. Hunt at the Packard factory on September 10, 1917.

No records remain of the actual completion dates of L-12 Exp. No. 2 through 5 engines. We have a test record of Exp. No. 2 dated September 21 and one of Exp. No. 3 dated October 16. We know that the first significant use of No. 4 was in the first flight of an Army DH-4 on October 29, 1917. Obviously, both engines had actually been completed a minimum of several days prior. We have a record of a test run of engine No. 5 on November 17, 1917.

Machining crankcase halves at the Packard factory in early 1918. (Courtesy of the Smithsonian)

It would seem that completion of the series of prototype eights was pushed much faster than the twelves. From the dates now known, it would appear that all six eights were completed by the time the second twelve was. The earliest record of L-12 Exp. No. 2 is a test recorded on September 21 with low-compression pistons. The engine was noted as having "standard 3 port intake headers." Output was 360 HP @ 1,700 RPM. Output of Exp. No. 1 on August 25 with low-compression pistons at the same RPM had been 337 HP.

In the early part of the report of the history of Packard's WW I aircraft production efforts submitted to the BAP by O. E. Hunt, Packard's Chief Engineer, Mr. Hunt made the following significant statement:

> *"During the early part of September continual development work on the twelve-cylinder engine resulted in improvements which raised its horsepower output from 315 to 400. Improved intake header design and increased valve lift were the principal factors in this improvement. During the latter part of September and the early part of October the drawings for the engine were undergoing changes in Washington in order to incorporate in them the improvements suggested by the 50-hour test and modifications in the oil pump and propeller hub directed by the military experience of our Allies. Drawings were consequently not furnished for production until October."*

POWER IMPROVEMENTS IN THE L-12

The question continually comes up as to just what was done to increase the power of the Liberty 12 and raise it from 315 or 320 HP to 400 HP. O. E. Hunt gives the answer in the above quote. **"Improved intake header design and increased valve lift were the principal factors in this improvement."**

Actually the reason for the question stems from misinformation provided by those people associated with the Liberty program more than from any knowledge of the actual power available from the first version of the engine verses its later output. Because we have available documentation of the actual power obtained during the first tests of the engine and the output of engines 2 and 3 of September 21 and October 16, we know that the Liberty 12 Army (high-compression) version was capable of 430 HP at 1,700 RPM when the first flight of a Liberty 12 was made on October 21, 1917.

Although other changes in carburetion and various "tweaking" of the design gave additional minor improvements in power output, by the first of October 1917, just six weeks after the first L-12 ran, output of this engine had reached the level it maintained for the remainder of its life.

The misinformation stems from statements made by various officials during actual production. As various changes in design were required to prevent mechanical failures in early versions of the engine, they were explained as necessary because the stepped-up power of the engine resulting from improvements caused failures in components not designed to withstand these new powers. This was later corroborated by the statement made in the official history of the production of aircraft by the Army following the war.

Actually there was some truth and some lies. The end result was the same but the truth was that the engine was capable of about all the power that was ever required of it and for which it was rated from the very start of its use in an aircraft. What changed was what was required of it.

It was to be an engine that was rated at 335 HP @ 1400 RPM and have a maximum output of 410 HP @ 1700 RPM. (See pages 36-37 and 52.) It was expected that it would normally be operated at its rated output and seldom pushed to maximum. Its mechanical design would allow reasonable reliability if operated within these parameters.

The change in power that required strengthening various parts to withstand the increased strain was not an increase in power ability but an increase in power demand. The engine was "rated" at 335 HP @ 1400 RPM. When they found that required normal power increased to say 375 HP, they changed the "rating" to 375 HP @ 1500 RPM. Eventually it was raised to 440 HP @ 1800 RPM. As can be seen on the power chart on page 51, the engine was capable of such power from the start. It was just not demanded of it and the engine was not designed to withstand such output on a continual basis.

As such demands were made, the engine had to be strengthened to withstand them. However, the official statement was that as power requirements became greater the engine was redesigned to produce them and resulted in failures which had to be prevented by improved parts. Actually no improvements in power ability were made, just increased demand. The "official story," so to speak, was stated as follows:

> *"The Liberty engines as originally manufactured was of approximately 330 horsepower. Information received from overseas led to the conclusion that a higher horsepower would be desirable. Therefore, after about 300 of these engines were in production, the engineers stepped up the horsepower to 375. Under the additional stresses induced, certain parts of the engine, particularly the crank shaft, required strengthening. This was done and several hundred engines were delivered at this horsepower. Then the final and definite information arrived from abroad to the effect that if an engine of 400 horsepower or more were produced, the United States would lead the combatant nations in size and power of engine during 1918 and 1919. The engine was therefore stepped up by the engineers to 440 horsepower. This enormous increase in power necessarily required an increase in strength of practically all the working parts of the engine. This required changes in a very large percentage of the jigs, tools, and equipment used in manufacture. It also required certain metallurgical changes in some of the parts, which in turn involved the development of new and better methods of producing steel for them. These changes in equipment were required in the engine plants, the parts plants, and the forging shops; in like manner they affected all manufacturers throughout the production line, including producers of raw metal."*

The above is quoted directly from *United States Army Aircraft Production Facts* (pages 22-23) written by Col. G. W. Mixter and Lt. H. H. Emmons at the direction of the Assistant Secretary of War, Benedict Crowell, and published in January of 1919. It was in turn paraphrased and reproduced, on page 277, in the Secretary's official report on the overall munitions efforts of the war, *America's Munitions 1917 - 1918,* which was published later in 1919. Variations of this same story may be found in many following publications, including the official *BAP History*.

We will see the changes referred to above, plus a great many not specifically mentioned, as we progress through this chapter. These changes were made to either strengthen parts, speed manufacture, or reduce manufacturing costs. Many changes were made to ease tolerances initially thought necessary to insure required quality and later found to be unnecessary, time-consuming and costly.

ONE DESIGN CHANGE

Although there were many changes made in the engine during its manufacture, these were not "design changes." Quoting again from the Mixter/Emmons report of 1919: "There was but one fundamental change in design in this engine after it was laid out in May, 1917. This was to change the oil system from the so-called scupper feed to pressure feed. Either system worked properly on the engine, but the latter system is foolproof while the former is not. The latter system was therefore substituted.

"As the engine was used in service and manufacturing progressed, it became evident that some of the limits might be changed, and these changes were made. This is common practice in all manufacturing establishments. It has always been so on automobile motors and will always be so on any manufactured product. As manufacturing processes are developed and as experience is gained, changes in limits are always expected and are always made, both to expedite production and improve quality."

Many significant changes were made in early engines. They involved, almost exclusively, Packard engines prior to Mnfg. No. 1050 and Lincoln engines prior to Mnfg. No. 200. These involved parts which were not interchangeable with earlier designs as well as those which were. Other manufacturers were unaffected by these early revisions in design and tolerance limits as by the time they started production things had stabilized considerably. There were a few complaints about changes in parts which affected machine setup, notably from Trego, but generally late production problems stemmed from late parts deliveries from vendors, loss of manpower to the draft and difficulties in obtaining the required production tools and machinery.

Packard had built the prototype engines and had rightly assumed they would get large production contracts. Although contracts with the first four producers were all let at about the same time, it was felt that Packard was farther along in production preparation and thus could be expected to be well-ahead of the others in first output.

PACKARD WAS THE FIRST PRODUCER

They were, but a deluge of early specification changes immediately made it clear that they would not be able to meet the contract production schedule. Because they had also planned to be a producer, and had made considerable preparation in that direction, Lincoln was expected to begin production shortly after Packard.

Since Packard and Lincoln were the most affected by the various early changes related to the engine, they made a point of documenting them in their post-war report to the BAP for inclusion in their history. Packard was the most affected, the most criticized for slow or substandard output and thus the one to document the changes in the most detail.

They mentioned the number involved in three different documents and detailed them in the latter. The first was a letter to Col. Montgomery acknowledging the receipt of the $2,500,000 advance the government had provided. This letter was dictated on **January 19th** and typed on January 21, 1918. In it was the statement: "Our contract was signed September 4. **Not expecting any, or at least many changes, we did not begin to keep a record of them until September 23. Since that date, however, which was less than four months ago, we have a record of over eight hundred changes (big and little) that have been put through.**"

On the left is L-8 Exp. No. 1 with original nose plate. In the center is L-8 Exp. No. 2 with no nose plate. Both have single-type crankcase strengthening webs. Also note the relocation of the front vertical through bolts and the fact that neither has a thrust bearing oil cup. On the right is the Buick-built BX No. 5 with thrust bearing oil cup and double crankcase strengthening ribs. (Author's collection)

EARLY CHANGES VERSUS ENGINE PRODUCTION

Record of completed engines		Changes in parts design made & notes	
Initial design completed	— June 6		
L-8 Exp. No. 1 completed	— July 3		
		July -	Nose plate deleted.
L-8 Exp. No. 2 tested	— July 25		
		Aug. -	Single to double webs.
L-12 Exp. No. 1 completed	— Aug. 12	Aug. 17	346 HP @ 1,700 RPM high-comp.
L-8 Exp. No. 3 completed	— Aug. 20		
L-12 Exp. No. 1 50 hours test	— Aug. 25	Aug. 25	337 HP @ 1,700 RPM low-comp.
L-8 Exp. No. 6 tested	— Sep. 10	Sep. 18	245 HP @ 1,700 RPM low-comp.
		Sep. 20	New carburetor and headers.
L-12 Exp. No. 2 tested	— Sep. 21	Sep. 21	360 HP @ 1,700 RPM low-comp.
		Oct. 8	High lift cam.
		Oct. 10	Change propeller hub to demountable.
L-12 Exp. No. 3 tested	— Oct. 16	Oct. 16	414 HP @ 1,700 RPM high-comp.
		Oct. 16	Add thrust bearing oil cup.
L-12 Exp. No. 4 flight	— Oct. 29		
		Nov. 13-14	Strengthen rods and rocker levers.
L-12 Exp. No. 5 tested	— Nov. 17	Nov. 17	Crankcase walls thickened and ribs added.
L-12 Prod. No. 1 completed	— Nov. 19		
L-8 Exp. No. 3 tested	— Nov. 27	Nov. 27	276 HP @ 1,700 RPM high-comp.
L-12 Prod. No. 3 completed	— Dec. 1	Dec. 4	Start of break-down tests.
		Dec. 5	Light crankshaft replaced by heavy
		Dec. 6	Navy pistons added to production.
		Dec. 28	Heavy crankshaft replaced by 2nd design & full pressure oiling adopted.
L-12 Prod. No. 36 completed	— Jan. 1		
		Jan. 28	1st use of Navy engine plates.
L-12 Prod. No. 116 completed	— Feb. 1	Feb. 8	End of break-down tests.
		Feb. 15	2nd design of heavy con. rods.
L-12 Prod. No. 230 completed	— Mar. 1		
		Mar. 16	Pressure oiling adopted to remaining stocks of light crankshafts. The original intention was to not switch over until remaining stocks of light crankshafts were used up.
L-12 Prod. No. 399 completed	— Apr. 1		

Notes: 1- Output/cylinder was the same in the Oct. 16 L-12 Exp. No. 3 test and the Nov. 27 L-8 Exp. No. 3 test.
2- The change of a part design happens somewhat in advance of the time the new part appears in an assembled engine. In most cases, existing stock would be depleted before new design parts were used on engines.

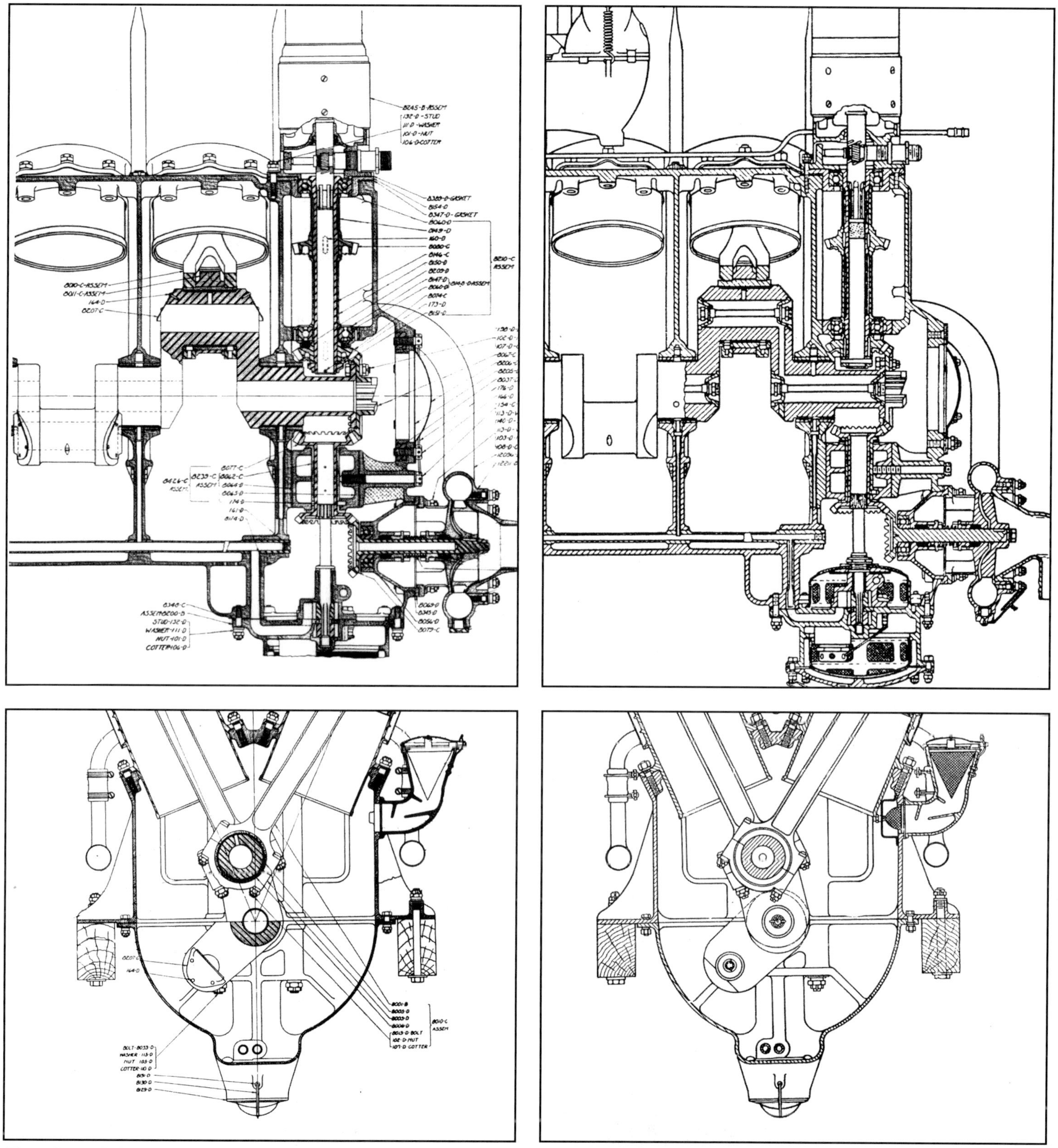

The above left sectional drawings are of the L-12 with scupper-type rod journal oiling as shown in drawings of about November 1, 1917. On the right is shown the later full-pressure oiling system in a drawing probably dating from mid-1918. (Courtesy of the Air Force Museum)

The next instance was in a letter of **February 7** to Capt. (later Major) Heaslet, at that time chief of the Detroit District and in control of Liberty production. In it they made the following statement: " In its general essentials the motor is excellent. We haven't lost faith in it for a minute, **despite the fact that our records show that considerable changes of all kinds — over nine hundred of them — have been made since September 23rd, when we began recording them."**

And last, in their **May 31, 1918** letter to the Honorable William L. Frierson, Assistant Attorney General, in response to an official inquiry into production lags, they made the following statement:

> *"Attention has already been called to the fact that improvements in design were rendered possible by experience gained with the first motors, produced by us at great expense and largely hand made. Recognizing that modifications in design, and specifications of material and test, should not be opposed if they tended to improve the performance of the motor, we cheerfully waived the provisions of Article X of the contract to our disadvantage, and undertook to absorb, as rapidly as they came to us from the Signal Corps, all engineering changes.*
>
> *"The effect of many of these on tools as well as delay in production which they necessitated may be gauged when I say that* ***up to February 9, 1918, 1,022 mechanical changes had been incorporated in the original design."***

In this case, the principal changes and their effects upon production were listed thereafter. The first listed as "Alteration #1" was dated 9-20-17. In prior statements regarding changes, they had said their first records of changes dated from 9-23-17. In this case, earlier changes are listed, the **ending date is 4-26-18**, and the total goes up to 1,022. Keep in mind that absolutely every change in specification, design of part, gauging requirement, inspection procedure, test procedure, etc. was counted. Each one had its effect on production, some minimal and some catastrophic.

There followed a ten-page list of items titled **Summary of Principal Mechanical Changes by Groups.** They were listed as "Alteration #1" through "Alteration #184" and dated 9-20-17 through 4-26-18.

I will list some of the more significant changes. The changes listed are, for the most part, direct quotes. My added information or emphasis are in bold type. The vast majority of these changes applied to Packard-built engines only. Lincoln was the only other company affected by any significant changes, and they were few by comparison to Packard. The following builders had a stable design from the start.

EARLY CHANGES

Alteration #1 9-20-17

1- Standardization of Delco electrical apparatus. This delayed production to the extent that the first engines were shipped without standard equipment. **(That is, they were shipped with early design equipment.)**

2- New design of carburetor and intake headers. Required new patterns, tools and gauges by both the carburetor manufacturer and Packard. **(Note the improved power accomplishments cited above by O. E. Hunt.)**

3- Crankcase upper-half drain tube change caused scrapping of old tools and gauges and building new ones.

4- Crankshaft and rod bearing babbitt anchoring changed.

5- Thickness of valve head increased.

6- Piston compression changed from 18.9% to 18%. **(This also resulted in a slight increase in power.)**

7- High tension cable terminal design change.

Alteration #3 10-8-17

High Lift camshaft. Required changes in dies, tools and test fixtures. **(Note the improved power accomplishments cited above by O. E. Hunt.)**

Alteration #6 10-10-17

Demountable propeller hub. Changed both propeller hub and crankshaft and related tools, etc.

Alteration #7 10-16-17

Addition of propeller hub thrust bearing oil cup. This added parts and required a new design upper crankcase half.

Alteration #8 10-16-17

Redesigned oil pump required all new parts for the pump except the gears and required new design lower crankcase half.

Alteration #20-21 11-13-17

Camshaft rocker levers strengthened.

Alteration #25 11-14-17

Connecting rods strengthened. This necessitated scrapping all rods machined up to that operation, leaving Packard without anything and requiring venders to scrap all forgings and restrike all dies. This greatly hampered production.

Alteration #27 11-17-17

Crankcase walls thickened and timing marks added. The crankcase was originally designed to have a wall thickness of 1/8" and 5/32"; it was found necessary to change the thickness in order to get the metal to flow in that size of casting. **(Also additional strengthening ribs were added at this time.)**

Alteration #29 11-17-17

Cams widened making necessary the scrapping of dies, tools and gauges.

Alteration #36 12-05-17

Original light-type crankshaft cancelled and superceded by heavy-type crankshaft. **(This was a manufacturing change only. Old type crankshafts were to be used until stocks were depleted, principally in Navy engines. Some remaining stocks of these light crankshafts would later be altered for full-pressure oiling. The change occurred in the engines at about Mnfg. No. 300. This actually became known as the medium-type crankshaft as it was strengthened again by Alteration #63.)**

Alteration #44 12-06-17

Intake header flanges slotted. **(This was done to better facilitate assembly and maintenance by allowing the headers to slide on and off their mounting studs from above.)**

Alteration #47 12-03-17

Ceased drilling holes in screws to save weight. **(Deemed too expensive and time-consuming for the small amount of weight saved.)**

Alteration #50 12-03-17

Oil pump intake screen added. **(The old and new designs can be seen on the sectional drawings on page 87.)**

Alteration #53 12-06-17

Addition of flat-top pistons for the Navy. **(Although low-compression pistons had been used in a number of test engines prior to this time, they had not been adopted as standard for Navy engines. These pistons were 20.5% compression.)**

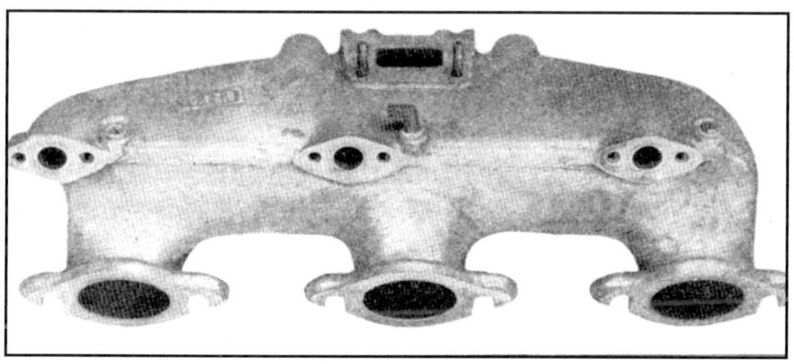

Intake header after the mounting flanges were slotted. (From the 1927 manual)

Alteration #57 12-08-17

New design water pump impeller.

Alteration #63 12-28-17

Crankshaft strengthened second time and changed to full pressure lubrication. **(Packard started using this design at about Mnfg. No. 1100 because stocks of old-style crankshafts were used until exhausted, though many were modified for pressure lubrication. All other manufacturer's engines used heavy crankshafts and full-pressure lubrication from the start. The use of full-pressure lubrication had been a desire of Vincent from the beginning and early tests vindicated this. Charles Vincent, brother of J. G. Vincent and a Packard experimental engineer, who was working with E. J. Hall on the break-down tests, wrote to J. G. Vincent with a report on a 20-hour test of Exp. No. 5 in which, among other things, he strongly advised full pressure lubrication of the rod bearings. In response, J. G. Vincent asked O. E. Hunt to have parts built to convert one engine to this system. Test results verified the advisability of this change and it became official with this alteration.)**

Alteration #64 12-28-17

New design piston pin retainer. **(Change occurred at about Mnfg. No. 225.)**

Alteration #71 12-30-17

Crankshaft and connecting rod bearings grooved.

Alteration #90 1-30-18

Issuance of Navy engine plate. Reduced orders for Army type and start of use of Navy plates on Navy engines. **(An example of an early Navy engine with an Army plate is Packard Mnfg. No. 43 with an Army plate with U. S. No. 17485 and Navy No. 32.)**

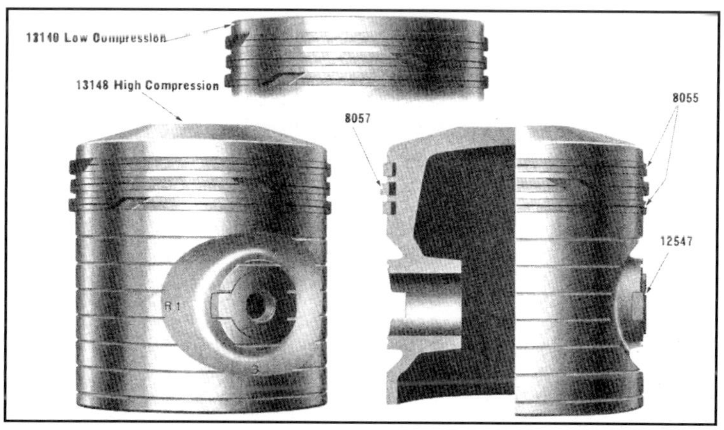

Standard Army pistons (bottom) and Navy piston (top). The new type piston pin retainer is illustrated. (From the early 1918 manual)

Navy engine data plate as adopted on 1-30-18. Note that this is from Packard engine Mnfg. No. 3922 and is Navy No. 02825. It is marked to the left of the MNFG. No. as H. C. (high-compression). Also note that it is a "Liberty Mod. A" and not marked as a "Model N." The plate was made of brass. Final inspectors stamp is to the right of the serials. (Author's collection)

Alteration #93 2-04-18

Carburetor. Settings of choke and jets changed to reduce horsepower.

Alteration #104 2-15-18

Connecting rods. Superceded by new design of greater strength. **(This became known as the heavy-type rod. The change occurred at about Lincoln Mnfg. No. 200 and Packard Mnfg. No. 520-550.)**

Alteration #118 3-02-18

Larger connecting rod bearing to go with new heavy-type connecting rods.

Alteration #120 3-05-18

Crankshaft bearing design changed. Old type not suitable for pressure lubrication.

Alteration #121 3-06-18

Light crankshaft pressure lubrication adopted to go into effect at once. **(Original intent was to go into effect after stock of light shafts was used up. Any light crankshafts remaining in stock were altered to full pressure lubrication. This change affected many Packard engines between Mnfg. No. 301 and 1100.)**

Alteration #126 3-08-18

Change in crankshaft oil hold closure from pressed-in-type plugs to detachable plug and stud method. **(This latter method is shown used on the pressure oil system shown on page 87.)**

Data plate of Packard-built Exp. No. 5 L-12. The circle in the lower left would contain the stamp of the final inspector and the rectangular area to its right would be stamped with the date of acceptance. The plate was made of white metal. This design was the standard for Army engines. Note there is no acceptance stamp and no date. (Author's collection)

Alteration #128 3-10-18

Generator driving shaft. Method of retaining lower gear nut changed from cotter to taper pin. **(This was started at Packard Mnfg. No. 275.)**

Alteration #133 3-18-18

Carburetor for low-compression engines added.

Although there were further changes after this time, they became fewer and less significant in nature. Most later changes affected tolerances, inspection and test procedures.

LABOR SHORTAGES

As difficult as the various design changes were making production, shortages of skilled labor and labor turnover were adding tremendously to the problem. There was a very heavy demand for skilled labor in the Detroit area. This resulted in considerable competition in the labor field and a subsequent escalation of wages. This in turn resulted in a very high rate of turnover — workers quitting one employer to work for another for much increased wages. In addition, losses to the draft were high.

Lincoln, as well as all of the other makers of Liberty engines, employed a large number of women during the war. Manpower was scarce and they found the female workers were very adept at skills requiring a fine touch. Here they are gauging pistons and valves to make sure they meet specifications. (Author's collection)

PACKARD

Packard reported labor turnover for their tool departments was astounding to say the least. These were all skilled workers who were highly paid and in much demand. As of September 1, 1917 their total employed tool workers was 595. Between that date and June 1, 1918 824 tool workers quit and 866 were hired. As a result, they had 674 employees in that category at the end date.

Put in simpler terms, they had an average of 641 employees in this category at any given time and an average of 92 quit every month and had to be replaced. That is about 14.5% of the work force leaving each month.

LINCOLN

Lincoln was just building up their facilities and work force when they received their contract in September of 1917. They did not complete the assembly of their first engine until February 4, 1918.

As of August 1, 1917 they had a total of 52 employees. By February 1, 1918 they had 977. By July 1, 1918 they had a total of 5,600. During the first six months of 1918 their average turnover was in excess of 20% and in the month of April it was 30%. How would you like to carry on a high-pressure manufacturing business where the government was breathing down your neck continually for faster output and better quality while 25% of your work force was walking off the job every month and the product was in a continual state of change?

This is what the early manufacturers were facing. Packard had the most pressure and problems to contend with and Lincoln was second in line.

PACKARD REQUESTS RELIEF

The pressure at Packard became so great in late January of 1918 that President Macauley was moved to request relief.

On February 7, 1918 he sent a letter to Capt. J. G. Heaslet who was in charge of Liberty production in the Detroit District. In that letter he went into the general history of Packard's involvement in, and resultant problems with, Liberty engine production in some considerable detail. He then requested that three changes in policy be made regarding those things required of Packard so that they might get on with production in a more normal manner.

Much of the relief requested was granted immediately and most of the remainder occurred within the next 60 days or so in the natural course of events. Things then settled down considerably at Packard. 1- They wanted to be relieved from having to conduct break-down tests. Break-down tests were terminated the next day, on February 8, 1918. Some testing did continue, but within the next 2 to 3 months the test location was moved to what became known as the Saxon Test Plant and later as the Liberty Engine Experimental Plant. This was located in the new Saxon automotive plant in Detroit which had been very recently built by Saxon but not yet occupied. 2- They wanted the experimental engineering and manufacture to be moved elsewhere. They were relieved of much of the experimental engineering by the creation of the Liberty Engine Experimental Plant mentioned above. They probably still did most of the parts manufacture that was required. 3- That they cease the manufacture of those parts they had to make by hand and await the receipt of final tools. They got little or no relief from this requirement.

However, they did begin to receive much of their final tooling about this time and, in addition, change requirements began to drop sharply as the design settled down. Just when Packard had begun to feel that there seemed to be no end to the changes, they began to taper off rapidly.

Because the contents of Macauley's letter serves well as a compact history of early Packard production, I will reproduce it in its entirety below.

February 7, 1918

AM-L

Captain J. G. Heaslet
United States Signal Corps
876 Woodward Avenue
Detroit, Michigan

Dear Sir:

We ask your careful consideration of the facts hereinafter related, as the matters involved are of very great importance to us as manufacturers of Liberty motors.

To present this matter it is necessary to go back some months and relate the facts that have led up to the situation which today is bothering us. When contracts were let with manufacturers of the Liberty motor, four firms were originally given orders to proceed with the work: The Lincoln Motor Company, Nordyke & Marmon Company, The Trego Motors Corporation, and the Packard Motor Car Company. Subsequently, contracts were let to the Ford Motor Car Company and the General Motors Company.

By reasons of our being equipped to build twelve-cylinder motors, and on account of the fact that we had so much to do with the design of the Liberty motor, and had built eleven of them by hand by order of the Signal Corps, we made an earlier start towards manufacture than any of the others.

Right there we made a fundamental mistake when our self-interest only is considered. From a selfish point of view, we should have followed the course being followed by all the other manufacturers of Liberty motors, and produced no parts until all the final, permanent tools, gauges and fixtures were ready and proved out. But it developed before we had gotten well started, that changes would be necessary in the design. It was not, in fact, until December 10th that Vandykes supposed to be final were delivered to us. As a matter of fact, there has never been a time until now when numerous changes were not in process.

When Major Hall was sent out by Washington to cooperate with us in getting into production, one of the first things he did was to strongly advise break-down tests of the Liberty motor, and under his general instructions we began to make preparations for these break-down tests about November 14th and they have been continued unremittingly since then. As expected, these break-down tests, or tests to destruction, have indicated features of the motor that could be improved and strengthened, and these tests show the wisdom and necessity of the numerous changes which have been made and are still being made, because as you know, the last features of the Liberty motor are still not standardized. In its general essentials the motor is excellent. We haven't lost faith in it for a minute, despite the fact that our records show that considerable changes of all kinds - over nine hundred of them - have been made since September 23rd, when we began to record them.

As you are familiar with those changes, there is no need to refer to them in detail, but since the bearings, among other things, are not in approved form, we can see that other changes are yet to come.

In starting on Liberty motor work and facing the numerous changes that had to be made, it became evident that some of the six manufacturers would have to do the experimental manufacturing, not only in order to standardize the processing, but also in order to assimilate and try out and test the changes which experience indicated were necessary. Packard, being ahead of the other manufacturers, undertook this experimental work, the experimental manufacturing and the break-down tests; and we have, since the beginning, undertaken to carry on concurrently all this various experimental and test work simultaneously with an effort to manufacture the motors.

But to carry out this program it was necessary for the pioneer manufacturer - the Packard in this case - to proceed with manufacture from temporary tools, pending the receipt of the final permanent tools. And we have been manufacturing Liberty motors to date, and are still doing so, wholly or in part, from temporary tools and by hand work. As no one of the contractors for

Liberty motors had an organization and experience in the production of aircraft motors, it was necessary the we undertake this experimental manufacturing with green men and break them into the work.

The work we have undertaken, in the way we have gone about it, has been a necessary evil. Some one of the six contractors had to do it in order that the motor might be perfected, and in order that five out of six plants should be saved the experimental manufacturing work. But we have found, and are finding that the experimental manufacturing from other than final tools was just about as satisfactory as might be expected. Producing with temporary tools, with green men, under the necessity of assimilating hundreds of changes, had resulted in a great deal of scrap, and has been very unsatisfactory and expensive to this company.

So long as what we were doing, and the way we had to go about it, were appreciated as a sacrifice on our part, to the good of the cause, we were willing to continue. But it appears we are subject now to criticism that our work is not up to the standard it would have reached had we waited until changes were all through and our tools in final, definite condition.

No manufacturer in the world could have proceeded under the conditions we have been working under and produce high-quality parts - of the quality that we shall be able to produce when our final tools are in and working.

Not only that, but our factory has been used as an educational institution for Government inspectors and representatives of all kinds. We have at the present moment 142 of them, and the number has reached 160. These largely were sent to our factory for preliminary education and training, and later on sent to the other factories producing Liberty motors. We were glad to make an educational school of our factory, because the thought that has been uppermost in out minds was to do that thing which would be to the best interest of the country, and the most valuable in quickly standardizing the engineering and manufacturing of the motor, and aid in getting all of the six manufacturers onto a producing basis. But the burden has become too onerous. Instead of reflecting to our credit, it apparently is turned to our discredit that the parts turned out are not to that standard which it is easy to maintain from finished tools and experienced organization.

Therefore, with your permission, we want to institute three changes:

1st: That we be relieved immediately from further break-down tests;
2nd: That the experimental engineering and experimental manufacturing hereafter be done at some other plant than ours, for we feel that we have carried more than our share of the burden;
3rd: That we cease the manufacture of these parts which we have produced wholly or in part from temporary tools or hand methods, until our final tools are received and tried out.

Our primary object in asking for these modifications of the present situation is that the public generally does not realize the conditions under which we have been and are working, and our reputation as manufacturers is suffering in consequence. Too, we feel that all the handicaps of the preliminary engineering and manufacturing should not be borne by us; others should share in it; they should not have so far the preferred situation. We still feel that we are willing to proceed as we have been, if the good of the cause requires it; but we believe that we have carried the matter so far now that we can be relieved without harm, because we have built and supplied, more or less by hand methods, all the motors necessary for the advance needs of the Government, and all that are necessary to standardize, up to this point, the engineering and manufacturing of the motor.

All the contracts with the manufacturers provide that the latter do not have to accept changes except for embodiment in the second lot succeeding the lot current at the time the change is decided upon. But we

Female workers drilling cylinder flanges with a multiple drill at the Packard factory. (Signal Corps photo from the Author's collection)

Liberty Break-Down Tests

December 4, 1917 to February 8, 1918
Conducted in the Packard Motor Car Co. plant

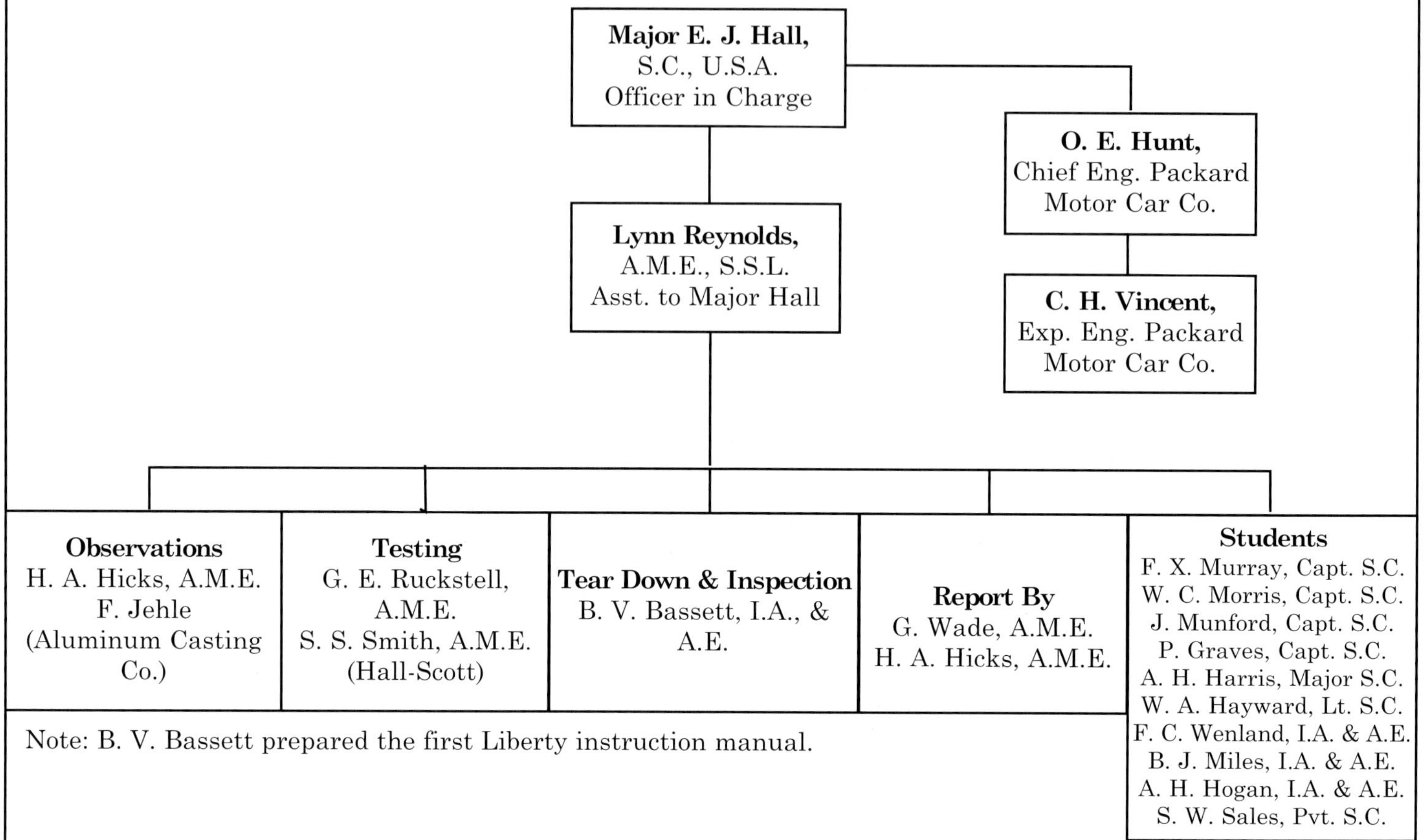

These tests, which were designed to run the engine at full power continually until a failure occurred, were run on Packard engines Exp. No. 5 and production No 12. When a failure occurred, the failing part was analyzed and, if found to be weak in some area, was redesigned to increase its strength. If the problem was the result of faulty lubrication, it was corrected by increasing oil flow by whatever means necessary or changing bearing clearances. If bearing failure was caused by inadequate material, new bearing material was selected. Hall listed 27 changes in his final report. They ranged from a different mesh on the oil pump screen and different make of spark plugs to strengthened crankshaft and rods. Many changes involved lubrication (8), including the change to full-pressure lubrication. A different design propeller hub was suggested and baffle plates were added to breathers.

have constantly waived this provision, and if necessary, will waive it hereafter, though we are able to manufacture scarcely two motors alike - the result of which is the evils hereinbefore referred to.

Please imagine yourself for a time in our position, and determine if we are getting quite a square deal; and then determine, please, what relief we can have.

We will have this letter delivered by special messenger, in order that, if possible, it may have your early attention and action.

Yours very truly,
Alvan Macauley
President

At this point in time, February 7, 1918, Packard had built approximately 120 production engines. Lincoln had completed its first engine on February 4 and no other manufacturer had assembled any. Lincoln, of course, had contended with some of the latest changes since that company had obviously started work some time prior. They did, however, benefit greatly from Packard's experience and were able to start production with mostly up-to-date design parts while Packard had to use up stocks of obsolete design parts to either prevent wastage or keep up output, or both.

Lincoln's factory facilities had been well along when their production schedule was set (see page 67) but it was unrealistic even so. Their schedule had called for the

Plant No. 2 of the Nordyke & Marmon Company of Indianapolis, Indiana. It was here that the Liberty engines were built. From an N & M company publication. (Author's collection)

Crankshaft and rod assembly during buildup at Nordyke & Marmon. (Courtesy of William Lewis)

Top: Crankcase assembled with cylinders in background above, intake manifolds on shelf to left and water manifolds on lower shelf to left. Left: Complete engine on assembly stand. Right: Complete engine on wood stand. The cooling system has just been flushed to prepare for shipping. All taken in the Nordyke & Marmon plant. (Courtesy of William Lewis)

Teardown for inspection after endurance tests at Nordyke & Marmon. Most of the major assemblies of the engine can be seen in these two views. (Courtesy of William Lewis)

Outdoor test stand at Nordyke & Marmon. Club propeller mounted. Note the exhausts exit to open air with no headers. (Courtesy of William Lewis)

Outdoor test stand at Nordyke & Marmon. In this case it is equipped for torque reaction measurements and with a detachable dynamometer mounted on a rolling carriage. Note the hand crank starter. This view shows the engine with a club propeller mounted and the stand set up as a torque stand. (Courtesy of William Lewis)

PRODUCTION AND SHIPPING RECORD OF LIBERTY TWELVE ENGINES

Special runs of test engines, geared engines, four, six and eight cylinder engines are not included.

MANUFACTURERS	Packard Motor		Lincoln Motor		Ford Motor		General Motors Buick		General Motors Cadillac		Nordyke & Marmon		Trego Motors	
Date	assembled	shipped	assembled	shipped	assembled	shipped	assembled	shipped	assembled	shipped	assembled	shipped	assembled	shipped
First Motor	11/19/17	12/4/17	2/4/18	3/26/18	5/23/18	6/14/18	5/14/18	6/16/18	5/24/18	6/15/18	6/15/18	7/12/18	5/--/18	6/1/18
1917 November	2													
December	33	22												
1918 January	80	39												
February	114	70	5											
March	169	128	26	15										
April	300	309	133	94										
May	502	483	107	127	8		13		8				1	
June	686	575	351	344	66	53	87	80	60	50	2			1
July	880	681	533	484	400	182	139	132	138	94	29	16		
August	659	650	851	761	550	519	265	203	44	106	59	42		
September	655	628	556	687	705	811	273	193			75	59		
October	580	903	1111	1050	1242	1059	498	558			271	308		
November	732	634	940	818	969	1201	390	408			253	143		
December	458	572	1301	1109	10	125	251	167			290	241		
1919 January	346	415	586	1011			362	404			21	191		
February			290	368				132						
March			14	23				1						
last motor	1/28/19	1/31/19	3/8/19	3/15/19	12/2/18	12/24/18	1/28/19	3/1/19	8/-/18	8/-/18	1/4/19	1/15/19	(see note)	
Total	6500	6500	6500	6500	3950	3950	2278	2278	250	250	1000	1000	(see note)	

GRAND TOTAL REGULAR PRODUCTION L-12 ENGINES FOR THE AIR SERVICE-- 20,479 L. M. McEnerling 3/17/19

Note: Trego production under Air Service contract is shown in this chart because they originally were contracted to produce 500 engines for the Air Service. However, after a single engine passed acceptance tests the remainder of the contract was transferred to Ordnance. Just how many of the 499 engines remaining in the contract were actually delivered to ordnance is not currently known as the Air service and the BAP no longer were responsible for Trego's production and stopped keeping records of it. Ordnance built 100 tanks using this engine. Trego production terminated at the Armistice and thus would have lasted five months at most. Considering their production problems and small capacity, it is not likely that they managed a total output of more than about 250 engines.

assembly of the first engine in November and that did not occur until three months later. Lincoln was also under pressure to produce, but nothing like that which was applied to Packard.

NORDYKE & MARMON

Nordyke & Marmon was scheduled to assemble their first Liberty engine in January of 1918. Setup of Liberty production facilities was to begin upon completion, or near completion of their contract for 1,000 Hall-Scott A-7a engines. The A-7a contract was originally scheduled for completion in October of 1917, but this was not to be. They delivered their first A-7a engine under that contract in November and their last in March of 1918.

As the pressures of A-7a production eased and both space and machinery for Liberty production became available, Liberty production got under way. As noted prior to this, Liberty design was very stable by this point and Nordyke & Marmon had few problems to contend with in this regard. Their first engine was assembled on June 15, 1918.

The initial contract called for a production capacity of 800 engines per month to be approached in the fourth month of production and reached in the fifth. Their actual production was 75 in the fourth month and 271 in the fifth. In their report submitted to the BAP history compilers they noted that at the signing of the Armistice they had 265,000 feet of factory space devoted to Liberty production with another 100,000 square feet of the automotive factory committed. With that total they would have had a capacity of 750 motors per month had the war continued.

As has been already noted, and will be further commented upon as we progress with production history, getting production up to expectations was a constant struggle. Officials of the Bureau of Aircraft Production (Equipment Division until May 20, 1918) were doing everything they could toward that end and in the summer of 1918 they came up with the idea of having a competition between the manufacturers each month. They would assign a production quota to each builder and award a trophy to the manufacturer who exceeded its quota by the greatest amount. The winner would be the CHAMPION LIBERTY ENGINE BUILDER of the month. These quotas were assigned based upon current abilities of each builder along with recent past-production history. Production of October, 1918 was the first contest month. As you will be able to see by comparing actual production for the month of October, 1918 with original contract expectations, the quota were much more realistic than were the original contract expectations.

October became not only the first month this honor was competed for, but the last. The Armistice was signed on November 11, 1918 and that ended the competition.

Nordyke & Marmon won first honors by having 246% of their quota pass acceptance and become ready for shipment. Originally all contracts were to be completed by June of 1918 and thus all contractors were to have been producing at maximum capacity by that time, and thus would have been at that capacity in October. The original expectations (noted as "contract"), production during the month prior, the assigned quota and the actual production for each manufacturer for the month is noted below and forms an interesting study.

It is immediately obvious that production was expected to improve dramatically at both Nordyke & Marmon and GM as their quotas were a very heavy increase over the prior month production.

Packard's 900th engine for October, 1918. Note it is mounted on Packard's unique rail transporting system. (Author's collection)

CHAMPION LIBERTY ENGINE BUILDER
October, 1918

Mnfg.	contract	Sept.	quota	actual	percentage
N&M	800	59	125	308	246%
GM	N/A	193	325	558	171.7%
Lincoln	1900	687	826	1050	127.2%
Ford	1000	811	900	1059	117.7%
Packard	1200	628	825	903	109.5%

In spite of the fact that Packard came in last in the competition, they still celebrated their accomplishment and made sure it was recorded by photographing the 900th engine completed during October and publishing it in the January, 1919 issue of their house organ, *The Packard* magazine.

By looking at the production figures of the manufacturers, it is obvious that the contest had rather dramatic effects. Although the Armistice caused a slowdown, the effects of the increased production of October remained obvious thereafter.

Another incentive to the manufacturers and their employees was instigated in the form of a publication called the *Aircraft Production News*. Volume 1 Number 5, dated November 14, 1918, carried the news of Nordyke & Marmon's winning of the Champion Liberty Builder trophy and the plant celebration that resulted.

Although the lead article was the announcement that Germany had just signed the Armistice, the majority of the paper was filled with articles already written regarding past-production, the grand reputation the Liberty engine had gained and various admonitions to keep production ever on the increase.

Quotas had already been set for the month of November and were announced on page seven. Although production continued for some months, the pace slackened considerably. Time was taken by Nordyke & Marmon for a big celebration and parade on Saturday

November 2 and, of course, all manufacturers took a day to celebrate the signing of the Armistice. November production thus lagged considerably that of October. As a matter of interest, however, quotas were set as noted below.

CHAMPION LIBERTY ENGINE BUILDER

November 1918 Production Quotas

Mnfg.	Oct. prod.	Nov. quota	Nov. prod.
N&M	308	375	143
GM	558	600	408
Lincoln	1050	1125	818
Ford	1059	1375	1201
Packard	903	875	634

I have never seen an explanation of why Packard's quota was set so low for November.

At times like these patriotism runs high and the more that can be done to foster it, the better for production results. Among the producers of the Liberty engine there was set up the United States Liberty Motor League (U. S. L. M. L.). All workers and management were encouraged to join and receive their membership card and lapel button. The league sponsored various social events and even organized its own band.

Indianapolis and the Nordyke & Marmon plant were the scene of a great celebration on November 2nd. The workers rallied at the plant in the morning for a celebration and preparation for a later parade in town. The downtown streets began clearing at noon and 3,700 N&M workers gathered by departments with banners flying and spirits high. The parade began moving at 2:30 PM, accompanied by many dignitaries in automobiles.

Aircraft Production News

Vol. 1 DETROIT, MICHIGAN, NOVEMBER 14, 1918 No. 5

"I NEED THOSE MOTORS!"

That is what General Pershing is telling every man and woman at work on the Liberty Motor program today. He needs 'em; all right, we'll build 'em and we'll keep on building 'em till he tells us he doesn't need any more of 'em.

Front page of the November 14, 1918 issue of the Aircraft Production News. (Courtesy of Melvin Mathias)

The *Aircraft Production News* described it as follows:

"Through the central part of the city, around the famous monument and to the square next to the post office led the line of march. There the city had erected a review stand, the central space reserved for workers."

"To the front stepped Mayor Jewett. The Marmon victory was an achievement, he declared, typical of Indianapolis spirit and skill. Governor Goodrich proclaimed the Indianapolis workmen were fighting with the same spirit Indiana soldiers were showing in France. Major Heaslet, commander of the Detroit District, told how disappointed the Detroit factories were when the banner went to Indianapolis, and congratulated the patriotic men and women of the Marmon plants. Lieutenant Emmons, chief of engine construction, told how Liberties were going into tanks, flying boats and planes, and how insistent was the government's need of more."

Marmon President Walter Marmon spoke for the officers and workers of the company in accepting the honor.

The above picture is from the November 14 issue of the Aircraft Production News captioned as follows: "Here is your banner, Mr. Marmon," remarked Lt. Col. J. G. Vincent, Father of the Liberty motor, as he disembarked from a Liberty-motored de Havilland plane at the Indianapolis Speedway Saturday morning, after a fast overland trip. (Courtesy of Melvin Mathias)

Above: Old Glory and the Champion Liberty Motor Builder banner fly over the Marmon plant. (Courtesy of Melvin Mathias)

Left: Visiting notables, ready for the parade. From left to right, L.V. Cram, Maj. Charles F. H. Johnson, Charles E. Buck, Lt. Col. J. W. Downey, John Henry Smith, Lt. H. H. Emmons, U. S. N., Chief of Motor Production, Maj. J. G. Heaslet, commanding the Detroit District, H. J. Saladin, Capt. Ralph J. Moore, Engine Inspection, and T. P. Myers. (Courtesy of Melvin Mathias)

"Actual presentation of the banner which had arrived that morning by a Liberty plane in charge of Lt. Col. Jesse G. Vincent, was made by A. A. Landon, Chief of the Production Division, to a committee of Marmon workers consisting of Miss Amy Bosma, rocker arm department; G. Watkins, motor assembly; E. J. Berry, crank case; T. J. Conway, production; and George Pasway, cutting."

TREGO MOTORS CORPORATION

Very little was recorded about Trego production. They submitted no final report for inclusion in the last BAP history and even their total production is currently unknown, although it is estimated at about 250 engines of their original contract for 500. This condition exists primarily because early in their production history, their contract was transferred to the Ordnance Department with their output to go into tanks. This action removed record keeping from the Air Service and, thus far, I have been unsuccessful in tracking down any remaining Ordnance records.

As noted prior to this, Trego Motors Corporation received a contract on September 6, 1917 and advances to set up production facilities on January 9 and March 15, 1918. A number of letters were received from Trego complaining about frequent changes in specifications which were delaying production and seemed to them unnecessary. Lt. Emmons replied in a letter of April 6, 1918 advising them to stop complaining and to go ahead with production on the same basis as the other companies.

Trego continued to have problems and to complain and finally, on June 6, 1918, Lt. Emmons stated at an APB meeting that the Liberty motors produced at the Trego plant were not of the same standard as those produced elsewhere, and that in his opinion those motors, while not entirely satisfactory for airplane requirements, would be quite adequate for tank purposes. Mr. Potter then stated that by direction of the Chief of Staff he had refused to furnish any Liberty motors to the Ordnance Dept., which was therefore seeking other sources of supply. He suggested that use of Trego engines for that purpose would possibly avoid possible competition with Ordnance for new sources of motor manufacture.

The board thereafter, "Resolved that a recommendation be made to the Director of Aircraft Production that, subject to the consent of the Chief of Staff, the Trego contract be transferred to the Ordnance Dept." The recommendation was accepted and approved and the transfer was made. One engine, Mnfg. No. 2, U.S. No. 7402 was accepted by the Air Service on June 1, 1918. All following Trego engines went to the Ordnance Dept.

The Air Service engine was never installed in an airplane but it remained in inventory for a number of years until it was sold as surplus in the 1920s and purchased by the Auto Engine Works Inc. of St. Paul, MN. It was then converted for marine use. The engine survives as a marine unit and today powers a boat belonging to wooden boat collector Rick McGraw.

Originally it was thought that there might be a number of changes in the engine for use in tanks, but in the end there were few. These changes and the uses the Liberty was put to in tanks will be discussed later in Chapter 6.

Trego Motors Corp. is listed in the New Haven tax books only in 1918, 1919 and 1920. Its 1918 value was listed at $208,450; 1919 at $210,675; and 1920 at $1,600.

Frank Trego first shows up in New Haven in 1917 as living at the Taft Hotel. The following year he lives at 458 Norton St. and by 1921 he has moved to 1592 Chapel and has his office in the Liberty Building at 152 Temple, room 511. That building was constructed in 1902 and remains today.

Motor Magazine listed the company as a specialty engine maker for automobiles in 1920 (and no other year). It would appear that Trego had aspirations to enter the engine building business following the war but could not make a go of it. Although this auto engine was listed as available and pictured (see page 105), no commercial

Data plate of the only Trego L-12 engine accepted by the Air Service. The engine survives in the collection of Rick McGraw. (Courtesy of Rick McGraw)

automobile manufacturer was ever listed as having used it in production.

Col. Vincent wrote an article titled *The Liberty Aircraft Engine* which was published in the *SAE Transactions Part I Volume XIV of 1919* in which he discussed in-depth the history of the development of the Liberty engine. In the discussion following it was a comment by Frank H. Trego which, since it is virtually the only recorded statement by Trego regarding the Liberty engine that the author has ever found, is reproduced in its entirety below.

"F. H. Trego: — The production of our plant was finally transferred to the Ordnance Department. Engines to be used in the large tanks called for in the Allied tank program required some few changes in the design, such as the incorporation of a flywheel and the welding in of a

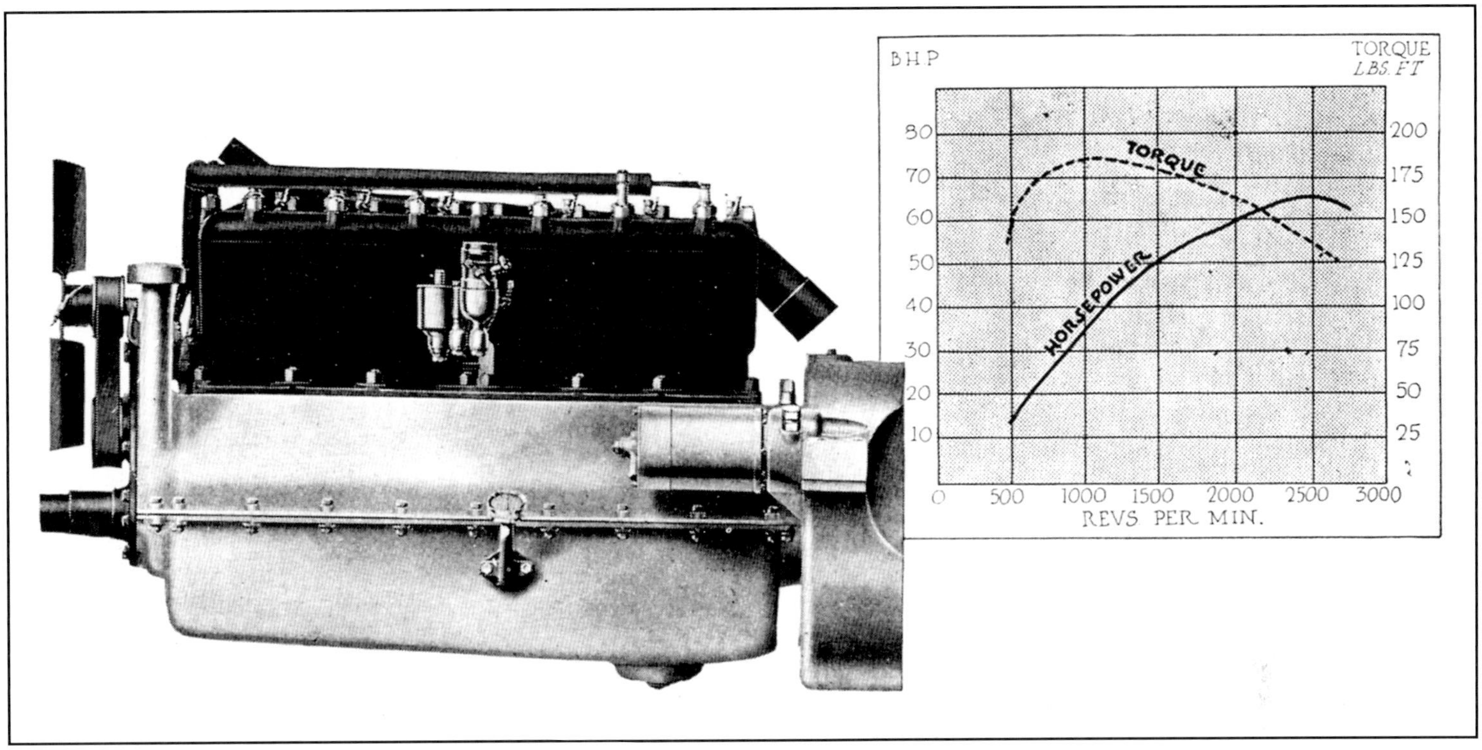

Trego automobile engine model 100 as pictured in the 1920 MOTOR Magazine. This was a six-cylinder L-head engine which developed 65 HP @ 2,500 RPM. Bore and stroke were 3 1/2 x 5 and it weighed 600 pounds. The distributor was mounted on the right front and the exhaust was on the right side of the engine. The carburetor was on the left. This was a specialty engine intended for the use of "assembled" auto makers but is not known to have been used by any. Trego's post-war entry into the engine-building business was short-lived and unsuccessful. (Illustration courtesy of the Hearst Corporation)

tie between the intake and exhaust elbows to prevent the cracking of the jacket at the top, and this we do in regular production. In connection with the testing of the engines for tank service, the United States Ordnance Department endeavored to comply with the testing specifications as suggested by the British Tank Service, and one of these tests called for four runs of 50 hours each with wide open throttle, on dynamometer, at 1400 r.p.m., all parts of the engine being carefully measured before and after each test. Outside of the cracking of jackets, which was later corrected by welding in of a tie, we found a tendency of the babbitt to break away from the bronze backing of the connecting-rod bearings in a spot about as large as a ten-cent piece, directly opposite the oil-hole in the crankshaft at the instant of the explosion of the cylinder carrying the plain-end rod. The hole in the crankshaft at this point has a depression on each side of it, leaving the babbitt unsupported for a considerable area and owing to the fact that the plain-end rod which runs on the back of the bronze bushing has 0.005 to 0.0065-in. clearance we decided that a rapping action was set up by this rod on the outside of the bronze and that this continual rapping directly behind the unsupported portion of

L-12 on test in the Trego plant in 1918. Trego production was allocated for use by the Ordnance Department for tanks after the acceptance of only one engine by the Air Service. (Air Service photo)

the babbitt had a tendency to break it away from the bronze. To eliminate this difficulty we had a shaft made up with an oil-hole drilled at a point 30 deg. in advance of the top dead center when the plain-end rod piston is at the outer end of its stroke. Thus a flow of oil on the center line portion of the bearing was assured before the explosion took place. This also provided complete support for the babbitt under the plain-end rod at the instant of the explosion. Tests of this shaft seemed to prove out the-ory, and we now consider it good practice, even with the single rod, as it assures an uninterrupted bearing surface under the rod at the instant of explosion. We also omitted the depressions on each side of the drilled hole and simply broke the sharp edge of the hole. After a test of 46 hours under full load at 1400 RPM the bearings were found to be in excellent condition."

FORD

In initial contract negotiations Ford agreed on a delivery schedule in which the first engines were to be delivered in April of 1918. 200 were to be delivered that month, 800 the following month and full production of 1,000 per month reached in June. That rate was to be maintained until the delivery of the final (5,000th) engine in September of 1918.

The chart of actual production on page 100 shows that Ford actually delivered its first engine on June 14, 1918, some two months behind schedule. As we have already seen, failure to have first deliveries on-time was the norm. All the producers seemed to have been either overly optimistic in their ability to obtain the required machinery and get things moving or succumb to the government's insistence that this schedule had to be met and could be met. In the case of Ford, and any following producers, the schedules laid out in the initial contracts were a bit more liberal than they had been for the first four.

Ford was by far the largest volume producer of goods (automobiles in this case) of any of those contracted to build the Liberty and probably could have been the largest volume producer of the Liberty as well. As can be seen from the initial production schedules shown on pages 66-69, that was not the original plan. Lincoln was to reach the highest monthly production (1,900), Packard the second (1,200) and Ford the third (1,000). This plan was revised in mid-1918. The APB contemplated awarding Ford a contract for an additional 7,000 engines and thus requested that they gear up for a rate of 2,500 engines per month.

Ford's history report to the APB following the war gives us some idea of the obstacles they had to overcome in their changeover to Liberty production.

When Ford signed their contract they were producing automobiles at the rate of 3,500 per day. However, producing an engine of the size and complexity of the Liberty was quite a different matter than building a Model T. The retail price of a Model T touring was $360 and it weighed 1,400 pounds. The purchase price of the Liberty was (by the time Ford got into production) $5,000 and it weighed about 810 pounds.

As Ford put it, "Its size and design did not lend itself to manufacturing equipment used for the Ford car. Evidence of this is established by the fact that we were able to use only 987 Ford machine tools out of 14,000 used on motor car production. Over 1,500 new machine tools had to be purchased at a time when the whole machine tool industry was flooded with orders, due to the great demand made upon it by contractors of war material."

This was the same story told by all of the manufacturers of the Liberty engine, and probably by any wartime manufacturer that required any new production tooling.

Ford's original estimate of Liberty production floor space requirements was 350,000 square feet. Ultimately 550,000 square feet were required. This necessitated tearing out several thousand machine tools and re-arrangement of over 50% of the plant equipment to allow an entire building to be devoted exclusively to Liberty engine production.

NO LABOR PROBLEMS AT FORD

Ford noted in their post-war report. "Being the first large Government contract placed with the Ford Motor Company gave the Liberty motor contract many advantages. The shop organization consisted of the very best mechanics, who, preliminary to starting production, were constantly impressed with the importance of the undertaking and the responsibility incumbent upon them in the undertaking. Men placed in charge of this work were given to opportunity to fly in an airplane so as to bring home to them the great danger in which an aviator is placed if any part of the motor should be defective.

"The selection of subcontractors was not made strictly on competitive bids. The ability to produce A-1 material was the qualification we sought primarily."

In marked contrast to almost all other Liberty builders: "No labor troubles were experienced because it was not necessary to hire any labor. The curtailment in the production of our regular product gradually released enough help for the production of Liberty motors."

The number of men employed in the production of the Liberty was 675 on the first of February of 1918. It rose steadily from that number to a peak of 11,288 on November 1 of that year. The Armistice was signed on November 11 and by the first of the next month the number of employees assigned to Liberty production had dropped to 826.

Although Ford was not plagued with labor problems, it did suffer the same delays caused by lack of tooling, changes in specifications and changes in design as did all other producers. Their comments regarding inspection requirements pretty much duplicated those made by the other producers, but were so well-put that I felt it would be beneficial to quote them.

"Government inspection was rigid and production was greatly hampered due to the fact that most of the Government inspectors were not properly trained. The selection of Government inspectors was not made strictly in accordance with their ability to act as inspectors. In many cases, men appointed as Government inspectors never had had machine shop experience. As a general rule, they were a very bright class of men but they had to get their experience, and they naturally got it at the expense of the contractor."

FORD'S PRODUCTION ENGINEERING CONTRIBUTIONS

Ford made three major contributions to Liberty production beyond the fact of being a major producer themselves. These contributions were, as a matter of fact, probably the equivalent of adding a major producer.

CYLINDER PRODUCTION

Initially, cylinders were manufactured by turning them from solid billets of steel. This method resulted in huge wastage of material because a cylinder of forged steel of some six inches in diameter had to be bored and then machined to an inside diameter of five inches and then be externally machined to final size. It then had its head manufactured and attached by welding followed by the welding-on of its mounting flange. The process was also very slow and would no doubt be a controlling factor in overall production.

As mentioned on page 47, Ford engineers developed a method of manufacturing cylinders from sections of steel tubing. The APB first mentioned this development on July 3, 1917. On August 7 Col. Deeds wrote to Ford noting he had reports that their process had proven successful and that they were now prepared to manufacture cylinders. He told them that the APB had allocated them 200,000 cylinder forgings and a potential schedule of deliveries of 20,000 per month running from September 1917 to July 1918.

Ford began to set up for production and notified Col. Waldon on August 13, 1917 that they expected to produce 1,000 cylinders a day in about three months if they could get the materials. It was on this date, the reader may recall, that L-12 Exp. No. 1 first ran. Ford would not sign their L-12 production contract until two and a half months later.

In this new process, a section of high-carbon drawn steel tube was formed into a cylinder by a series of steps which take an open-ended section of pipe cut at an angle as shown below as (1) and, through a series of operations, closes the end as shown in (2). Further operations result in the shape shown in (3). At this point the mounting flange shown in (4) is produced by an upsetting process. The rough forging then goes through two turret lathe

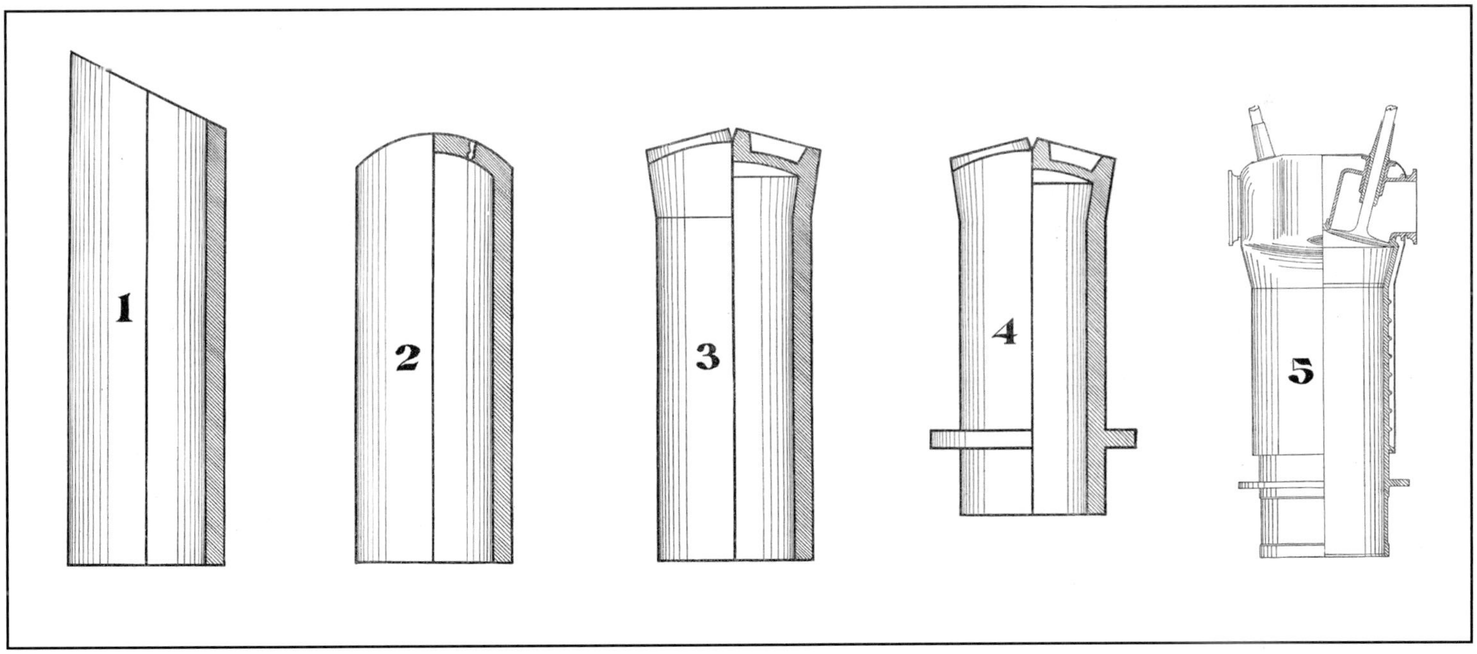

The process of producing a Liberty cylinder as depicted in the August 1919 version of the Liberty manual.

operations to remove stock from the inside and outside of the cylinder. The head and flange are finished by milling.

The valve elbows are separate steel drop forgings machined to shape and electrically welded to the openings in the cylinder heads immediately after they are machined. Water jackets of 16-gauge pressed sheet steel in two pieces are next welded to the outside of the cylinder by oxy-acetylene process. This last process was generally done by female workers. The process was a delicate one and women were found to be better adapted to this type of work than men. The resulting cylinder appears as (5).

BRONZE BACKED BEARINGS

The next important manufacturing achievement by Ford was the development of a special process to make bronze-backed babbitt-lined bearings, and the special application of these bearings to the crankcase and connecting rods. It was evident early in the testing process that bearing failures were likely to become a serious problem. Ford research produced an improved bearing which was made by them and supplied to all manufacturers.

WELDING OF INTAKE-EXHAUST ELBOWS

The third manufacturing improvement was the development of the electric butt welding of the inlet and

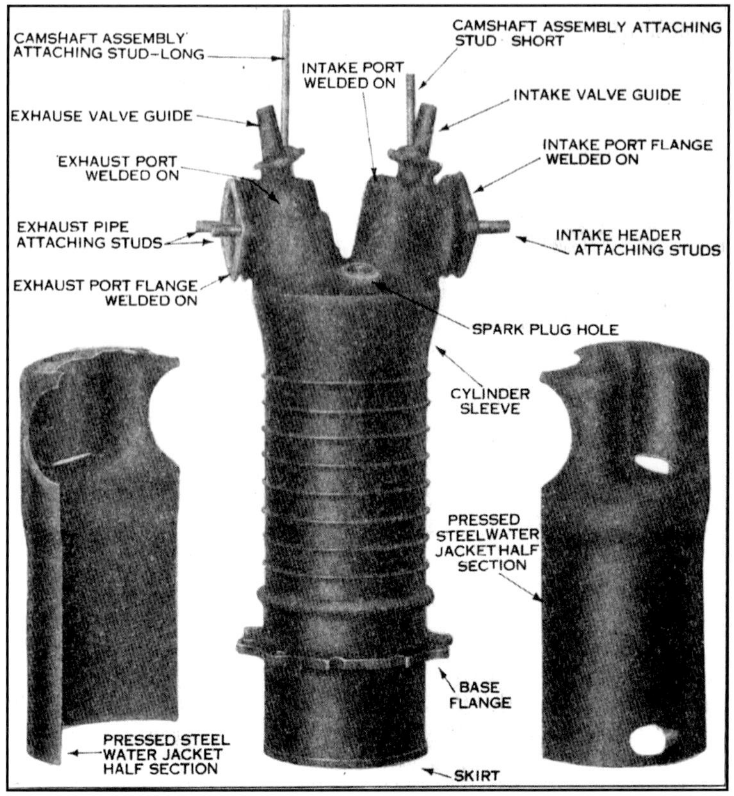

Cylinder assembly as pictured in the 1927 Liberty manual. (Courtesy of Curt Erickson)

Ford L-12 data plate. Enlarged from an Air Service photo. (Courtesy of the Air Force Museum)

3exhaust elbows to the top of the cylinder forging. Prior to this change, these were welded-on with an oxy-acetylene system which resulted in heat distortion of the head which took considerable time and effort to correct in each cylinder produced.

The contemplated second contract for 7,000 engines was signed on September 16, 1918. By that time Ford's total production from the initial contract had reached about 1,300 engines. Production capacity continued to be increased and on the day the Armistice was signed (November 11, 1918) Ford assembled 75 engines (a rate of 2,250 engines per month).

GENERAL MOTORS

Even though General Motors did not receive its first production contract until December 11, 1917, they were involved in the Liberty project almost from the beginning. The Lelands were among the first engineers called upon for advise when Packard was building the prototypes and Cadillac built the rods and rocker arm assemblies for them. In addition, a number of Cadillac engineers helped with the initial layouts and drawings, including D. McCall White, Cadillac Vice President. Walter Chrysler of Buick also was consulted early in the design process.

Women workers finishing cylinders by welding on the sheet steel water jackets, probably at the Ford plant. (Author's collection)

Cadillac supplied about one-third of the connecting rods for Packard's production engines up to May of 1918. From that point in time they concentrated on producing parts for their own Liberty production. The first mention of the General Motors contract on November 15, 1917 noted the engines were to be built at the Buick plant but later, on February 2, 1918, that both Buick and Cadillac were expected to come into production in May. The February 2 BAP note proved to be quite accurate as Buick assembled their first engine on May 14 and Cadillac their first on May 24.

General Motors' initial order for 1,000 Liberty 12 engines was increased to 2,000 on January 21, 1918. Both Buick and Cadillac were tooling up for production, Buick at their Flint, Michigan plant and Cadillac at their Holbrook Avenue plant in Detroit. Because of the extreme shortage of tool making facilities in the Detroit area, Cadillac manufactured many of their own tools as well as supplying some to Trego Motors, Nordyke & Marmon, Willys-Overland and others.

The complete 223,000 square feet of the Holbrook plant was devoted to Liberty production. In addition, a heat treatment facility, an employee restaurant, test sheds, tear-down and assembly departments and other facilities were added. Cadillac noted in their history that it was just 150 days after they started work on tooling that the first Liberty engine was completed and tested at the Holbrook plant.

It was originally intended that production under the General Motors contract was to be divided equally between Buick and Cadillac. Each concern was to assemble half of the engines. The first contract for 1,000 engines had been assigned A.S. numbers 37503-38502 and the second 1,000 A.S. numbers 43771-44770. By the time production was getting under way in the two locations, the decision had been made to assign the first block on A.S. numbers to Buick and the second to Cadillac.

Such parts production as was undertaken by the two divisions (Liberty parts were produced by a multitude of companies, not all of which were contracted engine producers) was set up as was deemed most expedient. That is, each plant produced those parts for which it was best suited. Cadillac manufactured all bearings, connecting rods, cylinders with jackets, all oil and water pumps, all intake headers and camshaft drive assemblies.

In late July of 1918 a decision was made within GM to redistribute Liberty production between the two divisions. The BAP had decided at their July 17 meeting that the demands for Liberty engines by the Allies would total at least 50,000 engines by July 1, 1919. They felt that the best way to meet this demand would be to give additional orders to existing producers under such conditions as would produce the most incentive for the largest possible production and enable manufacturers to commence preparations to do so rather than attempting to develop new sources of supply.

This message was conveyed to the current contractors and resulted in some early changes in preparation for the proposed contracts. The contracts would not be executed until the September-October time period.

The most observable changes took place at General Motors. The Holbrook plant was operating at capacity and different arrangements within the factory would have to be made if manufacturing requirements were increased. The most pressing product, so far as floor space was concerned, was cylinder production. The expected increase in contract production mandated more space at Holbrook for cylinder production as well as for the other parts and assemblies made there. Also, in anticipation of increased production, Ford placed an order with Cadillac for 10,000 intake headers.

The result of these changed requirements was the decision to move all General Motors Liberty engine testing and assembly to Buick's Flint plant and have Cadillac continue as a parts builder only. The switchover occurred at about the end of August of 1918. Cadillac shipped its last engine, A.S. 44021, Mnfg. No. C-250 just before the end of that month.

Buick had completed about 415 engines of its initial 1,000 by that time. They continued using their originally assigned A.S. serial series (37503-38502) until 1,000 engines had been completed. These were assigned Mnfg. Numbers B-1 through B-1000. Buick then picked up the remainder of the A.S. serial block originally assigned to Cadillac at 44022, assigning them Mnfg. numbers beginning at B-1001 and continuing to B-1750. General Motors received an additional contract for the 3,000 Liberty 12 engines which was assigned to Buick. Not all were completed however, as we shall see.

Data plate of Cadillac L-12 Mnfg. No. C-250. This image is an extreme enlargement from a full rear view of the engine taken for a test report and is thus not very sharp. Note, however, that the MNFG. BY designation is General Motors and not Cadillac. The manufacturer can only be told by the prefix of the Mnfg. No., C for Cadillac and B for Buick. (Courtesy of the Air Force Museum)

Cadillac's Holbrook plant where the Liberty was produced. (Courtesy of William Lewis)

R. H. Collins, President and General Manager of Cadillac. (Courtesy of William Lewis)

Crankshaft die for the Liberty Eight at the Buick plant in Flint. Air Service photo. (Author's collection)

THE SECOND CONTRACT SERIES

The additional contracts mentioned on July 17 were to be negotiated at a price (for twelve cylinder engines) of not to exceed $4,400 each, including profit. These figures were based on studies of costs by Packard. In its history of Liberty production as published in the BAP history, Lincoln notes that this last set of contracts were executed at a price of $4,000 per engine and that they were written as renegotiations of the remaining deliveries due under existing contracts along with contracting additional engines.

The actual contracts were executed with the various producers from August through October of 1918. Records of contracts of this period are often confusing. Most have been tracked down in the records by Bob Casari. (See acknowledgements and bibliography.) As a general rule, these contracts were given both order and contract numbers. Each carried its own date and the earliest date would seem to indicate the earliest point at which there was some type of formal acceptance of the proposal.

The earliest date may be associated with either the order number or the contract number and there seems to be no rhyme nor reason to which it will be. (All known orders and contracts are listed in Appendix A.) It can be noted in Appendix A when several of the original contracts were renegotiated at lower prices, but all of these changes do not seem to have so far surfaced.

Orders for an additional 34,000 Liberty 12 engines were placed in September and October of 1918. Before these new orders were placed, on-order engine totals stood at 22,500 and actual deliveries only about 7,500. The total ordered would now be 56,500.

These new L-12 orders were placed as follows:

Contractor	order #	date	number
Packard	730369	9/13/18	6,000
Ford	730407	9/16/18	7,000
Lincoln	730584	10/7/18	11,000
Nordyke & Mar.	730585	10/7/18	2,000
Willys-Overland	730520	10/15/18	5,000
GM (Buick)	30071-4	10/22/18	3,000
		Total	34,000

THE LIBERTY EIGHT IS FINALLY ORDERED IN QUANTITY

During this period it was also decided to place large orders for the Liberty 8. The answer to just why this decision was made at this time requires a detailed look at a number of related circumstances. This will be done in Chapter 5 and I hope will provide the answers to a number of long-standing questions about the design and production of the Liberty 8. I will give a brief look at these reasons here and present them in detail in Chapter 5.

For a start, let us refer to the July 17, 1917 APB board note on page 64 where Major Bolling suggested that, because of a foreseen lack of Hispano-Suiza engines, the Liberty 8 had to be looked to for this need. This was almost a year prior, but yet would seem to still apply. The 300 HP Hispano-Suiza Model H would be the engine for which the Liberty 8 would serve as a replacement. Up to this time, the only order for the Model H Hispano-Suiza was for 3,000 engines with Wright-Martin on November 20, 1917. By July of 1918 only 4 had been delivered.

In addition to powering aircraft for which the Hispano-Suiza H was to be used, four airplanes intended to be powered by the Liberty 8 were also in development; USB-2 (Bristol), USP-2, Pomilio FVL-8, and Victor D-8.

Two avenues were taken to cover this situation. On July 22, 1918 order 730083 was placed with Pierce-Arrow for 5,000 Hispano-Suiza Model H engines and on August 20 order 30011-A-1 was placed with Wright-Martin for an additional 2,000 Hispano-Suiza Model H engines. On August 20, 1918 order 730179 was placed with Willys-Overland for 3,000 Liberty 8 engines. On August 5, 1918 order 730155 was placed with General Motors (Buick) for 5,000 Liberty 8 engines and on August 20, 1918 order 730223 was placed with the Winton Motor Co. for 1,000 Liberty 8 engines.

Lincoln L-12 data plate from an engine in the collection of the Duxford Air Museum in England. (Courtesy of Jamie Fairchild)

Buick was expected to be the first to get into production because they were a current producer of the Liberty 12. The other two firms were not expected to deliver any engines until well into 1919. Obviously the end of hostilities which came about on November 11, 1918 would, in most cases, terminate these contracts before any deliveries could be made.

These contracts resulted in a total of 9,000 Liberty 8 engines on order. Very few would be delivered before the Armistice caused cancelation of these contracts. I want to point out here that most historians have stated that the very small number of Liberty 8 engines produced was a result of cancelations of contracts due to problems encountered with the engine. The engine did suffer from more vibration than most others of its size, but that was not a cause of any contract cancelations. (This subject will be discussed later.) Requirements for this size engine simply did not materialize until this late date and thus no production engines were contracted. The end of the war ended production, not problems with the engine.

Liberty 8 orders were placed as follows:

Contractor	order #	date	number
GM (Buick)	730155	Aug 5	5,000
Willys-Overland	730179	Aug 20	3,000
Winton Motor	730223	Aug 20	1,000
		Total	9,000

These Liberty contracts might seem to be contrary to the aforementioned desire to seek no new producers, but instead to increase the output of current ones. The Buick order would not be in question in this regard, but those with Willys-Overland and Winton might.

Willys-Overland had not previously produced Liberty engines, but was an early producer of aero engines, having been contracted to build 5,000 Curtiss OX-5 engines on September 9, 1917 and given a contract for an additional 2,000 on February 12, 1918. By the end of July, 1918 Willys had delivered about 5,259 OX-5 engines and was producing at a rate of about 1,000 engines per month. They had certainly proven their ability as an aero engine builder. The need for the OX-5 training-size engine was close to being satisfied and any further orders would be modest and aimed at replacement. Thus Willys production facilities would be available for 1919 and thereafter production of Liberty engines.

The contract with Winton is not so easily explained. Winton had been building automobiles since 1896 and had built 2,458 in 1916. They had already taken on some contracts for war goods, but not for building aero engines. It appears that the BAP felt justified in getting this company set up as a modest producer in 1919 and thereafter.

The BAP was getting production geared up in contemplation of a continuation of the war well into 1920. Thus, at this point 56,500 L-12 and 9,000 L-8 engines were on order for a total of 64,500 Liberty engines. This only accounts for standard engines. As we will see in the next chapter, a number of experimental engines were either built by the Engineering Division or contracted in small lots to others during the war.

Buick L-8 data plate MNFG. NO. BX 12. By the time production of the Liberty 8 got under way, it was known that only a small number of engines would be completed as possible prototypes for future production. Therefore, Buick chose to number them with a BX prefix, with the "X" indicating experimental, instead of the usual B only. Also note that this plate carries Buick's name as well as General Motors, which was not done with the L-12. This engine is in the Air Force Museum. (Author's photo)

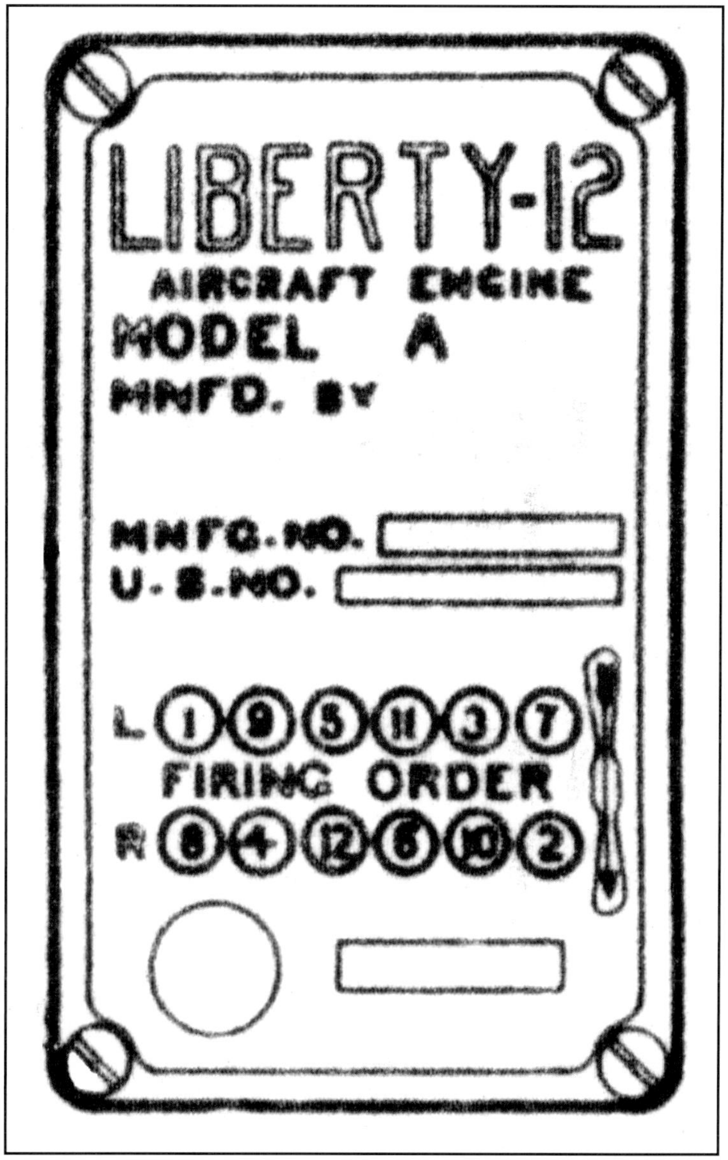

Above is the engineering drawing for the standard Liberty data plate as originally laid out for the L-12. The blank below the "MNFG. By" to be filled in with the name of the builder. Comparing this illustration with those of the actual plates of the various builders reproduced throughout the book will show that the only one who deviated at all from the standard was Trego (see page 104), and they made only slight changes. (Author's collection)

Marshal Foch and General Pershing. Signal Corps photo. (Author's collection)

mies. — There is now no longer any possible hope of forcing peace on the enemy. — The situation grows more desperate every day and may force the High Command to grave decisions. — "

The Chancellor yielded to this pressure and sent a cablegram to President Wilson, through the Swiss Government, on October 6th which read as follows:

> *"The German Government requests the President of the United States to take in hand the restoration of peace, acquaint all belligerent states with this request, and invite them to send plenipotentiaries for the purpose of opening negotiations. It accepts the program set forth by the President of the United States in his message to Congress on January 8 and in his later addresses, especially the speech of September 27, as a basis for peace negotiations. With a view of avoiding further bloodshed, the German Government requests the immediate conclusion of an armistice on land and water and in the air.*
>
> *Max, Prince von Baden,*
> *Imperial Chancellor"*

There followed a series of communications between the United States State Department on behalf of President Wilson and the German Government making clear the

THE END OF THE WAR

This last series of orders was placed from late July through mid October. Although the war was going much better for the Allies at this point than it had a few months earlier, it was still expected to continue for some time. Early October was, however, a turning point in the war. The Allies knew things were not going well for the German forces and for the German government, but they did not realize how bad it was until a sequence of events which started with an October 3, 1918 letter from German Marshal von Hindenburg to the German Chancellor von Baden which contained, in part: "The High Command insists on its demand of Sunday, September 29th, for the immediate forwarding of an offer of peace to our ene-

Armistice celebration, Place De La Concorde, Paris. November 11, 1918. Signal Corps photo. (Author's collection)

conditions under which the United States and her Allies would consider an armistice intended to lead to peace.

As a result of this, all these communications were transmitted to the Allied Governments on October 23, 1918 to obtain their views regarding proceeding. Their replies were in the affirmative and on November 5, 1918 the State Department replied to the German Government that Marshal Foch (French commander and Commander-in -Chief of the Allied Forces) had been authorized by the Government of the United States and the Allied Governments to receive properly accredited representatives of the German Government and to communicate to them the terms of an armistice.

American Army commanders, and General Pershing in particular, had received information through their Intelligence Section that the Germans had sued for an armistice, but they had no first-hand knowledge of the communications between Washington and Germany that followed. The war was pursued with even more vigor in order to keep the Germans on the offensive and put them in the weakest position possible while negotiations were going on.

As you can see, the last of the new Liberty contracts had been finalized prior to the November 5 affirmation by the Allies of a desire to proceed with armistice negotiations.

Although many of the military leaders, including General Pershing and Marshal Foch, felt the Allies should drive for unconditional surrender rather than an armistice, the political decision was to negotiate for an armistice, even though this would allow a better position for the German Government in the long run.

From November 5 until November 10 negotiations continued (as did also the fighting) and at 11:00 PM on the 10th a message was received from the German Government authorizing its delegates to sign the Armistice. The message was decoded and the documents drawn up and signed just after 5:00 AM on the 11th. Word was then sent out to all armies that fighting would cease at 11:00 AM on that day, November 11, 1918.

THE TERMINATION OF THE LIBERTY CONTRACTS

Upon the signing of the Armistice, the government was anxious to cancel aircraft contracts as rapidly as possible. Those producers who were a part of a very large organization which looked forward to returning to production of civilian goods as rapidly as possible were willing to wind up Liberty production as rapidly as possible.

The last Packard Liberty 12 leaves the assembly line on January 28, 1919. It had not yet been tested and accepted but the proud event was recorded for posterity. Quite surprisingly, the event was recorded witnessed by a government inspector, a shop foreman and an assembly worker, rather than by various high company and government officials. (Courtesy of NAHC)

FORD

Referring to the production chart on page 100, one can easily see which companies these were. Ford had contracted for a total of 12,000 L-12 engines but halted production on November 20. They assembled their last engine, number 3950, on December 2 and it passed final inspection on December 24.

GENERAL MOTORS

General Motors was also ready to get back to civilian production, although they did not appear to be as anxious as was Ford. They had contracted for 5,000 L-12 and 5,000 L-8 engines and, as previously noted, had made significant changes to facilitate increased production. Their wind-down was less hurried. They assembled their last L-12 on January 28, 1919 and delivered it on March 1. This was their 2,528th engine. Thus they completed their first two contracts for 1,000 engines each, and built 528 of the most recent contract for 3,000.

Production was stopped on their L-8 contract on November 14. By then, 15 engines had been assembled.

The Bureau of Aircraft Production had been actively working on the Liberty 8 design for some time (see the next chapter) and was thus quite interested in the early production engines. For this reason, they tested several of the early production engines, some of them even before they passed final acceptance by the Air Service. This has

Data plate of Buick L-8 BX 5 marked U. S. NO. -E-.
(Courtesy of William Lewis)

left us with records indicating that it took some time to decide just how to handle the above mentioned 15 engines.

General Motors had assigned Mnfg. numbers to their production L-12 engines using prefixes with the numbers which indicated the manufacture; B for Buick and C for Cadillac. Even though the data plates of the L-8 engines now carried the name of the builder (Buick) in addition to General Motors, they still followed this practice. However, it appears that by the time these engines were assembled and had data plates applied, they decided they would cancel the original contract and consider them as experimental rather than production engines. To so indicate, they prefixed these engines' MNFG. NO. with "BX."

By late November they had not decided how they were going to take care of the contract logistics. At least five of these engines had been subjected to testing up to this point, and their records always indicated Mnfg. numbers but no U. S. number. They had already decided that the originally assigned series of U. S. numbers (57209-66310) would not be used and evidently thought that they would have no assigned U. S. numbers. To that end, early engines were stamped in the U. S. NO. location with "-E-" for experimental.

Engine BX 5 was used in power plant report 67 and was photographed from several angles. The rear view shows the data plate clearly marked MNFG. NO. BX 5 and U. S. NO. -E-.

The ultimate decision was to write a second contract to cover the 15 engines. On January 27, 1919 order 730810, Contract 5463 was finalized and U. S. numbers 94472-94486 were assigned to the engines.

From the test which documented the data plate of BX 5 and from several earlier tests involving BX 2, 4 and 6, it is evident that initially these engines were not assigned U. S. numbers. In later tests, however, BX 12 and 14 are listed with U. S. numbers 94482 and 94485. In addition, the Air Service carried out a complete engine and aircraft inventory by U. S. number (now referred to as Signal Corps number) at the direction of House Resolution 190 dated August 1, 1919 and it was reported on December 3, 1919.

All 15 Buick L-8 engines were reported by the new assigned U. S. numbers. 94473 (BX 2) and 94482 (BX 12) were at McCook Field and the remaining engines were all at Little Rock. Thus, by this time, all 15 engines had their data plates stamped with the new series of U. S. numbers.

WILLYS-OVERLAND & WINTON

The Willys-Overland L-8 contract was canceled on November 13 and all production stopped. No engines had been completed.

Very little information has surfaced regarding the Winton contract. It has been found in the records listed as order number 730223, dated August 20, 1918, but so far no contract information has surfaced. It is thus possible that the actual contract was not consummated prior to the Armistice. Whatever the case, it was, or would have been, canceled immediately after the Armistice with no production completed.

NORDYKE & MARMON

Nordyke & Marmon had initially contracted to produce 3,000 L-12 engines and received a contract for an additional 2,000 in the second series. By the Armistice they had assembled about 550 engines of the first contract. The contracts were canceled and production stopped on December 7, 1918. An agreement was reached whereby Nordyke & Marmon would complete 1,000 engines total. The last of these was assembled on January 4, 1919 and delivered on January 15.

PACKARD

Packard had completed about 5,000 engines by the Armistice. It was agreed that the company would complete and deliver a total of 6,500 engines and the remainder of the second contract would be canceled. They assembled their last engine on January 28, 1919 and delivered it on January 31.

LINCOLN

The Lincoln Motor Company had been incorporated solely for the production of Liberty engines and had built two factories strictly for that purpose. This put them in a unique position as compared to the other producers (other than perhaps Trego). Although much of it was financed by the government, they had far more investment to become a producer than did any other builder. Cutting of production abruptly would have much more financial consequences for Lincoln than for the other builders. Thus they were afforded the most leniency of the builders.

Lincoln initially contracted for 6,000 L-12 engines and received a second contract for 11,000 additional engines. Although they had only assembled about 4,100 engines by the Armistice, they were also allowed to complete 6,500 for delivery. They assembled the last of these on March 8, 1919 and delivered it on March 15.

The second contract had been made, in the case of Lincoln, with some special clauses which, in effect, split the contract into two parts. As mentioned prior, this contract was also a modification of the first one in that all engines which remained to be delivered under the old contract as well as those to be delivered under the new, were at a fixed price of $4,000 each.

A second clause was included because of Lincoln's special situation. All engines remaining to be delivered from the first contract would be non-cancelable as well as an additional 3,000 of the second contract, making a total of 9,000 engines that Lincoln was guaranteed to build and sell. In addition, a 40% depreciation of the plant and facilities was to be allowed. This was done because the entire plant had been built for the production of this engine and, if such depreciation plus a certain amount of business was not guaranteed, an abrupt end of the war could result in the company facing the possibility of bankruptcy. This scenario could be the result of having to pay off the loans

Lincoln records the completion of its final Liberty 12. The picture was probably taken on March 15, 1919. In contrast to Packard's recording of their last engine, the Lincoln Liberty is pictured with company executives. Left to right: Henry M. Leland, president; Charles Martens, foreman of the test field; William H. Ebelhare, factory superintendent; William Guy, foreman of assembly; Paul W. Abbott, foreman of inspectors; Ernest E. Sweet; Wilfred C. Leland, vice president and general manager; William T. Nash, secretary and treasurer; LeRoi J. Williams, counsel and assistant general manager. (Courtesy of NAHC)

that were necessary to build and equip the plant, which could amount to more than the profits gained from production plus the after-war value of the facilities provided to build them.

The factory and production facilities had been built at inflated wartime prices and would have to be liquidated at post-war deflated prices. Thus one can easily see the potential of disaster if provisions were not made to circumvent it.

The last 8,000 engines of the second contract for 11,000 were to be cancelable if they were not needed. In addition, if the last 8,000 were not produced, additional depreciation was to be allowed.

When the war ended, it did not take long for the government to express a desire to end production short of the 9,000 engines. The Lelands, still as patriotic as when war broke out, acceded to the request and the figure of 6,500 was agreed upon even though it left their financial affairs in a precarious position.

Because the contract was non-cancelable (for the justifiable reasons stated above) a supplemental contract covering the cancellation was entered into. First, the rates of depreciation of the plant and equipment were modified to a flat rate of 55%. Second, it was agreed between the government and Lincoln that the profits that would have resulted from completion of the contract would have been not less than $2,700,000. Lincoln agreed to accept the sum of $1,000,000 and profits on certain inventory items in consideration for cancellation of the contracts at 6,500 engines.

The contract was signed for the government by no less than the Secretary of War, Newton Baker. Lincoln then paid taxes of $4,126,000 on its 1918 income.

By the end of March 1919 they had a quiet plant full of machinery which, for the most part, was unusable for any task they contemplated.

At first they considered the possibility of going into the business of building motors for other automobile manufacturers, but they shortly abandoned that idea thinking it best to go back into the business the Lelands knew best, the building of automobiles.

A completely new car was designed and the necessary preparations in plant and financing and dealer setup were made.

The reorganization was complete by January 1920 and new cars were expected to be available for sale by April. This goal was not met because of various unforeseen production problems.

In the middle of this, the Treasury Department served notice to Lincoln that an additional income tax of $5,725,673 was due on war profits.

Lincoln and government accountants again went through all the documentation involved and one by one the various items claimed by the government were dropped until finally the entire claim was withdrawn.

The new Lincoln was not on the market until September 6, 1920. By that time the 1920 selling season was over. In addition, a post war economic slump was just setting in. The Lincoln was a high-priced luxury car appealing to a small segment of buyers and was hit hard by the slump. They had bank loans due that they could not cover and were taking every available avenue to improve their perilous financial condition. They were not alone, however, as much of the auto industry was now in the mode of cutting prices in an attempt to boost sales and some makers were by now in financial difficulties.

The Lincoln Liberty factory in Detroit which was constructed in 1917. This picture probably shows it as it appeared in the middle 1920s. (Author's collection)

What Lincoln did not need now was just what they got on May 21, 1921 — a renewed Treasury Department claim for taxes on 1918 income, this time for $5,400,000. All indications are that a change of political administration plus a continued public clamor for a clampdown on what was considered as excess war profits led to this renewed claim.

The basic claim was that the depreciation allowed on plant and equipment following the war was, in effect, income and should be taxed. The matter was again reviewed between Lincoln and government accountants and as a result the government sent Lincoln a letter stating the claim was withdrawn and no tax was owed.

Matters continued downhill for Lincoln. The depression had continued and most of the automobile industry was hard hit. Several makers had gone into receivership and by November 1 Lincoln was trying to avoid the same fate.

On November 7, 1921 the government once again presented Lincoln with a tax claim, this time for $4,500,000. This claim resulted in a tax lean for that amount and, in effect, kept any potential financial saviors at bay. On November 8, 1921 Lincoln went into receivership.

The Lelands continued with whatever they could to keep the company in operation. Their best hope now was an interest in the company expressed by Henry Ford.

On December 29 the receiver announced that the federal government had found no fraud in the Lincoln income tax matter and the claim would probably be reduced to about $500,000.

On January 6, 1922 the federal government advised the receiver that the tax claim would be settled for $610,274. A little late, now that all hope for financial help to save the company was washed away by the last $4,500,000 claim.

On February 4, 1922 the Lincoln Company was sold at the receiver's sale to Henry Ford for $8,000,000. Although the Lelands would shortly be no longer associated with the company, in spite of initial promises to the contrary by Ford, Lincoln would continue in business to this day as a division of Ford and go on to build some of the finest luxury cars obtainable in this country.

Trego Motors displays a running Liberty engine on their float in the New Haven, CT November 16, 1919 Armistice Day parade. (Courtesy of The Whitney Library, New Haven Colony Historical Society, New Haven, CT)

An interesting historical note regarding the sale of Lincoln after receivership is that the legal portion of it was handled by attorney Harold H. Emmons, the same Lt. H. H. Emmons who headed engine production for the BAP during the war.

TREGO MOTORS

Most of Trego's history relative to production of the Liberty is recorded on pages 103-106 and little more can be added here. Their initial contract was for 500 L-12 engines. After acceptance of only one by the Air Service, their contract was transferred to the Ordnance Department for use in tanks. Although no records relative to the contract after that point have been found, their production rate, plus the number of tanks eventually built to use this engine, lead me to believe that a total production up to cancellation after the Armistice might be liberally estimated at 250.

Trego tried what Lincoln considered, entry into the market of building engines for automobile makers. The attempt was unsuccessful. The only attempt that I am aware of that they made to use their wartime Liberty production to advantage in postwar business ventures is illustrated by their use of one of these engines on a public parade float in New Haven, CT on Armistice day (November 16) 1919 as shown in the illustration on this page.

Thus ends the recording of the regular production contracts for the Liberty 12 and Liberty 8 engines. There were a number of experimental and other special purpose engines built during the war which we do not consider as production engines. These are covered in the next chapter.

NOTES:

Much information relative to those subjects that were covered in this chapter is contained in statistical compilations and their locations are listed below for quick reference.

1 - Production and shipping record by manufacturer listed by month. Page 100.

2 - Contracts proposed, issued and executed. Appendix A.

3 - Liberty production: by producer; by contract; by type; by U. S. serial number. Appendix B.

BUILDING THE ROLLS-ROYCE IN AMERICA

Initially, much thought was given to building combat-sized aero engines in this country in addition to those of training-size which were later built. The only foreign engine to have much consideration was the Rolls-Royce, as it was considered the best available in Europe. Col. Sidney Waldon compiled a brief history of the efforts of the APB and later the BAP to actually make this happen. I will quote this history extensively, as well as related entries from the APB minutes as I relate what actually happened relative to the attempt.

During the early summer of 1917 consideration was given to the production of Rolls-Royce engines in this country. On June 26, 1917, Mr. R. L. Montgomery (later Colonel) reported to the board that he had taken up the possibility of producing the Rolls-Royce engine in America with Lord Northcliff (head of the British Mission) who had recommended that Claude Johnson, manager of the Rolls-Royce plant in England, be asked to come to this country to negotiate for production of engines here. The board approved the suggestion. At this point in time the Liberty had been designed but the first prototype engine had not yet been delivered.

Although objections to this plan were raised a number of times, and several reasons given for them, including the complexity of the engines as compared to the Liberty, the difficulty in finding a suitable builder and the cost of producing the engines; serious consideration of the possibility continued almost to the end of the war.

The initial proposal was that we make arrangements with the Pierce-Arrow company to build the engines using competent assistance, samples, drawings, etc. provided by Rolls-Royce but Mr. Johnson preferred not to have anything to do with a company which, after the war, might derive any commercial benefit from an association with Rolls-Royce production.

On July 17 Mr. Montgomery reported that Rolls-Royce required purchase by the United States of either the Maxwell or Chalmers plant at Dayton for use for production. On July 18 Major Foulois (later Brig. General) recommended facilities be provided to build both the 270 HP Falcon and 375 HP Eagle VIII engines. Discussion which followed resulted in the feeling that production should start off slowly because of expected difficulties and because of the publicity the United States engine (later called the Liberty) would produce before the Rolls-Royce reached quantity production. The board felt it should not rely entirely on the U. S. engine. It was resolved that Mr. Claude Johnson should be informed that a plant would be provided for the manufacture of Rolls-Royce engines and authorizing him to bring over his men and equipment.

On July 19, 1917 Messrs. Johnson and Mr. Alan Royce (both of whom had recently arrived in this country) were present at the board meeting. Mr. Johnson said that they proposed to build a more-recent design engine than was now being built in England, anticipated no trouble providing the proper organization here, and expect to be able to deliver the first engine about January 15, 1918. The two gentlemen retired from the meeting and discussion followed. At this point it was felt that the engines would cost about $8,000 each.

Negotiations continued in which Rolls-Royce changed their plant requirements to one in the Pittsburgh area. In addition, there was great difficulty in agreement not only of a plant acceptable to both parties, but terms of a contract. These negotiations dragged along for months. By June of 1918 the board was getting information that our aircraft program would have to expand greatly. Even though there was much input against Rolls-Royce production now that Liberty production was now beginning to rapidly increase, Messrs. Ryan and Potter were in favor of a contract for 1,500 engines at $10,000 each as an alternative in case the Liberty ran into trouble. (By this time about 2,750 Liberty engines had been built and its price was about $5,000 each, soon to be reduced to $4,000.) A resolution for a contract proposal as put forth by Ryan and Potter was made but not ratified for lack of a quorum. By July 3, 1918 it had still not been ratified and Mr. Potter stated that after careful consideration he had decided not to enter into a contract with Rolls-Royce because: 1- Deliveries would not commence until February and the 1,500 engines would take a year to deliver. 2- The contract terms compare very unfavorably with those offered Liberty engine manufacturers, whose product has been enthusiastically received abroad. 3- Representatives of the British Air Service are unanimously agreed that the Rolls-Royce engine should not be manufactured in this country.

At this point the matter was finally dropped. Liberty engines were currently being manufactured at the rate of more than 1,000 per month and within four months, would reach 4,000 per month. The Liberty had proved to be as good or better than the Eagle VIII in most respects. It was easier to maintain, being less complicated and having interchangeable parts. The Eagle was a hand-fitted engine requiring very expert mechanics. For any serious repair work, it had to be returned to the factory and could not be field maintained. The Liberty was more powerful and had a better power-to-weight ratio. The Liberty was far cheaper to build and could be built more rapidly for much the same reason it was easier to maintain. Total production of the Eagle VIII in England by the end of the war was about 3,000.

The history of the Rolls-Royce contract negotiations as recorded by Col. Waldon were necessarily brief and, of course, tell the story as seen from the American side. The Rolls-Royce side may be seen in the excellent book by Ian Lloyd, *Rolls-Royce the Growth of a Firm*. This book, published in London in 1978 by The Macmillan Press, Ltd., devoted two chapters to the subject. The story Lloyd tells is little different from the above but much more detailed and allows one to see the issue from both sides of the negotiations.

Chapter 5

Wartime Development and Use of the Liberty

First let us try to wrap up as many of the remaining production changes as we can. Then I will document new engine designs as well as uses and proposed uses of the Liberty.

LATE PRODUCTION CHANGES

We are fortunate to have been left fairly detailed information regarding early changes required in parts, and manufacturing procedures and I have documented the more significant ones, from Alteration # 1 of September 20, 1917 through Alteration # 133 of March 18, 1918, in chapter 4. There were undoubtedly others after the 1918 date but their number and importance dropped rapidly beyond that point.

The BAP history carries little information relative to changes late in production but some information about them is contained in British records since most of their engine deliveries were from late production. I have copies of most of the British technical notes and correspondence related to their use of the Liberty from first receipts to the end of the war. The troubles reported by them and the fixes applied during this period are probably at least to some degree indicative of those of our own Air Service.

As a point of interest, first reports of specifications of the new Liberty, of course, reached the British long before their first engine did. In a report dated November 2, 1917 the "U.S.A. Liberty Motors" are described and specifications quoted and noted as being compiled from cablegrams received up to October 27, 1917.

OCTOBER 27, 1917 LIBERTY SPECIFICATIONS

Type	Disp. cu. in.	Weight in pounds	HP @ 1800 RPM	Weight per HP
8 cyl.	1099.6	560	263	2.131 lbs.
12 cyl.	1649.4	780	430	1.81 lbs.

These specifications would change to some degree before the engines actually went into service. The weight of the Liberty 12 would increase to about 820 pounds by the end of 1918 but its power would remain about the same as then quoted. The Liberty 8 would be proven capable of 283 HP @ 1,800 RPM by November 27, 1917, eventually be rated at 290 HP @ 1,700 RPM and capable of 300 HP @ 1,800 RPM. Its weight would be up to 650 pounds by 1919.

Once they had the Liberty 12 in-hand (the British never ordered nor did they acquire any Liberty 8 engines) they began to gain their own experience as to what to expect in the way of performance and failures. They had about 80 engines in-hand by the end of June 1918, about 200 by the end of July, 620 by the end of September and 1,074 by the end of the year.

The first reports of performance begin to appear in July of 1918, and the first showing any problems appeared on July 12 and 17. In this case it was **failure of camshaft bevel gears.** From the series of communications regarding this problem, it appears that it was not one of frequent occurrence. When it did occur, it indeed was a serious failure. By September 18 only one additional failure had been reported, this time it occurred at an American Seaplane Station at Killingholme, England.

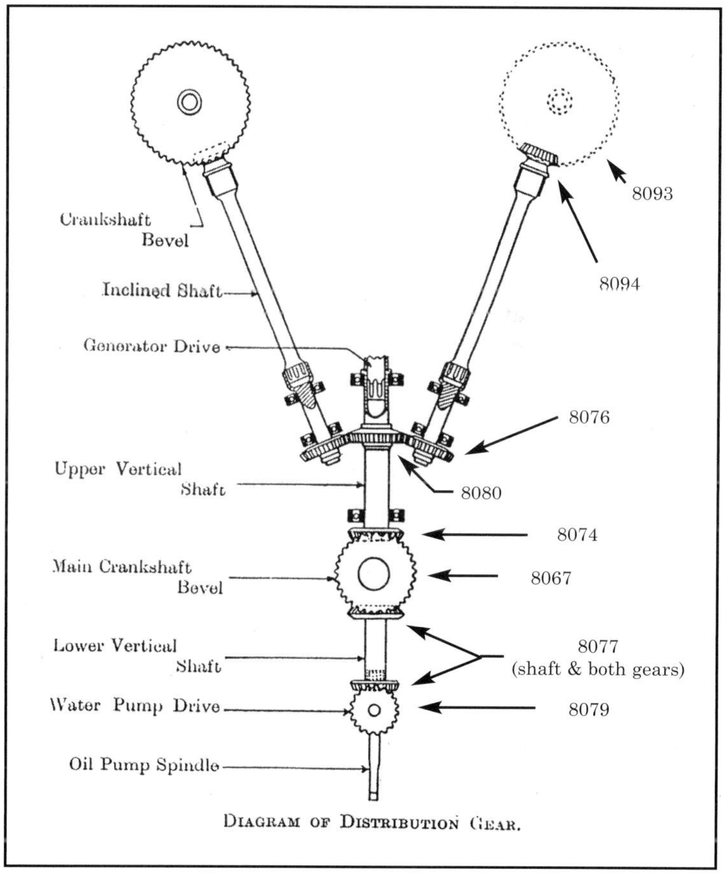

Diagram of the Liberty bevel gear train as depicted in the British Liberty manual of February 1924. (Part numbers added by the author. Author's collection)

The problem had been studied and it had been determined that it was one of design and should be solved with a new gear design. The British Director of Aircraft Equipment, Brig. General C. H. Whittington, wrote a letter to the General Officer Commanding, RAF British Armies in France on September 18 in which he made the following statement: "In view of the small number of cases in which failure of these gears has been reported, I am to inquire whether you are prepared to accept delivery of machines fitted with Liberty Engines with the old gears fitted. I am to add that it is considered that this is a reasonable risk to take." An affirmative response was received on the 19th.

In the mean while it would appear that the British took it upon themselves to produce some modified gears to use as replacements until new gears could be introduced in manufacture. On September 22 the Air Ministry received a letter from the British Maj. General commanding the Independent Force, RAF noting acceptance of the above-mentioned proposal and suggesting extra spare bevel gears be sent with engines until the new gears were available. In it he also noted that two engines had suffered from breakage of the main drive bevel gear (Part No. 8067) after being fitted with modified "Generator Driving Shaft Lower Gear Wheel" Part No. 8074. The cause of the failure was attributed to bad fitting of this new gear to the main bevel gear resulting from either "bad workmanship" or the "teeth being improperly formed." Hopefully the use of these new gears was halted.

On September 24, 1918 General Whittington sent a letter to the above-mentioned General Officer Commanding noting that information had been received from the "Director-General of Aircraft Production" in America that new Gear Wheels would be fitted in all engines produced beginning with the following U.S. numbers of the different makes:

MAKE	U.S. No.
Packard	19808
Lincoln	24445
Ford	29720
Nordyke & Marmon	4402
Cadillac	43892
Buick	37733

It should be noted in connection with this change that this was not the "stub-tooth" gear change often referred to in connection with Liberty engines following the war. That was yet another design.

A comparison of the above numbers with the list of numbers versus factory ship dates in Appendix B shows that in all cases these numbers fall in late July or early August. It would be some time later before affected engines reached the British.

During the month of September trouble was reported on several engines with **fouling of plugs** which they attributed to excessive oil pressure. They relieved this problem by reducing the normal oil pressure from 30-40 pounds to 15-20 pounds at running speed. This was accomplished by drilling holes in the head of the pressure relief valve. This could reduce idle pressure to something between 0 and 2 pounds when it should not go below 5 pounds. Although the engines seemed to survive that low a pressure at idle, it was eventually found that any excessively low idle pressure was caused by too large bearing clearances, oil leaks, missing oil passage plugs or other problems. It helped considerably with the plug fouling problem and a similar fix was applied in America.

Pistons were fitted with three rings, all of which were compression-type. The bottom ring was changed to a scraper-type, which helped to some degree.

The addition of an oil drain groove in pistons and 10 drain holes in the groove in addition to controlling oil pressure was the final fix to the problem of fouling plugs. This, however, was not arrived at until after the end of the war.

A third problem reported during September was **over-heating of generators.** By the end of October it had been determined that about all overheating of generators was being caused by "mismanagement of control switches," that is leaving one or both ignition switches in the on position with the engine stopped. An appropriate instruction plate was posted in each cockpit for the pilot's guidance.

This completes, so far as I am aware, the mention of all significant changes which were initiated during the manufacture of the Liberty 12. A number of minor procedural changes were made, such as which nuts were to be safety wired and which were not, acceptable tolerance limits on various parts, etc. but these will not be enumerated.

BRITISH REPORT ON LIBERTY PERFORMANCE

On November 4, 1918 Brigadier General Brook-Popham, Commanding General Royal Air Force In the Field, wrote to the Secretary Air Ministry reporting on Liberty performance in general with the following comments: "We have just taken down one engine (607 Packard - AS 18030). It had had 95 hours running and the engine was in perfect condition. The only fault was one piston slightly scored." "Generally speaking so far there is little cause for complaint on the running of Liberty engines."

Date Line of Wartime Development and New Designs

Designs being made for **L-16. - November 6-12, 1917**

November 19, 1917- **First production L-12** completed.

First L-12C order for 2 samples. - **November 30, 1917**

December 27, 1917- L-12A being redesigned for **tank** use.

Two L-4B engines were ordered built by Hudson. -**December 31, 1917**

January 7, 1918 - First test of **Dixie Magneto** on L-12. This was followed by tests of a **Simms** on Oct 28 and a **Berkshire** on Nov 21. Testing continued into the early 1920s.

First test of an **electric starter** (Bijur) on a Liberty 12. - **January 9, 1918**

Mid-January 1918 - Three L-1 test engines, known as **L-1D** models, were ordered built by the Eng. Div. in mid January. The first was tested on Feb 9.

First **L-12C** is completed. - **Late January 1918**

February 12, 1918 - L-4B No. 1 delivered.

Second order for **3 L-12C** engines. - **February 26, 1918**

Order 5257 to Thomas-Morse for 2 **L-6** engines. - **March 5, 1918**

Vincent talked to Olley about building a geared L-12 of R-R type. At a conference on **March 14, 1918** it was decided to build 100 and contracts for test engines were initiated. **First official order was on April 29.** These became the **L-12B.**

April 12, 1918 - Test of **Gates turbo-supercharger**. Several other designs followed. **(Sherbony June, Fergus in June, Moss in June)**

Test of first design of **articulated rods**.- **April 18, 1818**

April 25, 1918 - Earliest **alternate piston designs** noted.

One **L-12E** ordered. - **May 7, 1918**

May 8, 1918 - Alternate **carburetor designs** first noted.

A shop order was initiated for 6 **L-8 engines** to standardize the design for production. (appx.) **May 15, 1918**

May 16, 1918 1,000th L-12 production engine delivered.

First L-6 received. **July 2, 1918**

July 3, 1918. The first of two **Hall-Scott L-4** engines tested.

Vacuum oil control. - July 16, 1918

July 17, 1918 - The first **L-12B** was delivered .

The first of two **Hall-Scott L-6** engines tested. - **July 25, 1918**

July 20, 1918. Order 20387 for 2,000 Bristol USAO-1 canceled.

5,000 L-8 ordered from Buick. - **August 5, 1918.**

August 12, 1918 - Wright-Martin order for 2,000 Hispano H engines.

3,000 **L-8** ordered from Willys and 1,000 from Winton. - **August 20, 1918**

August 27, 1918. - L-12A HP design and testing of a **500 HP Liberty 12.**

6 FVL-8 single-seat fighters around L-8 ordered. - **September 21, 1918**

October 16, 1918 - 4 Victor D-8 single-seat fighters around L-8 ordered.

Armistice November 11, 1918
(By November 29, 1918 15,572 production L-12 engines had been delivered.)

December 5, 1918 - L-24 X configuration Liberty built and tested.

L-2 T built and tested.- **December 1918**

December 28, 1918 - Inverted L-12.

The last production L-12 is delivered, 20,479 - March 15, 1919

WARTIME NEW LIBERTY ACCESSORIES — THE ELECTRIC STARTER

On October 27, 1917 Lt. Emmons wrote to Col. Vincent advising that an electric starter should be developed for use on seaplanes with Liberty engines, whether in use by Army or Navy. This followed closely the first flight of a Liberty 12 in a Curtiss HS-1 flying boat on October 21, 1917.

I am sure there were designs of starters of which I am not aware, but the earliest noted was designed by the Bijur Motor Appliance Co. and was a propeller end-type and evidently put into service in small quantities (probably during late 1918 or early 1919) with mixed results. No test reports of this early type have surfaced but it is mentioned in reports of the design which replaced it.

It was described there as having what we now call a Bendix-type engagement which engaged the movable drive gear with a ring gear mounted on the back of the propeller hub. It was mounted between the radiator and the engine and this presented a number of problems. First; because of being in the air blast from the propeller, dust tended to sift into the working parts of the starter and cause the engagement system to stick. Second; the design had no protection against back-fires or heavy overloads. Third; its location required the removal of the radiator for service or replacement. Fourth; a special radiator was required on machines so equipped. Fifth; it could not be used with a supercharger.

The only picture the author has found of this type is a front view as applied to the Navy NC flying boats of 1918-19. (See illustration at lower left of this page.)

Bijur was requested to design a new starter which would mount on the rear of the engine and, by so doing, thus remove many of the objections raised by the first design. In addition it was to provide protection against overloads and back-fires. Engagement was by the Bijur automatic shifting mechanism and the starter was automatically connected to the rear of the crankshaft during the time of starting only. When the starting switch (either direct or solenoid-type) is depressed the splined starter screw shaft moves into mesh with the internal splines in the hollow end of the crankshaft. The engine is cranked at 40-50 RPM using a motor-gear reduction of 115 to 1. Several experimental models were submitted for testing and after several minor changes were made the design was considered satisfactory for production.

In March of 1920 production order No. 530037 was issued to Bijur for 590 starters of this design. Six were delivered for acceptance test on January 15, 1921. Tests resulted in some failures and design changes were incorporated to

Bijur starter as applied to the 1917 Packard 1A-905 engine. In this case the engine was of geared-type. The inset shows the ring gear-type Bijur starter applied to the Navy NC-4 of 1919. (Author's collection)

Liberty equipped with the second type Bijur starter as pictured in the 1921 Engineering Division test. (Author's collection)

correct them. A second batch of six incorporating changes was delivered on June 1, 1921. These passed tests satisfactorily and the remainder of the order was released for production.

Both Bijur and several aero magazines made a point to note that the Liberty engines used on the Douglas World Cruiser aircraft of 1924 used this starter.

As was often the case, and remained so for decades, the Navy often pursued its own solution to a problem the Army was working on. In this case a design was submitted to the Navy by the Liberty Starters Corporation of New York City (with its plant located in Poughkeepsie, NY). According to several articles which appeared in magazines in early and mid-1919, the Navy placed an order "for a number of starters." The statement did not change between April and September of 1919 so it is difficult to judge either how many starters were involved in the order (probably sufficient for testing only) or whether tests resulted in any significant production order.

The drive gears were connected to the crankshaft by its internal splines just as was the bijur starter. In this case, however, the engagement was permanent and most of the gear train rotated whenever the engine was running. Only the starting motor and its first order gears were disengaged after starting was accomplished. This was done by a ratchet system. The constantly rotating gears acted as "a small flywheel" according to the description, and were treated as an asset. One would, as a practical matter, feel the benefits of the flywheel effect were negligible whereas the potential for gear and bearing wear in a constantly moving gear train was high.

Although there was no motor overload protection in this starter, backfire protection was afforded by the ratchet engagement system which would disengage the motor and part of the gear train should a backfire occur.

A decided advantage of this design so far as the Navy was concerned was the fact that it also provided the ability to hand-crank the engine from the rear should it be so desired or if the battery or starter motor failed. In addition, a provision was made to drive accessories from the device if desired. This was desired by the Navy because in most of their flying boats all engines were accessible from the rear for hand starting whereas prop starting on land or water would certainly be classified as difficult (if not impossible).

The hand crank ratio was 21 to 1. The motor-driven ratio was 240 to 1 and resulted in a cranking speed of 25 to 50 RPM.

The Navy determined very early in the war that some type of hand starter would be required for the Liberty

End and side views of the Bijur Liberty starter. This model was available for several other make engines as well. (Author's collection)

The combination electric and hand starter of the Liberty Starters Corporation. (Author's collection)

Liberty Starters Corp. ad in the March 8, 1920 issue of Aerial Age Weekly announcing availability on the commercial market of their new starter. It was designed for use by the Navy and an unknown number was purchased by them. Some were no doubt sold outside the military as the Liberty became available on the surplus market in 1919. (Author's collection)

engine because of the inaccessibility of the propeller for starting purposes on most aircraft on which they used the engine. Their initial design effort resulted in a unit mounted on the rear of the engine and named the Type H-16 as it was designed for use on that model aircraft. The Navy contracted its manufacture to at least two makers (Commercial Truck Co. of America and Lewis Truck and Axle Co.). Serial numbers on the plates of two of these starters indicate production runs of a minimum of 300 from each of these makers.

Another somewhat more unique starting system was that designed to use stored compressed air to drive a starter motor or to drive the engine itself. One which used an air-motor was developed by the Motor Compressor Company of Newark, NJ and, they claimed, was tested successfully by the Curtiss Airplane and Motor Corporation in 1916, the Hall-Scott Motor Company in 1916, the Navy on October 23-24, 1917 and the Army on May 18, 1918.

In this particular design the starter motor also acted as the air compressor which served to pressurize the air storage tank used to supply the motor upon starting. In addition it could be fitted with a secondary compressor used to pressurize the fuel tank if so desired.

In a late 1918 ad they claimed the starter was approved by and being made for Navy hydroplanes. Just how many the Navy actually purchased and put into use is debatable. I have never noted any in Army or Navy use nor have I seen any test reports issued by either service.

Another system of utilizing compressed air as a starting medium consisted of equipping the engine with an air compressor and storage tank and then using the stored air to drive the pistons directly. This was done by an air distributor which was mounted on the compressor. This distributor supplied air under pressure to one bank of cylinders timed in such a manner that air pressure is applied to cylinders ready to go down on a power stroke, thus turning the engine over, and this is followed by fuel rich air being supplied to cylinders going up in preparation for a power stroke. The regular ignition fires any cylinder that is on a power stroke in the usual manner. The air feed lines to each cylinder are supplied with one-way valves which prevent the escape of cylinder pressure.

Although this system will certainly work in theory, it requires that either the standard cylinders be altered to facilitate connecting the compressed air lines to one bank of cylinders or the use of a special spark plug which also serves as an inlet for a high-pressure air line to the cylinder. For this reason it certainly was never adopted by any of the United States services. It was shown in some aircraft reference books of the mid-1920s as applied to the Liberty and was noted as the Heywood high-pressure injection starter manufactured by the Heywood Starter Corporation of Detroit.

It was listed in various publications as distributed in the United States by the Detroit Air Appliance Corp. in the 1920s and Sky Specialties Corp. in the 1930s. The system was also used in England and could possibly have

At above left is Packard-built Navy engine MNFG. No. 1758 Navy No. 0859 equipped with a Navy designed hand starter. At right is a Curtiss H-16A showing an engine istallation with hand starter and crank. (Courtesy of the Air Force Museum)

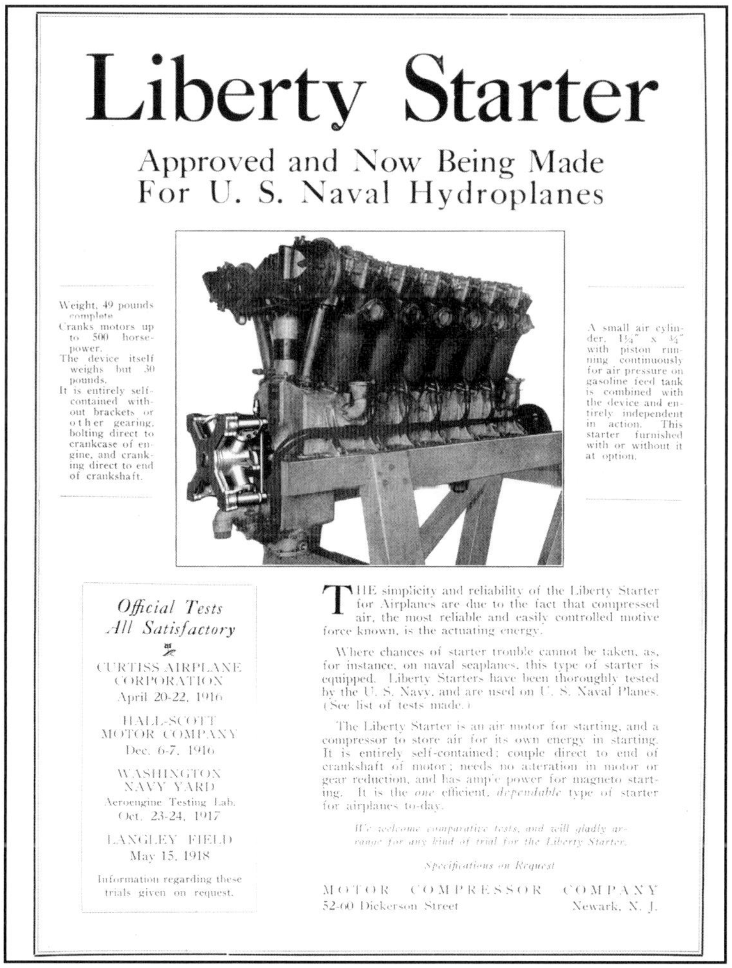

Liberty Starter

Approved and Now Being Made
For U. S. Naval Hydroplanes

Weight, 49 pounds complete.
Cranks motors up to 500 horse-power.
The device itself weighs but 30 pounds.
It is entirely self-contained without brackets or other gearing, bolting direct to crankcase of engine, and cranking direct to end of crankshaft.

A small air cylinder, 1¼" x ¾" with piston running continuously for air pressure on gasoline feed tank is combined with the device and entirely independent in action. This starter furnished with or without it at option.

Official Tests
All Satisfactory

CURTISS AIRPLANE CORPORATION
April 20-22, 1916

HALL-SCOTT MOTOR COMPANY
Dec. 6-7, 1916

WASHINGTON NAVY YARD
Aeroengine Testing Lab.
Oct. 23-24, 1917

LANGLEY FIELD
May 15, 1918

Information regarding these trials given on request.

THE simplicity and reliability of the Liberty Starter for Airplanes are due to the fact that compressed air, the most reliable and easily controlled motive force known, is the actuating energy.

Where chances of starter trouble cannot be taken, as, for instance, on naval seaplanes, this type of starter is equipped. Liberty Starters have been thoroughly tested by the U. S. Navy, and are used on U. S. Naval Planes. (See list of tests made.)

The Liberty Starter is an air motor for starting, and a compressor to store air for its own energy in starting. It is entirely self-contained; couple direct to end of crankshaft of motor; needs no alteration in motor or gear reduction, and has ample power for magneto starting. It is the *one* efficient, *dependable* type of starter for airplanes to-day.

We welcome comparative tests, and will gladly arrange for any kind of trial for the Liberty Starter.

Specifications on Request

MOTOR COMPRESSOR COMPANY
52-60 Dickerson Street Newark, N. J.

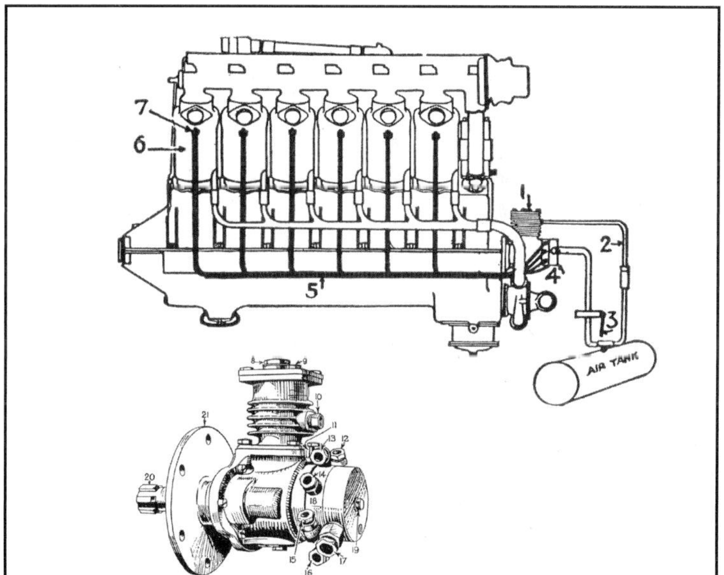

The Heywood air starter system of the 1920s as applied to the Liberty engine. Although available in the United States, it probably saw no military use. Possibly it saw some limited civilian use but even that is doubtful as electric starters would be far cheaper and more practical. The system was used briefly by the British on the WW II Nuffield Liberty. (Author's collection)

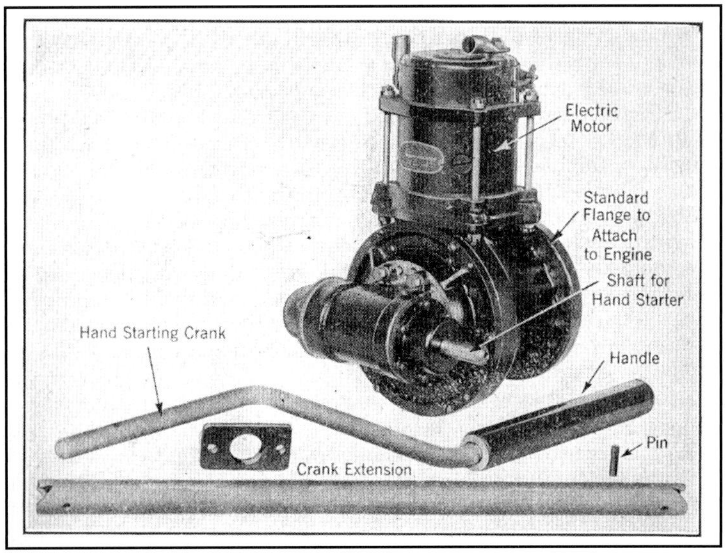

The 1924 Eclipse-Bijur combination hand/electric starter. (Author's collection)

British Lucus starter applied to Lincoln MNFG. No. 51 in the Duxford Air Museum in England. The driving motor has been removed. (Courtesy of Jamie Fairchild)

originated there. (It should also be noted that a 1917 reference has been found noting a "Christensen Air Starting System" which was described and illustrated in use on a Thomas-Morse aero engine. It used the same principal as the Heywood system.)

The Heywood air starter was used in 1939-40 on the British-built Nuffield Liberty which they used to power tanks. (See Chapter 11.)

For WW I use Lucas built an electric starter for use on the British Liberty engine. This was a gear reduction-type mounted on the rear of the engine with the motor mounted 90° to the engine crankshaft. The gear train was a combination of spur gears and a worm drive. A manual engagement of sliding dogs coupling to the end of the crankshaft was used in synchronization with contact of the motor starting switch.

Eclipse/Bijur inertia starter of about 1925. (Author's collection)

A detachable manual starting handle could be applied to the end of the worm drive shaft. Backfire protection was provided with both manual and motor starting.

Built in left- and right-hand models for mounting on twin engine installations, it could not be mounted on the engine as a complete unit but had to be assembled in sections after the main gear case housing was bolted to the engine. This design was in production before October of 1918 (the date of the official manual for it) and was thus available by the time the British received shipments of the engine in quantity.

By 1924 the Bijur Starter Company had been purchased by the Eclipse Machine Company of Elmira, NY and the 1920 Bijur starter had been improved by providing a new friction clutch overload release as well as the ability to manually crank the engine from either side. The Army tested this new starter and purchased some.

Not long thereafter Eclipse put on the market an inertia starter which could be either electric motor or hand crank-driven. In this design a flywheel was accelerated to a speed of 16,000 RPM by motor or hand and then the flywheel energy released by clutch to the engine with the ability to drive it four to five revolutions at a speed of 125 RPM. If this does not effect a start the flywheel can be again set to full speed and ready for another attempt in twelve to fourteen seconds by the electric motor or about a minute by hand.

The last design starter noted by the author as applied to the Liberty was the Coffman cartridge type as manufactured by the Standard Aero Starter Co. of Pittsburgh, PA. This type starter saw extensive use on a large variety

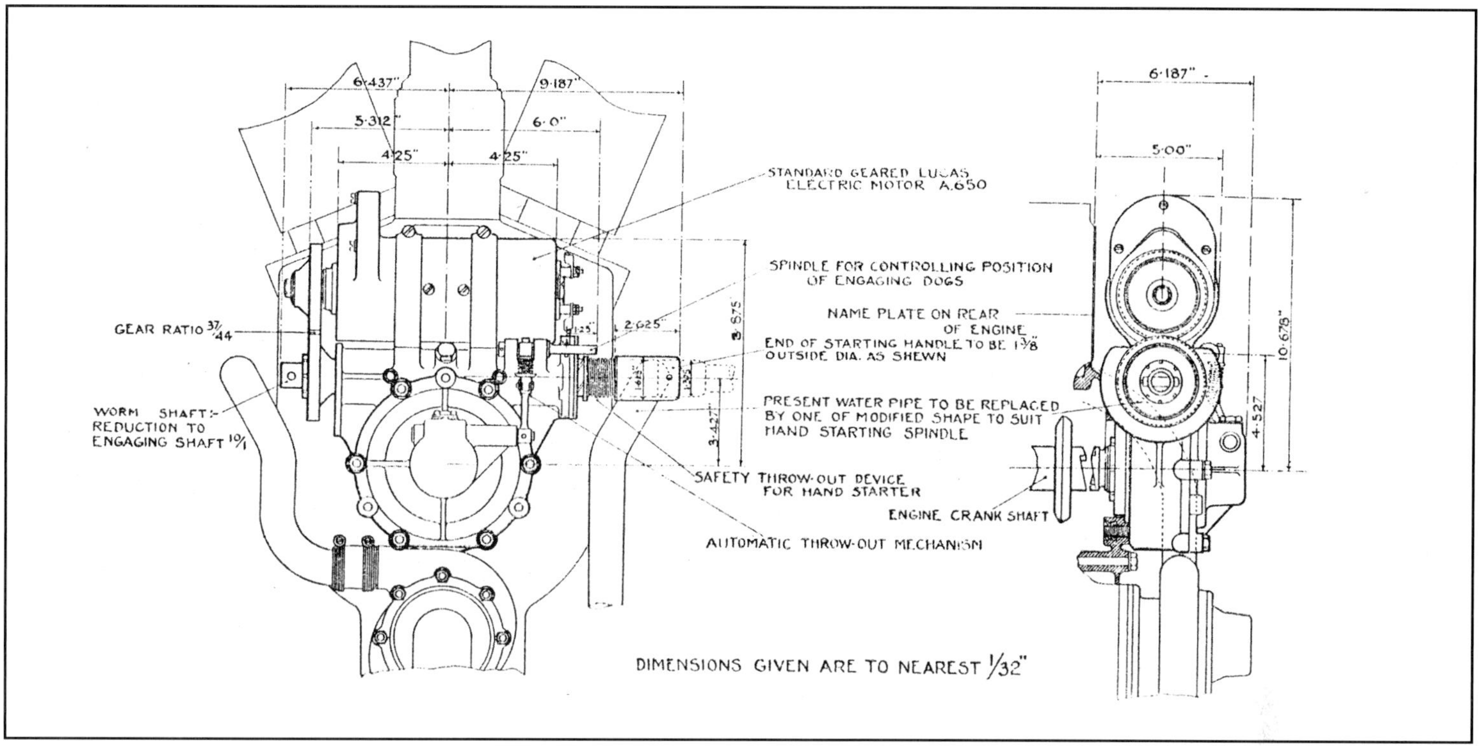

British Lucus starter of 1918 from the Ministry of Munitions manual dated October, 1918 and titled "The Starting Gear for U.S.A. Liberty Aero Engines." (Courtesy of the Air Historical Branch - Air Ministry)

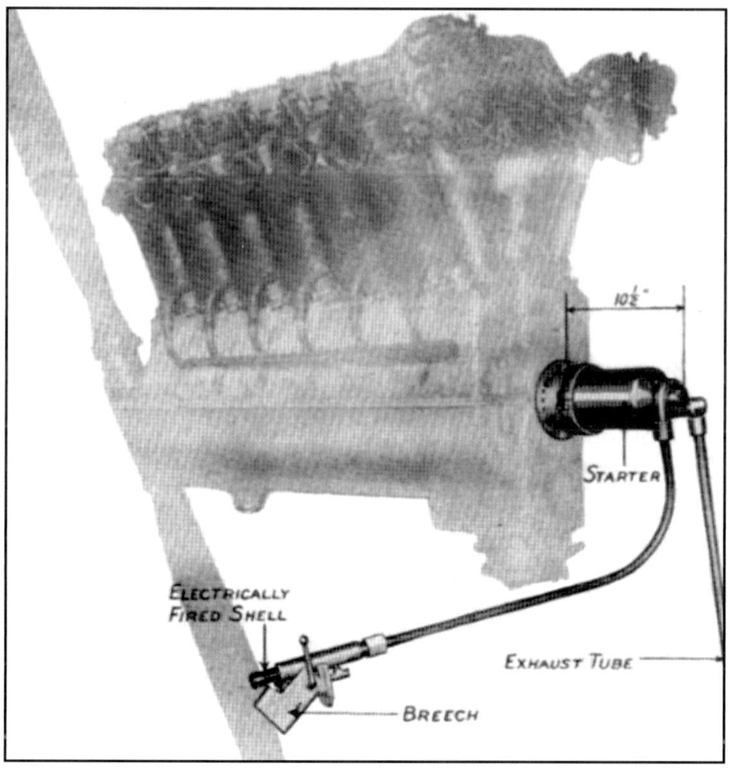

Illustration from Standard Aero Starter Co. ad of about 1931 for a Coffman starter shown applied to a Liberty engine. (Courtesy of Jamie Fairchild)

of aero engines from about 1931 all the way through WW II. The Liberty was an obsolete engine seeing its last few years of service when the Coffman was introduced and it is doubtful that many were so equipped.

THE MAGNETO

At the time the Liberty was designed there was no available magneto capable of operating a 45° V-12 engine. To be sure, the use of battery ignition was immediately questioned since the more common aircraft engine ignition system was magneto.

In addition to the question of which system was best so far as reliability and power output were concerned, was the question of whether DELCO had been favored over some magneto-producing company because of personal connections between DELCO's Charles F. Kettering and Col. Vincent. At least three different manufacturers eventually produced a workable magneto which the Engineering Division then tested.

The first was the Dixie Magneto produced by Splitdorf Electrical Co. This ignition was tested several times beginning in January of 1918 and continuing periodically until January of 1920. Results varied slightly depending upon the method and purpose of the different tests. The overall findings were that the Dixie and DELCO systems provided equal results "within the limits of experimental error."

The test report of October 4, 1918 was run on September 23 on an engine (Ford MFGR. NO. 144 U.S. NO. 29637) which had been equipped with both systems simultaneously with the ability to be switched from one to the other, thus eliminating any possibility of different engines giving different results and thus masking performance differences.

The output with each system was plotted on the same power graph and the resulting line was a single one although the different points can be seen. The calculated fuel consumption differed slightly, being 0.551 lbs. per H.P.H. (horsepower per hour) with DELCO and 0.541 lbs. per H.P.H. with Dixie. This slight difference was discounted by the reporting engineer with the following statement: "There is a slight variation in the economy of about .01 lbs. per horse power hour but these slight variations are no doubt entirely due to the personal factor entering into the manipulation of the test."

On October 2 the exact same test on the same engine was run using Simms rather than Dixie magnetos. The

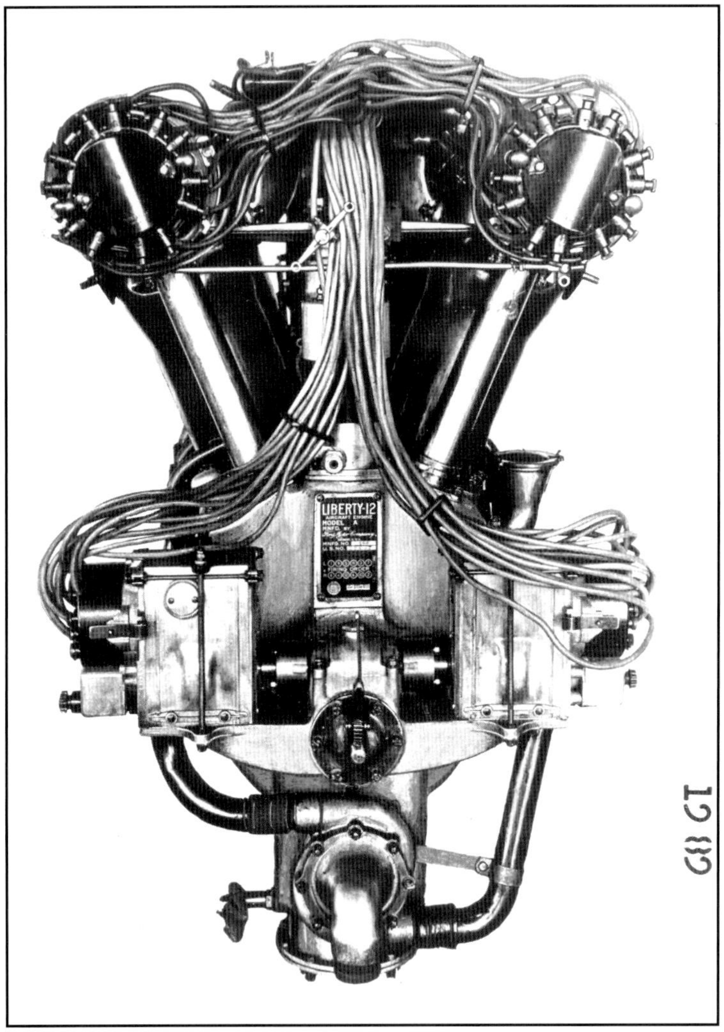

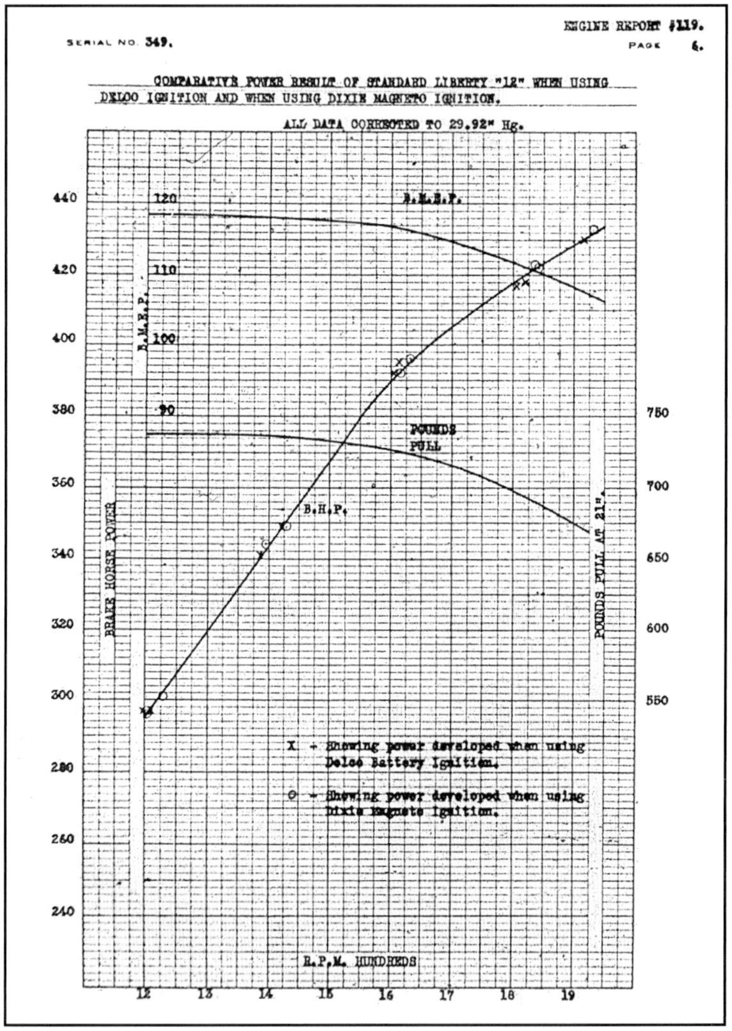

L-12 built by Ford MNFG. NO. 144 U.S. NO. 29637 equipped with both standard DELCO and special Dixie Model 1200 magnetos for comparison testing done on September 23, 1918. Acceptance date stamped on the data plate is 7-16-18. At right is the Simms magneto which was tested on the same engine on October 2, 1918. (Courtesy of the Air Force Museum)

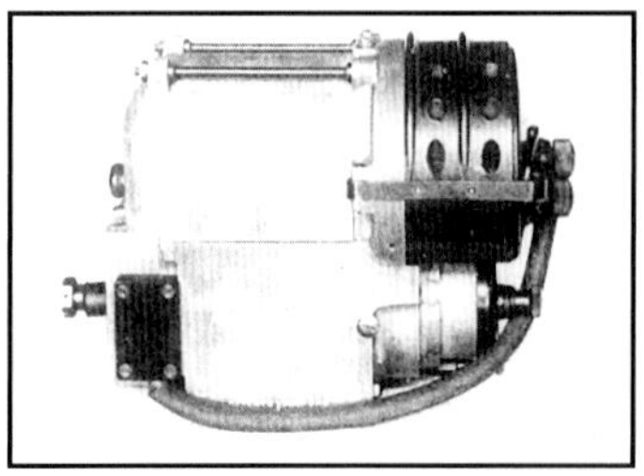

Power graph from the 1918 test of DELCO and Dixie ignition on Ford engine NO. 144. Power points are plotted for both ignition systems and were so close to the same that the curve was plotted with a single line. In a test run on the same engine under the same conditions but with a Simms rather than Dixie magneto the results were virtually identical and again a single line graph resulted (not shown). (Author's collection)

results were almost identical and a graph was again plotted in the same manner and again resulted in a single line on the graph. This time the fuel consumption was .555 lbs. per H.P.H with DELCO and .554 lbs. per H.P.H. with Simms.

In the February 13, 1920 test a single engine equipped with both ignition systems was again tested but this time a number of additional tests were run to see if either system was less liable to allow fouling of plugs under adverse conditions. Several runs with each system while pumping 1 gallon of oil per hour into the intake system both at idle and at full power. All conditions other than the ignition were the same for both tests. Even the same set of plugs were used. In this case we have no record of exact engine identification.

The end result of running an hour at 330 RPM while pumping a gallon of oil into the intake was a fouling of plugs sufficiently on the magneto system to result in only 11 cylinders firing with seven firing on one plug only. The same test on the DELCO system resulted also in only 11 cylinders firing with four firing on one plug only.

In the one-hour full-power test with the same oil injection, both systems ran the period with all plugs firing on all cylinders. An admirable performance indeed. Again a full-power graph was run with the results from both systems being the same "within the limits of experimental error." At least two full flight tests were also made with the two systems with about the same results. The 1920 tests seems to have put the matter to rest.

Nordyke & Marmon engine fitted with Dixie 1200 magnetos but no regular DELCO ignition. This photograph is dated 1920 but has no other notations. It is probably an engine used in flight testing. (Courtesy of Peter Bowers)

Standard Zenith 52 carburetor. Changes in altitude control were all interior. New production carburetors incorporated the changes and old ones could be easily altered in the field. (Author's collection)

INDUCTION SYSTEM CHANGES

The standard carburetor for the Liberty 12 was the Zenith Liberty. About the only work done on this carburetor after the best choke, jet and compensator sizes were settled upon was to redesign the altitude control. This was completed and tested and specifications issued on July 16, 1918. A description of the alterations as well as the results obtained was published in the September 1918 issue of the *Engineering Division Bulletin, Vol. 1 No. 4.* The changes were all interior and could be applied to existing carburetors in the field. New manufacture incorporated the changes. With these changes applied this became the Zenith 52 model.

A number of other carburetors were tested starting as early as May of 1918 and a few gave as good, or nearly as good, power and economy as the Zenith but none exceeded it to any appreciable degree. This held true even up to when the Liberty was phased out of active service in the early 1930s and certainly stands as a credit to the design. For this reason the Zenith 52 remained the U. S. standard carburetor for the Liberty throughout its service use.

On May 8, 1918 McCook Field tested Stromberg NA-D5 carburetors, which had been designed specifically for the Liberty 12, against standard Zenith 52 models on the same engine (Packard MNFG. NO. 316 U.S. NO. 17821). The results were almost identical. Power with the Zenith was 2 HP higher at 1200 RPM and 2 HP lower at 1800 RPM as compared to the Stromberg carburetors. Fuel economy with the Zenith carburetors was .4769 lbs. per HPH at 1600 RPM wide open throttle as compared to .4845 lbs. per HPH with the Strombergs. This was one of the cases where performance was not enhanced and thus the Zenith was retained.

THE CLAUDEL-HOBSON HC-7

On July 25, 1918 the Engineering Division at McCook Field (hereafter to be referred to as McCook Field) tested specially designed British Claudel-Hobson HC-7 carburetors for use on Liberty 12 engines. These carburetors were built by the Stromberg Motor Devices Co. of Chicago at the request of the British War Mission to be supplied on engines to be delivered to them.

Tests were carried out on Packard MNFG. NO. 714 U.S. NO. 18333 and were compared to standard Zenith carburetors. Results of the tests showed power output was better at higher RPM with the HC-7 but worse at low RPM, the equal point being about 1350 RPM. Altitude compensation was considered better on the HC-7 as compared to the old type Zenith 52 but equal to the new Zenith system mentioned above. Although not noted in this test, it was noted in subsequent ones that with the HC-7 carburetors the engine generally tended to stumble on acceleration when progressing through about 1200 RPM.

The HC-7 used a barrel-type throttle which required machining with extreme accuracy and each barrel had to be lapped into its respective seat. The disadvantage was, of course, more time required to manufacture and thus higher costs. An additional disadvantage was in maintenance as the entire carburetor had to be removed in order to change jets or floats.

Although a few of the early British Liberty engines may have been supplied with Zenith carburetors, most had HC-7 types or were later converted to that type. (By mid-July the British had received about 150 engines.)

CARBURETOR AIR INTAKE DESIGN

As originally designed, each carburetor has an individual elbow-type air intake scoop below and facing to the front of the engine taking in air over the propeller hub as it passed between the cylinder banks. These elbows had small sumps with fittings for pipes to drain to the outside any gas accumulating therein. This design was found to be prone to result in raw gas flowing out of the intakes and onto the top of the engine and then down into the interior of the engine cowl in cases of carburetor flooding and the aircraft at an attitude rendering the open intake lower than the bottom of the sump. Obviously this could present a considerable fire hazard.

In late July of 1918 design was begun on a new air intake manifold to eliminate this problem. The resulting manifold may be seen in the illustration on the next page. Finished and tested in September, it resulted in an enclosed manifold connecting the intakes of the two

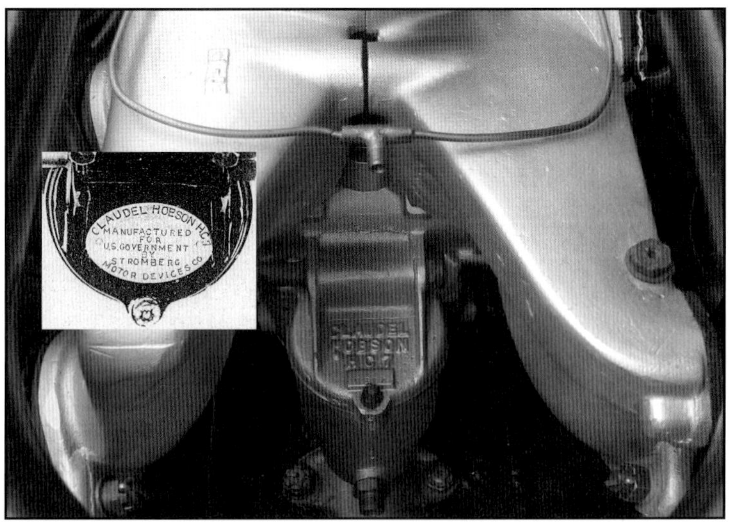

Claudel-Hobson HC-7 carburetor on Lincoln MNFG. NO. 51 SC NO. 23485. This is a British engine in the Duxford Museum in England. (Courtesy of Jamie Fairchild)

carburetors. A single center vertical intake stack fed air to the manifold and thus both carburetors. Any gas accumulating in the manifold would be carried to the outside by pipes attached to sumps at either lower end of the manifold. There was no way gas could spill out of the air intake. In tests, power was seen to increase slightly (379 HP @ 1,600 RPM with the original design and 386 HP @ 1,600 RPM with the new design) and fuel economy decrease slightly from .533 lbs. per H.P.H. to .540 lbs. per H.P.H.

The new design was accepted as standard by October or November and was applied to production engines as soon as parts required could be made available. A new water pipe connecting the two pairs of intake manifolds was also required in order for the pipe to clear the vertical intake stack.

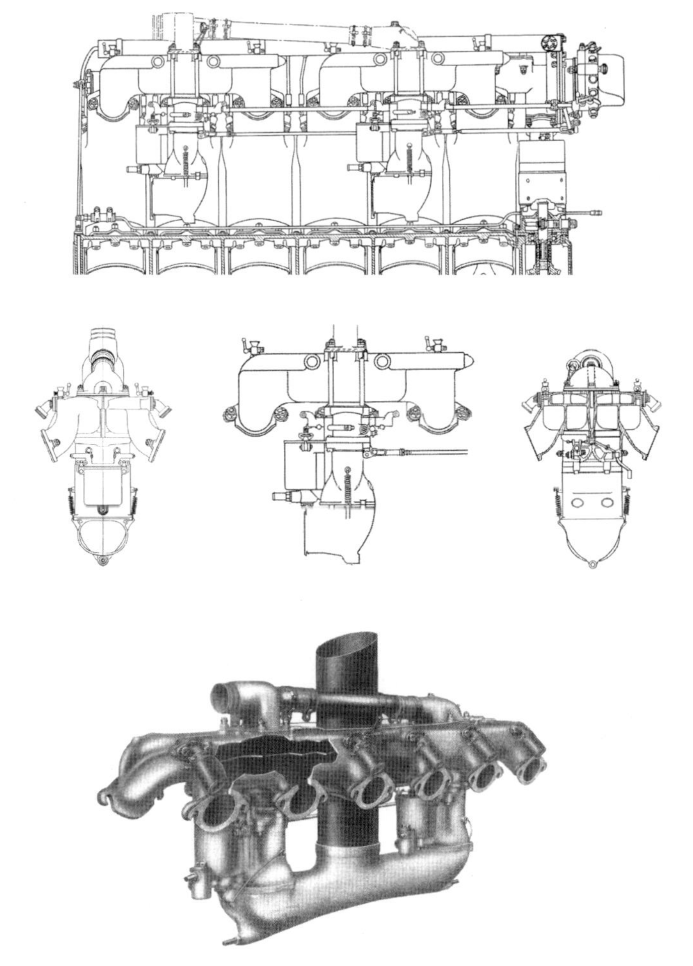

Above top is a line drawing showing a complete engine with the original-type individual open mouth air intakes on the two Zenith carburetors. Center left to right shows a single carburetor with the old design - front, side and rear views. Below is the new design. (Author's collection)

INVERTED CARBURETORS

Work began early on designing an inverted carburetor for the Liberty. Even though the fire hazard of the standard updraft design had been reduced considerably by the new common intake manifold with vertical stack just described, the problem of inaccessibility to carburetors mounted in that position had not. The latter problem could only be solved by putting the carburetors above rather than below the intake manifold. This would require downdraft carburetors.

Updraft was the norm and, for the most part, is today. Thus a downdraft carburetor would be considered inverted.

Earliest inverted designs were Zenith and appeared to have been a design collaborated between Zenith and the Engineering Division since weekly Engineering Division reports refer to their progress on drawing layouts as well as testing of engines equipped with the carburetor, which was known as the Zenith USD 52. Weekly reports indicate an engine was under test with these carburetors by the first week in October of 1918 and were completed by the second week of November.

A report (test serial No. 520) was also issued on 11/14/18 with the results of testing carried out on a pair of Venus inverted carburetors. These did not show promise and no more is heard regarding this model.

Result obtained from the new Zenith USD 52 was more promising and was issued in Vol. II No. 4 of the *Bulletin of the Experimental Dept.* of January 1919. In that report it was noted that after the design of the new common air intake system just described was completed and tested it was suggested that the carburetors themselves might be mounted in an inverted position above the manifolds with their air intakes projecting through the top cowling, thus producing an even shorter and more direct air passage as well as a further reduction of fire risk.

The carburetor was patterned after the US 52 as to principals of operation but resulted in a much different physical unit. Test results were quite encouraging. In general, power output was improved over the standard carburetor. Because of the increased complexity of the castings required for this new carburetor, it would be more expensive to produce than the US 52 but they felt this could be worked out.

Notes in following weekly reports indicate they immediately set about to alter the design to this end. However, it soon becomes evident from following reports and testing of other inverted carburetors that the inverted Zenith soon was replaced by an inverted version of the Stromberg NA-D5 known as the NA-L5. The inverted intake manifolds designed for the Zenith would serve either carburetor.

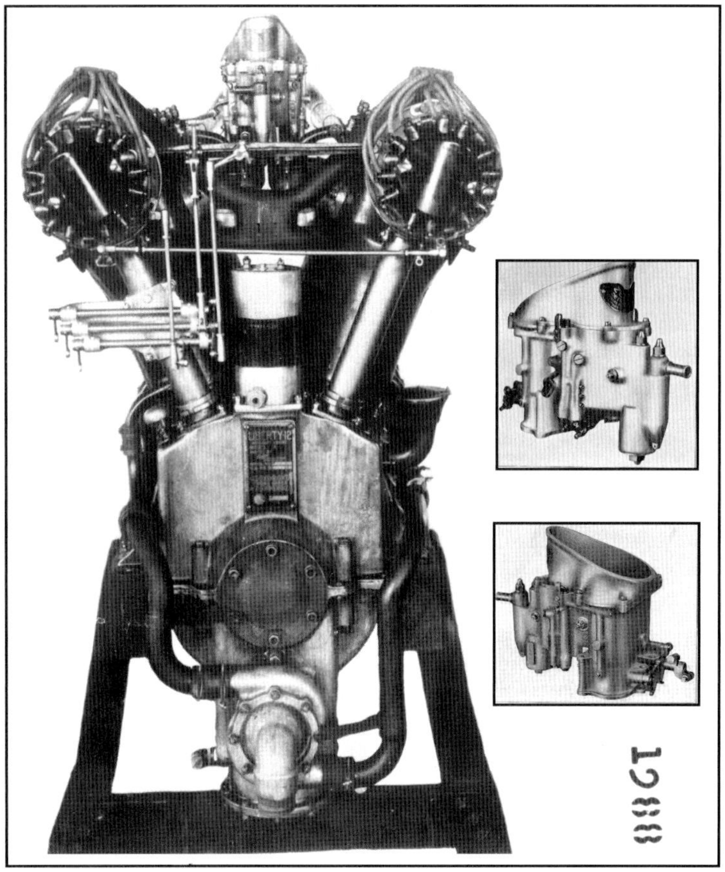

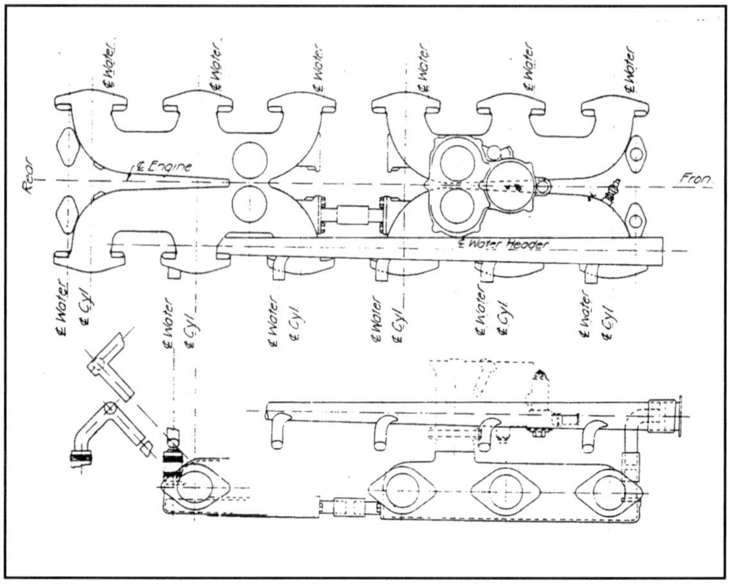

At left is Liberty 12 Exp. No. 5 equipped with the new manifolds and Zenith USD 52 inverted carburetors with the insets showing the carburetor from the sides. Courtesy of the Air Force Museum. Above is a line drawing of the new design manifolds required for their installation of inverted carburetors. The insets and line drawing images are from the test report published in Vol. 11 No. 4 of the Bulletin of the Experimental Department of the Airplane Engineering Division. (Author's collection)

The Stromberg model was not completed and submitted to testing until after the end of the war so I will cover it in Chapter 7. You will notice the intake manifold design is the same for both carburetor models.

VACUUM OIL CONTROL

The problem of plug fouling caused by excess oil pressure at low engine speeds was addressed early in this chapter and it was noted that it was controlled by drilling holes in the head of the pressure relief valve and the use of a groove and oil drain holes in the piston skirt. Although this fix remained in use throughout the life of the Liberty, others were tried. A number of different piston designs were tried, including cast iron types, but most were worse than the original design and were discarded.

Other methods of controlling oil pressure at varying speeds and engine loads were also devised. The best of these was a device to control oil pressure with intake manifold vacuum. The more throttle applied, the lower the intake manifold vacuum and the higher the output of the engine. Application of this manifold vacuum to a properly designed control valve was used to regulate oil pressure, reducing pressure under light load and increasing it under heavy load. During the progress of this design the control valve was applied to the oiling system in several places, but the most practical was found to be directly on the oil pump. Its use in that location required only slight changes to the oil pump and with little difficulty could be introduced rapidly into production.

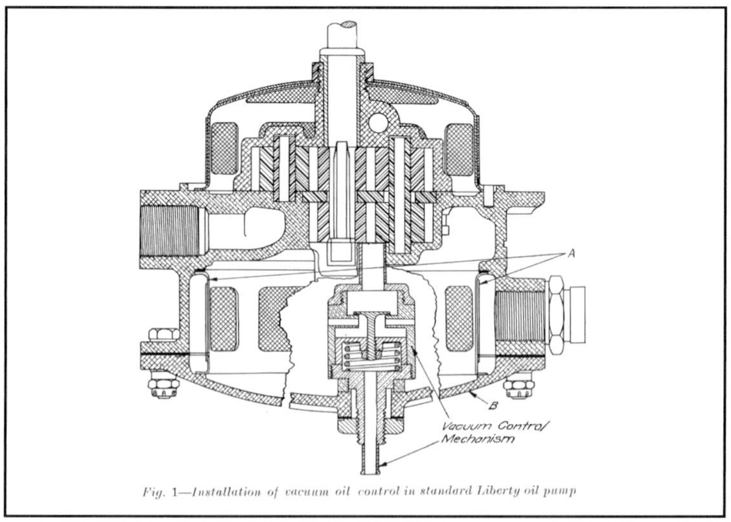

Sectional drawing of the vacuum oil control applied to the standard Liberty oil pump as pictured in the Bulletin of the Experimental Dept. Airplane Eng. Div. (Author's collection)

Development of this design did not reach the stage of a decision to officially adopt it until near the Armistice. A full report was carried in the September 1918 issue of the *Airplane Engineering Div. Bulletin* (Vol. I No. 4). Although there were other oil pressure regulation devices introduced in post war years, it is not currently believed the vacuum oil control was ever introduced into production. The 1927 issue of the Air Corps Liberty maintenance manual, a several hundred-page master manual of maintenance and overhaul, makes no mention whatever of it.

ARTICULATED RODS

Beginning as early as April of 1918 McCook Field began experimenting with the use of articulated rods on the Liberty and we know of two designs which were completed. A number of tests were completed and three were recorded in published documentation as Technical Reports serials 18, 35 and 469.

At the time these designs were initiated McCook Field was still working on improving both bearing reliability and structural reliability of the fork and blade-type rod. The articulated design passed 50 hour reliability tests by late in 1918 and test engines produced power and smoothness comparable with the fork and blade design.

Work on the early design rods was carried out on Packard engine Exp. NO. 2. This engine continued to be used for these rod tests into June of 1918.

The engine on which this finalized design was tested was Packard MNFG. NO. 77 AS NO. 17507. The engine produced a normal power curve with output of 424 HP @ 1,800 RPM. Fuel economy was .468 lbs. per HPH at 1,600 RPM, and 386 HP and oil economy of .0352 lbs. per HPH at the same output.

By the time this was accomplished the heavy design fork and blade rod came into use and bearing life had been materially improved. The need for a better alternate to the original type rod design thus evaporated. Although there were a few additional articulated tests after the war, interest rapidly disappeared.

The reader will shortly see, however, that this articulated rod design was used in the Liberty "X" or L-24 engine so the design effort was not entirely a waste.

SUPERCHARGING THE LIBERTY

McCook Field was interested in supercharging or "boosting" the Liberty fairly early in its life. The first test I have noted was on April 8 and 9, 1918 of the Gates "Turbo-Booster Blower." This device consisted of a turbine

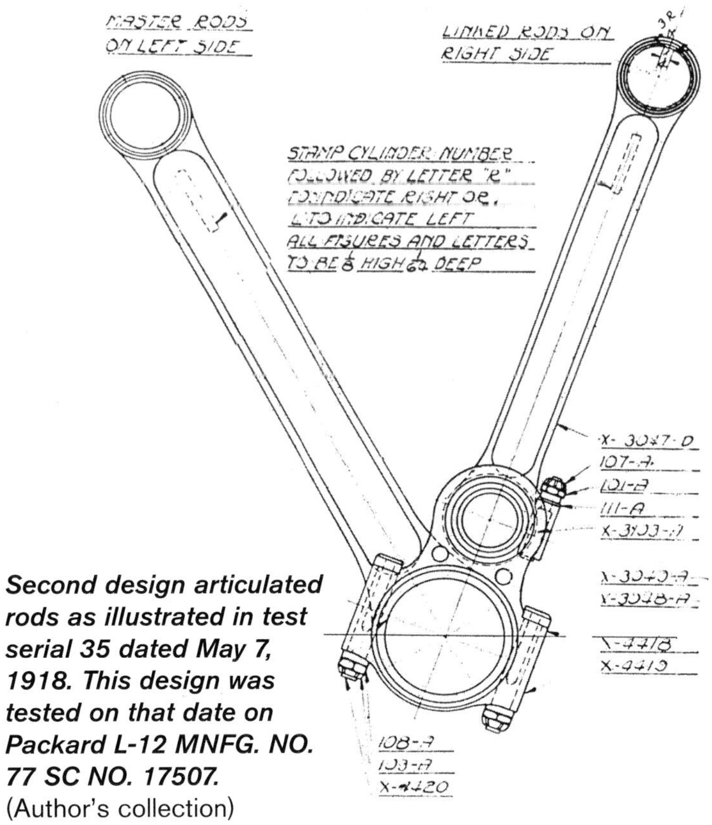

Second design articulated rods as illustrated in test serial 35 dated May 7, 1918. This design was tested on that date on Packard L-12 MNFG. NO. 77 SC NO. 17507.
(Author's collection)

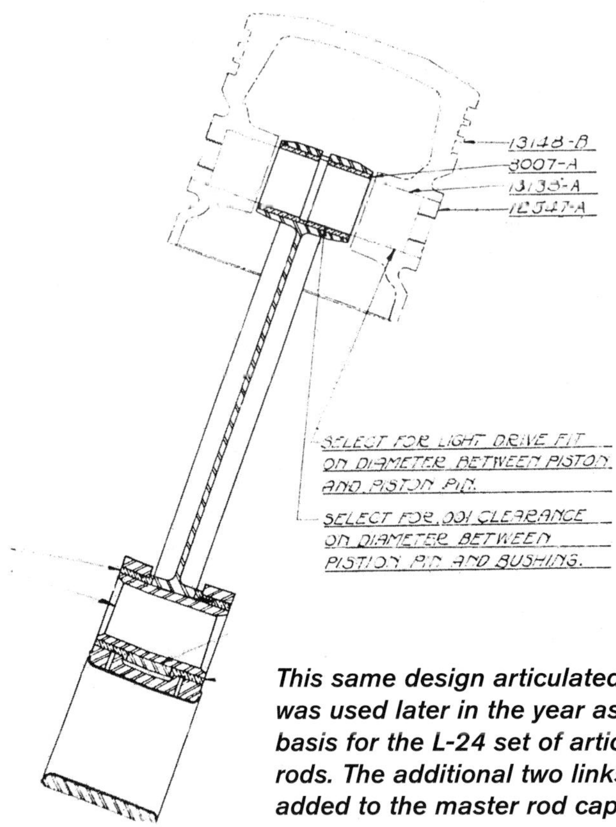

This same design articulated rod was used later in the year as a basis for the L-24 set of articulated rods. The additional two links were added to the master rod cap.

having a 10" diameter impeller driven by a 1 1/2" diameter pully which was in turn driven by four rubber belts from a driving pully of 10" in diameter. The latter was to be driven directly by the engine at engine speed.

Tests proved the capacity of the unit to be entirely too small to be of any use on the Liberty and, additionally, the belt drive, as one might suspect, was found to be unsatisfactory. Within a short time they became very hot and "showed signs of going to pieces."

This was just the first (the last for this particular device) of many tests of various supercharger designs and the subject is so extensive as well as interesting that I have devoted an entire chapter to it. The reader is referred to Chapter 9, *Supercharging the Liberty*, for extensive material on the subject.

ENGINE DESIGNS BEYOND THE TWELVE

The remainder of this chapter will be devoted to the various designs of Liberty engines beyond the standard Liberty 12 production engine. Although they certainly did not come into existence chronologically in order of the number of cylinders they may have had, I will present them in that order.

LIBERTY MODEL SUFFIX LETTERS

The one-cylinder Liberty was known as the 1D or L-1D, since "L" had already been accepted as the prefix to be applied to all Liberty design-based engines. I suppose this is as good a place as any to attempt to explain the model letter suffixes applied to the various Liberty types, if there is indeed an explanation.

If we go back to the origins of the name "Liberty" we find that as soon as the name Liberty was adopted to replace "U. S. A. Standardized Aircraft Engine" the model suffix "A" was applied. Thus we first had the Liberty 8A, or L-8A and then the L-12A. Although we have not yet covered them, there followed the L-12 models B and C. Ah-ha, one immediately says. It is obvious that they are going to follow a logical progression of model letters with each engine type (number of cylinders) as new models are designed within each type. With the Liberty 12 they seemed to have (almost) followed this rule.

Then why did they start with the L-1D instead of L-1A? Frankly, I do not know. There was by this time an L-12A, L-12B and L-12C. Did someone decide that the next Liberty should be a model D no matter what type (number of cylinders) it was? I doubt it, but it sounds good for a little while. You see, there was no L-12D and there was an L-12E

However, as soon as you start to look at the model letters assigned the other various Liberty types any overall logic one tries to apply seems to fall apart. Only in the case of the Liberty 12 did early test reports bother to even refer to a model letter, and then usually only when it was not an "A" model. The L-1 was usually called the "Single-Cylinder Liberty," the L-2 was the "Two-Cylinder Liberty," the L-4 was the "Four-Cylinder Liberty," the L-6 was the "Six-Cylinder Liberty," and the L-8 was the "Eight-Cylinder Liberty."

We find the L-1 referred to as the "Model D Single-Cylinder Liberty" in early documents, and then not always. No illustrations of the engine's data plate have ever been found, if it indeed had one.

The L-2 is usually referred to as the "Two-Cylinder Liberty" but occasionally as the "T-1" or "T-2." There were two of these engines built and the "T" probably made reference to "Twin" cylinder construction.

The L-4 is usually referred to as the "Four-Cylinder Liberty" but is sometimes called the "L-4B" in later documentation. Why model B? There has never come to light any indication that there was ever but one design. Only two of this model were built, both at the same time and both by Hudson. The one known survivor is missing its data plate and no early photos show its plate. I will go a little further into this when I cover the engine.

The early L-6 when referred to by model was always the Model A. The 1920 version built by Wright was the L-6B

The L-8 when referred to by model was always the Model A.

Only one L-24 was built and it was invariably referred to as the "Twenty-Four Cylinder Liberty." In some later documentation it was called the "Liberty X" engine. I have never seen it given a model letter assignment but if it had one it would be the A.

THE L-1D

Six Liberty one cylinder engines were built by the Airplane Engineering Department at McCook Field. The first noted in records were three produced under Order 32. The first of this three was tested during February 1-7, 1918. A second order, No. 70, followed shortly thereafter and these were finished by April 25. These single-cylinder engines were built strictly for test purposes and strictly for testing Liberty cylinders and directly associated parts. These included valves and their driving mechanisms, intake manifold design, pistons, rods, bearings and spark plugs as well as various types of fuels. These six engines were used in development by McCook Field engineering staff, where four were still listed on inventory as late as June 1921 and were probably actively used for some years after, since Liberty development continued for another 10 years or more.

L-1D from August 1918 test report. Note the flywheel at right, Liberty cylinder, throttle linkage and shaft coupling to dynamometer at right. (Author's collection)

The best way to acquaint the reader with the engine's construction and use is to quote from one of the published tests carried out by the Airplane Engineering Department in the summer of 1918. "In certain kinds of engine tests it is possible to use a single-cylinder engine just as satisfactorily as a multi-cylinder type of the same design, thus avoiding the added expense of setting up and running the multi-cylinder engine, and reducing materially the time required for overhauling, rebuilding or changing the engine between tests.

"A single-cylinder engine can be used just as well as an eight or a twelve, for instance, for most tests on pistons, piston rings, valve construction, compression ratio, valve timing, cam design, ignition timing, intake and exhaust gas action, manifold design, cylinder lubrication, spark plug tests, etc. In fact for some of these tests the single-cylinder test engine is actually better than the multi-cylinder type, because it is easier to detect slight variations in the running of the single-cylinder engine in a test than corresponding variations in perhaps one or two cylinders of an eight or twelve. A single-cylinder test engine which in effect is one-twelfth of a Liberty 12 except for the necessary changes in crankcase, crankshaft, camshaft, connecting-rod, etc., and the addition of a flywheel, was constructed and has now been put through a large number of tests with the object of determining the effect on the Liberty engine of various changes in design and construction, which have been suggested from time to time as the result of experience.

"The first test run with this engine was a standardization test in which every feature of design was identical with the Liberty 12. This was made for the purpose of obtaining a standard basis of comparison for subsequent tests. In this particular series of tests a set of 13 variables were tried in various combinations and the power output of the engine recorded at speeds of from 1,000 to 2,000 RPM with 38 combinations tried. The results of those tests are summarized in Table 1. It will be observed from Table 1 that the best combination of variables is that shown in test 15, consisting of the 1 5/8 in. choke carburetor, 10 deg. overlap camshaft and vertical crooked intake pipe. The performance curves obtained from the single-cylinder Liberty test engine with this combination of variables are given in figure 14.

"Referring to figure 14, it will be observed that at 1,800 RPM the single-cylinder Liberty engine developed 39 HP. Multiplying this by twelve would give 468 HP for

L-1D showing the special carburetor at left, L-12 distributor and its vertical drive. (Author's collection)

Table 1

Summarized Data of One-Cylinder Liberty Engine Tests

Letters indicating conditions of tests—refer to key below table for meaning of letters

Number of pounds pull above or below pull of standard test—(noted as test 1)

Standard basic test / Conditions described below			A	B	C	D	E	C D E	D E	A B C E	A C D E	B E	A B E	B D	B G D	B D E	A E	A C	C E	B C E
Test No.		1	2	3	4	5	6	7	8	9	10	11	12	13	14	15	16	17	18	19
r.p.m.		Lbs. pull																		
1000		77	2	0	0	2	1	1	5	4	5	—2	6	6	0	7	5	4	1	—1
1200		79	2	—1	—2	4	1	2	3	4	5	—3	5	6	—1	6	3	3	1	0
1400		79	4	—1	0	4	0	3	4	5	6	—2	5	8	1	6	3	4	1	2
1500		76	8	0	2	8	4	7	8	10	9	4	9	10	5	10	8	9	4	3
1600		77	9	—1	3	7	3	6	7	10	7	4	8	10	3	10	7	9	3	1
1700		78	6	—1	2	5	1	5	6	10	6	0	7	7	2	8	7	7	2	0
1800		78	4	—2	2	3	1	4	5	8	5	2	8	6	0	9	7	5	1	0
2000		76	1	—1	2	—4	2	4	5	9	6	—1	6	3	—6	5	7	5	2	—7
Total		620	36	—7	9	29	13	32	43	60	49	2	54	56	4	61	47	46	15	—2
Percent over test 1			5.8 %	—	1.4 %	4.7 %	2.1 %	5.2 %	6.9 %	9.7 %	7.9 %	—	8.7 %	9.0 %	—	9.8 %	7.6 %	7.4 %	2.4 %	—

Conditions	A B	H	J	B E K	A E L	A B E H	A B E J	B D E H	B D E J	B D M	B D E	B N E	N	D O	D E O	O	D E L	L	D L
Test No.	20	21	22	23	24	25	26	27	28	29	30	31	32	33	34	35	36	37	38
1000	3	—6	—4	0	4	0	3	2	2	5	4	1	—1	3	3	—3	2	—3	1
1200	3	—7	—6	0	5	—2	2	2	2	4	4	1	—3	2	1	—4	1	—4	2
1400	5	—7	—3	0	4	—2	1	2	3	4	5	2	—2	4	2	—6	2	—3	4
1500	9	—5	0	5	10	3	5	6	5	8	8	6	—2	7	5	2	6	0	7
1600	8	—4	—4	5	10	4	5	5	7	6	9	6	—1	5	6	2	6	1	7
1700	4	—5	—5	4	11	3	4	6	3	5	6	4	—3	3	6	2	6	—3	5
1800	4	—5	—5	3	5	3	4	3	4	3	7	4	—2	1	5	1	4	—4	0
2000	—2	—3	—2	2	7	2	3	1	1	0	4	4	0	—2	1	—3	0		—7
Total	34	-42	-29	19	56	11	27	27	27	35	47	28	-14	23	29	—9	27	-29	19
Percent over test 1	5.5 %	—	—	3.1 %	9.0 %	1.8 %	4.7 %	4.7 %	4.7 %	5.5 %	7.6 %	4.5 %	—	3.7 %	4.7 %	—	4.7 %	—	3.1 %

Key to Test Conditions.

Standard basic test (No. 1)—standard engine, 6 to 1 piston.
A 18 in. horizontal intake.
B 10 deg. overlap cams.
C Inlet valve guide cut away.
D Vertical crooked intake pipe.
E $1\frac{5}{8}$ in. venturi carburetor.
G Zenith carburetor—36 choke, 170 jet, 160 compensator.
H 5.5 to 1 piston.
J 6.5 to 1 piston.
K Water jacketed 18 in. horizontal intake pipe.
L Hollow ground cams.
M Zenith carburetor—38 choke, 170 jet, 160 compensator.
N Cylinder reversed—intake for exhaust.
O 15 deg. overlap cams.

All figures corrected to standard barometric pressure of 29.92 in. mercury.
All torque readings taken at 15.75 in. lever arm.

BULLETIN AIRPLANE ENGINEERING DEPARTMENT

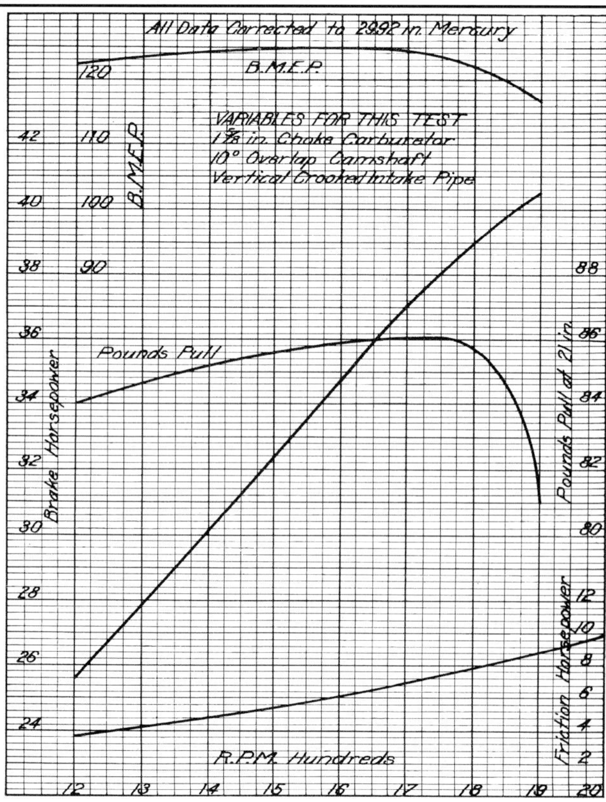

Fig. 14—Graphical representation of results of test No. 15

L1-D rear view showing water pump and starter as well as a good view of the distributor and carburetor. (Author's collection)

a 12-cylinder Liberty engine similarly fitted. Theoretically, the power output of the 12-cylinder engine should be about 25 HP more than this due to the fact that the friction losses of the 12-cylinder engine would not be twelve times as great as in the single-cylinder. This then would result in a theoretical availability of 493 HP from the like designed 12-cylinder Liberty at 1,800 RPM."

Although an engine with all these features and the resulting theoretical output was never put into production, a similarly designed one was tested shortly thereafter and did produce 495 HP at 1,800 RPM and 526 HP at 2,200 RPM. (See power graph and illustration of the engine.)

The single-cylinder Liberty engines were used for a number of years in the testing of various existing and new designs of Liberty engine components. In addition, however, they were soon shown useful in the study of flame propagation in engine cylinders as well as the design of anti-knock fuels.

Beginning in late 1919 Donald MacKenzie of the National Bureau of Standards and R. K. Honaman of AT&T conducted a number of experiments on this engine with a cylinder adapted to use a spark chronograph to record flame propagation.

Certainly just as important were experiments using this engine and a cylinder pressure recording device developed at the National Bureau of Standards by a number of scientists, the most noted of which was Thomas Midgley, Jr., who along with C. F. Kettering and others, had so much to do with the development of anti-knock fuels.

THE UNIVERSAL TEST ENGINE

It did not take long after the war ended for McCook Field to begin actively working on other than Liberty engines. The usefulness of the single-cylinder Liberty was by then well proven. However, the engine was designed for use specifically with a Liberty cylinder and could be used with no other engines.

The theory was not to be wasted on a single engine design and the Engineering Department came up with the Universal Test engine, the first of a total of four was built in late 1919. This was an extension of the single cylinder Liberty theory in that it could be adapted to a great variety of cylinder types, including the Liberty. It was used extensively in the development of the air-cooled cylinder, water-cooled cylinders of later engines, anti-knock fuels, spark plugs, etc.

It could accommodate any conventional-designed air or water-cooled cylinder with bores ranging from 4 to 8" and strokes from 4 to 10". One to four plugs could be accommodated with either battery or magneto ignition. Compression ratio could be varied through a wide range.

THE L-1 SIX CYCLE

This engine is somewhat of a mystery but, since it is known to have existed, I will mention it here. Perhaps some future historian will come across more detailed information.

Thus far, the only place I have found any mention of this engine is in the weekly progress reports of the Engineering Division and one test report, serial No. 594 of

Above left and right are views of the Universal Test Engine of 1920 fitted with a Liberty cylinder. Note the flywheel and multiple ignition system choices. In the center picture the cylinder is fitted with a Midgley cylinder pressure indicator used for investigation of knocking. (Author's collection)

December 17, 1918. I suspect the December 17 test represents the end report of work on this project as no more is heard of it. Unfortunately I have never been able to find a copy of the December 17 test report.

In the weekly report of September 21-28, 1918 we find the first mention of this project. "New tests have been started during the week on a 6-cycle single-cylinder Liberty engine." The October 5-12 report adds a little to the information in that it states "experimental work continues on the six-cycle single-cylinder Liberty test engine without a cylinder water jacket."

Testing continues to be mentioned in each weekly report through the remainder of October and the week of November 9-16. There is a lapse and then it is again noted as being tested in the November 23-30 report. That is the last report to mention this engine.

In my research I have only been able to find two references to a "six-cycle" engine. The first is a short description of a Salmson design in Glenn Angle's 1921 *Airplane Engine Encyclopedia* and also in his 1939 *Aerosphere 1939*, with a more complete description in the latter book. This 1915 Model 2M-7 engine was apparently a double-row 7-cylinder water-cooled radial using a single-throw crankshaft. It had a specially arranged gear providing a rotation of the cams at one-third crank speed. It made no further explanation of its operation.

The other reference is found in S. D. Heron's 1961 *History of the Aircraft Piston Engine*. He refers to a British project carried on during WW I about which he said he had few details. He said the design was developed by Alan E. Chorlton with government backing. His only knowledge of it consisted of the following statement: "I think the idea back of this project was that the two strokes involved in scavenging would be sufficient by means of increased BMEP to overcome the effect of what might be described as the "waste" strokes of the cycle." Supposedly it was a single-cylinder only project.

THE L-2

The earliest mention I have found of the L-2 Liberty was in the June 1-8, 1918 progress report. In it was the statement "The parts of the two-cylinder test engine are now completed and are awaiting assembly." By the following week the story had been altered slightly to read "Parts are now completely machined and delivered to the engine assembly department for the two-cylinder Liberty test engine with the exception of one housing cover which was re-designed and ordered from the foundry June 14. All of this work is promised by the 19th."

By the June 15-22 report the housing in question had been identified as the machine gun fire control housing and "should be finished in a day or two."

We hear no more about the L-2 until the report of October 12-19. It is obvious from the interceding reports that an overwhelming amount of experimental and testing work was going on at McCook Field during this period. It would seem that the L-2 had taken a back seat in the scheme of things and work on it had remained dormant for some four months. It finally rates a mention as follows: "Four 8-cylinder Liberty engines are being built from parts furnished by the factory. A 2-cylinder experimental engine of Liberty type is also being built." Finally all parts were ready, the time was ripe, and the L-2 was being assembled.

We hear again from the L-2 in the November 2-9 report when "A 2-cylinder experimental engine which has been under construction, was completed during this week." During the next week a dynamometer test of the engine was started.

It suddenly becomes evident that parts were being made for two of these engines when in the November 16-23 report it mentions that "one of the 2-cylinder Liberty engines was run in." And during the following week that "tests of a twin-cylinder Liberty-type engine with special headers and carburetors" was carried out.

This engine with special carburetors and headers continues to be mentioned and in the December 7-14 report it is referred to as twin-cylinder Liberty T-1. Thus the two L-2 engines were referred to as T-1 and T-2, the "T" being for "Twin" since they were now being called "twin-cylinder" Liberties.

L-2 No. T-1 as tested on February 20, 1919. Note that it is configured with outside the "Vee" carburetors. (Courtesy of the Air Force Museum)

L-2 No. T-2. Shown in what must be standard flying configuration as it is driving a propeller rather than a dynamometer, has conventional carburetion and appears to be cooled by a conventional radiator in the slip stream. With a power output of about 70 HP, would it have been considered for actual flight? (Courtesy of NASM)

Official tests reports were published as serials 609 in December of 1918 and 682 on February 20, 1919. It appears that this engine was found to be of limited use for testing, particularly after the end of the war and it became evident that there would be more than ample standard V-12 Liberty engines available for any type of test desired.

The February 20 report is the last I have found. Its object was to determine the power characteristics and differences in vibration in the two-cylinder Liberty engine between power impulses occurring 45° and 405° apart. The engine was described as wet sump and using special Zenith AS-8 single venturi carburetors with special 18" induction pipes with the carburetor on the outside of the "Vee." The results were rather inconclusive. The most significant statement being that the engines output under any of the configurations tried was some 10% less per cylinder than that of a standard L-8 or L-12. The engine under test in this configuration was T-1 and is pictured on page 140.

A photograph of what would appear to be engine T-2 was found in the NASM archives in what seems to be flying configuration. It is mounted on a test stand with a propeller of appropriate size to its power and cooled by a standard radiator. Although it certainly could have been flown, it is highly unlikely that it was.

L-4B No. 1 as restored and currently residing the the collection of the Air Force Museum. (Author's collection)

THE LIBERTY L-4B

The initial report to the Aircraft Production Board of May 31, 1917 by Hall and Vincent included a four-cylinder engine. It was to have a rated power of 110 HP and a maximum output of 135 HP. Weight was estimated at 375 lbs.

While writing his book *The Liberty Engine 1918-1942* for the Smithsonian Annals of Flight, Philip S. Dickey, III came upon the earliest reference I have noted regarding actual building of a four-cylinder Liberty. In the McCook Field letter files he found a letter to G. H. Brodie of Packard from Colonel Vincent dated November 3, 1917 in which Vincent directed Brodie to rebuild an L-8 into two L-4s.

At this point in time six experimental L-8s had been built and the last experimental L-12, No. 5, had probably just been completed.

It is not clear just why, at this early date, Vincent or anyone else would have felt there was good reason to build an L-4, particularly at the cost of one of the five usable L-8s in existence. It is not likely we will ever know. There is absolutely no evidence that this request was ever carried out. Colonel Vincent must therefore have retracted the directive.

Shortly thereafter, on December 31, 1917, however, the Engineering Division issued order number 1373 for the production of two L-4B engines by the Hudson Motor Car Co. Just why Hudson was picked is not clear. They were not contracted as producers of any regular models of the Liberty. There were, of course, a number of indirect connections between Hudson and various people associated with the BAP in one way or another. Col. Vincent had been Hudson's chief engineer from 1910 to 1912, Roy D. Chapin and Howard E. Coffin had been founders of the company and Chapin was its president when he went into government service. Howard Coffin became Chairman of the APB. Charles H. Vincent, currently experimental engineer with the BAP, had worked for Hudson from 1911 to 1912, been offered the opportunity to purchase a substantial block of stock in the company in 1911, left briefly and returned from 1914 to 1916 before then going to work for the Packard Company.

The contract was probably offered to Hudson purely because they had the facilities available. Hudson never received any aero engine production contracts other than this minor one.

The first engine was delivered to the Airplane Engineering Division Experimental

L-4B No. 1 exhaust side as photographed in the Wright Field Museum sometime during the 1930s. Note that the data plate was missing from its normal position even then. (Courtesy of the Air Force Museum)

Department in Dayton for testing on February 12, 1918. It was reported that Hudson had not run the engine prior to delivery but had jacked it in at 120 RPM for 9 hours.

The test reports seem to indicate the engine was probably supplied with no intake manifold and no carburetor. For testing purposes the Engineering Division then made up an intake system and a makeshift carburetor consisting of 1/2 of a standard Zenith L-12 model modified as best as possible to adapt it to the L-4. Weight was listed as 398 lbs. without intake manifold. (A June 11, 1918 report comparing this and the L-6 engine with a number of engines listed its weight dry at 400 lbs.)

The engine was then run on a dynamometer and after only five minutes at 700 RPM it burned out number 2 and 3 rod bearings. Upon disassembly a number of problems were found:

2 and 3 rod bearings had burned out and #1 was on the point of failure.
Inlet valves opened at top dead center instead of the standard 10° after top dead center.
One cam shaft gear bolt was found to be broken.
The upper cam shaft drive shaft had no end-play.
All bearings, including the main bearings, had only .001" clearance.

The engine was reassembled with .002" bearing clearances and all other problems corrected. It was then submitted to power tests and a power graph generated with the Zenith carburetor. Carburetion was found to be very poor and power output was likewise poor. Maximum output was at 91.5 HP @ 1,200 RPM and fell off dramatically after that point.

A special Schebler carburetor with 1 1/4" venturi was then tried with much better results but still below what was expected of the engine. Maximum power of 102.1 HP was recorded at 1,400 RPM. The test engineer noted the engine had excess vibration above 1,500 RPM. (Four-cylinder engines were noted as being prone to a high amount of vibration, so this comment may not have been significant.)

The power curves for both of these tests were included in the report and are reproduced herein. They are certainly not representative of the best potential of the engine. With proper carburetion the engine should give its peak power at about 1,800 RPM and produce close to 140 HP at that speed.

Unfortunately only the two earliest test reports have been found. Thus the true potential power of the engine has not been observed, at least not under controlled conditions. The June 11, 1918 report mentioned above listed its power as 135 HP @ 1,700 RPM and maximum power as 147 HP.

Rear view of L-4B No. 1 showing the generator, water pump and battery ignition system. (Author's collection)

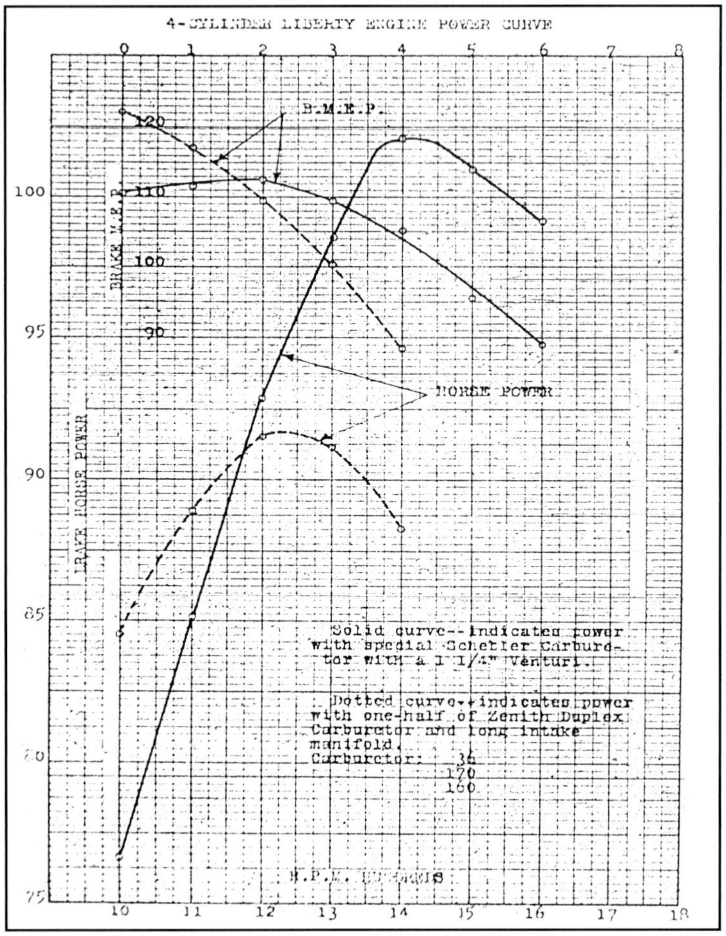

L-4B No. 1 power graph from test serial No. 7 with the results with both carburetors shown. This graph is not to be taken as the norm for this engine, however. (Author's collection)

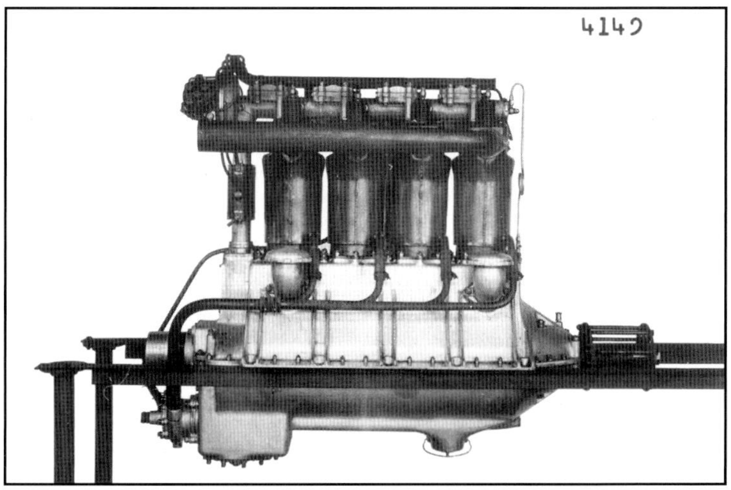

The above three images are believed to be from test serial 38 dated 6/29/18 and of L-4B No. 2. You will note that a different intake manifold and carburetor are installed and an exhaust manifold is in place. It should also be noted that the oil pump is missing. (Courtesy of the Air Force Museum)

Interest in the L-4 was certainly not high. Both Curtiss and Hall-Scott engines in its power range were already in production and active use in training aircraft. For this reason only two L-4 engines were ever built and interest in them was limited at best.

Officially published Technical Reports of the Engineering Division list L-4 tests as follows:

Test Serial	Subject	Date
7	Performance of L-4 No. 1	2/23/18
8	L-4 water pump	3/4/18

Test Serial	Subject	Date
232	Test of 4-cyl starting motor	6/11/18
59	L-4 Westinghouse generator	6/12/18
38	Performance test L-4	6/29/18
64	Hall-Scott L-4 test	7/3/18
72	Performance test L-6	7/11/18
52	Hall-Scott L-6 test	7/25/18

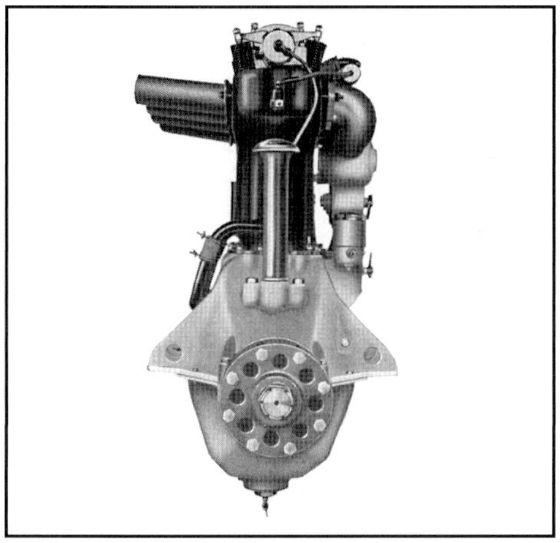

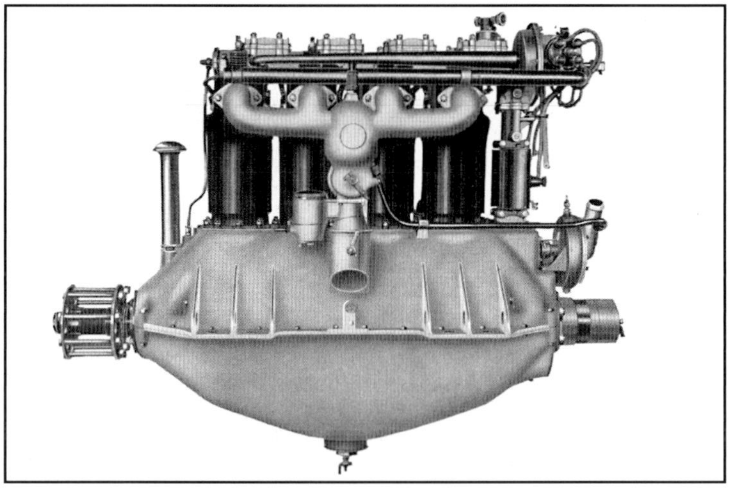

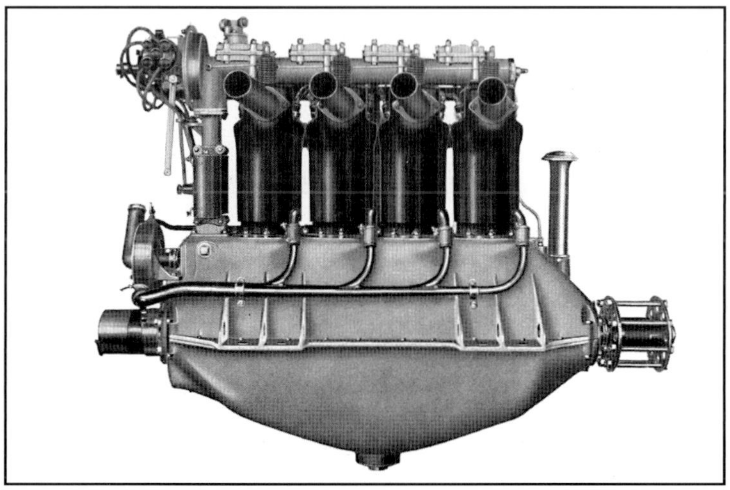

The above three images are of a Hall-Scott L-4 engine. This engine was submitted to the Air Service BAP for tests at about the same time the two Liberty L-4 engines were. Although offered to the Air Service, only the two test engines were purchased. It was also offered on the civilian market following the war but few were sold. (Courtesy of William Lewis)

They are listed in date-order for clarity and tests of Hall-Scott L-4 and L-6 and the first test of the Liberty L-6 are included for their historical significance. Tests 7, 8, 52 and 64 have been found. Test 38 has not been found but was probably carried out on L-4 No. 2. A series of photographs believed to be from that test have been found and are reproduced at left.

No evidence has been found indicating that either L-4B was ever actually flown. The June 11, 1918, report already mentioned was submitted to Lt. Col. Thurman H. Bane, head of the Technical Section of the Department of Military Aeronautics, by First Lt. T. Maynz. The report contained basic specifications and performance data on the L-4, L-6 and a number of foreign engines in the same power range. The final comment of the reports was: "It seems to me, however, advisable to install one or two Liberty 4's in a J.N.-4-D and try it out. I believe the Liberty 6 entirely too powerful for a training plane." Thus a recommendation was made to Col. Bane that the engine be flown and Col. Bane certainly had the authority to have it done, yet it remains that there is no evidence the recommendation was ever carried out.

Both Hall-Scott L-4s and both Liberty L-4s were still in Air Service inventory at McCook in December of 1919. Liberty L-4 No. 1 was transferred to the Wright Field Museum some years later and No. 2 was sold as surplus. No. 1 is still in the Air Force Museum today. No. 2 was listed for sale in the classified section of the July 27, 1929 issue of *Aviation* magazine and was reported in a private collection in the year 2000 but has so far not surfaced.

All references so far found relative to the two L-4 engines refer to them only as No. 1 or 2. This would be the Manufacturer's number. So far, no reference to an assigned AS number has been found. The norm was to stamp both the manufacturer's number and the assigned AS number on the engine's data plate. No known photograph of an L-4 engine shows an attached data plate, thus we cannot use this source to find what numbers, if any, were assigned.

AIRPLANE ENGINEERING DEPARTMENT SERIAL NUMBERS

All production aircraft and engines ordered by the Air Service were assigned numbers referred to as SC (Signal Corps), AS (Air Service) or US (United States) numbers. In the case of production contracts blocks of numbers were assigned to each contract to match the number of items ordered. Although the blocks were generally assigned in ascending order as the various contracts were chronologically issued, this was not a hard fast rule. The reader can see what numbers were assigned to the various production contracts listed in Appendices A and B. When these numbers were stamped on the engine's data plate in area usually titled "U. S. No." they were generally prefixed by SC (Signal Corps) on early engines and then AS (Air Service) on later engines, although in many cases no prefix at all was applied.

On about February 1, 1918 a special block of 100 numbers was assigned to the Airplane Engineering Division at McCook Field for use on experimental and special engines that they either built themselves or contracted to outside manufacturers. The block was 47828-47927 inclusive. The assignment of these numbers was totally under the control of the Airplane Engineering Division and their use continued until about January of 1919.

The assignment of these numbers to the various experimental or special engines was not controlled by any set of standard rules and records of their assignment are not to be found in any one location. Numbers were not necessarily assigned in the order in which production was ordered. Not only that but, as we will see, orders were changed or canceled so frequently that keeping assigned numbers in logical order might have proved somewhat of a problem.

To this we can add the fact that not all special engines manufactured under Engineering Division control were assigned numbers, in this series or any other. The one and two-cylinder Liberties are believed to fall in this category.

As I progress through the various experimental engines produced during this period I will note the numbers when they are known. I will try to refer to these as US numbers when they are mentioned, unless there is some reason to be specific about the actual prefix used.

HALL-SCOTT L-4

The reader will note in the preceeding list of tests that a Hall-Scott L-4 engine was tested and the report was published on July 3, 1918, just four days after the test of the second Liberty L-4. The test actually occurred on June 28 and was carried out on engine MNFG. NO. 3001 U.S. NO. 48065. The engine was one of two ordered, the first being U.S. NO. 48064 and probably MNFG. NO. 3000. The reader will note the U.S. number falls outside the experiment engine series. It and its L-6 Hall-Scott companions (which will be covered later) were purchased for testing after being offered by Hall-Scott rather than being built at the request of the Engineering Division.

U.S. NO. 48064 was transferred to the Wright Field Museum, was still inventoried there in 1958 and is probably now in the Air Force Museum inventory.

The Hall-Scott L-4 was basically an A-7 model using as many Liberty cylinder and cam parts as possible with them mounted on an A-7 type crankcase modified so as to accept the Liberty cylinders. The oiling system was of standard Hall-Scott wet sump-type but with full pressure on both main and crank pins.

The engine used standard Liberty cylinders and associated parts and "the same type of camshaft and housing construction" as noted in the test report. It was rated at 125 HP @ 1,650 RPM and actually produced 132 HP at that speed. Fuel consumption was found to be .598 lbs. per HPH.

The report was a short one and the test appears to have been made purely to report on the output of the engine. No comments on performance or construction were made other than to produce a power graph. All indications are that there was no interest in proceeding further with any new design four-cylinder engine, either Liberty or Hall-Scott. Although its weight was not noted in the test, later Hall-Scott brochures listed it at 380 lbs.

It is believed that only the two test engines were purchased. The engine and its companion Hall-Scott L-6 were offered on the civilian market following the war but very few of either model were sold. By 1923-24 Hall-Scott had abandoned the aero engine market and was concentrating on marine engines. The company was purchased by American Car and Foundry in 1925 and during the transfer of its assets no mention was made of an aircraft engine product line.

THE LIBERTY L-6A

The first mention found regarding the actual building of a six-cylinder Liberty was the issuance on March 5, 1918 of order number 5257 to Thomas-Morse for construction of two L-6A engines. This was more than two months after the order for two L-4 engines had been issued.

In an April 18 letter to Mr. Potter Col. Vincent made the following observations:

Liberty 4 - 2 sample engines built.
Liberty 6 - Thomas-Morse to build 2 samples.
Liberty 8 - 5 samples - could be manufactured on short notice. Hispano 300 is better design however.
Liberty 12 - Model A direct-drive in production.
2 geared Liberties built by Packard.
1 duplex Liberty 2-crankshaft job looks good.

On May 31 order 5257 was replaced by order CS-85 for 2 engines and spare parts. The first engine was delivered on July 2 and immediately test run and then put on a dynamometer to obtain power curves and fuel and oil economy figures. The results were reported in test serial 72 dated July 11. The second engine was completed a few days later and put on tests at the Thomas-Morse plant.

Liberty L-6A No. 5 SC 47843. (Courtesy of Peter Bowers)

In the mean time order CS-95 was issued on June 6 with the same company for an additional 12 L-6 engines. It seems that either the BAP or the Experimental Division was constantly changing its mind and either changing specifications or quantities, or both, on experimental orders and contracts. It continued here and we will see it do the same throughout the remaining wartime contracts for experimental engines. CS-95 was replaced on June 25 by CS-125. This order reduced the quantity ordered to 4 plus one set of spare parts.

The initial CS-95 order for 12 engines would have increased the total to 14 engines. One would certainly think that such a quantity indicated a need for more than just a small quantity for testing purposes. Logic would lead to the conclusion that someone intended to also supply a small quantity of aircraft suited to the engine. Obviously the plan was rapidly altered since the order was reduced less than three weeks later. Thus far, no plan for an aircraft built around this engine, during this period of time, has surfaced.

The weekly progress report of July 6-13 noted that the first of this second batch of four was currently being assembled and the balance were in progress. From this point there is no mention of the L-6 engines in the weekly progress reports until the October 5-12 report at which time they noted that CS-125 was completed and that the first engine was delivered to McCook Field on October 10.

This therefore marked the completion of a total of six engines. MNFG. NOs. were 1-6 and U.S. NOs. were SC 47839-47844. These were the only Liberty L-6A engines built during the war. Ten more were built by Wright in 1920 and I will cover these in Chapter 7.

On 5/15/19 test serial 829 was issued as the standard report on the L-6A. That test was run on Thomas-Morse MNFG. NO. 5 U.S. NO. SC 47843. These standard tests were quite detailed in listing specifications and output. In addition they illustrated the engine tested from all standard angles as well as disassembled and showing all basic assemblies and major parts. The illustrations from this test are reproduced in Appendix G and detailed specifications in Appendix D.

Specifications of the L-6A

Bore	5 in.
Stroke	7 in.
Number of Cylinders	6
Displacement per cylinder	137.41 cu. in.
Total piston displacement	624.46 cu. in
Compression ratio	5.42/1
Valves per cylinder	2, 1 intake & 1 exhaust
Carburetors	2 Zenith Model 55 AS
Ignition	2 Delco Model X 903 battery-type distributors
Engine weight dry	567.5 lbs.
Normal power	231 HP @ 1,700 RPM
Maximum power	240 HP @ 1,850 RPM
Weight per HP dry normal	2.45 lbs.

Liberty L-6A No. 5 SC 47843. (Courtesy of the Air Force Museum)

FLIGHT TESTS OF THE LIBERTY L-6

Interestingly enough, the first flight test of an L-6 engine was carried out in a civilian plane and must indeed be classified as a disaster. This disaster was in no way caused by engine performance, however. The story is so interesting that I relate it below in some detail.

THE CHRISTMAS BULLET

Both the BAP and the Experimental Airplane Division at McCook Field had numerous experiences during the war with proposals by supposedly qualified aircraft designers and engineers who were trying to sell their goods to the Air Service. Although these efforts took up some of the time of McCook engineers and other military personnel, usually it was limited since drawings and specifications would generally be required before any work paid for by the government would be authorized. In most cases the proposal would be found to be either for an item not desired by the Air Service or one which, by studying the submitted material, obviously would not operate as proposed.

On occasion, however, inventors were able to exert political pressure resulting in more than usual attention having to be paid to their proposal. Such was the case here.

In this instance Dr. William W. Christmas was the designer of the aircraft. Dr. Christmas was born on September 1, 1865 and received his medical degree from George Washington University in 1905. The Doctor's interest soon turned to flight in addition to his medical practice. He claimed he built and flew his first plane in late 1907 to early 1908. It was supposedly damaged in a crash and later purposely destroyed to prevent others from acquiring his design secrets.

There never seems to have been any verification of the actual building and flight of this first airplane. He built his second in 1911 at College Park, Maryland and his third was built in 1912 at Washington, D. C. by the Christmas Airplane Company and displayed at the aeronautical show in the Grand Central Palace at New York City in May of that year. Although there seems to be no doubt that the last two airplanes were actually built, any proof that either actually flew seems to be lacking. The Doctor, of course, claimed they all did.

Although the Doctor's interest in airplanes continued, he built no more airplanes until approaching the Continental Aircraft Company of Amityville, Long Island in early 1918. He placed an order for a small biplane of his own design which he called the "Christmas Scout." The name would later be changed to the "Christmas Bullet."

Some time later the Doctor initiated some indirect pressure on the Air Service to consider his new airplane design by having a Charles A. Collins write a letter to Senator James W. Wadsworth, Jr. of New York. In this letter Collins went to some lengths to impress the senator with the qualifications of Dr. Christmas as an airplane designer. He said that Christmas had been associated with Professor Langley, that he had built and flown a number of airplanes prior to 1913, that he was well financed and currently building both a single-place fighting scout and a three-place fighting machine of high-speed and asked for the loan of a Hall-Scott A-5a and a Liberty-12 to power them.

The senator forwarded the letter to Col. Vincent, commander of McCook Field and head of the Engineering Division. Vincent had been down this road before, so to speak, and could see the potential for problems with yet another "new and wonderful" airplane design from an unknown and unproven designer. He wrote directly to

The Liberty 6-powered **Christmas Bullet** ***powered by L-6A No. 1. This plane crashed during its first test flight in January of 1919, killing the pilot and destroying both the airplane and its engine. The plane was marked with military-type cockades on the wings and fuselage and given vertical rudder stripes by the builder to give it a military look. It was, however, not contracted for, purchased or tested by the Air Service.*** (Courtesy of Peter Bowers)

Collins indicating he was not particularly impressed with the Continental Aircraft Company as he had seen what appeared to be thus far its only product, an outdated pusher-type powered by a Hall-Scott engine, and found it to be of no use to the Air Service. He suggested that instead of sending the requested engines that he first send specifications of the two proposed airplanes for McCook engineers to study.

Whether this was done or perhaps the influence of the senator was effective, we cannot say. We do know that all further contact with the Air Service came through a new company formed by Christmas, the Cantilever Aero Company of Copiague, Long Island. This location was only a mile from the address of the not well liked Continental Aircraft Company (who it is believed actually completed the construction of the Christmas Bullet) and leads one to believe it was a probably a name only company formed to get the Continental Aircraft's name out of the picture.

Which ever was the case, by June 11, 1918 Dr. Christmas had been promised his two engines, a Liberty 12 and, substituted for the Hall-Scott, a Liberty 6. The L-12 was shipped on June 19 but the L-6 was delayed because none had yet been delivered to McCook Field.

The plane (or planes, as a second had been mentioned along the way) for which the L-12 was intended was apparently never completed. Whether or not the loaned L-12 was ever returned to McCook Field is uncertain.

On July 1 Christmas wrote Vincent that construction of the Scout was being delayed awaiting arrival of the L-6 engine. Vincent replied on July 15 that they would loan a Liberty 6 for use in the Scout on condition that Christmas would not attempt to fly the plane when completed, but ship it to McCook Field for evaluation, along with a second duplicate airframe for sand testing. Christmas was to wire Vincent accepting the terms before an engine would be sent.

Christmas replied by wire on July 17 as follows:

> *"Your letter received this morning. Conditions therein satisfactory to us. Am writing confirmation today. Please rush motor."*

A letter of confirmation dated the same day was shortly received by Vincent. He had L-6 No. 1 packed and shipped on July 24. It arrived at Continental Aircraft on July 31.

Vincent believed the Scout must be near completion and wired Christmas on August 26 to ship the first airframe without engine to McCook Field for structural tests and if it passed they would ship a set of instruments so the second plane could be completed for flight.

No airframe was sent to McCook Field and nothing more is found in the records regarding it until Christmas wrote to McCook Field on October 16 telling of his intentions to fly "three" Scouts extensively in order to demonstrate their advantages. This, of course, was in direct disregard for the agreement under which he obtained the engine.

On November 12 Lt. Col. Thurman Bane, head of the Technical Section, advised General William L. Kenly, OIC of the Division of Military Aeronautics, of Christmas' intentions and the condition of the engine loan and noted that Christmas should contact Lt. Col. Vincent.

Since this was the day after the signing of the Armistice, the Army's interest in the Christmas Scout (soon to be called the Christmas Bullet) went from near none to absolutely none. Their prime interest now was to try to insure the return of the loaned engine. The result was General Kenly met with officers at Hazelhurst Field on November 15

The Hall-Scott L-6-powered Christmas Bullet. This airplane was completed not long after the crash of the first Christmas Bullet. Dr. Christmas also proposed to demonstrate this plane to the Air Service but was refused. This plane also crashed on its first test flight and killed its pilot, probably in April or May of 1919. (Courtesy of Peter Bowers)

(the proposed flying site) and directed that the Christmas plane was not to be flown by any Air Service pilot until it had been sand tested at McCook Field.

From this point only a few details of the story have surfaced, and some of them are somewhat in doubt as they were told by Christmas himself at later Congressional hearings and were likely colored in his favor. On December 5 he took the plane to Hazelhurst, advised General Kenly that the plane had been tested and requested permission to demonstrate it. Permission was refused and he then took the plane to Central Park, Long Island on the 6th. Christmas later claimed that Kenly reversed his refusal and the plane was flown to Hazelhurst on the 7th, demonstrated and photographed there by government photographers. Absolutely no records of it or photographs have been found. Records show that Kenly wrote Christmas on December 11 indicating that if he wanted to fly the plane he should obtain a license from the Joint Army and Navy Board of Aircraft Cognizance to use a private field.

Not much more is known of the airplane until in mid-January of 1919 (exact date unknown) when Christmas attempted to have the company pilot fly it. He refused and was fired. After several attempts a pilot was found who agreed to fly it, an American named Cuthbert Mills who had just returned from overseas.

At this point a short description of the plane might be in order. This was a modified biplane whose upper wing had a span of about 28 ft. and the lower about 12 ft. Overall length was about 21 ft. The upper wing was flexible and drooped, taking on dihedral in flight. The wing supposedly swept backward slightly under load and the camber changed. The elevator design was a warping type in which the rigid trailing edge was bent upward or downward for control.

At what is believed to have been a private field, Mills climbed into the plane and taxied off for a short distance and climbed into the air and began to turn. The plane immediately shed its wings and the fuselage fell nose first into the ground, killing the pilot instantly and destroying the plane and its engine. Among the unfortunate witnesses was the pilot's mother.

Not easily discouraged it seems, Christmas completed a second airplane of this design and fitted it with a Hall-Scott L-6 engine. This plane was displayed at the New York Aero Show from March 1-15, 1919 with a placard reading:

> "CHRISTMAS BULLET - *safest, easiest controlled plane in the world. Speed 200 m.p.h. - 6 cylinder Hall-Scott 200 h.p. motor. Cantilever Aero Company, 1265 Broadway, New York City."*

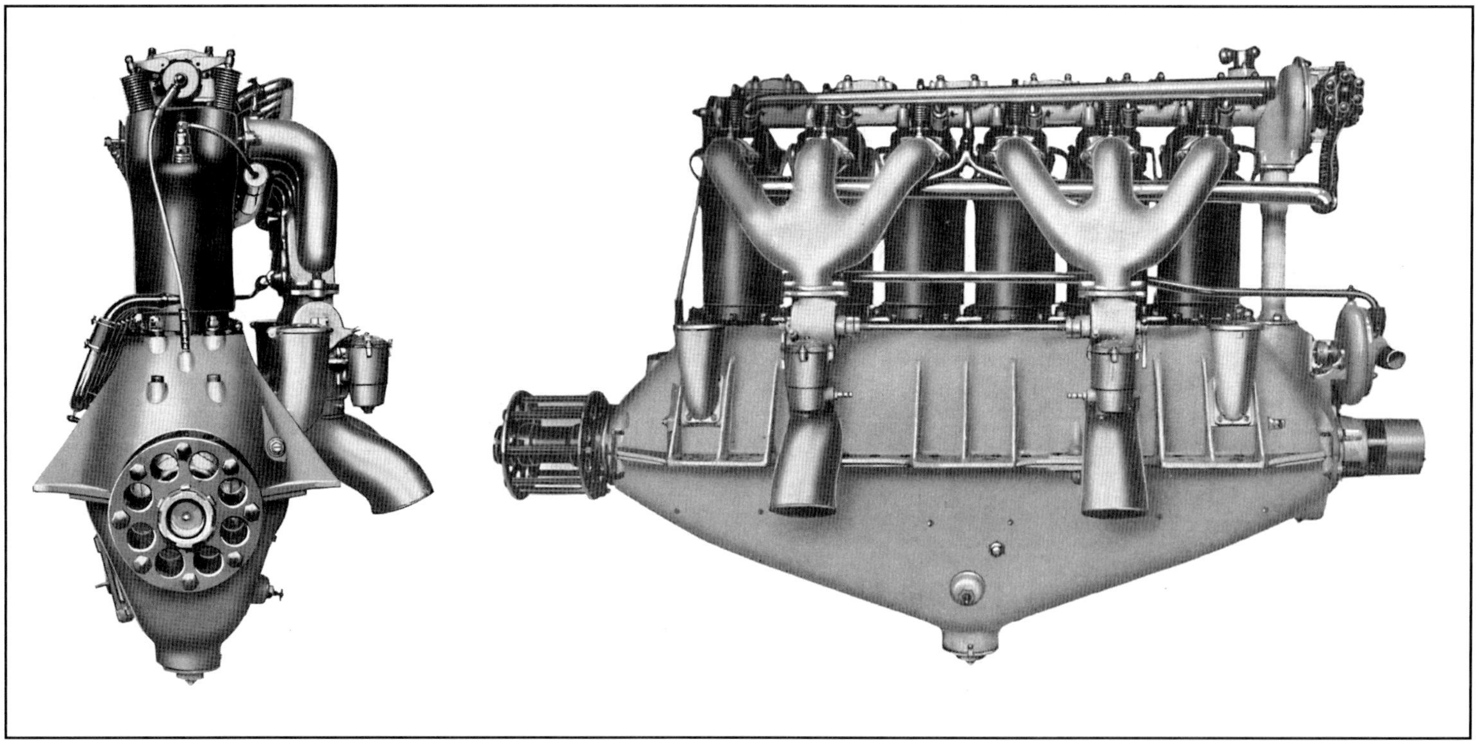

The Hall-Scott L-6 engine as pictured in their brochure of 1919. Six Liberty 5 x 7 cylinders were used thus giving the same displacement as the Liberty 6. Dry weight was 530 lbs. Two special Miller carburetors were used. (Courtesy of William Lewis)

Following the show Christmas made another attempt to have the Air Service observe a demonstration at an Army field, this time of the Hall-Scott-powered plane. He was refused.

A short while later he managed to hire Captain Allington Jolly of the Royal Air Force to pilot the second (and last) Christmas Bullet on a test flight. This first, and last, flight was a short one. Although the plane kept its wings attached while in flight, through an apparent lack of flight control the Captain flew it straight into a barn. The plane was destroyed and the Captain killed.

Although Doctor Christmas continued to claim great things of his airplane designs, and of the "Bullet" in particular, there is no evidence that he ever built another airplane of any type.

HALL-SCOTT L-6

Like the Hall-Scott L-4, the L-6 was an A-5 modified in design to allow the use of Liberty cylinders and associated parts as well as many Liberty parts as possible in the camshaft assembly and associated parts. The oiling system was wet sump with full pressure on main and rod bearings.

On June 25, 1918 Hall-Scott L-6 MNFG. NO. 2098 was tested at McCook Field. It was a brief test (serial 52 dated June 27, 1918) to determine the power output of the engine. The engine was rated by Hall-Scott at 200 HP @ 1,700 RPM but its output was noted as only 183 HP at that speed. The resulting power graph reflected this and indicated the engine peaked at 185 HP @ 1,800 RPM. No fuel nor oil consumption data was obtained. As brief as the Hall-Scott L-4 test was, this one was even more brief.

No U.S. number was listed in the above test. Records have been found assigning U.S. numbers 48064-65 to Hall-Scott L-4 engines and test serial 64 shows L-4 MNFG. NO. 3001 was assigned U.S. number 48065. U.S. 48064 was no doubt L-4 MNFG. NO. 3000. Records show U.S. numbers 48066-67 as assigned but not to what engines. Considering the facts as we now know them, assembling the puzzle parts we have would tell us that most likely there were two Hall-Scott L-6 engines purchased at this same time; MNFG. NO. 2098 tested in serial 52 was probably assigned U.S. NO. 48066 and a second Hall-Scott L-6 (likely MNFG. NO. 2099) assigned U.S. No. 48067.

Although we observe no further interest in the Hall-Scott L-4, L-6 MNFG. NO. 4519 was subjected to a standard engine test in test serial 1319 in June of 1920 and a 50-hour test serial 1383 on 9/17/20. It showed a lack of proper lubrication of the water pump bearings and burned out several during the series of ten 5-hour test runs subjected to. Fuel consumption was also noted to be high, averaging 0.565 lbs. per HPH. Power output was much better than in the test of No. 2098 above. Power at 1,700 RPM was 224 HP. At 1,800 RPM it was 231 HP.

Although no more published tests appear, the engine must have been shown some interest as during 1919 the Army had Lowe, Willard, and Fowler (L.W.F.) develop a twin-engine modification of the DH-4 for Air Mail use. The L-12 was removed and its location used for mail storage. Power was supplied by two Hall-Scott L-6 engines

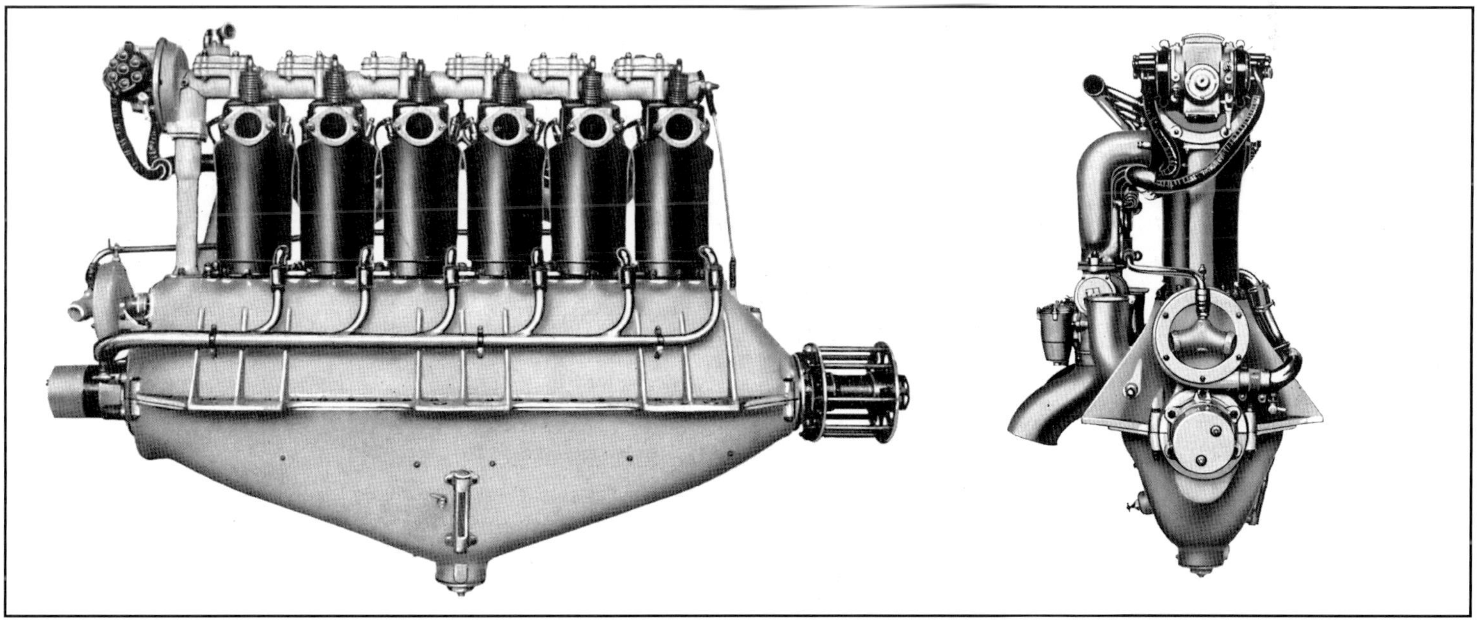

Hall-Scott L-6 as illustrated in their 1919 brochure. Ignition was 8 volt Delco. The generator and water pump were mounted on the rear and an oil sight gauge on the right side of the wet sump. The crankcase held 14 pints of oil. (Courtesy of William Lewis)

installed between the wings whose span was increased by 10 ft. Twenty of these were built for the Post Office and ten for the Army. This would have required 60 engines plus spares. These aircraft did not stay in service long and by November of 1925 a block of 50 of the engines were offered on the surplus market by a Detroit company who had purchased them from the Army.

One of the engines which were ordered for these aircraft has recently turned up in a private collection, Hall-Scott L-6 MNFG. NO. 4580 U.S. 94675, just 61 numbers from the 1920 test engine mentioned above.

In addition to the Twin DH-4 an Air Service photo of a Fokker D-VII with a Hall-Scott L-6 installed for testing has also been found as well as one of a Fokker D-VII apparently loaned to Hall-Scott for use in testing the L-6 as well as for advertising purposes.

Hall-Scott offered the L-6 and a slightly improved model L-6a on the commercial market probably before the end of the war. I have a copy of a nicely done brochure for the L-6a which could have been used for commercial sales if so desired. The reason it is of particular note is that it is receipt date stamped July 19, 1918 at the School of Military Aeronautics at Princeton, NJ. The closing remark of the brochure was "The Hall-Scott Company feel that this engine will meet an urgent necessity at the present time not only in advanced training machines, but also in single combat and scout machines." Obviously they were trying to sell the engine to the Air Service. How many were sold is not known, but these were the company's last aero engine models. They had retired from the aero engine market entirely by 1925.

Twin DH-4 powered by two Hall-Scott L-6 engines. Twenty DH-4 airplanes were converted to this configuration by LWF for the Post Office and ten for the Army. (Courtesy of Jesse Davidson/Peter Bowers)

FIRST ARMY FLIGHT TEST OF THE LIBERTY L-6

On March 29, 1920 Liberty L-6A MNFG. NO. 3 U.S. NO. 47841 was flight tested in Fokker D-VII P-127. The test was of both aircraft and engine but the pilot's observations were directed primarily at the performance of the airplane. It was noted as equipped as a single-seat pursuit, night attack.

Air Service Fokker D-VII No. 8323 McCook Field P-127 used in flight testing Liberty 6 No. 3 US 47841. (Courtesy of Peter Bowers)

The airplane was noted as handling well in all general areas. Service ceiling was 15,900 ft. and absolute ceiling was 17,500 ft. Speed at ground level was 120 MPH, at 15,900 ft. it was 92 MPH and at 17,500 ft. it was 76 MPH.

Engine performance was good and basic specifications were noted as follows:

Make - Liberty 6 Model A Thomas-Morse NO. 3
U.S. number 47841
Type vertical 6
Rated horsepower- 215 @ 1,700 RPM
Compression ratio- 5.42 to 1
Weight dry - 567.5 lbs.
Gas consumption - 0.49 lbs. per HPH
Oil consumption 0.0256 lbs. per HPH
Weight of water in engine - 21 lbs.

I have carried the Liberty 6 history to this point, which is well beyond the end of the war, primarily to show when the engine was first flown relative to production of the first series of engines. I will return to the L-6 in later chapters where I will cover the second batch of L-6 engines built by Wright and the remaining aircraft powered by the engine.

THE LIBERTY 8

As we know, the Liberty 8 was the first Liberty to be completed and the first to actually power an airplane. Packard was contracted to build the first six of this model and completed the last on or before September 10, 1917. The engine which was the first to fly was L-8A No. 3 in LWF-F AS 2505 on August 29, 1917. The airplane was later assigned McCook Field number P-8. It was kept in service for several months and then condemned and burned on April 5, 1918, probably following some type of damage.

Its engine however, Packard Exp. No. 3, continued to be used and was one of the engines which powered the first Bristol USB-2 Fighter P-37 in its series of flight tests from August through November of 1918. But that is getting ahead of the story.

The history of production orders for the Liberty 8 ties in very closely with the history of the Bristol Fighter which was proposed to be built in quantity in this country early in the war. The British Bristol fighter of which this was to be a copy with American power was the F2B which was designed in the fall of 1916 and was currently in production and in use as a fighting and reconnaissance machine. It was powered by a 190-210 HP Rolls-Royce Falcon engine. It was noted as having only fair performance but excellent maneuverability and fighting abilities.

In August of 1917 the A. E. F. recommended our use of the Bristol Fighter with the 200 HP Hispano engine. It was noted that the Bristol Fighter was fully tested but not with the Hispano engine but that the British Air Board had modified the design for this engine and that it is considered best for American production.

On August 14 a sample Bristol with an incomplete set of drawings left London bound for the U. S., reaching Washington on September 5, 1917. (Some historians say September 1.)

On September 4, 1917 the APB board met and recommended a contract with the Fisher Body Corp. for 1,000 Bristol Fighters powered by the Liberty 8 engine.

In the mean time Capt. (later Col.) V. E. Clark returned from Europe where he had been serving on the Bolling Mission and the decision was made to redesign the Bristol Fighter to use the Liberty 12 with Capt. Clark doing the work. This was apparently done because the Liberty 8 was not yet ready for production and the Liberty 12 was seen as being available far sooner. At this point in time the first set of production contracts for L-12 engines had just been issued. Packard was seen as an early producer. (The first production engine was completed by Packard on November 2.)

On October 30 the board passed a resolution recommending a contract with the Curtiss Airplane and Motor Co. for 1,000 Bristol Fighters to be powered by Liberty 12 engines. The next day, on October 31, the BAP recommended eliminating the order with Fisher for 1,000 Bristols with L-8 power. Since there remains no record of a contract being written or proposed for the first 1,000 Liberty 8 Bristols to Fisher, it seems that the BAP recommendation of October 31 stopped the Fisher contract before it became an official document.

Thus, from this point the Bristol Fighter proceeded as a Liberty 12 powered airplane. The sample British Bristol was sent to Curtiss on November 1, 1917 where it was used to help finish the required drawings for production. It later went to McCook Field where it was fitted with a Hispano-Suiza H 300 HP engine (the first U.S. production engine) and became P-30 with the designation of USB-1 (United States Battler -1). By November 18 it had gone through much flight testing and set an unofficial altitude record on that date. From that time it continued to see much use in active flight testing and performed admirably.

The Liberty 12 Bristol Fighter went through a great deal of problems in the attempt to get it into large-scale production. The contract for 1,000 was first renegotiated and replaced by order 20285 and then was replaced on January 10, 1918 with order 20387 for 2,000 airplanes. These airplanes were designated USAO-1 for observation, even though the original name used was "fighter."

The British Bristol re-engined with a Liberty 8, assigned McCook Field number P-37, known as the USB-2, and flight tested between August and November of 1918. It was initially powered by Packard Exp. No. 5 and later re-engined with Packard Exp. No. 3. (Courtesy of Peter Bowers)

The first production USAO-1 was delivered in March and a second in April. Thirteen were delivered in May and in that month an airplane was lost in a fatal crash during testing. June saw nine delivered and again a plane was lost in a fatal crash. Both of these crashes were caused by loss of fabric from wing panels, a problem which had been encountered in other Liberty-powered airplanes and corrected. In July a third fatal crash occurred, this time credited to pilot error. The earlier crashes had already set minds that the Liberty 12 was too heavy (actually it was only slightly heavier than the Rolls-Royce the British used in it) and too powerful for the plane and the contract was canceled on July 20, 1918 with a total of 27 airplanes delivered.

Mid-1918 brought some major changes in the aircraft program. The USAO-1 was already seen as in trouble in June and the USB-1 flight tests had been very successful.

That the Liberty 12-powered Bristol Fighter had been abandoned is evident from the following quote from the June 22-29 weekly report of the Engineering Division. "Decision has been made to return to the original program for the Bristol Fighter, therefore the work of completing drawings for the English type Bristol have been resumed and design work is being done for planes with the Hispano 300 and Liberty 8 motors."

As a result, a second sample Bristol Fighter was shipped from England, arriving in early July of 1918, and was re-engined with a Liberty 8 and assigned McCook Field number P-37. This plane was designated USB-2. The USB-1 with the Hispano-Suiza 300 had performed well. Now McCook Field wanted to see if the same plane with a Liberty 8 of about the same power would perform as well and give them two engine options.

I think quoting directly from the second BAP history regarding the expectations of these two airplanes would be quite informative at this point.

History of USB-1 and USB-2 (Corps & Army Observation)

"Owing to the disappointing outcome of the experimentation on the Bristol Fighter adapted to the Liberty-12 engine, it was decided that the Bristol should be put into production as nearly as possible like the original model received from England, but using the Hispano 300 HP motor, or the Liberty-8 cylinder type; the difference in the fuselage for the two being in the front end, to accommodate the different motors.

"The work along these lines was carried out by the Airplane Engineering Department, and it was expected that one of each type would be released for test by the middle of September, 1918.

"It was made clear the last of August by Major Carl B. Ely, of the Plane Production Department, that the official name of these planes was USB-1 (U.S. Battler #1) and USB-2 (U.S. Battler #2). He stated that it is particularly desired that the word 'Bristol' be not confused with 'Battler.'"

On September 5th, Colonel Vincent wrote:

"USB-1 - This machine is an exact copy of the English Bristol Fighter, with the exception that the McCook Field organization has redesigned the front end to make a good installation job of the 300 HP Hispano-Suiza engine.

"The preliminary machine of this type was obtained by taking an English built machine and installing a 300 Hispano engine by the "cut and try" method. This sample machine has been put through full flight tests by the Technical Section of D.M.A., and they have approved it for preliminary order for 1000 machines.

"USB-2 - A sample machine has been obtained by taking an English built machine and installing a Liberty-8 engine by the "cut and try" method. This machine has made some preliminary flights at McCook Field and is just ready to deliver to the Technical Section of D.M.A., for full flight test. It will be delivered the first good flying day. Very few additional drawings will be necessary from the USB-1 aside from engine mounting, cowling, radiator and gun location."

It was contemplated using as many of the fabricated Bristol parts as possible in connection with the manufacture of the USB-1 plane in order that these parts might not be a total loss. By the last of October Curtiss had started work on three USB-1s, the number of experimental planes to be built by Curtiss as placed by the Airplane Engineering Department.

The day of the signing of the Armistice a formal order was placed with Curtiss Airplane & Motor Corporation for 2,000 USB-1 planes, Order No. 720904. On November 16th this was cancelled, however, with nothing toward production started.

Pomilio FVL-8 Fighter No. SC 40081 at McCook Field. This was the second of six delivered (SC 40080-85) and was assigned P-69. It was powered by Buick-built L-8 MNFG. NO. BX 14 U.S. NO. 94785. (Courtesy of Peter Bowers)

In the case of the U.S. Battler models, USB-1 and USB-2, both were in serious consideration as production airplanes as early as June. The USB-1, P-30, had already proved that the Bristol Fighter with Hispano 300 power was a good performer. The airplane with L-8 power had been the original intention as seen by the September 4, 1917 recommendation of the APB. The USB-2 was expected to prove itself soon.

In addition to the two British Bristols which became known as USB-1, P-30 and USB-2, P-37, the Engineering Division ordered built in-house two additional of each model in mid-June. By the first week in July they had increased the order to four of each. It was on July 20 that Order No. 20387 for the 2,000 USAO-1 Bristol Fighters was canceled after only 27 deliveries.

In early September two additional models with each engine was added to the order. These were to have improved fuselage designs using 3-ply plywood construction and were known as the USXB-1 and USXB-2. None of these 12 prototypes were completed before the Armistice halted all work.

In addition to the two USB models, there was serious work being done to put single-seat fighters into production as soon as possible. To this end a number were designed between early summer of 1918 and the Armistice, several of which were ordered as prototypes for testing. Some of these were halted before any actual airplanes were completed and some had one or more prototypes built before the end of the war and in some cases limited production was eventually ordered following the war.

Proposed Fighters built around either the Hispano 300 or the Liberty 8

Model	Engine	No. built
ORENCO-D	Hispano 300	4 plus 50 prod.
T-M MB-3	Hispano 300	4 plus 50 prod.
VCP-1	Hispano 300	1
USP-1	Hispano 300	0
Pomilio FVL-8	Liberty 8	6
Victor D-8	Liberty 8	0
USP-2	Liberty 8	0

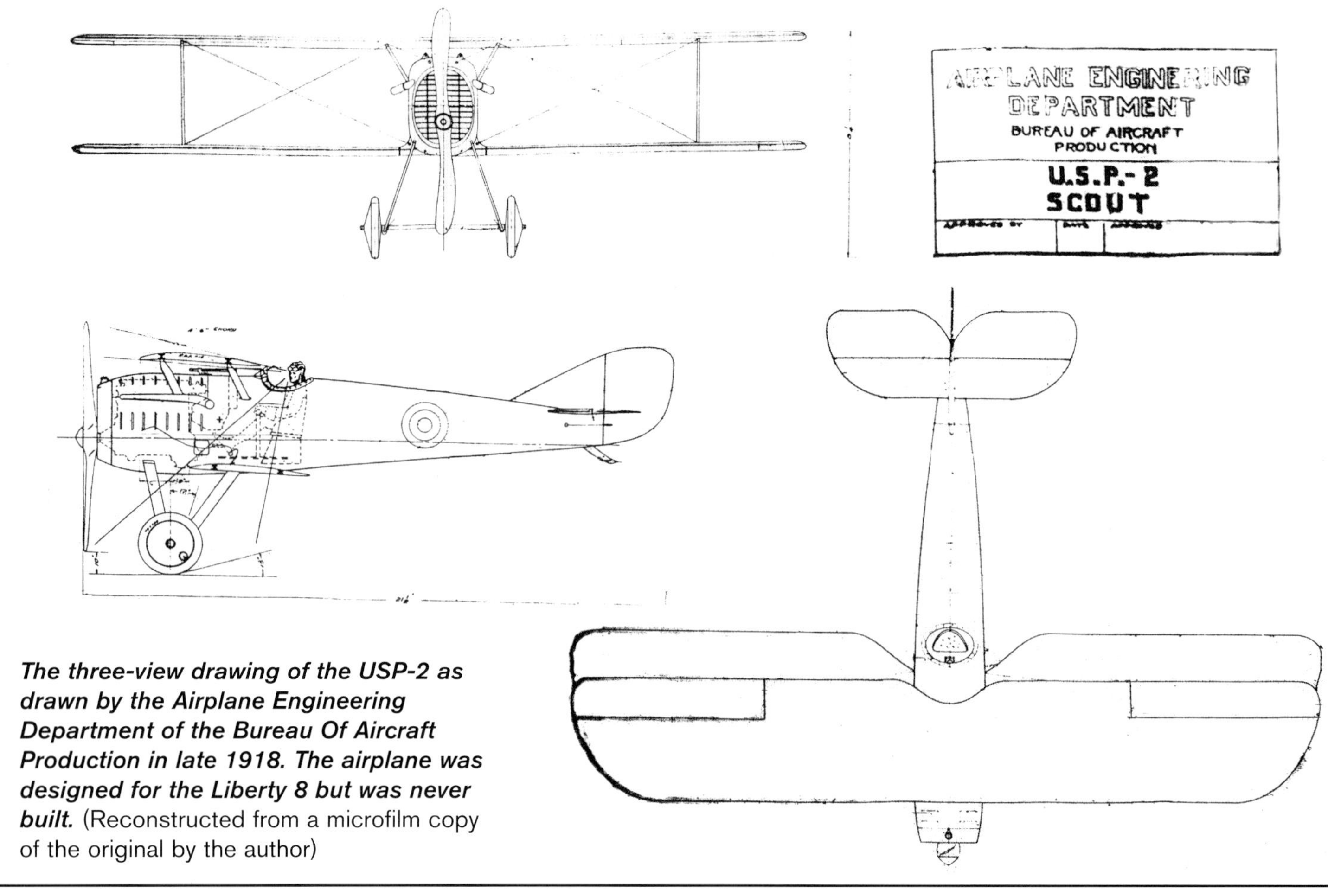

The three-view drawing of the USP-2 as drawn by the Airplane Engineering Department of the Bureau Of Aircraft Production in late 1918. The airplane was designed for the Liberty 8 but was never built. (Reconstructed from a microfilm copy of the original by the author)

Four prototypes of the Victor D-8 were on order but were canceled at Armistice. The USP (U.S. Pursuit) airplanes were Engineering Division designs of a single-seat pursuit adapted to either the Hispano 300 or the Liberty 8. They had not advanced far enough to be ordered built as test models but, had the war continued, they probably would have been.

Now we can see that it was felt there was good reason by June of 1918 to take steps to get both the Hispano 300 and the Liberty 8 into quantity production.

In the June 17, 1918 entry of Col. Vincent's diary is this note: "Held a conference this morning with Mr. Peters, Mr. Moorhouse and Mr. McCain about the standardizing of the drawings for the Liberty "8" which is going to be put into production." This entry confirms, of course, the entries seen in the weekly reports and also allows a closer estimate as to just when the decision was made to put the L-8 into production. With the above statement recorded on June 17, we can safely assume the decision was made only a few days before; let us say June 15 just to fix an estimated date.

Almost a year after the Bristol design was first considered for production, engine availability could be considered in an entirely different light. The Hispano 300 had been ordered into production at Wright-Martin back on November 20, 1917. After a long struggle it could be seen that they would soon have the engine in quantity production. The Liberty 8 had never been ordered up to this time, but the Liberty 12 was in quantity production by several manufacturers and thus any production problems associated with the Liberty could be considered pretty much resolved. The Liberty 8, once ordered, could be expected to be in quantity production fairly rapidly.

STANDARDIZING THE L-8 DESIGN AND ORDERING ENGINEERING DIVISION PROTOTYPES IN PREPARATION FOR PRODUCTION

On April 18, 1918 Lt. Col. Vincent wrote to Mr. W. C. Potter, Chief of the Equipment Division, updating him on status at McCook Field and noted, among other things, that 5 sample Liberty 8 engines could be built on short notice. He also noted that the Hispano 300 was the better choice, however. One must certainly find this an interesting and honest note, coming from the co-designer of the Liberty engine.

Shortly thereafter, final production design detailing of the Liberty 8 was initiated and by the week of June 1-8 the McCook Field weekly report noted the job was about half-done. In the June 22-29 report it was noted that the Liberty 8 would be completely standardized in a few days. In that same week four L-8 engines were ordered built in-house (Production order 737) and the Engineering Department promised completion by August 15.

JUNE 29, 1918 - ENGINEERING DIVISION PROTOTYPE L-8s ORDERED

US numbers 47854-55-56-57 were assigned to these first four engines. It would seem that things were still in a state of flux as the week of June 29 - July 6 report noted the L-8 design was nearly finished during that week and was sent to Detroit to be completed by the Detroit District Office. In that same report it was noted that the order had been increased to six engines and because of drawings not yet received, completion would be delayed to September 1. The additional two engines would extend the US numbers to include 47858-59.

ENGINEERING ORGANIZATIONAL CHANGES OF JUNE-SEPTEMBER 1918

Although I will not go into them in any great length. a number of details regarding the division of engineering design responsibility between the BAP and the DMA were agreed upon and laid out in detail during this period. They had some effect upon activities regarding the Liberty 8 and the Bristol Fighter P-37 in which it was tested. Being aware of some of these changes will make it easier to understand some of the occurrences I will be mentioning shortly.

On April 27, 1918 the Air Service Division of the Signal Corps was succeeded by the Aviation Division and Brig. General William L. Kenly became its commander. A few days later, on May 4, 1918, John D. Ryan was named Director of Aircraft Production and head of the Bureau of Aircraft Production which had succeeded the Equipment Division. On May 30, 1918 the Air Service was separated from the Signal Corps and the Air Service Division became the Division of Military Aeronautics. Very shortly after that, W. C. Potter, who had been assistant to Ryan, became head of the BAP and Ryan rose to second assistant Secretary of War and Director of the Air Service, thus heading both the BAP and DMA.

On about July 1, 1918 O. E. Hunt, formally Chief Engineer at Packard but now in the employ of the BAP was appointed Associate District Manager of the Detroit District office of the BAP and in full charge of Liberty Engineering. One of his chief assistants was Charles H. Vincent, the brother of Col. Vincent.

It was at about this time that the final work of design detailing of the Liberty 8 was moved to the

Charles H. Vincent's BAP ID card which allowed him entrance to all production and testing areas. (Courtesy of Roberta Vincent Mocabee)

Detroit District Office. Even so, much Liberty-related work continued to be done at McCook Field under the control of the Airplane Engineering Division and its Experimental Engineering section.

On September 15, 1918 the Detroit District office of the BAP set up its own Experimental Engineering Department to take over the direction of testing and interpretation of results relative to production Liberty engines. This department took over the recently completed Saxon automobile plant in Detroit to use for this purpose and it was shortly designated the Liberty Engine Experimental Plant. C. H. Vincent was appointed by O. E. Hunt to be "in charge of experimental engineering for the Engineering Department of the Bureau of Aircraft Production, Detroit District." In this capacity C. H. Vincent headed operations at the Liberty Engineering Experimental Plant.

Also during this period of time the Technical Section of the DMA was set up with headquarters in Dayton, Ohio. Its test location was Wilbur Wright Field, which was near to McCook Field. At that location it had the needed facilities to test aircraft and their engines with the purpose of requesting new designs or changes in designs, testing new designs and approving them for service use. The Airplane Engineering Division of the BAP having the actual design responsibility and the Production Division the production responsibility.

Now I will return to the history of the production of the Engineering Division sample L-8 engines and, shortly the full-production L-8 engines ordered from Buick and others.

Work progressed slowly on the in-house L-8 engines with the forecast completion date continuing to slip as time went by.

THE FIRST PRODUCTION LIBERTY 8 ORDER FROM BUICK FOR 5,000 ENGINES

On August 5, 1918 the first Liberty 8 production contract was signed with Buick. This was order number 730155 for 5,000 engines. At about this same date Cadillac ceased assembling L-12 engines and became a parts producer only. Buick was thereafter the only GM factory to build and test complete engines. They had sufficient facilities to take on the Liberty 8 in addition to the 12 and did so very rapidly.

In preparation for anticipated large production of USB-1 Hispano 300-powered "Battler" airplanes, contract 30011-A-1 was issued to Wright-Martin for 2,000 Hispano-Suiza 300 HP model H engines on August 12, 1918. (2,000 USB-1 airplanes were ordered from Curtiss, order 720904, on November 11, 1918.) To cover additional airplanes needing this engine, order 730083 was issued to Pierce-Arrow for 5,000 Hispano-Suiza 300 engines on July 22, 1918.

THE SECOND PRODUCTION LIBERTY 8 ORDER FROM WILLYS-OVERLAND FOR 3,000 ENGINES

On August 20, just two weeks after the Buick order, Willys-Overland was given an order for 3,000 Liberty 8 engines, order 730179. Willys was an early producer of engines for the Air Service, having received their first contract for 5,000 OX-5 engines in September of 1917. This order had been completed by mid-July. A second order for 2,000 had been added to it in February of 1918 and it would be completed by the end of October. Willys had a production capacity for the OX-5 engine in excess of 1,000 per month. The need for further production of the OX-5 was rapidly diminishing and Willys' proven ability to produce aero engines was going to be channeled to Liberty production. On October 15 an order for 5,000 Liberty 12 engines would be added.

THIRD PRODUCTION LIBERTY 8 ORDER FROM WINTON MOTOR CAR CO. FOR 1,000 ENGINES

As noted back on page 113, Winton seems to have been chosen as a builder just to try and insure adequate production facilities from late 1919 on. Order 730223 was written up for 1,000 engines and appears in several official listings, but only for a short time and then disappears. Evidently there was a change in official thinking regarding Winton as a producer and the order was never actually executed. All intentions in this direction were officially canceled after the Armistice.

By this time not a single Engineering Division L-8 had been delivered and the expected date of delivery of the first engine had slipped to September 15. Standardized drawings had, of course, been completed and were used for building the Engineering Division in-house engines and were delivered as well to the three contract builders.

As time slipped by, predicted delivery date of the first Engineering Division L-8 moved to September 25, then to October 5, then October 12 and was finally delivered on October 21. Initially this first engine was to have been delivered by August 15.

ENGINEERING DIVISION L-8 ORDER INCREASED TO A TOTAL OF EIGHT

In the midst of this, Production Order 737 for these engines was again increased by two on about October 1 bringing the order to eight engines. These last two engines were assigned U.S. numbers 47865-47866, the intervening numbers following the first six having already been assigned to other experimental engines.

THE POMILIO FVL-8 IS ORDERED

The Pomilio brothers had submitted plans for both the L-8 and L-12, the FVL-8 and the BVL-12. Both were accepted and order CS-216 was issued on September 21, 1918 for the construction of six of each model.

Two views of Engineering Division L-8 Exp. NO. 7 U.S. NO. 47854. This was the first of a series of eight engines built to be production prototypes. Note the lack of a data plate. Hand stamped in its location was "EXP NO. 7" "U.S. NO. 47854". (Courtesy of the Smithsonian)

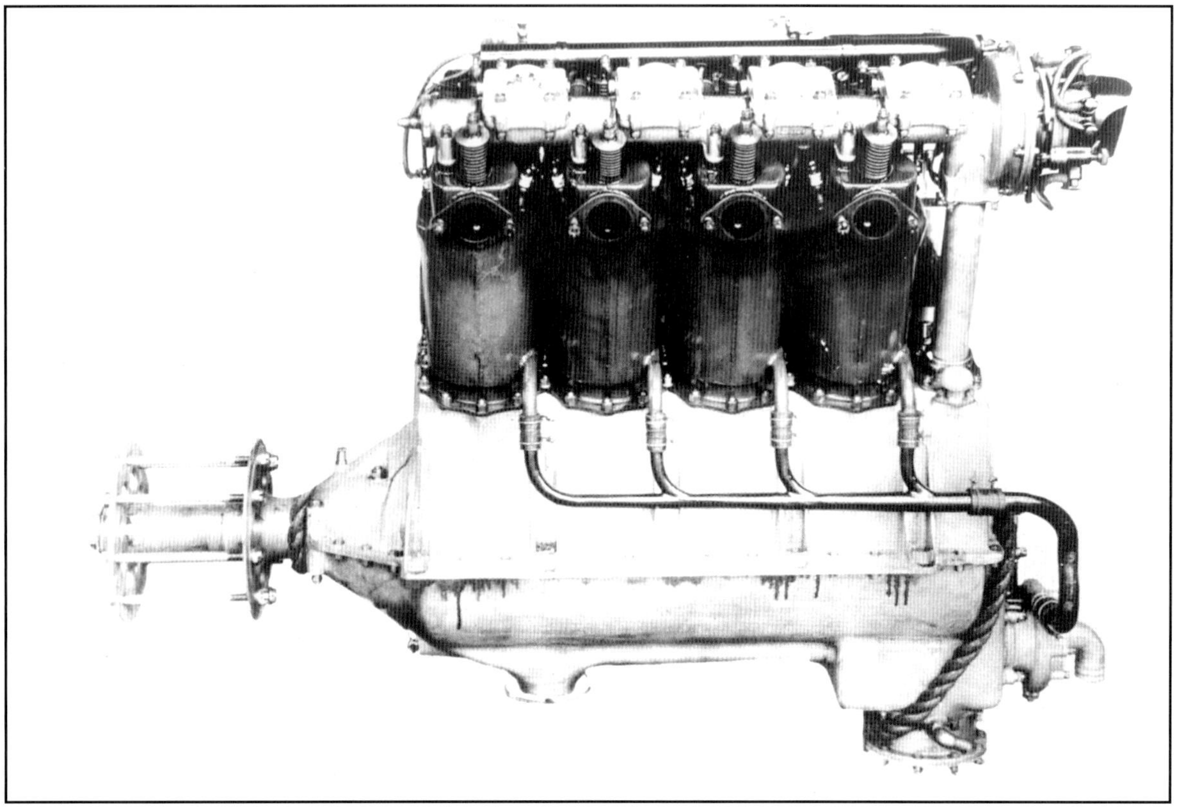

Two views of Buick L-8 BX 5. This is an early photograph with the U.S. number stamped -E-. Later it would be restamped 94476 when the new series was assigned. (Courtesy of the Smithsonian)

THE VICTOR D-8 IS ORDERED

Mr. Heinrich had submitted a single-seat pursuit design named the D-8 to be built by the Victor Aircraft Corp. It also was accepted and order CS-244 was issued on October 16 for four examples.

As I just noted, the first Engineering Division in-house standardized L-8 was delivered on October 21.

THE FIRST BUICK PRODUCTION L-8 IS DELIVERED

During the first week of November the first production Buick L-8 was delivered to C. H. Vincent at the Saxon Liberty Experimental Plant in Detroit. By November 14 Buick had completed a total of 15 engines and production was halted. **The reason: November 11, 1918 was the signing of the Armistice.**

The second and third Engineering Division L-8 engines were delivered during the week of December 14-21 and the remaining five by Feburary 1, 1919.

NOVEMBER 11, 1918 THE ARMISTICE

On November 13 the Willys-Overland Liberty 8 contract is canceled with no engines completed. Any plans laid to issue the proposed contract to Winton were also canceled. On November 14 Buick's L-8 contract was canceled with 15 engines completed. The Pomilio contract was allowed to continue. The Victor D-8 aircraft had not advanced notably and that contract was canceled November 13. Also on November 13 Wright-Martin's Hispano 300 order for 3,000 was reduced to 500, the remaining 2,500 being canceled. The Pierce-Arrow order for 5,000 of that model was also canceled. Order 720904 with Curtiss for 2,000 USB-1 Hispano-powered planes was canceled on November 16.

The Engineering Division in-house engines were produced by having most parts manufactured by outside vendors, the majority of which were already building Liberty parts for the various current producers. They then did any final milling or finish work required and final-assembled and tested the engines. Weekly reports tell of many instances of awaiting parts from various suppliers. The initial order was placed about June 25. The first engine was completed on October 21, almost four months later. It took another nine weeks to complete the remaining seven engines.

Buick received its order on August 5 and delivered the first engine about November 5, three months later. By ten days later, on November 14, they had completed 15 engines. Buick was currently producing the Liberty 12 at the rate of about 100 per week and would probably have reached a similar rate of Liberty 8 production within 90 days.

We have followed the history of when and why the Liberty 8 was put into production and can certainly see that the Armistice was the reason why production was stopped. But was there any other reason? The engine proved to have excessive inherent vibration, but did that have any bearing on its cancelation? Would it have eventually effected production considerations? Now lets look at when and how this vibration problem surfaced and try to come to some legitimate conclusions.

LIBERTY 8 BRISTOL P-37 FLIGHT TESTING

The second British Bristol received at McCook Field in early July, and assigned the designation USB-2 P-37, was having L-8 Exp. NO. 5 (Packard-built) installed during the first week in July. Later reports noted that the installation should be completed by August 5 and the plane ready to fly. The first flight was made on August 6, 1918.

Flight testing continued at McCook Field throughout August and into September, whenever weather permitted. No information has been found to indicate any problems during this period. Col. Vincent's September 4 note regarding this early testing period would seem to indicate he expected full flight testing at Wilbur Wright Field by the Technical Section of the D.M.A. to be successful.

From its arrival at Wilbur Wright things began to go down hill. It was flown there with L-8 Exp. No. 5 as power. Just how much testing was done with that engine installed, we do not know. We do know that on October 5 C. H. Vincent, Experimental Engineer at the Detroit District Office of the BAP telegraphed Major Hallett at Wilbur Wright that he was shipping L-8 Exp. No. 3 by express to be used for the completion of the full flight tests of P-37.

The letter that followed on the same day described the engine's current status and revealed some facts which will be of interest later in our story. This engine was, of course, built very early in the design process of the Liberty and would follow much the same change evolution we have seen in the Liberty 12 as it progressed through production and testing.

First its latest power curve was noted:

RPM	BHP
1410	256
1510	274.2
1600	289.4
1700	300.2
1820	311.6
1920	320.7

Gas consumption at 1634 RPM and 291.3 BHP was: .503 lb. per HPH

The outputs are interesting when compared to a power graph of the same engine reproduced on page 54 as plotted about a year earlier.

He noted that except for minor details, the intake and carburetor systems were up-to-date. In addition it was noted that the rods, pistons, etc., were up-to-date but the crankshaft was the old light design which had been fitted with up-to-date force feed oiling. All original engines were, of course, fitted with scupper oiling as the twelves.

His final note regarding the engine would come back to haunt him a month or so later. It is quoted below.

"This motor has seen a tremendous lot of hard service, and it would not be surprising if the crankshaft or some other old part were to give way most any time, but we hope that you will be able to get a general idea of its performance before it develops any trouble of this kind."

He ended by requesting to be notified the day prior to first flight as he wished to be present during some of the flights.

From this point a series of correspondence went back and forth between the Detroit District Office and Wilbur Wright which does not seem to show the Engineering Department of the Detroit District Office in a very good light and turns into a battle of wills.

October 9: WW (Wilbur Wright) writes that the engine was received without any safety devices (safety wires or cotter pins) and had to be put in the shop and completely gone through and safetied. The start of testing was therefore held up.

October 22: DDO (Detroit District Office) writes asking how the installation is coming as he is anxious to get a performance report on the engine at the earliest possible moment. He also noted that although he had offered to send a mechanic to aid in the installation, he had been unable to do so and had asked Col. Vincent of McCook Field to give him assistance.

October 25: WW (Major George. E. A. Hallett) wrote that since the outside nuts were not safetied, the OIC of the test section had the lower crankcase removed to be sure the inside was properly safetied (it was) and then the motor was installed. At this point it was found the engine control connections were not the same as those on Exp. No. 5 and some new connections had to be made. He followed by noting that when the plane was pulled out on the line and started the carburetor flooded. It was removed and taken to the shop for repair. The weather was bad that day but it was intended to test again on the first good weather day.

October 26: DDO writes that he was surprised the pilots refused to fly an unsafetied engine and that they seldom (exterior) safetied 50-hr test engines which were subjected to much harsher use and never had any problems and that they were considering cutting out many of the cotter pinning operations. He further noted his surprise at the carburetor problem as the engine was running perfectly when it left.

October 31: WW Motor Repair Division to Major Hallett. Reported that the float was reversed in the carburetor and that was the problem.

November 1: DDO writes that he received a wire from WW later yesterday after noon (October 31) that when started the engine vibrated so badly that the pilots refused to fly the plane until DDO sent someone to investigate. He said he had not been able to get a mechanic on a train that night but had requested Col. Vincent to send someone from McCook Field (only a few miles away) in the mean while.

He then went on to note that the 8 would, of course, produce much more vibration than the 12 and that the production engine with heavier crankshaft might be smoother.

He followed by suggesting possible causes of rough running such as missing and what to look for, etc. and noting the qualifications of the man being sent, Mr. George Buzaine.

Although the series of correspondence during this period does not mention it, it appears from a test report copy that test No. 115, a preliminary climb test, was carried out by Major Schroeder on November 2 from 10 to 12 AM checking climb from 1,000 ft. to 3,000, 4,000 and 5,000 ft.

After a couple of letters back and forth commenting on the various points brought up by the two locations, Major B. Q. Jones, CO of Wilbur Wright Field, wrote:

November 8: WW writes that Mr. Buzaine arrived, inspected the engine and declared nothing wrong. Up to that time no WW mechanic had been allowed to make any changes as it was understood the engine had been specially set up by DDO (Vincent) on the dynamometer and nothing should be altered. After that their mechanic looked it over carefully and found the backlash on cam drive gear of the left cylinder bank was very excessive and caused timing on that bank to be well out. After correcting the engine ran much smoother and flight testing was resumed.

In most of these letters back and forth there are numerous comments on each side's comments. When one suggests that perhaps operation under certain conditions result in this or that kind of operation, the other responds politely that they are "well aware" of those facts, but—

November 12: DDO writes responding to WW letter of the 8th and in it mentions "We are also glad to advise that our first production engine is now under test, and that it is showing a very satisfactory performance." This tells us that Buick No. 1 L-8 must have been delivered to the DDO between November 4 and 12, 1918.

November 15: Major R. W. Schroeder, OIC Testing Section at WW Field wrote Col. Jones a letter regarding the Bristol Liberty with five recommendations: "1- Owing to the very poor design of the engine bed, it is recommended that all tests on this machine, with this style of bed, be discontinued. 2- The Liberty motor which is now in this machine, is in very bad shape, as the attached letter will verify. 3- We have six propellers to be tested out on this machine, and I am sure that the motor will never hold up to a good enough standard, to make a good comparative test. 4- Captain Boeree has information that there is an improved engine bed being used in England, and also says there is a set of drawings at McCook Field showing same. 5- Major Hallett, Captain Boeree and Captain White also agree that this is a very dangerous bed."

As a result of this recommendation from Major Schroeder in addition to all the prior problems related to testing P-37, Col. Jones wrote a letter noting the following:

November 15: WW Lt. Col B. Q. Jones - (promoted since November 8) to DMA Technical Section at Dayton. This was a memo noting in detail the work done and changes made in the P-37 Bristol since the new motor (Exp. NO. 3) was installed. The performance, or lack of it, was beginning to become a serious headache to Col. Jones and Maj. Hallett and they were going to officially document it to the right people along with their recommendations. Every change was noted, down to repacking the water pump and any alterations to the plane, such as water pipe location changes, etc.

Among the things mentioned were all we have seen in the prior letters plus many not there. Additional running and flight details were include, however, and will be noted. His comments were dated as to action and are so noted below in abbreviated form.

October 9: Motor installed in plane. It was then returned to the shop and safetied.

October 25: Motor run and had a carburetor float problem.

October 26: Machine tried out on the ground. Ran well at high speed but would not idle without missing and had considerable vibration between 1,200 and 1,300 RPM.

November 2: Motor running fairly satisfactorily. (He did not mention any flight time but test report 115 indicated 2 hours flown in climb tests.)

November 4: Plane flown for 46 minutes and engine run 10 minutes on the ground. Vibrated badly between 1,200 and 1,300 RPM. Smoothed out fairly well at 1,500 RPM. Maximum RPM 1,575.

November 6: Motor run on ground and vibrated badly. Too much backlash in left camshaft drive gear.

Total time on motor three hours and thirty minutes. (This seems to include the two hours on November 2 not mentioned in the letter.)

November 14: "Major Hallett and Captain White each went up in the plane to observe the engine vibration. They believe that the running vibration of the engine is very severe and that the metal fittings in the front-end of the fuselage cannot last very long under such vibration.

"It is believed that the vibration of the Liberty-8 motor is a serious argument against its use. It has been observed that the 300 HP Hispano motor runs very satisfactorily, even on this same mounting, in the Bristol Fighter. It is believed that the 45-degree Liberty-8 will prove impractical, due to its high vibration.

"THE ENGINE MOUNTING: It is the opinion of this office that the engine mounting in the Bristol Fighter is entirely impractical, for either the Liberty-8 or the Hispano 300, because it is not rigid enough. When the Hispano 300 is run on seven cylinders, simulating one missing cylinder, the top of the radiator cap swings side-ways as much as two inches; obviously there is a great danger of crystallizing and breaking the engine mounting under these conditions. When the Liberty-8 is allowed to become a little too cool it "kicks" the engine mounting exactly the same, even though all eight cylinders are firing.

"In a letter from Mr. C. H. Vincent, in the Detroit District Office, Bureau of Aircraft Production, dated October 5th, he states, in paragraph 6: 'this motor has seen a tremendous lot of hard service, and it would not be surprising if the crankshaft or some other old part were to give way most any time, but we hope you will be able to get a general idea of its performance before it develops any trouble of this kind.' The Test Pilots at this Field do not care to risk having a crankshaft break and seize, in an engine mounting so weak as that on the Bristol Fighter, it is feared that should the propeller stop very abruptly, as often occurs when a crankshaft breaks, that the entire engine would be twisted off the nose of the machine.

"RECOMMENDATIONS: It is recommended that test flights of the Bristol Liberty eight be discontinued until a better Liberty-8 can be furnished, which will not be in danger of breaking a crankshaft. It is further recommended that the Liberty-8 not be used in the Bristol machine, because it would necessarily be a short lived machine if subjected to such severe vibration."

The last letter preserved in this Wilbur Wright file on the P-37 while at its location contains a November 20, 1918 memorandum to the Testing Section over the signature of the Technical Executive, 1st Lieutenant William G. Marvin. "1- In accordance with authority from the Technical Section, no further flight tests will be made in the Bristol Fighter with Liberty-8 engine. 2- Lieutenant Colonel Jones desires to fly this ship within the next couple of days and immediately after his flight it will be dismantled. By order of Lieut. Col. Jones."

SOME CONCLUSIONS

It is evident from the history of the P-37 flight tests that it was not until November 15 that the vibration problem was considered serious enough to halt testing on P-37 until a better engine bed was available. By that time the Armistice had been signed (November 11) and all L-8 contracts had been canceled (between November 13 and 14).

No reference has so far been found indicating concerns about excessive vibration of the Liberty 8 prior to the testing of P-37 with L-8 Exp. NO. 3 by the DMA. Of the six L-8s constructed up to that time, five had been tested extensively (NO. 1 was not run.) NO. 3 had powered the first flight in L.W.F. -F. L-8 NO. 5 had flown P-37 far more hours at McCook than it was flown with NO. 3 at Wilbur Wright Field. As related on page 53, the L.W.F. was flown more than 10 hours between August 29 and September 2, 1917 with L-8 NO. 3 and no doubt a number of additional later flights.

There is no doubt the engine actually had a high amount of inherent vibration. The question is why that fact had not been noted as a potential problem before the testing of P-37. The answer probably lies in the fact that all prior flight testing had been carried out in a single

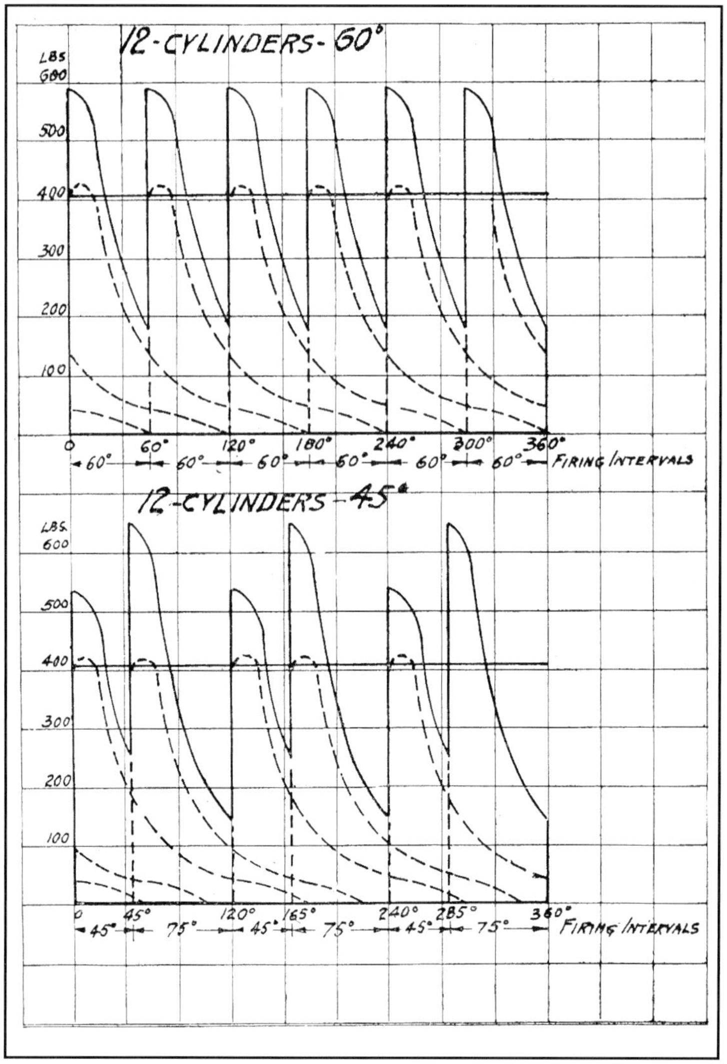

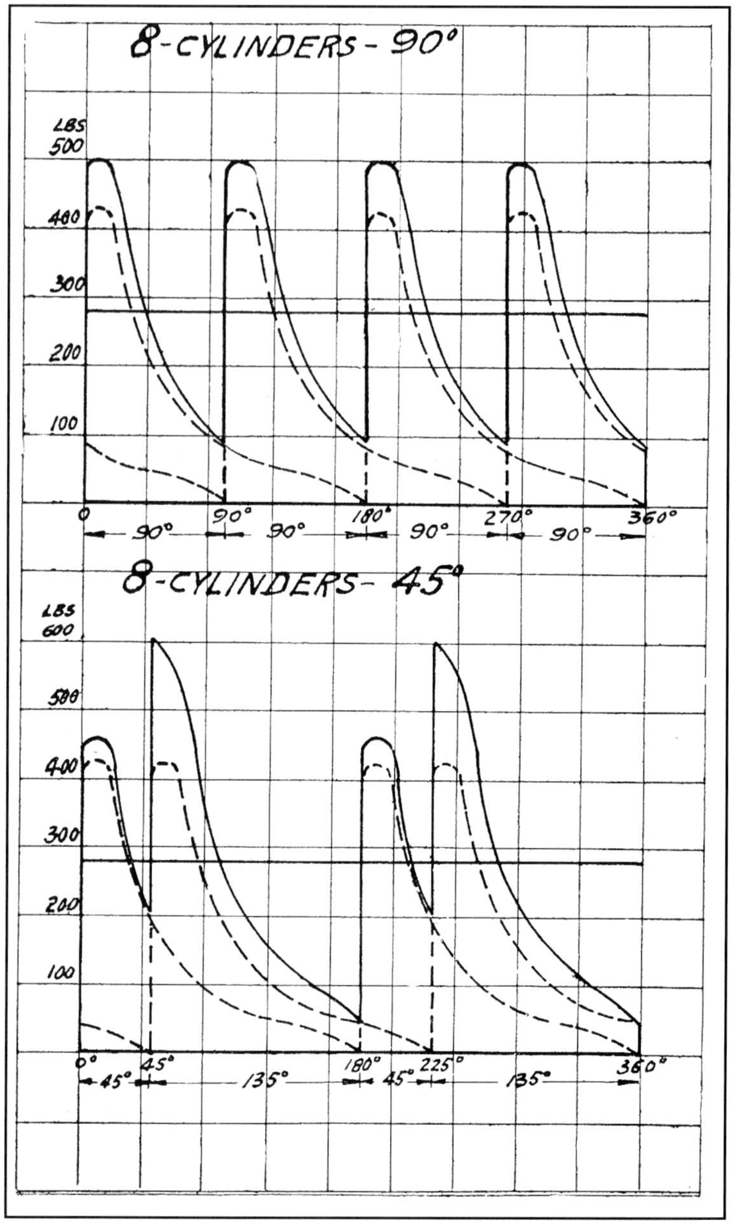

A power impulse chart provided by Charles B. King in the comments to Col. Vincent's 1919 SAE article on the Liberty. It points out clearly the irregularity of these impulses on a 45° V-type engine, the V-8 being much more affected than the V-12. (Courtesy of the SAE)

plane which no doubt had a much stronger engine bed than did P-37 and thus no one noted any unusual vibration. All other testing was done on the very solid engine bed used with dynamometer and torque stand testing.

Had the war continued the matter would certainly have gotten serious attention. How rapidly the Hispano 300 had become available would have had a big impact on later L-8 production decisions, as would a successful redesign of the Bristol engine bed.

Would production have eventually been halted if the war had continued? I think probably so; certainly short of current contract completion. The controlling factors would have been how fast the Hispano 300 became available in quantity and how successful they were in providing exceptionally strong engine beds in the airplanes the L-8 was to power. When the FVL-8 was flight-tested in 1920 with Buick's L-8 NO. BX 14, the pilot noted that the engine mounting gave trouble because of the inherent vibration of the Liberty 8 engine.

On April 18, 1919 Buick production engine NO. BX 5 was subjected to tests at the Power Plant Section of the Engineering Division in order to produce a "Standard Engine Report," the purpose of which was to both evaluate the engine and document its statistics and performance. The test was carried out by Major George E. A. Hallett, now chief of the power plant section, the same Major Hallett who had experienced so many problems with L-8 Exp. NO. 3 at Wilbur Wright Field just five months before.

The engine performed well and had no part failures. Power output was good, as was oil and gasoline consumption. Hallett's comments in conclusion were quite complimentary. He considered its general design as very good in that it was a neat and compact design and well adapted to mass production at minimal costs. He particularly commented on the improved intake header over that of the original model. This header was designed at the Detroit District Experimental Plant by C. H. Vincent.

He noted its power-to-weight ratio was very good for an engine in its class and its compactness, narrow width because of the 45° included cylinder bank angle and short length adapted it well to airplane installation.

On the negative side were various comments regarding shortcomings relative to accessibility of the carburetor, oil strainer and some other parts with the engine mounted in an airplane.

After a generally good report in all other regards, he concluded with a one sentence summary: "The Liberty eight is considered unsatisfactory for actual service due to its excessive vibration."

This inherent vibration was caused by the combination of 8 cylinders and a V angle of 45°. The chart reproduced on the prior page shows diagrammatically the firing impulses of both 45° and 90° eight cylinder engines. It is easy to see that this firing irregularity would result in vibration. This diagram, along with the one showing the power impulse pattern of a 60° versus 45° twelve, was provided by Charles B. King along with comments reproduced in the discussion following Col. Vincent's SAE article on the Liberty published in the 1919 *Transactions Volume XIV Part 1.*

Had the engine been laid out as a 90° engine it would have been quite a good engine. I believe that the Liberty family of engines certainly represented the state of the "proven" art for the day. They were mechanically sound designs, any model of which would have given good service. I believe the only major mistake made in laying out the original design criterion was the decision that low head resistance was a concern so serious as to let it control the included angle of the V-type models. This was the reason for the selection of a 45° angle. It caused some amount of vibration in the twelve but because of the large number of cylinders involved, it was never a serious problem. In the eight it was. If the twelve had been a 60° engine it would have been a much smoother one. Had the eight been a 90° engine it would have been as good a performer as the twelve.

This was Col. Vincent's pet theory at this time. He many times expressed the concern that engines must be designed with low head resistance in order to give the aircraft they powered the best performance. Of the co-designers, it was Vincent who pressed for this included angle. We must therefore lay at his doorstep the responsibility for what was probably the only significant design blunder of the Liberty. Even so, it was only significant for the Liberty 8. The Liberty 12 suffered little adverse effects from the 45° angle.

THE LIBERTY 8 IN SUMMARY

There were a total of 29 Liberty 8 engines produced. The first six were built at Packard between July and September of 1917. No more L-8 engines were ordered built until late in 1918 when first eight were ordered built by the Engineering Division and just a few weeks later production orders were given to Buick, Willys-Overland and Winton. The Engineering Division eights were delivered between October 21, 1918 and December 31, 1918. The first Buick engine was delivered about November 5, 1918 and the last before cancelation, number 15, was completed on November 14 and officially accepted in February of 1919.

The original production order for the 5,000 Buick engines was number 730155. US numbers were to be assigned to these engines from the block 57209-66310, probably the first 5,000. With cancelation coming so early in production, it was decided to treat the engines as experimental and write a new contract to cover them. This was order 730810 issued on January 27, 1919. A new block of US numbers were assigned to these 15 engines, 94472-94486. Buick would normally assign MNFG numbers to their engines prefixed by the letter "B" followed by numbers beginning with 1 and progressing upward. In the case of the L-8 the letter "B" was followed by an "X" for experimental. Thus the Buick L-8 engines were numbered BX-1 through BX-15. This numbering decision was made after the engines had actually already been completed and thus early photos showing the data plate of some engines show them marked differently. Official final tests and acceptance stretched into February of 1919.

MNFG.	Number built	U.S. serials	Mfg. serials
Packard	6	EXP	1-6
Eng. Div.	8	47854-47859	EXP 7-12
		47865-47866	EXP 13-14
Buick	15	94472-94486	BX-1-15

THE GEARED LIBERTIES

The earliest official mention of geared Liberty engines that I have found are in the March 7, 1918 minutes of the APB board meeting. In a report given by General Cormack (British representative), Commander Briggs and Captain Wheatley after a recent investigation of the technical and production status of the Liberty was the following statement: "The power as developed was adequate for the time, but for the next year's requirements development of the geared motor should at that time be started at one of the manufacturing organizations, and preparations made for production."

It did not take long for steps to be taken toward putting this suggestion into action. On March 14, Col. Vincent had Maurice Olley, from Rolls-Royce, and Commander Briggs in his office for a conference about geared construction. As a result, Col. Vincent told Olley that he would like to build a sample Liberty with Rolls-Royce-type of gearing. At Vincent's request, Olley said he would send over a set of drawings of the Rolls-Royce design. The Rolls-Royce design was of the epicyclic type.

Col. Vincent then submitted a report on the geared Liberty to the board which was presented by Mr. Potter on the 18th. This presentation resulted in a resolution noting that since it was probable that some form of geared Liberty would be required in large numbers, particularly for night bombers, and that experience with that type has been largely in Great Britain and Italy, the Chief Signal Officer and Chief of the Equipment Division be requested to ask Col. Vincent, Commander Briggs (Great Britain) and Captain D'Annunzio (Italy) to sit as a committee to recommend the best type of geared Liberty motor to be constructed. It was further recommended that the Chief Signal Officer take steps to provide additional production facilities for manufacture of these geared engines without interfering with production of the present type.

Allison epicyclic gear set mounted in its gear case with propeller hub and ready to mount on the engine.
(Courtesy of Rolls-Royce / Allison)

On April 23 it was requested that Mr. H. M. Crane (Wright-Martin Aircraft Corp.) be substituted for Captain D'Annunzio on the committee.

On April 26 Col. Vincent reported to the board that he recommended equipping the Liberty with the Rolls-Royce type epicyclic gearing and that it would weigh about 900 lbs. so equipped. The next day he notified the board that he was sending blue prints of the new parts to be used.

Col. Vincent had initiated action on an epicyclic design immediately following the March 18th meeting. Assigned to the task were A. Moorehouse of the Engineering Division and W. R. Griswold, a Packard engineer who may have, by this time, also come on board as an Engineering Division engineer.

The two engineers submitted a report on the design to Vincent which, unfortunately, was not dated. It included drawings of the resulting design which were dated April 30, 1918. The result of that study was expanded somewhat and published in Vol. 1 No. 2 of the *Bulletin of the Airplane Engineering Department* in July of 1918 and the two principal drawings from it are reproduced at right.

The design was explained in detail and in so doing it was pointed out that the R-R gears were designed for less power (360 HP) delivered at greater speed (1,800 RPM) than that required of the Liberty, which was to be 410 HP @ 1,700 RPM. Although it would seem not to be a great difference, from a design standpoint, it was. Because of various stress factors involved and the required results, the Liberty gear design was significantly different from the Rolls-Royce as can be seen in the drawings. The reduction of both the Rolls-Royce and Liberty gears was 1.66 to 1.

THE LIBERTY 12B

The epicyclic Liberty 12 became known as the L -12B although in a majority of the official references it is referred to as the epicyclic-geared Liberty. The standard Liberty is the L-12A.

On April 29 the Engineering Division issued order No. CS-24 to Packard for 3 epicyclic-geared Liberty engines. CS-24 had been immediately followed by order CS-34 to the Allison Experimental Company of Indianapolis, Indiana for the 3 epicyclic gear sets required. On May 8 CS-24 was replaced by order No. CS-43 for 6 engines of the same type. The engines would be shipped to McCook as would the gear sets. The Engineering Division would then assemble the complete engines. US numbers 47833-47838 were assigned to these engines.

During the first week of May, board minutes noted that Willys-Overland was preparing figures on its ability to produce geared Liberties and Locomobile was also being considered as a potential builder. Neither company was ever contracted as a producer, however.

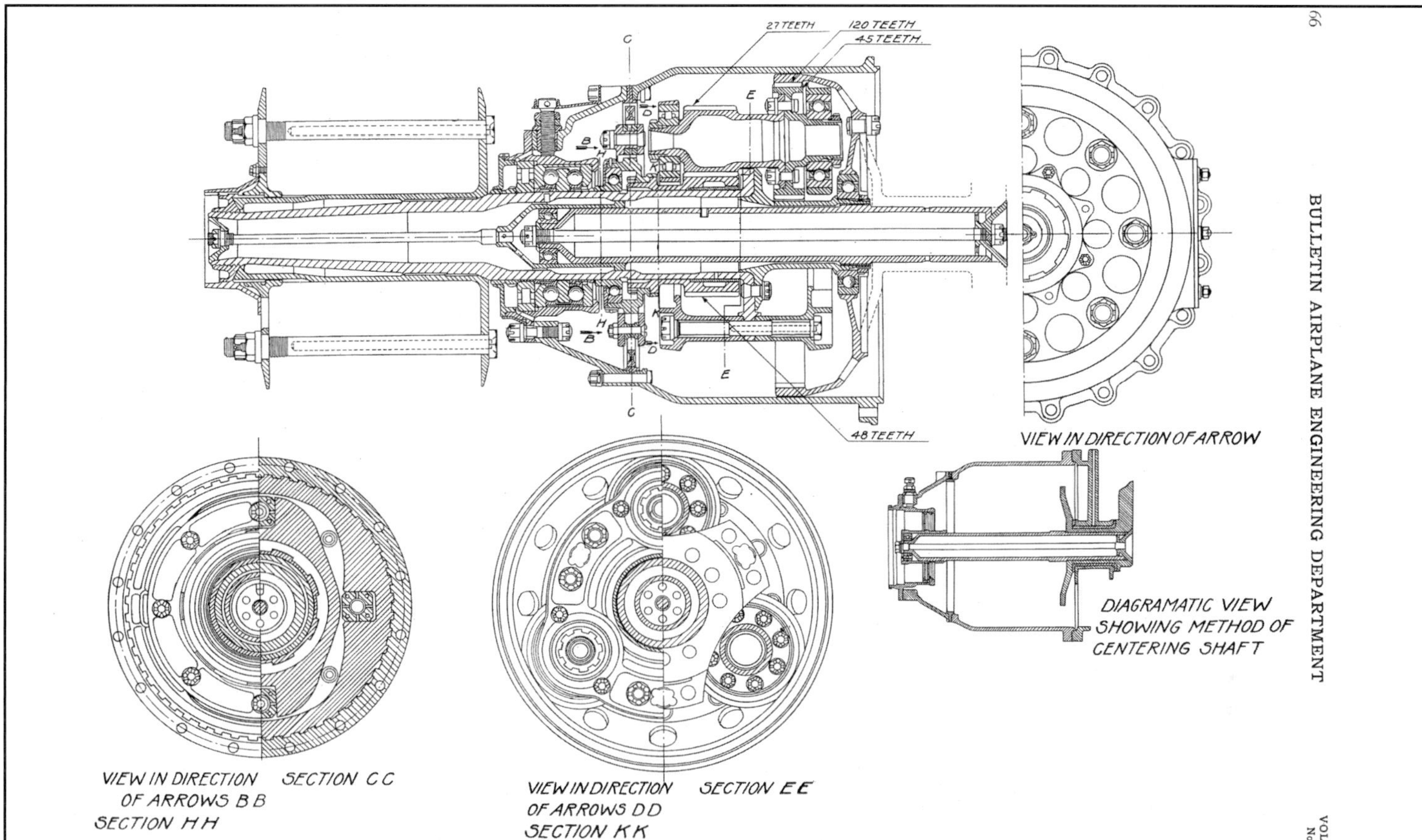

Fig. 15—Assembly drawing Rolls-Royce epicyclic reduction gear

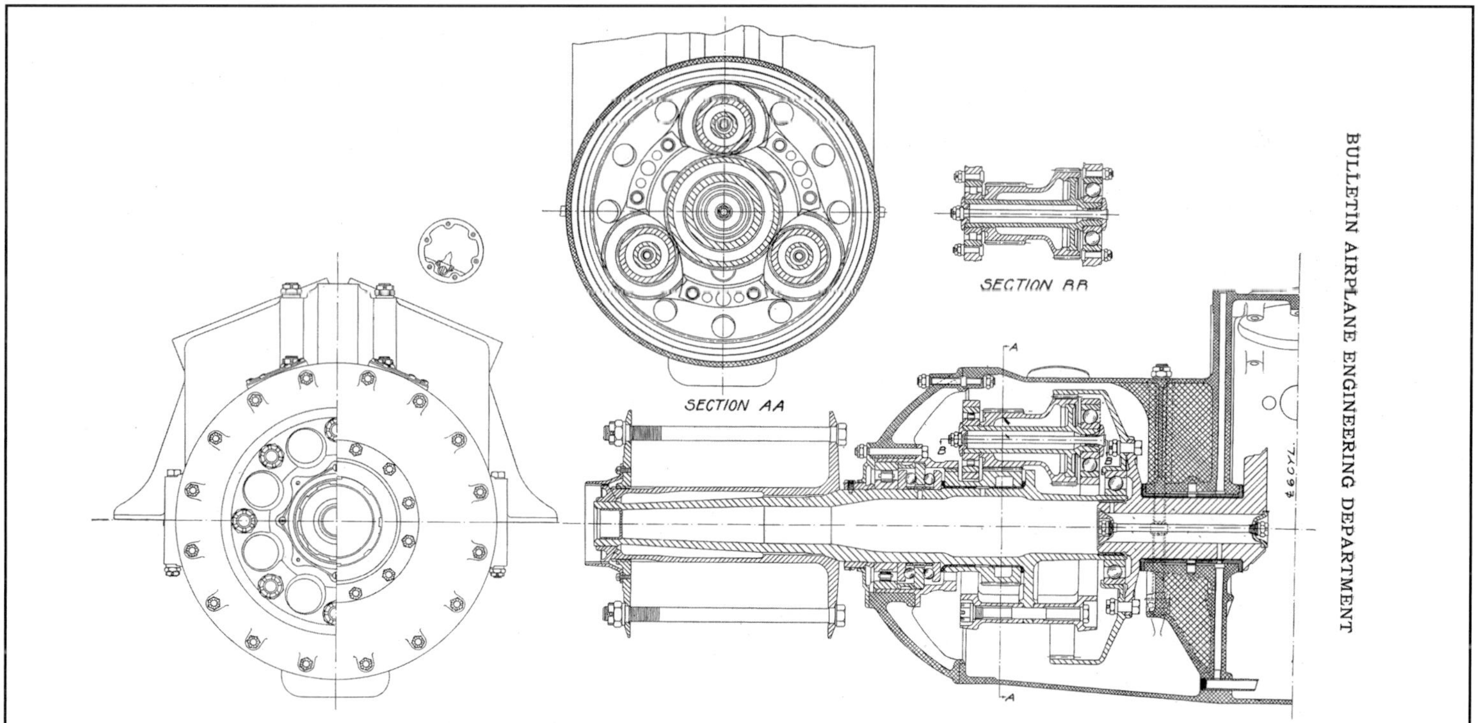

Fig. 16—Details of design of Liberty epicyclic reduction gear

On May 18, 1918 a conference was held in the New York Biltmore Hotel (room #640) of the complete committee set up to consider the subject of "Geared vs. Direct Drive Engines." Those present are listed below:

Commander Briggs	English Admiralty
Ensign Breese	U. S. Navy
Major B. D. Gray	Production Dept.
Mr. M. Olley	Rolls-Royce, Ltd.
Mr. Charles Day	Standard Aircraft Corp.
Mr. H. M. Crane	Wright-Martin Aircraft Corp.
Lt. Col. J. G. Vincent	Airplane Eng. Dept.
Mr. A. Moorhouse	Airplane Eng. Dept.

The meeting resulted in the recommendation that 100 of these engines be built as quickly as possible. This was presented to the board and eventually resulted in BAP order number 730187 (contract 4469) being placed with Packard for 100 engines on August 12, 1918.

The pending order was no secret as Mr. N. H. Gilman of Allison called Col. Vincent on June 28 relative to getting an order for 100 sets of epicyclic gears. Vincent told him he would write Mr. Kellog about the matter. Allison did get the order.

The 6 engines ordered by the Engineering Division would serve as prototypes for testing and finalizing of production design.

The July 13-20 weekly report notes for the first time that 3 epicyclic gear sets will also be received from the Willys-Morrow Company of Elmira, N. Y. There is no explanation as to why this order was first mentioned at this date since it was no doubt placed not very long after order CS-34 to Allison to provide the remaining 3 sets of gears required for the 6 engines ordered. Also for some unknown reason the order number of the Willys-Morrow gears is never given.

Progress of the six prototype engines is followed rather closely in the weekly reports of the Engineering Division with the first being received on July 17. Allison had promised the first of the gear sets by July 13 and then continually slipped the ship date until the first was not shipped until about August 13.

In the meanwhile, Willys-Morrow delivered its first gear set on July 31. It was installed on engine Exp. No. 1 and the next week power curves were begun on Exp. No. 1 engine. The Willys-Morrow gear set burned out a bearing after 20 minutes of running. A new bearing was installed and the engine was put on endurance tests.

One of the six prototype L-12B engines built by Packard between September and December 1918. The close up of its data plate at left shows it with no MNFG. NO. or U.S. NO. yet stamped. The data plate used was a standard L-12A plate with the "A" painted over and no model "B" designation replacing it. It is probably MNFG. NO. 1 U.S. 47833. (Author's collection)

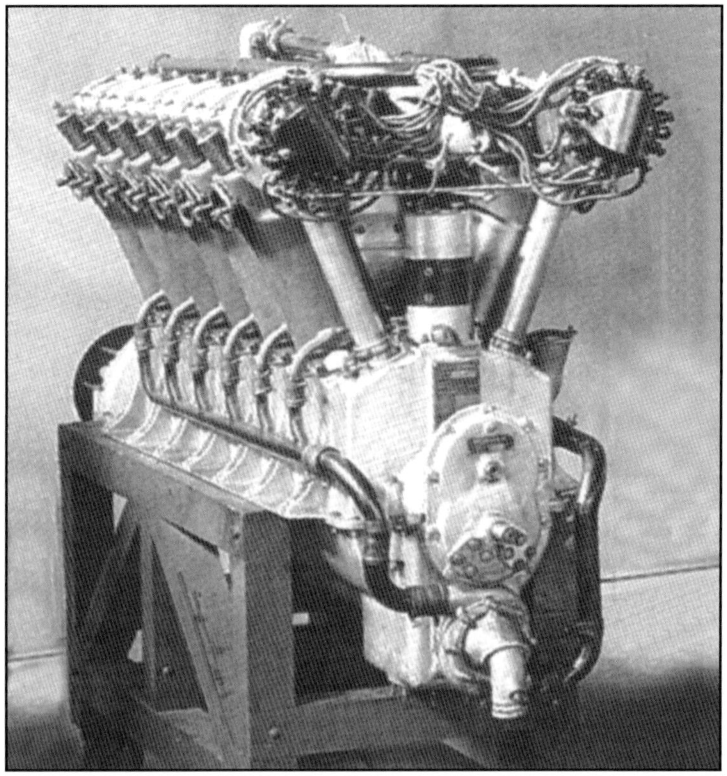

Above at left are pictured several L-12B engines of 200 contracted by the Navy in 1919 to be built by Packard with epicyclic gear sets built by Allison. The engines were assembled and tested by Allison at their Indianapolis plant and the left group are shown on the assembly floor at that location. At right is a rear view of one of these engines showing the Navy hand starter with which they were delivered and the Navy-type data plate. (Courtesy of Rolls-Royce/Allison)

By the end of August three engines had been received along with all three Willys-Morrow gear sets and one from Allison. One of the Willys-Morrow sets was returned as defective. The final two Allison sets were received on September 6. On about September 10 an additional three gear sets were ordered from Allison on order CS-203. This made a total of six engines and nine gear sets. It is apparent from the running commentary in the reports that the weak point was definitely the gear sets and they were trying to have extra sets to keep the testing progressing. The second set of three gear sets ordered from Allison was delivered in late January and early February of 1919.

By the end of September five engines had been received and the sixth was finished but being held at Packard per O. E. Hunt's direction. Two of them, Exp. No. 2 and 4, were shipped to the Navy. The sixth engine was delivered on November 26, 1918. By December 5, Exp. No. 6 had also been transferred to the Navy. The Navy's expressed need for geared engines to lift their multi-engine flying boats under full load was a prime factor in the development of the geared Liberty.

I note here that the design of epicyclic-, or planetary-type reduction gears results in the output, or propeller, shaft rotating in the same direction as the engine crankshaft. The Liberty 12B reduction ratio was 1.66 to 1. In the case of the Liberty, and most other aircraft engines, this is called right-hand rotation. This is in reference to the engine as seen from the rear in that the crankshaft rotates to the right or clockwise. Spur type reduction gears are generally designed with only two gears and this results in the propeller rotating in opposite direction of the crankshaft, or left-hand, counter-clockwise as seen from the rear. The result is, of course, that the propeller must have its blades pitched for opposite rotation. With this in mind, one can determine from an aircraft picture which direction the propeller rotates.

Throughout the fairly detailed history carried in the Engineering Division weekly reports we find that the engines present no difficulties, it is the gear design which is being developed and were constantly requiring parts replacement. New design repair parts begin to be ordered from Allison only and Willys-Morrow drops from the picture. It would thus seem apparent that the Allison gears were considered superior. This is born out by the fact that from this time on Allison was contracted by the Air Service and private engine builders as a designer/producer of high-power reduction gears.

In the weekly report entry for the last week in December 1918 it was noted that a power curve was run on Exp. No. 3 in preparation for installing it in an airplane. The name of the plane was not given and thus far we do not know what, if any, airplane it may have been installed in.

Testing with the six engines built by the Engineering Division was continued by both the Army and Navy and interest in the design remained high for some years to come. However, with the end of the war on November 11 came shortly thereafter, on December 4, cancellation of the Packard contract for 100.

As you will see in a later chapter, the model was eventually reordered in small quantity and used in a number of different airplanes. Some of the engines from that order are pictured on the prior page. They were externally the same as the original six.

So far as I know, there are no surviving engines of this first order.

THE LIBERTY 12C

Unfortunately records of the order and production of the spur-geared Liberty 12C are sketchy at best. Robert Casari's very diligent years of research uncovered the following material relative to the orders for this engine. It was ordered from Packard on an experimental basis and although one would have expected it to have been assigned US numbers within the Engineering Division's assigned block of 47828-47927 set aside for this purpose, no documentation has so far surfaced to indicate any were assigned.

The earliest dated order was No. 842 of November 30, 1917 for 2 examples. That order was replaced by order CS-5 of April 4, 1918 for the same number of engines. Order CS-5 was canceled on June 4, 1918 with no complete engines delivered and those parts which had been finished were ordered shipped to McCook Field.

This order was for two spur-geared engines with plain bearings. The fact that the design was for engines with plain bearings in the reduction gears seems to have been the determining factor in the cancelation of the order. A few reports of problems with early spur-gear engines have surfaced which might be related to failures of the plain bearing design as reports of tests of the second design, using ball-bearings, indicate the gears are very rugged in design and perhaps even excessively large for the power involved, and that the ball-bearings and their spacing and lubrication were quite satisfactory.

We do not know the cause of the cancelation of the first order for two engines, but we must assume that they either had serious problems with it or the second design, the order for which was placed on February 26, was seen as the better one.

Order number 4857 was placed with the Packard Motor Car Co. for three spur-geared Liberty 12C engines with ball bearings supporting the gears on February 26, 1918. The first of these engines was actually completed in late January of 1918. Col. Vincent was at the Packard plant that week on business and noted in his diary that he inspected a completed engine on February 10. On Monday the 11th he again looked it over, this time after O. E. Hunt had it torn down for inspection after a test run.

Here we probably have a case, and this was not very unusual during this period, of an unofficial order being placed some time before the paperwork was actually completed and "everyone had signed on the dotted line." In this particular case the first engine was completed before the order was official. I will note at this point that in some later published documents this engine was said to have been designed by "the Engineering Division" and produced by Packard. The Engineering Division in this case was no doubt Col. Vincent (he also designed the Liberty 12E which I will cover shortly). The reduction gear used on this engine was a scaled-up version of that used on the Packard 1A-905 (also called the 905-3) designed just before the war by Col. Vincent.

By mid-March the second engine was completed. On March 28 it was reported as running on the dynamometer and that the third would be complete in about 30 days. Thus by about May 1 all three engines were finished. (No. 3 was tested on May 11.)

Liberty 12C No. 1 on a stand at the Packard factory. Note the use of the early-type air intake system. The picture was probably taken in February of 1918. (Author's collection)

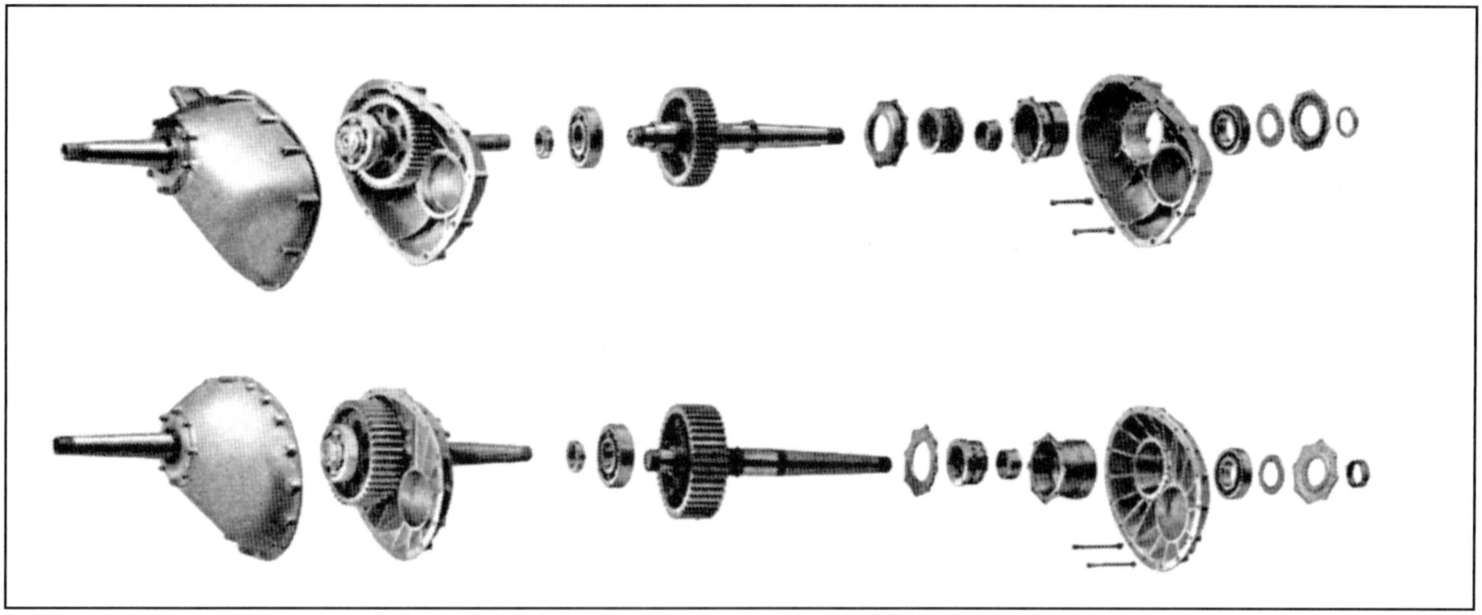

Above are pictured the reduction gears of the Packard 1A-905 (top) and the Liberty 12C (bottom). One can see that the designs are essentially the same except for the difference in gear ratios and overall size to accommodate a 285 HP engine in the first case and a 420 HP engine in the second. (Author's collection)

In the March 14 Aircraft Production board, at the suggestion of Col. Deeds, it was recommended that Mr. Potter discuss with Packard the use of their facilities in the production of the geared Liberty, the tests of which would soon be completed at McCook Field.

This inquiry was in preparation for possible series production but no such contract was forthcoming. Although the few surviving records of L-12C tests indicated no problems relative to the design or operation of the spur-gear drive, the first reduction-geared Liberty chosen for production was the epicyclic Liberty 12B, and it was given to Packard. The inquiry only noted "geared Liberty motor" and not which design.

Flight-testing was initiated during May and by the first week in June it was reported that the USAC-1, by this time referred to as the USC-1, was being flight-tested

This is a Packard factory photograph titled machine shop crankcase planing and dated January 1918. In the center is the upper half (upside down) of the L-12C No. 1 crankcase and slightly to the left can be seen the front gear case. (Courtesy of NAHC)

Thomas-Morse MB-2 U.S. 25806 powered by L-12C No. 3 using a left hand prop and photographed at McCook Field, probably in late December of 1918 or early January of 1919 . There seems to be no evidence that this plane was ever flight tested. (Courtesy of Peter Bowers)

with the L-12C. By the second week in June a four-blade propeller was being designed for use of the L-12C in the Thomas-Morse MB-2. That design work was reported as completed during the first week of August.

The flight-testing of the USC-1 apparently did not go particularly well, whether because of airplane characteristics or engine shortcomings is not entirely clear. The engine was still considered an acceptable one and well-suited for series production as late as mid-1920 but the airplane was never flight-tested beyond mid-1918 so far as is known, even though the epicyclic Liberty 12B of the same power could have easily been substituted for the 12C and the airplane further tested if the design had been considered good but the engine bad. I think we must therefore come to the conclusion that the flight characteristics of the USC-1 were found to be wanting and the design was abandoned in favor of better ones.

The L-12C-powered USAC-1 U.S. No. 40045 during one of its test flights at McCook Field during the summer of 1918. For reasons not precisely known, this design was dropped from consideration and only a single example was built. (Courtesy of Robert Casari)

At least the first two, and possibly all three, L-12C engines were built using the early light-type crankcase and light crankshaft. No. 2 suffered failure of the center main bearing support web during an early test and they somehow repaired it and went on with further tests until it shortly broke away entirely and sent pieces through the side of the crankcase. During the last week of June a rebuild was started on the engine and included a new, heavy crankcase. By the end of July the rebuild was finished and the engine was being prepared for a 50-hr. endurance test. During the first week of August the weekly report noted the engine was being torn down for inspection following the test.

Also during that week the design work on the four-blade propeller for use of the L-12C in the Thomas-Morse MB-2 was completed and work begun on building it. It might be noted at this point that by the first week of August the L-12B had not been yet run. Gear sets were just being installed in the first two engines and L-12B No. 1 would not be subjected to endurance tests until the second week of August.

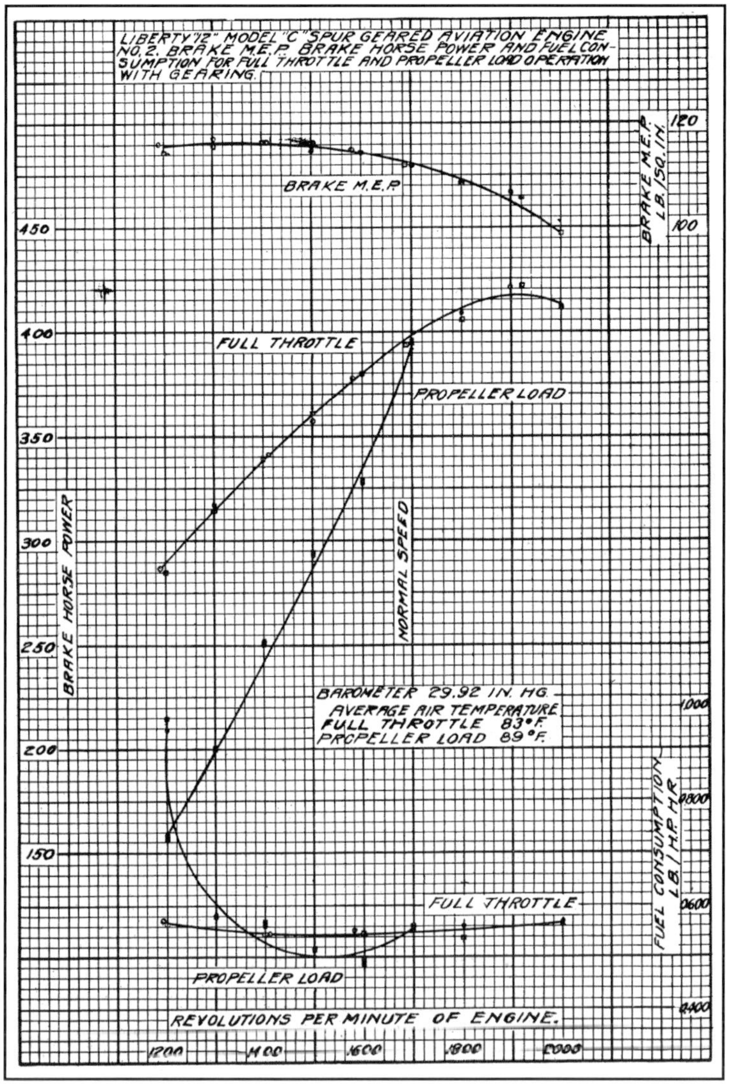

L-12C No. 2 power graph from a test run probably made in late 1918. (Author's collection)

A Thomas-Morse MB-2 fuselage had been delivered to McCook Field for sand tests sometime in October and failed under a load factor of 2 ½, which was noted as quite unsatisfactory. Two of this model were purchased and they were assigned US numbers 25805 and 25806. The usual practice was to subject the first aircraft to sand tests and flight test the second. Such was the case here. After the failure of 25805 some redesign was done resulting in structual changes in the fusalage of the second machine. In addition, at the same time the wing design was modified.

It was then reported (and pictured from three angles, one of which is reproduced at left) in Technical Orders No. 4 of January 1919 that the "Thomas-Morse MB-2 Geared Liberty" — "Second machine has been delivered to McCook Field for flight tests." Nothing further was said regarding the plane. I have found no record of its being flight-tested. Perhaps the sand test failure combined with the ending of the war on November 11 finally ended any further consideration of the airplane.

There seems to be little further work done with the L-12C until the Engineering Division issued McCook Field report 1283 on June 7, 1920. This report was titled *Report on performance and design of five representative geared aviation engines.* The object of the tests and resulting report was given as "To make available to designers a single report embodying complete descriptions of American-built geared engines with data on performance tests."

The report covered the following engines in detail and named them just as I note below:

Model "C" Hispano-Suiza
Model "D" Hispano-Suiza
Model "C" Liberty
Model K-12 Curtiss
Epicyclic-geared Liberty

Note that the L-12B is called the Epicyclic-geared Liberty and not the Model "B" Liberty. This is another example of Epicyclic being more often used than Model B.

L-12C NO. 1 on a stand in the Packard plant, probably taken in February of 1918. (Author's collection)

Front view of L-12C NO. 1 on a stand in the Packard factory in February of 1918. (Author's collection)

L-12C with gear case removed showing construction. Note the late type vertical air intake stack. (Author's collection)

<u>L-12C statistics</u>

Weight	966 lbs.
Crankshaft torque	1,225 ft. lbs. @ 1,300 RPM
Power	420 HP @ 1,900 RPM
Reduction ratio	1.50 to 1
Propeller torque	1,922 ft. lbs. @ 1,300 RPM
Propeller gear	42 teeth
	11" diameter
Crankshaft gear	28 teeth
	7 ½" diameter

In the description of the engine it was noted that it was supplied with the old-type light crankcase and crankshaft and that the propeller hub was slightly heavier in construction and larger in diameter than the standard Liberty hub. In the description it was noted that the hub weighed 29.5 lbs. as compared to 26.0 lbs. for the standard Liberty hub.

The report noted that the engine was satisfactory with regard to production manufacturing and that its reduction gear was of simple and rugged design well suited to production. Performance of that particular engine (No. 2) was not satisfactory in that it would only produce about 95% of the output generally achieved with a standard Liberty. They attributed this to the poor condition the engine was in upon receipt for testing. As a matter of fact the center main bearing support had failed on a previous test and been repaired. It failed completely during the running of a full power curve without gearing and threw the lower bearing retainer out through the side of the crankcase. Since part of the tests used to make up this report were on the several engines which were run prior to the time of the report and then used here for input, it is entirely possible that the report used for this engine (No. 2) was the report from the June 1918 test mentioned on page 172. Considering the duplicate nature of the problem, on the same engine, as well as the fact that the June 1918 rebuild included a new heavy crankcase and the engine as described in this test started with a light crankcase, I would say the early test was used for this report.

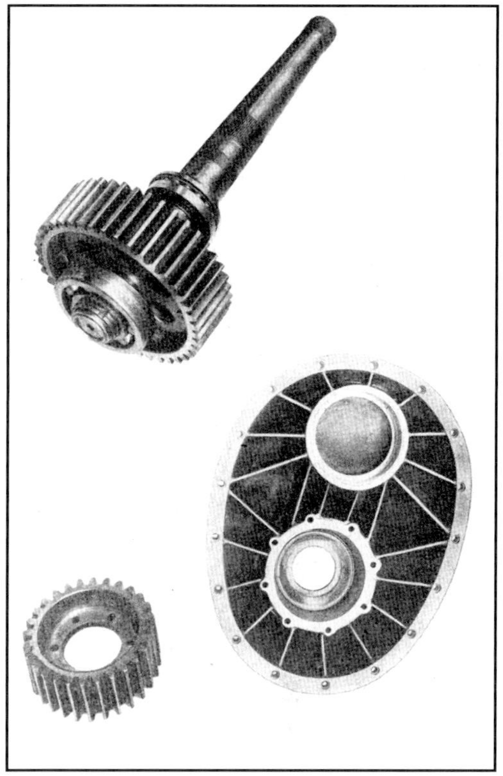

L-12C reduction gears and front gear case removed. (Author's collection)

Two views of the Packard experimental engine assembly room taken in January of 1918. L-12C No. 1 can be seen in both views. On the wall in the top picture is a January 1918 calender. (Courtesy of NAHC)

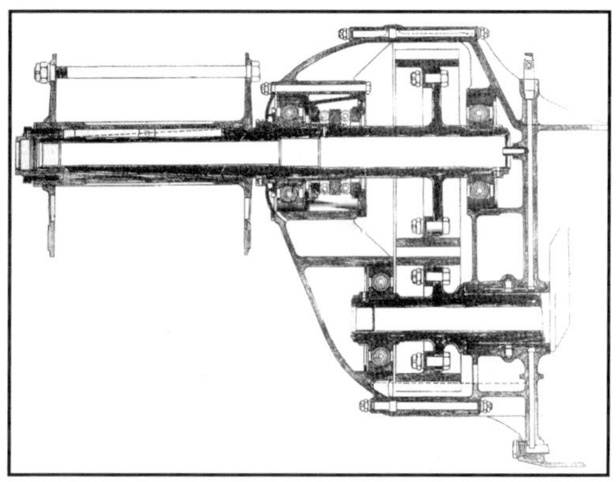

Sectional drawing of the L-12C reduction gear. Lubrication was introduced to the gear box from the main system through the rear of the hollow propeller shaft. From there it was distributed to all three ball-bearings and the thrust bearing and then supplied the compartment made up by the gear case. This retained oil at a level high enough to allow the lower spur gear to dip in it and thus lubricate the gears. A drain was provided back into the main engine sump at a point to provide a proper oil level in the gear sump. (Author's collection)

This is the last published report on the L-12C that I have found. Interest in a reduction-geared Liberty appears to have moved entirely to the L-12B from this time on.

Only three engines were ever built and none are known to have survived.

THE LIBERTY 12E

By late 1917 serious consideration was being given to the possibility of designing a Liberty engine adapted to being used with an automatic cannon firing through the propeller hub. On January 4, 1918 Lt. Emmons cabled General Pershing that experiments were being made as to the possibility of using a cannon with Liberty motors.

After receiving an inquiry from Col. Vincent, Lt. Emmons wrote him on January 18 that there were no drawings of the Puteaux 37 automatic cannon in existence, but a Puteaux 37 simi-automatic was to be sent over from Paris by Capt. Malony of the Ordnance Department. On the 21st Emmons sent a follow-up cable to France regarding the matter. On January 30 Col. Vincent noted in his diary that "I arranged with Capt. Hughes to keep right after information from Europe on the 37 mm automatic cannon."

All of this was in preparation for designing what was to become known as the Liberty Cannon Engine, and eventually as the Liberty 12E. This design had no doubt been ongoing as by February 10 the engine was in the process of construction, Col. Vincent having noted in his diary on that date that he "inspected various parts of the new Liberty cannon motor Mr. Hunt is building."

On February 28, 1918 order number 4978 was issued to the Packard company for the construction of two Liberty Model E engines. The engine was already designed and the first one being built when the contract was issued.

Col. Vincent was the designer of this engine and that fact was noted several times in various documents of the period. In a letter to Mr. Potter dated April 18 Vincent noted that the "duplex Liberty 2 crankshaft job looks good." This is the earliest official mention of the use of two crankshafts in this engine that I have seen, although it is obvious that such was the case from the initiation of the design. It also seems to have bestowed upon the engine a name which stuck, the "Duplex Liberty."

Liberty 12E No. 1 manufactured by Packard and the only example completed. This engine survives in the Air Force Museum collection. (Courtesy of the Air Force Museum)

As we can see, the first engine was nearly complete by mid-April. On May 7, 1918 the order was rewritten as CS-40 and quantity ordered increased from two to three and a wood model was added to the contract.

The June 1-6 weekly report indicated the wood model was already complete and the first engine was close and expected to go on test in about two weeks. This prediction, as we will see, was somewhat premature and the first (and only) engine was not delivered until after the end of the war.

In the June 15-22 report we find out that the hold-up on this engine was the lack of crankshafts. Completion was now predicted to be about July 1. Subsequent reports continued to slide delivery dates until the July 20-27

Liberty 12E No. 1 from the old Wright Field inventory photo collection. This photograph probably dates from the late 1930s and shows the lack of care shown some of the collection. The distributor head is missing on one bank and the water manifolds have been removed from the pump and laid between the cylinder banks. The rear cover for the cannon tunnel has been removed as has the lower right accessory cover. The latter shows the internally splined end of the right crankshaft which could be used for a starter drive. (Courtesy of the Air Force Museum)

report said the first engine would be ready to ship on August 1. The next mention, on August 10-15, it said the first engine "promised for shipment by August 1 has been held-up by the epicyclic motor order." By September 28-October 5 the word was that "The first motor on this order will be ready in about three weeks." The next week's report notes it is (still) expected in about two weeks and that it was ready except for crankshafts. During this period the cause of the hold-up is always crankshafts if any specifics are mentioned. The October 12-19 report promises delivery on October 26.

Nothing more appears in the weekly reports until November 16-23 when it notes: "Order CS-40 for three special Liberty cannon engines. On November 13, contractor was advised to discontinue all work on this order and submit accounts as soon as possible." The Armistice was signed on November 11 and most experimental contracts for which no immediate use was seen were canceled. At about that same time the order from Packard for two LePere-designed LUSAC-31 aircraft to be powered by this engine was also canceled.

Rear view of L-12E No. 1. (Author's collection)

OLD WRIGHT FIELD MUSEUM INVENTORY

A small museum was started in late 1918 at McCook Field to preserve a few of the early aircraft and engines. When McCook Field was closed in 1927 some of the material was transferred to the old Wright Field section of Wright-Patterson. The museum was maintained at that location until a few years after the end of WW II when it was transferred to a new location set up at Patterson Field. In the late 1960s the museum was moved to its present location at the old Wright Field portion of Wright-Patterson Air Force Base. The museum's engines were inventoried and photographed during its first stay at the old Wright Field location and although now scattered within the museum's archives, they often surface. When they do they can usually be identified as such because they are numbered to match the inventory list that was generated at the same time. Two of the illustrations used here are from that file. The Liberty 12E was item number 41 on the inventory which included a total of 81 engines, many of which are still in the Air Force Museum inventory.

ONE ENGINE AND ONE WOOD MODEL DELIVERED

At this point the first engine was either finished or so close to it that the decision was made to do so. It was delivered to McCook Field on November 26. No record of the US number assigned to this engine has been found, if indeed one was assigned. The wood model was received there on December 2.

The only official test I have found record of was Test No. 593 done at McCook Field on December 7, 1918. Unfortunately I have been unable to locate a copy and thus we have little statistical data on this engine model. Because it is based directly on the L-12A with only those modifications necessary to allow use of a 37 mm cannon firing between the cylinder banks and through the propeller hub, we can estimate these specifications fairly closely and I will do so below.

Throughout the references to this engine in the weekly reports it is always referred to as the "Liberty Cannon engine," yet in the final entry of December 7-14 is the note "The test of the Liberty cannon engine, model C1, was completed —." This is the only use of this model number I have seen. Perhaps its use stemed from the fact that it, as did the L-12C, used spur reduction gears.

Several photographs of the engine have been located. The best and most detailed, however, are Packard factory photos of the wood model. This model was an exceptional piece of work and is very detailed and the finished product appears to follow it very closely. Unfortunately the wood model did not survive. But the engine did and is now in the Air Force Museum collection. It is in storage and in rather poor condition but nothing major is damaged or missing and it is thus quite restorable and hopefully the museum will do so at some future time.

This engine was designed for the use of a 37 mm cannon firing through the propeller hub. This was accomplished by modifying the basic L-12A, but only as necessary to accomplish the end result. Whenever possible the design allowed use of standard L-12A components.

Standard Liberty cylinders, pistons, valves, oil pump, water pump and carburetors were used. The angle of the

Liberty 12E No. 1 from the Old Wright Field inventory photo collection. Note the gear case with the bearing caps of the two crankshaft gear bearings. The reduction ratio was about 1.6 to 1. (Courtesy of the Air Force Museum)

All pictures on this page are of the wood model of the L-12E built by the Packard Motor Car Co. and were taken by Packard in their factory. The wood model was a very close duplicate of the single engine built.
(Courtesy of NAHC)

Packard factory photo of the L-12E wood model taken in the Experimental Airplane Department of the plant.
(Author's collection)

cylinder banks was set to allow use of the L-12A distributors and cams. Two crankshafts were used in order to move the cylinder banks outward and allow a cannon barrel to pass from the rear through the crankcase and out through the hollow propeller hub. Each bank of cylinders utilized pistons mounted on Liberty 6 rods and driving their individual crankshafts. Each crankshaft had a small drive gear on the forward end which drove the larger central gear which in turn drove the propeller shaft. This resulted in a gear reduction ratio to the propeller of about 1.6 to 1.

Although very little is known regarding detailed specifications of this engine, we do know from surviving correspondence from C. H. Vincent to Col. Vincent that in early May Packard was building intake headers to fit special 2 3/16-inch bore double venturi Stromberg carburetors to be used on this engine. This might lead us to believe that the L-12E was expected to produce more power than the standard L-12A, but we have no solid evidence of that and I will therefore assume that the engine was rated at the same power as a standard Liberty 12.

C. H. Vincent was experimental engineer at Packard at this time working with O. E. Hunt on both experimental and production Liberty matters. On May 24, 1918 Hunt wrote a memo to C. H. Vincent directing him to handle the break-down testing of the Duplex engine.

Estimated statistics of the L-12E

Weight	1,000 lbs.
Power	420 @ 1,700 RPM
Reduction ratio	1.60 to 1

THE HIGH-POWER L-12

The building of this engine was evidently initiated by a suggestion of C. H. Vincent to Col. Vincent on May 6, 1918. In a memo to Col. Vincent from C. H. Vincent dated May 6, 1918 was the following:

> *"Confirming our conversation of yesterday, it is my opinion that the special intake manifolds to fit the 2 3/16 inch bore carburetor which Packard Motor Car Co. is now bringing thru for the duplex, in connection with the present cam shaft No. 2, incorporating 1/2" lift, and hollow ground cams, would enable us to bring a standard direct drive Liberty up close to five hundred horse power, particularly if it were fitted with the double venturi Stromberg carburetor, which I have asked Mr. Hunt to order for the duplex. I believe that such a motor will go thru a fifty hour test under French specifications in good shape, fitted with the latest type crank shaft and connecting rods, and there is no question but that it will stand up all right in the air. My opinion is that this is the quickest and most practical way of producing the five hundred horse power motor and with the speed which such power will give us, in the LePere ship for instance, it would be perfectly feasible to turn it up to 1800 or 1850 RPM.*
>
> *"Since the expense of testing out this proposition would be comparatively small, requiring only a few special parts, I would suggest giving the Packard Motor Car Co. an order for sample motor."*

High-power Liberty 12 developed in late 1918. This engine, Packard NO. 316 U.S. 17821, developed 526 HP @ 2,175 RPM and was run as high as 2,300 RPM. Camshafts were reversed, this reversed the location of the intake and exhaust valves and allowed mounting the carburetors on the outside of the "V" and the exhaust stacks inside. (Author's collection)

Col. Vincent acknowledged the memo with the note that "I am putting this matter up to Major Marmon." At this point in time Major Marmon was heading the Experimental Engineering section of the Airplane Engineering Division under Col. Vincent.

Evidently the idea was approved by Major Marmon and in late May of 1918 order CS-82 was issued to The Packard Motor Car Co. to build a high-power Liberty 12 engine, the object being to obtain at least 500 HP by modifying the standard L-12A.

As we shall see, the objective was easily obtained. A series of modifications were tried with each having some effect on output.

The engine selected for the tests was Packard No. 316 US 17821. It was chosen simply because it was an engine on which a number of minor tests on carburetors, pistons, rods, etc. had already been run and was already assigned to testing purposes.

It had been given a standard power test on April 17, 1918 and had produced 420 HP @ 1,800 RPM. Following that test it had been installed in a DH-4 for flight tests but an unusual "knock" was observed in the engine before the first flight and the engine was removed for examination. It was torn down and carefully examined but nothing was found wrong. It was then reassembled and used on a number of carburetor tests.

On May 8 it was put back in standard condition (the same as on the April 17 test) and again a power curve run. This time the output at 1,800 RPM was 431 HP indicating, as they put it, the engine was now in excellent condition. The engine was then used for several more tests.

It was shortly after this time that the decision was made to try to modify a standard L-12A in such a way as to produce 500 HP and order CS-82 was issued. This engine was available and had recent power curves already documented and so was chosen as the test vehicle.

The first thing tried, of course, was to install the best case set of changes which had recently been tried on the L-1D engine. Although about 495 HP @ 1,800 RPM was expected, only 442 HP was obtained, an increase of only 2 1/2%.

Faith was not quickly lost, however, and it was eventually found that the best combination for the single-cylinder was not necessarily best for the twelve. There followed a number of further tests, using mainly various adjustments and alterations suggested as a result of the tests with the single-cylinder engine, which eventually resulted in an output of 510 HP @ 2,175 RPM.

526 HP @ 2,175 RPM OBTAINED

The final test on August 26, 1918 was to run the engine with the same modifications as the 510 HP test but on fuel consisting of 80% gasoline and 20% benzol. Output was now 526 HP @ 2,175 RPM.

The final report on this effort was published in the September 1918 edition of the *Bulletin of the Experimental Department of the Engineering Division*. In it they made a special note as follows: "In this test the engine was run up to a maximum speed of 2300 r.p.m. to determine the peak of the horsepower curve. This was the highest speed at which the engine was operated during any of the tests, and it is worthy of note that although the Liberty engine was not designed for such speed, no serious defects developed on account of the highly increased stresses to which the parts were subjected at these high speeds."

This was not the end of the testing of this engine, however. The contract was evidently an open-ended cost-plus affair in which testing was continued as long as it was felt that new information was to be gained.

Work continued on this contract and brief notes so indicating continued to occasionally be seen in the weekly reports, including endurance runs, until during the week of November 16-23 (the week following the Armistice). In that week's report we find: "Order CS-82 for fitting out and adjusting a Liberty 12-cylinder engine to develop 500-hp. and test at that power to destruction. Contractor instructed to discontinue all work on this order and to submit accounts as soon as possible."

C. H. Vincent was in control of this project from its initiation but we cannot say if he continued to be until the end as during this period he became an employee of the BAP and left the Packard plant and went to the Saxon facility. (See page 158.)

O. E. Hunt wrote C. H. Vincent the below memo on April 24, 1918:

"I am handing you herein a copy of a program on Liberty engine development work that Mr. Anderson and Mr. Reifel are working to now. I wish you would read this over at once and let me have any suggestions that you may have in connection with it.

"I want you to take off my hands completely, the matter of handling break-down testing of the Duplex, and the development of a 500 h.p. standard Liberty motor. I will, of course, keep in touch with you from day to day as the work progresses, but I want you to issue all orders to Mr. Anderson and Mr. Reifel in connection with additional parts and material that you will need to get a real answer on these jobs. Also, to follow up the actual testing and see that any additional parts that you will need are promptly manufactured. I think the most convenient way for you to keep me educated will be to check over your orders with me before they are released to the shop or laboratory."

Table 1

Chronological Table of Series of Liberty-12 Tests which led up to Final Result of 526 hp. at 2175 r.p.m.

Period of Tests: May 8, 1918, to Aug. 26, 1918.
Manufactured by: Packard Company.
Manufacturer's No.: 316.
Oil: Veedol Airplane Engine Oil No. 1.
Gasoline: 68° Baume.

Test	Date	Horsepower at 1800 r.p.m.	Maximum horsepower obtained	Maximum torque obtained pounds	Special features
A	April 17, 1918	420	420	715	Standard Liberty-12-A. No special features.
B	May 8, 1918	431	431	730	Standard Liberty-12-A. No special features. (Engine given inspection and probably in better adjustment than in Test A.)
C	June 9, 1918	442	442	777	1. Special 6 to 1 compression pistons—3 rings. 2. Camshaft, standard lift, but exhaust closing 10° after top dead center, and inlet opening at top dead center. Other timing standard. Right and left camshafts interchanged. 3. Benz type intake headers coupled to exhaust parts; short exhaust pipes connected to inlet parts. 4. Four standard Zenith carburetors fitted, having 36 choke, 165 jet and 170 compensator. One half of each double carburetor blocked off. 5. Connecting-rod clearance cut down to 0.00175 in. for forked rods and 0.003 in. for plain-end rods.
D	July 4, 1918	436	482	760	Same as Test C except: Four Schebler model A carburetors with $1\frac{13}{16}$ in. chokes replaced the Zeniths. (This was the first dynamometer run of a Liberty-12 at a sustained speed of over 2000 r.p.m.)
E	Aug. 4, 1918	450	485	774	Same as Test D except: Air intakes of the Schebler carburetors cut away so as to leave an unobstructed passage through the chokes.
F	Aug. 17, 1918	466	510	800	Same as Test E except: Frictional losses reduced by removing lowest of the three standard rings from the pistons. New heavy type connecting-rods with standard clearances. New semi-heavy crankshaft. Special 12-volt Delco ignition.
G	Aug. 21, 1918	468	510	810	Same as Test F except: Intake headers wrapped with heat-insulating asbestos tape. Special "AC" combination mica and porcelain plugs, type A-1 fitted.
H	Aug. 26, 1918	496	526	858	Same as Test G except: Fuel used: Combination of 80 percent gasoline and 20 percent benzol.

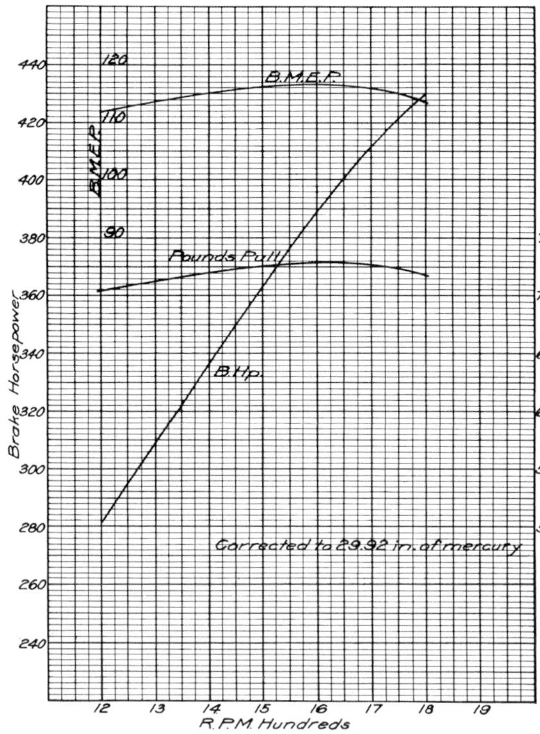

Fig. 3—Liberty-12 No. 316 after inspection and running as a standard engine. Carburetor setting: choke 36, jet 165, compensator 170. Outlet water temperature, 170° F

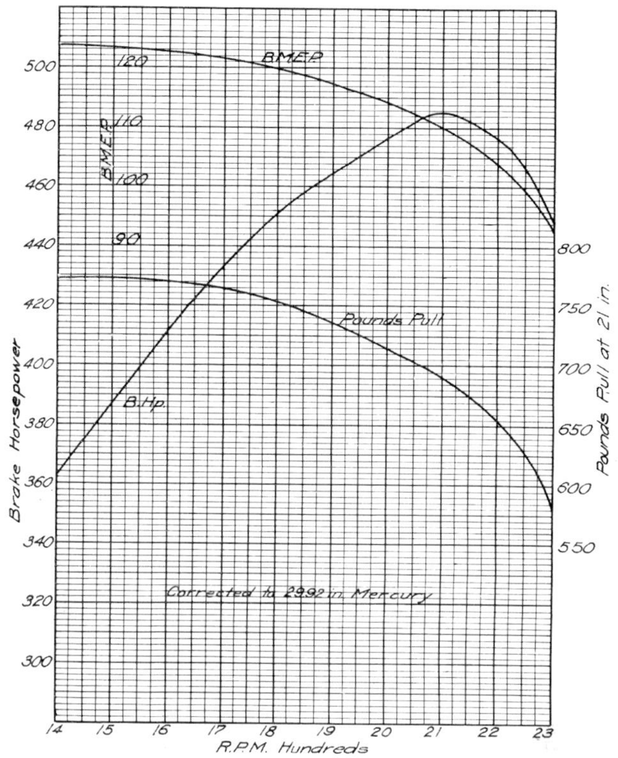

Fig. 6—Fifth test of Liberty-12 fitted as noted in test E in the table on page 29

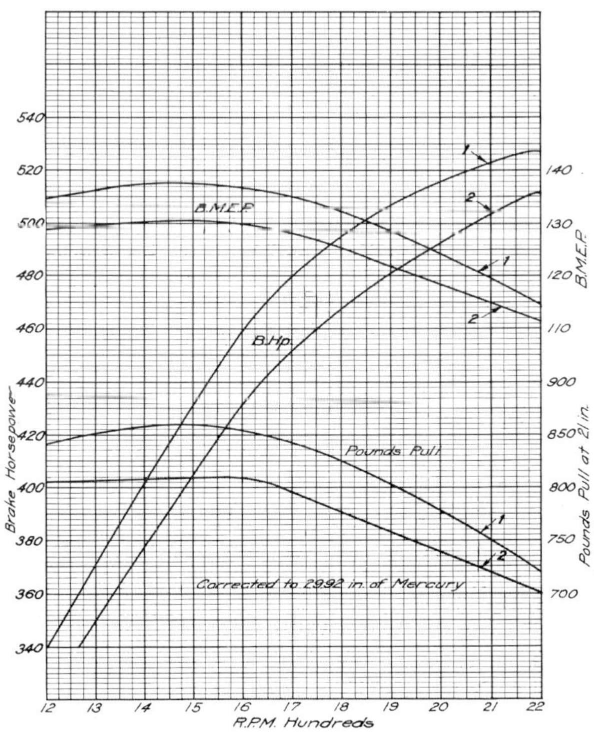

Fig. 7—Performance curves of Liberty-12 No. 316 fitted as noted in test G on page 29. Curves marked No. 1 were obtained with special fuel which is a mixture of 80 percent gasoline and 20 percent benzol. Curves No. 2 are the results with standard 68° gasoline only

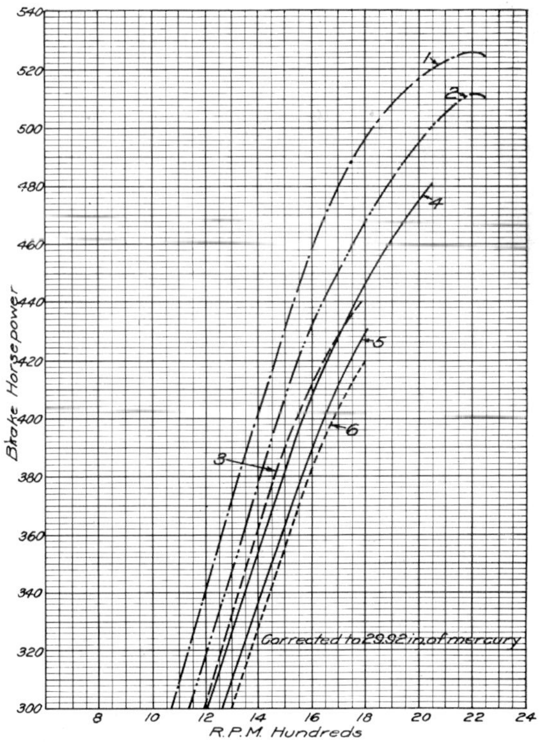

Fig. 8—Composite plotting of curves of No. 316 Liberty-12, showing development in horsepower from 420 at 1800 r.p.m. (curve 6) to the final result of 526 at 2175 r.p.m. (curve 1)

It would thus appear that he controlled the development of the the High-Power Liberty until it had accomplished it main purpose, and the testing of the Liberty 12E until he left to take over the Experimental Engineering Department of the Detroit District Office of the BAP in September of 1918.

THE TANK LIBERTY

The Liberty Tank, or Ordnance, engine will be covered here, but with only enough of the basic history behind it to give the reader an overall view of what led to the design and use of this version of the Liberty.

As soon as the United States became involved to the war, the Army became interested in the possible use of tanks. Since we had little experience in this field, a fact-finding team was sent to Europe to try and determine what tanks then available could be best used by this country. The French Renault FT-17 light tank was recommended but no available heavy tank was found to be suitable. The mission's recommendation was therefore to create a joint Inter-Allied Tank Committee to come up with a suitable design.

Such a committee was formed and came up with a new design based mostly upon prior British experience. The idea was to combine British experience with American production abilities. The end-result was the Mark VIII tank.

On December 26, 1917 a set of plans and a full-scale wood mock-up were completed and approved. The idea was that the British would supply the armor plate, structural components, weapons, ammunition and track shoes

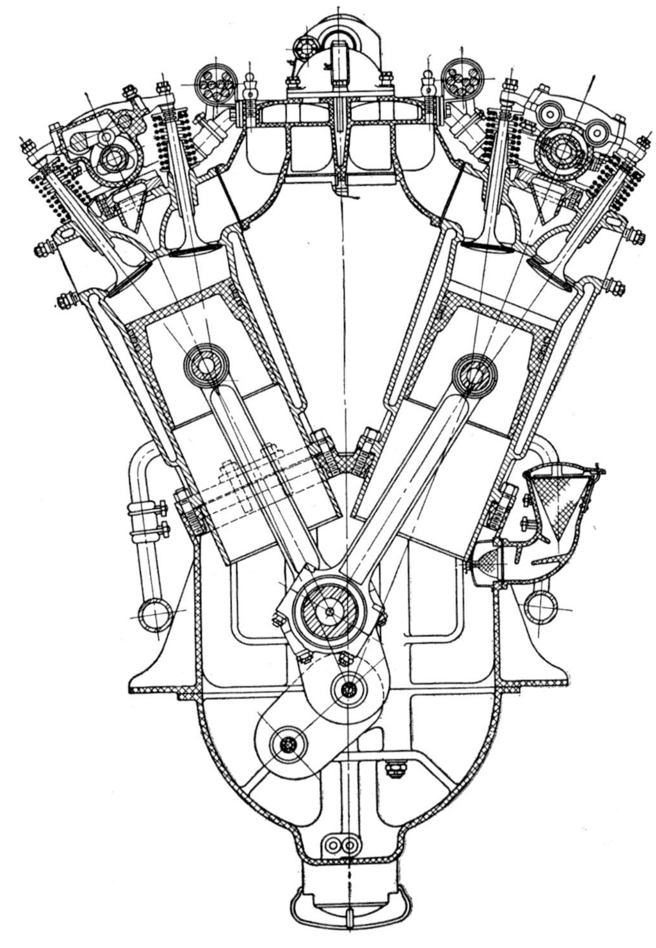

Sectional view of the proposed Liberty 12 Tank Engine. The only noticeable differences are the use of cast iron cylinders, the 90% rotation of the sump cover lock spring, and exhaust port bolts, and use of Navy pistons.

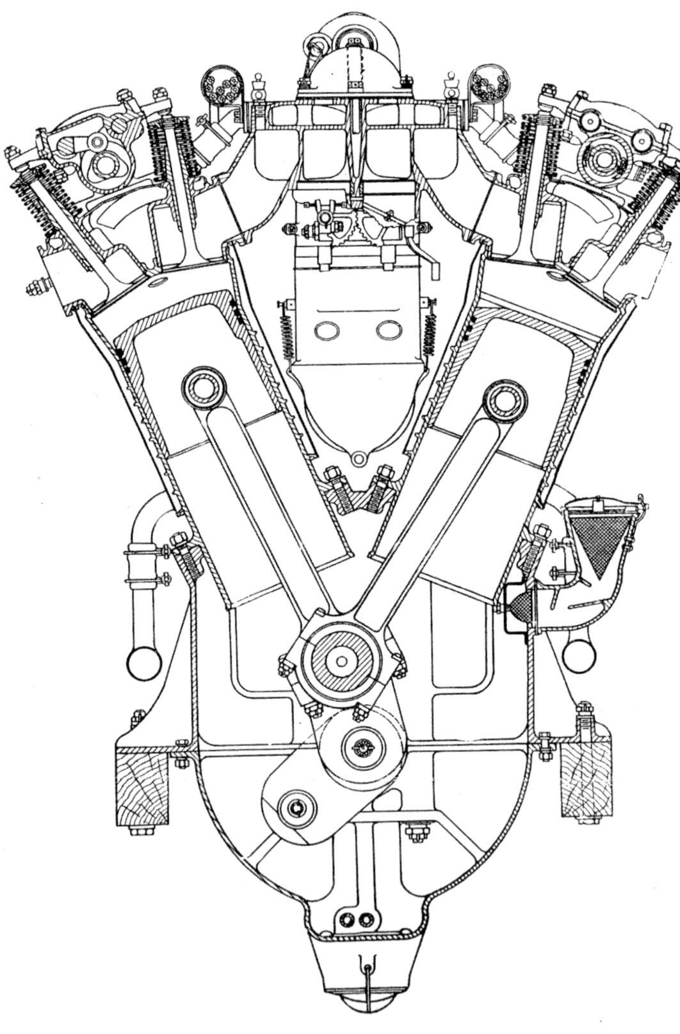

Sectional view of the standard Liberty 12A. (Both views from the author's collection)

and the Americans would supply the engine, brakes, suspension and running gear. The initial proposal was for the construction of 1,500 such tanks.

At this point in time, the Liberty had barely reached production status. Everyone concerned had great hopes, of course, that the engine would very soon be available in massive quantities. In reality, by the end of December Packard was the only maker that had actually assembled an engine, and by the last day of that month only 22 had passed final inspection and been shipped.

The minutes of the Aircraft Production Board immediately reflect receipt of this report. On December 27, 1917 we find the note: "It was stated that the Liberty engine was being redesigned for use in tanks, and that 1,500 would be required for this purpose."

On February 19 the board discussed "the use of the Liberty Motor in tanks and it was felt the engineers of the Ordnance Department should submit their redesign of the motor for such use to the Engineering Department of the Signal Corps, and the Chairman was asked to advise the Ordnance Department to that effect."

The changes recommended in the engine itself were minimal and, so far as we know, only involved redesigning the cylinders so as to use cast iron rather than the steel cylinders with welded-on sheet steel cooling jackets, rotating the exhaust manifold bolts on the cylinders by 90°, rotating the sump cover lock spring by 90°, and the use of Navy low-compression pistons. The remaining changes involved only accessories and the addition of a flywheel, clutch and transmission.

Although these were the changes recommended, in the end standard low-compression Navy engines were used and the cast iron cylinders were never put in production. Throughout the following three or four months the proposed needs of the Ordnance Department, both overall and as to a monthly program, are found in various official documents. Also several tests were run by the Engineering Department on low-compression engines to gather data for Ordnance. As yet, no evidence has surfaced indicating the cast iron cylinder was ever put into production nor that any changes were made to the lower crankcase to accommodate a 90° rotation of the sump cover lock spring.

THE BRITISH TANK MISSION

Early in 1918 the British sent a mission to this country to gain knowledge of our ordnance designs and to learn more about the Liberty engine as related to tank use. The Assistant Controller of Workshops, a British Major whose name we, unfortunately, do not know, on April 2, 1918, wrote a six-page report on the Liberty engine which revealed some rather interesting facts.

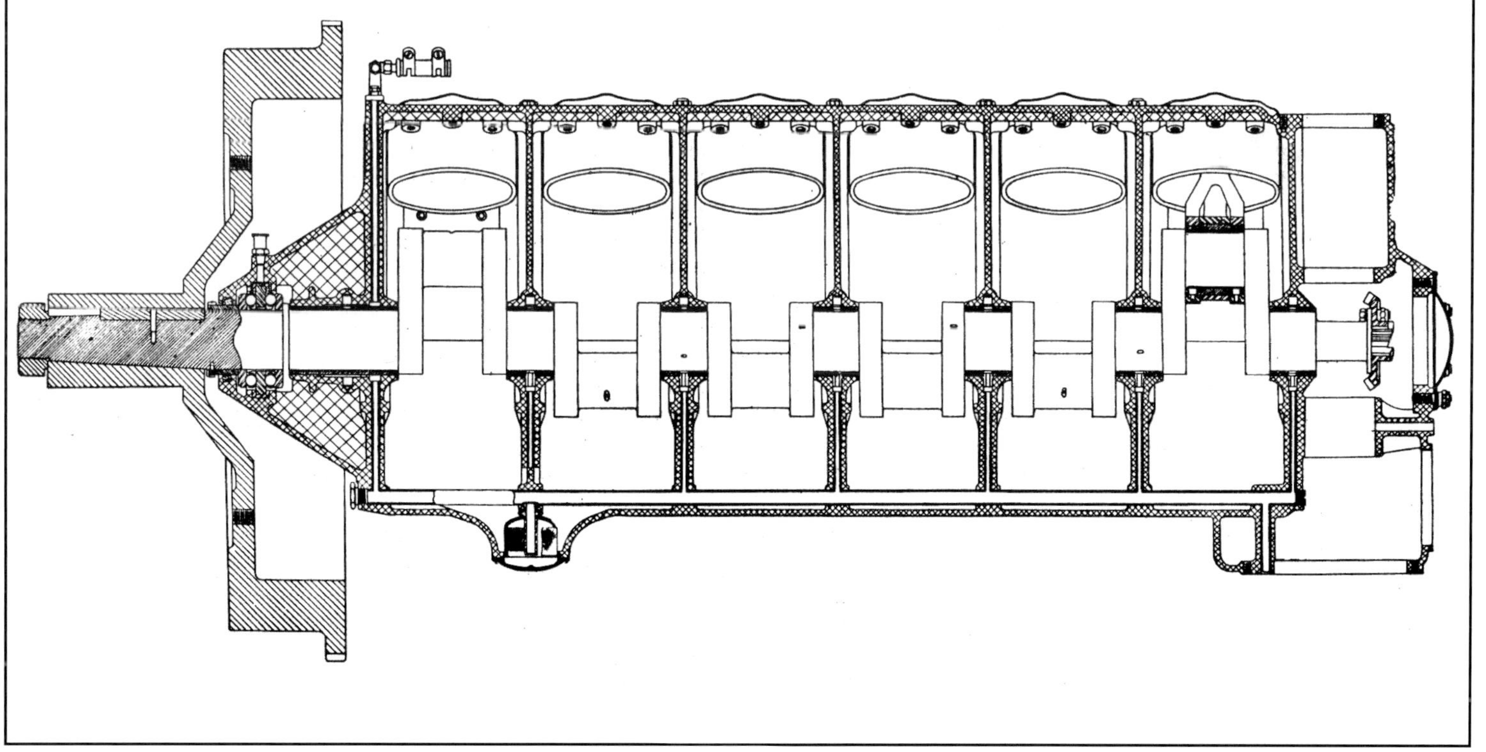

Side sectional of the Liberty 12 Tank Engine. The only change from the standard Liberty 12 is the addition of a flywheel with ring gear. (Courtesy of Jim Dillon)

A Liberty engine on test for the Mark VIII tank in the Ordnance Department. It has standard cylinders and shows no external difference from a standard aircraft version. If it has the welded on tie between the intake and exhaust elbows, it is hidden behind the camshaft housing. (Courtesy of Jon Hornbostel)

He described the engine in some detail and then made some interesting observations while comparing it with the British Recardo 6-cylinder 150 HP tank engine which it would replace as potential power for the Mark VIII tank. The first thing noted was the weight of the engines. The Liberty weighed between 796 and 804 lbs. ready to pack for shipment. The Recardo weighed 2,688 lbs. (The clutch and flywheel, as applied to the Recardo, weighed an additional 672 lbs.) There was therefore a 1,884 lb. advantage in the use of the Liberty which the Major did not take lightly.

The Liberty intended for use was the low-compression, or Navy model. Its power and torque were quoted in a chart with listed speeds of from 1,200 to 1,800 RPM. The Major said "As regards power output — apparently no official tests have been made below 1200 R.P.M. — there are a number of HP and Torque curves in existence, the figures of which vary somewhat — below are results obtained from a certified Laboratory Report of the Airplane Engineering Department, McCook Field. These figures appear conservative and are lower than the majority of those that I have seen."

The lowest power figure of 1,200 RPM was of concern because the plan was to use the engine at a rated power of 300-325 HP. This equates to a maximum of about 1,300 RPM. Performance from just above idle speed to 1,200 RPM would be of interest for an engine driving a land vehicle though of little use in aircraft applications.

The figures given for the low-compression version are quoted below. Torque is given in lbs./ft., which was used in early Liberty tests. Most later tests gave this figure in lbs. at 21 inches.

LOW-COMPRESSION LIBERTY OUTPUT

RPM	HP	Torque
1200	302	1205
1300	325	1190
1400	344	1185
1500	365	1165
1600	386	1150
1700	401	1120
1800	399	1040

Toward the end of the report he listed the advantages of the Liberty engine as:

a- Large production.
b- Light weight.
c- High efficiency.
d- Ease of obtaining spares and immense repair facilities in France due to standardization by American Flying Corps.

He ended the report with: "In conclusion I strongly recommend that every effort be made to utilize the output of the Liberty for tank work."

As we follow the various entries relative to the Ordnance engine it rapidly becomes apparent that aircraft use, both by America and its Allies, take precedence over Ordnance use.

On April 13, 1918 Mr. Potter presented at the board meeting an estimated schedule for delivery of Ordnance Liberties as follows: April 50, May 75, June 200, July 400, August 400, September 375, for a total of 1,500.

A cable from General Pershing dated April 16 put the brakes on any progress that may have been made up to this point. It said, in part, "Allied nations must be provided with Aircraft materials even if American program is delayed. English requests and also those of France and Italy should be given priority. Use of Liberty engine in tanks is not approved at present."

The April 23 Aircraft Production Board notes include the following communication from the Adjutant General to the Chief Signal Officer: "No Liberty engines produced prior to July 1st will be furnished the Ordnance Department."

The complexion of the matter changed considerably when on June 5 Lt. Emmons (head of engine production) reported to the board "that the Liberty motors produced at the Trego plant were not up to the same standard as those produced elsewhere, and that in his opinion such motors, while not entirely satisfactory for airplane requirements, would be quite adequate for tank purposes. Mr. Potter stated that by direction of the Chief of Staff he had refused to furnish Liberty motors to the

Ordnance Department, which therefore was seeking other sources of motor supply, and suggested that development of the Trego plant by that Department would avoid possible competition for new sources of motor manufacture."

TREGO BECOMES THE ORDNANCE LIBERTY PRODUCER

The following resolution followed: "RESOLVED, that a recommendation be made to the Director of Aircraft Production that, subject to the consent of the General Staff, the contract for the manufacture of Liberty Motors by the Trego Motors Corporation be transferred to the Ordnance Department." The General Staff approved the change and the contract was transferred to Ordnance.

The Air Service had accepted only one engine from Trego, MNFG. NO. 2 U.S. NO. 7402. The original Air Service contract was for 500 engines. The remaining 499 of that contract were transferred to the Ordnance Department. In conjunction with this change, all records of production and acceptance for this contract were also transferred to the Ordnance Department. Unfortunately I have been unable to uncover these records. As a result, all I can do is estimate the number of engines the Ordnance Department probably received before the Armistice of November 11 no doubt caused the cancellation of the remaining undelivered engines. Considering the amount of time remaining before November 11 and the probable production rate of Trego, I would doubt more than 250 were delivered.

There are no indications that those engines produced for the Ordnance Department were different in any significant regard from the standard aircraft engine. Since the transfer of the contract to Ordnance resulted in so few surviving records beyond that point in time, I will reproduce in its entirety what F. H. Trego said regarding the engines in the discussion section of a SAE paper on the Liberty given on February 4-6, 1919 by Col. J. G. Vincent.

COMMENTS OF FRANK H. TREGO REGARDING THE LIBERTY

F. H. Trego: "The production of our plant was finally transferred to the Ordnance Department. Engines to be used in the large tanks called for in the Allied tank program required some few changes in the design, such as the incorporation of a flywheel and the welding in of a tie between the intake and exhaust elbows to prevent the cracking of the jacket at the top, and this we do in regular production. In connection with the testing of the engines for tank service, the United States Ordnance Department endeavored to comply with the testing specifications as suggested by the British Tank Service, and one of these tests called for four runs of 50 hr. each with wide-open throttle, on dynamometer, at 1400 r.p.m., all parts of the engine being carefully measured before and after each test. Outside of the cracking of the jackets, which was later corrected by the welding of a tie, we found a tendency of the babbitt to break away from the bronze backing in the connecting-rod bearings in a spot as large as a ten-cent piece, directly opposite the oil-hole in the crankshaft at the instant of the explosion of the cylinder carrying the plain-end rod. The hole in the crankshaft at this point has a depression on each side of it, leaving the babbitt unsupported for a considerable area and owing to the fact that the plain-end rod which runs on the back of the bronze bushing has 0.005 to 0.0065 in. clearance we decided that a rapping action was set up by this rod on the outside of the bronze and this continual rapping directly behind the unsupported portion of the babbitt had a tendency to break it away from the bronze. To eliminate this difficulty we had a shaft made up with an oil-hole drilled at a point 30 deg. in advance of top dead center when the plain-end rod piston is at the outer end of its stroke. Thus a flow of oil on the center line portion of the bearing was assured before the explosion took place. This also provided complete support for the babbitt under the plain-end rod at the instant of explosion. Tests of this shaft seemed to prove out theory, and we now consider it good practice, even with a single rod, as it assures an uninterrupted bearing surface under the rod at the instant of explosion. We also omitted the depressions on each side of the drilled hole and simply broke the sharp edge of the hole. After a test of 46 hr. under full load at 1400 r.p.m. the bearings were found to be in excellent condition."

Thus we see that, beyond some perhaps debatable improvements in bearing oiling, the only change was to weld a tie between the intake and exhaust elbows on top of the standard cylinder so as to strengthen the cylinder at that point. A similar alteration was used on some engines by the Navy after the war.

ACTUAL TANK PRODUCTION

A mild steel prototype tank was shipped from England to the United States in July of 1918. After arriving here an engine and transmission was fitted and the tank tested. Parts for 100 tanks were completed in England by the time the war ended but no more had been assembled as complete tanks.

These 100 sets of parts would be shipped to the United States after the war and the completed tanks built here. I will cover the design and history of these tanks in more detail in the following chapter.

As a final note on the wartime prospects of using the Liberty in tanks, Major G. A. Greene of the British Tank Corps requested McCook Field to test a low-compression Liberty with carburetors mounted outside the "V." The reasoning behind this request was that putting the carburetors in this location for tank use would allow better access to them than having them in the conventional position within the "V." It also allowed the carburetors to be mounted so as to have the float chamber and jet at right angles to the engine. This would allow greater climbing angles without as great an effect on the float chamber.

The test was carried out on November 18, 1918 on Packard Exp. No. 5 equipped with the same intake system used on 500 HP engine No. 316 covered a few pages back. Low-compression pistons were fitted and the camshafts were reversed as on No. 316. Power graph was plotted from 600 to 2050 RPM and showed that a low-compression so equipped produced 20 HP more output at 1,800 RPM than did the same engine with high-compression and standard carburetion. This, of course, followed the same pattern set by the 500 HP tests.

THE LIBERTY 16

The Liberty was built in so many configurations that it might seem that no possible variation of cylinder layout could have been missed. Actually, it appears that this was an "almost" cylinder configuration. There is no evidence that a sixteen

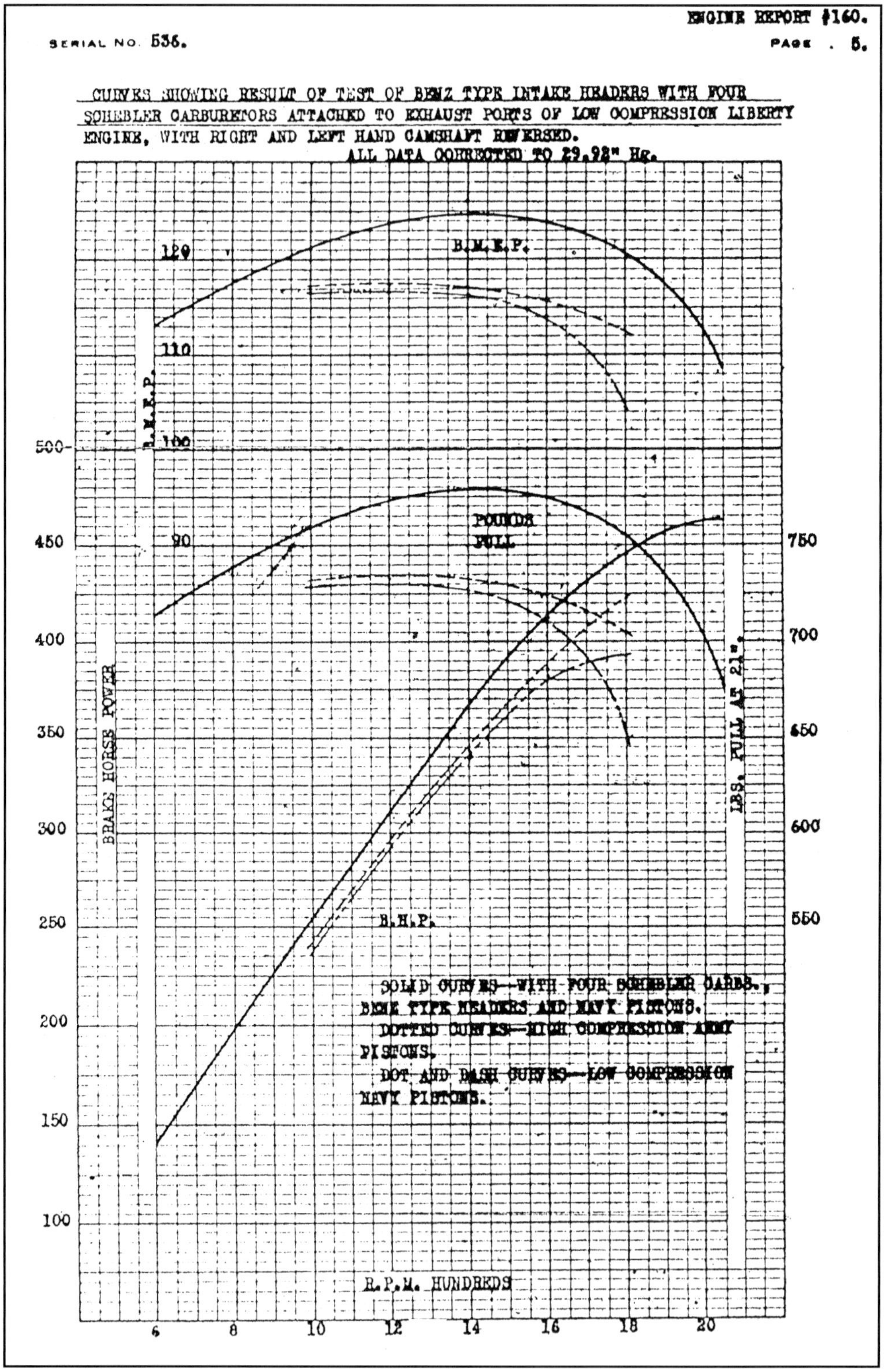

Power graph of Exp. No. 5 equipped with LC pistons and outside carburetors and manifolds of the same design as the 500 HP High-Power engine. On the same graph are plots of the engine with standard carburetion and both HC and LC pistons. This graph is also quite unique in that it is the only one I have seen plotted as low as 600 RPM. (Author's collection)

cylinder Liberty was ever built, but it is quite possible that one was designed.

In the early Liberty history prepared by Potter, Benet and Kent in June of 1918 is a note that between November 6 and 12, 1917 designs were being developed for a 16-cylinder Liberty. I have found no other mention of such a configuration except an entry in the BAP minutes dated September 11, 1918 stated "Commander Atkins stated that the Navy was strongly in favor of the early development of the 16-cylinder Liberty motor."

If we try to interpret this quote as literally as possible, then one could very well assume there was a completed design since the statement said development of "the" 16-cylinder, not "a" 16-cylinder.

This is pure conjecture, of course, but it does perhaps give some hope that somewhere in the World War I Air Service files are the designs of a Liberty 16 just waiting to be discovered.

What layout would they use? The two most likely possibilities are a V-16 and an X-16. The easiest to accomplish would be the X-16. Captain L. M. Woolson of the Engineering Department was, at about this time or very shortly thereafter, developing an X-24 Liberty which was constructed by joining two upper L-12 crankcases. This same procedure could have been used to build an X-16 from two upper L-8 crankcases and, unless the extra height was objectionable for some reason, was probably the layout chosen.

I restate, however, there is no evidence this engine was ever designed or built.

THE LIBERTY 24, OR X-24

This engine was proposed and built at the suggestion of Col. Vincent. The task was assigned to Captain Lionel M. Woolson, a test engineer assigned to the Engineering

Liberty 24. This engine was conceived by Col. Vincent and designed and built by Captain Lionel M. Woolson of the Engineering Division in November-December of 1918. The engine was basically two upper crankcase halves of a standard Liberty 12 bolted together with required modifications in accessory locations and oiling and water routing. (Courtesy of the Air Force Museum)

Division under Col. Vincent. Woolson received his engineering education at St. Paul's School in London. He worked for the Bijur Motor Appliance Co. from 1912 until entering the Army Air Service in 1917 as a Lieutenant. He was assigned to McCook Field and shortly became engineer in charge of engine testing.

We do not know the precise date the decision was made to build this engine, but by the last week in November the necessary oil and water pipes were being fabricated in the Engineering Department at McCook Field. About a week later it was reported that power curves and gasoline consumption tests had been completed. This would indicate the engine was completed and tested by December 7, 1918.

Although construction of the engine was fairly simple and straightforward as compared to the design of a completely new engine, it certainly must have taken a month or two from conception to execution, since some adapters had to be designed and built and special articulated rods had to be constructed.

With this in mind, we can see that the design was conceived prior to the Armistice and was in the midst of execution when the war ended. The decision was made to finish the project and build and test the engine, but upon completion of the first test the project was ended. The results were considered important enough to publish in the December 1918 issue of the *Bulletin of the Experimental Department, Airplane Engineering Division.* We therefore have excellent documentation regarding the design and performance of the engine. The project ended at this point, however, and the engine was no doubt scrapped shortly thereafter.

THE DESCRIPTION AND TEST

I will, for the most part, use direct quotations or edited versions thereof, from the published report of this engine as the most accurate and complete source of this description.

Although the tests were not very conclusive, "it is believed they point out the possibility that a 24-cylinder engine could be developed which, if run at not to exceed 1400 to 1420 r.p.m., would develop about 700 hp. and give a reasonable length of service. Inasmuch as the dry weight of the complete engine including propeller hub flange and without water and oil was 1385 lbs., this would give a dry ratio of about 1.97 lbs. per horsepower."

The above statement is certainly interesting in the point made and, considering that the Engineering Division not long thereafter (late 1920 through 1923) designed and had built by Packard a 700 HP engine called the W-1, it is a wonder that they did not further pursue the Liberty 24 avenue. The basic statistics of the two engines make an interesting comparison:

Statistics of the Liberty 24 and Eng. Div. W-1

	Liberty 24	Eng. Div W-1
Type	**X-24**	**W-18**
Displacement	**3,300 cu. in.**	**2,832 cu. in.**
Power output	**700 HP @1420**	**700 HP @ 1700**
Weight	**1,385 lbs.**	**1,770 lbs.**
P/W ratio	**1.97 lb/HP**	**2.53 lb/HP**

The report went on to note that operated at a running speed considerably lower than the regular operating speed of a standard Liberty would result in both long life and the ability to use a relatively large slow-speed propeller without the need of a reduction gear. In addition, "An engine of this type could no doubt be built quickly with very little extra tooling or production delay, and it seems likely that it could be adapted to a plane without very much trouble, although of course, it would require a very deep fuselage."

In spite of these initial observations, there is no evidence that this Liberty configuration was ever seriously considered thereafter.

CONSTRUCTION OF THE ENGINE

"Crankcases: The upper halves of two standard Liberty 12-cylinder crankcases were bolted together, the mating surfaces being lapped-in in the regular manner. Extra long main bearing bolts were used, extending through both halves of the resulting crankcase. The outer supporting ledges of the half in the inverted position, were planed off just enough to provide clearance for the engine bearers, the weight of the powerplant being supported in the usual way on the extension edges of the upper half of the crankcase.

"Connecting-rods: Articulated connecting-rods, of the same type as those tried experimentally in several Liberty-12s, were used in the 24-cylinder powerplant, as a number of these rods were on hand from previous tests. Each master rod was fitted with a special cap incorporating bosses to take two extra link rods, in addition of course to the link rod already attached to one side of the master rod. Fig. 5 illustrates the design of the special caps used, while the appearance of these rods, consisting of one master and three link rods, may be seen in Fig. 8. The master rod of each set was used in the lower right cylinder.

"After assembling the pistons to the rods, the stroke of the four members of one set was measured, and it was found that there was but very slight variation.

"Lubrication system: A standard oil pump, the same as used on the 12-cylinder Liberty, was adapted to this engine. On account of the fact that the 24-cylinder engine had only the same number of crankshaft and connecting-rod bearings as the twelve, it was believed that the capacity of the regular oil pump would be sufficient for the 24-cylinder under normal running conditions. As a matter of fact, no lubrication troubles were experienced during the tests, showing that the capacity of the standard oil pump was sufficient for the larger engine.

"The pump was driven off the vertical drive shaft at the rear or ignition end of the engine, and was mounted on the lower half of the crankcase in the position usually taken by the ignition generator when the crankcase is right side up. A special adapter, as shown in Fig. 4, was required to locate the pump in this position.

"On account of the fact that the ends of the cylinder barrels project about three inches into the crankcase, a form of sump, which is found to be ample, was provide in the lower half of the case. The general layout of the lubrication system for the 24-cylinder engine is shown in Fig. 7, the oil pump being located as at D and J on the lower half of the crankcase. B is the oil return pipe from the lower crankcase sump, while C is the pressure delivery pipe running along the bottom of the crankcase and discharging into the seven vertical oil holes leading to the crankcase main bearings. A portion of the oil reaching the propeller end bearing of the crankshaft was led to the crankshaft housings of the upper rows of cylinders through pipe H, just the same as in the Liberty-12. An extra pipe G was provided in the 24-cylinder engine to care for the crankshafts of the lower cylinder banks in the same manner.

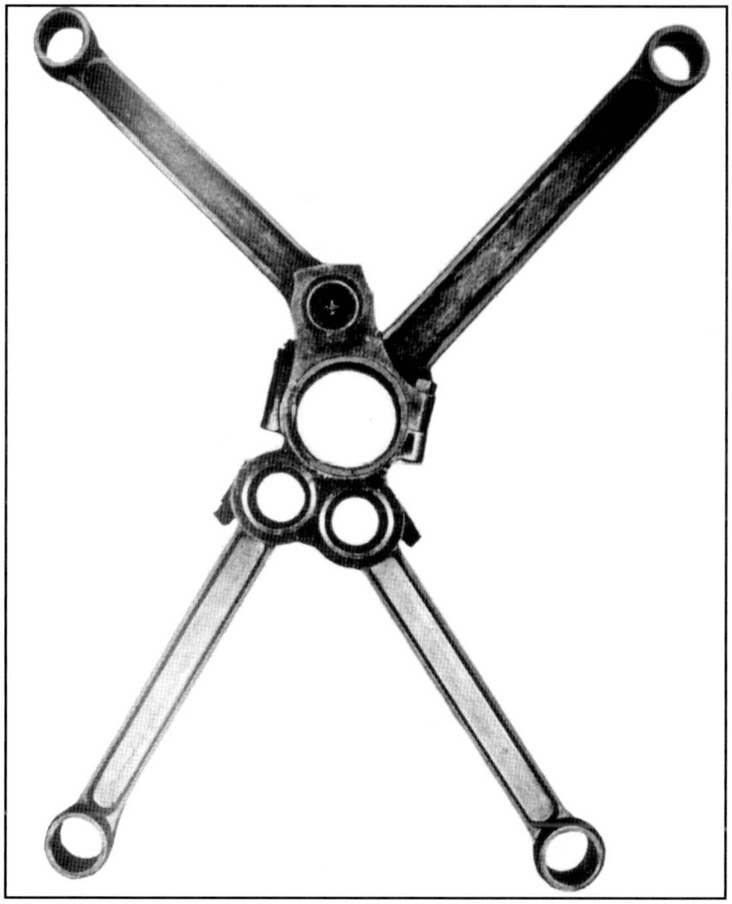

Fig. 8 One set of articulated connecting-rods, consisting of one master and three link rods. (Author's collection)

"Cooling system: Only one standard Liberty-12 water pump was used, which was located as shown at A in Fig. 7, and driven direct off the rear or ignition end of the crankshaft. It was fitted with a special impeller which gave a greater capacity than the present standard type used on the twelve.

"Cooling water from the pump was sent to the twelve upper cylinders in the regular way through distributing manifolds running along either side of the crankcase. Circulation in the twelve lower cylinders was in just the reverse direction to that in the upper set, however, on account of the inverted position of the lower members. To accomplish this, the inlet and outlet connections simply were reversed, the water entering those which would be outlet connections with the cylinders right side up, and discharging from the jackets through the standard inlet openings.

"Carburetion: The carburetors for the twelve upper cylinders were fitted in the regular way, with standard intake headers. Similar headers also were used for the twelve lower cylinders, but it was of course necessary, in order to have the carburetors right side up, to change the method of fastening them to the inlet manifolds. To do this, a set of U-shaped connecting pipes or adapters was made, as shown in Fig. 9. These were attached in place of the carburetors to the lower intake headers, and the carburetors themselves were bolted to the other ends of the adapter pipes. The location of the carburetor adapters may be observed in Fig. 6.

"Ignition system: Standard Delco ignition, the same as used for the Liberty 12, was utilized on both the upper and lower halves of the 24-cylinder engine, although there was no interconnection between the upper and lower sets. As in the standard 12-cylinder motor, the left distributor in the upper set was connected to the front spark plugs in all twelve of the upper cylinders, while the right upper distributor was connected to the rear spark plugs in the same set. Similarly, the lower left distributor served one set of plugs in the twelve lower cylinders, and the right lower distributor the remaining set." The firing order of the 24 cylinders is shown in Fig. 11.

"It will be noted that two upper cylinders fired in sequence, then two below and so on. This was necessary on account of the position of the cylinder banks in relation to the crankshaft. The upper rows of cylinders were 45° apart, as in the Liberty-12, so that the angle between the

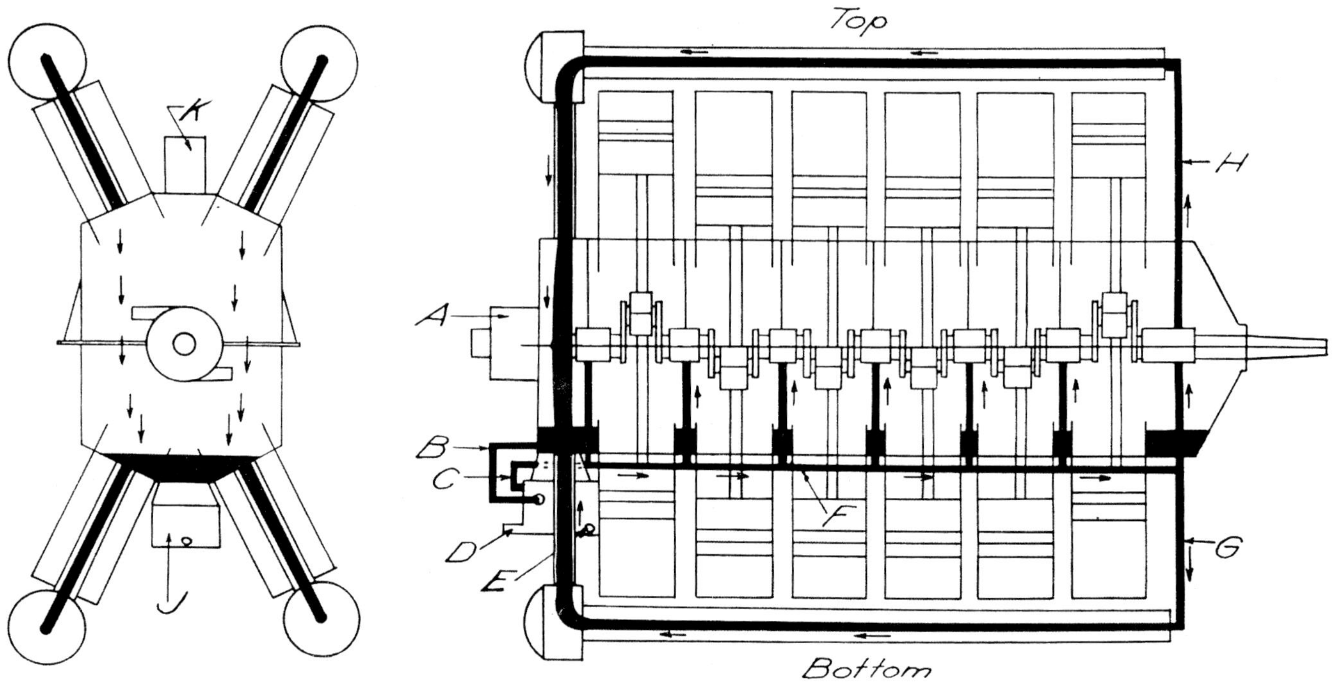

Fig. 7. Diagram of oiling system, showing how projecting sleeves of lower cylinders acted as stand-pipes, making possible the use of the cylinder side of the lower crankcase as a sump.

upper and lower cylinder banks on the same side of the engine was 135°. As the crankshaft was the standard six-throw type, *(Author's note: A special heavy-duty crankshaft designed by C. H. Vincent was used in this test.)* with crankpins 120° apart, it will be apparent that when Nos. 1 and 6 crank arms, for instance, were in the firing position for the upper left cylinders, Nos. 3 and 4 throws, 120° ahead, were 75° past the upper right cylinders, with 60° still to travel before reaching firing position for the lower right cylinders. However Nos. 1 and 6 crankpins only had to travel 45° to reach firing position for the upper right cylinders, or 15° ahead of Nos. 3 and 4 throws for the lower set. The firing sequence between the four rows of cylinders was, therefore, upper left, upper right, lower right and lower left, and the gaps between power impulses were 45° and 15°, as compared to 45° and 75° for the Liberty-12.

"The timing range for the 24-cylinder engine was the same as for the twelve, and the upper and lower distributors were synchronized exactly the same as the left and right units in a 12-cylinder motor."

(Note that all figures mentioned in the quoted text were not reproduced here.)

Tests: The engine was mounted on the test stand of a standard dynamometer in the McCook Field test laboratory which had recently been set up. It was purposely built in such a way that two dynamometers could be connected in tandem by a shaft to provide sufficient power

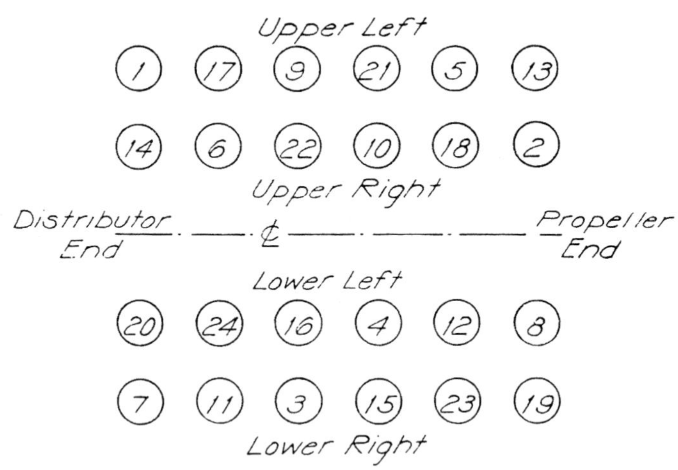

Fig. 11. Diagram showing firing order of 24-cylinder Liberty.

absorption for testing of very large engines. This was the first instance of this feature being used.

The first trial consisted of a 2-hour run-in after which a short power test was run with readings taken under full load at two speeds, 1,185 RPM and 1,325 RPM. Output at those speeds were 585 HP and 650 HP respectively. The engine functioned quite well and did not smoke excessively from either upper or lower cylinders.

Fig. 6. Propeller end of the Liberty 24. The adapter tubes for the lower set of carburetors can be seen between the lower banks of cylinders. (Author's collection)

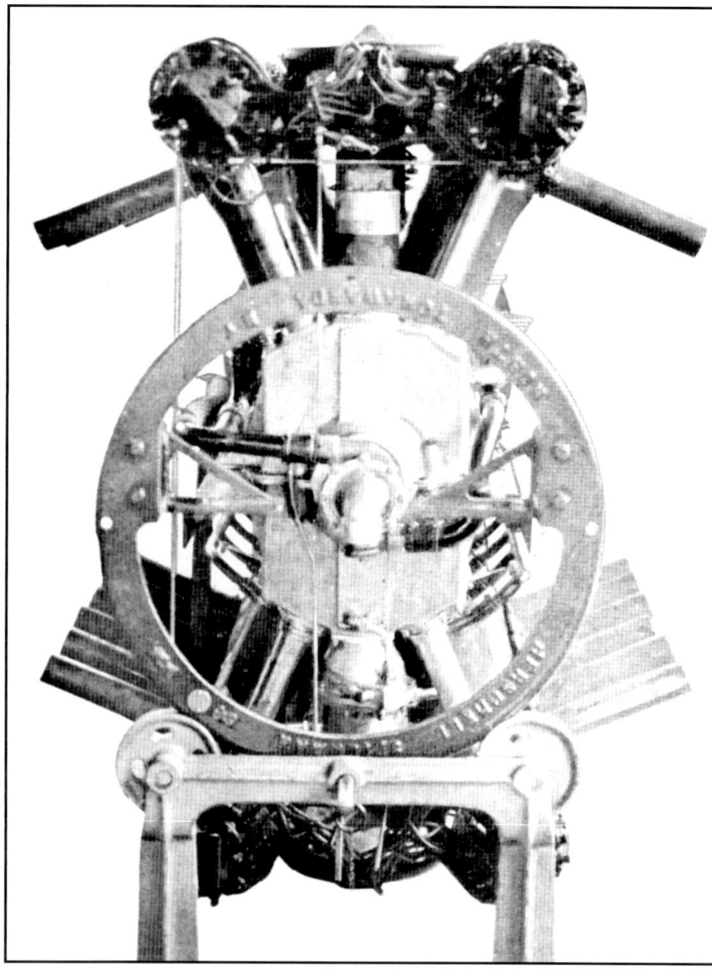

Fig. 3. Rear end view of the Liberty 24. The oil pump adapter can be seen mounted in the generator position of the lower half of the crankcase. (Author's collection)

Lionel M. Woolson
1888-1930

Mechanical and Aeronautical Engineer.
He was born in Los Angeles and received his engineering education at St. Paul's School in London.
Worked for the Bijur Motor Appliance Co. from 1912 to 1917, during which time he was issued 14 patents.
He joined the Air Service as an engineer at McCook Field and rose in rank from Lt. to Capt. during 1917-1918.
Woolson was an accomplished airplane pilot as well as a designer of aircraft engines. After the end of the war, Woolson went to work for Packard as Engineer in charge of tests.
He continued to work at Packard for the remainder of his life and became their chief aeronautical engineer.
He was the principal designer of Packard's aero and marine engines from about 1923 until 1930.
He designed their 1500 and 2500 series of aero engines as well as the X-2775 and the DR-980 Diesel aero engine which was the first Diesel to power an airplane and which set the world heavier-than-air non-refueling endurance record in 1931 which stood until December 23, 1986.
He was a very prolific inventor and by the time of his death had been issued a total of 191 patents.
He died on April 23, 1930 in an aircraft crash near Attica, New York.

"A second test was run later, at which time a power curve was obtained, together with a gasoline consumption reading at 1,400 RPM." The power curve is reproduced in Fig. 12, and for comparison another curve is shown in dotted lines, representing the amount of power which theoretically should be obtained from the engine, on the assumption that it should develop double the power of a 12-cylinder Liberty. In fact, it would be reasonable to expect the power to exceed that amount as the 24-cylinder engine would have less losses to friction.

It can be seen from the plotted power curves that the output obtained during the full power curve test did not reach expectations. The initial two-point test, also plotted on the graph, did produce expected results, however. The report noted this fact and their explanation was that the lower 12 cylinders had their carburetor intakes very close to the floor because of the lack of clearance resulting from the very deep engine. This restricted air flow and also allowed intake of accumulated exhaust gasses which had dropped to floor level.

The X-24 Liberty is tested in the McCook Field dynamometer test lab. Note the right dynamometer is connected by a shaft to the left one, making up a double unit. The engine mount position of the right dynamometer is open. (Author's collection)

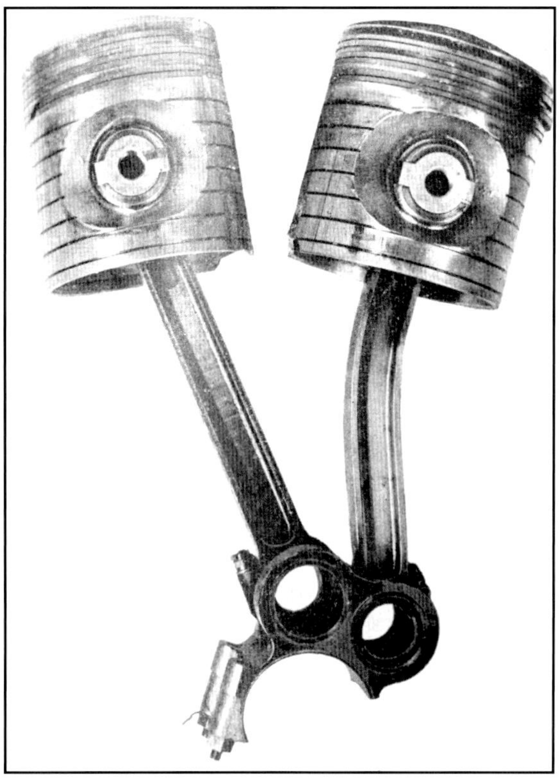

Fig. 13. Two auxiliary link rods which broke in the test, showing how the failure occurred in the special cap on one side and in the master rod on the other. (Author's collection)

All of this no doubt had detrimental effects and did, to some degree, reduce output. It is of interest, however, that the initial two-point test did produce the expected output. It is also noteworthy that the engine had just had its power recorded at 1,782 RPM (last plot on the curve) when the connecting rod of No. 1 cylinder failed and ended the test. Between the last two plots on the curve (1,617 and 1,782 RPM) the curve was expected to remain fairly flat, yet it dropped off considerably. This could well indicate the pending rod failure was already causing power loss. Now back to some direct quotes.

"Immediately after obtaining the power reading at 1,782 RPM, the engine failed, the connecting rods of No. 1 cylinders at the rear of the engine giving way and piercing the aluminum crankcase, which of course, made it necessary to discontinue the test. It was found on tearing down the engine for inspection, that the connecting-rod had broken on one side of the crank-pin, and the connecting-rod cap on the opposite side."

It was demonstrated that there was no reason to fear excessive spark plug fouling in the lower cylinders and it was not necessary to change any plugs during the tests.

"As regards vibration, the engine ran quite smoothly at speeds above 1,000 RPM, but below this point it had some tendency to oscillate around its center."

Examination of the failed connecting-rod cap showed crystallization from poor material or lack of heat treatment. Teardown showed bearings to be in good condition.

In conclusion it was stated that if put in production the engine could be constructed quickly and with little extra tooling and that it should have a useful life comparable to other types.

No further testing was done on an "X"-type Liberty but the lessons learned regarding the relative ease with which the engine had successfully been operated in the inverted position were not lost.

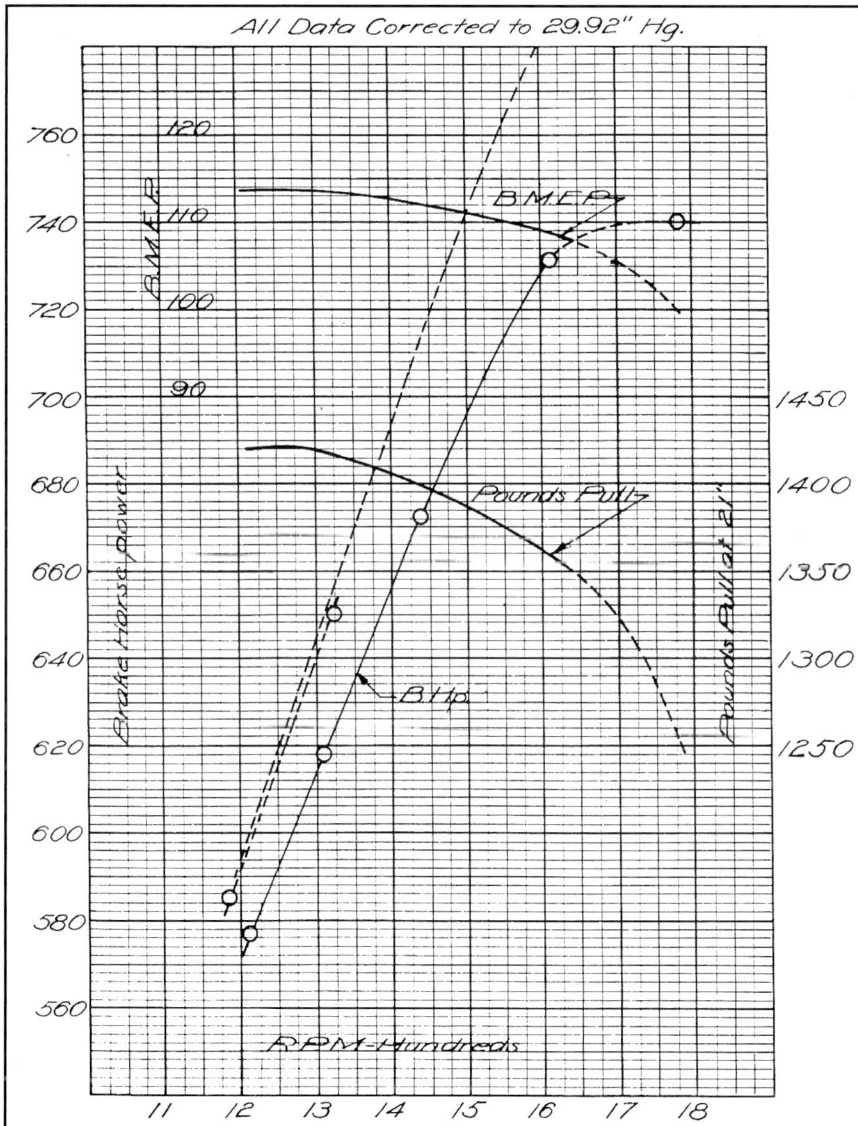

Fig. 12—Power curve obtained in test of 24-cylinder engine. The long dotted curve indicates the power which theoretically should be obtained, and the short dotted curve just below it shows the power actually obtained in the first test

THE INVERTED LIBERTY 12

During the week of December 28-January 4, 1919 it was noted that "A Liberty 12 engine was overhauled for an upside-down test." Obviously the X-24 Liberty success had sparked continued interest in operating the Liberty in the inverted position. Some testing continued and on February 20, 1919 and official test report serial 680 was issued as "Test of Inverted 12-Cylinder Liberty Engine."

This testing took place on February 5 and 7, 1919. The conclusions arrived at were that inverted operation would be practical but steps would be necessary to reduce oil consumption and exhaust valve overheating. The project was shelved until early 1923. I will cover the subject in depth in *Chapter 8, Allison and the Liberty*.

DELIVERY OF LIBERTY ENGINES TO THE ALLIES

The reader is directed to Chapter 3 for the story of the initiation of Liberty engine contracts by the Allied nations.

THE BRITISH LIBERTIES

The initial British order was for 3,000 engines and was executed in February of 1918. Only about 100 engines had been accepted by the Air Service by that date but production was expected to soar to grand heights within mere months. Thus deliveries were scheduled to begin with the first 10 in March, reach 700 per month in September and deliver the final 720 of the 3,000 in October. As we know from the story of actual production told prior to this, output of engines was not nearly up to predictions. Deliveries to the Allies were thus scaled back.

As already noted, General Cormack and Commander Briggs of the British staff suggested at a March 7 board meeting that facilities be set up in England to assemble and test Liberty engines, and complete sets of parts be shipped for that purpose. It was requested that shipments be divided in half and shipped on two different steamers to avoid total loss in case one steamer did not make it. Although the idea was seriously considered for some time, it was ultimately discarded.

After being notified that deliveries would not be up to earlier expectations, the British replied by cable "Referring to my cable of yesterday, please see Aircraft Board at once and explain the extreme gravity of situation as regards supply of Liberty motors in this country. Such motors are now embodied as integral parts of British program and are intended for very valuable new models of aeroplanes for specific important air operations which must be carried out period July-September. It is vital to allied air operations that whatever Liberty motors may be produced, some should be put at once to the most valuable use. The doubts expressed as to deliveries in your 960 have dismayed us extremely, —"

Evidently General Pershing had also been advised of "the gravity of the situation" and was swayed by it as at the April 16 board meeting "Col. Montgomery stated that the Chief of Staff had been advised by General Pershing that Allied nations had to be fully provided with raw and finished aircraft materials, even at the expense of temporarily delaying the American program; therefore the English request of Liberty engines of the estimated output of July 1st should be given priority, as well as those of France and Italy; —" It was further stated that 734 engines were allocated to the Navy.

The effects upon British deliveries were immediate, as can be seen to the right, but by this time the French and Italians had only expressed a desire for engines, they had yet to sign a firm contract.

The Adjutant General wrote the Chief Signal Officer on April 23 giving the following directive: "You will allocate Liberty engines produced prior to July 1st as follows: (a) To the British Government 980 (b) To the Navy Department 734. You will be informed at a later date as to the number of Liberty engines which are produced prior to July 1st that will be allocated to the French and Italian Governments."

Although it is not clear just how many were on firm contracts at this time, an additional 5,000 engines were eventually ordered for a total of 8,000.

British deliveries, as accurately as can be determined, are listed below. Deliveries from March to the end of the war come from the official British history *The War In The Air,* published in a series of six volumes by the Imperial War Museum between 1922 and 1938.

ACTUAL BRITISH DELIVERIES

Month	Num.	Accumulated total
March	10	10
April	20	30
May	175	205
June	225	430
July	620	1,050
August through Nov. 11	Priorities changed and deliveries all but halted until the end of the war.	
After Nov. 11	1,202	2,252

By July 1, some 2,389 Liberty engines had been shipped from the producing factories. 430 went to England and if the Navy got their 734 that left 1,225 for other users. Production rose rapidly after that time. The British got another 620 engines in the month of July and this actually put them 70 engines ahead of schedule.

From that point, however, priorities of allocation changed and the British received no more Liberty engines until after the end of the war.

British DH-9a in use by the RCAF. (Courtesy of Peter Bowers)

They had planned a great deal of their day bomber production as the DH-9a aircraft powered by the Liberty 12 engine and continued with further engine orders to power them. Another 2,500 were contracted for, and verbal orders given for, an additional 2,500, for a total of 8,000 engines. The BAP promised deliveries would restart in January of 1919 at the rate of 750 per month during the first-half of the year. We currently have no records as to post-war British deliveries, but we do for the French. French deliveries began very rapidly following the war and during February, 1919 reached 750 per month. It is thus likely that the 1,202 needed to complete British deliveries were in British hands by the end of January, 1919.

Data plate from a very late Packard Liberty. MNFG NO. 6425, U.S. NO. 71680 with an acceptance date of 2-21-19. (Courtesy of John F. Mac Arthur)

From the cover of the first issue Liberty manual of 1917.

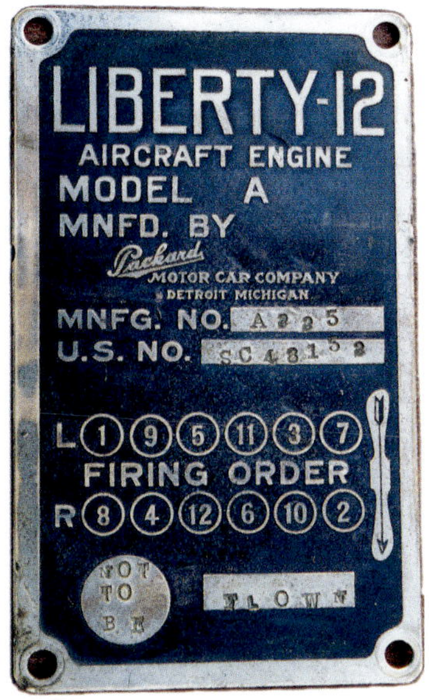

Data plate from Packard School engine MNFG NO. A225, U.S. NO. 43152 marked NOT TO BE FLOWN. (Courtesy of John Ulrich)

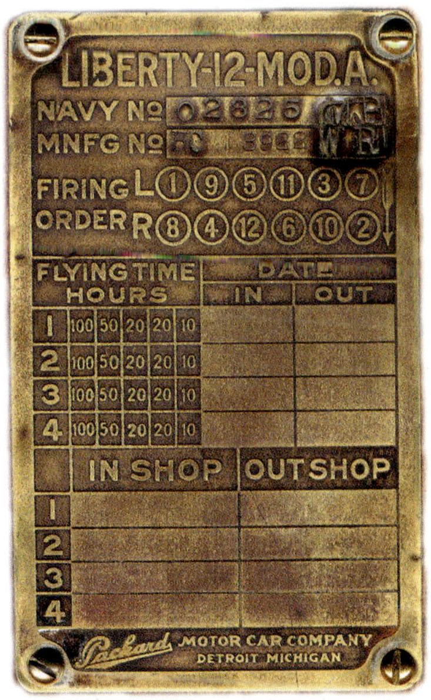

Navy Liberty data plate. The Navy plates were made of brass and of a different design from the Army plate. This one is from Packard MNFG NO. 3922, Navy NO. 02825. (Author's collection)

Data plate from Buick L-8 MNFG NO. BX-12, U.S. NO. 94482. (Author's collection)

Data plate from the only Trego Liberty accepted by the Army, MNFG NO. T2, U.S. NO. 7402. (Courtesy of Jim Aamodt)

L-4 MNFG NO. 1. The engine is in the Air Force Museum collection and the only example currently known. The right view shows its battery-type ignition and generator. (Author's photos)

L-4 MNFG NO. 1. From the Air Force Museum collection. (Author's photo)

L-6 MNFG NO. 4 equipped with late design intake manifolds and magneto ignition. From the Air Force Museum collection. (Author's photo)

L-6 MNFG NO. 4. Magneto ignition and Nelson machine gun synchronizer are clearly visible. (Author's photo)

L-6 MNFG NO. 4. Buick L-8 MNFG No. BX12 is in the background. (Author's photo)

L-8 MNFG NO. BX-12. This engine is in the Air Force Museum and until BX-15 was found in 1985 (see below) it and EXP NO. 1 were the only known surviving L-8 engines. (Author's photos)

L-8 MNFG NO. BX-15. This last Buick L-8 built was sold surplus following WW I, converted to marine use and installed in this boat, Liberty The Second. The boat capsized during a 1922 race and went to the bottom of Pennsylvania's Conneaut Lake. The boat was found and raised on July 13, 1985. Both the boat and engine were restored to operating condition. At the right is the boat and engine as photographed the day it was raised. At the left is the boat and engine as it appears today. (Courtesy of Dan Styduhar)

Two views of a Nuffield Liberty Mark V engine from the Mike Dennis Roberts collection. Note the distributor heads have been removed and placed on the cam housings. (Courtesy of Jamie Fairchild)

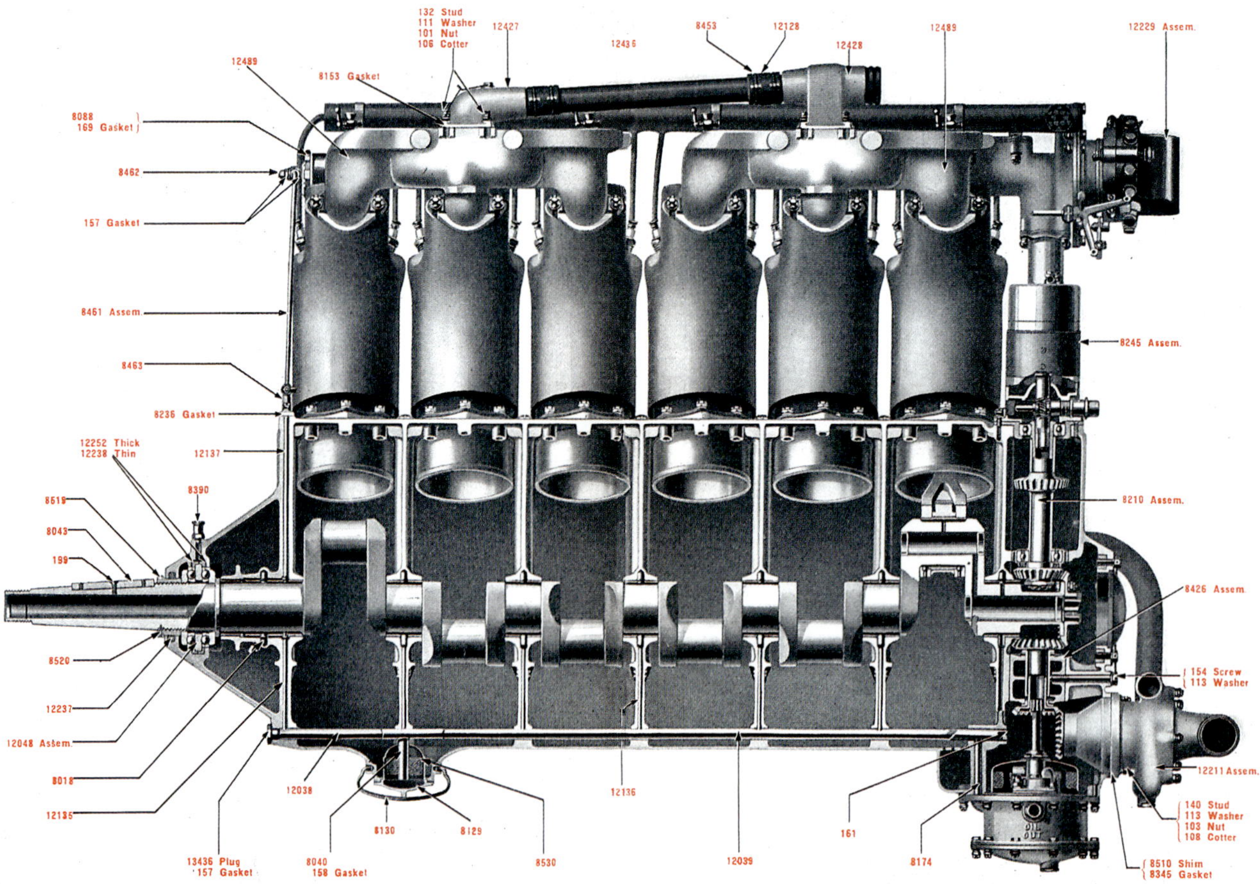
132 Stud
111 Washer
101 Nut
106 Cotter
12427
12436
8453
12128
12428
12489
12229 Assem.
12489
8153 Gasket
8088
169 Gasket
8462
157 Gasket
8461 Assem.
8245 Assem.
8463
8236 Gasket
12252 Thick
12238 Thin
12137
8390
8210 Assem.
8519
8043
199
8426 Assem.
8520
154 Screw
113 Washer
12237
12048 Assem.
12211 Assem.
8018
12038
12136
140 Stud
113 Washer
103 Nut
108 Cotter
12135
8130
8129
161
13436 Plug
157 Gasket
8040
158 Gasket
8530
12039
8174
8510 Shim
8345 Gasket

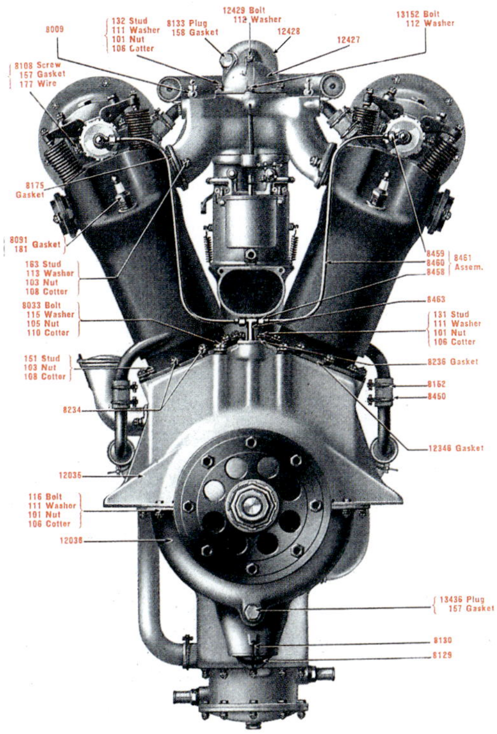

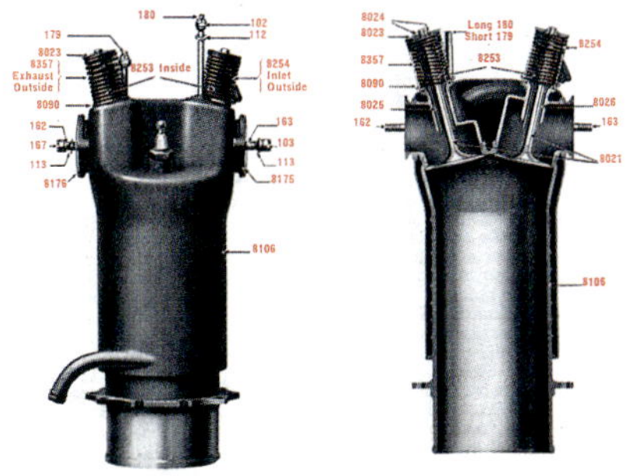

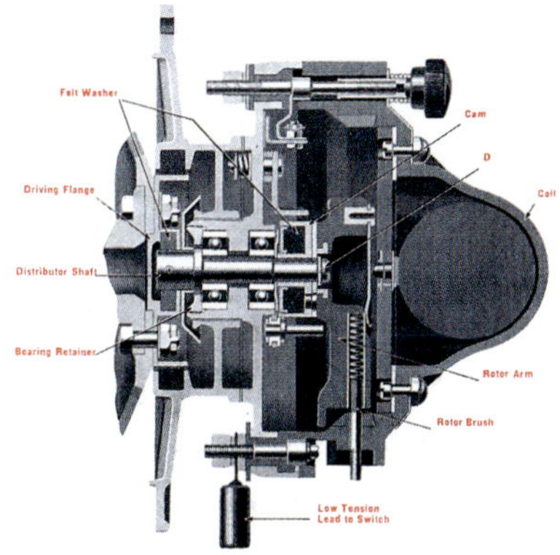

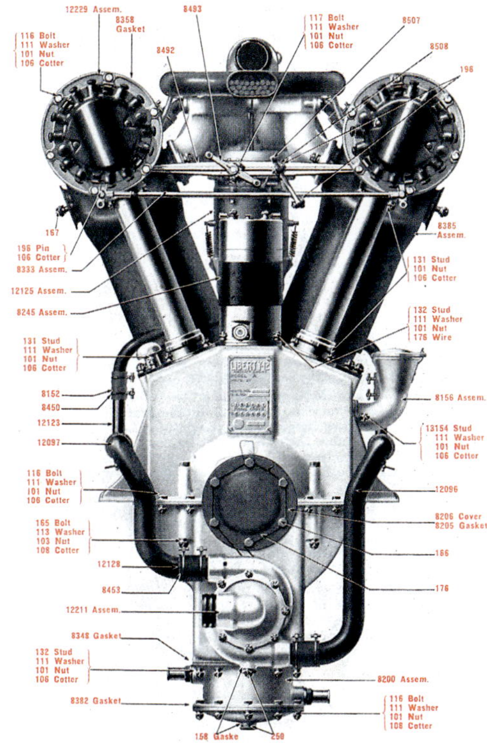

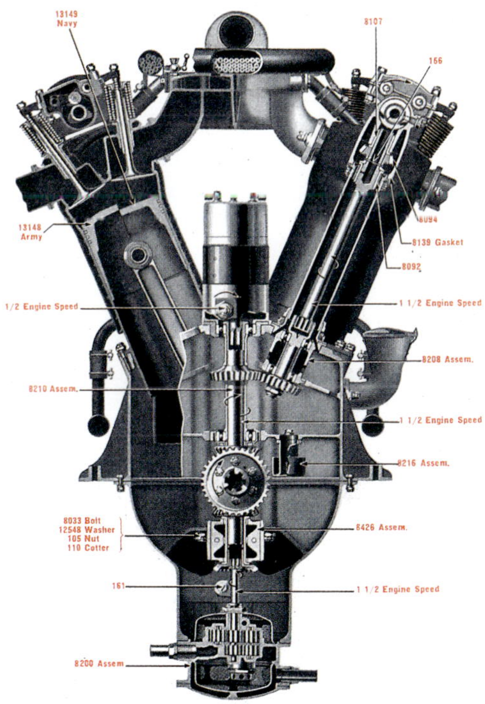

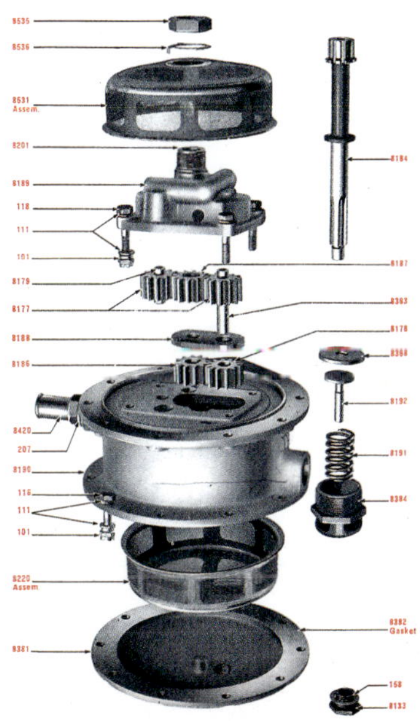

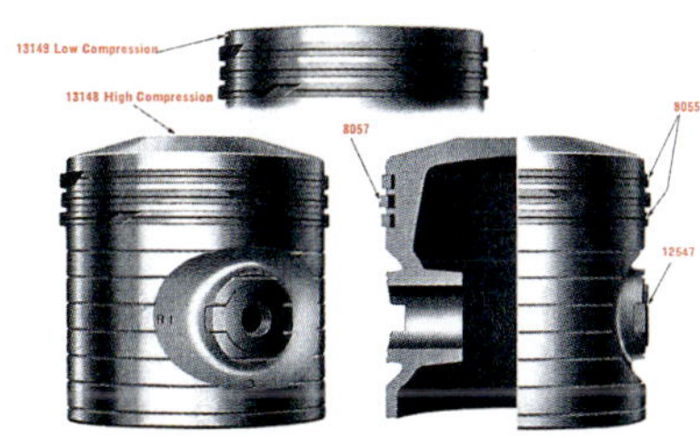

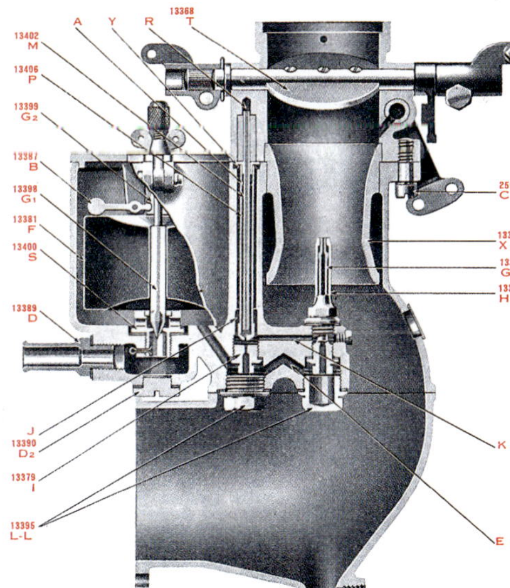

On the facing page and above are a number of sectional and exploded views of the Liberty 12 as they appeared in the first official manual produced in late 1917. These illustrations not only show much of the details of construction but they also show part numbers (in red) of many of the parts. (Author's collection)

This is the last Liberty engine flown by the Army Air Service. The engine is a Lincoln with original acceptance date of 12-3-18. It was rebuilt to current standards in 1927 and assigned AC 27-926. It was reconditioned again on 8-22-28 by the Allison Engineering Co. and installed in AC 27-204 on 10-3-33. Its final service was carefully recorded and upon completion of its last flight, which was on 9-13-35 from Fort Clark, Texas to Kelly Field, Texas in Douglas C-1C Ambulance AC 27-204, the engine was then removed and transferred to the Wright Field Museum. The Allison rebuild plate is on top of the nose and the relocated original data plate on the right side of the nose. It is currently in the Air Force Museum. (Author's photos)

The only surviving Liberty B engine known to exist, this engine is in the Air Force Museum. It is unique in several respects in addition to being the only remaining example of a rare model of the Liberty. It is equipped with a Form F supercharger for high altitude use and the nose cap has been modified to carry what appears to be control linkage for a variable pitch propeller. The use of this type propeller would maximize the abilities of both the geared and supercharged features of this engine over the wide range of altitude in which it would operate. The data plate (shown in the inset) calls this an Epicyclic Geared Liberty Engine (not Liberty 12B), as was the most common practice of the Air Corps. The rebuild date was 3-3-27 and the manufacturer's number was 215-G. Only 206 of this type are currently known to have been built so the author cannot explain a number nine higher. The crankcase of this engine is stamped with a Navy inspector's mark and is one of the 200 built under Navy contract in 1919-1920 and later transferred to the Army.
(Courtesy of John Leonard)

At left is a cut-away Russian M-5 Liberty displayed in the Russian Air Force Museum at Monino. (Courtesy of Harry Woodman) ***At right is a standard Buick-built Liberty on display in the Museum of Flight at Seattle.*** (Author's photo) ***The only difference noted by the author is the supply of oil to the thrust bearing directly from a pressure oil line on the M-5 whereas the standard Liberty uses an oil cup.***

Above: Capitol marine conversion of a Liberty 12. Courtesy of Walter Hill. Below: Capitol marine conversion of the only Trego-built Liberty 12 accepted by the Air Service. This is the only know surviving Trego Liberty. (Courtesy of Richard McGraw)

This double-V 24 cylinder Liberty was designed and built as a marine engine. Built with four banks of six Liberty cylinders, the center "V" operates as a standard Liberty 12. The two outer banks each use a standard Liberty Zenith carburetor with adaptor manifolds. Rods are of articulated type and the crankcase is unique. (Courtesy of Warren Greatbatch)

Above is a Gar Wood TS-26 500 HP marine conversion of a Liberty. Notice the use of four outside carburetors just as did the 1918 Engineering Division 500 HP engine. Courtesy of Sierra Boat Co. Below is one of two Gold Cup marine engines built by Gar Wood for the 1927 race. This engine was built on a Liberty crankcase but virtually all other components were of Gar Wood design. (This engine is in the collection of Craig Magnusson)

By December of 1918, 885 DH-9a aircraft were reported in service. Some of this model were powered by the 350 HP Rolls-Royce Eagle VIII engine but most by the Liberty 12. In addition, the DH-10 was in development and was powered by two Liberty 12 engines.

The DH-9a with Liberty power was reported in active service on the Western Front in R.F.C and R.A.F. Squadrons 18, 99 and 110, and Naval Squadron 5 (later 205) before the war ended.

As can be seen from the list of deliveries, more than half of the British total were delivered after the war ended. The remaining 1,202 engines were purchased as partial fulfillment of contract agreements and brought the total to 2,252 engines.

As will be noted in Chapter 10, the British put these engines to good use in the years following the war. Just why the number 2,252 of a total of 8,000 ordered was settled on we do not know. According to the brief history written by Col. Edgar S. Gorrell, official historian of the A.E.F., the British paid $16,589,718.90 as settlement for these 2,252 engines and all spare parts and other equipment associated with them. If the total is divided evenly among the total engines, this comes to $7,366.66 per engine. Obviously the material delivered in addition to the engines was considerable, as the official cost of the engine listed by the British was £1,215. With the exchange rate in 1918 of $4.76 per £, that comes to $5,783.40. This figure seems reasonable in light of the additional cost of packing and shipping to England and the fact that early engines cost the US government $5,600 each with the price being reduced to $4,400 late in the war.

THE FRENCH LIBERTIES

The French purchased 6 engines for testing in early 1918 and finally contracted for 250 engines in July. Before the war ended another 2,650 engines were contracted and, in addition, verbal contracts were placed for 6,000 more. This resulted in firm executed contracts for 2,906 and additional commitments raising the total to 8,906.

After the initial 6 engines sent for testing, no direct shipments were made to the French by the BAP until about October of 1918. Between then and November 11, 1918 283 engines were shipped directly and 122 from stocks already in France under AEF control, making a total of 405 officially delivered by the Armistice.

The final settlement for delivery of Liberties to the French was made late in 1919 and was for a total of 3,575 engines plus 3,310 sets of spares.

Although no details of the dates of post-war shipments to the British have surfaced, this is not true of the French shipments. From a surviving memo of about January 15, 1919; *(It is unfortunately undated, but its content dictates its composure at very close to January 15, 1919.)* we know the dates and quantities of the final shipments to the French along with the quantity of spares, their costs and contract cancellation costs.

British Engine Cost

ENGINES

Type.	*Normal H.P.*	*Number of Cylinders.*	*Price.* £	*s.*	*d.*
Adriatic (B.H.P.)	230	6	1,089	0	0
Atlantic (Galloway)	500	12	2,073	10	0
Anzani	100	10	550	0	0
Arab (Sunbeam)	200	8	1,017	10	0
Adder (Hispano Wolseley W.4b)	200	8	946	0	0
B.R.1	150	9	643	10	0
B.R.2	200	9	880	0	0
Beardmore	120	6	825	0	0
Beardmore	160	6	1,045	0	0
Clerget (7.Z)	80	7	484	0	0
Clerget (9b) and (9bf)	130	9	907	10	0
Clerget (11eb)	200	11	951	0	0
Curtiss (Ox. 5)	90	8	693	10	0
Dragonfly (ABC)	320	9	1,072	0	0
Eagle II, III, and IV (Rolls-Royce)	250	12	1,430	0	0
Eagle V (Rolls-Royce)	275	12	1,721	10	0
Eagle VI and VII (Rolls-Royce)	350	12	1,919	10	0
Eagle VIII (Rolls-Royce)	360	12	1,622	10	0
Falcon I, II, and III (Rolls-Royce)	190	12	1,210	0	0
Falcon IV (Rolls-Royce)	225	12	1,617	0	0
F.I.A.T. (A.12 Bis.)	300	6	1,617	0	0
Gnome	80	7	430	0	0
Hawk (Rolls-Royce)	75	6	896	10	0
Hispano Suiza (French)	200	8	1,004	0	0
Lion (Napier)	400	12	1,897	10	0
Liberty	400	12	1,215	0	0
Le Rhône	110	9	771	10	0
Le Rhône	80	9	620	0	0
Monosoupape	100	9	696	0	0
Maori (Sunbeam)	250	12	1,391	10	0
Pacific	600	12	2,145	0	0
Puma (B.H.P.)	230	6	1,089	0	0
R.A.F. 1a	100	8	522	10	0
R.A.F. 3a	200	12	1,210	0	0
R.A.F. 4a	150	12	836	0	0
Renault	80	8	522	10	0
Salmson	200	9	968	0	0
Viper (Hispano Wolseley) W.4a	200	8	814	0	0

The above list of prices of British aircraft engines is extracted from a Ministry of Munitions "Priced Vocabulary of Aircraft Supplies, 1918-1919." It includes a few early post-war and experimental engines as well as those produced and purchased by the British during the war. Prices are, of course, in English pounds. The exchange rate in 1918 was $4.76 per £. The price of the Liberty was thus $5,783.40, the R-R Eagle VIII $7,720.72 and the Napier Lion $9,029.72. The cost of any others of interest may be calculated using the stated exchange rate.

Four hundred and five engines, with spares, had been received by November 11. Between that date and January 1, 1919, an additional 1,095 were delivered, bringing the total to 1,500 engines and a like number of sets of spares. From that time deliveries were made at a set rate of 750 per month until the agreed-to number of 3,575 were in French hands. Spare parts deliveries continued also, but at reduced rates.

ACTUAL FRENCH DELIVERIES

	Number		Accumulated Total	
Month	Engines	Spares	Engines	Spares
Oct.-Nov.	405	405	405	405
Nov.-Dec.	1,095	1,095	1,500	1,500
January	750	690	2,250	2,190
February	750	600	3,000	2,790
March	575	530	3,575	3,310

Documentation available indicates the planned settlement as of about January 1, 1919 was to be a delivery of 4,875 engines and 3,615 sets of spares. The plan was changed, probably within the next 30 days, and the final settlement was as noted above.

Exactly where these engine shipments came from cannot be ascertained. Some were shipped from the Airplane Spare Parts Depot No. 2 which was a section of the Air Service Production Center No. 2, a huge AEF aircraft facility located at Romorantin, France. We have specific documentation of 428 shipped from this location to various French military locations as well as directly to French aeronautical production plants between September 8 and November 25, 1918. ASPC No. 2 records note shipment of 90 engines to British forces. An additional 1,536 engines remained on hand at that location which could have been used to supply either British or French delivery requirements. Sufficient new engines had been shipped to various locations in England and France for AEF use to supply remaining requirements needed to fulfill shipments to the British and French.

FRENCH USE OF THEIR LIBERTY ENGINES IN BREGUET 14 TYPES

In 1917, the French government decided to standardize aircraft production on a limited number of types. One of these was the Breguet 14 that was to be used both in the day-bomber and army-cooperation roles. By July of 1918, the Breguet 14 was being produced by eight firms. The bombardment B2 version was produced solely by Michelin whose maximum production was about 130 airplanes per month, barely enough to sustain 15 bombardment escadrilles. The remaining manufacturers produced the army-cooperation A2 version.

A French Breguet 14A2 with Liberty 12 power. Current information indicates most of the Liberty 12 engines purchased by the French and used by themselves in aircraft were used to power Breguet 14A2 or B2 airplanes. (Courtesy of Peter Bowers)

The availability of Renault motors was less than required for all Breguet production. By mid-1918, it was necessary to use the less-desirable Fiat A-12 motors for Breguets supplied to escadrilles that were then being re-equipped. Out of 58 army-cooperation escadrilles, 26 were equipped with Fiat motors. Moreover, the American Air Service could not obtain enough B2 types to maintain their one Breguet-equipped bombardment squadron. The anticipated availability of large numbers of Liberty motors offered a solution to the limitations imposed by the shortage of Renault motors.

Accordingly, two Breguet A2s and a B2 (4360) were fitted with Liberty motors. These tests became the basis for series production when sufficient Liberty motors were available. The following tables indicate actual deliveries of Liberty motors to the French in the last quarter of 1918 in as much detail as we know.

As noted above, between sometime in October and November 11, 405 engines were delivered to the French. From what we currently know it would appear that 283 were shipped direct from the BAP and 122 from the Airplane Spare Parts Depot No. 2, Air Service Production Center No. 2 at Romorantin.

We have no way of knowing specific destinations of the direct shipments but the destinations of those shipped from ASPC No. 2 are known and are as follows:

Pre-Armistice, Sept. 8 - Nov. 11, ASPC No. 2 French Liberty delivery locations

Manufacturer	Number
Bellanger	9
Michelin	58
Schmitt	0
SIDAM	0
Breguet	50
depots	5
Total	122

The remaining 306 engines recorded as shipped from ASPC No. 2 to the French between November 12 and 25 were as follows:

November 12-25 ASPC No. 2 French Liberty delivery locations

Manufacturer	Number
Bellanger	29
Michelin	142
Schmitt	26
SIDAM	4
Breguet	1
depots	104
Total	306

**Estimated Production of Airplanes By European Manufacturers For Delivery To A. E. F.
Extracted from a BAP document dated September 27, 1918.**

		Estimated Production								
			1918				1919			
Type	Engine	Aug	Sep	Oct	Nov	Dec	Jan	Feb	Mar	Total
Breguet	**Renault 300**	**60**	**45**	**27**	**27**	**27**	**21**	**6**		**213**
Breguet	**Fiat A12 240**	**66**	**18**	**18**	**18**	**18**	**12**	**6**		**156**
Breguet	**none**			**150**	**200**	**200**	**250**	**250**	**250**	**1300**

Author's note: This chart specified the engine as none under the last Breguet line entry. This represents the production estimate of the 1300 Liberty-powered machines ordered in May of 1918. Since the chart was dated September 27, 1918, it would be safe to assume the numbers noted in August and September were already delivered. Indications are that deliveries for October and beyond were not made.

The French converted Breguet B2 No. 4360 to Liberty 12 power and tested it at Villacoublay on September 20, 1918. The engine used was Packard U.S. NO. 17657. Another Liberty-powered Breguet B2, No. 6463, was shipped to the Air Service Engineering Division and assigned U.S. NO. 94097 P-148.

Although the tables indicate that Liberty motors were supplied to several firms, only three are known to have entered into series production, these being Breguet, Michelin, and Bellanger. Other firms may have anticipated using the Liberty motor but there is no evidence that any such aircraft were ever produced.

Breguet: The Breguet firm produced a Liberty-powered version known as the Ap2. This was intended as a long-range fighter to provide protection to bombers. An American fact-sheet on this type states: "In November 1918 there were 50 of these planes under construction at the Breguet factory at Villacoublay, and were substantially the same as the model equipped with the Renault motor except for very minor changes necessitated by the use of the Liberty motor...this plane was never put into quantity production and it never reached the front." The number of 50 is consistent with the number of motors delivered to Breguet. Known serial numbers fall in the range 9060 - 9101. At least one Renault-powered machine was recorded in this group. However, it is very likely that all 50 were constructed. One such machine, Ap2 9089, was photographed with the markings of escadrille R.246. Therefore, the type did enter French service after the war. Two of these machines, Ap2 9095 and Ap2 9101, were recorded in the USA on 12 July 1919. Ap2 9095 was at McCook Field and was transferred to the Post Office Department on 4 April 1921; Ap2 9101 was wrecked at Bolling Field on 17 August 1920.

Michelin: In May 1918, the Americans placed an order for 1300 Breguet B2Ls with Liberty motor. One prototype, B2L 4360, was completed in August and tested at Villacoublay in September. It appears that series deliveries did not commence until about February 1919. The earliest known example was B2L 12536, one of 10 or more furnished to the Romanian government. The highest number is 13021. Two machines each may have been furnished to Poland and Japan. One example was in service with French escadrille Br.134. None in this series are known to have been furnished to the Americans. One could speculate that Michelin produced about 100 B2Ls.

Bellanger: The Bellanger firm commenced delivery of Fiat-powered Breguet 14 A2s in July 1918 and continued production at the modest pace of about 25 per month. However, they were able to garner a contract to produce Liberty-powered B2Ls and, at their normal production rate, would have delivered the first of this series in about February 1919. A photograph taken inside the Breguet factory at Villacoublay shows early-production Bellanger-built B2Ls undergoing final assembly alongside of Breguet-built A2s. Apparently, the Bellanger firm relied on the facilities of Breguet for this final stage of assembly. In any case, the Bellanger machines are painted with American roundels indicating production on behalf of an American contract. Ironically, only one such machine appears to have been accepted by the US Air Service. This was B2L 6463 that was brought to the USA where it became AS 94097. It was first recorded at McCook Field on July 18, 1919.

Somewhat later, it was reconditioned, given the project number P-148, and in September 1920 used to test new camouflage schemes. But there is no record it was ever given a performance test. Known Liberty-powered B2Ls fall in the ranges 6463 - 6522 and 6831 - 6986. Thus, it seems likely that Bellanger production of B2L aircraft exceeded 200 by some margin. Interestingly, Bellanger-built B2Ls in French service continued to carry their original American markings into the early 1920s, for example B2Ls 6469, 6488, and 6522 of the early series. Airplanes in the later series carried French markings indicating that the French continued to use the Liberty motor after the cancelled American contract had run its course.

In summary, it is very likely that total French production of Breguet types with Liberty motor was not less than 350 airplanes. Thus, approximately 10% of French Liberty motors were actually used in airplanes. Certainly a large number would have been used in aircraft production had the war continued into 1919. To place this in perspective, about 4,000 Breguets were produced in the last nine months of 1918. The potential would easily have been to use all of the delivered Liberty motors in the expanded 1919 aircraft production program.

The research into the French Breguet 14 use of the Liberty and the ASPC No. 2 Liberty engine files were provided by Alan Toelle. Without that information this section on the French use of the Liberty would have lacked much of the detail included.

We currently have no information regarding the specific destination of any other of the French Liberties and this, of course, represents the majority. It would seem that with the exception of those few noted in Appendix F as used in French aircraft on an experimental basis, the majority were sold as surplus.

EUROPEAN TRIP OF JOHN D. RYAN SEPTEMBER 7 - OCTOBER 6, 1918

Since it may have had an effect on the use of the Liberty engines delivered to the British and French before the Armistice, as well as the total numbers finally delivered, I will relate briefly what was agreed between those nations and Mr. Ryan relative to Liberty engines during this official trip.

In August 1918, Mr. Ryan was appointed Second Assistant Secretary of War and Director of the Air Service, with general authority over the Department of Military Aeronautics and the Bureau of Aircraft Production.

Breguet 14A2 delivered to the Engineering Division, powered with a Liberty 12, assigned US No. 94097 and P-148. (Courtesy of Peter Bowers)

Mr. Ryan was requested by the Secretary of War to accompany him to Europe in order that he might become familiar as quickly as possible with the condition of the Air Service abroad, both as to personnel and equipment. With knowledge of Air Service and AEF needs he could also make needed decisions regarding the supplying of materials to France and England. This decision was brought about not only by the fact that for the first time there was one head over both the Air Service and the BAP, but because beginning in September there could be seen for the first time a large enough supply of Liberty engines to be able to furnish not only our own needs but also those of our allies.

The party left America on board the *USS Northern Pacific* on August 31, 1918 and arrived at Brest, France on September 7. Over the next several days many fields and facilities were visited and most US military officers with control and knowledge of matters relative to the Air Service and its supply, including, but not limited to the following: Major General Patrick, Chief of the Air Service Aeronautical Expeditionary Forces; Colonel Dunwoodie; Brigadier General B. D. Foulois; Chief of Air Service Zone of Advance; Colonel Gorrell, representing the Air Service at General Headquarters; Colonel Allen, in command of ASPC No. 2 at Remorantin; and others were consulted at length.

British representatives, including Mr. Winston Churchill, British Minister of Munitions, Lord Weir, Air Minister, and Sir Arthur Duckham, Master of Aircraft Production, were consulted. French representatives Premier Clemenceau, Mr. M. Luchere, Minister of Munitions, and others were consulted. The following agreements were reached and are quotes from the 1920 BAP history.

French Agreement

"A definite agreement was reached to give the French 1500 Liberty motors for shipment from the United States before January 1, 1919, all to be consigned to the Air Service Aeronautical Expeditionary Forces to be delivered to the French Government, with the provision that no deliveries were to be made for planes that were not actually ready for motors. After January 1, 1919, 750 motors per month for the first six months were to be shipped to the French direct, on contract to be formally entered into by representatives of the French Government with the Bureau of Aircraft Production in the United States, on condition that the French were to furnish to the Aeronautical Expeditionary Forces certain planes and engines in such proportions that we were to exchange horse power for horse power. In other words, calculating the Liberty motor at 400 HP, 750 motors a month called for 300,000 HP, and the French, therefore, were to deliver to us single seater fighter planes and engines for them that would require a like total horse power of motors. It was calculated that the French would, therefore, deliver to us about twice as many motors in planes during the period covered by the proposed contract, as the planes they were to furnish equipped with motors would average not over half the power per motor of the "Liberty 12"."

By looking at the actual delivery figures on the prior pages, it can be seen that the delivery schedule agreed to was followed through the month of February and into March. Just what the contract actually executed stipulated, I do not know. But it must have had some "teeth" in it to have resulted in the sale of this number of engines after the Armistice when one would surely believe the French would rather not have acquired more engines. It would also appear that the engine for airplanes trade would not be adhered to as payment was later settled in cash.

If the initial agreement was followed until the time of the Armistice, and that would be a logical assumption, then possibly 400 French airplanes were powered by Liberty engines prior to the Armistice, most being Breguet 14 types. Alan Toelle's specific research noted above appears to verify this. Deliveries beyond the Armistice were, I feel sure, under no restrictions as to the use they were put to. Current knowledge indicates very few Liberty engines were installed in French aircraft other than the Breguet types already mentioned. The reader is directed to Chapter 10 and Appendix F for what information has so far surfaced regarding eventual uses the French engines may have been put to.

The final settlement for 3,575 engines and 3,310 sets of spares came to $21,272,250.00. Early agreements were to provide 100% spares but according to the January 15, 1919 memo this number would be reduced but the expense of cancellation of the unwanted spares would be 30% of the value of those cancelled. The value of spares had been set at $500 per set. The cancellation penalty was therefore set at $39,750. This was added to the $1,655,000 value of the spares delivered for a total of $1,694,750. Subtracted from the total settlement of $21,272,250 this results in the engines costing $19,577,500, or $5,476.22 each.

BRITISH AGREEMENT

Although an attempt was made at a firm agreement with the British regarding Liberty deliveries, none was possible. The notes of the 1920 BAP History in this regard are interesting, however, and are quoted here.

"The next day, September 29, Mr. Ryan and Mr. Stettinius conferred with Mr. Churchill, Lord Weir, Air Minister, and Sir Arthur Duckham, Master of Production, and a long conference was held in the evening at the Savoy Hotel. Upon Mr. Stettinius's request that he be allowed to use the allotment of Liberty engines that we were ready to make to the British, in exchange for tanks, which the Ordnance Bureau was anxious to get from the British and which General Pershing had told Mr. Ryan he wanted very badly, 1,000 Liberty motors, in addition to the 986 that had been contracted for in the United States (shipment to be made by the 1st of January, 1919), were definitely promised in exchange for 375 "Mark VIII" tanks."

(The reader will note this figure differs slightly from the 980 engine allotment referred to on page 198.)

"Mr. Ryan and Mr. Stettinius were assured by Mr. Churchill that the trade was satisfactory and that the tanks would be delivered within a reasonable time, but it was ascertained upon further investigation the next day that it was out of the question for the British to furnish us any such number of tanks and that no "Mark VIII" tanks had yet been produced successfully, so that after several further conferences Mr. Ryan left Mr. Stettinius the 1,000 Liberty motors to be delivered to January 1st, and 4,500 additional in the first six months of 1919, Mr. Stettinius to make the best trade he could, getting tanks in return."

As seen from the prior discussion of British post-war settlements for Liberty engines, the delivered engines were paid for in cash. Ultimately 100 sets of parts for Mark VIII tanks were shipped to the US in January of 1919 but it would appear that an actual trade was not consummated.

Caproni Ca. 5 U.S. NO. 42119. This machine was 85% completed by Standard and then turned over to Fisher for completion. This was a prototype airplane which would have served for both American and Italian production. (Courtesy of Peter Bowers)

ITALIAN DELIVERIES

The story of the proposed Liberty engine orders of the Italians revolves entirely around the proposed building of one or more models of the Italian Caproni bomber. These three-engined bombers were designed by Count Gianni Caproni. There were three basic models with which we are concerned here, the Ca. 3, Ca. 4 and Ca. 5. The Ca. 4 was a triplane and the other two models were biplanes.

Examples of these types were shipped to the United States from Italy, arriving here between August of 1917 and February of 1918. Although other models were considered for a time, ultimately the Ca. 5 model fitted with 3 Liberty 12N (low-compression) engines rated at 330 HP were chosen to be manufactured for the Air Service as heavy bombers for night use. (The Italian Ca. 5 was powered by three Fiat A-12 Bis engines rated at 300 HP.)

Actually the Caproni orders involved airplanes for both the United States and Italy. In both cases it was intended to build the airplanes here and ship them to Europe disassembled. At first, assembly was proposed to be done in Italy but that was soon considered to be impractical because of the need to subsequently transport them back to France for battle use. Next, it was decided to build an assembly plant in France and by the end of the war, some progress had been made in that direction.

The process of setting up for production here proved the be excruciating. It was initially thought that the Italians would turn over to American officials both sample airplanes and complete drawings so that production could be started. In the end, that was far from the case. Count Caproni retained ownership and control of the sample airplanes which had been sent here early in the war and, in addition, failed to deliver complete drawings. Because of this, preparations for production dragged and as a result little happened until almost the end of the war.

Four prototypes were ordered built by the Standard Aircraft Corporation of Elizabeth, New Jersey, but only the first of the order was actually delivered by that company. The contract was transferred to the Fisher Body Corporation of Detroit, Michigan after continual problems with Standard. The Standard Ca. 5 was completed in July of 1918. The first Ca. 5 completed by Fisher was delivered to McCook Field for sand test in November.

With only one airplane completed by Standard and the second of the initial order for four completed by Fisher, the original order was canceled and an order for an additional three placed with Fisher. These last three Ca. 5 airplanes were delivered by Fisher between January and April of 1919.

After various initial proposals, order 20035 for 500 Air Service Caproni Ca. 5 airplanes was placed with the Curtiss Airplane and Motor Corporation on October 8, 1917. Because of problems obtaining access to and control of the Italian sample machines and also finished plans, this order was canceled on November 17 and replaced by order 20285 for 50 dated November 26, 1917. Because of the cancellation of a portion of this order which was for Spad fighters as well as the inability to get the needed information required to initiate production, this order was never officially executed. No production Caproni Ca. 5 airplanes were ever delivered.

Any Caproni Ca. 5 airplanes built under contract for US Air Service use would have been powered by Liberty engines already under contract and would have been delivered by the BAP directly to the manufacturer for installation or shipment with the knocked-down airplane.

In late 1917 the Italians had expressed a desire to have 2,000 sets of Caproni Ca. 5 parts built in this country for delivery to Europe along with 3,000 Liberty 12 engines. It would appear that initially only 1,000 of the airplanes would be Liberty powered, since three are required for each plane. By mid-February of 1918 the general plan had been modified to 1,000 sets of Ca. 5 parts for Italy and 1,000 sets for the United States. As noted above, no actual contract had been executed because of the extreme difficulties encountered in getting production set up.

On February 19, 1918, the APB was informed by the Army General Staff that it had been decided not to increase the number of planned bombing squadrons and to equip all those authorized with the British-designed Liberty-powered Handley-Page 0/400 models then under construction. Thus with this new information in-hand, production for the United States of the Caproni Ca. 5 was dropped.

The Italian request for 1,000 Caproni Ca. 5 airplanes and 3,000 Liberty 12N engines was still an active issue in spite of the fact that they had so far not executed a formal order. In preparation for the airplane order, an order was placed with the Curtiss Airplane and Motor Corporation for 500 on July 7, 1918, and another for 500 with the Fisher Body Corporation on July 8, 1918. These orders only served to prepare the way to production and would be executed only when a formal contract was executed with the Italians. This never happened and the orders were canceled after the Armistice. Because the airplanes were never officially ordered, neither were the 3,000 Liberty engines meant to power them.

Thus the Italians only purchased 5 engines for testing. In late March or early April of 1918, 3 high-compression and 2 low-compression engines were delivered.

The complete story of the plans for production of the Caproni bomber in this country is a very complex and interesting one. I have told it here with extreme brevity, only covering enough to give the reader a good idea of just why the Italians intended to order Liberty engines and why that order never materialized. Robert Casari

Lined up at the Dayton-Wright air field are 151 DH-4 airplanes ready for delivery to the Air Service. Except for the lead airplane, none have had their wings mounted. The photographer's shadow can be seen in the right foreground. (Courtesy of Peter Bowers)

required some 70 pages to tell the story in detail in Volume 2 of his *U. S. Military Aircraft the World War I Production Program*. The reader is directed to that publication for the only detailed story of the WW I Caproni production that I am aware of.

LIBERTY 8 CONSIDERED AS A POWER PLANT

Also of interest is the fact that during the late part of the war when the possibility of U. S. use of the Caproni Ca. 5 was still an active issue, the Liberty 8 engine was also considered as a possibility for its power and planning for its potential use was taken into consideration when the first L8 orders were placed. The Production Planning Report of July 12, 1918, allocates the Liberty 8 to the Caproni Night Bomber starting in November with 75 engines, rising to 300 engines in January of 1919 and continuing to the end of the

LePere LUSAC-11 airplanes in final assembly at the Packard Motor Car Co. in Detroit. Dated September 19, 1918, this photograph shows five airplanes. (Author's collection)

chart in June. The total L-8s so allocated through June of 1919 was 2,377. During the same period 1,982 were allocated to "Bi-place Pursuit" of "type not determined."

US MILITARY USE OF THE LIBERTY IN PRODUCTION PLANES DURING THE WAR

The Army made plans to build, or actually built, a number of experimental airplanes which were powered by some version of the Liberty engine. I have covered all intended for the L-4, L-6 and L-8 in the sections in which I told the story of the production of those models.

PRODUCTION L-12-POWERED ARMY PLANES FOR OBSERVATION AND FIGHTING

The Liberty 12 was envisioned as the most widely used of the various models. The DH-4 was built in the largest numbers and saw the most actual use. Wartime production was 4,841. The DH-9 was an improved version of the DH-4 and it was planned to switch production over to that model at the earliest possible moment. A number of factors slowed this switch and by the end of the war only prototypes of the DH-9 had been built and no production ever started. In the next few years following the war, a large number of DH-4 airplanes were converted to DH-4B models, a configuration which incorporated the more significant improvements which made up the DH-9.

The Bristol F2B fighter was chosen in August of 1917 as a good design to produce in this country. Although initially to be powered by the Liberty 8, by October a decision had been made to change the powerplant to the Liberty 12. The first order (for 1,000) was placed in November of 1917 with a designation of USAO-1. By January of 1918 an order for 2,000 of this model was in place. After a series of problems getting the plane into series production, the first example was delivered in March of 1918. By July some 25 of the model had been delivered and testing had resulted in three fatal crashes. These, in combination with serious doubts as to the wisdom of mating the Liberty 12 with the Bristol airframe, led to cancelation of the production contract after 27 planes had been delivered. The story is related in more detail beginning on page 153.

The LePere LUSAC-11 (LePere United States Army Combat, type 11) was designed by Captain Gene LePere, an engineer with the French air mission to the United States. This was a two-place bi-wing combat airplane designed for either pursuit or observation work. It was found to be strong and agile and after construction of six prototypes and 25 production samples, the type was ordered into serial production. On September 7, 1918, an order for 1,000 was placed with the Packard Motor Car Company, the company that had built all prior examples. An additional order for 500 was placed with Brewster on October 17, 1918. Both production orders were canceled just after the Armistice with no deliveries made.

Although a number of other airplane designs fitting this category were considered and some prototypes built, no other designs were ordered into production before the end of the war.

Curtiss H-12L. One of a total of 19 of this model built by Curtiss. A-766 in the foreground and A-776 in the background. (Courtesy of Peter Bowers)

MULTI-ENGINED DESIGNS

Two Liberty-powered designs were considered as night bombers, the Caproni Ca. 5 and the Handley-Page O/400. Both were to be built in this country and delivered to Europe unassembled. The three-engine Caproni Ca. 5 was discussed under Italian engine deliveries and we see that although production of as many as 2,000 was contemplated, orders for only 1,050 were in place at the end of the war. These were cancelled immediately following the Armistice and ultimately only five prototypes were built.

The two-engine Handley-Page O/400 met with more production success. Standard was contracted to produce 500 of this model, 50 were to be delivered assembled in this country and 450 to England as unassembled parts to be finished there. Before the Armistice resulted in cancellation of the contract, seven airplanes had been delivered here and 100 sets of parts shipped to England. I do not believe any of the British parts were assembled into complete airplanes.

Although other multi-engine designs were considered and some prototypes built, no other designs were contracted for series production during the war.

PRODUCTION L-12-POWERED NAVY PLANES

The naval air service was in about the same condition when the United States entered the war on April 6, 1917 as was the Army. On that date they had 54 aircraft on inventory: 45 seaplanes, 6 flying boats, 3 land planes and one lighter-than-air craft. At the signing of the Armistice, on November 11, 1918, they had 2,107 aircraft on inventory: 695 seaplanes, 1,170 flying boats, 242 land planes and 15 lighter-than-air craft.

Although the Navy initially showed no interest in the Liberty engine, that attitude changed very rapidly.

Although the Navy had been making some progress toward modernizing and enlarging their Air Service during the period immediately preceding our entrance in the war, advances in that direction came much more rapidly thereafter. On May 14, 1917, the Navy estimated their airplane requirements from July 1, 1917 to July 1, 1918 as 250 training planes and 250 service types. By May 23 their estimated needs had changed considerably. In a report of the Joint Army-Navy Technical Aircraft Board to the Secretaries of War and Navy was the following statement:

"It is estimated that the needs of the United States Navy for heavier-than-air aircraft during the coming fiscal year, in order to equip the Navy Coast Stations and to train 330 pilots, will be as follows:

300 School seaplanes with 100 H.P. engines.

200 Service seaplanes with 150-250 H.P. engines.

100 Speed Scouts with 100-150 H.P. engines.

100 Large seaplanes with 200-400 H.P. engines.

The most satisfactory School seaplanes at present are the Curtiss N-9 and the most satisfactory Service seaplanes are the Curtiss R-6 type. The other classes are not sufficiently developed by any firm to form an idea as to what firm has the best product."

On May 31, 1917, at the joint meeting of the Aircraft Production Board and the Joint Army-Navy Technical Aircraft Board at which Vincent and Hall first presented their Standardized Aircraft Engine design, "Navy members stated that the proposed engine would be merely for military battle and not adapted to Navy needs." Obviously the Navy had no current vision of what designs might surface for the proposed 100 Large seaplanes with 200-400 HP engines.

Curtiss introduced the 200 HP Curtiss V-2-powered HS-1 patrol flying boat in early 1917 and the Navy shortly became interested in its use as their Service seaplane. The first Liberty 12, Exp. No. 1, passed its 50-hour test on August 24. By this time the Navy had begun to take interest in the Liberty. Performance of the HS-1 would be improved considerably by the extra power which could be provided by the Liberty 12. In perhaps a dual-purpose trial, L-12A Exp. No. 3 was shipped to the Curtiss plant at Buffalo, New York and installed in a Curtiss HS-1 in place of the V-2 and flight-tested on October 21, 1917. The test was very successful.

The Navy had already accepted the Curtiss V-2 powered Curtiss H-12 as a Large seaplane and eventually 20 were purchased. An improved model of it, known as the

Curtiss H-16 powered by two Liberty 12N engines. 274 of this model were produced, 150 by the NAF and 124 by Curtiss. From the Rex Beisel collection. (Courtesy of Peter Bowers)

H-16, was introduced in early 1917 and the British had purchased a number and had designed an improved hull for it. By the time the Navy became interested in this improved version, the Liberty appeared to be a proved engine.

All of these events seemed to have come together about the end of October 1917 and resulted in the building program recommendation by the Joint Army-Navy Technical Board and approved by the Secretary of the Navy for the procurement of 1,185 HS-1L (HS-1 with Liberty power) and 235 twin-Liberty engined H-16 airplanes. At this point the Navy had thus replaced the 200 HP Curtiss V-2 with the 330 HP Liberty 12N as power for their Service seaplanes and Large seaplanes. Engine requirement would be 1,655 L-12N engines plus spares. An allocation of 2,289 such engines was requested for delivery as required for anticipated production of these airplanes.

The Navy would go on to procure other types with Liberty power as the war progressed, but the HS-1L and H-16 were the first types in production. The reader is directed to Appendix F for a complete list and illustrations.

PACKARD BECOMES THE LIBERTY 12N PRODUCER

Packard was going to be the first producer able to deliver any engines and was thus the first builder of the Navy Liberty 12N. With the exception of seven 12N models built by Lincoln in March and April of 1917, Packard remained the sole supplier of Navy type engines, either to the Navy or any other users. Packard shipped its first 12N models sometime in January of 1918. From that time until about June 1, 1918 Packard completed 650 12N models out of a total production of about 1,200 engines.

Early production output was more heavily 12A models but as the Navy began to get airplane production under way, both at commercial companies contracted and at the Naval Aircraft Factory (NAF), and Lincoln production came on line in March to help supply 12A needs, Packard output became predominantly 12N models.

Out of a total Packard production of 6,500 standard Liberty 12 engines, 5,324 were 12N models and 1,176 were 12A models. It remains, therefore, that Packard produced only about 626 12A models after June 1, 1918.

McCook Field staff photo taken late in 1918. Standing, left to right – Mr. Frank Caldwell, Mr. O. H. Skinner, Mr. F. R. Roper, Captain Blee, Mr. Wahl, Mr. Roche, Lt. Colonel J. G. Vincent (Commander), Lt. Colonel H. C. Marmon, Captain Blood, Mr. Sprug, Mr. Simmons, Mr. Riehle, Captain Philip Wyman, Mr. Moon, Mr. Peters, Lt. Spencer and Captain L. M. Woolson. Seated, left to right – Mr. R. A. Loomis, Lt. McMahon, Mr. Caleb Bragg, Sgt. Alexander Klemin, Mr. G. L. McCain, Mr. Larkin and Mr. E. L. Nelson. (Author's collection)

BULLETIN OF EXPERIMENTAL DEPARTMENT, AIRPLANE ENGINEERING DIVISION VOL. II. No. 1.

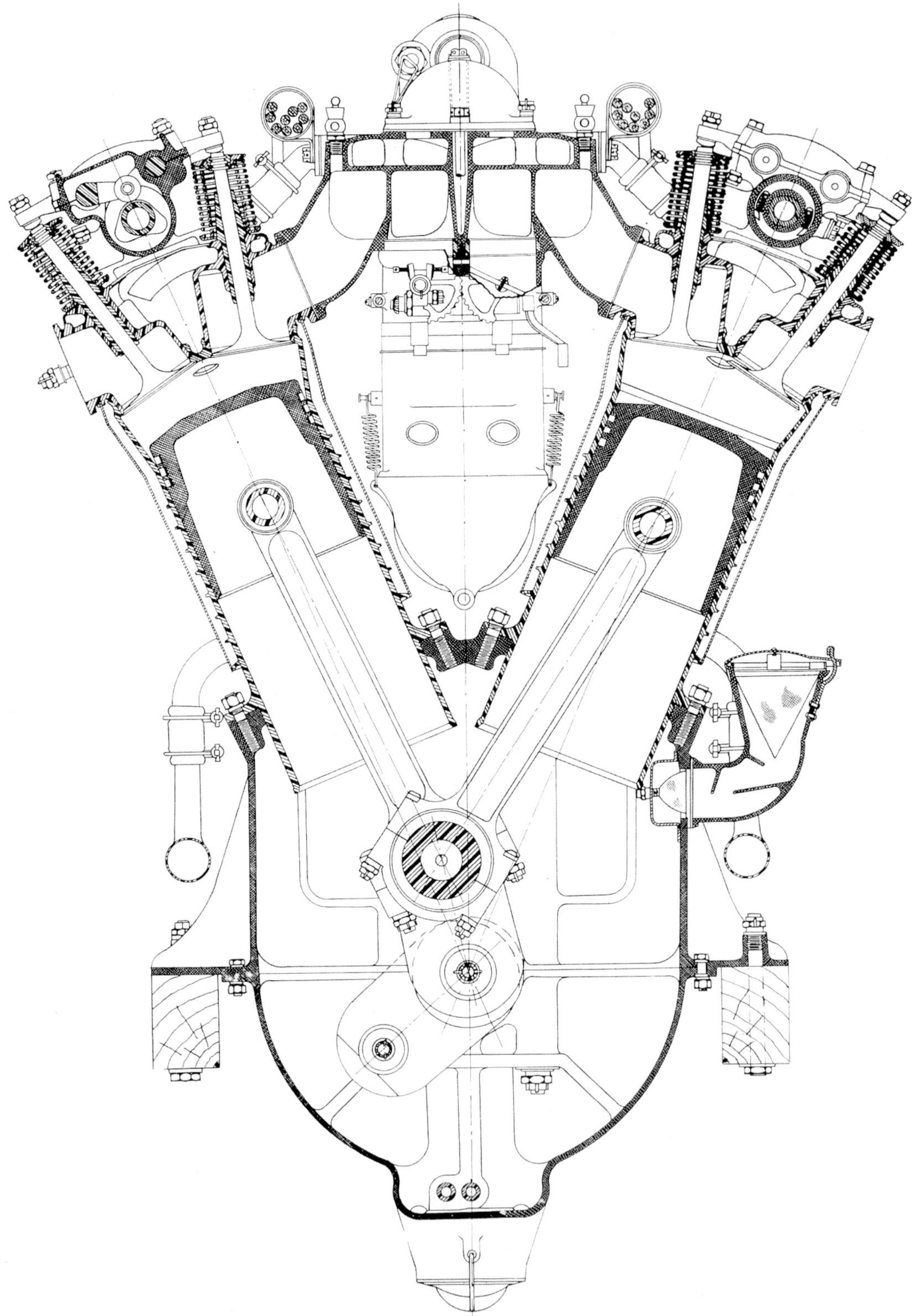

—Transverse section through cylinders of Liberty-12

October 1918

BULLETIN OF EXPERIMENTAL DEPARTMENT, AIRPLANE ENGINEERING DIVISION VOL. II. No. 1.

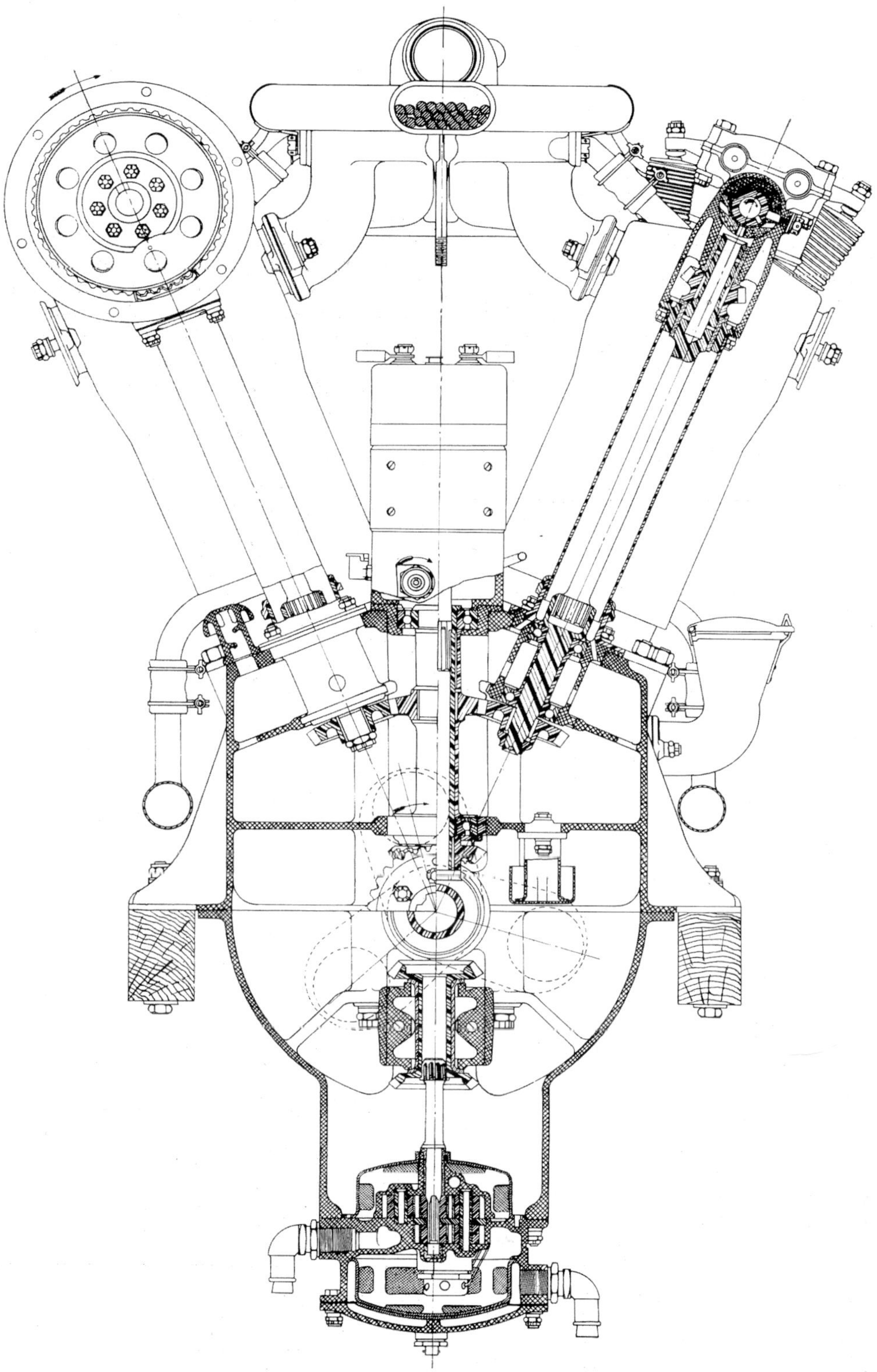

Transverse section of Liberty-12 through timing gears

October 1918

BULLETIN OF EXPERIMENTAL DEPARTMENT, AIRPLANE ENGINEERING DIVISION VOL. II. No. 1.

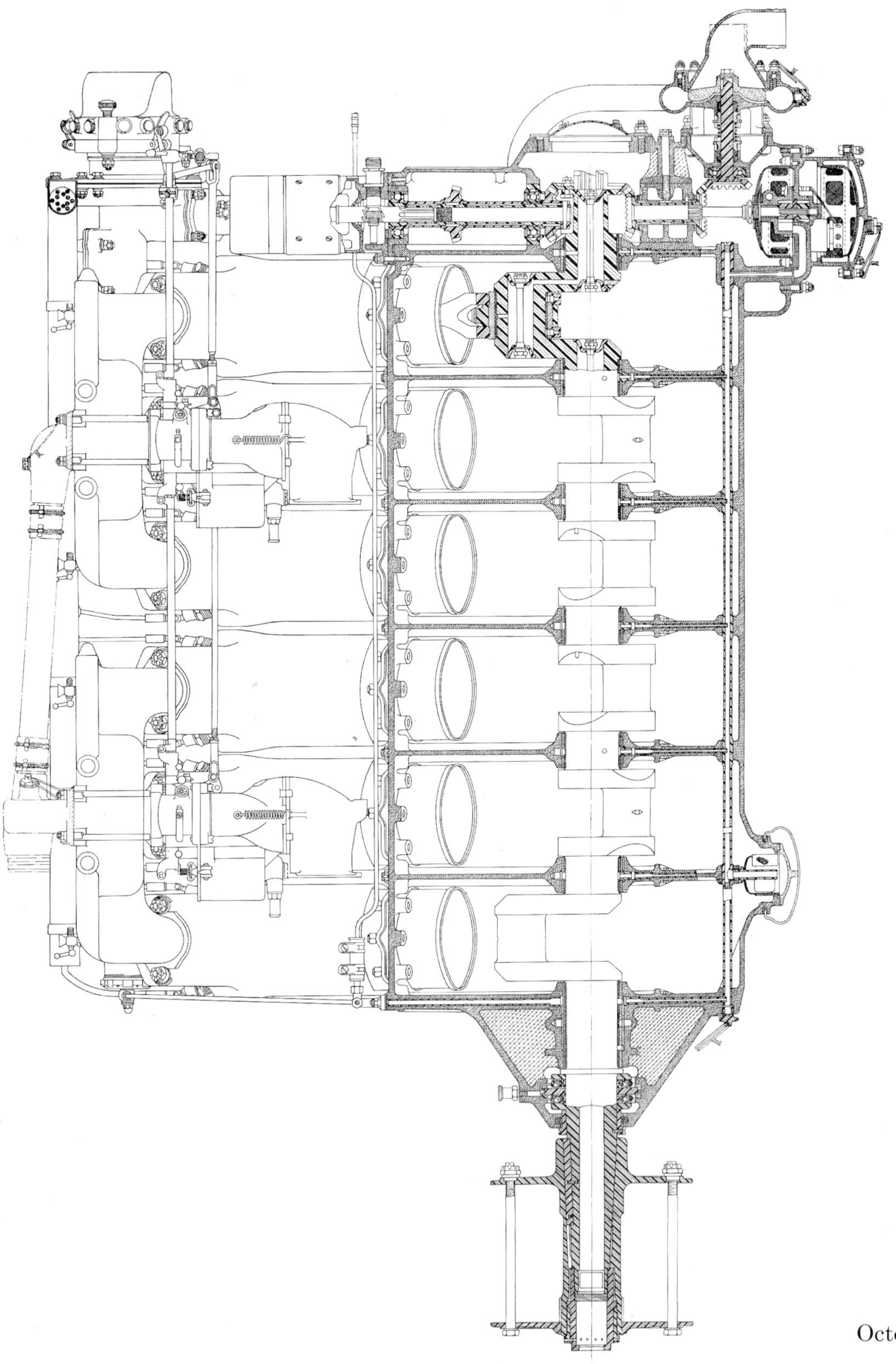

—Longitudinal sectional drawing of Liberty-12A engine

October 1918

Chapter 6

United States Tank Use From 1917 - 1939

We will soon discover that use of the Liberty to power tanks of the United States was very limited. We will find in Chapter 11, however, that this country's limited use of the Liberty for this purpose led, eventually, to Liberty production by Russia and Great Britain for such use. Indeed, total production of Liberty engines in these two countries (about 14,600) would equal about 65% of the WW I production in the United States (21,386).

The reader is referred to the text of the prior chapter starting on page 186 for a detailed history of the selection of the Liberty to power the Mark VIII Heavy Tank. The tank ultimately became known as the *Liberty Tank* (or sometimes the *Allied Tank* because of its joint design).

In the fall of 1917 the Army sent a fact-finding mission, headed by Major James Drain and Major Herbert Alden, to Europe to try and find what current Allied tank designs might be suitable for use by our services. The Renault FT-7 Light Tank was found suitable but no heavy tank was acceptable. Their report recommended that the Americans and British jointly develop a heavy tank based on British past experience and American production capabilities.

That recommendation was carried out with the creation in November of 1917 of the Joint Inter-Allied Tank Committee. Initially the committee was headed by the British and Americans with French participation in an advisory capacity only, the French having decided to pursue their own designs. Co-Chairmen of the committee were Secretary of the Royal Navy Air Service Landships Committee, Lt. Colonel Albert Stern, and U. S. Army Major James Drain.

Work on the new design started immediately. The Tank Committee directed Lt. G. J. Rackham to develop detailed drawings. The general design followed the rhomboid shape of its British predecessor, the Mark V. (The Mark VI was designed and built as a wooden mock-up but never put in production. Three examples of the Mark VII were built for tests but the Mark VIII was soon settled as the selected new design for the great offensive planned for 1919.) That design was enlarged and improved upon in many details.

Even though the Mark V tank would prove its worth as a battle vehicle, it would also bring to light many features which needed changes. The Mark VIII design tried to improve upon many of those found lacking. The Mark V's six-cylinder Recardo 150 HP tank engine was one of the items to be improved upon if possible. (See page 188 for details.) The American Liberty was chosen as the best available power plant. Its power was more than double that of the Recardo and its weight advantage was more than 1,800 pounds.

Other improvements included an engine compartment separated from the fighting compartment by a steel bulkhead provided with access hatches. The fighting compartment was then provided with an overpressure ventilation system to keep heat and fumes from the engine compartment from most of the crew. The mechanic was not so fortunate, however, as his station was in a chair beside the engine.

Plans were completed and a full-scale wooden mock-up built. Approval for the design was given on December 26, 1917. The plan was for the British to supply armor plate, structural components, weapons, ammunition and track shoes and the Americans would supply the engine, brakes, suspension and running gear. It will be noted, however, that in the few tanks actually completed and put in use by the two nations the British used their own armament and the Americans used theirs.

Initial plans were to build 1,500 tanks. Components assigned to the two countries would be fabricated in their respective locations and shipped to France where an assembly and test plant was to be built at Neuvy-Pailloux.

The plant was started in early 1918 but progressed slowly and was still not ready for use by the time of the Armistice in November of 1918. Probably in connection with the decision to build the assembly plant in France, the French changed their former advisory position to one demanding a share of the Mark VIII tank production.

In July of 1918 a mild steel prototype was completed in England. It was then shipped to the United States where the engine and transmission were added and the tank put on trials.

By November of 1918 the British had completed 100 sets of parts. The Americans had no doubt completed at least that number of Liberty engines at the Trego plant and assigned them to tank use. The remaining parts to be furnished by the Americans were probably completed or nearly so. The overall production schedule was initially effected primarily by the unavailability of Liberty engines for this use. The Americans refused to release any Liberty engines for tank use until mid-June of 1918. This kept progress from being made as there was little impetus to stockpile parts or complete an assembly plant if there were no engines available. This attitude would certainly

Mark VIII tanks under construction in July of 1919. Some idea of the size of these tanks can be ascertained by comparing their size with that of the six-ton Renault Light Tank seen between the first and second Mark VIII. (Courtesy of Mike Kendall)

This Rock Island Arsenal photograph of December 5, 1919 shows the status of construction of Mark VIII No. 5. (Courtesy of Jon Hornbostel)

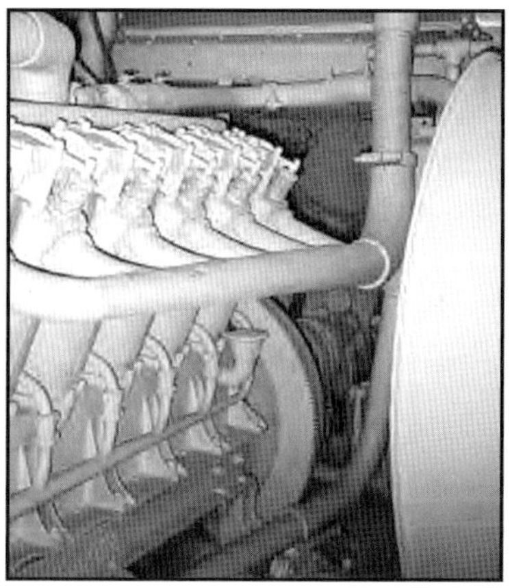

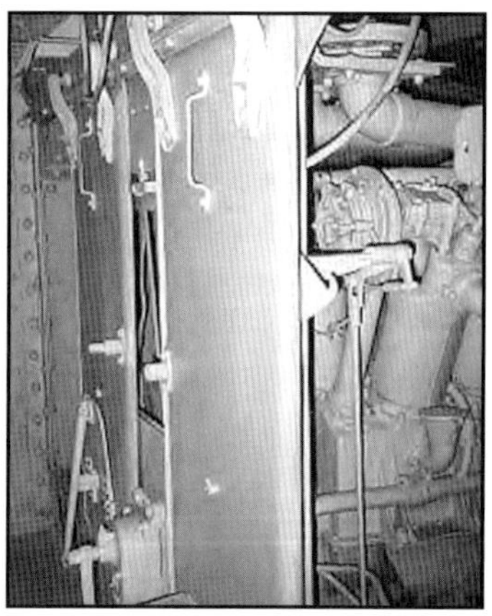

A group of Mark VIII tanks being serviced during exercises. Their large size is once again apparent, this time by comparing them to the service personnel. The Mark VIII never fought in war but served as a training vehicle from 1919 into the early 1930s. 90 of the original 100 were then retired to storage and, except for two retained for museum display, were transferred to the Canadians for training use under Lend-Lease in 1940. It was thus once again an active military machine during WW II. (Courtesy of Mike Kendall)

The above picture is from the Patton Museum showing a Mark VIII in service. The 6-pounder normally mounted in the sponsion is absent, perhaps explained by the very late date of 1935 with which the photograph was marked. (Courtesy of Jon Hornbostel)

At left are two views of the Mark VIIIs Liberty engine as seen through the access hatches. Note the hand starter accessible on the fighting compartment side of the division bulkhead. (Courtesy of Jim Hensley) have to be followed with considerable caution as once engines were released for tank use they could certainly be expected to arrive in large numbers.

The Armistice was signed on November 11, 1918 and the British and French withdrew from the Mark VIII tank program on November 27. The Americans then negotiated the purchase of the 100 completed sets of parts from the British and shipped them to Rock Island Arsenal, Illinois.

In addition to these sets of parts, the British later assembled seven Mark VIIIs at the North British Locomotive Company at Glasgow, Scotland. The first completed is said to have been powered by a Rolls-Royce aircraft engine. The remaining six were powered by a new Recardo 12-cylinder tank engine of 300 HP which was created by mounting two of the 150 HP Recardo tank engines on a common crankcase. Later, the original tank was re-engined with a 300 HP Recardo.

The principal differences between the British and American versions of the Mark VIII were power source and the fact that the machine guns of the British version were .303 cal. Hotchkiss whereas the Americans used .30 cal. Browning. Both used the same 6-pounder cannons.

In January of 1919 the parts were shipped to Rock Island Arsenal and Colonel Jordan, Arsenal Commander, was directed to begin the manufacture of 100 complete tanks. During February the mild steel prototype arrived at Rock Island and was immediately sent to Savannah Proving Grounds about 50 miles north of that location. Tests resulted in recommendations for some changes in the exhaust and cooling systems.

Production was started at Rock Island on July 1, 1919 and the first tank was completed on January 5, 1920. Progress after this time was fairly rapid and the last (100th) tank was completed on June 5, 1920.

With the exception of a handful of experimental types, this, 15 three-ton Ford and the 950 six-ton Model 1917 Renault designs, built by various companies in the U. S. under Ordnance direction during 1918-1919, formed the the basis of the U. S. Tank Corps into the early 1930s. Light and medium tank development was carried out on a rather limited basis because of the usual chronic shortage of funds. As long as there was a strong potential need for the Mark VIII Heavy Tank in the advent of war, it was kept as some kind of standard.

Heavy Tank Mark VIII General Data (100 built)

Length	**34' 2 1/2"**
Width	**12' 5" (sponsons out)**
Height	**10' 2 1/2"**
Combat weight	**81,000 lbs.**
Engine	**Liberty low comp. 338 HP @ 1,400 RPM**
Power/weight ratio	**7.8 HP per ton**
Gear box	**Planetary, 2 forward, 1 reverse**
Speed	**6.5 MPH**

Armament

2 6 pdr. (2.24")	
5 .30 cal. Browning MG	
Main armor thickness	**16mm (0.63")**
Secondary armor	**6mm (0.236")**

Rock Island Arsenal sectional drawing of the Mark VIII. This exceptional drawing shows clearly the engine and drive train. Courtesy of (Jon Hornbostel)

This unlabeled photograph is from the Patton Tank Museum and shows a Liberty engine connected to a drive system which appears very much like that used on the Mark VIII tank. This is probably a picture of a test stand used to develop or refine the design of the Mark VIII drive system. (Courtesy of Jon Hornbostel)

The Mark VIII was maintained as a service tank and modifications to affect improvements in safety and performance were made throughout its life, which continued to the middle 1930s. These changes included such items as additional fire extinguishers, relocated fuel lines and tanks, new design intake air manifold, new cooling water header tank, etc.

In August of 1927 the 2.24" (6-PDR) tank gun, Mark II, was approved as substitute standard for manufacture and issue to the Army. This cannon was used in no other Army equipment but was so declared to facilitate possible manufacture in case of hostilities requiring active use and possible further manufacture of the Mark VIII tank, which itself had been declared as substitute standard. It was noted in the Ordnance Committee minutes that "It is hoped that the tank development will progress with sufficient rapidity so that the actual manufacture of Mark VIII tanks will not be required." Fortunately, this was the case, although, as noted above, the Mark VIII remained in service use for several years beyond 1927.

THE LIBERTY INTERNALLY AND EXTERNALLY AIR-COOLED ENGINE

Early in 1932 Mr. Neils Anderson approached the Ordnance Department with a proposal for the construction of an internally and externally cooled Liberty engine of his design. Although we know very little about the early history of this project, it is evident that a great deal of design work had been done prior to the time Anderson brought the idea before the Department. What Anderson wanted was the loan of a Liberty engine, which he would then modify to his specifications and submit to standard S.A.E. test on a dynamometer at the Aberdeen Proving Grounds. If the tests proved satisfactory, the engine would be installed in a Mark VIII tank and given performance tests.

The intention was to allow Anderson access to a free Liberty engine on which to try his design, provide a testing facility at no cost to him and a suitable vehicle in which to prove its worth should the block test indicate the design

The Anderson internally and externally air-cooled Liberty 12 under test at Aberdeen Proving Grounds on August 8, 1932. From this angle the standard cam assembly can be seen as can the taller-than-normal intake manifolds. (Courtesy of the Air Force Museum)

had good promise. This, of course, would save Anderson a considerable amount of cash investment and provide a tank to test the engine in. The cost to the Ordnance Department would be modest, although even modest funds were very hard to come by during this period of depression, and the government might benefit in the end by having access to an improved engine for use by either the Ordnance Department or possibly the Air Service.

The Artillery Division, Manufacturing Service, recommended on May 23, 1932 that the Ordnance Department cooperate with Mr. Anderson in this project and on May 26 the Ordnance Committee authorized it. The engine, US NO. 90558, one of the last batch of 528 built by Buick, was supplied from Air Service stocks and sent to the Heat Controlled Motor Company where it was modified to incorporate Anderson's changes. It was then sent to Ft. George G. Meade, Maryland where the Ordnance officer cooperated with Anderson in the preparation of it for block tests. Upon completion of block test at that location the engine was shipped to Aberdeen Proving Grounds where it was subjected to a complete engine test in accordance with standard S. A. E. procedure as set forth in the Society's 1931 handbook, page 101.

By 1932 the Air Service and the Allison Engineering Company had done a great deal of work on air-cooling the Liberty and had built a number of them for use in airplanes. Anderson's design was quite different from any prior efforts in that air-cooling was also applied internally. It also used more standard Liberty parts than did the Allison version. Its design was described in some detail in the first partial report (Ordnance Program No. 5163) dated August 25, 1932. The below description is extracted from that report.

A- The standard cylinders were replaced by air-cooled ones designed to utilize the standard valve drive train of the original engine. Bore and stroke remained standard Liberty and thus displacement was the same as the standard Liberty 12.

B- New pistons were of increased length in order to accommodate air tubes over which the pistons slide. Four air tubes were installed in the crankcase at the bottom of each cylinder, two in front of the connecting rod and two behind it. The pistons were constructed so that the four air tubes at the bottom of each cylinder fit into four passages in the inside of the piston. An effective air seal was formed by rings set in grooves in each air passage. The under side of the piston heads were finned parallel to air flow. Compression ratio was 6:1.

C- The connecting rods and cylinder walls were increased in length to accommodate the increased length of the pistons.

D- The exhaust valves were made with a hollow stem with a tube in the center running within ¼ inch of the valve head for internal cooling. The outside chamber of the valve stem had four ports which registered with a counter-bored section in the valve guide, which in turn was connected to the air-to-valve manifold. The tube which was inside the

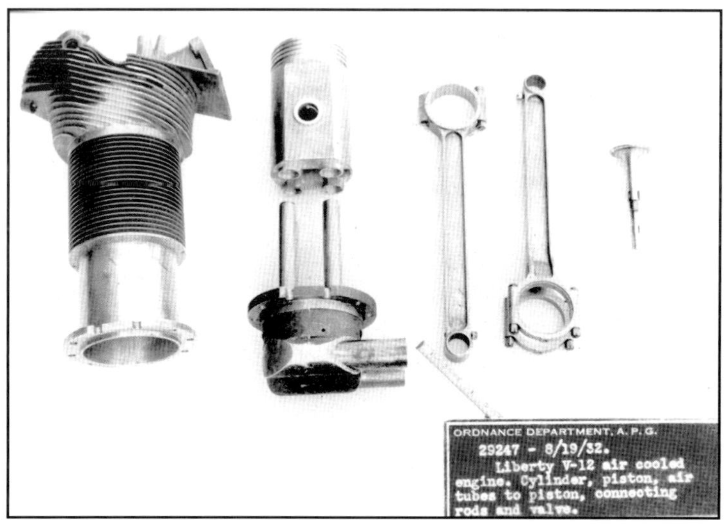

The air-cooled cylinders, special air-cooled piston with its four feed tubes and their manifold, the extra length rods and the special air-cooled exhaust valve of the Anderson air-cooled Liberty. (Courtesy of the Air Force Museum)

valve stem had four ports above the valve guide which permitted heated cooling air to exhaust into the atmosphere.

E- Air for the internal-cooling system was furnished by a small centrifugal type fan mounted to the rear of the engine and belt-driven from the Cardan shaft. In the valve cooling system the air was carried from the fan through tubes to a distributing manifold from which leads were taken to individual valves. The air was led to the counter-bore in the valve stem and from there through the ports into the outside chamber of the valve stem, down the stem, across the valve head, into the inner tube and out the exhaust ports which are above the valve guide. In the piston cooling system air was carried from the fan by air tubes to a distributing manifold from which leads were taken off to each cylinder. These leads connected with the two front air tubes at the bottom of the cylinder. The air was then led to the two front passages in the piston, up around the inner side of the head and down the two rear passages to the rear tubes and thence exhausted into the atmosphere.

F- The external air-cooling was accomplished by means of a large centrifugal type fan mounted on the front end of the engine and driven from a shaft counter-bored into and extending out from the crankshaft. This fan forced air externally around the cylinders, which were individually enclosed in conventional type air-cooling fins.

G- Special intake headers were provided which used two standard Zenith US 52 carburetors set up with 160 main jets, 170 compensating jets and 36 venturi.

Anderson air-cooled Liberty under dynamometer test on July 14, 1932. The flywheel and starting ring gear are easily seen. Because of the excessive back lighting from windows, it is difficult to see the details of the air-cooled cylinders. (Courtesy of the Air Force Museum)

The tests were run on August 8 and 9, 1932 and, because of the 6:1 compression ratio, 87.1 octane aviation gas was used. No detonation was observed while using this grade fuel. Ethyl gasoline with an octane rating of 74 was then tried and considerable detonation was experienced while under heavy loads. This fact was noted as a shortcoming if it was intended to operate the engine at near full loads. It is interesting that no suggestion was made to reduce the compression ratio to avoid having to use aviation gas, an item not easily found where tanks are used.

The engine performed well under the limited testing and the following conclusions were documented:

A- That the internally and externally air-cooled Liberty engine develops a greater horse power than was expected.

B- That the engine cools adequately on the block.

C- That the thermal efficiency is high due to high compression ratio made possible by internal air-cooling.

D- That the necessity of using special blend gasoline in this engine of octane rating higher than government issue is a disadvantage from a military standpoint.

The test results were found to be encouraging and after being reviewed by the Ordnance Board it was officially recorded in the Board minutes that "It is considered desirable to determine the suitability of this type of engine for use in the heavy tank. This procedure, it is believed, will determine whether engines of this general character will be satisfactory in other similar combat vehicles." This gave the official confirmation that the engine had so far met of the original intent of the program. It

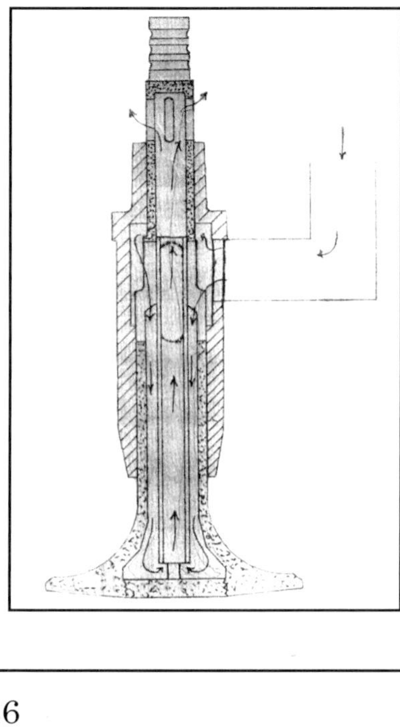

At left is a diagram of the air-cooled exhaust valve of the Anderson Liberty 12. (From the test report and courtesy of the Air Force Museum)

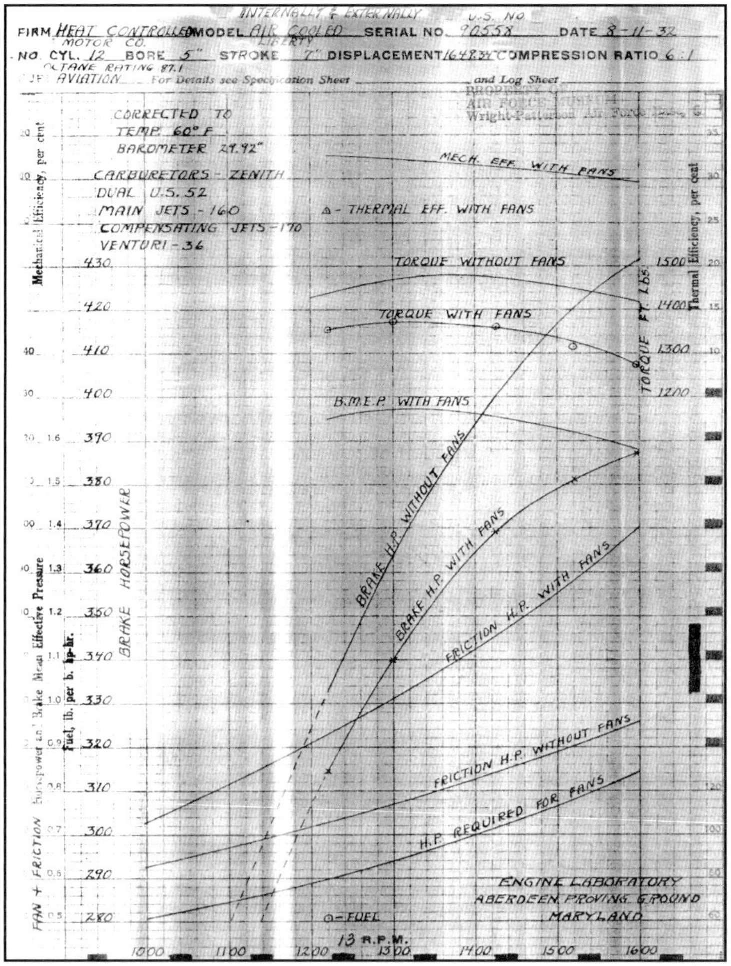

Power graph generated at Aberdeen Proving Grounds for the Anderson air-cooled Liberty 12. (Courtesy of the Air Force Museum)

The Anderson air-cooled Liberty seen from the large blower end. Cooling shrouds are not in place. (Courtesy of the Air Force Museum)

was estimated that preparing a Mark VIII tank for installation of the Anderson engine would cost about $500.00, and that installation and test of the engine would cost an additional $500.00. The tank used would be designated while so equipped as "Tank, Mark VIII E1."

A tank was selected, its Liberty engine removed and the necessary modifications made to accept the Anderson Liberty. The Anderson engine was then installed and the tank prepared for a short period of testing. It appears that the board minutes estimate of costs was used as an actual limit of funds which could be expended on the project because several minor modifications to the installation were noted as desirable during the test but were not done because of a lack of funds. Almost all were associated with changes in handling of the exhaust of cooling air or heat shielding baffles. After an initial test run, six modifications were noted as being desirable but not done because "at this time a check of the funds revealed only $50.00 available." It was then decided to run the vehicle under various load conditions until test funds were exhausted. Even without these changes, however, cooling of the engine presented no problems during the tests.

The tests were run at Aberdeen Proving Grounds from August 29-31, 1932. The final day's run was documented in detail and is interesting enough to warrant inclusion below.

PM	
1:30	Engine started in Mark VIII tank at machine shop.
1:45	Left machine shop for Aviation Field Course - speed 4 to 5 mph.
1:50	Stopped at sand course, Aviation Field, to attach M1921 Medium (23-ton) Tank as a towed vehicle.
1:59	Started with 23-ton tank as towed load using brakes to hold engine RPM down to 1400. In other words the internally and externally air-cooled engine was running at 1400 RPM under maximum load - speed 4 to 5 mph.
2:01	Stopped - brakes on towed vehicle overheated.
2:05	Started with towed vehicle in low transmission and high gear planetary - engine RPM held at 1400. Speed 4 to 5 mph.

2:14	Stopped to tighten brakes on Mark VIII tank.
2:18	Started - towed tank in low gear transmission and high gear planetary. Speed 4 to 5 mph.
2:23	Stopped - man temporarily injured when open door struck a tree and banged close on him.
2:25	Started - same conditions as above.
2:35	Stopped at sand course to disconnect towed load.
2:37	Started - return to machine shop.
2:50	Returned to machine shop.

Recapitulation:
Engine run time - 1 hr. 20 min.
Vehicle running time - 43 minutes

The engine suffered from no overheating and performed well. The final report noted the engine held good promise as power for various ordnance vehicles but would, of course, require further testing and refinement before becoming acceptable as a production engine. It made specific mention that air-cooling had many advantages over water-cooling if perfected, and noted that even though the engine with attached fans and cooling shrouds was larger than the standard Liberty, the overall space taken was less because of the removal of the water cooling accessories.

The February 16, 1933 Ordnance Committee notes carry the cancellation notice of the project. It was noted that the Proving Ground reports indicated that with further development this type of engine can probably be made to function satisfactorily, "however, it is the opinion of the Subcommittee (on Automotive Equipment) that developments of this character are functions of industry and should not be undertaken by the Ordnance Department."

It was then ordered that the engine be turned over to Mr. Neils Anderson as part of the agreement, the Mark VIII tank be restored to original condition, and the designation "Tank, Mark VIII E1" be cancelled.

EARLY UNITED STATES TANK HISTORY

It is certainly not the intention of this work to attempt any detailed documentation of early United States tank history. However, I feel that at least an overall view of what this country did in the way of tank development, how the effort was organized and what was accomplished from our first interest in 1917 until the end of active use of the Liberty engine as tank power in the late 1930s is essential to understanding what part the Liberty played.

Much of the general information which is presented below is extracted from the Minutes of the Ordnance Committee Meetings, of which I have copies from 1919 to 1933 and beyond.

"Tanks were first introduced experimentally in the World War by the British about 1916. The purpose was to restore mobility to the offensive and to combat machine guns used on the defense. During 1917 and 1918 such progress was made that all the belligerents adopted some form of tank and had ambitious plans for the use of tanks in 1919, had the war lasted. The tanks were characterized by their principal short-comings; slow speed and poor reliability. During the World War the Ordnance Department built a *light* (known as the six-ton) tank based on the French Renault tank and a *heavy* (forty-ton) tank based on a tank designed by the British and known as the Mark VIII tank. This tank was fitted with a modified Liberty airplane engine. A few 3-ton Ford tanks were also built. A limited number of American tank units were engaged in actual battle in the latter part of 1918. After the Armistice the Ordnance Department was engaged in completing a number of the above types, as follows:

Ford 3-ton Tank	15
Six-ton (American Tank)	950
Mark VIII Tank	100

A very small expenditure for development work on tanks was made during the fiscal year 1919, but considerable work was done on gathering data. The development of the medium tank (intermediate in weight between the light and heavy tank) properly dates from the fiscal year 1920, although data had been accumulated and preliminary studies made by the Ordnance Department, which indicated the desirability of the development of a type of medium size."

Although the War Department had appointed a board (the Westervelt or Caliber Board) to use the experience gained during the war to make recommendations for future development of artillery, no such steps were taken toward guiding the future development of the tank. The wartime Tank Corps was brought under the infantry in 1920 and all tank development left in the hands of the Ordnance Department.

"The Ordnance Department used as a basis for its post-war tank development policy a memorandum dated 1 August, 1919, by Major R. E. Carlson, Ordnance Reserve Corps, and a more complete report of about 150 pages by the same author dated 16 March 1921, recommending development of a fast medium tank. Major Carson was, during the World War, assistant to Colonel James A. Drain, Ordnance Reserve, the American member of the Anglo-American Tank Commission. The lack of a War Department board to lay down a development policy became increasingly serious as the post-war tank development proceeded. Although attempts were made from time to time to supply this deficiency by various boards of limited scope and by policy letters from the Adjutant General and from the Chief of Infantry, the development of tanks may fairly be said to have suffered from the lack of a definite and fairly consistent policy."

It was their decision that the Tank Corps would utilize the above noted light and heavy tanks then in-hand and direct development toward a new medium tank. In the mean while the light and heavy tanks on-hand would be maintained and upgraded as seen necessary to keep them in service.

MEDIUM TANK M 1921

The initial work order approving the design and manufacture of an experimental medium tank was issued on June 13, 1919. It was to be similar to the British Medium D and have a speed above twelve miles per hour. On August 16, 1919 Major Carlson went to Camp Meade, Maryland, as an official representative of the Ordnance Department, to discuss with Colonel S. D. Rockenbach (then Chief of Tank Corps and formally Brig. General and Chief of AEF Tank Corps) the medium tank recommended by the Ordnance Department.

"The work was carried out in close cooperation with the Chief of the Tank Corps and agreed with the principals laid down in a paper prepared by its then chief, Brig. General S. D. Rockenbach, entitled *Tanks Function in Relation to Design* (dated 18 August, 1919). This paper, which was virtually the Tank Corps' statement of policy, said in closing '— push design and construction of an 18-foot long tank, weighing not over 18 tons, power 10 HP per ton of weight, speed 3 1/2 miles to 12 or more miles per hour, armor protection against armor-piercing rifle bullets; cannon, quick firing, not less than 2.28 or over 3 inch cal., two machine guns, 18 inch ground clearance, flotation 4 1/2 pounds, one man drive, self starter, good vision, cruising radius 60 miles one fill, with the equipment generally of the Mark VIII'."

After some changes in detail, one of which was that protection was to be upgraded to resist .50 cal. armor-piercing rounds at close range, the Ordnance Board formally approved the specifications of the new Medium Tank M 1921 so that detailed design work could proceed. After the detailing of the design and building of a full-size wooden model, it was recommended that two pilot models (Medium A Tank) be built. At this point the engine had been chosen as a 6-cylinder 250 HP Murray and Tregurtha (a modified marine engine) which would provide 12 1/2 HP per ton of weight. Length was to be 21' 6", width 8' and weight now 20 tons (approximately).

MEDIUM TANK M 1922

At about the time this approval had been granted it came to the attention of the Ordnance Board that the British had a promising idea for a cable-type suspension system for a medium tank. The Tank Corps Technical Board reviewed this design and recommended that a pilot tank incorporating it be built. The Ordnance Committee approved this and ordered that the second of the two pilot models recently approved be designed and built with this type of suspension and designated Medium Tank M 1922. Thus in the summer of 1920 both the Medium M 1921 and the Medium M 1922 were put under construction. This coincided very closely with the completion of the last of the Mark VIII tanks.

The M 1921 pilot model was completed at Rock Island Arsenal in December of 1921 and delivered to Aberdeen Proving Ground on February 20, 1922. It carried a four-man crew and weighed 41,000 pounds loaded. The Murray and Tregurtha engine was rated at 220 HP @ 1,200 RPM but was found to actually deliver only 195 HP at that speed. Rated power would give a power-to-weight

Medium Tank M 1921. This tank began life with a Murray and Tregurtha engine, was repowered with a Liberty upon that engine's failure, then used to test a new Packard V-8 tank engine, the 1T-1330, and then had its Liberty reinstalled. It retained the Liberty until it was retired in the mid-1930s. (Courtesy of Jon Hornbostel)

Medium Tank Model 1921
(1 prototype built)

Length	21' 5"
Width	8'
Height	9' 9"
Combat weight	41,000 pounds (eventually 46,000 pounds)

Engine

original	Murray and Tregurtha 195 HP @ 1,200 RPM
second	Liberty 12 338 HP @ 1,400 RPM
tested	Packard 1T-1330 200 HP @ 1,200 RPM

Power/weight ratio

original	9.5 HP per ton
second	16.5 HP per ton
tested	9.8 HP per ton

Speed

original	10.1 MPH
second	20+ MPH

Armament

1 6-pounder (2.24") - 2	.30 cal. Browning MG
Main armor thickness	25.4mm (1")
Secondary armor thickness	9.53mm (3/8")

ratio of 10.7 HP per ton but actual output resulted in 9.5 HP per ton. Maximum speed was 10.1 MPH with the engine turning at 1,200 RPM.

Tests were carried out at the proving grounds from February to December of 1922. At mileage 281 it was shipped to Camp Meade, Maryland for service-testing by the Tank Corps. It was under test at that location from February 10, 1923 until October 1 of that year, when the Murray and Tregurtha engine broke its crankshaft. The decision was made to replace the engine with a Liberty 12 of the type used in the Mark VIII tank. Following that the remainder of the service tests were completed and the tank returned to Aberdeen on October 22, 1923. It remained at Aberdeen and was used for testing not only of the basic design but numerous new design tank accessories for a number of years. By mid-1925 it had traveled a total of more than 1,200 miles and by the end of 1926, 2,128 miles. It was used, for example, as late as 1932 as a towed load during the testing of the air-cooled Anderson Liberty in a Mark VIII tank.

The failure of the Murray and Tregurtha engine to perform well and, it was claimed in the Ordnance Committee notes, its lack of power led to an extensive survey of available commercial engines which would fit the requirements. None were found and on November 8, 1923 it was decided to have the Packard Motor Car Company (one of the builders of the Liberty) design and build a motor to Ordnance specifications.

THE PACKARD 1T-1330 TANK ENGINE

The contract was awarded on February 11, 1924 and the single engine ordered was delivered in June of 1925. In early 1926 it was installed in the M 1921 tank for testing. The Ordnance Board considered this engine very successful and reasonable in price ($33,000) for a single-unit design and production. The engine was an overhead valve 60° V-8 of 1,330 cubic inch displacement with a bore and stroke of 5 ½" x 7". Rated power was 200 HP @ 1,200 RPM. Dual Delco ignition was used. An adjustable governor (1,200 RPM to 1,500 RPM) was provided and the engine was capable of about 250 HP at 1,500 RPM.

As we shall see, the M 1921 tank's successor was the Medium Tank T1 and it was the tank for which the Packard 1T-1330 engine was intended. The T1 was completed at Rock Island Arsenal in May of 1927 and the Packard engine removed from the M 1921 and installed in

Packard 1T-1330 tank engine operator's-end view. Note that all accessories, including distributors, are grouped at this end of the engine. Notice the similarity of the dual ignition switch to the Liberty design. (Author's collection)

the T1 at that time. The M 1921 then went back to Liberty power for the remainder of its useful life.

The Packard 1T-1330 remained in the T1 throughout its testing period and was considered a very good engine by the Ordnance Department. In 1932 the T1 was re-engined with a Liberty but the reason was not because any fault had been found with the Packard. The fault lay with the Ordnance Department in not ordering the design for more power. Had they requested an engine of 300-350 HP it would probably have become the standard tank engine for a number of years. Instead, it gained the honor of being the first purpose-built American tank engine and the only one until production of the V-8 Ford tank engine of WW II, all others having been adaptations of marine, industrial or aircraft engines.

The M 1922 tank was completed at Rock Island and shipped to Aberdeen on March 1, 1923. Testing there soon showed that the new cable type suspension was susceptible to rapid wear. In addition, the Murray and Tregurtha engine had the same under-power and unreliability problems as when used in the M 1921. Although the lack of reliability of the Murray and Tregurtha engine was noted in the Ordnance records, there seems to be no record of it ever being replaced. It continued to be used in testing various accessory improvements at Aberdeen until August of 1926 when, with 854 miles on it, the tank was retired to the Ordnance Museum.

MEDIUM TANK T1 AND T1E1

It did not take long to decide that between the M 1921 and M 1922 Medium Tanks the M 1921 was the better

Packard 1T-1330 tank engine left-side view showing the gear-driven 36" cooling fan. Both electric and hand starter were provided. (Author's collection)

design. However, a firm policy regarding tank development had yet to be laid down. To that end, on May 23, 1921, the Chief of Ordnance requested that an approved War Department policy and tactical requirements for all tanks be laid down. As a result, the War Department, A. G. O. on April 3, 1922 published such a policy. Included was the requirement for Medium Tanks as follows: "The medium tank not exceeding 15 tons in weight, thereby bringing it within the limits of average highway bridges, the capacity of railroads and the limit of 15 tons placed by the War Department on the medium pontoon bridge." The Ordnance Department was very satisfied with the M 1921 Medium Tank as were all using services. Their only objection to it was the lack of performance and reliability of the Murray and Tregurtha engine. This shortcoming they felt could be overcome by the development of a new engine (the Packard 1T-1330). The M 1921, however, weighed some 23 tons and the basic design could not have its weight reduced to the 15-ton restriction and maintain the required armor protection. A deciding factor in choosing the 15-ton limit was the carrying capacity of pontoon bridges (along with regular bridges, etc.).

On November 1, 1924, the Medium Tank T1 was ordered from Rock Island Arsenal. The first work was, of course, to make a complete set of design drawings based on current requirements and studies so far carried out with the M 1921.

A turning point was reached when on June 23, 1925 the M 1921 Medium Tank successfully crossed a Model 1924 medium pontoon bridge at a weight of 23 tons. It was at that point decided to amend the tank development policy and continue development of the M 1921 with improved design components installed and tested in it with the idea of later putting all the successful features in the T1 which had been put on order in November of the prior year. In the mean time work had been started in 1924 on the design of a new 15-ton Medium Tank which was at first known as the M 1924 and would later become the T2. Further work on the 15-ton tank was deferred pending the recommendation of the Chief of Infantry.

Plans were finished and the T1 completed in May of 1927. By January 6, 1928, it had completed its shop test, Ordnance test and Tank Board test. The Tank Board report noted that the T1 "is in general an excellent vehicle and gives promise of being the most satisfactory tank constructed by the Ordnance Department to date." In addition it was noted that the highest average speed of the tank was 11.28 MPH (at the 200 HP rating; it was 14 MPH at the 250 HP rating). Although this was some improvement over the 10.1 MPH speed of the M 1921 with its original Murray and Tregurtha, it was still not considered satisfactory. Perhaps the speed of in excess of 20 MPH obtained by the M 1921 after its change to Liberty power had some

Medium Tank T1 (T1E1) pictured at Rock Island Arsenal not long after completion. Originally powered by a purpose-built Packard 1T-1330 tank engine and re-powered with a Liberty in April of 1932. (Author's collection)

influence. The report did make a point to note that the Packard engine was "rugged and dependable" and that it produced the power required by the design contract.

The Ordnance Committee recommended the T1 be standardized on January 24, 1928 and that standardization was approved by the War Department on February 2, 1928. For reasons currently unknown the War Department reconsidered the matter and withdrew the standardization on March 30, 1928. Although the power contracted for in the Packard engine had been considered adequate in early 1924, they probably realized that that was no longer the case and that it would not be wise to standardize an under-powered tank.

Both the M 1921 and T1 Medium Tanks remained in service into the early 1930s. The M 1921 had been powered with a Liberty most of the time since October of 1923. Performance was considerably enhanced by that installation but the increased power did reduce the life of the drive line which was not designed to withstand the much higher power.

The T1 was also re-powered with a Liberty in April of 1932 and became the Medium T1E1. This gave the 43,900-pound tank a power-to-weight ratio of 15.4 HP per ton and increased its top speed to 25 MPH. As with the M 1921, it also increased drive line failures.

Medium Tank Model T1 and T1E1
(1 prototype built)

Length	**21' 6"**
Width	**8'**
Height	**9' 4 ½"**
Combat weight	**43,900 pounds**
	Engine
original	**Packard 1T-1330**
	200 HP @ 1,200 RPM
	250 HP @ 1,500 RPM
second	**Liberty**
	338 HP @ 1,400 RPM
	Power/weight ratio
original	**9.1 HP per ton (200 HP)**
	11.4 HP per ton (250 HP)
second	**15.4 HP per ton**
	Speed
original	**11.28 MPH (200 HP)**
	14 MPH (250 HP)
second	**25 MPH**
	Armament
1 6-pounder (2.24") - 2	**.30 cal. Browning MG**
Main armor thickness	**25.4mm (1")**
Secondary armor thickness	**9.53mm (3/8")**

COST OF TANK DEVELOPMENT

The Ordnance Committee made a special project of documenting a brief history of the development of the

Medium Tank T2 Built in 1929-30 by James Cunningham, Sons and Company. Although this tank was designed by the Ordnance Department, it was the first medium tank not built by them. (Courtesy of University of North Dakota, Patton papers and Jon Hornbostel)

medium tank in the minutes of their August 22, 1929 meeting. They had done the same for the development of the light tank in the minutes of their August 1, 1929 meeting and the same for the history of the Christie wheel-track type vehicles in their March 14, 1929 minutes. In the later case the history included much information on Christie-designed wheel-track vehicles built by or for the Ordnance Department which were other than tanks. These three items provide invaluable historical material and were of great aid to the author in this chapter.

The following two paragraphs provided such an interesting insight to the development of the medium tank that I have quoted them below.

"The cost of the medium tank development (exclusive of the Christie tank, but inclusive of all accessories and minor projects) over the nine-year period covered in this history (F. Y. 1920-28) was $982,675. This money was used for design, construction, tests, repairs, alterations and all other work in connection with the development of the tanks and accessories discussed herein. It was also used to develop many improvements in standard tanks. It should be borne in mind that development and research work are always very expensive due to (1) the individual nature of each job (2) the high-class personnel required and (3) the special and high quality of material and equipment used.

"From the above it may be seen that *approximately* $100,000 per year has been spent on medium tank development, including many accessories thereto and minor projects. Of incalculable value have been the experience and data gained, which give the necessary technical foundation for a type of equipment which is entirely new. It may safely be said that the experience gained in tank development in the last nine years accounts to a considerable extent for the almost immediate success of the new Light Tank, T1E1, delivered to the Experimental Mechanized Force by the Ordnance Department in the summer of 1928. A great deal has also been accomplished in the improvement of existing standard tanks."

The committee went on to summarize the accomplishments since 1919. Many were listed but I will include below only those items which seem the more important to this study.

A-Built three medium tanks.

B-Built a medium tank much better than any previous tanks (Model T1).

C-Accumulated a great deal of data and experience on tank design, a new field with no technical history to fall back on.

D-Prepared the ground for the quick and successful development of the light tank, T1E1, in 1927-1928. (Author's note: The light tank history quotes the cost of development and production of eight vehicles based upon the new Light Tank T1 at $187,535. These included 1 pilot chassis, T1 (later completed as a cargo carrier), 4 Light Tanks T1E1, 2 light Cargo Carriers T1E1 and 1 Light Tank T1E2. This would average about $23,442 per vehicle. If one took the total cost of the medium tank project of $982,675 and only credited it with producing three medium tanks the average cost per tank would be $327,558. The Ordnance Committee wanted to make sure it was known that there were many benefits gained from the medium tank project besides the production of three tanks.)

E-Developed tank accessories.

F-Developed many new improvements to existing standard tanks. (6-ton and Mark VIII). (Author's note: In the light tank history it is noted that consideration was given to updating modifications to the 800 existing 6-Ton Tanks (On October 21, 1926). Costs were estimated at $10,050 per tank for a total of $8,400,000 and it was decided not to make these changes. The more interesting information contained here is that there were still 800 of the original 950 6-Ton Tanks in service. We have figures on the Mark VIII tank indicating that all 101 built (100 production and one prototype from England) were still serviceable. In 1940 90 of these were sent to Canada and one, or possibly two, went to museums.

MEDIUM TANK T2

The origin of the Medium T2 tank design was noted on page 231. It was to be the 15-ton type called for in the April 3, 1922 War Department policy and some work had been done on the design while the main efforts toward a Medium Tank had been directed to the 23-ton types. When the order was issued on March 11, 1926 to continue development of the Medium M 1921 that order also contained a directive to actively pursue development of the M 1924 15-ton medium. Design work advanced but actual construction was held up until 1929 by lack of funds. The final design was based upon that of the T1E1 Light Tank mentioned above. The T1E1 Light Tanks were built by James Cunningham, Sons and Company who also received the contract to build the new Medium T2.

This design was quite different from the Medium T1. It carried a four man crew and the engine, a Liberty, was front mounted whereas the Medium Tanks all had rear-mounted engines. Loaded weight was 31,200 pounds and the 338 HP Liberty engine gave it a power-to-weight ratio of 21.7 HP per ton. The result was a top speed of 25 MPH which was limited to 20 MPH by a governor to extend the life of the running gear.

Hailed as the best tank yet designed by the Ordnance Department, the continued lack of funds prevented the type from being ordered into production. As a result, the one T2 built was used for testing and development well into the 1930s.

Medium Tank T2
(1 prototype built)

Length	**16'**
Width	**8'**
Height	**9' 1"**
Combat weight	**31,200 pounds**
Engine	**Liberty 12**
	338 HP @ 1,400 RPM
Power/weight ratio	**21.7 HP per ton**
Speed	**25 MPH**

Armament

1 semi automatic 47mm cannon	
1 .50 cal. Browning MG	
Main armor thickness	**22.2mm (7/8")**
Secondary armor thickness	**6.35mm (1/4")**

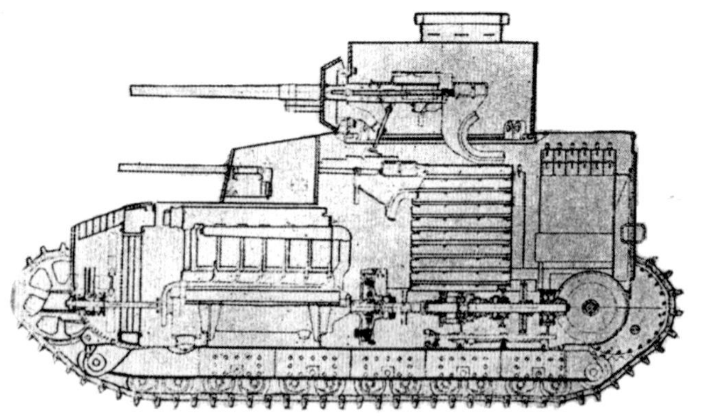

Sectional drawing of the Medium Tank T2. Note the Liberty engine located in the front. (Author's collection)

THE FINAL DAYS OF THE WW I 6-TON TANK M1917 AND THE MARK VIII HEAVY TANK

The 6-Ton M1917 tank was not, of course, Liberty powered but its long and relatively useful life does provide an interesting bit of history. As noted above, of the 950 built during WW I 800 were still in service in October of 1926. Improvements were still being proposed as late as 1932. In April of 1933 the Ordnance Department officially declared they would no longer consider any major modifications to this tank. From about 1925 until 1935 it was classified as Substitute Standard light tank. In April of 1935 its classification was changed to Limited Standard. In April of 1937 they were officially classified Obsolete and ordered stripped of their armament and their turrets removed and stored for possible future use. The hulls were to be either scrapped or expended as targets.

The number of these tanks still in service at that time is not known to me but in December of 1939 it was requested that those turrets in storage be removed to make the space available for other uses. 100 turrets were to be retained in open storage for use as targets at Aberdeen Proving Grounds and the remainder scrapped.

The Mark VIII was still being updated as late as 1935. On April 16, 1935 its classification was changed from Substitute Standard to Limited Standard. On April 9, 1936 it was recommended that the requirement for a Heavy Tank be dropped from the Ordnance Book of Standards and on May 7 it was approved. On June 16, 1939 the Chief of Field Service recommended the Mark VIII be classified Obsolete and on September 7, 1939 that was officially done. The quantity on hand was noted as 101; value per unit was $45,052; condition was serviceable; estimated scrap value $500 each. The on-hand quantity of 101 would indicate that all the 100 production as well as the one prototype sent from England had survived.

Although scrapping was authorized it evidently was never done as 90 are known to have been given to the Canadian government for use in training in 1940 and one sent to the Aberdeen Proving Grounds Museum.

CHRISTIE WHEEL-TRACK TANK HISTORY

Although it would be some time before a Christie wheel-track tank would be built using the Liberty engine for power, the concept and design which would eventually lead to one began just after the end of the war. As noted above, the early tanks were characterized by their principal shortcomings; *slow speed and poor reliability*. In a paper written by Major General Charles P. Summerhall to Brig. General Rockenbach, commenting on the latter's paper on tank lessons of the war, presented before the Army Conference in the AEF, was stated: "The organization should include trucks to move tanks to reach rendezvous positions near the jumping off line. I have been much impressed at the useless wearing out of tanks by having them drive long distances to their point of deployment."

Actually this statement led in two directions. One was to build tanks of a size limited to that which could be transported by truck and the other to build a tank which could transport itself rapidly without undue wear. This would lead to the Christie theory, which was to build a tank that could ride on wheels to transport itself on roads at rapid speeds until it reached the battle area and then be easily converted to tracks for across country use during battle.

J. Walter Christie was president of the Front Drive Motor Company of Hoboken, New Jersey. He had approached the Ordnance Department sometime in 1916 with a design for a wheel carriage for a 3" A. A. gun. They evidently felt it had some merit and ordered a pilot model built on 12/22/16. Testing produced enough confidence to lead to a production order for four units on 12/13/17. This was the beginning of a long and not always smooth association between Christie and the Ordnance Department.

In early 1919 Christie proposed a combination wheel and track-laying self-propelled 155 G.P.F. carriage to the Ordnance Department and it was received well enough for him to proceed with its building. On August 25, 1919 it was noted in the Ordnance minutes that it was going to be shipped to Aberdeen Proving Grounds for tests. The formal contract for it was dated 12/29/19.

Although later comments in the Ordnance Department minutes indicate they did not have the same opinion, demonstrations of this gun carriage greatly impressed the Chief of the Tank Corps, Colonel (later Brigadier General) Rockenbach. The result was that the Ordnance Department ordered a preliminary study of "wheel-track" vehicles (tank-type) on July 31, 1919. On November 14, 1919 the following request from Rockenbach was officially recorded in the Ordnance Department minutes: "Mr. Christie has made a preliminary study which is satisfactory, and it is urged that he be allowed to go ahead with all energy possible."

This resulted in an order for study, design and manufacture of one Christie tank with the Front Drive Motor Co., of Hoboken, New Jersey on November 22, 1919. Specifications were drawn up and, after examination, were modified on June 8, 1920.

On June 15, 1920 the Ordnance Department purchased the non-exclusive rights to existing Christie wheel-track and all other patents related to vehicles or parts thereof or arts, processes or machinery used in the

At left is the M 1921 Christie tank at Rock Island Arsenal shown on wheels. At right it is under test at Aberdeen Proving Grounds on tracks. The tank weighed 25,305 pounds and had armor ranging from 1/4" to 3/4" thick and carried a crew of 4. It was 18' 2" long, 8' 6" wide and 7' 1" high. Maximum speed with its 120 HP Christie engine was 15 MPH. It carried two 30 cal. machine guns, one mounted in each 1/4 front, and one 57 mm gun mounted in the front of the hull but not yet in place in the early picture on the left. (Photos from the Rock Island Arsenal Museum via Jon Hornbostel)

production thereof and the rights to any future patents or developments growing out of them. In consideration for this he was paid $100,000 and was to receive a royalty of $5,000 on each vehicle built. Considering the fact that no proven vehicle had been given any lengthy test, this would seem to be a very unusual expenditure, particularly considering the statement on page 233 that, exclusive of the Christie tank, over the nine-year period from 1920 to 1928 slightly under $100,000 per year was spent on medium tank development, the actual total being $982,675. As you will shortly see, from 1917 to 1924, eight years, $995,000 was spent on Christie contracts and patents.

The tank was built and demonstrated at the Christie plant in January of 1921. A report on this test along with some requests by Christie resulted in several authorized changes. These were made and the tank finally delivered to Aberdeen Proving Grounds on February 5, 1921. The tank was powered by a 120 HP six-cylinder Christie engine.

This tank underwent testing at Aberdeen Proving Grounds until April 21, 1921 at which time they were suspended at Mr. Christie's request. He wanted to make several changes in the tank which were approved by the Ordnance Committee on May 10, 1921.

It was returned to Christie and modified and returned to Aberdeen by March 28, 1922. Testing was resumed and continued until October 24, 1922 after which it was shipped to Camp Meade for service tests. These tests were begun on February 8, 1923 and continued until March 20, 1923.

In its original configuration this tank is generally referred to as the Christie Model 1919, or M 1919. As it was modified and finished its trials, it is known as the Model 1921 or M 1921. Although considered a medium tank it was called the Christie M 1921 or Convertible Tank M 1921 which kept it from being confused with the Medium Tank M 1921.

Throughout the period that work was being done on either design, modification or testing of this tank the Ordnance Department had great difficulty working with Mr. Christie and getting him to follow contract requirements. In addition, the tanks performance was found wanting in numerous areas. In a 1921 report by Colonel Moody, Chief of the Tank, Tractor and Trailer Division, some 15 "major specific troubles" were cited which he felt should be corrected before attempting any development of guns, turrets, armor or other items dependent on the chassis for their effectiveness.

On July 10, 1924 it was ordered that the tank be removed from further testing and returned to Aberdeen Proving Grounds and placed in their automotive museum.

As a final note regarding the Christie M 1921, the Ordnance Committee minutes carried the following comment: "Summing up the above, it may be said that whatever merit there is in the wheel and track idea was largely offset by the poor engineering, and poor execution (materials and workmanship) in the specific vehicle tested." Although this might seem to lay-to-rest any future dealings with Mr. Christie and his wheel-track tanks, it did not.

To add further reason to stay clear of Mr. Christie the board followed that comment with a brief history of the large series of purchases and tests of gun carriages designed and produced by Mr. Christie. I will not go into the comments in great detail except to say that not a single gun carriage passed its tests. All were rejected by the using services (none of which were the same as the using service of tanks).

The amount of money spent on these designs, especially considering the fact that from the beginning none passed muster, is rather remarkable. Perhaps a little light is shed on this by a statement found in a note written by Major General William J. Snow, former Chief of Field Artillery, to the Chief of Staff dated March 29, 1927. Under the subject heading "Research and Development Work for the Field Artillery under Christie Mount" said, after summarizing this project briefly: "In addition, the continuing work on Christie mounts was not done without some political pressure on the Army. It ceased when funds began to be curtailed." This statement is made more specific by reference to a letter from a U.S. Senator indicating Mr. Christie was seeking in 1922 a specific appropriation for continuing work on his projects. The General was referring specifically to gun mounts but the the statement applies to all the Christie contracts.

For informational purposes, a complete list of the Christie contracts up to that time was included and is reproduced below.

Expenditures on Christie Contracts and Patents 1917-1924

Date	Quantity	Material	Amount
12/22/16	1	Pilot 3" A.A. Wheel Carriage	$7,880
12/13/17	4	Production 3" A.A. Wheel Car.	60,000
10/30/18	1	Exp. Carriage for 8" Howitzer. (Salvage parts used for below 155 Carriage.)	86,960
11/22/19	1	Tank (First construction.)	67,000
12/29/19	1	First 155 G.P.F. Combined Wheel & Track-Laying Carriage	67,500
3/13/20	3	Carriages for 155	150,000
3/26/20	2	Carriages for 75-105	95,000
6/1/20	1	Design & Mnfg. of 4.7" A.A. Carriage.	83,000
6/15/20		Purchase of Christie patents	100,000
10/3/20		Repairs & Alt. to the pilot 155 Carriage.	6,000
8/25/20	1	Design & Mnfg. of 3" A.A. Carriage	78,000
6/15/21		Tank rebuild	15,000
6/15/21	1	Trailer (Apx.cost)	15,000
8/27/21		Furnish & install 4 speedometers on 4 155 Carriages	300
8/18/21		Overhaul & repair one 155 Carriage	5,400
9/23/21		Repair & modify 75-105 Carriage	2,200
1917-1924		Cost of testing all vehicles	155,760
		Total costs	**$995,000**

The above history was documented in some considerable detail in the Ordnance Board Minutes of March 14, 1929 under item 7522. Also part of that same item was the recording of a directive of the Secretary of War through the Adjutant General's Office to the Chief of Ordnance dated February 19, 1929, that in cooperation with the Chief of Cavalry the M 1928 Christie mentioned below be procured and equipped as an armored car. It was directed that $15,000 will become available from the 1930 Ordnance Appropriation on July 1, 1929 for the purchase. At this point it would appear that the Ordnance Department was none too happy with the prospect of having to again work with Mr. Christie and was going to leave a good paper trail to fall back on when it again did not work out.

CHRISTIE M 1928

Christie was continually working on new designs or modifications of his existing ones and the rejection of his M 1921 by Ordnance led to the design of a new wheel-track design which he felt would overcome the objections raised to the M 1921. During this period he reorganized his Front Drive Motor Company as the U.S. Wheel Track Laying Corporation located in Rahway, New Jersey. The new design was completed as his M 1928 and a prototype demonstrated before Army representatives at Fort Leonard Wood and Fort Meyer, VA on October 28, 1928.

This new vehicle weighed 17,200 pounds and was powered by a 338 HP low compression Liberty engine giving it a power-to-weight ratio of 39 HP/ton and an extremely impressive performance. Its much larger engine (the M 1921 was powered by a 120 HP Christie

Christie tank M 1928. This new Liberty-powered design was demonstrated to the Army by Christie and subjected to tests by the Army. The Ordnance Department contemplated purchasing it but never did. (Courtesy of the Rock Island Arsenal Museum via Jon Hornbostel)

engine) and extensively redesigned suspension gave it vast improvement in both speed and agility. Those observing were so impressed that the Chief of Cavalry convinced the Secretary of War to give the above directive to procure the M 1928 Christie.

In February of 1929 Congress had approved a 1929 appropriation of $250,000 for the Ordnance Department to purchase T1E2 Light Tanks. In March the Chief of Infantry had requested those funds be used for procurement of M 1928 Christies for infantry use. On January 9, 1930 (OM item 8095) the Secretary of War approved the request.

Just prior to this, on January 2, The Secretary of War withdrew the prior approval to purchase the M 1928 chassis with Armored Car revisions. It has been reported this was done because Christie would not quote a price acceptable to the Ordnance Department. In any case, it laid the groundwork for an alternate plan, the procurement of one or more M 1928 derivatives. It would appear that it was felt that some significant changes in the M 1928 design were desirable and they should be incorporated in one or more newly produced machines.

CHRISTIE M 1930 (THE M 1931 PILOT)

On January 22, 1930, Christie presented the Ordnance Department with a proposal for supplying a modified version of the M 1928 with a price of $26,000 each if purchased in a quantity of 5 or $25,000 each in a quantity of 10. In the February 13, 1930 OM item 8095 is a recommendation that 6 Christie tanks be procured with 1929 funds (the $250,000 mentioned above).

THE RUSSIAN M 1930 ORDER

At some time during the month of January (the exact date is not known) a Russian delegation visited Christie to examine his new tank. News of the spectacular performance of the M 1928 had come to their attention and they were very interested. The Russians liked the tank and on April 28, 1930 signed a contract with Christie to purchase two M 1930 tanks to be delivered no later than four months after the signing of the contract, or by September 1, 1930. (We shall see that the delivery deadline was not met.) The design was most likely the same as the current proposed design of the pending U.S. order.

The price agreed to was $60,000 for the two tanks, $4,000 for spare parts and $100,000 for patent and production rights and technical assistance. The $100,000 patent rights figure was based upon what Christie had gotten from the U.S. Government for similar rights in this country. Although the Russians already had the Liberty in production in the form of their M5 engine, I have found nothing to indicate they shipped their own engines to Christie for use in the tanks. By 1930, surplus brand-new Liberty engines could easily be found in this country for as little as $400 each. The cost of a new M5 plus shipping from Russia would probably exceed that figure.

Nothing was done with the January 22 prices quoted by Christie and on April 10, 1930 the Ordnance Department issued an RFQ for 6 Christie tanks and one armored car. (Note that the Russian contract was signed on April 28.) Christie responded to the somewhat different specifications with prices of $35,000 each for the tanks in quantity of 6 or $38,000 each in quantity of 5. He refused to accept the design changes imposed by Ordnance for the Armored Car and there were no other responses so the authorization to procure the Armored Car was revoked. After some extended discussion regarding Christie's response, the proposed contract for 6 tanks was never executed.

On June 19, 1930 the Chief of Ordnance recommended the proposed order for 6 tanks be reduced to one pilot for testing before ordering the remaining 5. The Chief of

Infantry wanted the order to stand at 6 but the Secretary of War sided with Ordnance and reduced the order to one, still out of 1929 funds. On June 19 it is also decided to designate the tank as Medium T3.

The Ordnance Department had not had a history of smooth sailing with Christie and it was evident this had not changed as on June 26 Chief of Staff General Summerhall ordered the Chief of Ordnance to get over his problems with Christie and sign the contract for one Christie tank or risk loosing all of the $250,000 funding for infantry tanks. (The fiscal year for these funds ended on July 1, 1930, and any unused funds would revert to the Treasury on that date.)

Christie T3E1 No. 2 at Aberdeen Proving Grounds. The photo is dated January 1932. (Courtesy of Jon Hornbostel)

The contract was signed on June 28, 1930 for one M 1930 (with turret) with the cost being $55,000 plus $7,000 for testing and engine modifications. (The Army supplied the engines for all ordered Christie tanks.) This tank would become the M 1931 pilot and be designated as the Medium Tank T3. On July 1 the remaining $188,000 of the infantry tank allocation was returned to the Treasury.

The Liberty engine as fitted to the Christie tanks. This was a standard low compression V-12 Liberty rated at 338 HP @ 1,400 RPM. It was supplied with a 12 volt battery/ generator ignition system then current on the military Liberty. Electric starter was mounted on the transmission and a Navy H-16 type hand starter was mounted on the rear of the engine. An eight-blade 30" fan is mounted on the shaft end of the engine just behind the flywheel. The pipes on top of the engine are connections to the cooling system. (Courtesy of R. P. Hunnicutt)

The agreed-to delivery date was September 1. Its actual delivery was on January 19, 1931, some four months late. Christie was continuing to live up to expectations. During this period he was also building the two tanks ordered by the Russians. Ordered as M 1930 tanks, they appear to have been delivered close to the same design as the American prototype (but without turret). They were also scheduled for delivery in September but were delivered in late December for shipment to Russia, where they arrived in January of 1931. The Russian Christie is covered in detail in Chapter 11.

THE POLISH M 1930 ORDER

The Polish military had also been attracted by the publicized performance M 1928 and had approached Christie regarding possible purchases. The Polish purchasing mission arrived in Washingtom in March of 1930 and asked the U.S. government if they had any objections to the Polish government purchasing Christie tanks. After finding there were none, the mission negotiated a contract with Christie for the purchase of one M 1930 gear-driven tank for $30,000 plus an additional $3,500 in spare parts. They also secured an option to purchase the rights to produce the M 1930 in Poland for $90,000. The contract must have been executed about the first week in June 1930. $10,000 was paid on deposit with the remainder payable on delivery, which was to be in four months.

The Polish military deligation arrived at the Christie plant in early October of 1930 to take delivery. They found the tank was not yet completed and that Christie's

Pictured above is F Troop, 1st Mechanized Cavalry in February of 1934. The center four vehicles are all Combat Car T1s. Notice that the all-machine gun car is on the right, and all three others are still equipped with 37 mm guns. (From the Patton Museum via Jon Hornbostel)

attitude about the matter was somewhat hostile. He refused to deliver the tank unless they paid the $90,000 for manufacturing rights immediately.

They would not agree to that requirement and the tank was not delivered. They brought legal action against Christie and, according to Edward Christie (Walter Christie's son) the deposit was eventually returned. The one tank, which was in the process of construction, was completed and evidently delivered as No. 2 tank of the seven soon to be ordered by the U.S. Ordnance Department.

After delivery of the M 1930 on January 19 it was submitted to extensive testing at Aberdeen Proving Grounds. It was returned to Christie for modification on March 27, 1931, and then returned to Aberdeen for further testing. The contract called for the delivery of the tank (after acceptance) and a set of drawings for $55,000. After the completion of acceptance tests, Ordnance offered Christie $54,000 for the tank and a set of drawings. Christie replied with a refusal to sell the tank if not given assurance he would get an order for additional tanks. Ordnance would not agree to any such commitment and withdrew the offer. The tank was returned to Christie and not paid for. This would result in the return of the remainder of the $250,000 allocation to the Treasury.

This tank was retained by Christie and was sold to the British government in late 1936, the story of which is covered in Chapter 11. He was quite happy to sell it to anyone by then as he was in deep financial trouble and had been trying to get Ordnance to pay the $55,000 he claimed they still owed for it, without success.

CHRISTIE M 1931 PRODUCTION CONVERTIBLE MEDIUM TANK T3 COMBAT CAR T1

It appears that the Army was not easily discouraged as on June 18, 1931 Christie's U.S. Wheel Track Laying Corporation was given a contract to built seven Medium T3 Tanks (ref. Office of the Chief of Ordnance, 451.25/3758) with 0.188-inch nickel-steel alloy interior body panels (rather than armor). Exterior armor ranged from 1/2 to 5/8 inch. Total price was $241,500.

By September 17, 1931 it had been decided that 5 of the 7 would be furnished as tanks and designated Medium T3 Tanks and 2 furnished as combat cars and designated Combat Car T1. The only physical difference between them was to be the armament. The tanks were to be furnished with a turret mounted 37 mm and .30 cal. machine gun combination; in the combat cars the 37 mm was replaced by a .50 cal. machine gun. The tanks were to be assigned to the infantry and the combat cars to the cavalry and the technical designations of the vehicles as "tank" or "combat car" was done to satisfy the legal requirement that all tanks be assigned to the infantry.

The first of the seven was delivered to Aberdeen for tests on October 9, 1931 and the seventh during the month of March 1932.

By the end of 1931 the division between the infantry and cavalry had been changed to 3 T3 Medium Tanks and 4 T1 Combat Cars. The vehicles were assigned numbers from 1 to 7. The M 1928 and the M 1930 (M 1931 pilot) were chain drive to the rear pair of bogie wheels when traveling without tracks. Six of the seven production versions had the same design, the seventh (vehicle No. 2) was furnished with gear drive instead. No. 2 was undoubtedly the undelivered Polish tank.

Nos. 1, 3, 4 and 5 were assigned to the Mechanized Cavalry as Combat Cars T1 and Nos. 2, 6 and 7 to the infantry as Medium Tanks T3.

Medium Tank T3
Combat Car T1
(Christie M 1931)
(A total of 1 prototype and 7 production vehicles were built. In addition, 2 were sold to Russia and 1 to Great Britiain, all without turrets.)

Length	17' 5"
Width	7' 3 ½"
Height	7' 4"
Combat weight	22,220 pounds
Engine	Liberty rated 338 HP @ 1,400 RPM The engine was not governed and was capable of 400 HP @ 1,700 RPM
Power/weight ratio	30.6 HP per ton (at rated)
Speed @ 1,400 RPM	46.7 MPH on wheels 27.2 MPH on tracks

Armament

1 37 mm gun (Medium T3)
1 .30 cal. MG (Medium T3)
1 .50 cal. MG (Combat Car T1)
1 .30 cal. MG (Combat Car T1)

Main armor thickness	15.9 mm (5/8")
Secondary armor thickness	12.7 mm (1/2")

MEDIUM TANK T3E1

Because of its unique drive design, No. 2 was designated as T3E1. It probably arrived at Aberdeen for testing about November of 1931. By early 1933 it had been decided to convert its on-wheels drive from gear to the chain system used by the others. The February 2, 1933 OM (Ordnance Minutes) item 10434 directed that Medium Tank T3E1 be converted to chain drive and after accomplishment its designation would change to T3 No. 2 and the designation E1 be removed from its name plate.

COMBAT CAR T1

The first two combat cars arrived at Fort Knox, Kentucky in the spring of 1932 and the last two during the summer. The last one was modified by the Ordnance Department to convert it to all machine gun armament, the other three were delivered with 37 mm guns the same as the T3 Tanks.

COMBAT CAR T1E1

On May 19, 1932 OM item 9837 carried an extensive list of modifications submitted by the Subcommittee on Automotive Equipment to be applied to one of the Combat Car T1s as a test before they are applied to the remaining vehicles. These changes were

Christie Medium Tank T3 viewed from the front. (Author's collection)

aimed at improving the T1 and also with a view to their use in any new design.

Although the list was quite long and detailed, it made no mention of changes to the chain drive used for road travel without tracks. The combat car on which the modifications were to be made was to be designated Combat Car T1E1. Some historians state that the modifications included changing the road drive system from chain to gear drive as the Convertible Medium Tank T3E1. Perhaps that item was added to the list after May 19. If that change was indeed applied then there were, for a while at least, two Christies with gear drive.

If the change was indeed made one would wonder how long the feature may have remained on this Combat Car since it was removed from the T3E1 about nine months later. This writer seriously doubts that feature was applied.

COMBAT CAR T1E2

On June 2, 1932 OM item 9876 noted the authorization to temporarily install a 120 HP Cummins Diesel engine in one of the Combat Car T1 vehicles while its Liberty was removed for overhaul. The idea was to see what benefits the relatively constant torque over a range of some 1,000 RPM might be. It was recognized that the power of this engine was not adequate but since it was to be supplied free by the manufacturer and they felt some good engineering data could be gained. The combat car was to carry the designation T1E2 while so powered. On July 7, 1932 OM item 9958 noted the cancellation of the project. The conversion never took place.

COMBAT CAR T1E3

On July 21, 1932 OM item 9958 carried the following item: The American-La France and Foamite Corporation suggested the use of their V-12 30° 753.9 cu. in. 246 HP @ 2,800 RPM engine in Combat Car T1. The engine weighed 1,600 pounds, which is about 700 pounds more than the Liberty. The engine and required alteration of gear ratios of third and fourth gears would be supplied free with the understanding that if the tests proved satisfactory the engine would be purchased at a cost of $1,500 and the gear modifications would be paid for. The proposal was accepted. The combat car would become T1E3 when so equipped.

On October 20, 1932 it was reported that Combat Car T1E3 was at the American-La France plant at Elmira, New York for completion of the engine installation and should be returned by the end of the month.

The August 31, 1933 OM item 10953 reported that the tests of the new engine in Combat Car T1E3 had not proved satisfactory. Performance was not adequate and the engine had actually failed. It was directed the the original Liberty be reinstalled and the designation returned to Combat Car T1.

MEDIUM TANK T3E2

The T3/T3E1 was the last Liberty-powered tank of U.S. design but not the last U.S. Christie design. By late 1932 The Ordnance Department had approved a set of specifications for an improved version of the Christie Medium T3 tank. By November 1932 plans had been approved for having 5 examples built. The new design was to incorporate a number of improvements in the design of the tank itself, including increasing its length to 18' 9", width to 8' and height to 7' 8". This allowed seating the driver and a gunner side by side, increasing the size of the turret and adding two more crew. The Liberty, now easily recognized as obsolete, was to replaced by a more modern engine. The new tanks were to be the Convertible Medium Tank T3E2.

Christie was offered the contract but evidently balked at the changes required. He had refused to build the combat car specified in the January 22, 1930 RFQ primarily because of a requirement to increase the size of the fighting compartment. Perhaps the increased dimensions of this new design was a factor. The contract was, instead, awarded to the American-La France and Foamite Company.

The original proposal was to use a 625 HP Curtiss Conqueror engine supplied by the contractor but on January 31, 1933 this was changed to a 450 HP Curtiss D-12 to be supplied by the Ordnance Department (acquired from the Air Service).

After extensive testing a number of modifications were applied to the five tanks and they were then redesignated as Convertible Medium Tank T3E3.

MEDIUM TANK T4

The last of the Christie-based design American tanks were the Medium T4 and T4E1. These were built by Rock Island Arsenal in 1935-1936. Sixteen T4s and one T4E1 were built. All were convertible types. Power was now by a 268 HP Continental 7-cylinder aircraft engine adaptation.

THE FINAL DAYS OF THE LIBERTY-POWERED T-3 TANKS AND T-1 COMBAT CARS

May 28, 1936 marked the beginning of the end for the Liberty-powered Convertible Medium Tank T3 and the Combat Car T1. OM item 12888 carried an item titled

"Pilot Combat Cars T1, and Medium Tanks T3, Convertible - Disposition." The following is excerpted from it:

Seven Christie convertible-type vehicles were produced in 1931. Four of these vehicles were equipped as Combat Cars T1 and assigned to the Mechanized Cavalry at Fort Knox, Kentucky. The remaining three were equipped as Medium Tanks T3 and assigned to the Infantry at Fort Benning, Georgia, for test purposes. These experimental pilots have been in service for approximately five years and have well served the purpose for which they were procured. The cost of maintenance has been extremely high for all seven vehicles. Reports compiled by Field Service show that the combat cars at Fort Knox have averaged $3.60 per mile of operation for maintenance and the tanks assigned to Fort Benning have averaged $2.63 per mile, making an average for the group of $3.18 per mile of operation. Maintenance parts have to be specially made which results in unusually high costs.

There followed the note that the Infantry had been supplied five of the more modern T3E2 types and the even later design T4 was now in production and are being delivered by Rock Island Arsenal. It was further noted that the new Combat Car M1 was now being supplied in quantity. It then went on to describe the need to conserve limited maintenance funds for use on new standard tanks and combat cars and noted that to that end it was essential that maintenance costs on worn-out experimental pilot vehicles be reduced to the absolute minimum, and that:

"These pilot Combat Cars T1 and Medium Tanks T3 be withdrawn from service at an early date. It is the opinion of the Subcommittee (on Automotive Equipment) that parts from one or more of these old Christie vehicles should be used for maintaining the balance in service to avoid procurement of new parts.

Recommendations:

a-That as individual Medium Tanks T3 and Combat Cars T1 become unserviceable parts be taken from these vehicles to maintain a fewer number of vehicles of this type in serviceable condition.

b-That as the individual Combat Cars T1 and Medium Tanks T3 become wholly unserviceable due to use of parts for maintenance of the other vehicles or otherwise, these unserviceable vehicles be placed on I. and I. reports with a view to their condemnation in accordance with existing regulations.

c-That after condemnation these vehicles be used locally as targets."

All seven vehicles were considered as still in service as of about the first of June 1936 but the numbers remaining no doubt reduced fairly rapidly thereafter. Probably all had been removed from active use by the time the Mark VIII Heavy Tank was declared obsolete in 1939.

The Liberty engine had powered the Army's tanks and combat cars with good service from 1917 to 1939.

The Christie M 1930 - 1931 Tanks Chronology

Date		
1930		
April	29	Russian contract for 2 M 1930
May		
June	xx	Polish contract for 1 M 1930
	28	U.S. contract for 1 M 1930
July		
August		
	29	Russian M 1930 delivery due
September	1	U.S. M 1930 delivery due
October		
	15	Polish M 1930 delivery due
November		
December		
	29	Russian M 1930 delivered
1931		
January	19	U.S. M 1930 delivered
February		
March		
April		
May		
June		
	18	U.S. contract for 7 M 1931
July		
August		
September		
October		
	9	first U.S. M 1931 delivered
November		
December		*Seven M 1931 were delivered to Ordnance during this period. No. 2 is believed to have been the defaulted Polish M 1930 updated to a M 1931.*
1932		
January		
February		
March	xx	last U.S. M 1931 delivered

Polish/Christie default

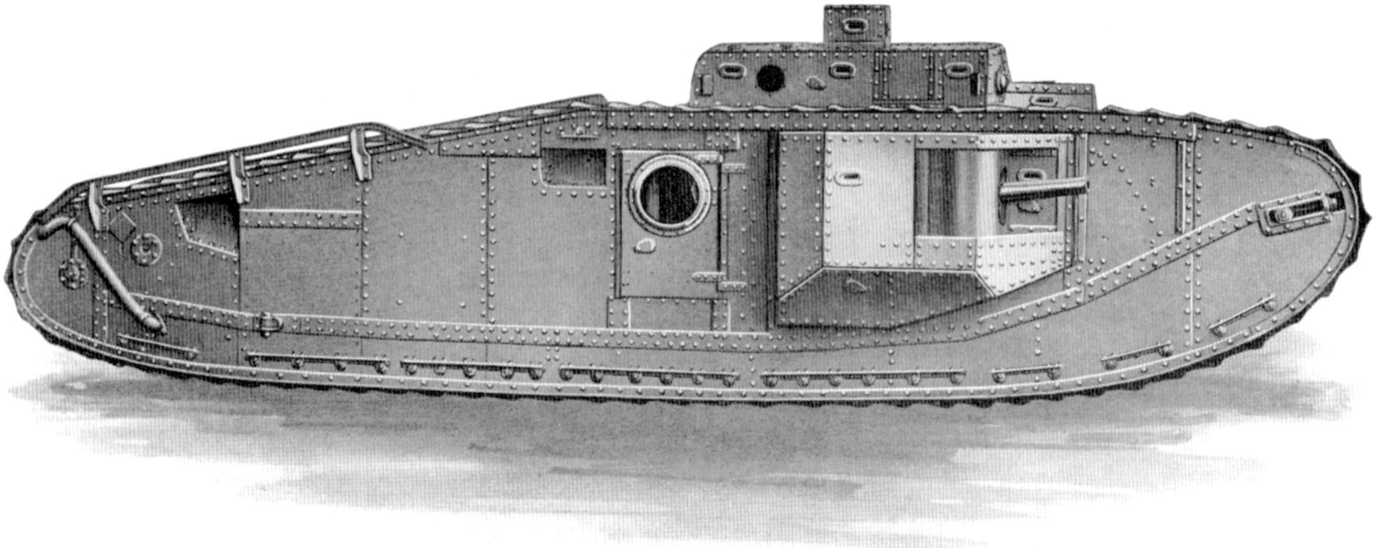

RIGHT SIDE VIEW OF 35-TON TANK, MARK VIII.

(Views from the November 15, 1918 Mark VIII Tank Manual)

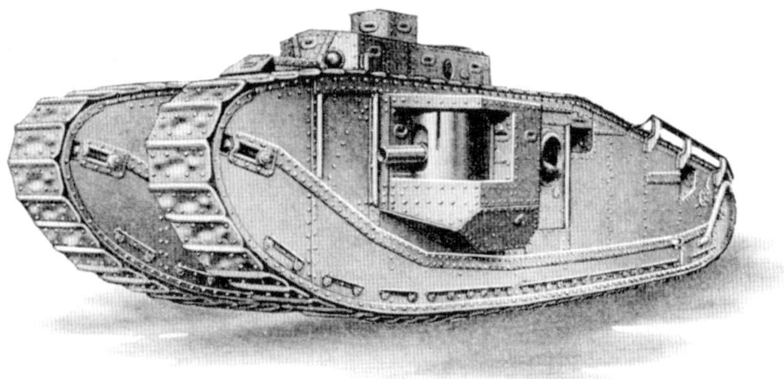

THREE-QUARTER FRONT VIEW OF MARK VIII

Chapter 7

Post-War Engineering Developments and New Designs

In this chapter I shall cover those engineering changes and new designs which followed WW I, most of which were initiated by the U.S. government. These were primarily done under the control of the Army Air Service although the Navy was also a user and initiated a few of its own modifications.

Several Liberty-based new engine designs were initiated by various people and companies. Some of these were intended for government use, the aim being to either extend the useful life of the multitude of Liberty engines held in government stocks or to extend the range of service to which the engine could be applied, usually by building a more powerful engine based on Liberty components. A few were designed to produce a high power engine for civilian use around the proven and easily available components of the Liberty.

In a couple of cases these new designs or modifications result in a subject so extensive that I have chosen to dedicate a complete chapter to each. Thus I shall touch on them only briefly in this chapter.

The first is the story of Allison Experimental/Allison Engineering. Allison's first connection with the Liberty appears to begin with the building of production tooling for Nordyke & Marmon in late 1917. Allison then began contracting to build Air Service-designed modifications of the Liberty, the first being contract CS-34 of early May of 1918 for the building of epicyclic reduction gears for the L-12B. From that time until the early 1930s Allison was involved in both the building of modified Liberty engines and the rebuilding of Liberty engines under government contract. They also initiated a few designs of their own. This subject is so extensive that I have devoted Chapter 8 to it.

The second subject is the supercharging of the Liberty. This modification was initiated by the Engineering Division of the Air Service in June of 1917 when they delegated Mr. E. H. Sherbondy to investigate the supercharging of aero engines. The original idea was to be able to retain sea level atmospheric pressure to the carburetor at high altitude so as to maintain sea level power at those altitudes. Since it was to be the principal fighting engine of the Air Service, the Liberty was chosen as the engine around which these designs were based. By the fall of 1917 the General Electric Company and their principal turbine engineer, Dr. Sanford A. Moss, had been requested by Dr. William F. Durand of the National Advisory Committee for Aeronautics to also pursue the subject of supercharging the Liberty engine.

Sherbondy's investigations resulted in several designs for both turbo and gear-driven superchargers. Only the turbo type was actually built and tested. They were of limited success and the Sherbondy effort was terminated shortly after the end of the war. The Moss design was also for a turbo-type supercharger but was more successful and, with only a short interruption, was continued for a number of years using the Liberty engine as its base. The basic design developed for the Liberty was carried on and became the basis for the very successful turbosuperchargers used on many of the aero engines of World War II.

There were also gear-driven centrifugal superchargers and Roots-type superchargers built for the Liberty. All of these will be covered in Chapter 9, *Supercharging the Liberty*.

ENGINEERING DEVELOPMENTS BY OR AT THE DIRECTION OF THE ENGINEERING DIVISION

—

CRANKSHAFT AND RODS

Crankshaft design went through an evolution of three designs during the production phase but no changes were made thereafter. The original design was known as the "light" type and was built for scupper rod journal oiling. It was replaced with what became know as the "medium" type by alteration #36 on December 12, 1917. The "medium" weight crankshaft was replaced by the "heavy" type with alteration #63 on December 28, 1917. The "heavy"

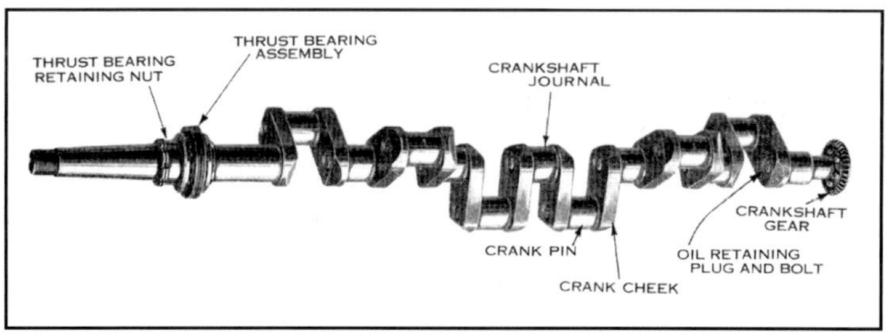

Crankshaft as pictured in the 1927 manual. Note that this crankshaft is not counterbalanced. That feature would not see any extensive use until well after the Liberty was designed. Even the WW II Nuffield Liberty built by the British would use this same design, without counterbalance. (Author's collection)

type was the first to be manufactured with full pressure lubrication. Only Packard engines would be originally fitted with either "light" or "medium" crankshafts and many of these were later altered to full-pressure lubrication.

The 1927 maintenance manual noted: "There were three different-sized shafts used, the first two being lighter than the present standard shaft. All crankshafts that do not measure 3 5/16 inches across the cheek at the pin shall be replaced by those meeting this dimension. Check the shaft for cheek measurement at the pin."

Rod design had been settled by the end of regular production and by 1927 the only notes in the overhaul manual were regarding early rods which were made for early light bearings being obsolete. Such rods were to be disposed of. These rods (both plain and forked) were identified by their inside measurements of 2.844 and 2.750 inches respectively.

I noted on page 135 under wartime development that articulated rod designs were developed during the war and that some additional interest was shown shortly thereafter. That interest was sparked by the 50-hour test report of the L-12A 1921 Model reported in test serial 1844 on 1/24/22. The actual test was run between June 17 and July 1, 1921. Captain George E. A. Hallett, Chief Power Plant section, made the final report of the test and among his comments were some regarding the rod bearings. First he noted that they showed much cracking as a result of the extended running at high power levels. He followed this with the following comment: "This cracking of the white metal lining is probably a matter of fatigue." (He was, of course, correct in his assumption and the cause, and thus babbitt failure, was corrected by the development of the less-flexible, steel-backed bearing in 1923.) "This fatigue, it was thought, is caused by the present non-rigid support of the forked connecting rod construction. It is believed that in an articulated rod, a more rigid support of the bearing would be possible, thereby, eliminating much of the bearing flexure with its resultant failure. In view of this inherent weakness, and the satisfactory results obtained with the articulated connecting rod in such engines as the Curtiss C-12, Model W-1, (a W-18-type engine designed by the Engineering Division and built by Packard) Rolls-Royce and Napier Lion, it seems advisable to investigate the articulated connecting rod for the Liberty Twelve engine."

Although some small amount of work was done on further testing of articulated rods for the Liberty, it was soon halted. Captain Hallett's report was published on January 24, 1922 and within a year Allison had made significant progress on their steel-backed bearing which would reduce the flexing of bearing shells to an acceptable level and be a much more cost effective cure than complete replacement of rods with those of new design.

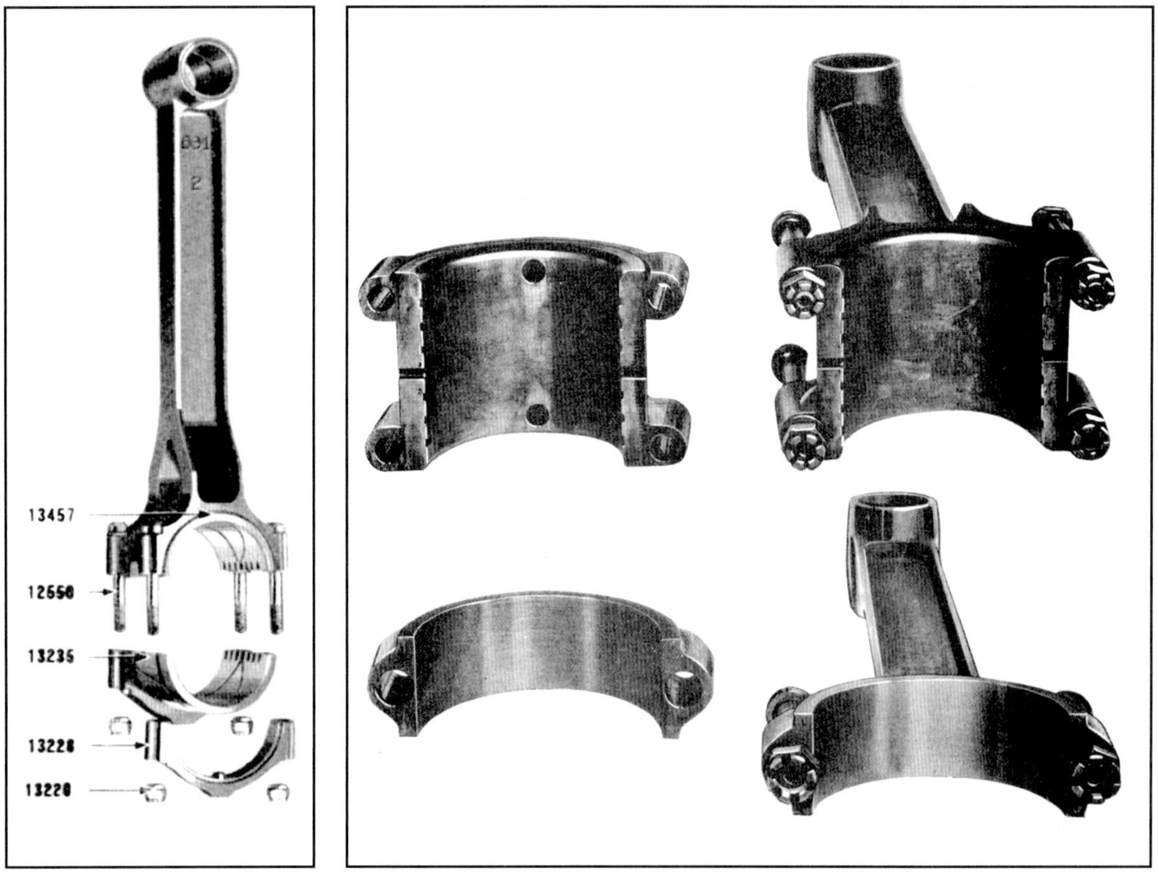

At left is illustrated a rod with grooved-type bearing used in early 1918. At right is a pair (both fork and blade) of rods with plain surface-type bearings. (Author's collection)

BEARINGS: The initial work on bearing development was begun during the breakdown tests carried out at the Packard plant under the direction of Major E. J. Hall (see page 94). It was during this period that a number of basic features regarding bearing design were decided upon. Main bearing

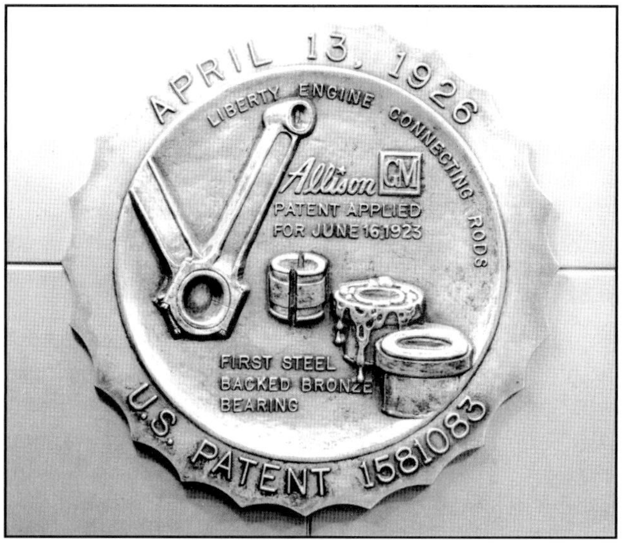

An impressive plaque cast by Allison in 1954 commemorating the the steel-backed bearing design which did so much to gain the company both reputation and business in the 1920s and 1930s. (Courtesy of Rolls-Royce / Allison)

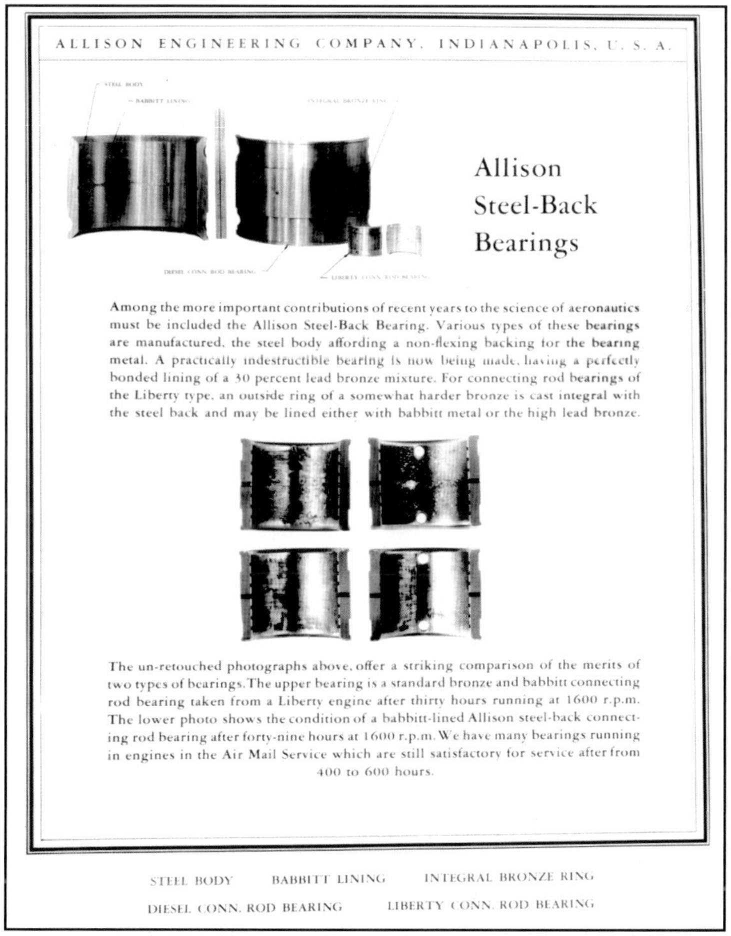

A bearing ad run by Allison in the late 1920s. (Courtesy of Rolls-Royce / Allison)

position stabilization was to be accomplished by use of a single central locating dowel in each half of the bearing. The lower dowel would be hollow and also be the bearing oil feed from the lower crankcase oil manifold. The big end rod bearing would be stabilized on the lower half only and by a single dowel in the center of each of the two rod caps.

Bearings would have bronze backs with babbitt lining. Initially it was thought that the thicker the babbitt the better would be the wearing characteristics but testing showed that bearing thickness was better applied to the bronze backing to give strength and prevent deflection or distortion in the bearing as such movement would cause the babbitt to crack and chip off the backing.

Both grooved and plain bearing surfaces were tried and initially grooves were thought to be best as they would distribute the oil better on the total bearing surface. Early bearings were thus grooved but later thinking on the matter resulted in the adoption of plain surfaces.

The method of stabilizing the bearings as well as the oil source to lubricate them is well-illustrated in the color section on page 202, and on the sectional drawing on page 220. The bearings themselves are illustrated on this and the prior page.

Although a few other methods of manufacturing the bearings were tried, those made by Ford seemed to give the best results as they appeared to have better babbitt retaining characteristics.

Also of significant help was the change from scupper to full-pressure oiling of the rod bearings which occurred early in production (see page 89).

ALLISON STEEL-BACKED BEARINGS

Although the Engineering Division continued to test bearings made of various materials, little progress seems to have been made until about 1923. In early 1923 Allison engineer Norman Gilman perfected a process whereby babbitt and bronze lining could be successfully applied to steel bearing shells. Babbitt was used on the interior surface which contacted the crank journal and bronze on the outer surface which contacted the blade rod. The steel backing had twice the strength and three times the stiffness of the bronze backing formally used for bearings. Flexing of the bronze had been the principal cause of bearing surface deterioration and the ability to apply bearing material to the much stronger steel backing greatly reduced that problem.

Allison had been involved in assembling and testing Liberty engines since their first contracts to assemble and test the L-12B epicyclic engines built in 1920. They were in constant communication with the Engineering Division and kept abreast of Liberty engine performance.

Application for a patent on the manufacturing process was made by Allison on June 16, 1923. After some considerable delay caused by initial doubts by the Patent Department that the process was new to the art, patent No. 1581083 was finally granted on April 13, 1926.

Samples of the bearings were submitted to the Engineering Division for tests in the summer of 1924 and 50-hour endurance test on a high-compression (6.5:1) Liberty engine showed them to remain in excellent condition. The Air Service shortly thereafter accepted the bearing for use on both main and rod bearings of the Liberty.

Allison later claimed that time between overhauls of the Liberty were raised from about 200 hours to 600 hours by these bearings.

At about this time Allison was contracted to convert a number of L-12A engines to inverted configuration. All converted by Allison were fitted with their steel-backed bearings, as were later inverted air-cooled Liberty engines they built for the Air Service. Both of these models will be discussed in detail later.

Although by the end of 1924 Allison bearings were available in quantity for Liberty rebuilding, there is no evidence that the Air Service switched to their exclusive use nor that they made any attempt to retrofit any engines in current use. I will cover the subject of rebuilding in detail later.

The May 1927 manual *Liberty-General Instructions For Maintenance* covers the repair and rebuilding of the Liberty engine in detail. In the sections on main and rod bearings it is noted that (by 1927) steel-backed bearings had been installed in a number of engines.

Bearings which were not damaged and which met fit specifications were to be reused after cleaning, whether or not they were steel-backed types. The exception was early type rod bearings with curved oil grooves, which were to be replaced if found. Steel-backed rod bearings could be identified by the oil holes in their centers. No method of identifying steel-backed main bearings was given although slightly different clearances were called for when reassembling. They were probably checked for magnetic properties.

GEARS: Occasional failures of teeth of the various bevel gear drives in the train from the rear of the crankshaft continued to be reported in spite of the earlier redesign during initial production reported on page 122. This resulted in a second redesign which became known as the "stub-tooth gears." The new type gears were applied to all in that train. This redesign was accomplished in early 1921.

Repair directions in the 1927 manual indicate, however, that it was permissible to replace damaged gears in pairs with stub-tooth gears without replacing the complete set of gears in the train so long as this was not as part of a class III overhaul (a complete overhaul). In the case of a class III overhaul, if not already done, all gears must be replaced with stub-tooth type.

ACCESSORIES

OIL PUMPS: The original standard oil pump carried a part number for the assembly of 8200 and it was therefore known as the "8200-type" pump in all following documentation. This pump was modified in two different ways to facilitate regulation of oil pressure. The first was the vacuum oil control developed in late 1918 and described on page 134. This design is not believed to have ever been put into production but was mentioned in the 1920 manual as available for use to relieve high oil pressure at low speeds. It required a new lower pump cover and oil strainer and the replacement of the old pressure relief valve with the vacuum control valve. It evidently saw very limited use as it was not mentioned in the 1927 maintenance manual.

An oil pressure control method that did see extensive use was introduced in about 1920. In this system the oil pressure relief valve head of the standard 8200 type pump is drilled with four holes with a No. 52 drill. Doing so would limit maximum oil pressure to about 30 pounds and thus relieve excessive oiling. If, however, bearing clearances became excessive or oil leaks developed somewhere in the system this could result in pressure being too low at low engine speeds. The 8200 pump with this

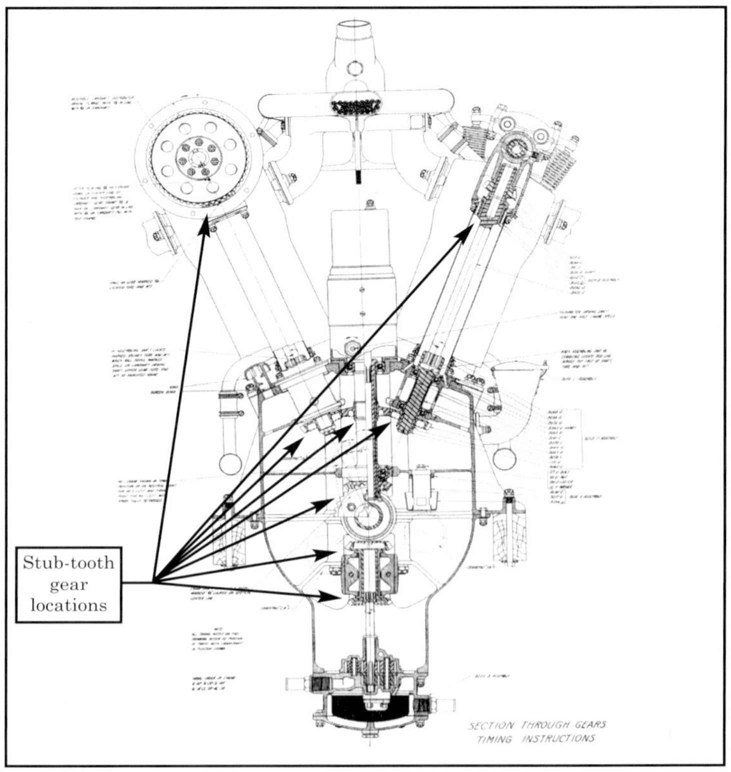

type of pressure control continued in use until the Liberty was removed from service.

CENTRIFUGAL OIL CLEANER: An improved oil filter which showed great promise was introduced in late 1920. This was a centrifugal oil cleaner which was installed on the 8200 pump (and later on the A-1 pump as well) in place of the standard bottom cover and oil strainer.

The theory was that unwanted particles suspended in the oil could be removed by centrifugal force by a whirling bowl containing the oil. This bowl was mounted on the bottom of the pump and spun a 1 1/2 times engine speed. The resulting force at 45 gravities was sufficient to cause most suspended particles to be deposited against the outside surface of the spinning bowl. They would periodically be removed and thus the oil, which would normally be changed after about 5 hours service, could be used indefinitely, or so it was thought. In actuality, although it did appear to extend oil life the oil would still have to be changed periodically. This device was installed on a number of 8200 oil pumps and tried in service.

A-1 AND A-2 TYPE OIL PUMP: In mid-1922 a new design oil pump was made available. (The new design was announced in the October 1922 Technical Bulletin No. 29.) This new pump was installed as required, that is, at the discretion of servicing organizations. All in-service engines were not retrofitted with it.

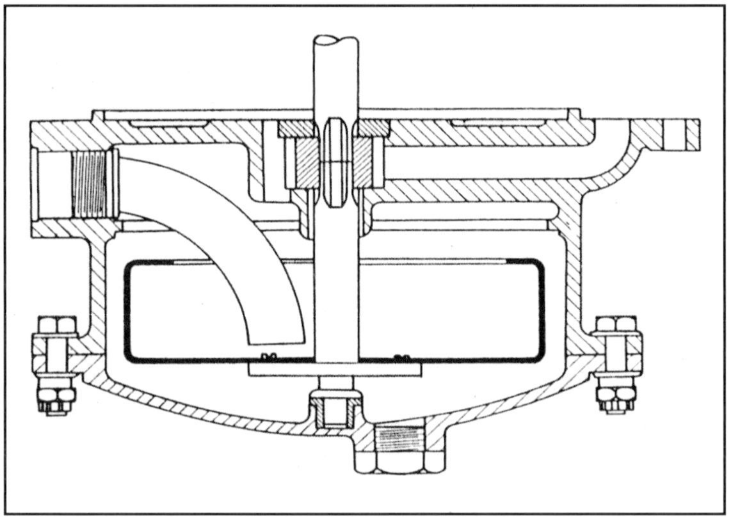

Centrifugal oil cleaner as applied to the 8200-type oil pump. (Author's collection)

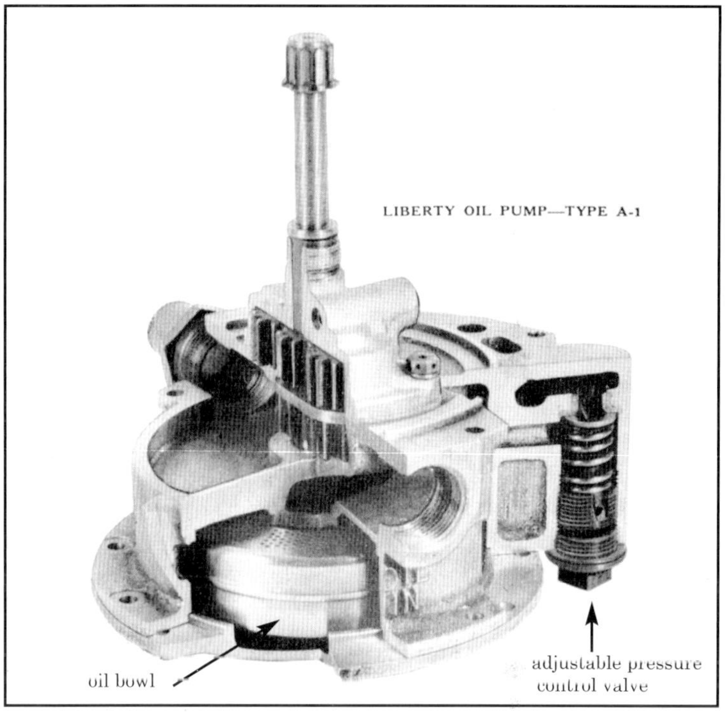

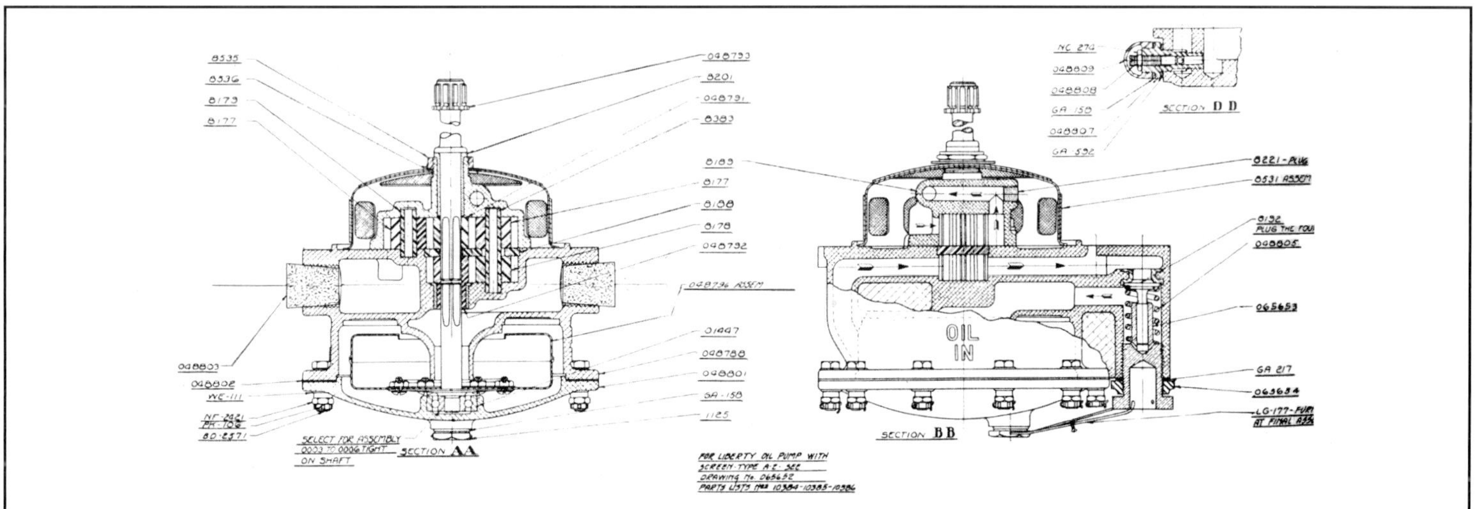

LIBERTY OIL PUMP—TYPE A-1
(Dwg. 048787)

The new design was produced in two versions which were designated as the A-1 and A-2 Liberty oil pumps. In both types the design of the inlet and outlet connections were changed from the old standard in that they were enlarged as it was found the old fittings were not large enough to pass the necessary oil flow at low temperatures. In addition, the old pressure relief valve of the 8200 type was replaced with externally adjustable pressure control valves, one for high pressure and one for low.

In addition, the A-1 model was equipped with a centrifugal oil cleaner. The A-2 model lacked this feature.

Even though it was thought by this time that the centrifugal oil cleaner had proved its worth there seemed to still be some reason to believe that its use was not always advisable. This is evidenced by the fact that the new design was issued with or without the cleaner. This fact was later verified by the instructions issued in the 1927 manual to remove the cleaner from all A-1 pumps during overhaul, replace them with A-2 lower covers and strainers, and remark them as A-2 models.

A-3-TYPE OIL PUMP: By 1925 a third design known as the A-3 was introduced. This pump had the large inlet and exit connections of the A-1/A-2 as well as the adjustable high and low pressure feature but used as many of the old 8200 pump parts as possible. As noted above, the A-1 was phased out of use by the late 1920s. The 8200, A-2 and A-3 all continued in active use until the Liberty was discontinued as a service engine.

PISTON OIL GROOVES AND HOLES: As I noted on page 122, excessive piston oiling and thus plug fouling was controlled by reducing oil pressure and providing oil drain holes in the pistons. This piston modification was tried in 1921 tests and was thereafter directed to be applied during overhauls "where facilities are available."

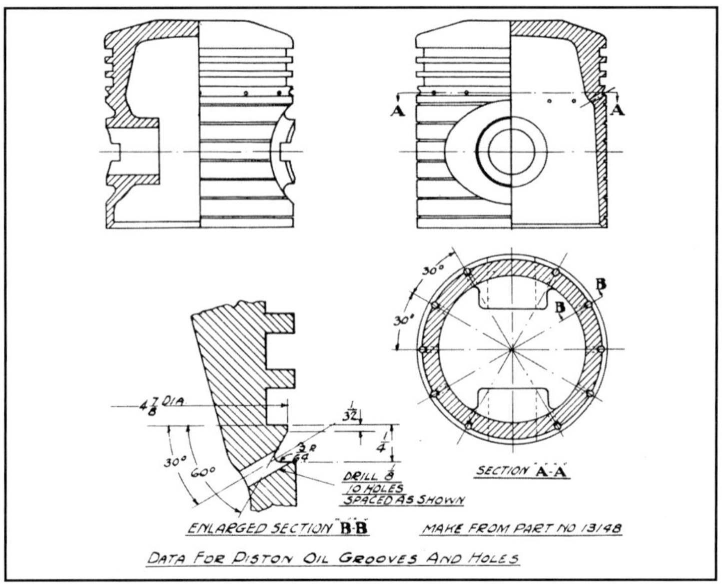

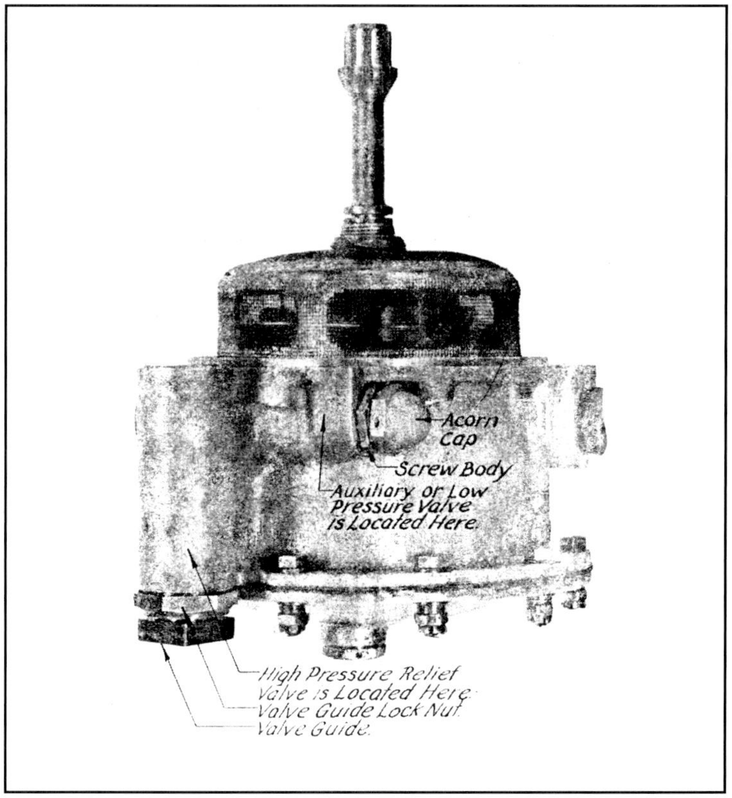

A-3-type oil pump showing the locations of the low and high pressure adjustments.

WATER PUMP: In about 1926 the standard water pump was redesigned in order to incorporate an additional support bearing. This changed only the bearing container, bearing, packing gland and spring. New parts were to be used in repairs and overhauls when parts became available. At this same time a new stainless steel shaft with removable gear was also introduced and was to be used for replacements.

ELECTRICAL SYSTEM: Probably the earliest change which resulted in any modification to the electrical system was the early use of the electric starter. This began before the end of the war and I covered the entire evolution of the Liberty starters in Chapter 5 even though it continued into the late 1920s.

Use of electric starters rapidly pointed out the need to increase the capacity of the original 8-volt electrical system whose primary requirement was only to operate the ignition system of the engine. The engine was provided with an 8-volt generator and regulator and lead/acid battery. The generator could provide a maximum of 10 amps charge to the battery should it be almost completely discharged. Under normal conditions the battery would provide voltage to the coil-type ignition upon hand-starting with the generator providing the source once the engine was started, as well as recharging the battery as necessary.

Should the generator cease to function during flight, the engine would run for a considerable time on the battery alone and allow safe return of the aircraft to a landing field.

Electric starters required more power than the original 8-volt battery could supply. Starters were thus designed for 12-volt operation and larger capacity batteries. By 1921 the first change had been made with the introduction of the Leece-Neville 1-L 12-volt, 25-amp generator to replace to old 8-volt Delco one.

By 1927 a series of various different generators had been introduced into use. The 1-L was supplemented by the 2-L of 50 amp capacity and, in addition, Leece-Neville introduced the updated versions of 25 and 50-amp capacity known as the B-1 and C-1 models.

In any installation in which a 12-volt generator is installed, the battery is changed to a 12-volt one, a new regulator is provided and the ignition switch is replaced with one having an additional dropping resistor and a higher capacity ammeter. The remainder of the ignition system remained the same.

ADDITIONAL ACCESSORIES: A number of additional accessories were attached to the Liberty during the years in which it was in use. These were not an original part of the engine and affected its operation only indirectly. Some were used on engines other than the Liberty, fuel pumps being a good example. In many early aircraft installations fuel was supplied to the engine by pressurizing the fuel tank. This proved to be a considerable fire hazard since any leak in the fuel system resulted in fuel under pressure being pumped out at the point of the leak.

BELLOWS FUEL PUMP: Beginning in 1921 a bellows-type pump, built in dual form for mechanical backup, was introduced as a positive-type pump. This pump was driven from the spline at the rear of the crankshaft and was also used on the Wright Model H engine.

ENGINE-DRIVEN C-5 GEAR-TYPE FUEL PUMP: The C-5 pump is one of a series of gear-type fuel pumps which started with the C-80. This was a McCook Field design started in October of 1920 and intended to become the standard fuel pump for all engines of about 500 HP and under. The design was completed in January of 1921, and designated as the C-1. Drive adapters were designed to allow its use by the Packard 1A-1237, which was the latest pursuit engine, and the Liberty 12.

C-1: Two of these pumps were built and tested in February of 1921. Seven more were built and the first one was flight-tested on a Packard 1A-1237 in the PW-1 with good performance. In May of 1921 some minor design changes were made and two additional pumps were ordered built with them incorporated. After these eleven units were built, the C-1 was replaced by a significantly changed design.

C-2: The new design was designated the C-2 and 15 were ordered and received in January of 1922. With some minor changes in design an additional 10 were ordered in June of 1922 and received in July. An additional 40 were ordered and delivered in January of 1923.

C-3: While the C-1 and C-2 pumps were in development a larger version, designated as the C-3, was designed for use by engines in the 500 to 1,000 HP class,

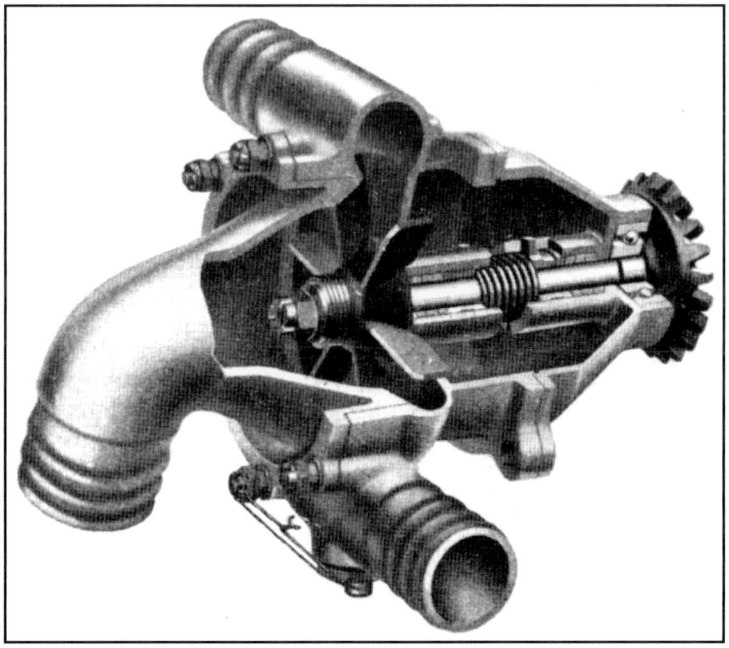

Water pump with original single-bearing-type container and impeller mounted. (Author's collection)

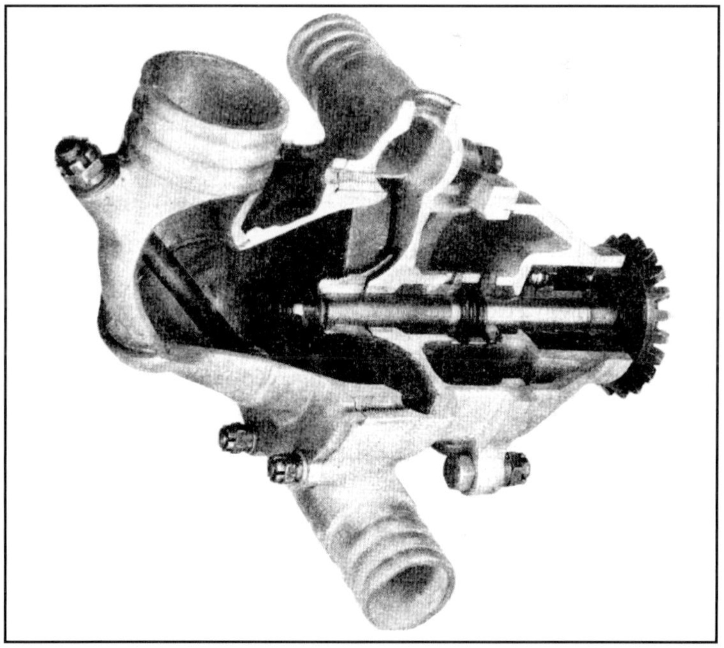

Water pump with a two-bearing-type container. Shown with adjustable type inlet. (Author's collection)

in particular the W-1A. Two were built for testing in February of 1923. In March of 1923 15 additional, with minor changes, were ordered.

C-4: The C-4 pump is of the same design as the C-2 but adapted to being wind-driven for aircraft using engines with no prevision for an engine-mounted pump.

C-5: In February of 1923 it was discovered that under some combinations of conditions the C-2 pump did not have sufficient capacity to operate a Liberty 12 at idle. Capacity was increased by adding 1/8" to the face of the gears. In addition, gear hardness was increased and the gears were changed to stainless steel. This new design was designated as the C-5 and 10 were ordered. In April of 1923 100 were put on order and it became the standard model for up to 500 HP engines.

GASOLINE TANK PRESSURE PUMP: For those applications which used a pressurized tank to supply fuel to the engine, a single-cylinder air pump could be attached to the engine at the same location as the C-5 pump. It could also be driven by the spline at the rear-end of the crankshaft like the bellows pump.

NELSON MACHINE GUN CONTROL: Soon after the Liberty 12 was, in fact, in series production, McCook Field was established and its engineering team in operation, work was begun on the design of a machine gun synchronizer to be used in conjunction with the Liberty engine and the Marlin machine gun. Designed at McCook Field under the direction of A. L. Nelson, the device was initially know as the McCook Field Gun Control. It soon took on the identification of the Nelson Gun Control.

The design, with various improvements, continued in use for a number of years and was capable of controlling machine gun fire through a four-blade propeller.

FUEL AND PRIMER LINES AND C-1 FUEL STRAINER: By the mid-1920s, the earlier strainer which was attached to each carburetor at the fuel inlet had been either replaced by or augmented by the C-1 type strainer placed in the main fuel line, usually mounted on the left side of the engine just forward of center. The last Liberty flown pictured on page 204 is fitted with the C-1 strainer at this location. Fuel line, strainer and primer line layout is shown below.

OVERHAUL

Overhauls were carried out by a number of entities ranging from individual units to contracted companies such as Allison Engineering. Initially the using units controlled when an engine would be overhauled and to what extent parts were replaced by new ones under the overall guidance of the issued manuals. Many individual Directives and Technical Orders were issued under the authority of the Air Service/Air Corps to control the

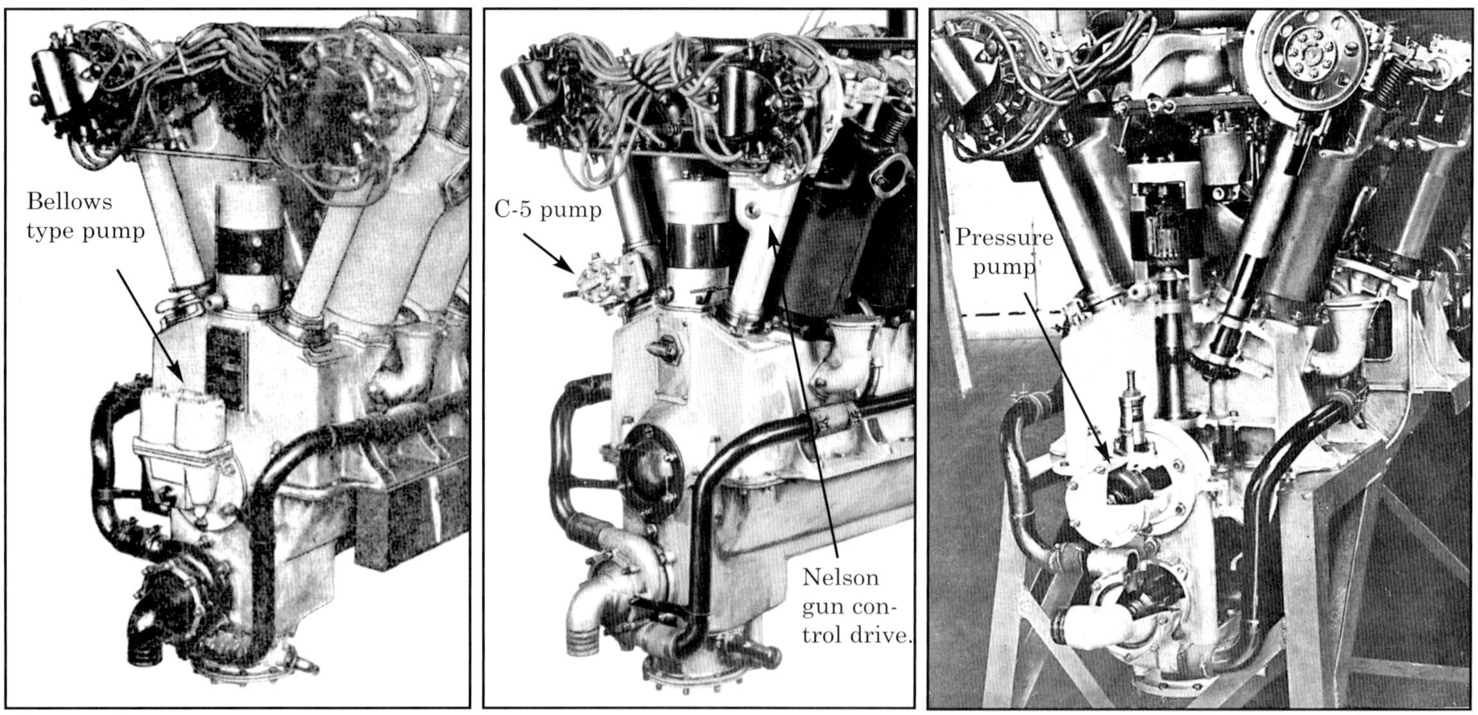

Shown above are some of the accessories mentioned as applied to the Liberty engine. The C-5 pump and drive could be applied to either cam drive but was usually seen on the left one as pictured. The Nelson machine gun drive could also be applied to either side but is usually found on the right as pictured in the center photo with a late design drive. (Author's collection)

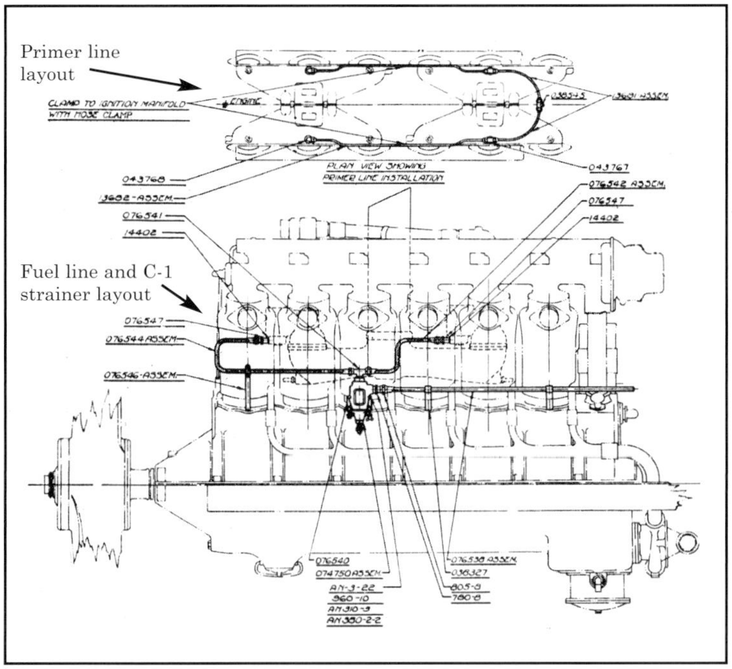

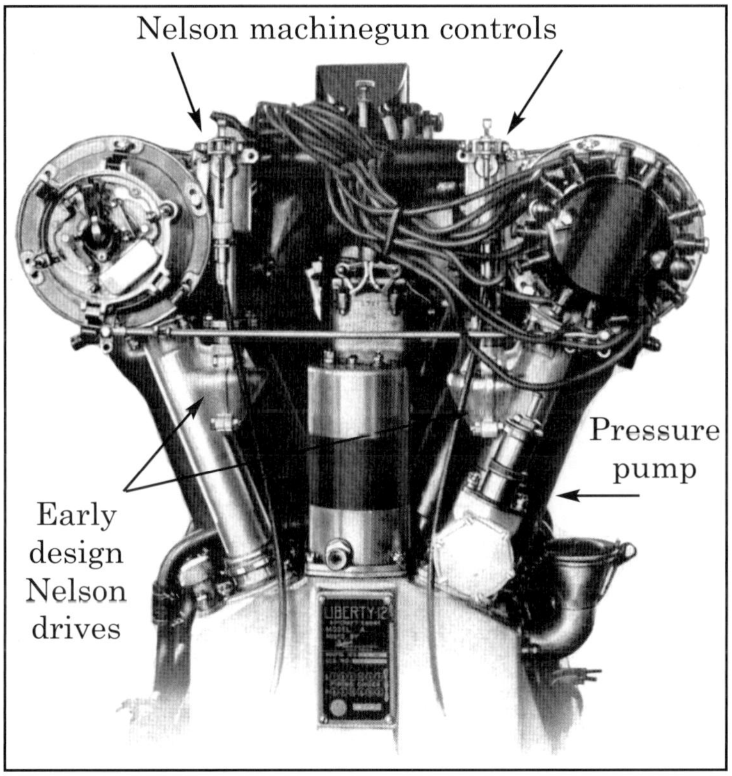

Above is shown a dual Nelson machinegun control with an early design drive of mid-1918. The left distributor head is removed to better view one of the controls and sits atop the engine. In addition, a pressure-type fuel pump is shown driven from the right camshaft's driveshaft. The illustration is from the June 1918 edition of the Bulletin of the Airplane Engineering Department. (Author's collection)

rebuilding of service engines, announce new design replacement parts, etc. The Liberty engine was not officially removed from Air Corps service until 1934 and I am sure changes were still being made in the required overhaul procedures up to the very end. However, the general requirements probably changed little after the late 1920s. I have a copy of *Engines and Parts Liberty-General Instructions For Maintenance Technical Order No. 021-5* dated May 1927 which gives an interesting detailed description of classes of overhauls, what they are composed of, how engines are marked following them, etc. Since the information answers a great number of questions regarding markings found on surviving engines as well as how overhaul was controlled, I will quote extensively from the manual.

General rules of overhaul from the May 1927 Technical Order No. 021-5

Classification of repairs: Terms to be employed for the different classes of work required on engines are as follows:

a. Class I - This includes operations which are generally considered minor repairs, such as replacement of camshafts and cylinders, and the installation of special equipment, such as gun-synchronizing drive units, self starters, new oil pumps, new water pumps and fuel and priming lines.

b. Class II - This applies to repairs generally considered top overhaul, such as the removal of camshaft assemblies, intake-header assemblies, cylinder assemblies, etc., for the purpose of replacing and lapping valves; replacing piston rings and pistons; removing carbon, etc.; and does not include any work performed on an engine requiring the separation of the crankcase halves.

c. Class III - Complete overhaul.

d. All work required in Class I and II is hereby defined as minor overhaul.

e. All work required in Class III is hereby defined as major overhaul.

Marking the crank case and name plate - *a. Letters "X," "Y," and "G" on the name plate.* - The letter "Y" is used as a symbol indicating that all pistons in the engine have had the oil-drain grooves and holes machined in them, and a No. 9 Woodruff key has been installed in the generator drive-shaft lower ball-bearing container. The letter "X" was used as a symbol indicating that all gears bearing the marks "G," "GW," "3-M," "D-S," and a letter "R" inclosed in a circle, had been removed from Nordyke & Marmon engines. (These were the markings to be found on gears used in Nordyke & Marmon engines later found to be sub-standard and ordered replaced.) The letter "G" is used as a symbol indicating the engine is equipped with stub-tooth gears.

It is directed that the letters "Y" and "G," as explained in paragraph a., be stamped on the engine name plate directly after the Air Corps serial number. For example, the number will then appear 9999 Y. G. Due to the installation of stub-tooth gears in all engines during Class III repairs, the letter "X" shall no longer be used. (It appears that any campaigning of Nordyke & Marmon engines to remove rejected gears would cease and the mark "X" would no longer be used since any Class III overhaul would replace all such gears if any remained in the engine.)

b. Recording Class III repairs on crank-case - To prevent the possible loss of, and errors in, the records of Class III repairs, and to provide an immediate reference, it is directed that the record be stamped, with five-thirty-seconds-inch letters and numerals, on the propeller-end vertical face of the upper-half crankcase, just below No. 6 left cylinder. Class III repairs, completed by air intermediate depots or other organizations, shall be indicated by the name or initials of the activity performing the work and the date it was completed. The initial reconditioning or repairing of an engine taken from long-time storage, either by a contractor or an Air Corps activity, shall be considered as one Class III repair, and recorded as such. Engines reconditioned by contractors, bear a reconditioning plate, in addition to the standard engine name plate, and the date of such reconditioning, together with the Air Corps order or purchase number, shall be stamped on the crankcase of such engines, at the time of recording the next Class III repair.

Data plate from Packard-built Liberty MNFG NO. 6425 U.S. NO. 71680. This is a very late manufacture engine being number 6425 of 6500 Packard engines. It carried the "Y" and "G" markings, but stamped below rather than following the number because of a lack of space. The reader will also note the "L" prior to the number. It has been noted on a very limited number of other engines (not build by Packard) but no explanation of its meaning has yet been found. (Courtesy of John MacArthur)

(1) *Examples of markings:*

816-23—1-21-24
S. A. A. I. D.—5-15-25
F. A. I. D.—8-10-26
Langley—6-2-27

(2) This record shall not be made for any repairs other than Class III.

Note: In the above examples 816-23 would be a number identifying a field unit. S. A. A. I. D. indicates San Antonio Air Intermediate Depot. F. A. I. D. indicates the Fairfield Air Intermediate Depot near Dayton, Ohio. Langley indicates Langley Field, Virginia.

Four Class III repairs permitted: It has been the policy of Air Corps activities to perform Class III or major overhaul repeatedly on the same engines, regardless of the number of proceeding overhauls, with the result that some engines have received as many as 10 to 12 Class III repairs. This naturally has required extensive replacement of parts, developing a shortage of cylinders, crank shafts, connecting rods, main and connecting-rod bearings, and several others. To relieve this shortage, no engine shall be given Class III repairs more than four times, and only repairable new engines which have been prepared for long-time storage, and those, the crankcase markings of which indicate that they have had but three Class III repairs or less, shall be shipped to air intermediate depots or turned over to engineering departments for overhaul.

Disposition of certain engines: *Engines having light-type crankcases.* - The complete overhauling of engines having light-type crankcases has not proven advisable. It is, therefore, directed that no such engines be given Class III repairs, or shipped from any activity without specific orders from the Field Service Section.

(1) The light-type crankcase can be distinguished from the heavy-type on an assembled engine, by checking, with calipers, the thickness of the horizontal cross web or boss, on the lower-end of No. 1 crankshaft-bearing supporting web. This cross web or boss is approximately one-fourth inch in thickness in the light-type case and three-eights inch in the heavy. The inspection can be made through the crankcase center, gear-end opening where the starter is usually installed.

(2) No general inspection of engines in stock shall be made to determine which engines have light-type crankcases. This inspection shall be made when engines are being selected for Class III repairs or for shipment. When found, such engines shall be tagged with an identification tag bearing these words, "Light-type crankcase — not suitable for overhaul," set aside, and prepared for storage as given in the following paragraph b. These engines shall be reported to the Field Service Section, in the regular monthly "Consolidated Engine Report," on the reverse

side of which is written this note: "There are on-hand (give number of engines) engines which have been inspected and tagged as having light-type crankcases."

a Engines not to be repaired. All engines having had Class III repairs four times or more, and those found with light-type crankcases, shall have the following listed parts removed and returned to stock, then be prepared for temporary storage, in compliance with Technical Order No. 02-1-1, and stored, awaiting instructions from the office of Chief of Air Corps for further disposition:

Air-gap-type distributors, part No. 005269.
Gear-end starter.
12-volt generator.
Oil-pump assembly, type A-2, part No. 065652 or A-3, part No. 069030.
Mica spark plugs, B.G. types and Mosler M-1.
Carburetor air-intake-manifold and pipe assembly, part No. 14409.
Fuel-pump assembly, type C-5.
Camshaft lower drive-shaft and fuel-pump-drive installation, parts Nos. 070105 or 073126.
Camshaft and gun-control-drive-housing assembly, part No. 068546, with drive.
Gun-control drive-shaft assembly, part No. 068548.
Fuel-line strainer and fuel-line installation, drawing No. 0151815.

b Packard and Lincoln engines of certain serial numbers. Engines bearing Packard name plates and numbering from 1 to 1000, inclusive, and those bearing Lincoln name plates and numbering from 1 to 200, inclusive, are apt to have light-type crankcases; if not, they shall receive, during repairs, special inspection for the following nonstandard parts:

Main bearings with end oil grooves.
Crankshafts with oil scuppers.
Light-type connecting rods.
Thin-type connecting-rod bearings.
Light-type pistons and piston pins.

When such parts are found they shall be replaced by standard parts or the engines shall be surveyed and the serviceable standard parts reclaimed.

c Engines taken from long-time storage. Each engine taken from long-time storage shall be completely disassembled, cleaned, inspected, and reconditioned before being used for flight. Under no circumstances shall these instructions be deviated from.

NAME PLATE RELOCATION

During any overhaul or repair, at a location having the necessary equipment available, it was directed that the name plate be removed from its normal location, the holding screws reinserted in their holes and a 1 1/2-inch inspection hole be drilled and tapped in the old plate location.

The hole was closed by inserting an oil vent connection assembly. This was done to "provide a means for inspecting gears assembled in the upper-half of the crankcase and to vent the oil tank."

The name plate was relocated to the right-side of the engine nose as shown in the drawing reproduced on this page.

RECONDITIONING AND UPGRADES BY THE AIR SERVICE AND CONTRACTORS

The earliest that large-scale rebuilding has been noted is 1921. During this fiscal year some 275 Liberties were reconditioned by Air Service Depot facilities. During 1922 1,000 were done. No records of large-scale reconditioning have surfaced for 1923. Perhaps the 1,000 of 1922 gave them a stock in storage from which to draw. In 1924 another 600 were reconditioned by Air Service Depots.

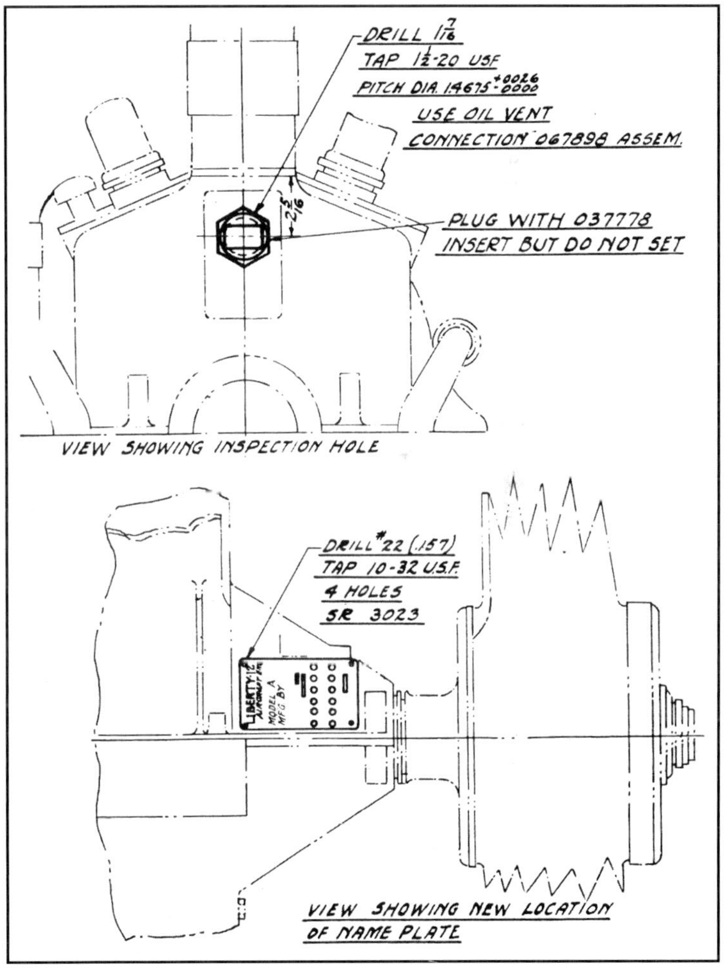

Instructional drawing for relocating the name plate to allow an inspection hole to be drilled in the rear of the upper crankcase. (Author's collection)

Beginning at least as early as 1924 the Air Service contracted, probably by putting out bids, the reconditioning and upgrading of Liberty engines. The earliest data I have is for the "remodeling" of 500 engines which was let during 1924 to The Steel Products Engineering Corp. of Springfield, Ohio. By December 31, 1924 it was reported that 200 had been completed, the remaining 300 being delivered during 1925. Considering the date, these were probably being brought up to "1921 Model" standards.

From that time until as late as 1931 the Air Service/Air Corps continued to contract large lots of engines to be reconditioned. All seem to have been done by either The Steel Products Engineering Corp. or Allison Engineering of Indianapolis, Indiana.

During all this time some rebuilding was still being done by the Air Service (and Navy as well, although not much data has surfaced regarding their activity in this area) at their own facilities. All Class I and Class II repairs were, of course, done at government facilities. Contracted large-scale remodeling and reconditioning would all be Class III repairs. It would appear from surviving records of this work that the supplied engines would come from either new stocks still held in storage or engines from the field which had seen service and were in need of rebuilding but which, in some cases, may have already been rebuilt before and brought up to current standards. Thus we find records of both "remodeled" and "reconditioned" engines with unit costs of engines done during the same fiscal year being different for the different type of work.

Although I cannot say for sure, since there are no verifying records, it would seem logical that contracted large-scale rebuilding would include the use of Allison steel-backed bearings since they were available beginning in 1925. Allison records show they sold to the Air Service 1,300 Liberty sets in 1929 at about $65 per set. This amounted to only about 10% of the average unit rebuild cost so it would be foolish not to have included this upgrade.

It is obvious from the totals indicated in the following chart of contracted rebuilds that by 1925 the Air Service was testing the feasibility of having quantities of engines rebuilt under contract and by 1928 was no longer carrying out their own large-scale Class III rebuilding. That seems to be verified by the comments published in the October 1928 Material Division Activity report. In it was a comment on the progress of the contracts with Allison Engineering and The Steel Products Engineering Corp. for a total of 2,090 reconditioned engines (fiscal year 1929) along with a note on Liberty engine overhaul equipment: "All depots have been instructed not to dispose of equipment used in connection with the overhaul of Liberty engines, as it may again be necessary to do overhaul work at depots."

A Steel Products Eng. Co. ad from the September 1931 issue of* Aviation Magazine. *Note that they indicated they had just completed a contract for "over 1500" Liberty engines. (Author's collection)

LIBERTY 12 MODELS

(See chart on page 288)

Over the years the Liberty was a service engine it was assigned several model names. Names were assigned to indicate what improved features had been incorporated in the engine. The first new model name assigned was the 1921 Model.

1921 MODEL

The Engineering Division at McCook Field completed a 50-hour test on 1921 Model Nordyke & Marmon engine MNFG. NO. 661 AS NO. 5041 on July 21, 1921 and filed the official report serial No. 1844 on January 1, 1922.

As officially designated a 1921 Model, the engine carried the following features and modifications from the original 1918 production standard:

Standard Zenith carburetors with automatic altitude control;
Standard pistons with oil grooves and drain holes.
12 volt ignition with buzzer starter;
A-1-type oil pump with centrifugal (Stryker) cleaner;
Gear-type fuel pump; (This must have been a C-1)
Oil tank vent (and gear inspection hole);
Stub-tooth gear train.

I have already mentioned the change to a 12-Volt ignition system but did not, at the time, mention the buzzer starter modification. This was a device designed to start engines mounted so the prop could be manually turned but needed only be moved sufficiently to turn the engine slowly and pull a primed charge into the cylinders. Once charged, the device could fire a spark plug of a cylinder which was in firing position.

Steel Products Eng. rebuild plate.

This was accomplished by using a special rotor in one distributor head and a special auxiliary high tension coil with battery voltage fed to its primary by a starter button and vibrator type circuit interrupter. Its secondary (high voltage output) is fed to the proper spark plug by the special rotor.

With this device the engine could be turned slowly (ignition switches off), by either ground crew or the pilot himself, one or two revolutions to prime the engine. The pilot would then turn on the ignition and push the starter buzzer button which would fire a cylinder which was in proper position. This would turn the engine over and start it. Although this system worked fairly well it was never installed on engines as a standard, particularly after the use of electric self-starters became common.

1923 MODEL

The use of this model designation has only been seen in connection with the engines used in the Douglas World Cruisers. These engines were specially rebuilt at McCook Field to provide engines for the airplanes, and an additional 15 to be distributed around the world at strategic locations for use as spares. These engines were 1921 Models with the following changes:

The cylinders had reinforced heads designed to reduce the incidence of cracked and leaking water jackets by reducing flexing of the cylinder. The reinforcing was accomplished by welding an external plate between the valve port elbows.

Table of known large-scale Air Service rebuilding and contracts for the rebuilding of Liberty engines. Contracted unit costs are stated where known.

Fiscal year	No. remodeled	Unit cost	No. reconditioned	Unit cost	Comments
1921			275		Air Service Depots
1922			1,000		Air Service Depots
1923					none known
1924			600		Air Service Depots
1925	500	$587.76			Steel Products Eng.
1926					none currently known
1927					none currently known
1928	167	$736.99	558	$712.41	Allison Engineering
1928			235	$636.00	Steel Products Eng.
1929	105	$532.43	1,240	$594.39	Allison Engineering
1929			745		Steel Products Eng.
1930	1,500	$450.40			Allison Engineering
1930	1,500				Steel Products Eng.
1931	90	$509.85			Allison Engineering
Totals	**3,862**	**avr. $505.65**	**4,653**	**avr. $631.59**	

3,660 Allison - 2,980 Steel Products - 1,850 Air Service - Grand total - 8,515
(Average costs reflect contracted work only.)

Engines were equipped with Bijur electric starters with auxiliary hand start feature.
The latest type C-5 fuel pumps were installed.

This, of course, was only slightly different from the standard 1921 Model. The principal change was the use of specially reinforced cylinders, a practice that never became standard.

1926 MODEL

This is the latest year designation I have found applied to the Liberty, and only rarely. In this case I have found no specific documentation as to what changes from the 1921 Model may have been incorporated. I do not believe the use of the reinforced cylinder head of the 1923 Model was ever adopted as a standard change and do not believe it would have been used on 1926 or later models.

Those changes which were likely incorporated in addition to the features of the 1921 Model were:

Use of Allison steel-backed bearings;
Use of the C-5 type fuel pump;
Use of the A-3 type oil pump;
Use of the late two-bearing water pump.

TIME BETWEEN OVERHAULS (TBO)

One would think that this subject would be easy to find information on in regard to the Liberty but the author has not found it to be the case. I do believe enough information has been found to be able to put to rest some of the myths that have circulated regarding this subject.

I looked through my material relative to TBO for the Liberty. I have the March 1918 Army Liberty manual, the August 1919 Army Liberty manual, the 1919 Navy manual and the 1927 Army manual issued to major service units.

Specific information did not surface for the later years for the Liberty but here is what I found.

The March 1918 manual (only 250 engines had been built when this manual was published) says disassemble, inspect and repair as necessary at 50 hours. This is also about the same time that the Navy experienced some problems and issued a similar order for a 50-hour checkout.

The August 1919 Army manual says disassemble, inspect and repair as necessary at 150 hours.

The 1919 Navy manual says the same as the 1919 Army but expands a bit by noting just why 150 hours was picked. It notes a specific series of tests, which were run to determine just when overhaul was advisable, which were a set of 160-hour in-air tests run on two production engines by the Army in May through July of 1918. The engines were Packard MNFG NO. 719 and 801. They were flown in two different Curtiss R-4 airplanes in a manner thought to closely duplicate likely military combat-type usage. Normal maintenance and minor repairs were made between flights. The conclusion was that 160 hours between tear downs for overhaul would be considered safe because tear downs of both engines showed them to be in good enough condition to expect a number of additional safe operating hours and therefore setting a 150-hour standard would leave a more than ample margin of safety.

The several hundred page 1927 Liberty manual, strangely enough, tells how to overhaul every nut and bolt on the engine but does not give a TBO.

When the Air Corps was doing a study in early 1927 on the cost of operation of Liberties verses the Curtiss D-12, they calculated three overhauls and a life span of 320 hours. This equates to 80 hours between overhauls. They used 120 hours between overhauls for the D-12. Both figures were very conservative and may not have been realistic. More detail may be found at the end of this chapter.

An article published in the October 25, 1926 issue of *Aviation Magazine* about the aircraft of Western Air Express Co., which was flying the Air Mail with Douglas Mail Planes with Liberty power, stated their current TBO for the Liberty was 200 hours. The article was written by Col. Thurman H. Bane, former chief of the Engineering Division, having succeeded Col. Jesse G. Vincent following WW I.

I have a large collection of early engine manuals and some give specific TBO for their engines. (Although I have many Wright radial engine manuals dating from 1932-1944, they do not give any specific recommendations for TBO.) The recommended TBO given in these factory manuals make an interesting study when compared to those stated for the Liberty. I list them below in chronological order along with Liberty recommendations.

Factory recommended TBO

1918 Liberty 50 hours
1918 Curtiss OX 50 hours
1918 Wright Hispano-Suiza 50 hours
1919 Liberty 150 hours
1920 Curtiss C-6 100 hours
1923 Curtiss D-12 100 hours
1926 Liberty 200 hours (Air mail report)
1927 Curtiss D-12 200 hours
1927 Wright Tornado T-3A 300 hours
1927 Wright Tornado T-3B 300 hours
1927 P&W Wasp 300-400 hours
1928 P&W Wasp 300 hours
1928 Wright Cyclone 200-300 hours

1928 Curtiss Conqueror 200 hours
1928 Packard 3A-1500 & 3A-2500 200 hours
1929 Curtiss Challenger 200 hours
1929 Wright Whirlwind 200-300 hours
1929 P&W Wasp 300-400 hours
1930 Wright Whirlwind 200-300 hours
1930 Wright Cyclone 350-500 hours
1930 Gipsy (85 HP) 300 hours
1931 Packard DR-980 Diesel 350-500 hours
1932 P&W Wasp & Hornet 300 hours
1932 R-R Kestrel 400 hours
1932 R-R Buzzard 400 hours
1934 P&W Twin Wasp 300 hours
1934 Gipsy Major (130 HP) 750 hours
1935 P&W Wasp & Hornet 300-350 hours
1936 P&W Twin Wasp 300-350 hours
1937 P&W Twin Wasp 300-400 hours
1938 Gipsy Queen III (200 HP) 750 hours
1939 P&W Twin Wasp 300-400 military, 400-500 civil
1940 P&W Twin Wasp 300-400 military
1941 P&W Wasp & Hornet 300-400 military, 400-600 civil
1942 P&W Twin Wasp 350-400 hours
1943 P&W Twin Wasp 350-400 hours
1944 P&W Wasp 350-400 hours

As a note of interest, Allison claimed that Liberty engines in Air Mail service equipped with their steel-backed bearings gave good service to between 400 and 600 hours.

This gives you an idea of what was about standard for TBO of the period. As can be seen by the above figures, 200 hours TBO was common on current design engines to as late as 1930.

Another interesting study is specific fuel consumption of the period with the Liberty compared to other aircraft engines from WW I to WW II. Such figures are most commonly stated as fuel consumption in pounds per horsepower hour. In other words, the quantity of fuel used by the engine to produce one horsepower for one hour of operation.

Factory reported Specific Fuel/oil Consumption

1917 Hall-Scott A-7a (100 HP)
fuel: 0.58 lbs./HPH
oil: 0.035 lbs./HPH

1917 Curtiss V-2 (200 HP)
fuel: 0.54 lbs./HPH
oil: 0.03 lbs./HPH

1918 Curtiss OX (90 HP)
fuel: 0.60 lbs./HPH
oil: 0.03 lbs./HPH

1918 Liberty 12 (420 HP)
fuel: 0.496 lbs./HPH
oil: 0.032 lbs./HPH

1920 Curtiss C-6 (160 HP)
fuel: 0.50 lbs./HPH
oil: 0.030 lbs./HPH

1921 Liberty 1921 Model (420 HP)
fuel: 0.47 lbs./HPH
oil: 0.016 lbs./HPH

1923 Curtiss D-12 (400 HP)
fuel: 0.53 lbs./HPH
oil: 0.015 lbs./HPH

1926 BMW VI (500 HP)
fuel: 0.485 lbs./HPH
oil: 0.033 lbs./HPH

1927 Curtiss D-12 (460 HP)
fuel: 0.50 lbs./HPH
oil: 0.010 lbs./HPH

1928 Curtiss Conqueror (625 HP)
fuel: 0.52 lbs./HPH
oil: 0.015 lbs./HPH

1929 Curtiss Challenger (170 HP)
fuel: 0.55 lbs./HPH
oil: 0.015 lbs./HPH

1929 American Cirrus (90 HP)
fuel: 0.54 lbs./HPH
oil: 0.010 lbs./HPH

1930 Wright Whirlwind R-975 (320 HP)
fuel: 0.55 lbs./HPH
oil: 0.025 lbs./HPH

1930 Wright Cyclone 1750 & 1830 (535 & 605 HP)
fuel: 0.55 lbs./HPH
oil: 0.035 lbs./HPH

1930 Lycoming R-680 (210 HP)
fuel: 0.55 lbs./HPH
oil: 0.025 lbs./HPH

1931 Packard DR-980 Diesel (225 HP)
fuel: 0.46 lbs./HPH
oil: 0.04 lbs./HPH

1934 Curtiss Conqueror (650 HP)
fuel: 0.535 lbs./HPH
oil: 0.025 lbs./HPH

1936 Continental A-40 (40 HP)
fuel: 0.75 lbs./HPH
oil: 0.025 lbs./HPH

1936 Jacobs L-5m (285 HP)
fuel: 0.53 lbs./HPH
oil: 0.018 lbs./HPH

1941 Guiberson T-1020 Radial Diesel (250 HP)
fuel: 0.415 lbs./HPH
oil: 0.021 lbs./HPH

1942 Packard Merlin V-1650-1 (1325 HP military)
(150-hour test of production engine No. 5)
fuel: 0.528 lbs./HPH
oil: 0.0082 lbs./HPH

The 1921 Model Liberty was equipped with pistons with oil grooves and drain holes and gave noticeably better performance in oil consumption than the earlier Liberties and compared quite favorably with other makes of the period.

It can easily be seen that fuel economy did not change from 1918 to 1942 (if you do not count the Diesel engines). Oil economy improved by about 33% from 1918 to 1936, but by 66% by 1942.

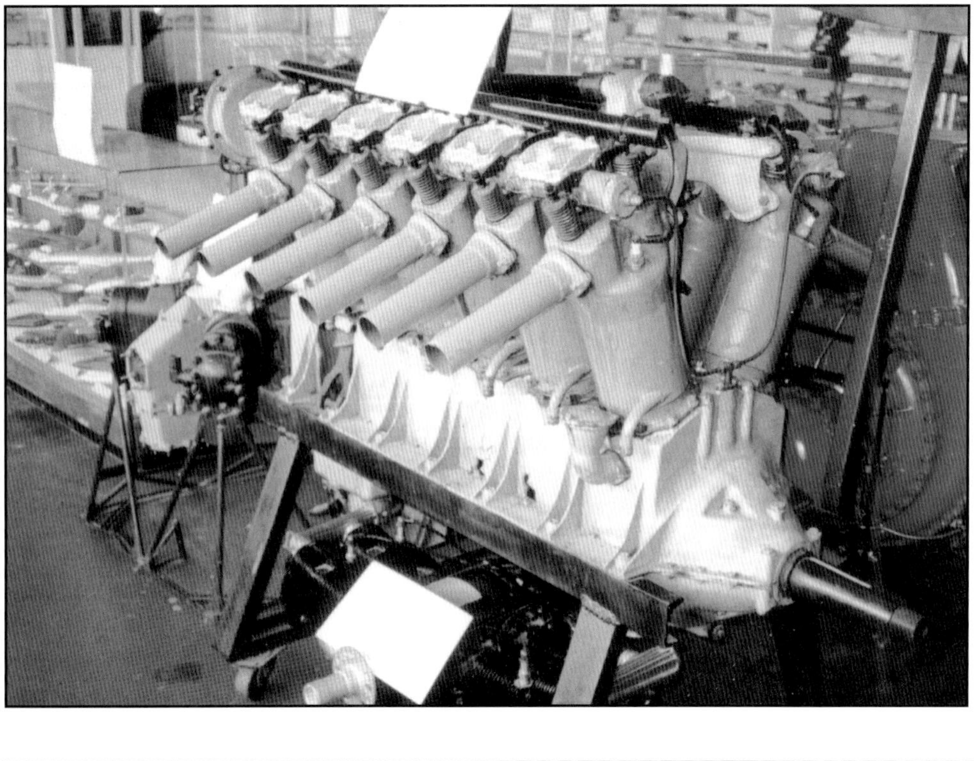

Above and left is an example of a failed rod bearing after which the engine threw a rod. In this case the forked rod bearing end went through the side of the crankcase, knocked the rear oil filler off the side of the engine and lodged sitting on the mounting flange. As can be seen, a few loose pieces put small holes in the side of the crankcase forward of that location. The engine probably came home, if that was not too far, on the front ten cylinders. (Courtesy of Joe Freeman)

A Form D turbocharged Liberty with honeycomb aftercooler mounted in a DH-4B. This picture is from the March 15, 1926 issue of Aviation Magazine *and was accompanied by a very descriptive caption which follows: "Two airmen, testing new engine equipment, recently had a miraculous escape when their engine was torn to pieces 12,000 ft. in the air. A new type propeller, which was being tried out, suddenly burst. One piece of the propeller blade flew clear, while the other, still attached to the engine, whipped around with terrific force, breaking the crankshaft just back of the front two cylinders. The propeller and hub, front of the crankcase, crankshaft, two connecting rods and pistons were torn loose and fell clear of the airplane. Flying fragments from the engine ripped open the front of the fuselage, and tore the wing covering in many places. The airmen succeeded in regaining control of the machine and made a safe landing, while the front part of the engine fell in a farmer's yard over ten miles away." Such an incident might either make an airman change his profession or decide he was impervious to destruction.* (Author's collection)

Another case of a propeller failure on a Form D turbocharged Liberty. This DH-4B was testing a "Hart" variable pitch propeller when it seized and rotated the control plate. This in turn rotated the drive assembly which was bolted to the crankcase of the engine. This rotating force tore the front section off the crankcase. The propeller departed the airplane and the pilot glided it to a landing. Not quite so dramatic as the occurance at left but surely one to make a pilot think twice before his next test flight. (McCook photo 13757) (Courtesy of the Air Force Museum)

Did the Liberty engine ever fail under test or in flight? Certainly it did but it was not a common occurrence. Ignition failures which caused the engine to quit in flight were almost unheard of. The Delco system was efficient and reliable and because of the totally dual system employed, a total failure of ignition was indeed rare.

Early engines did have reported gear failures which resulted in the engine quitting in flight, but more often than not the engine kept running with teeth missing from a gear and the failure was determined after landing and investigation of the rough or noisy running of the engine.

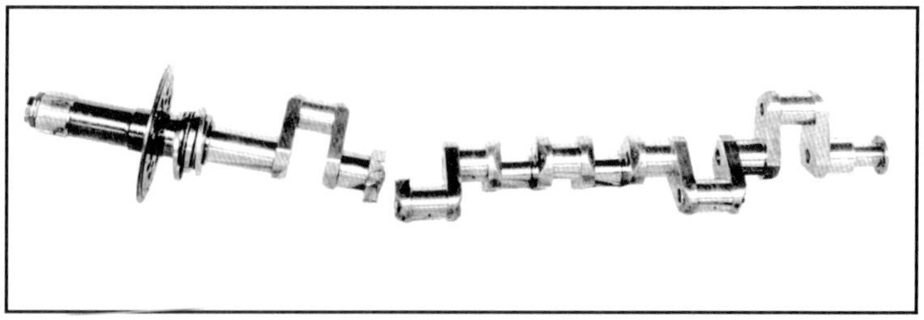

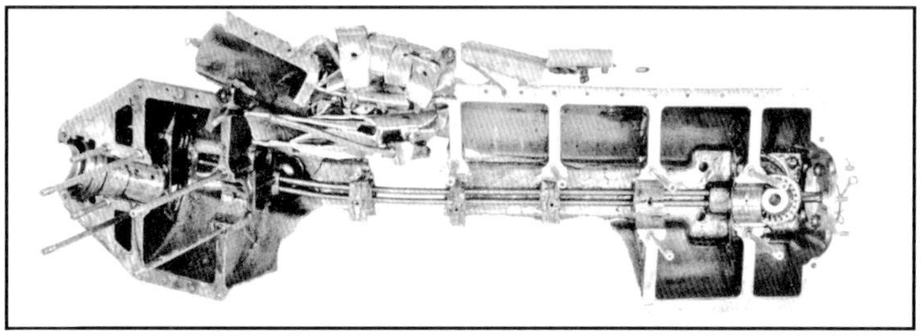

The above illustrations show the crankshaft, broken at the second cheek of No. 5 throw, which is just to the rear of No. 6 main bearing.

The illustrations on this page are all of Packard MNFG NO. 147 after it broke its crankshaft during a high-power test run on August 16, 1918. After the crankshaft broke, the forward section, held by main bearings 6 and 7, remained in place in the crankcase. The rear section was forced downward, pulling main bearings 3, 4 and 5 off the upper crankcase and driving the lower-half of the crankcase down with it. The movement of the crankshaft is apparent by noting that its center line is normally at the point the two halves of the crankcase join. The engine was almost severed into two seperate sections. The condition of the various parts and assemblies was noted and the list of those items still usable was small. (Courtesy of the Air Force Museum)

Main bearing failures were rare. Rod bearing failures which caused drastic engine damage happened, but not frequently. Engines were known to continue running long enough to get the plane back to the field after throwing a rod if the field was not far away.

A big advantage of airplanes of the period was that most could be landed dead stick on a road or in a field if the engine failed or they ran out of fuel.

The failures that could prove most drastic were a broken crankshaft or loosing a propeller blade. The latter is not an engine failure, of course, but will rapidly cause one. We seldom see such drastic failures pictured and since it is really interesting to see the results, I did so here.

On page 260 is pictured an engine which threw a rod but may have come home running minus the rear two cylinders, if home was not far away. On page 261 are illustrated two catastrophic in flight failures which instantly disabled the engine but still did not cause a crash. In these cases the failure destroyed the engine but was a propeller rather than engine failure. The end result was about the same as if a broken crankshaft had been the origin of the problem.

On the page at left are pictures taken from test report D-34 dated August 24, 1918, which originated at the Liberty Engine Experimental Plant in Detroit. This was a failure report after a medium-heavy crankshaft broke 19 hours and 56 minutes into a series of short runs at high power, most being about 15 minutes, to test various fuel types and mixtures. The failure occurred 2 hours and 23 minutes into a full-power test at 1,600 RPM.

The report stated: "Engine failure was due to a broken crankshaft of the medium-heavy-type. This appears to have occurred under excessive torsional strain created by intermittent running at full load with mixtures of varying richness. This failure further demonstrates the inadequacy of the medium-heavy-type of crankshaft for the Liberty engine." The engine was Packard MNFG NO. 147 (U.S. NO. unknown) with Navy pistons.

FURTHER DEVELOPMENT OF THE L-6

Other than the twelve, the Liberty 6 was the only engine of the original series to see further development after the end of the war. The eight had undergone refinement, a second series of prototype engines built, and series production ordered before the ultimate conclusion was reached that the engine was subject to such inherent vibration as to make its use questionable.

The reason the six had received no further attention was the lack of need for an engine of its size. Once the war ended more attention was directed to possible uses for it. By November of 1919 development work had been started

Thomas-Morse L-6A NO. 4 U.S. 47842 now in the Air Force Museum collection. This engine is equipped with all the features added on the Wright L-6B engines built in 1921; namely compact Mercedes-type intake manifolds, Stromberg NA-S5 carburetors, Dixie magneto ignition and new-type gun synchronizers (only the right one is mounted). (Author's collection)

Wright L-6B U.S. 94577 installed in Engineering Division TW-1 P-200 AS 64317. The new Mercedes type intake manifolds and Stromberg NA-S5 carburetors are clearly visible. The gun synchronizers cannot be seen. The left Dixie magneto sits on its platform just behind the rear carburetor. Clearer views may be seen in the marine conversion chapter. (Courtesy of Peter Bowers)

on the L-6. The original Thomas-Morse engines were modified and tested in preparation for modifying the design and contracting a second small batch of engines. The decision was made that 10 engines of the revised design would be built and US numbers 94577 - 94586 were assigned to them. I have not yet found a record of the contract let to Wright Aeronautical Corp. for the production of the engines but it must have been finalized between December of 1919 and perhaps February of 1920. I say this because I do have records of Packard contracts for 1A-744, 1A-1116, 1A-2025 engines assigned AS numbers 94551-94560 dated November 4, 1919 and for 1A-1116 engines 94598-94600 dated October 5, 1920.

THE WRIGHT-BUILT L-6B

The engines were not delivered until 1921. In the mean time final decisions were being made as to just what changes from the original prototype engines would be incorporated when actual production of the second batch got under way.

As related earlier, the first Army flight-test of an L-6 was of U.S. 47841 MNFG NO. 3 in a Fokker D-VII P-127 AS number 8323 on March 20, 1920. The plane was equipped

The larger illustration shows the location of the data plate of the Wright L-6B engines. It was placed on the left side of the upper crankcase below the rear carburetor. The inset shows the plate from engine No. 94584. This small batch of engines were still considered experimental and used a generic Air Service data plate which had blank locations in which to stamp the AS number, engine type and maker's name. (Courtesy of Keith Brayer)

as a single-seat pursuit, night attack. Performance of the airplane and engine were found to be good. Maximum speed at sea level was 120 MPH and absolute ceiling was 17,500 ft. at a speed of 76 MPH. Among the technical specifications of the engine was the fact that it used two Stromberg type NA-S5 carburetors and Delco ignition. The Thomas-Morse engines had all originally been equipped with 2 Zenith U.S. AS 55 types. The Stromberg NA-S5 would be used on the Wright contract engines also.

During this period extensive testing was being carried out on three intake manifold designs utilizing the Stromberg NA-S5 carburetor. The first was the standard Liberty 6 type, the second was the Mercedes type, and the third was the Benz type. Tested were power output, fuel economy and general operation. Another feature considered in the tests was compactness of the overall installation. Power output was slightly higher with the Benz type, a result also found when tests were being conducted with the high-power L-12 back in 1918. This manifold, however, had no water heating which resulted in poor warm-up characteristics. It was also the least compact of the three. Performance with the Liberty and Mercedes types was quite comparable but the Mercedes type was more compact. The only drawback of the Mercedes type was the fact that the front and rear manifolds, unlike the Liberty type, were not alike, requiring two different parts. In spite of this fact, the Mercedes type was decided upon and the Wright-built series of engines were so equipped. These tests were done on Thomas-Morse engine No. 4 and No. 5.

The next flight test was done on DH-4B P-173 AS 64009 using Thomas-Morse No. 5 US 47843. Shortly thereafter a flight test was conducted in Curtiss PN-1 P-198 AS number 63276 using Thomas-Morse No. 3 US 47841. Just how much flight testing was done with this airplane is not clear. It was retained on inventory for some time, being surveyed on January 14, 1926.

The PN-1 was a night pursuit laid out and designed by the Engineering Division to use the Liberty 6 engine and built by Curtiss. A contract was let to Curtiss for two of this model as Eng. Div. order number 223 on February 28, 1920 but only one was completed.

The original Thomas-Morse L-6A engines were used to test the new features to be incorporated in the new Wright-built engines. No. 4 engine was equipped with all the final changes and tested. It was retained in that final configuration and is now in the Air Force Museum collection. Those changes were: 1-The use of Dixie magneto ignition; 2-The use of new design intake manifolds with Stromberg NA-S5 carburetors; 3-The use of improved machine gun synchronizers.

The Wright-built engines differed from the modified Thomas-Morse engines primarily in the magneto drive system utilized. The Thomas-Morse engines used a bolt-on

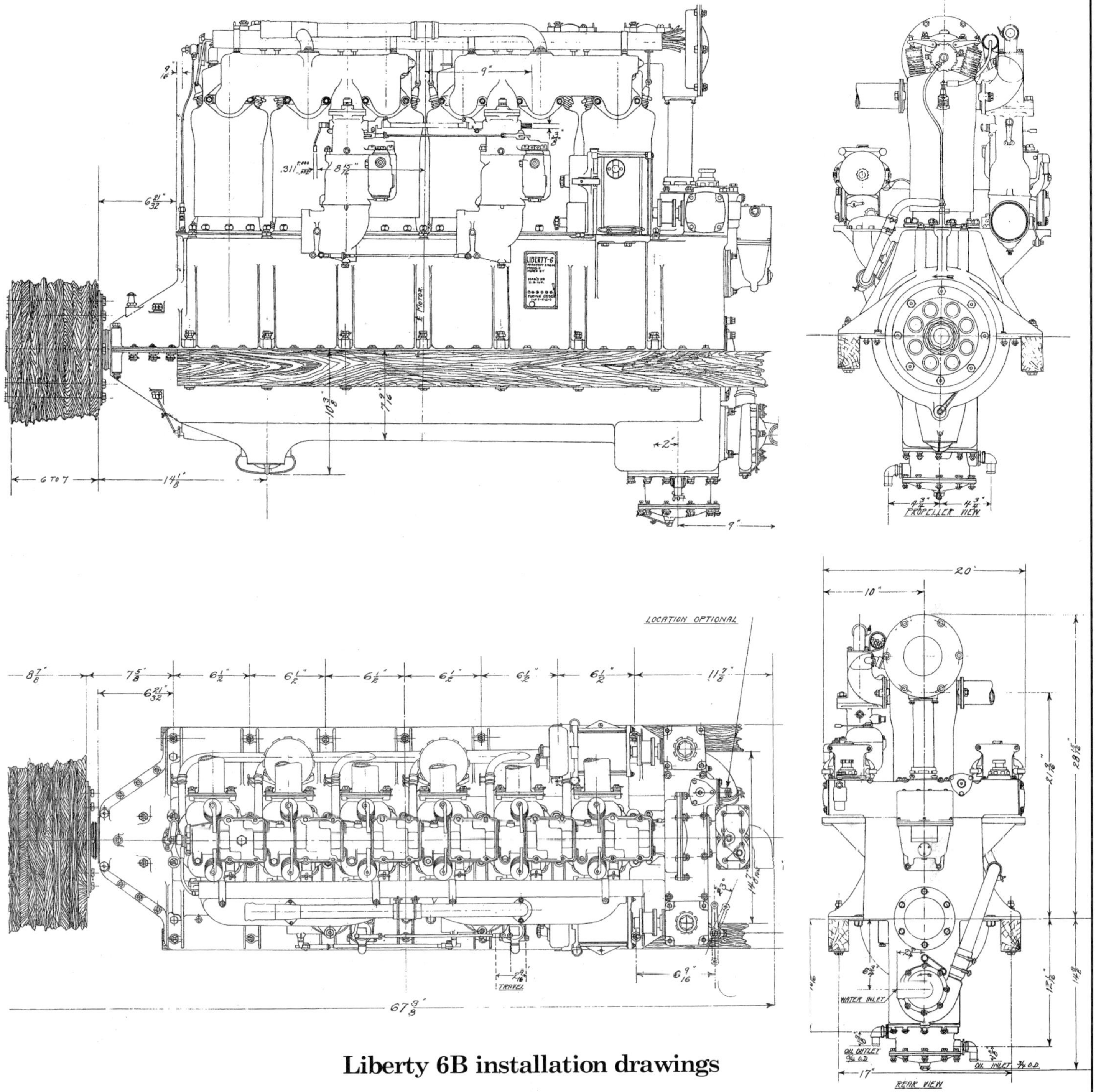

Liberty 6B installation drawings

Installation drawings of the Wright-built Liberty 6B. All features which were new to this model can be seen in these drawings. They were: 1-The Delco ignition was replaced by a pair of Dixie magnetos mounted on shelves cast as part of the upper crankcase on each side of number 6 cylinder. They are driven by a double right angle bevel drive from the cam drive vertical shaft. The use of a magneto no longer made a generator mandatory so its drive location is capped. 2- New Mercedes-type intake manifolds and Stromberg NA-S5 carburetors. 3- A new auxiliary drive position was added on the rear of the upper crankcase where the data plate formerly was. A bellows-type fuel pump was mounted in this location. 4- The data plate was moved to the left side of the engine just below number 5 cylinder. (Courtesy of NASM) right-angle drive driven from the rear of the crankshaft. The Wright engines had a double right-angle drive built into the engine which allowed the magnetos to be mounted on two shelves cast in the sides of the upper crankcase half. Although not generally referred to by a name other than Liberty 6, the Wright-built engines were officially designated as the Liberty 6B.

By mid-1920 the Engineering Division had begun design work on a new training plane around the Liberty 6. Assigned the designation TW-1 (training-water-cooled, first design) order number 2947 was issued in early 1921 for construction of two examples by the division at McCook Field. They were assigned AS numbers 64317 and 64318. 64317 was assigned McCook Field number P-200 and first flight-tested in early 1922 with the first Wright-built L-6 (US 94577) as power. Both airplanes were completed and powered by Liberty 6 engines but the second was later re-engined with a Packard 1A-1237. Among the comments in the test report of the P-200 flight was "This airplane stalls very slowly and does not have a tendency to drop off on the wing. This is a good quality from the standpoint of safety. It is very stable laterally and longitudinally. However, the performance is not considered entirely satisfactory for this type of training plane." Perhaps this is the reason behind the second example's later 1A-1237 power (a 300 HP engine versus the 220 HP of the Liberty) and the fact that the design was not carried further. P-200 was surveyed on January 14, 1926.

Technical Order 24 of April 1922 reported that the new Liberty 6 had just passed the standard 50-hour test and "as a result, a few minor modifications will be made in the ten (10) engines of this type now on hand. These modified engines will be used in airships." This is an interesting statement considering the fact that none were so used. There was, however, some reasoning behind the statement because the Air Service did have hopes of building a large lighter-than-air service. It was, however, a constant uphill struggle for allocation of funds.

In spite of this, 1921 was a particularly good year as compared to some others. Some 2% ($773,519) of the Air Service appropriation was allocated to LTA (Lighter-Than-Air). This included $195,000 to purchase the Italian semi-rigid ship *Roma,* about which we will see more later, as well as funds to begin the design and building by Goodyear of the TC class non-rigid airships. TC-1 through TC-3 were ordered in 1922 and TC-4 and TC-5 in 1923. As actually constructed these five airships were powered by two 150 HP Wright I engines. Considering the statement made in very early 1922 regarding the L-6B being held for LTA use, it is probable that at that time they were seriously considered for powering the TC series airships.

Two L-6 engines were transferred to the Navy and installed in two planes of a new series of five seaplane gunnery trainers built by the Naval Aircraft Factory. Assigned numbers A-6344-6348, the first two were powered by Liberty 6 engines, probably the new Wright engines. The next two were powered by 200 HP Aeromarine T-6 engines and the last by a 180 HP Wright-Hispano E-4. Assigned designations were TG-1 through TG-5, with the Liberty powered versions being TG-1 and TG-2, A-6344 and A-6345.

A total of 16 Liberty 6 engines were built, six L-6A types by Thomas-Morse and ten L-6B types by Wright. Three are known to have survived, Thomas-Morse US 47842 in the Air Force Museum and two Wright-built engines, US 94581 and 94584, both of which are now in the Kermit Weeks Air Museum in Florida. The Liberty 6 engine is known to have powered in flight the following aircraft:

LIBERTY 6-POWERED AIRCRAFT

Plane	Engine	Comments
Christmas Bullet	US 47839	1918
Fokker D VII	US 47841	1919
DH-4B P-173	US 47843	1921
PN-1 P-198	US 47841	1921
TW-1 P-200	US 94577	1922
TW-1 US 64318	US 94578?	1922
TG-1	US 945??	1922
TG-2	US 945??	1922

FURTHER DEVELOPMENT BY WRIGHT THE WRIGHT MODEL M

It would appear that the Air Service decided, at about the same time they picked Wright to build the redesigned L-6B, that they wanted to determine the possibility of improving the Wright 300 HP Model H engine by incorporating in it the cylinder and valve construction of the Liberty engine. To that end they contracted Wright to build such an engine using Wright Model H S.C. 13449 as a basis.

The engine differed from the design of the Model H only in the incorporation of Liberty-type cylinders, valves, valve train and its drive shafts. All other components were from the original Model H engine. Thus the bore and stroke and displacement were the same (5.5" x 5.9" and 1,127 cu. in.) as was the rating of 300 HP @ 1,800 RPM.

The Air Service generally refers to the engine as the 300 HP Hispano-Suiza with steel cylinders but Wright assigned it the designation Model M. Design and fabrication must have started in late 1919 as the engine had been completed and was submitted to a standard test on February 16, 1920. The test was assigned McCook Field

Pictured above are, at the top-left, a side and rear view of the Wright Model H engine. At the top-right are the same views of the Wright Model M engine. At bottom-left are views of standard Liberty cylinders. At the bottom-center is a top view of the Wright Model M. At bottom-right are views of a Wright Model M cylinder. (Author's collection)

Test Serial 1209 and was published as Air Service Information Circular 64 on May 29, 1920.

The following quotations from that report probably serve best to describe the object of the design and the results of its test.

Object

"The Model M was designed to determine the possibility of improving the Hispano-Suiza 300-horsepower engine by using the cylinder and valve construction of the Liberty engine. Better cooling of the cylinders and valve gear, larger valve area, easier production, and cheaper maintenance were the results desired."

Test method

"The engine submitted for test was the only one of its type built. The apparatus used and the method of conducting the test was that employed in making a standard engine test. The following successful runs were made:

One full-power run, 1,400-2,200 revolutions per minute.

One friction horsepower run, 1,400-2,200 revolutions per minute."

Observations

"The first two runs which were made with this engine were successful from an operating standpoint, but the horsepower was considerably lower than that of the average Model H engine. On continuing the test, two failures occurred, both due to exhaust-valve failure. In the first, the head of a valve broke off and imbedded itself in the piston, injuring the valve, piston and cylinder beyond repair. This is shown by the photograph on page 13. When replacement had been made, another run was attempted and another exhaust-valve failure occurred, the head again breaking off and shattering into several pieces through contact with the piston, which was also seriously damaged."

Tear-down inspection

All exhaust valves appeared to have been excessively hot. Exhaust valve stems had lengthened to the extent that there was no valve clearance. There was no other evidence of wear or pending failure of any part of the valve mechanism. All cylinders and pistons except those damaged by valve failure were in good condition with the exception of badly carbonized exhaust valve seats.

Analysis of engine

"*Design.*- This type of cylinder construction has several advantages as compared to that used in the Model H. Better cooling is provided by having the water in direct contact with the cylinder walls as well as the provision which is made for direct contact of the water with the cylinder head and valve seats. The cylinders are individually removable for repair or replacement. Greater valve-opening area is provided with these cylinders and the valve springs and stems are air-cooled. On the other hand, the valve gear has the disadvantage of being more complicated and the camshaft housing being more liable to oil leakage.

"*Production.*- One of the best features of this cylinder construction is the simplicity of design and ease of manufacture in large quantities. The valve gear is also simple and easily produced.

"*Performance.*- A complete test of this engine was impossible, due to successive valve failures, but on no run did the engine show as high power nor as good fuel economy as is the case with the Model H engine. Apart from the valve failures, no other trouble was encountered. The low power output, high fuel consumption, and valve failures, however, are not inherent in steel cylinder construction since many other makes of engines employ it with success. The manifold and cam design may contribute to the loss in power and economy, while the valve failures may be due to cam design or valve material.

"The savings in weight with this type of cylinder construction is appreciable when the weight of the water is taken into consideration.

"The steel cylinder valve gear is more accessible for adjustments of valve tappets, since no special tools are required and it is not necessary to remove any cover for adjustments, as is the case with the model H."

Total dry weight of the Model M was 623 lbs. That of the Model H was 632 lbs. Weight of water carried in the Model M engine was 38 lbs. The weight of water carried in the Model H was 57.9 lbs. This gave the Model M a wet weight advantage of 29 pounds. This works out to be a 4.6% reduction in weight for the same rated power. Such a gain was considered very significant indeed.

Although the cylinders of the Model M were of Liberty-type, the different bore, stroke and crankcase mounting required a totally new structure. Liberty cylinders could not be modified for this use. Also, even though of similar design as the Liberty 8, the camshafts, housings and drive shafts of that engine could not be modified for use as the center-to-center cylinder dimensions were different by 3/16".

As seen from the above, two valve failures resulted in the destruction of what were evidently the only two spare cylinders available. It appears that even though the opinion reached was that the model would not be difficult to develop into a viable engine, interest waned rapidly and there is no evidence any further development was carried out.

THE L-12 DIRIGIBLE ENGINE

On July 29, 1921 McCook Field completed a series of tests on a standard Liberty 12 with the object of determining what choke would be required to limit the output of the engine to 280 HP if operated at 1,400 RPM for use on

LTA craft. Zenith 52 carburetor settings required to accomplish this were found to be 25 mm choke, 1.10 mm main jet and 1.30 mm compensating jet. The final recommendation was that the Liberty 12 with these settings become the standard for dirigible use and it was used when ever its power output was appropriate.

The same tests were run on a Liberty 6 engine with a single Zenith 52 carburetor adapted to its manifolds with the expected result of half the power output of the Liberty 12, that is 140 HP @ 1,400 RPM. Apparently this test was run just to verify the expected output as one half that of the L-12. No recommendation was made for any use of a Liberty 6 so modified.

However, as noted on page 266, Technical Order 24 of April 1922 brought up the possibility of the Liberty 6B being used in airships.

THE *ZODIAC* RN-1

The stated reason for determining the requirements to get this power from Liberty 12 engines was the desire to

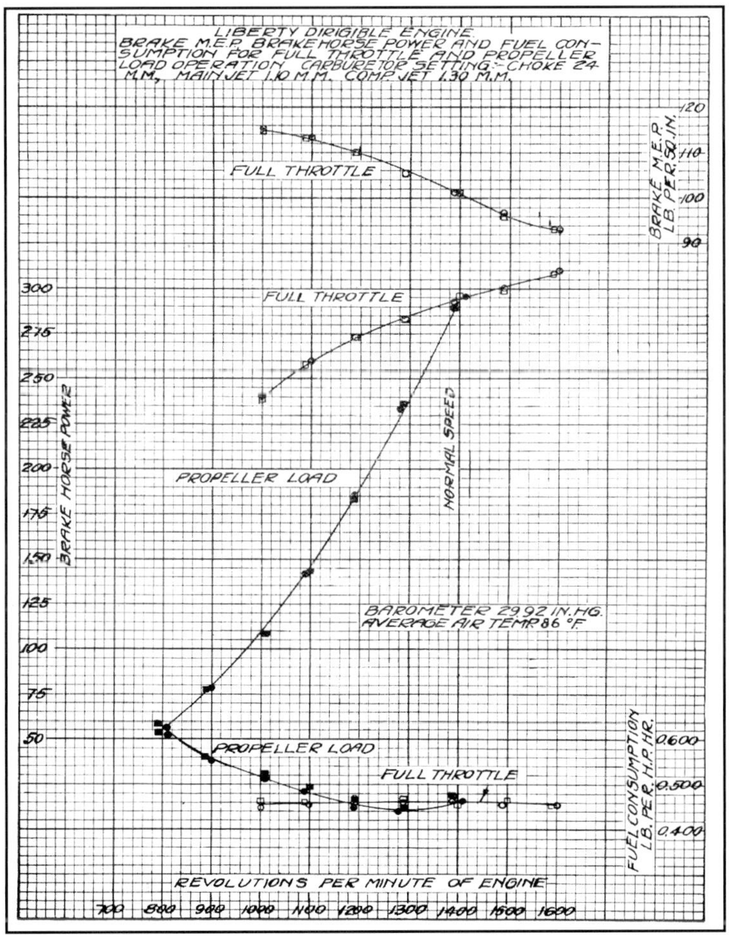

Power graph of L-12 engine set up for use in Lighter-Than-Air craft. Generated on July 29, 1921. (Author's collection)

The Zodiac RN-1 over Scott Field in about 1922. This aircraft was 264' long, 49.5' in diameter, had a volume of 326,500 cubic feet of hydrogen which gave it a total lift of 21,027 pounds. It was powered by 2 Liberty 12 engines. (Courtesy of Peter Bowers)

replace 280 HP Renault engines then used in Army dirigibles with Liberty engines. Actually there was only one airship in Army inventory that was powered by 280 HP Renault engines, and that was the RN-1.

The RN-1 was a French *Zodiac* non-rigid airship built in France in 1919 as part of a contract for a reported 13 such airships by the US Navy during WW I. Of the thirteen only nine were actually delivered to the United States. Of these, the Navy sold one to Goodyear (which it never erected), eventually transferred two to the Army and scrapped the remaining six. Of the two Army airships the first was scrapped at Langley Field in 1919. The second was erected there in late 1919 and was the only one of the nine to ever see service. Originally it carried the French designation ZD-1 which was then changed to ZDUS-1. After being refurbished and the installation of two Liberty L-12 engines to replace the 280 HP Renaults, it was designated as the RN-1.

In 1923 the Liberties were replaced with Packard 1A-1237 engines. This was done to try out the new Packard design which was rated at 300 HP. At this rating they produced the same power as the derated Liberties and saved a total of about 250 pounds of weight. The RN-1 was put in storage in 1924 and eventually scrapped.

THE *ROMA*

The Army had great interest in large airships and wanted very much to acquire rigid airships but was restricted from doing so by a joint Army-Navy agreement approved by the Secretary of War in February of 1920. That agreement allowed that type of airship only to the

The Roma near Langley Field on December 17, 1921. This aircraft was 420' long, 82' wide, 92' high, had a volume of 1,193,000 cubic feet and a useful load capacity of 39,600 pounds. It crashed on February 21, 1922 killing 34 of the 45 men aboard. It was powered by six Liberty 12 engines. (Courtesy of Peter Bowers)

Navy. Thus the Army had to be satisfied with semi-rigid types as the largest it was allowed.

The Army ordered the Italian semi-rigid airship *Roma* on June 9, 1920. The type T-34 airship had taken its maiden flight on March 19, 1920. After a number of test flights with both Italian and American crews it was dismantled and shipped to the Langley Field, Virginia where it was again erected. It made its first flight in America on November 5, 1921.

This was a hydrogen-filled airship of 1,193,000 cubic feet capacity. It was 420' long, 82' wide and 92' high. It had a useful load capacity of 39,600 pounds. The original engines were six Italian Ansaldo V-12 types of 450 HP each. These proved so troublesome that they were replaced with six Liberty engines. Documents indicate the Liberties were kept at the normal 400 HP rating in order to maintain the performance given with the original engines.

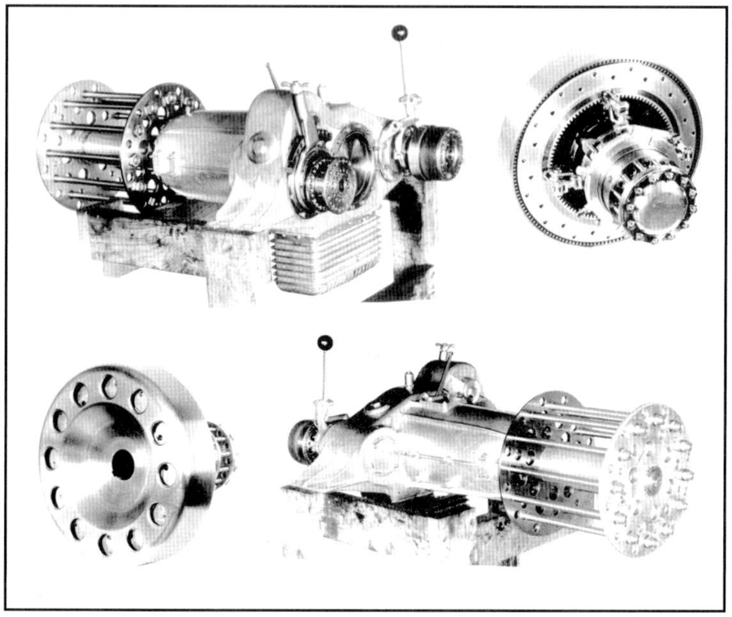

Views of the actual reduction and reverse gear and the disk clutch. (Courtesy of the Air Force Museum)

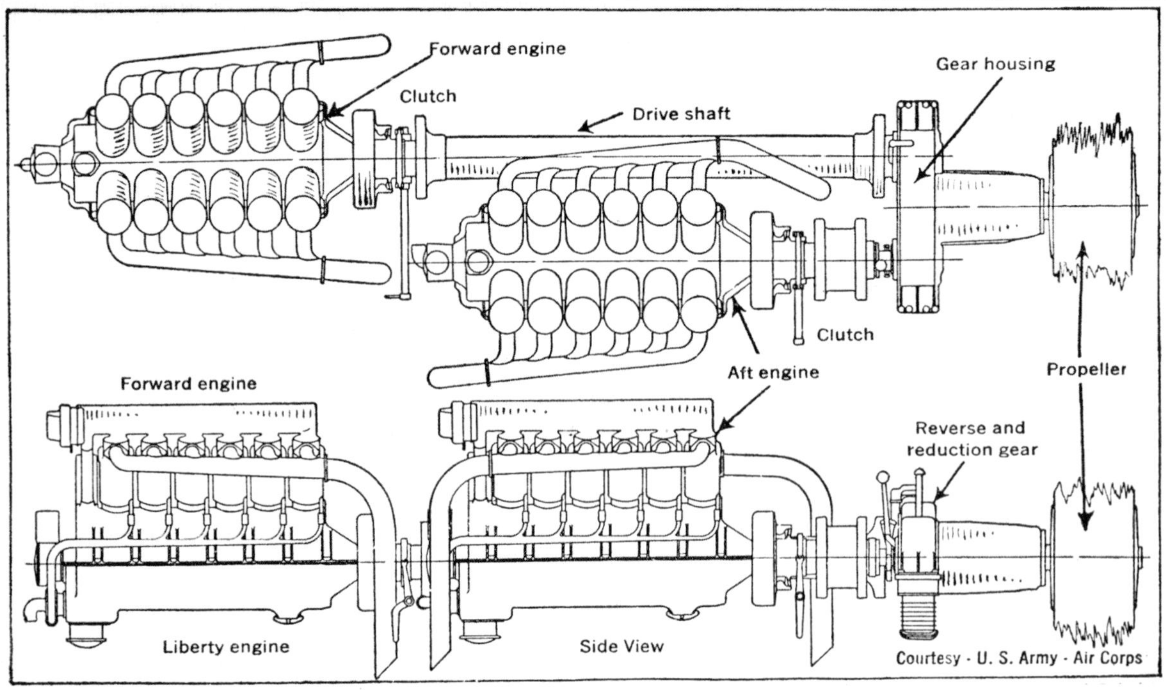

Engine car layout of the RS-1 Liberty engines which drove a single 17 ½' propeller either singlularly or in pairs. The drive system was designed and built by the Twin Disc Clutch Co. of Racine, Wisconsin. (Courtesy of the Air Force Museum)

The RS-1 in flight on May 13, 1926. (Courtesy of Peter Bowers)

This airship met a tragic end when on February 21, 1922 it crashed on the Norfolk Quartermaster Depot 25 minutes after departing Langley Field. The hydrogen-filled craft caught fire and was totally destroyed. 34 of the 45 men aboard were killed in the crash.

The investigation which followed resulted in Congress refusing to fund any further airships with hydrogen as a lifting medium.

THE RS-1

The RS-1 was the first semi-rigid airship designed and built in the United States. The project was initiated in 1922 and became a joint effort of the Air Service and the Goodyear Tire and Rubber Company of Akron, Ohio. At the time the project was conceived, Italy was the only country with any success in the construction of this type of aircraft and they were very reluctant to release any detailed information. As a result, a great deal of new research and design had to be done here before a final plan was completed.

A contract was signed with Goodyear on October 13, 1922 and actual construction was started at the Goodyear facility in Ohio in 1923. The ship was delivered to Scott Field in early 1925 where it was erected. It was inflated and subjected to numerous tests and was taken on its maiden flight on January 8, 1926.

This ship used helium as a lifting medium. It was 282' long, 74 ½' wide and 66' high and had a capacity of 719,500 cubic feet resulting in a useful load of 12,138 pounds.

Power was supplied by four 300 HP Liberty 12 engines in dirigible configuration, resulting in a total available output of 1,200 HP. There were two power cars, each having a pair of engines driving a single 17 ½' propeller through a special transmission designed and built specifically for this purpose. The basic layout of the transmission was probably done by the Engineering Division. The contract to have it built was given to Goodyear along with one to build the airship itself. Goodyear subcontracted the job to the Twin Disc Clutch Co. of Racine, Wisconsin. The design work was done by Thomas Farwick. This company had designed for Goodyear the friction clutches used in the Army airship AC-1 with two 130 HP Aeromarine U6D engines. Three units were built, one for testing and use for any refinements required, and two for installation in the RS-1. The first unit was delivered to McCook Field and passed its 50 hour test in the summer of 1925. The remaining two units were delivered directly to Goodyear.

The design provided 12 to 5 reduction gearing to a single 17 ½' pusher propeller being driven by either two 300 HP Liberties or by a single one. The transmission could reverse the direction of the propeller but one engine had to be disengaged during this operation. Either or both engines could be disengaged by its own disk clutch.

The engines were mounted in staggered form. Each had its own disengaging clutch. One drove a transmission input through a short shaft and the other through one long enough to clear its adjacent engine. (See diagram on page 270.)

In 1927 the Liberty engines were replaced by 2A-1500 Packard engines. The Packards were lighter than the Liberties by some 150 pounds each. The normal rating of the Packards was 500 HP @ 2,000 RPM and 600 HP @ 2,500 RPM. They would produce about 400 HP at the 1,550 RPM at which speed the Liberty dirigible engines were throttled to produce 300 HP. Whether the Packards were throttled to reduce their output I do not know. The 17 ½' prop was replaced with one of 16' diameter.

The airship was taken out of service in November of 1928 because of envelope porosity pending its replacement but, because of economic conditions, a new envelope was never installed and the airship was eventually scrapped.

AEROMARINE LIBERTIES

The Aeromarine Plane and Motor Co. of Keyport, N. J., which had been in the aero engine business since at least 1914, in June of 1921 submitted to the Air Service a design for a 24-cylinder engine built around the basic Liberty design using as many standard Liberty parts as possible. According to their report, the object was to "provide a suitable power unit for some of the larger types of land planes and sea planes whose range of operation requires great carrying capacity."

THE AEROMARINE AL-24

The Engineering Division was currently developing the W-1 series of engines, an 18-cylinder "W" configuration

Below are two views of the Aeromarine AL-24 engine submitted to the Air Service for consideration for use in large bombing and seaplanes, the same purpose for which the Engineering Division was developing the W-1 engine. The AL-24 was a 24-cylinder engine built around Liberty components in a fan, or double V-12 configuration. The outer banks were 45° and 35° separated them. A sectional view is at right. This was one of four 24-cylinder Liberty engines known to have been built. The first was the Engineering Division X 24. The last two were designed for marine use and will be covered later. One of the two had the same configuration as the AL-24. (Courtesy of the Air Force Museum)

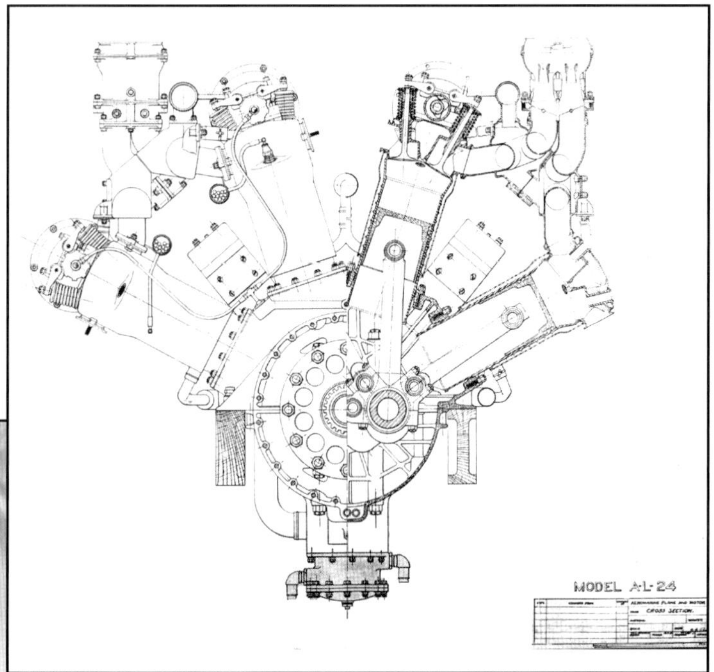

with a bore and stroke of 5½" x 6½" giving a displacement of 2,832 cu. in. and a rated output of 700 HP @ 1,700 RPM. Total weight of the engine was 1,720 pounds, or 2.46 lbs./HP. Their first engine had been tested in April of 1921. The Aeromarine engine was no doubt presented in response to the interest of the Air Service in engines of this class.

It is not clear whether or not the Air Service actually contracted Aeromarine to build a test engine, but it appears unlikely. Even so, an engine was actually built by Aeromarine and photographs of it were submitted to the Air Service as part of a presentation package of the design submitted on June 18, 1921. The project life appears to have been short as the April 1922 Air Service Technical Order No. 24 reported a summary of Test Report No. 1777, which must have been issued in the last quarter of 1921.

The summary was brief and is quoted as follows: "Design of the 24-cylinder Liberty Engine - Submitted by Aeromarine Plane & Motor Co. Report serial No. 1777. This is a discussion of the drawings, design calculations, and specifica-

At the above-right is the Engineering Division W-1A which was built by Packard. The photo is of a late engine (1922) now in the Air Force collection. Above are front and rear views of the Aeromarine AL-24 built by that company in 1921 and submitted to the Air Service for consideration. They were turned down. (Courtesy of the Air Force Museum)

Engineering Division W-1A and Aeromarine Liberty AL-24 comparison

	W-1A	AL-24
Crankcase	cast Alum.	cast Alum.
Height	51¾"	46 13/32"
Width	39 7/16"	51 3/16"
Length	90 15/16"	69 7/8"
Weight	1814 lb.	1492 lb.
Cylinder const.	forged with sheet metal water jackets	forged with sheet metal water jackets
Number of cyl.	18	24
Bore	5.5"	5"
Stroke	6.5"	7"
Displacement	2779 cu. in.	3387 cu. in.
Rated output	700 HP @ 1700	850 HP @ 1650
Wt/pwr ratio	2.59 lb/HP	1.76 lb/HP
Valves/cylinder	4	2
Ignition	quad magneto	dual battery generator
Carburetors	6 Stromberg NA-S6	4 Stromberg NA-L5
% use of existing parts	N/A	75% standard Liberty

tions on a proposed 24-cylinder Liberty submitted by the Aeromarine Plane & Motor Co.

"The design calls for a 24-cylinder engine having four rows of six Liberty cylinders each, at an included angle of 45, 35, and 45 degrees between rows. The bore and stroke was to be 5 by 7 inches respectively, piston displacement of 3387 cu. in., compression ratio 5.42:1, and horsepower 800 at 1650 RPM. It was proposed to use as many Liberty engine parts as possible.

"The proposal was rejected as it was deemed inadvisable to build the engine according to this particular design."

There was no reference to any actual tests of the engine, just a curt "no thanks." The engine was indeed built and from its appearance and the detailed specifications submitted, it would seemed to have been a very good design to achieve the proposed power using existing and fairly well proven-components. It was obviously proposed to do the same job the Engineering Division W-1 was to accomplish.

The detailed specifications submitted by Aeromarine allow some comparison with the W-1 which present an interesting study and might lead one to the conclusion that the Air Service was prejudice toward its own project and was not about to abandon it in favor of an outside proposal. On paper, at least, the AL-24 was the better engine overall, particularly considering they both used the same separate cylinder, exposed valve train system and proposed no significant new concepts of design. The AL-24 was certainly a clean design.

DEVELOPMENT OF THE STROMBERG NA-L5 CARBURETOR

As related in Chapter 5 on page 133, during late 1918 the Engineering Division began work on inverted carburetor designs for the Liberty. The standard Zenith US 52 was redesigned, in collaboration with the Zenith Co., for use in inverted position, inverted in this case meaning downdraft since the normal position of that period was updraft. Although the results were promising, Stromberg submitted a proposal for an inverted version of their NA-D5 carburetor which they initially called the NA-5L, and eventually the NA-L5.

By mid-1920 this design had been fairly well developed and was submitted to testing, the report on which was published as Test Report No. 1340 on July 31, 1920. It was noted that performance was superior to that given by standard Zenith carburetors, that it provided much better accessibility and less fire danger and was recommended for service use on the Liberty after some improvement in fuel consumption under propeller load.

By July of 1922 a few minor changes had been made in the design. They issued instructions for the use and adjustment of the carburetor as Air Service Circular No. 314 and by now were stating that it "had been especially designed for use on the Liberty 12 with supercharger installation." This was not the initial intention but the

Above are illustrated a pair of Stromberg NA-L5 carbs mounted on a Liberty 12. The lower two illustrations show a single carburetor mounted on its pair of intake manifolds to feed six cylinders, three from each barrel. (Author's collection)

***November, 1919 Packard ad for the 1A-1650 as it appeared in* Aviation Magazine.** (Author's collection)

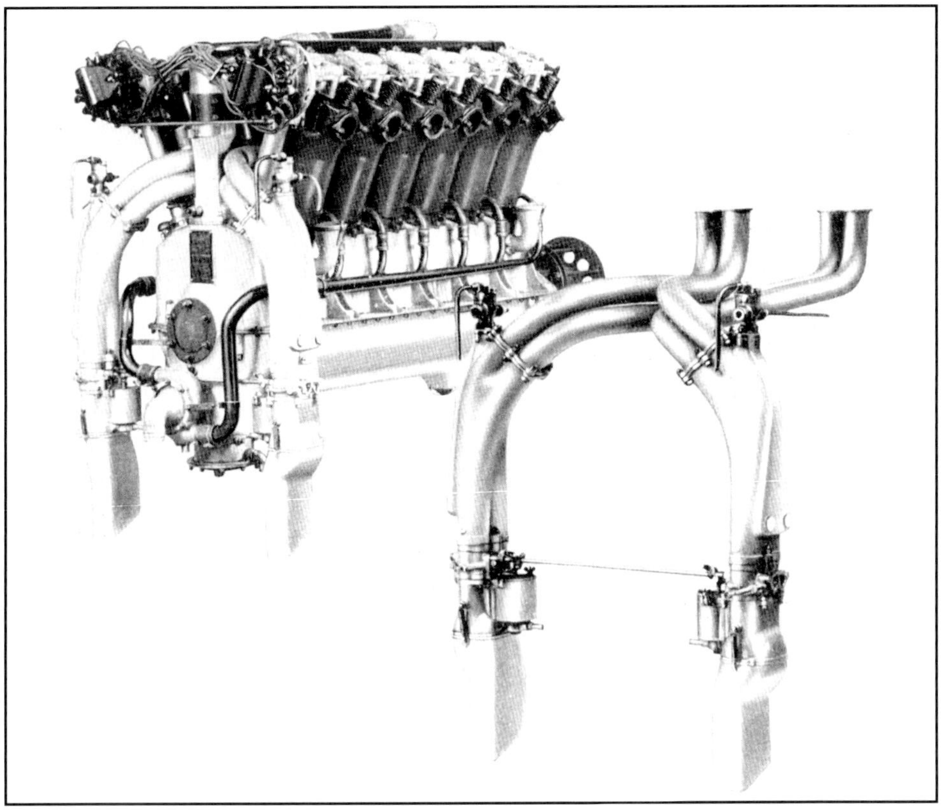

Packard 1A-1650 engine with "Fuelizer" attached and detached. The 1A-1650 was a Liberty as manufactured by Packard during the war but now carrying a Packard name. (Author's collection)

decision had been made that it would not be worth while to change the intake system of all service engines to use these improved carburetors. However, they were much better adapted for supercharger use than the standard Zenith and were therefore adopted for that purpose.

By October, 1922 and Test Report No. 2092, a few additional changes had been made resulting in the NA-L5A type and 150 of this design were ordered under contract No. 7334 of 1922. Report 2092 came to some somewhat different conclusions regarding the NA-L5 series of carburetors. Now the conclusion was "Even with these modifications, however, the NA-5L carburetor is not as satisfactory as the Zenith U.S. 52 and is not recommended for general use." "The General Electric superchargers, Form "B" and Form "D", are designed to fit the Liberty engine with Stromberg inverted carburetors, and it will be necessary to use such carburetors with these superchargers."

THE PACKARD 1A-1650 AND THE FUELIZER

As one might expect, Packard put on the postwar aero engine market in late 1919 a slightly redesigned Liberty twelve. This new commercial Liberty became Packard's model 1A-1650. An advertisement of October 1919 noted "For the first time Packard offers this noteworthy engine on a commercial basis." "Only a few of the first-run of these motors remain unsold. Telegraph orders advised."

Displacement was 1,649.5 cubic inches with a bore and stroke of 5 x 7 inches. Rated horsepower was 420 @ 1,750 RPM, and weight was 860 pounds. Mechanically it was almost a duplicate of the Liberty and was most likely made up by using parts remaining from military contracts. Some, if not most, of these parts came from Nordyke & Marmon Co. of Indianapolis whose military order for 3,000 L-12A engines had been reduced to 1,000 at the end of the war. In the minutes of the Packard board of May 5, 1919 is found the following entry:

> *"Management was authorized to purchase from Nordyke & Marmon Co. of Indianapolis, for $100,000.00, their stock of finished and semi-finished Liberty engine parts, together with jigs, fixtures and gauges necessary to complete the semi-finished stock, this material to be used by the Engineering Department in the construction of special motors."*

Considering the pending surplus status of the Liberty engine, it is not likely another production run was ever made although the engine continued to be listed as available until 1922. Estimated production is 70 to 90 engines.

The engine made its first public appearance at the New York Aeronautical Exposition which opened at the 71st Regiment Armory in New York City on March 6, 1920. Displayed were Packard's new 1A-744, 1A-1116 and the 1A-1650 engines, along with the fuelizer carburetor system for the 1A-1650 described below. Packard noted that the 1A-744 and 1A-1116 engines were not yet in production but the 1A-1650 was available.

THE PACKARD FUELIZER

A novel invention never available on any other aero engine was offered on the 1A-1650, the Packard "Fuelizer." This was a preheating device for the fuel mixture which was invented by Capt. Lionel Woolson. It was covered by patent 1,550,136 issued on August 18, 1925. Application for

Packard LePere LUSAC-11 AS 42148 being equipped with a Liberty 12A with Packard "Fuelizer" for testing. One of the carburetors can be seen through the access port and in the inset. (Courtesy of Peter Bowers)

it was filed on December 23, 1920. Packard used a version of this "Fuelizer" in their automobiles with success, notably in the 3-35 Twin Sixes and later in their Single Six cars.

Essentially this device took advantage of the pressure difference on either side of the carburetor throttle to shunt some of the mixture into a heating chamber and igniting it with a separate spark plug. The resulting hot gases were routed back into the intake manifold along with the unshunted mixture, and thus heated it. The more open the throttle the lower the pressure differential and thus less shunting and less preheating took place.

The Fuelizer feature did not become available until about January 1920. Price of the engine was $4,500 without Fuelizer, and $4,800 with Fuelizer.

It is doubtful that many Fuelizers were produced and sold. The Air Service purchased one and installed it on a Liberty engine and mounted the engine in DH-4 "P-135." This combination was tested between June 7 and July 1, 1920. Results of those tests were not particularly good and in many cases the engine operated better without the device than with it. Air Service engineers came to the conclusion that much of the trouble was caused by the cooling effect on the intake mixture as it traveled up the long pipes between the Fuelizer and the intake manifold. Their opinion was that the theory was good and if it worked as designed they would like it very much. However, their conclusions as stated in their Test Report Volume II No. 134 dated November 20, 1920, was: "It is concluded that in its present form the Packard Fuelizer applied to the Liberty engines is unsatisfactory for service use."

This engine equipped with Fuelizer was also tested on Packard LePere 42148 but nothing further came of the design so far as aircraft were concerned.

THE PACKARD LIBERTY-BASED VICTORY AUTO ENGINE

Both Hunt and Vincent no doubt retained an active interest in the Liberty engine design for some time after

An experimental six-cylinder Packard auto engine of about 1920 based upon Liberty cylinders. Packard called this their "Victory" engine. They used the transmission of their "Third Series Twin Six" auto engine. The engine was never put into production but if it had been used as a replacement for the 90 HP "Twin Six," its 200 HP would have had no rivals in the automotive world. (Author's collection)

returning to Packard at the end of the war. One, or both, of the two also contemplated the possibility of utilizing the basic Liberty design in a new line of automobile engines. Although no specifications are now known, several photographs of an experimental six-cylinder automobile engine captioned on the back as the "Victory" engine have been found. It appears to be an automotive version of a Liberty six with a single Packard automotive carburetor. Displacement would thus be 825 cubic inches and output probably about 200 horsepower. Considering its intended use, it was likely de-stroked to reduce its size somewhat and allow higher RPM. Had this engine replaced the then-current 90 HP Twin Six of about half the displacement it would have certainly set the luxury car market "on its ear" so-to-speak. The decision to progress further never came about and the engine was built on an experimental basis only.

Hall-Scott Motor Co. also considered building and selling to automotive manufacturers a six-cylinder engine based on their prewar 5" x 7" cylinder aero engines. Although surviving photographs indicate they built a prototype, no production of what they called their "aviation-type automotive engine" is known to have resulted.

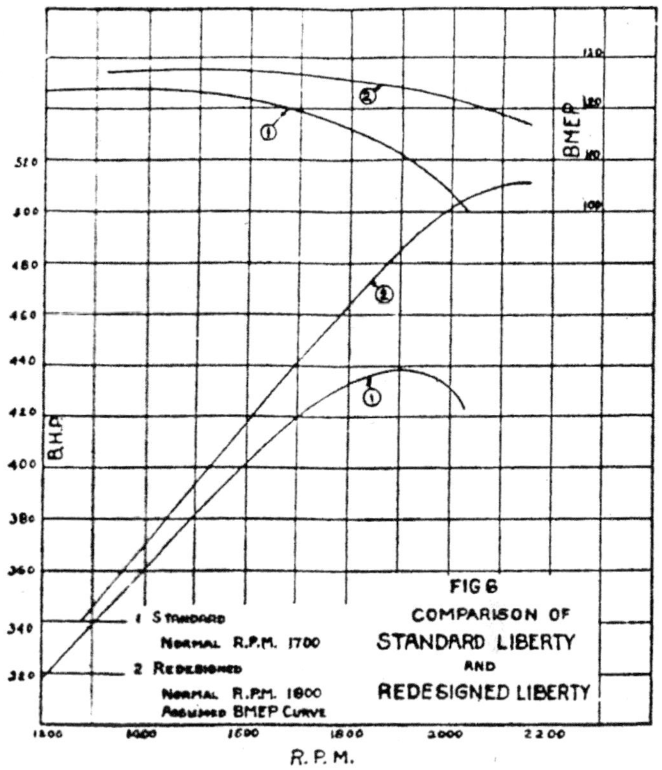

Navy-generated power graph of the proposed aluminum block Liberty designs.

THE NAVY'S LIBERTY IMPROVEMENT EFFORTS

In May of 1923 Lieutenant Bruce G. Leighton of the U.S. Navy released an extensive article (with Navy sanction) titled *Recent Developments in Aircraft Engines in the Navy*. This article was published by *Automotive Industries* magazine in their May 1923 issue and covered six and one half pages.

He prefaced the article with the statement: "In classing this development work as work done "in the Navy" or "by the Navy", I refer to work done under Navy supervision and control, and paid for principally from Navy appropriations." He further noted that he was covering engine developments in the most broad terms because the great number of items prohibited great detail, and that the names of the various organizations employed in this work would not be named for obvious reasons. He also noted that, for the most part, these developments were carried out by private firms at the expense and direction of the Navy and it was the Navy's intention in so doing to maintain a healthy private aircraft industry which would be needed in times of emergency to support the country's military needs.

Recent developments in air-cooled engines were covered and the Navy's great expectations for the type were expressed. Within a few years, of course, the Navy had moved totally to air-cooled engines and abandoned water-cooled types altogether. That was not to be practical for some time, of course, and in the mean time they pressed hard for further development of the water-cooled engine.

Wright was developing the T-2 under Navy subsidy and it was compared to the Liberty. The T-2 was designed to mount in Liberty engine barriers.

The Wright E-2 was also being further developed at Navy request resulting in the E-4 model. He stated that the average life between overhauls for the Wright E-2 was 101 hours against an average for the Liberty of 72 hours. Even though the Navy Liberty manual of 1919 called for an average time between overhauls of 150 hours, they appeared to always be more conservative in actual practice.

The Navy had just recently initiated a future requirement of having the standard 50-hour acceptance test replaced by one of 300 hours. The first engine to pass that required test had been the Aeromarine U-8D, followed shortly by the newly designed Packard 1A-1551 dirigible engine. The Navy's intention was to raise all their service engines to that capability. The Liberty was one of their major engines and they would like to see its reliability raised significantly, if possible. If not, it might be replaced by the Wright T-2.

Three projects were initiated relative to the Liberty. All three were to redesign the complete upper section of the engine by replacing the individual Liberty cylinders with aluminum cylinder blocks. This would get rid of the inherently leak prone welded sheet-metal water jackets, allow better intake manifolding and carburetion and replace the exposed valve gear with an enclosed design. The aluminum cylinder blocks with their top-mounted cam assemblies acted to greatly strengthen the crankcase and thus the entire engine structure.

THE WRIGHT MONOBLOC LIBERTY

The Navy was actively working with several aero engine builders to come up with new designs. Among them were Wright, Aeromarine and Allison. Wright and Aeromarine built aluminum block aircraft engines supported by the Navy, and Allison was doing extensive rebuilding and had built 200 L 12Bs for the Navy as well as new gear design work for both the Army and Navy. All were requested to develop new cylinder blocks for the Liberty.

In early 1923 Wright announced the development of a new cylinder construction for the Liberty 12. This information was released to the various aviation magazines and printed in Wright's house organ, *The Wright Engine Builder*, Volume II No. II of May 1923. It was carried in the May 7, 1923 issue of *Aviation* magazine, for example. The article started by noting the occurrence of water jacket failures of Liberty engines currently in use. This failure was probably the most common encountered with the Liberty and, although it rarely resulted in a catastrophic engine failure, was a constant maintenance problem. I quote directly from the company publication all pertinent facts stated:

"To remedy these defects, the Wright Aeronautical Corporation, of Paterson, N. J., has developed a new cylinder construction for Liberty engines similar to that which has been so successfully used on their Model "E-4" and "T-2" engines. This construction was illustrated in the July 13, 1922, issue of *Automotive Industries*, and consists of cylinders in blocks of threes, with aluminum combustion chambers, and both valves and spark plugs seating in aluminum bronze bushings shrunk into position.

"Besides overcoming the above mentioned difficulties, the engine is considerably strengthened by the block cylinder construction and cam boxes. Moreover, the valve gear is completely enclosed, and there is, therefore, no oil leakage on the top of the engine. The original power of the Liberty engine will be increased approximately 10%, and it is perfectly possible to operate the engine for six times the period possible with the old cylinder construction, without a top overhaul." Claimed output was thus about 450 HP at 1,700 RPM. If the T-2 design was followed, the design would have used four valves per cylinder.

The *Automotive Industries* article referred to was a detailed one on the new Wright T-2. That article was strictly on the T-2 and made no mention of the Liberty cylinders.

The above composite image was generated by the author using Wright T-2 cylinder blocks to replace the standard individual cylinders of the Liberty. The Navy instigated such a project with Wright, Allison and Aeromarine. The Aeromarine-converted engine was assembled and photographed and is shown below but no photograph of a Wright or Allison conversion has surfaced. To see just what the Wright engine might have looked like, the author generated the above image.

The official photo of what the Aeromarine Plane and Motor Co. called their N.A.L model. Probably it stood for New Aeromarine Liberty, but possibly Navy Aeromarine Liberty. This engine mounted 2 newly designed 6-cylinder aluminum cylinder blocks in place of the original individual cylinders. 2 Stromberg NA-L5 carburetors were used. (Author's collection)

No photograph of a Liberty with this cylinder conversion has so far surfaced, thus we cannot positively say Wright actually built them. Considering the fact that the project was instigated by the Navy, it seems very likely that test units were built. Their article certainly indicated that the cylinder blocks had been not just designed, but actually built. They were mentioned in *Jane's* from 1924 to 1926. **However, I have put together a composite photograph using Wright T-2 cylinder blocks and a Liberty crankcase to let us see just what such a conversion may have looked like.** I did so to see just how close to the Aeromarine Model N.A.L. Liberty, which was built, it might look.

The Navy showed no further interest in the Wright project, and Wright did nothing more related to the Liberty.

THE AEROMARINE N.A.L.

The description given by Lieutenant Leighton more closely described the Aeromarine design than that of Wright in that it mentioned a six-cylinder aluminum cylinder block, whereas the Wright design used two three-cylinder blocks connected by a single cam housing serving the two units. Both designs served to accomplish the same purposes, although, by virtue of its single-piece construction, the Aeromarine was somewhat more rigid. Carburetion was by two Stromberg NA-L5 carburetors.

The Aeromarine conversion was described and illustrated in *Jane's* from 1924 to 1926. The same 450 HP output was noted as Wright claimed for their conversion. The aluminum cylinder blocks had inserted steel liners and the bore and stroke were standard Liberty of 5" x 7". It was noted that provision had been made to allow an increase in bore to 5¼". A smart decision which would allow an increase of displacement from 1,649 cu. in. to 1,818 cu. in. and an output of almost 500 HP.

Specifications listed were:

Bore	5 inches
Stroke	7 inches
Compression ratio	5.4 to 1
Overall length	70 inches
Overall width	30 inches
Height above crank center	30 inches
Rated HP	450 HP @ 1,800
Weight dry	915 pounds

THE ALLISON ALUMINUM LIBERTY

Allison was also requested to design an aluminum block Liberty. The engineering paper for it has survived in the Allison archives. It is written as a design project and since we have no other documentation and no known photograph of such an engine survives we cannot say whether an example was built or not.

As described, it would have an external appearance much like the Aeromarine N.A.L. The Allison description follows.

"The cylinders are cast aluminum, two blocks of six. The block castings are bolted to the standard Liberty case, using the same hold-down studs as the standard engine. These cast aluminum blocks rigidly bolted to the crankcase greatly stiffen it and prevent the weaving of the crankcase which is so detrimental to the main bearings. Steel cylinder liners are screwed into the aluminum castings at the cylinder heads, and a rubber ring is used to retain the cooling water at the lower portion of the cylinder. The cooling water is in contact with the cylinder liners.

"The combustion chamber is of aluminum with aluminum-bronze or cast iron valve seats, either cast-in or pressed-in, as is found best.

"Two intake and two exhaust valves per cylinder are used. The valves are tulip-shaped, 1 5/8" clear diameter, and have a 13/32" lift. The valves are actuated by rocker arms from a common camshaft on the top of each cylinder block. The cam contours have been developed so that the maximum accelerations are moderate, even at a speed of 2000 RPM, and at the same time permit a considerable period of dwell at full opening. The relatively light weight of the small valves and forged duraluminum rocker arms makes it possible to use moderate spring pressure, reducing the load and wear on the valve actuating mechanism. The camshaft and rocker arms are entirely enclosed and all moving parts are thoroughly lubricated. It is possible to remove the entire valve gear assembly from a cylinder block as a unit. The camshaft is driven by standard Liberty gears in the usual manner."

The engine was designed to produce a rated 450 HP at 1,800 RPM at a dry weight of 10 to 14 pounds less than the standard Liberty but the new cylinder blocks held more water which resulted in a operating weight about the same as the standard Liberty.

"Two types of carburetor equipment are indicated. One uses two double Zenith carburetors, as in the standard Liberty, except that the manifolding is slightly different. On account of the short distance from the carburetors to the cylinders it is not considered necessary that these manifolds be water jacketed.

"A single installation is also shown. This single carburetor feeds all twelve cylinders through a special designed manifold and throttle developed by the Allison Engineering Company. A similar installation has been used on the Allison Twelve Marine engine and found extremely satisfactory. The carburetor is of the sliding venturi type, similar to that used in the Allison Twelve Marine engine. Since compensation in this type of carburetor is accomplished by floating the venturi on the air stream, the position of the venturi depending upon the velocity and weight of the air, no altitude adjustment should be necessary for ordinary flying. The manifold in this installation is water jacketed."

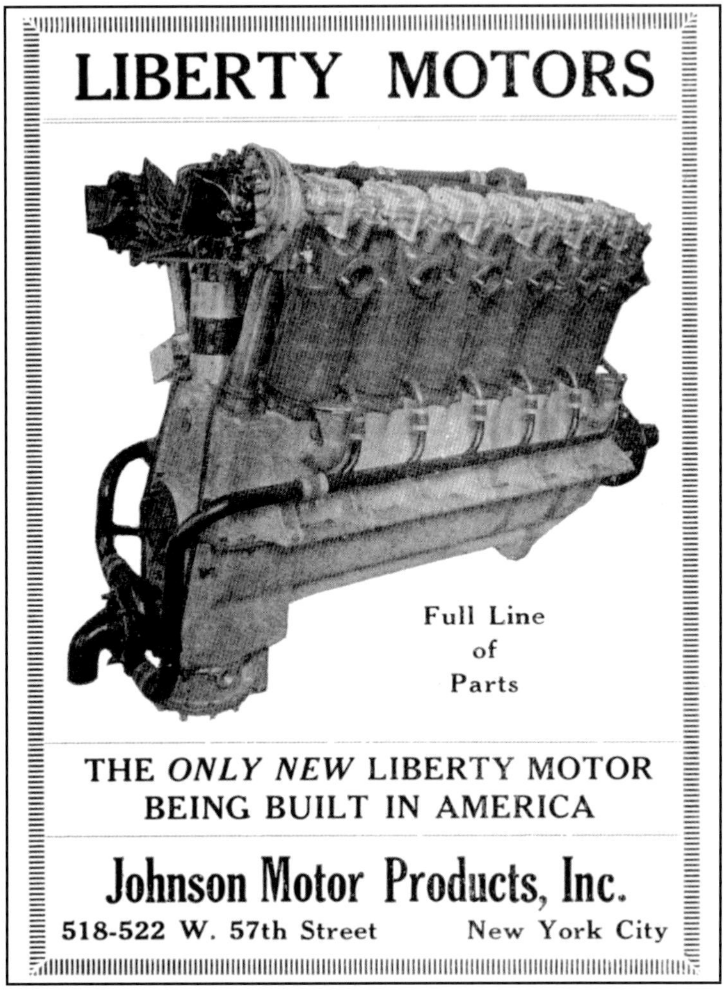

Above is the Johnson Motor Products ad which began to appear in prominant U. S. aeronautical magazines in early 1924 and continued for a few months, to then be replaced by various classified type ads until the 1927 ad on the next page ended the series. (Author's collection)

Cooling water was carried to the cylinder blocks from the standard Liberty pump by manifolds bolted directly to the aft end of the two cylinder blocks. From there the water flowed across the heads and out the forward ends and to the radiator.

"All parts of the engine are standard Liberty, including camshaft drive mechanism and the ignition, excepting cylinder blocks, valves, valve actuating mechanism, camshaft and inlet and exhaust manifolds.

"The work of converting the standard Liberty to the Allison Liberty can be done at any flying field or repair depot, as no machine work is required on the standard Liberty parts."

The Navy let the projects die at that point and purchased neither Allison, Wright or Aeromarine conversions for use on their considerable stock of Liberties. Although I have not yet been able to track down the actual contracts between the Navy and these companies, I would think the Navy paid for the design and production of the company's efforts and would therefore claim ownership to the end result. That must have been the case with the Aeromarine engine as it was purchased surplus and marine converted by Capitol.

MORE BUILDERS OF NEW LIBERTIES

On page 256 I reproduced an advertisement by The Steel Products Engineering Co. of Springfield, Ohio in which they noted that they could furnish Liberty engines of the latest Air Corps specifications. These would be rebuilt engines, of course, but would be equivalent to new. In the years following WW I a number of dealers and rebuilders would advertise surplus Liberty engines for sale either new, used or rebuilt. Some of the more significant of

The above ad for new manufactured Liberty engines appeared in aeronautical magazines for a brief period in 1919. A few engines may have been assembled from parts. (Author's collection)

Very early ad for surplus Liberty engines placed in the November 10-17, 1919 issue of* Aerial Age Weekly *magazine. (Author's collection)

these dealers will be noted. I have noted only three companies that actually claimed to be manufacturing new engines. They were Packard, the Liberty Engine and Foundry Co. of Seymour, Connecticut, and Johnson Motor Products, Inc. of New York City.

Packard actually built new engines from new parts, either currently on-hand or finished on their own machinery. The other two companies were almost certainly selling engines either purchased new as government surplus or assembled from a combination of new and used parts.

Packard's 1919 entry into the market was followed shortly by the Liberty Engine and Foundry Co. As was the case with Packard, the Liberty Engine and Foundry's offering disappeared about as rapidly as it appeared. Its offering was an attempt to benefit from the reputation of the Liberty as a preeminent military aero engine. In this case, however, it is highly doubtful that the proposed manufacture of these engines ever actually happened. The Air Service first offered surplus engines for sale during 1919 and it may well have been that this company hoped to purchase and resell these engines - the availability of the Liberty surplus on the open market could have been what killed the project. Which ever was the case, by the end of 1919 the Liberty Engine and Foundry Co. had disappeared from the scene.

The only other company I have found that claimed to manufacture new Liberty engines was Johnson Motor Products, Inc. of New York City. Beginning in 1924 this company placed a series of ads in various aeronautical magazines announcing themselves as producing "The Only New Liberty Motor Being Built In America." By 1925 their display ads had been replaced by blocked classified ads offering Liberty parts and services and new engines, but no longer claiming themselves as manufacturers. In this case also, it is very doubtful the company ever actually manufactured a Liberty engine. The next ad of significant size I have noted was a half-page blocked classified in the March 15, 1926 issue of *Aviation* magazine for six new Liberty Twelve models with all improvements on a special of $2,000 each. These were no doubt new surplus engines or engines assembled from new parts. Johnson placed a full-page ad in the May 10, 1926 issue of *Aviation* magazine noting "The once in a life time opportunity! A $750,000 stock of Liberty material, all new and unused and manufactured under Air Service Specifications by the Ford Motor Company will be sold at 20 cents on the dollar as we have gone into another model of motor, where we cannot use this material. If disposed of in one lot, we will include a complete set of drawings and assembly tools for $150,000." It further stated "This material is the same as that built into JOHNSON LIBERTY MOTORS used by Sikorsky, Huff Daland, Fokker, the Canadian Department of Defense and the Wilkins Arctic Expedition Fokker

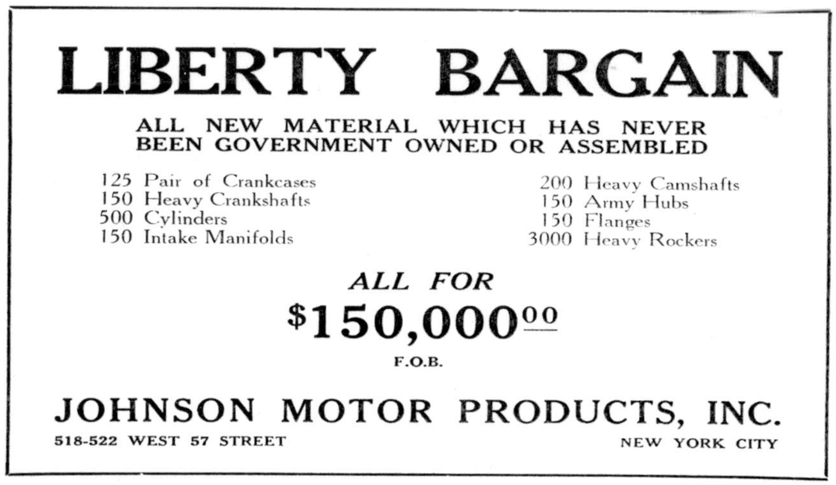

This 1927 ad was the last Liberty related; one placed by Johnson Motor Products, Inc. (Author's collection)

Surplus Disposals by the Army Air Service
From November 11, 1918 to April 20, 1922
Compared to the Air Service inventories of 1919-1921

Airplanes	1919-21	Sold	Remaining
Avro 504K	9	4	5
Aeromarine MI	?	3	?
Breguet	5?	15	?
Curtiss JN4A	1919 inventory listed a total of 3458 JN of all models. Exact matches to those on sale list not given.	57	2820 remained of all Curtiss JN models.
Curtiss JN4B		14	
Curtiss JN4C		347	
Curtiss JN4D		216	
Curtiss JN6H		4	
Curtiss R4L	31	10	21
DH4	2463	193	2270
Fokker	56	12	44
Martin Bomber	7	6	1
Heinrich Scout	1	1	0
L.W.F	62?	85	0?
L.V.G.	3	3	0
Newport	36	22	14
Rumpler C4	?	10	?
Standard D (seaplane)	1	1	0
Standard H3	9	8	1
Standard J1	1357?	1386	0?
Standard JR1B	3	3	0
Standard S4	1	4	0?
Sopwith 1A2	24?	27	0?
Spad	26	2	24
Thomas Scout	?	6	?
Total		2439	

Engines	1919-21	Sold	Remaining
Anzani 50 hp	7	5	2
Anzani 90 hp	105	18	87
Astro-Daimler	2	1	1
Beardmore	35	32	3
B.M.W.	35	3	32
Bentley Rotary	?	5	0?
Bugatti	33	11	22
Clerget 130 hp	28	17	11
Curtiss VX	12	7	5
Curtiss V2	2	1	1
Curtiss V 203	70	9	61
Curtiss V2C3	10	1	9
Curtiss V2C10	8	6	2
Curtiss OX5	7605	4634	2971
Curtiss S	1	1	0
De Dion-Bouton	?	2	0?
Duesenberg	2	2	0
Fiat	389	379	10
Gnome 90 hp	1919 inventory listed 258 Gnome of all types.	11	210 remain of all types.
Gnome 120 hp		12	
Gnome 165 hp		20	
Gnome 132 hp		5	
Hall-Scott A5	28	28	0
Hall-Scott A5A	50	38	12
Hall-Scott A7A	2083?	2114	0?
Hall-Scott A7	1	1	0
Hall-Scott Liberty 6	2	1	1
Hispano-Suiza A	1733	322	1411
Hispano-Suiza E	150	25	125
Hispano-Suiza I	762	36	726
Hispano-Suiza H	1429	122	1307
Hispano-Suiza 220 hp	24	8	16
Isotta-Fraschini	5	5	0
Lawrance A3	379	155	224
Le Rhone 80 hp	748	254	494
Le Rhone 120 hp	127	44	83
Le Rhone 300 hp	?	2	0?
Liberty 12A	**10915**	**312**	**10603**
Liberty N	**1002**	**55**	**947**
Liberty 8-cyl.	**27**	**11**	**16**
Liberty Experimental	**?**	**13**	**?**
Lorraine-Dietrich	10	1	9
Maybach	7	6	1
Mercedes	133	107	26
Oberurdel Rotary	1	1	0
Opel	6	6	0
Peugeot	5?	9	0?
Renault	159	98	61
Sturtevant 5A	116	113	3
Thomas	135	19	116
Wright 8	?	1	0?
Wright D6	5	1	4
Total		9090	

I had available three different 1919 inventories taken within a six-month period and one of June 4, 1921, all of which listed both aircraft and engines by specific model and type. As seems to be characteristic of Air Service inventories, none totally agree. The same model names for both engines and aircraft are not necessarily used. This leads to uncertainty as to exactly what item is being referred to. Items are noted above from the inventories with a ? because they either were not specifically listed or because the maximum number listed was less than later listed as sold. In most cases, the four inventories consulted each showed a different count for the same item. If that were the case for an item on the sold list, I picked the highest count of the four inventories.
The report of sales noted that the airplanes and engines, plus spare parts and miscellaneous equipment also sold during the period, had a estimated war-time cost of $29,441,138 and had brought $2,337,567 or approximately 8% of their original cost.

Plane "Alaska"." It would appear that Johnson must have supplied at least a few engines to these builders.

Johnson Motor Products also sold Capitol marine conversions under their own name in the 1925-26 era. (See Chapter 12.)

Less than a year later, in January and February of 1927, Johnson placed a series of large blocked classified ads offering a large quantity of new Liberty parts for sale as a lot. (See the reproduction on page 281.) This may well have been the same lot of parts offered in the May 10, 1926 ad. After this time Johnson Motor Products, Inc. seem to have gotten out of the Liberty engine business. I have come across no other companies which claimed to actually build new Liberties, only sellers of new surplus ones.

SURPLUS SALES

The last subject of this chapter is, appropriately, what happened to all those Liberty engines built for, but never used by, the American military services.

By the end of the war a total of 57,962 Liberty 12 and 9,029 Liberty 8 engines had been placed on order for a total of 66,991 engines. The abrupt and unexpected early end to the war prevented production from starting on most of the late series of contracts, which covered about 43,000 of the total ordered. As a result, only 21,121 8- and 12-cylinder engines were delivered by the time all production ceased at the end of January of 1919. Of these 5,822 were delivered to our Allies. The remaining 15,299 engines (almost all of which were 12-cylinder models) included 342 school engines and 29 eight-cylinder models, mostly for experimental use. Thus there were about 15,000 useful engines, all twelve-cylinder models, remaining under the control of U. S. military services.

About 5,000 engines had been distributed to aircraft manufacturers and installed in new planes for the Air Service. This left about 10,000 Liberty engines distributed at various locations throughout the United States and AEF locations in Europe plus, of course, those engines operating in aircraft which had not been destroyed in action, accidental crashes, or otherwise. By December of 1919 6,600 were in storage at the Little Rock Supply Depot and another 1,400 at various other depots across the country. The remainder were at various fields and stations and installed in aircraft.

As we have seen in the prior discussion of Liberty rebuilding and upgrading, the Air Service did not follow a "use up and throw away" policy. With the number of reserve engines in stock one could see how it could be a temptation but it was never adopted as a policy. With in excess of 8,500 engines rebuilt between the end of the war and 1931, it is evident that they were not drawing new engines from storage and disposing of used engines needing rebuilding. It is also obvious from instructions issued regarding the number of Class III rebuilds allowed that a significant number of engines went through as many as four or more rebuilds. Currently known records of engines remaining on-hand over the same period reinforce this conclusion.

Most of the following discussion will relate more to Air Service stocks and disposition of Liberty engines than to the Navy's actions in this area, primarily because a great deal more statistical information has survived relative to Army Liberties than Navy ones. Of the 15,000 American engines, about 4,200 went to the Navy, of which about 2,200 had been installed in production aircraft built by or

AIR SERVICE
COST OF OVERHAUL
JULY 1925 - DECEMBER 1926

Liberty

Number	Total cost	Average cost per engine
109	$51,863.29	$475.81
159	103,631.43	651.77
47	29,650.89	630.87
315	$185,145.61	$587.76

Curtiss D-12

Number	Total cost	Average cost per engine
16	$7,589.12	$474.32
11	6,705.82	609.62
27	$14,294.94	$529.44

OPERATING COST PER RUNNING HOUR FOR THE LIFE OF AN ENGINE

Liberty

Initial cost	$2,500.00
Three overhauls @ $587.76	1,763.28
Total	$4,263.28
Cost per hour with 320 hours life.	$15.16
Gas (24.52 gal/hr @ 15 ¾ cents/gal.)	3.85
Oil (7 qt/hr @ 31 cents/gal.)	.54
Cost per operating hour	$19.55

Curtiss D-12

Initial cost	$8,000.00
Four overhauls @ $529.44	2,117.76
Total	$10,117.76
Cost per hour with 600 hours life.	$16.86
Gas (20.2 gal/hr @ 15 ¾ cents/gal.)	3.18
Oil (3.25 qt/hr @ 31 cents/gal.)	.25
Cost per operating hour	$20.29

for the Navy. This left the Navy a stock of about 2,000 Liberty engines in storage.

It did not take long for those in authority to realize that there was a surplus of Liberty (and many other types of aero engines as well - but we are only concerned here with the Liberty) engines above possible Army requirements. Initial requirement estimates of 1919 were set unrealistically high at 10,000 engines. Even that figure allowed the sale of large numbers as surplus. A policy was adopted that serviceable surplus engines would be sold only to United States citizens for use in commercial or civil aeronautics or to educational institutions. None would be sold to foreign governments without approval of the Department of Air Service.

On April 17, 1919 the War Department announced that 1,000 new engines had been placed on sale and that bids should be submitted to the Sales and Salvage Section, Army Air Service, Washington, D. C.

The Navy was not far behind and announced their first surplus sale on September 15, 1919. The offered items included 219 aircraft with engines and also a group of 441 engines. The aircraft and engines had been assigned an appraised value and any bids received at or above that value would purchase, any below would be considered but a sale not guaranteed. The aircraft included 83 HS-2L, 52 H-16 and 20 F-5L types, all to be equipped with new Liberty engines. This would account for 227 new Liberty engines. None of the 441 engines were Liberties.

The November 10-17, 1919 issue of *Aerial Age Weekly* carried the first commercial ad for surplus Liberties I have noted. New-in-the-crate engines were priced at $2,750. The prices at which new and used Liberties were offered varied greatly over the following years. I have compiled an average price list and posted it on page 286.

From this time on, both the Army and Navy periodically put both airplanes and engines out for bids. In most cases, minimum acceptable bid prices were set. Of all the various types of engines they put on the market, the Liberty was always priced the highest. By no means all engines offered for bid were purchased and it took a long time to significantly reduce the stock of surplus engines.

Fortunately, I have found an Army Air Service record of disposals up to April 20, 1922 which lists the exact number of specific models of aircraft and engines sold between the end of the war and that date. It is worthy of reproduction with comparison against the Air Service inventories of these items from 1919 to 1921. As the reader will note, after offering the first 1,000 Liberty engines for sale in April of 1919 with more offerings no doubt made between that time and April of 1922, only 391 Liberty engines had been individually sold. Aircraft were generally sold including their engines. Thus the 193 DH-4, 6 Martin Bombers and 10 Curtiss R4L aircraft accounted for an additional 215 engines. This, of course, does not account for any Navy surplus sales.

The ready market was for the smaller airplanes and engines which were adapted to small airplanes, thus the Standard J1, Curtiss JN, Hall-Scott A7, Hispano-Suiza and Curtiss OX5 were big sellers.

Both the Army and Navy continued to offer the Liberty as surplus but seemed to keep the prices high enough to prevent flooding the market. The Army was initially somewhat reluctant to dispose of too large a quantity of Liberties as they considered them as a potential source of supply in the event of a national emergency requiring an immediate buildup of air power. That feeling became less and less prevalent as the Liberty became more and more obsolete.

In January of 1924 Delos C. Emmons, Chief of Production Engineering at McCook Field noted in a letter to Fairfield Air Intermediate Depot, repository for some 4,000 surplus engines of all types, about 500 of which were Liberties, that although the large stock of Liberty engines "is our only valuable asset in aeronautical supplies" "it is more than apparent that if a large number of Liberties are sold it would give a great impetus to commercial aviation."

This view was expanded upon and passed on to the Chief of the Air Service with a recommendation that 5,000 Liberty engines should be sold or otherwise disposed of within the next four years. At that time the on-hand inventory of Liberty engines was in excess of 11,800. By 1929 (four years) it had been reduced 3,044 to 8,756. By 1931 it had been reduced 5,521 to 6,289.

To give the reader an idea of how sales progressed the report of the Chief of the Air Corps for the period from July 1, 1925 to June 30 1926 is of interest. General Patrick reported that 867 new and 926 used Liberty engines were put up for sale during that period. Of these 156 new and 426 used were sold. In addition, 200 new engines were transferred to the Coast Guard "for use as marine engines in prohibition enforcement work." Left for disposal at the end of the fiscal year were 511 new and 500 used Liberty engines.

In the mean while, it was becoming widely recognized that the longer the Liberty remained acceptable as power for not only airplanes already in service but continued new designs, the longer it would be before our military aircraft became totally up-to-date. As it was, we were falling rapidly behind other military forces in the world. This resulted in the initiation of a number of studies aimed at determining the costs of operating Liberty engines verses a comparable modern type with which the Air Service had wide experience. The Curtiss D-12 was that engine. This study was initiated in early 1927 and used overhaul costs for the two types for the period July 1925 through December 1926.

The last Liberty flown by the Air Corps on September 13, 1935. After landing at Kelly Field, Texas in a Douglas ZC-1C Ambulance airplane it was removed and shipped to the Wright Field Museum where it was installed in this specially-built case for display. The above is the official Air Corps photograph after installation. It remained in that case for a number of years but has now been removed and is displayed on an open stand in the Air Force Museum at Wright-Patterson Air Force Base. See the illustrations on page 204. (Courtesy of the Air Force Museum)

As shown in the table on page 283, this study included 315 Liberties overhauled during that period and 27 Curtiss D-12s. An interesting note in this study is that the service life of the Liberty was set at three overhauls and about 320 hours whereas the May 1927 instructions in *T.O. No 021-5 General Instructions For Maintenance* stated that four overhauls were allowed. If the reader will review the information on page 254 he will recall that by 1927 the Class III overhaul requirements stated that a Liberty was now allowed only four Class III overhauls. The first one had to be applied when an engine was removed from storage to be put in service. Therefore a new, stored engine had its first of four overhauls before it was put in service. After that point it could receive three more, thus the use of three overhauls to calculate operating life. Also note on that page that prior to setting four as a limit, to preclude using excess parts, engines would sometimes be overhauled ten to twelve times during their life.

D-12 life was calculated at four overhauls and 600-hour life.

Using a little arithmetic gives a TBO of 80 hours for the Liberty in the case of this study. The D-12 time of 600 hours and four overhauls results in a TBO of 120 hours for that engine. Both figures are lower than recommendations for their type during that period and are thus quite conservative.

The initial cost of the Liberty was set at $2,500. Why that figure was chosen is not recorded. The lowest contract paid for new L-12 engines when they were in production was $4,000 each. Perhaps the price picked was the current value of a new Liberty on the civilian market. In other words, its value, not its cost.

Using the figures they assigned, the operating cost per hour for the Liberty over an assumed life of 320 hours was $19.55 per hour. The operating costs per hour for a D-12 over an assumed life of 600 hours was $20.29 per hour. Even though the study seemed to justify the use of the Liberty, at least on a cost of operating and overhaul basis, the Air Service claimed they had based no new designs around the Liberty since the observation competition held in the summer of 1924 and expressed themselves as satisfied that no new observation types would be needed for two or three years. In the mean while the current stock of Liberty engines was sufficient to last for the foreseeable future, the average consumption of them being about 500 engines per year. At that time (early 1927) there were slightly more than 9,000 Liberty engines on-hand.

NO NEW LIBERTY-POWERED AIRCRAFT

Finally, on July 1, 1929, the Air Corps officially prohibited the use of the Liberty in any new aircraft. The 1932 annual report of the Chief of the Air Service noted that they were still supporting about 950 Liberty engines at that time. A number of Liberty-powered aircraft were in active service and they lived on for another five years.

THE LAST LIBERTY FLOWN BY THE AIR CORPS

Finally, in 1935, the last active Liberty-powered airplanes were removed from service. The order to phase out those Liberty-powered aircraft remaining in service was issued late in 1933.

The significance of the Liberty's contribution to the Air Service was recognized and orders were issued in May of 1934 to the last remaining squadrons with them in service to keep meticulous records of the flights of any Liberty-powered airplane which might be the last flown at their station.

That aircraft was Douglas Ambulance Airplane ZC-1C Air Corps 27-204 powered by a Lincoln-built L-12 which was originally accepted on December 3, 1918 and reconditioned in 1927 and then again in 1928 by Allison Engineering Co. It was installed in the Douglas on October 3, 1933. It made its last flight on September 13, 1935 from Fort Clark, Texas to Kelly Field, Texas.

The engine was then removed and sent to the Wright Field Museum. It remained at that location until the new Wright-Patterson Museum was built. It now resides in the Air Force Museum at Wright-Patterson Air Force Base.

AVERAGE PRICES AT WHICH LIBERTY ENGINES WERE OFFERED FOR SALE
(compared to the average price of a new Ford sedan of the same period)
1919-1931

Year	New	Used	New Ford price
1919	$2,775		$525
1920	2,700	$1,400	675
1921	2,700	1,400	510
1922	2,600	1,350	443
1923	2,500	1,200	380
1924	2,500	1,195	375
1925	2,000	1,250	375
1926	1,850	1,200	380
1927	1,300	975	380
1928	1,150	700	460
1929	600	450	525
1930	400	275	580
1931	350	250	590

LAST NAVY USE OF THE LIBERTY

The Navy was not quite so interested in recording their last use of the Liberty as was the Air Corps. Although they too had relied heavily on the Liberty, they had been quicker to migrate to more modern engines. They had more rapidly embraced the air-cooled engine as the aero engine of the future, and they had a far smaller stock of Liberty engines to phase out of use. Even so, a number of Liberty-powered Navy types remained on inventory into the late 1920s and two types beyond 1930. Those two were both built during WW I. The last Curtiss H-16 was removed from inventory during May of 1930 and the last NAF F-5L in January of 1931. The OL-2 and OL-4 Loenings were also in service into that era but the precise dates are more difficult to ascertain because they are grouped with the remaining OL series which went through the P&W-powered OL-9 which was last in inventory in July of 1938.

PRICING OF THE SURPLUS LIBERTIES

As already noted, government Liberty engines were, at least during the first few years, put on sale on a bidding process with minimum acceptable prices set. Although there were exceptions made, they usually accepted only bids at or above their set prices. This initially kept prices up but sales limited. As the market drove prices down lower prices were set. The Depression of 1929 dramatically effected the price of everything, the Liberty included. By the time the effects of the Depression were easing in the late 1930s the Liberty was more obsolete than ever and its prices never rebounded.

In the chart above I have noted the average retail prices new and used Liberty engines were offered at in advertising of the period. As a matter of interest I have also listed the retail price of a new Ford sedan for the same years.

SOME OF THE MAJOR DEALERS IN LIBERTY ENGINES

An early dealer in surplus Liberties was, of course, Johnson Motor Products, Inc. of New York. They assembled new motors and offered large quantities of parts as well. Another dealer was the Detroit Marine-Aero Engine Co. of Detroit. This was a company owned by the famous boat racer and industrialist, Gar Wood. They purchased large quantities of surplus aero engines of all types and also sold parts for them. Although they did sell unmodified stock engines, their principal business was the conversion of aero engines to marine use. I will discuss this company in some detail in the chapter on marine conversions.

The Johnson Airplane & Supply Co. of Dayton, Ohio claimed to be the largest dealer in aeronautical supplies in the country and sold aero engines and aircraft of all kinds. The Netherlands Aircraft Manufacturing Co. of New York City was also a large dealer as was the Yackey Aircraft Co. of Forest Park, Illinois. F. T. Holliday of Indianapolis, Indiana offered new Liberties for sale for several years, including several L-6 engines converted for marine use.

During the 1920s the Crawford Airplane Co. of Venice, California offered Liberty 6 and 12 and other makes for sale, both new and used, as did the Floyd J. Logan Airplane Co. of Cleveland, Ohio.

Among the claimants to having the largest stock of airplane motors in the United States (including Liberties) was Nicholas-Beazley Airplane Co. of Marshall, Missouri. Also a large dealer, as well as being a company that converted the Liberty to marine use, was the Vimalert Co. of Jersey City, New Jersey.

The above listed companies which dealt in Liberty engines is representative but certainly does not include all that handled major quantities of them. Almost all of these companies dealt in many types of surplus military aero engines. The Liberty was not their only product by any means.

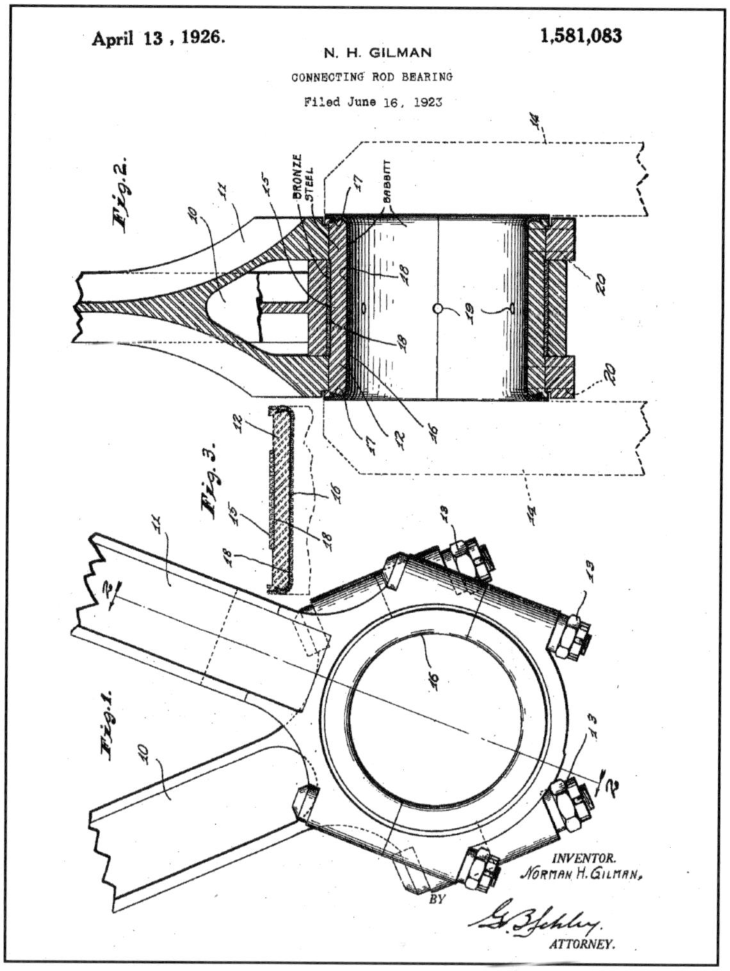

The Allison steel-backed bearing patent. This patent was taken out by Allison engineer Norman H. Gilman and assigned to the Allison Engineering Company of Indianapolis, Indiana. Although applied for on June 16, 1923, it was not issued until April 13, 1926. It is shown applied to Liberty connecting rods and described as applied to that engine. The patent claimed it increased the life of the Liberty engine several fold. It could be applied to other engines, not just the Liberty, and was applied to many types, aircraft and otherwise. Production and sale of this type of bearing became a significant income producer for Allison for many years thereafter. (Author's collection)

MORE POST-WAR DEVELOPMENT

Although I have presented a large amount of material regarding Liberty developments following the war, there remain several subjects so major that I have set aside two additional chapters for them. The first will cover various Liberty projects initiated, for the most part, by the Engineering Division but contracted to the Allison Engineering Co. for further development and execution. It will also document any other Liberty-related projects undertaken by Allison on their own initiative. That chapter, *ALLISON AND THE LIBERTY*, follows this one. In it are covered the inverted water-cooled, the air-cooled and the geared Liberties as well as any other projects with a Liberty connection.

And last will be the subject of *SUPERCHARGING THE LIBERTY*. That chapter will cover all efforts at supercharging the Liberty. A great deal of effort was expended in supercharging the Liberty engine, not so much to improve upon the Liberty in service use, although this was an initial aim, but mostly to develop the supercharger itself. The Liberty was the engine upon which the initial designs were based and was a convenient and suitable "work horse" upon which to continue its development.

LIBERTY MODEL NAMES GIVEN IN
MODEL DESIGNATIONS OF USAF AIRCRAFT ENGINES

Up until 1926 military aero engines were known by the names assigned to them by their manufacturers. At that time an official form of designations was introduced. With only minor modifications, this basic system remained in use until the end of the piston engine era. The basic system is explained as follows:

1- A type prefix is assigned based on cylinder layout as follows:
  H= two-row inline
  L= Single row inline
  O= Opposed
  R= Radial
  V= Double row V
  W= Three row
  X= X configuration
2- A number indicating the engine's displacement to the nearest 5 cubic inches.
3- A model suffix, originally a letter but from 1933 onward a number with odd numbers assigned to the Army and even assigned to the Navy. Prior to WW II the type letter was sometimes accompanied by one or more additional letters to further describe the engine, such as "I" for inverted, "G" for geared or "S" for super charged.

The Liberty thus went through all of these systems of designation. Unfortunately it caused more confusion than clarification in the case of the Liberty. This came about because by the time of the introduction of the standardized system in 1926, the Liberty was an obsolete engine still in use but not in production. The assignment of the suffix numbers of 1933 had to have been done from purely a historical viewpoint as by that time there were only a small number of Liberty engines still in service and they were being rapidly phased out. This suffix assignment was probably not done until a comprehensive listing of all USAAF and Navy engines, which included reciprocating, turboprop, jet and rocket types, was put together just after WW II. It was titled *Model Designations of USAF Aircraft Engines* and its stated purpose was "to serve as a basic reference identifying and briefly describing each approved designation. — This publication includes all prime movers designated for use in USAF Aircraft from the inception of the engine designation system (in the mid-nineteen-twenties) to date." Thus it did not include any early WW I engines which were no longer in service in 1926, but it did include the Liberty.

The originator evidently attempted to generate suffix numbers and even older model names for prior variations of the Liberty from historical records. Given the massive amount of material he was working with (the document contained some 80 pages listing an average of 10 engine models per page), and the method he must have used it is no wonder that his resultant list of Liberty variations was less than accurate. Unfortunately the resultant document is likely to be accepted as "official and thus accurate" even though that is not the case.

The aforementioned list gave basic information on all engines which included a description of it and a listing of the aircraft in which it was used. All listings of the Liberty erroneously gave its bore and stroke as 5.4 x 6 inches, which is the bore and stroke of the Packard Merlin V-1650. The correct measurements are, of course, 5 x 7 inches. A weight of 905 pounds was given to all models listed, which included two geared models as well as the standard direct drive. Although 905 pounds may have been correctly applied to a standard model with accessories attached, it was not correct for either geared engine listed. The assigned designations of V-1650 with suffixes of 1, 3, 5, 7, 9, 11 and 13 were applied. The Model designation column attempted to list each engine by USAF and manufacturer's name. Thus each V-1650 model also carried a manufacturers model name. Names applied were: V-1650-A, V-1650-D, V-1650-E, V-1650-F, V-1650-HD, V-1650-N and IV-1650. I will comment on each assigned model individually.

USAF Listing with comments

USAF model	Manufacturer model	Description	Author's comments
V-1650-1	V-1650-A	reconditioned engine	Strictly speaking, V-1650-A was never used as a designation; it would have been Liberty 12A or L-12A. All Liberty engines in service by 1926 were reconditioned, so that is correct.
V-1650-3	V-1650-D	reconditioned engine	The aircraft listed as using this engine carried standard L-12A engines although probably rebuilt to 1926 or later standards. This list is the only place I have noted a "D" model Liberty 12.
V-1650-5	V-1650-E	equipped with 2:1 spur gear with single reduction	The brief description given does not match any spur reduction gear engine built. The L-12C had a 1.5:1 ratio, the single reduction Allison had a 1.67:1 ratio, the two-speed Allison had a 1.54:1 and 2.03:1 ratio. The 2.03:1 is close enough not to argue with except that the engine with it was a two-speed not single reduction. The Liberty cannon engine was spur-geared but had a 1.6:1 ratio. It was called the L-12E by Packard. The two-speed engine was also called the Liberty "12E" when listed in the Old Wright Field Museum inventory. I am therefore at a loss to say what engine this entry refers to. There were no entries in the user aircraft column for this or any following Liberty models.
V-1650-7	V-1650-F	equipped with epicyclic 5:3 reduction gear.	The epicyclic reduction model was known as the L-12B. This is the only place I have seen "F" used.
V-1650-9	V-1650-HD	equipped with heavy crankshaft	All Liberties in use by 1926 had heavy crankshafts. This is the only place the author has seen "HD" used.
V-1650-11	V-1650-N	reconditioned engine	Listed as standard engine with HC pistons. The "N" model is generally used for Navy low-compression engines.
V-1650-13	IV-1650	inverted air-cooled	The inverted 1650 engine was water-cooled, not air-cooled, and the air-cooled engine was a V-1410.

In conclusion, I feel I must consider this listing of Liberty models as basically unreliable and confusing.

Two views of the same early configuration Allison-converted Buick-built inverted water-cooled Liberty. Its data plate is shown in the inset. The MNFG. NO. appears to be B327 but its U.S. numbers cannot be read. Note that it is marked General Motors Corp., like the Cadillac plate on page 110, but does not carry the Buick name. Also note the design is slightly different from the Cadillac plate (different location of "MNFD BY").

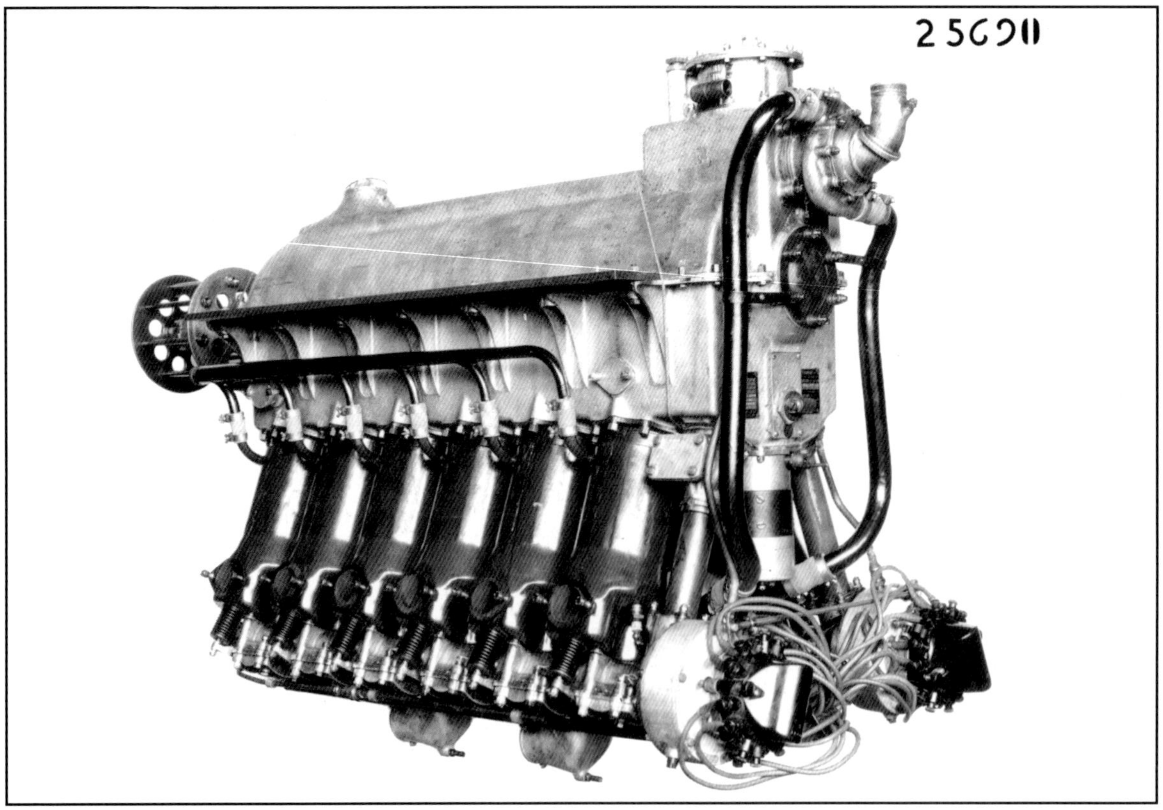

Chapter 8

Allison and the Liberty

On September 14, 1915, the Indianapolis Speedway Team Company was formed with James A. Allison as its president. This company was the predecessor of the Allison of today. Fifty-five year old James A. Allison, President and principal owner of Allison Engineering Co. died on August 4, 1928, after a brief bout with bronchial pneumonia. The executors of the estate immediately began to look for potential purchasers of the company who would continue the business in Indianapolis. Interest was limited and bid prices relatively low. An interested party was Eddie Rickenbacker. Rickenbacker had only some nine months before (on November 1, 1927) concluded the purchase for $700,000 of the Indianapolis Motor Speedway. He had been interested in the purchase of the Allison Engineering Co. and approached Allison in that regard and was told by Allison that he would not sell but instead was offered the chance to purchase the Speedway.

With the mild interest shown he purchased the company with a high bid of $90,000. The purchase was completed in late 1928. Rickenbacker noted in his autobiography that he had no more than taken possession when he realized that he had neither the time to run it properly nor the finances to carry out the expansion it would need in the immediate future.

The Fisher Brothers Investment Trust was currently buying into the aviation industry and showed an interest. As Rickenbacker put it, they agreed to take the Allison Engineering Co. "off his hands." They became the official owners on January 1, 1929. The Fisher Brothers Trust was a holding company set up in 1924 to handle the Fisher family proceeds from the sale of the Fisher Body Corporation to General Motors in a 1916 - 1926 deal. Fisher sold Allison to General Motors only 90 days later. Allison was acquired from General Motors by Rolls-Royce in 1995.

James A. Allison, Founder and president of Allison.
(Courtesy of Rolls-Royce / Allison)

This, of course, is Allison history in its most brief form. To understand just why and how Allison moved from an auto racing enterprise to a manufacturer of precision machinery, we must add considerably more detail.

James Allison's father, Noah, had built a thriving printing and publishing business in Indianapolis, Indiana, but passed away in 1890 at an early age (44) leaving his business to his wife and three sons, one of whom was, of course, James. The Allison family, owners of the Allison Coupon Company, was thus prominent in Indianapolis business circles. By 1904 James Allison was vice president of the enterprise.

Another prominent business man of the area was Carl G. Fisher. Fisher was an entrepreneur who had started his business empire in the bicycle business. He soon shifted to the automobile and by 1904 had an Indianapolis dealership which sold a number of brands, including Oldsmobile, Packard and Stoddard-Dayton.

Fisher had become interested in bicycle racing during the period he was in the business and when he switched to the automobile so did his racing interests. By 1901 he was not only an avid spectator whenever the opportunity to watch any kind of auto race presented itself, but was a race car owner and driver as well. In 1904 he had a special race car built by the Premier Company of Indianapolis.

Although just how they met is not recorded, Fisher and Jim Allison were close personal friends by 1904. They shared the interest in automobile racing although during this period Fisher seems to have been the only one of the pair to actually own race cars.

The scene is thus set for a couple of significant events which followed within the next decade. Carl Fisher and Jim Allison were close friends, were both affluent, and had a keen interest in the automobile and in racing them. Fisher was more the entrepreneur and Allison the planner. (Carl Fisher, by the way, was in no way connected with the Fisher brothers of Fisher Body Company fame who later purchased the Allison Company.)

Norman H. Gilman, Allison executive and engineer from 1916 - 1936. (Courtesy of Rolls-Royce / Allison)

The first of these very significant events occurred in 1904. The story of just what actually led up to it varies considerably, depending upon whose version one reads. Percy C. Avery had purchased the French patent for a system of storing acetylene gas, a highly flammable gas which was then used for lighting buoys and lighthouses. It was also used to some extent for automobile lights but in that case it was generated during its use by a system which dripped water into a container of calcium carbide and burned the acetylene gas as it was generated. The key here was being able to store quantities of the gas for easy use. The light produced by a good acetylene light was far superior to anything else available at the time for automotive use.

Avery had exhausted his resources in an attempt to develop a safe method of compressing the gas to be used in his storage device and was looking for investors willing to join him. His contact with Fisher and Allison yielded the necessary investment interest. The two men raised the necessary capitol to purchase half-interest in Avery's patent and finance the building of their first plant located at Pennsylvania and 28th Streets in Indianapolis. The new company was incorporated as the Concentrated Acetylene Company and their product named "Prest-O-Lite" for its ability to provide an instant flame.

Acetylene gas is highly explosive and as a result a number of rather destructive disasters occurred in various manufacturing plants over the next few years. Eventually their causes were found and overcome and the business prospered by leaps and bounds. Fisher, Allison and Avery became quite rich. In 1906 Avery withdrew from the business and Fisher and Allison changed its name to the Prest-O-Lite Company.

Eventually the Union Carbide Company began to show an interest in purchasing the company. Although initial offers were refused by the partners, they finally accepted an offer of $9,000,000 for the company in 1913. This would represent approximately $60,000,000 in today's dollars. By this time both of these men were already millionaires from prior holdings and earnings from their businesses.

Now lets drop back a few years to 1908. As the automobile became more popular so did racing them. This was done, for the most part, either as road races carried out on town and country roads mapped out as a course for a particular race, or on county fairgrounds horse race tracks. The fairgrounds tracks were becoming too short for the speeds of the cars and road racing was becoming too dangerous.

Some were considering and beginning to build tracks specifically for auto racing. In 1908 Fisher and Allison obtained an option to purchase a large block of farm land northwest of Indianapolis with the intention of building a three-mile racetrack on it. They invited three mutual friends to join them in the venture: Arthur Newby of the National Motor Vehicle Company, Frank H. Wheeler of the Wheeler-Schebler Carburetor Company, and Stoughton Fletcher, an Indianapolis banker.

The venture was to become the Indianapolis Motor Speedway Corporation. However, before the final incorporation Fletcher pulled out of the venture and Fisher and Allison took over his share, bringing their holdings to 30% each with Newby and Wheeler each holding 20%.

The articles of incorporation were filed on February 9, 1909 and construction was begun in earnest. The original design as a three-mile track was altered to a two and one half-mile to allow room for future grand stands around the outside of the track. The first races were held in August of 1909. The original surface was compacted dirt and rock with oil treatment and did not fare well under racing conditions. This rapidly led to the decision to pave the track with brick. This job was completed by the end of the year and thus was born the "brickyard."

Several races were held following the completion of this project. Public response to these races led to a decision by the owners that a single 500-mile race each year with a large purse would be the best use of the facility. Thus was inaugurated, beginning on May 30, 1911, the Indianapolis 500 International Sweepstakes race - the Indianapolis 500 we still run today.

Obviously the four men who started the Indianapolis Speedway were all very interested in racing. Arthur Newby, principal owner of the National Motor Car Company had race cars built at his factory and formed a race team with Johnny Aitken and Joe Dawson as two of his drivers. Joe Dawson drove one of the Nationals to victory in the 1912 event. No Nationals were entered in the 1913 event. Sometime in 1914 Johnny Aitken approached Allison with the idea of starting a race team. Allison liked the idea and enlisted his friends Fisher, Newby and

Wheeler. They joined together and formed the Indianapolis Speedway Team Company, which was incorporated on September 14, 1915.

A shop for preparing and maintaining race cars was set up in a rented building on Georgia Street in downtown Indianapolis. By late 1916 it had become evident to Allison that a location so remote from the track did not work efficiently. It was decided that Allison would take over the Indianapolis Speedway Team Company and build a small shop near the track in Speedway. Aitken would become General Manager. Norman H. Gilman, recently assistant superintendent of the National Motors plant would shortly come to the Indianapolis Speedway Team Company as chief engineer and superintendent. A 20-man machine shop with all the latest equipment was established.

ALLISON EXPERIMENTAL CO.

On February 23, 1917 the company was reorganized and the name changed to the "Allison Experimental Company."

When America entered WW I on April 6, 1917 Allison announced that there would be no more races at the Speedway until the end of the conflict. He also stopped all race car preparation at the Speedway Team shop and told Gilman to keep the shop and its staff in-tact and seek any war work requiring the precision machining capabilities of which they were capable. With Allison's financial resources behind the shop there would be no problem with keeping it in operation until work was found. Jim Allison's association with the company was supportive but, for the most part, remote. It would shortly be Norman Gilman who would run the day-to-day operations of the organization.

There seemed to have been no difficulty in finding war-related contracts. Principal among them seems to have been several related to caterpillar-type vehicles. In addition they received many contracts to build machine tools, jigs and various other precision equipment.

ALLISON AND THE LIBERTY DURING WORLD WAR I

The first Allison contracts known to be related to the Liberty were to build gauges and machine tools for the Nordyke & Marmon Company of Indianapolis. The first of these orders must have been for Hall-Scott A-7a tooling as they signed a contract to produce 1,000 of this model on June 21, 1917. On September 7, 1917 they signed an additional contract to produce 3,000 Liberty engines. Various delays kept them from delivering the first A-7a until November of 1917, even though initially it was planned to have that contract completed by the end of October of 1917. Their last A-7a was delivered in May of 1918. Although manufacture got under way some time earlier, their first Liberty was delivered on June 15, 1918. Although we have no specific dates of this tooling job, the history of what was occurring at Nordyke & Marmon during this period would indicate that Liberty tooling would probably not have been required until January of 1918 at the earliest. There are vague references in some Allison historical papers of the assembly of two Liberty engines for Nordyke & Marmon during this period, perhaps as instructional or master engines. If this actually happened, the parts would probably have been supplied by Packard and parts vendors supplying Packard since Packard had completed their first production engine in November of 1917.

Except for the building of the tooling and possible assembly of the two engines mentioned above, Allison's first Liberty-related job was the building of 3 epicyclic gear sets for the newly designed Liberty 12B beginning in April of 1918. The reader is referred back to page 166 of Chapter 5 for the detailed story of the development of the epicyclic-geared Liberty 12B.

The Willys-Morrow Company had also been given an order for 3 sets of these gears. The 6 sets were to be used in 6 engines to be built by Packard. Allison delivered the third gear set from the initial order for three in September of 1918. On about September 10 an additional three sets were ordered from Allison making a total of nine sets for the six experimental engines being built by Packard.

The last engine was delivered by Packard on November 26, 1918, and the last of Allison's gear sets was delivered in January or February of 1919.

Pending the expected good results of testing the six prototype Liberty 12B engines ordered from Packard, an order for 100 production engines was placed with them on August 12, 1918, order No. 730187 contract No. 4469. The Armistice of November 11, 1918 brought its cancelation on December 4, 1918. Allison had received the contract to build 100 sets of epicyclic gears for these engines which was also canceled.

THE POMILIO LIBERTY SPUR GEAR

This was probably only a design exercise. But since it appears in Allison historical material I think it should be mentioned here. Perhaps some future historian will come across some mention of a proposed Pomilio airplane of later 1918 which was to use it and this bit of material will add something to his knowledge.

Pomilio was an Italian aircraft company, based in Turin, which began operations in this country very late in WW I, about August of 1918. They set up facilities in Indianapolis, Indiana and submitted aircraft designs to the Army Engineering Division for possible purchase. Two were accepted, the FVL-8 and BVL-12, and contracts signed with Pomilio to produce six of each type. The FVL-8 was a single-seat fighter built around the Liberty 8A and the BVL-12 a day bomber built around the Liberty 12A. When the war ended only a few months later, Pomilio had completed only one of each type and was already in financial difficulties. These two were delivered to McCook Field along with all parts so far completed for the remainder of the contracted 12 airplanes. Ultimately the remaining 10 were completed by McCook Field personnel.

During this short period Pomilio submitted a proposal for an additional bomber design. Before it could be seriously considered the war ended. Thus we are left, so far, with only a reference to the design but no hard facts relative to it. However, Allison later built a "Pomilio-Type Liberty Gear." This gear was a stand-alone gear box designed for use on a standard Liberty 12 engine. We have photographs of the final product and at least four different engineering type references to it. In one describing its design in some detail is the statement that it "is a slight modification of a gear designed by Pomilio." The reference in Engineering Division documents to this proposed additional Pomilio bomber is the only thing we have to tie this Pomilio-designed Liberty gear to a possible original use.

During this period of time the Pomilio company was in financial difficulties and one would not think in a position to contract Allison to do more than a basic design study, much less actually build the gears. There is no evidence the Air Service contracted the design. They had not approved the design of any Pomilio airplane to use such an engine. No photograph of Pomilio gears has been found. Under normal circumstances a company of Allison's size certainly would not risk the funds required to do such a design and construction project on speculation. The reference in the "Pomilio-type" stand-alone gear material indicates the gear was designed by Pomilio and nothing indicates that it was built. There is also nothing to indicate Allison carried the design any further until they used it as a basis for the stand-alone gear.

This brings us to the end of the war with Allison's Liberty-related work up to this point consisting of the building of tooling for Nordyke & Marmon and the possible assembly of two Liberty engines for that company, plus the building of six sets of epicyclic gears and a canceled order for 100 more sets. And, of course, the remote possibility of having built their first reduction gear, the Pomilio Liberty gear.

ALLISON AND THE LIBERTY AFTER WORLD WAR I

By the end of the war Allison had expanded from a 20-man operation to a good size company with a staff of 100 plus a temporary complement of as many as 150 draftsmen. In spite of reports in some Allison histories that during the war several hundred Liberty engines were built in-house, *that did not happen.* These statements are invariably the result of erroneously thinking references to engines built just following the war, which would have been the contract to assemble and test 200 epicyclic Liberties which follows, were speaking of engines built during the war.

ALLISON BUILDS 200 LIBERTY 12B ENGINES

Of the six prototype L-12B engines built very late in the war, three were immediately transferred to the Navy. Although the Army interest at that time was minimal, they would eventually try a few of this model in some special aircraft. They would, as we will soon see, also generate an air-cooled version. Initial interest was, however, by the Navy. After testing the prototypes acquired from the Army, in early 1919 they placed an order for 200 engines.

The manner in which this contract was executed was a little unusual but certainly logical considering the circumstances. The Navy contracted Packard to manufacture the engines as finished sets of parts without the epicyclic gears. Allison was contracted to build the gear sets and assemble and test the engines. In addition they would do any development work necessary on the gearing since there was as yet no long-term experience on the design.

Allison did not have the machinery necessary to produce complete engines. Packard, of course, did and had produced the prototypes and thus had the crankcase patterns. The crankcases were, of course, different from the standard Liberty. Thus it was a relatively easy task for them. The Packard board minutes of March 29, 1919 contain the following entry: "The Management was authorized to execute a contract with the United States Navy Department, to manufacture 200 finished crankcases, and crankshafts and other parts for epicyclic-gear-driven Liberty motors, at a fixed price of $1,354.37 per set. The crankshaft forging to be supplied by the Government." The total contract price would have been $338,592.50.

An additional note of interest is found in the Packard board minutes of May 5, 1919, some five weeks later. "The management was authorized to purchase from Nordyke & Marmon Co. of Indianapolis for $100,000.00 their stock of finished and semi-finished Liberty engine parts together with their stock of jigs, fixtures and gauges. To be used by the Engineering Department for construction of special motors."

In Chapter 7 I documented Packard's production of a commercial Liberty they called their 1A-1650 model. They publically announced its availability in late 1919 and first showed it in early 1920.

Assembly area at Allison Plant 1 during completion of the Liberty 12-B contract of 1919-1920. At least 38 assembled engines are pictured along with perhaps an additional 4 in process, and half a dozen bare crankcases in the left rear of the room. Below is a factory photo of an engine. Note that these engines have the early type carburetor intake system rather than the late vertical stack and are equipped with Navy hand-starters and Navy data plates. (Courtesy of Rolls-Royce / Allison)

It is not difficult to tie these three events together. Packard received an order for 200 sets of Liberty parts in March, purchased a large stock of Liberty parts from Nordyke & Marmon in May and announced a commercial Liberty a few months later. I am sure they did well on the ventures.

With the number of engines involved, fulfilling this contract carried into 1920. Testing revealed two problems; 1-the failure of planet gear bearings, and 2-lack of adequate lubrication of the gears caused by insufficient capacity of the standard Liberty pump when required to carry the additional load of supplying the epicyclic gears.

The first difficulty was solved by finding a supplier of higher quality bearings. The second was solved by redesigning the original pump whose capacity was increased by 44% by changing the design of the pump gears. In this manner most of the original pump parts were retained.

The Navy was evidently a bit over optimistic in their estimated requirement for this engine and never did put them all to use. They did a number of comparison tests in F-5L aircraft between the L 12B and L 12N powering the same airplane with first direct-drive and then geared Liberty engines and comparing performance. Performance was definitely improved by the geared engines. F-5L A-3682 and A-3787 were tested in this manner in late 1922 and the results documented in NACA Report 116 of October 1922.

In a Navy report made public in June of 1922 they noted that Navy aircraft had been operating successfully on a daily basis for the past year with geared engines. It would seem that although the geared Liberty was to prove the worth of the more efficient, slower speed propeller allowed by gear reduction, ultimately the Navy used more modern geared engines when they made the change.

The Army requested transfer of 100 surplus Navy Liberty 12B engines to their inventory in February of 1925 for use as air-cooled engines. The requisition was approved on March 1, 1926. As we shall soon see, Allison would get some of these engines back and rebuild them into VG-1410 air-cooled Liberties.

THE LIBERTY REVERSE GEAR

In 1920 Allison Experimental Company designed a reverse gear for use on a 300 HP Liberty 12 dirigible engine. Although not stated in the engineering papers which described it, it was probably designed early in the ZR-1 *Shenandoah* Navy project when that ship was intended to be Liberty powered. Later, of course, the Navy changed their mind and contracted Packard to design and build the 300 HP 1A-1551 engines which ultimately drove it.

This transmission was designed to transmit 300 HP at 1,400 RPM in forward and 200 HP in reverse. All gears were 1.5" wide and were of 10/12 pitch. Pinions had 55 and 60 teeth, countershafts 35 and 38 teeth and idlers 40 teeth. Reversing was controlled by a toothed clutch. The transmission was mounted in an aluminum case.

Little more is known of its design and there is no current evidence that it ever became more than a design exercise before the plan changed to using the new Packard 1A-1551 engines with entirely different design transmissions, which Allison also designed and, in this case, built.

THE 1920 ARMY REORGANIZATION ACT

Among the many changes issued by the above act was the establishment of the US Army Air Service which took over control of the Army's flying service from the Signal Corps. In conjunction with this change also came a change in the system of assigning serial numbers to both aircraft and aircraft engines. Aircraft numbers had begun at 1, it being assigned to the first Wright Model A purchased in 1908. Numbers had progressed upward to a high of 68592 plus a special block of 1919 - 1921 experimental procurements of 94022 - 94112.

Engine serials had also started at 1, it being assigned to the Wright 4-cylinder engine which powered the first Model A airplane. Engine serials had progressed into the 95000 range when the new numbering system was initiated.

The new system numbered both airplanes and engines in accordance with their procurement within each fiscal year. The first fiscal year for which this was in effect was Fiscal Year 1922 which began on July 1,

Design features of the L 12C reduction gears as compared to the Allison Single-Speed Spur-type

	L-12C	Allison
Reduction ratio	1.5:1	1.67:1
Diametral pitch	4	9/7
Number of teeth		
Propeller gear	42	70
Crankshaft gear	28	42
Pitch diameter		
Propeller gear	10.4375	10
Crankshaft gear	6.3125	6
Tooth width	3.375	3.5
Engine weight	1000 lb	974 lb

Above and above right are two views of the 1921 Allison Spur-Gear Liberty. Only one of this type is known to have been built but it was used to test the use of gear reduction in aircraft from 1921 until at least 1925 and possibly later. The engine is not believed to have survived. (Courtesy of Rolls-Royce / Allison)

1921 and ended on June 30, 1922. The first airplane was thus numbered 22-1 the second 22-2 etc. The same system was applied to engines.

THE ALLISON SINGLE-SPEED SPUR REDUCTION GEAR

The basic design work on this engine was done by the Engineering Division at McCook Field. Contract number 51411 was then let to Allison in late 1920 or very early 1921 as the low bidder. This engine was an updated Liberty 12C with Allison doing the final design and production. Design power was 475 HP. The reduction ratio of the 12C was 1.5:1 and this new engine was 1.67:1.

Specific design characteristics are listed in the accompanying chart and although it can be seen that the physical sizes of the gears are similar, the gear case designs are very different and the two engines are easily distinguishable. In

LWF-T3 transport. Only one example of this type aircraft was built. It was tested with a standard Liberty 12A, Allison single-speed spur reduction gear (shown here) and epicyclic-geared Liberty 12B. The gear case can be seen in the enlarged inset. (Author's collection)

Allison Single Mesh Liberty Reduction Gear. Below left is an extreme enlargement of the data plate seen in the above image. The herringbone design of the gears is seen at the upper left. The spring coupling is in the lower gear and the propeller shaft coupling is at the lower right. The propeller hub and its bolts are at the far right. Some of the components can be compared with like units used in the stand-alone epicyclic gear box pictured on the following page, notably the spring coupling and the propeller shaft coupling. (Courtesy of Rolls-Royce / Allison)

addition, the Allison engine used their special design spring coupling between the crankshaft and the drive gear. This coupling would flex slightly under impulse load and thus absorb most of the torsional impulses imposed by the crankshaft and relieve stresses otherwise imposed upon the gear train. This spring drive was used by Allison on all following designed gear trains that the author is aware of.

All current information indicates that engine was assigned AS number 94783.

The engine was probably completed in early 1921 and is believed to have been initially test-flown in DH-4B P-137 AS 63600. It was also tested in the LWF-T3 P-328 AS 23-1220 transport in 1924 and possibly other aircraft as well. As late as December of 1924 Allison was making various modifications to this engine under Engineering Division contract. The Air Service continued to maintain an interest in geared engines but built no more of this type.

ALLISON EXPERIMENTAL CO. BECOMES ALLISON ENGINEERING CO.

AE Co.

Allison was still working on this project when they again went through a reorganizing effort and changed their name to the Allison Engineering Company on January 4, 1921.

ALLISON STAND-ALONE SINGLE MESH REDUCTION GEAR - POMILIO TYPE

Shortly after receiving the contract to build the Spur-Gear Liberty, the Engineering Division let another contract

Allison designed and built stand-alone epicyclic reduction gear for bolting directly to a standard Liberty engine. It used the Allison spring coupling designed for use in the stand-alone spur reduction gear built for the Engineering Division. (Courtesy of Rolls-Royce / Allison)

to Allison (No. 51579) for a Liberty reduction gear. This contract was let under Engineering Division Order number 54-21.

In the Allison engineering paper describing the features of the design is the statement "This is a gear of the single reduction type especially designed for the Liberty engine, and is a slight modification of a gear designed by Pomilio." Since no mention is made of such a "Pomilio design" having been actually built, only designed, I feel I must assume it was indeed only a design submitted by Pomilio, probably for the reasons stated earlier in reference to the Pomilio bomber proposal.

This was different from the Allison Spur-Gear Liberty in several very significant respects: first, it was a stand-alone gear box designed to be attached to a standard Liberty 12A engine on either extended engine barriers or special rails attached to the engine; second, the reduction gears were very different in design, being double herringbone rather than straight-cut spur gears. The reduction ratio was 1.82:1. The weight of this gear installed was 185 pounds.

Only three units were built. I have found no record of it being flight tested and no example is know to have survived.

The building of this "Single Mesh Reduction Gear" led almost immediately to the design and building of another stand-alone reduction gear, this time using the planetary gear developed for the production Liberty 12B.

THE ALLISON STAND-ALONE LIBERTY EPICYCLIC GEAR BOX

Allison either thought the epicyclic reduction gear for the Liberty held some promise as a saleable item and thus designed and built them as a private venture, or the Engineering Division contracted its building. Currently, I cannot say which since I have found no record of a contract covering it.

This stand-alone gear box carried a similar gear train as the Liberty 12B but was designed to transmit up to 475 HP rather than the 400 HP of that engine. It was described in their engineering paper *A. E. Co. Liberty Planetary Reduction Gear* dated April 28, 1921 as follows: "This gear is a planetary type reduction gear complete in its own case, built to be driven by any standard Liberty engine. It is designed to transmit a maximum of 475 H. P. at 1800 engine revolutions with a reduction of nearly 5:3. The use of this gear increases the distance from the front cylinder to the propeller by 14 3/4" over that of the standard Liberty engine."

Design features of Allison epicyclic gears

	L-12B	Stand-Alone
Reduction ratio	1.667:1	1.669:1
No. of teeth in annuls	135	153
No. of teeth in planet wheel	45	45
No. of teeth in planet pinion	30	33
No. of teeth in sun gear	60	75
Diametral pitch of teeth	12/14	12/14
Pitch diam. of annuls	11.25	12.75
Pitch diam. of planet wheel	3.75	3.75
Pitch diam. of planet pinion	2.50	2.75
Pitch diam. of sun gear	5.00	6.25
Pressure angle	20°	20°
Width of gears sun/annuls	1.688/1.125"	1.75/1.25"

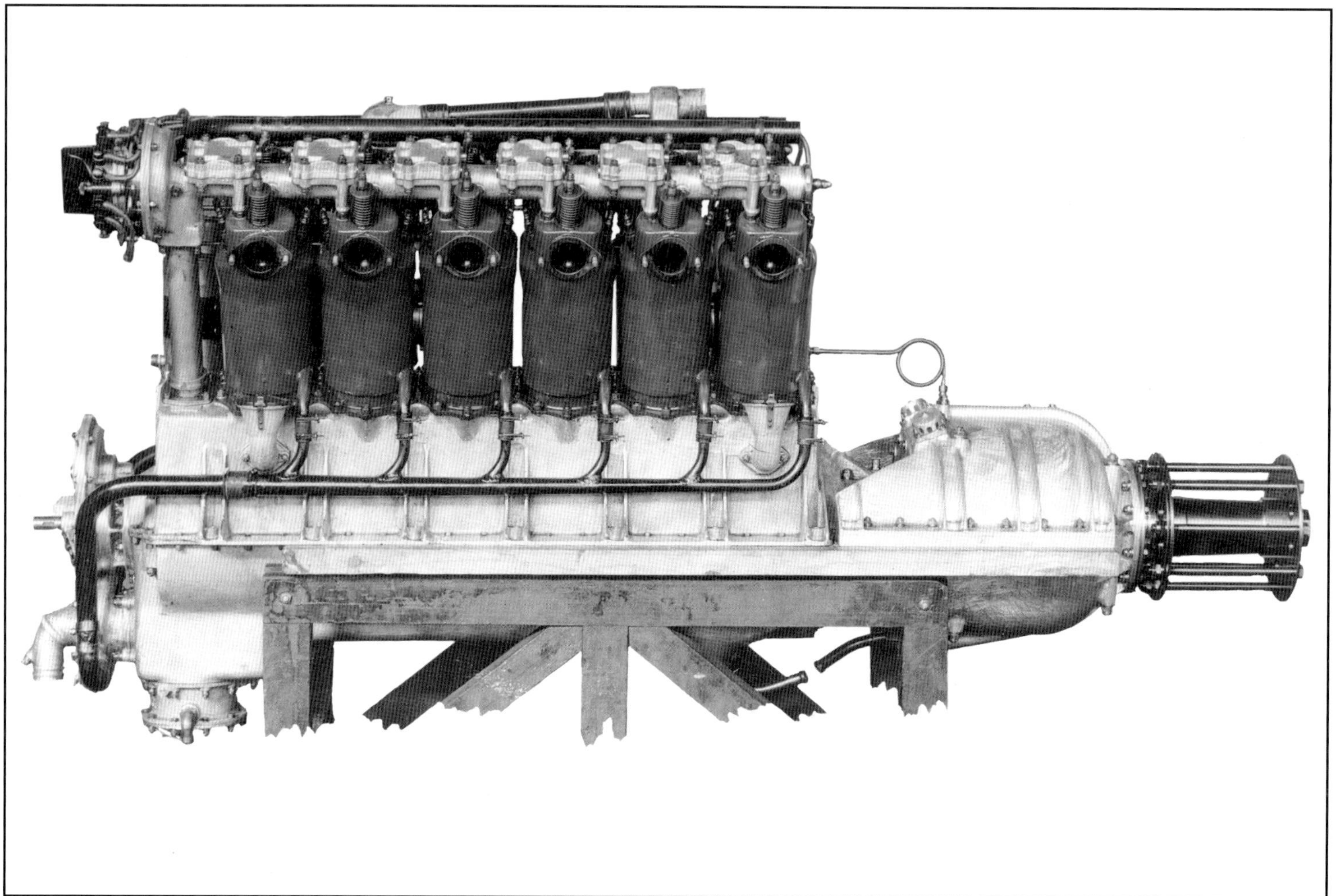

Allison Stand-Alone Epicyclic gear installed. Note the transmission is mounted on rails which are bolted to the engine. Allison records indicate they sold 20 of these units. (Courtesy of Rolls-Royce / Allison)

The Allison engineering reports on this project compare the design with that of the standard planetary geared Liberty and note the use of the same spring coupling used in the "Single Mesh Reduction Gear" contract number 51579.

Use of the spring coupling device was a significant improvement in design. Not long after this, Allison proposed to the Army and Navy that they install this spring coupling in the original Liberty 12B engines (which did not have this feature). Whether or not they did is not currently known.

Although the reduction ratio was similar to the standard Liberty 12B, the actual gearing was different and the unit was constructed for the transmission of more power. It weighed 240 pounds installed.

Allison records indicate 20 of these units were sold but no records have surfaced to indicate whether these were civilian or military sales. It would therefore appear that their anticipation of profitable sales was realized.

LIBERTY ANGULAR DRIVE TRANSMISSION

Sometime in the 1922-23 era Allison designed, at the Navy's request, a rather unique transmission to be used to drive two propellers from a single Liberty 12 engine. Although an engineering paper describing the transmission has survived, it is not dated. This fault is shared by most Allison engineering papers. Just why Allison almost invariably failed to date these papers is beyond the author's imagination. They certainly did not have future historians in mind when they were written.

The paper described it as follows: "This transmission is designed to transmit the power of a Liberty engine to two propellers located outboard, above and slightly forward of the engine.

"The engine drives a bevel gear through a flexible spring coupling. Two bevel pinions engaging with the gear transmit the power through tubular shafts to two outboard gear boxes. The design is intended for tractor propellers."

Allison two-speed spur reduction gear Liberty as it appeared in the Wright Field Museum in the 1930s where the engine was assigned number 49 in the museum inventory. Although most of the information on the data plate seen on the rear of the engine cannot be read, the large identification lines can be read as Allison Engineering Co. Although the engine obviously survived into the 1930s, its current location is unknown. (Courtesy of the Air Force Museum)

The mechanical description was as follows:

Main Gear Box:	
Engine Horse Power	450 HP
Engine speed	1800 RPM
Speed of drive shafts	2455 RPM
Weight	162 pounds
Outboard Gear Boxes	
Propeller horse power	225 HP
Propeller speed	1300 RPM
Weight each	115 pounds
Drive Shafts	
Shaft horse power	225 HP
Shaft speed	2455 RPM
Assembled weight each	38.5 pounds

The total weight of the transmission is 469 pounds and the added weight to the Liberty motor is 449 pounds.

The author has not found an airplane designed to use this power system nor do I currently believe such an airplane ever got past the design stage. One might think it would be a biplane with the engine mounted in the fuselage behind the pilot with the propellers located between and forward of the wings.

This was an interesting design which never got beyond the design stage.

ALLISON TWO-SPEED SPUR-GEAR LIBERTY ENGINE

In about January of 1923 the Engineering Division contracted Allison to refine the design of a shiftable two-speed spur reduction gear Liberty, the basic layout of which their engineers had laid down during the later part of 1922. The reasoning behind the design was clearly stated in an article carried in the February 11, 1926 issue of *Automotive Industries*.

"To make possible the full utilization of the power developed by supercharged engines at high altitude, the U. S. Air Service has built an engine of the Liberty type with a two-speed transmission. At the ground the low speed is used, the propeller turning slowly in the denser air, while the engine runs at rated speed. At higher altitude, as the engine speed increases, the pilot shifts to high gear, increasing the speed of the propeller in respect to the engine speed.

"The application of a super-charger to an airplane engine brings up unusual conditions of propeller design and operation. The supercharger maintains constant engine power with increasing altitude, while the power absorbed by the propeller at constant speed decreases with the decreasing air density. The engine speed is thus

DH-4B P-139 modified to accept the Allison Two-Speed Spur-Gear engine for high altitude testing. Seen here equipped with a 12', four-blade, left-hand propeller and special landing gear for ground clearence of the extra large propeller, radiator below the fuselage and GE Form F side turbo on the engine. (Courtesy of Peter Bowers)

allowed to increase, and would soon reach an excessive figure. Variable pitch propellers have been used, in which the blade angle is increased to compensate for the decreasing air density and so maintain the engine at rated speed at all altitudes. Propellers of this type have not been particularly successful, however; hence the engine with two-gear reductions was built."

Design Features of the Allison Two-Speed Liberty

	Low Speed	High Speed
Reduction ratio	2.03:1	1.54:1
Diametral pitch	6/8	6/8
Number of teeth	36/73	43/66
Pitch diameter	6.0/12.16	7.16/11.0
Tooth width	3.125"	2.625"
Pressure angle	20°	20°
Clutch	sliding tooth	
Total engine weight	975 lbs	

So here we see the reasoning behind the design. The variable-pitch propeller referred to was, at the time, in its design infancy and was yet to become a reliable piece of equipment. Not many years later, however, it was perfected and, along with the supercharger, allowed the piston engine to power aircraft to operational altitude of 30,000 feet and more. In the mean while the two-speed propeller drive was tried as a possible alternate.

Although the idea would seem to have some merit, and did in fact actually work, one can easily see from the following description of the mechanical design, that its use was, at best, not very practical.

"Two sets of hardened nickel steel gears are used, mounted on ball and roller bearings. The jack shaft, also mounted on ball and roller bearings, is connected to a regular Liberty crankshaft by a flexible spring coupling." Obviously this is a slight misstatement as the Liberty crankshaft was not "standard" but was shortened and adapted to the Allison spring coupling. "The driven gears float on the propeller shaft. Between the two gears the shaft is splined, and a sliding tooth clutch transmits drive through either set of gears. The gear shift mechanism is connected to a lever in the pilot's cockpit. To shift the pilot closes the throttle and with a short jerk of the lever changes from one speed to the other." This description of the shifting operation was, obviously, an over simplification. One does not change between two gear ratios connected to their drive by a one-tooth clutch by simply jerking the shift lever. The Allison description of the shifting requirements noted that a special propeller speed indicator would be desirable in order to be able to correlate engine speed to propeller speed after moving the clutch to neutral position and before shifting to the desired gear, thus avoiding clashing.

Gear lubrication was by the engine pressure oil system. The same extra-capacity oil pump developed for the single-speed spur-geared Allison Liberty was used on this engine. Oil from the camshaft oil lead is pumped into the gear-drive shaft and propeller shaft and then distributed by centrifugal force to the gears and bearings. From there it drains into the gear case and from there through a drain hole into the engine sump.

The engine was assigned AS number 23-244 in the new numbering system. As previously explained, this would indicate the 244th engine of the fiscal year 1923.

The engine was completed in late 1923 and during the first part of 1924 was equipped with a side supercharger and flight-tested in a DH-4B equipped with a four-blade 12-foot propeller and extra high landing gear designed to allow ground clearance for the extra length propeller.

Testing continued into at least 1926, by which time development of the controllable pitch propeller had advanced sufficiently to show it was to be the more practical solution to the problem.

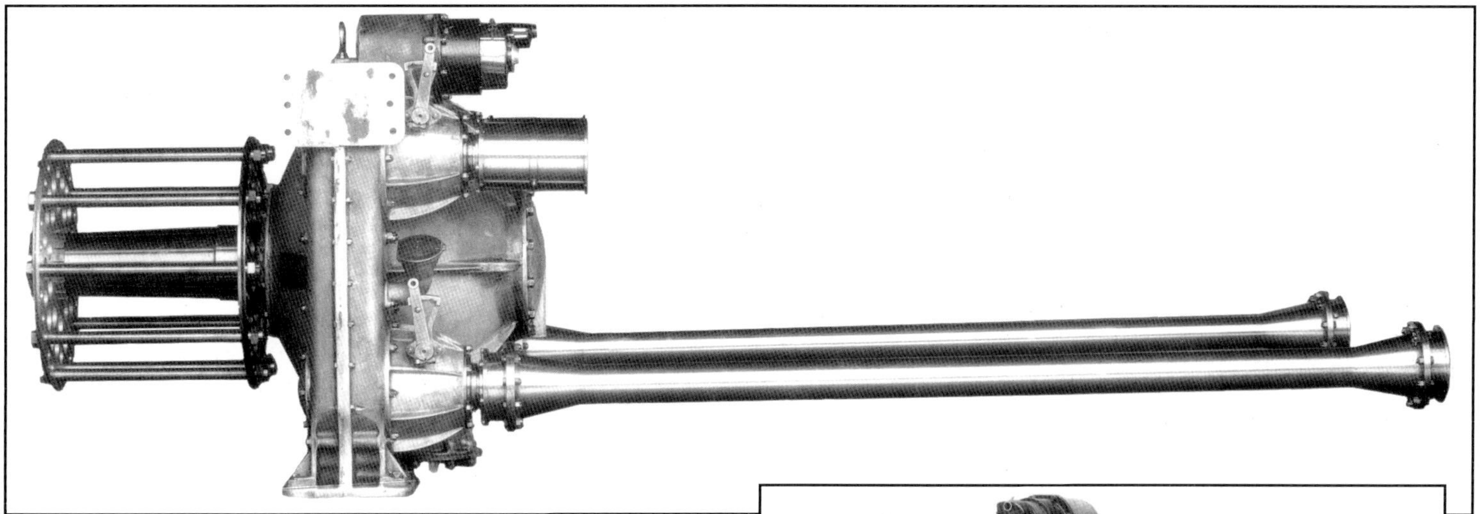

Allison Four-Liberty Gear Box. The 600-watt generator can be seen on the top of the unit and the disconnect levers of the left two engines are visible in the top view and all four in the right view. The disc clutches and flywheels were mounted on the engines and thus are not seen here. The gear housing is vented with a Liberty engine breather seen in the left-center of the unit. (Courtesy of Rolls-Royce / Allison)

THE ALLISON FOUR-LIBERTY GEAR BOX

In the summer of 1923, a short time after the Engineering Division contracted Allison to build the Two-Speed Liberty they contracted them to develop and build a transmission which could connect any one or more of four standard Liberty engines to drive a single propeller. Any engine could be connected or disconnected from the drive by its own sliding-tooth clutch.

During this period the Engineering Division had a high interest in providing high power sources for long-range heavy bombers. The source could be in the form of large engines, multiple engines or methods to power a single propeller by more than one engine. Remaining Air Service documentation indicates that it was to be for a long-range night bomber utilizing a centralized power plant which would be dimensionally larger than the Barling Bomber. Had development continued the plane would probably been assigned the designation NBL-3

The reduction ratio was an extremely high 2.94:1 which resulted in a propeller speed of only 577 RPM with the engine's operation at a rated speed of 1700 RPM. Initial design specifications had been for a ratio of 1.89:1 which would result in a propeller speed of 900 RPM at that engine speed, but those were changed before actual construction.

The transmission was built into a large rectangular aluminum case which carried a central large spur gear which drove the forward-mounted propeller. This gear was driven by four pinions grouped around the spur gear at the four corners of the case. The spur gear had 212 teeth and the pinions each had 72. Each pinion was driven by a standard Liberty 12 engine through a sliding tooth clutch and an Allison spring coupling. Two of the engine drives were also supplied with Allison multiple disc friction clutches so that either could be used to initially start the propeller.

All engines were equipped with electric starters. The two engines without disc clutches were equipped with flywheels weighing 53 pounds each. On the other two the 121 pound clutch assemblies acted as flywheels. So equipped, engines could be started and stopped at any time while disengaged, brought up to the speed (within 50 RPM) of the engine or engines driving the propeller and then engaged with its sliding tooth clutch. The transmission was lubricated with a dry-sump type oiling system and was provided with an external oil tank and its own oil pump which acted as both a supply and sump pump. A 600-watt generator was mounted on top of the transmission case and was driven from the spur gear.

Each clutch was controlled by an individual lever which allowed any one or more engines to be engaged or disengaged from driving the propeller. Since all engines were mounted within the fuselage of the aircraft, any engine could be disengaged and stopped to allow maintenance in flight.

The total weight of the unit without the generator was 875 pounds, 146 pounds of which was drive shafts, coupling and adapters.

The transmission was subjected to and passed its 50-hour acceptance test in the Allison factory during July and August of 1924. The test consisted of ten five-hour runs of four hours at 90% power, and one hour at 100% power. The unit passed the test with no significant problems. Over the test period, however, the Liberty engines suffered from a number of cylinder water jacket leaks and several plugs were changed because of fouling.

The test was run driving a 15'5" club to absorb the power. The size of the proposed operational propeller was not stated in the test report but it would have to have been very large. Although the Air Service had some basic

The first inverted Liberty 12A under test at McCook Field in February of 1919. The engine ran well and during its brief test period the only problems noted were excessive heating of exhaust valves and plugs and high oil consumption. (National Archives photo courtesy of Kim McCutcheon)

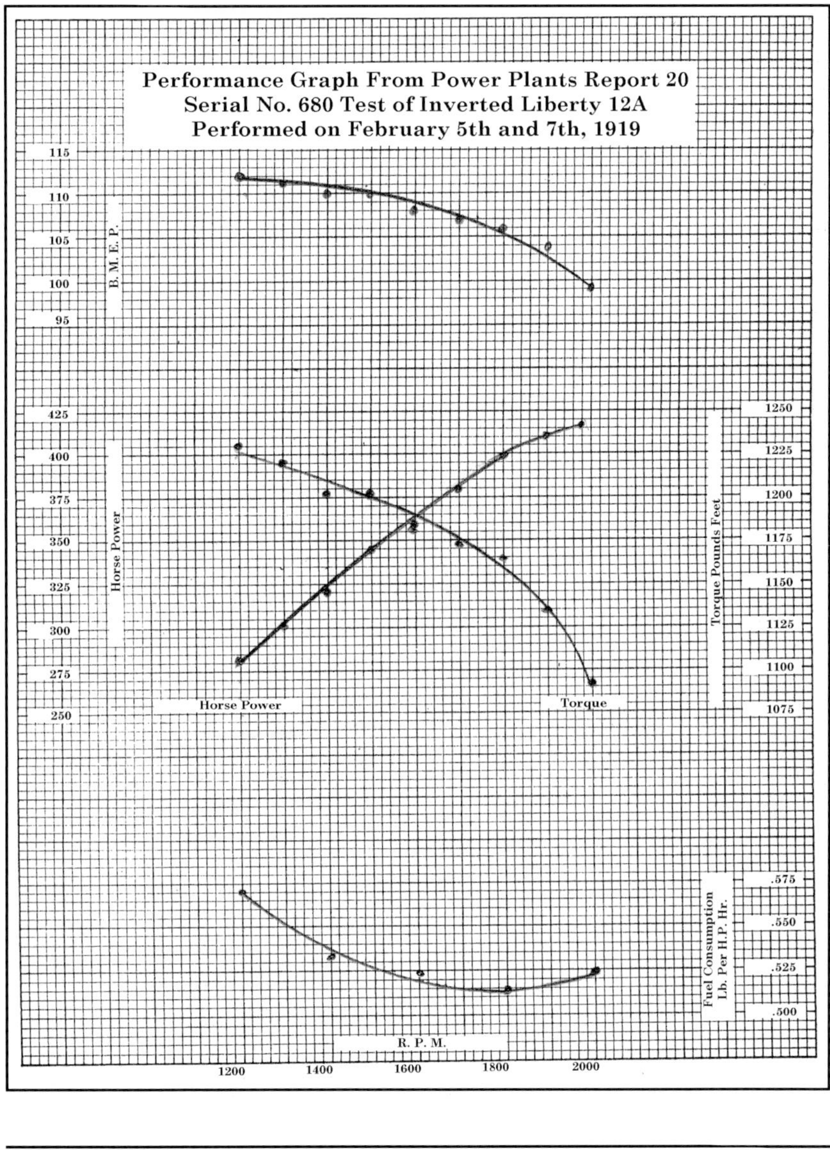

Rear view of the first inverted Liberty 12A under test at McCook Field. Re-routed water lines can be noted in this and the above photograph. (National Archives photo courtesy of Kim McCutcheon)

aircraft in mind when this transmission was contracted, the author has never seen any design information regarding it and such an airplane was never built.

THE ALLISON ALUMINUM LIBERTY

In late 1922 or early 1923 the Navy initiated the development of a major improvement of the Liberty engine in the form of replacement cylinders. One of the most significant improvements in water-cooled aeronautical engines to arrive in reliable form since the Liberty was designed was the aluminum cylinder block with its overhead valve system enclosed from the elements.

Although the Liberty was being made more reliable all the time, and would make another significant improvement when the Allison steel-backed bearing came into use, it would always suffer from leaky sheet metal water jackets on its individual cylinders, and wear and oil leaks from its exposed valve train. It was these two shortcomings the Navy was trying to eliminate.

The Navy contracted with three of the four aero engine companies it was currently doing business with to design new aluminum cylinder blocks for the Liberty. These were Wright, Aeromarine and Allison. Just why they chose not to also get Packard working on the project is not known. Perhaps they did and Packard was not interested.

I have already covered the subject and the designs these three companies came up with in the prior chapter, but will give a brief summary of the Allison effort here as well.

Surviving Allison documents indicate a design was made. Nothing has so far been found to indicate an engine of this design was actually constructed. I will not discount the possibility as it is known that Aeromarine did build an example and it is strongly suspected that Wright did. There is a photograph of the Aeromarine engine but so far none have been found of either a Wright or an Allison engine.

The surviving Allison design paper describes the engine as having the individual cylinders replaced by six-cylinder cast aluminum banks which utilized wet-type steel cylinder liners screwed into the aluminum casting at the cylinder heads. Cooling water was retained at the lower end of the liner by rubber rings. Combustion chambers were aluminum with aluminum-bronze or cast iron valve seats, either cast in or pressed in. The cylinder banks bolted directly on a standard crankcase.

Two intake and two exhaust valves per cylinder were used. These were driven by a single overhead cam operating forged aluminum rocker arms. The cam assembly was driven by the standard Liberty drive train.

Carburetion was to be either by the two standard Zenith carburetors or a single large carburetor of similar design as the one used on the Allison Twelve marine engine of 1919-20. The intake manifold used with the Zenith carburetors would not be water jacketed as Allison noted their short design would not require it. If the single Allison marine type were used the manifold would be jacketed.

Standard Liberty water and oil pumps and all other accessories would be used. It was noted that any Liberty could be converted to an Allison Liberty in any flying field or repair depot.

For more details of the design the reader is referred to page 278. In outward appearance the engine would probably look much like the Aeromarine "N.A.L." pictured on page 278 but with no carburetors showing and with two exhaust ports per cylinder rather than one.

THE INVERTED WATER-COOLED LIBERTY

The history of inverting the Liberty begins, of course, with the development of the X-24 engine. This was a project proposed by Col. Vincent and assigned by him to Captain Lionel M. Woolson, test engineer in the Engineering Division. It was probably initiated in early November of 1918 because it was completed and power curves run during the first week of December. I described the engine and its testing in some detail in Chapter 5 (see pages 191-197).

The engine suffered a rod failure which ended the test. The Armistice had been signed on November 11, right in the middle of this project. It was carried to completion because it had been well under way. However, the need for early development of an engine of this power was no longer pressing. Thus the lessons learned in the test were duly noted and the X-24 project abandoned.

Probably the most significant lesson learned was that the Liberty lent itself well to operating in the inverted position. Although the need for a 700-800 HP engine waned with the end of the war, the idea of inverting the Liberty seemed inviting enough to cause the Engineering Division to almost immediately overhaul a standard Liberty 12A engine and make it ready to test on a dynamometer.

A STANDARD L-12A IS RUN IN THE INVERTED POSITION IN 1919

The following modifications were made to the engine. Other than being mounted on the dynamometer cradle upside down, it was in standard configuration.

1- The water circulation system was reversed. Water flow from the pump was piped to the lowest point on the engine - the intake headers - and admitted through the elbow, which in the standard assembly is the water outlet.

The water inlet from the cylinders was through the water manifold which, normally, is the water inlet.

2- No crankcase breathers were used because of the narrowness of the dynamometer cradle, but the sump cap at the propeller end of the lower case was removed to allow the engine to breathe at this point.

3- The inlet headers were standard, but "inverted" Zenith carburetors were used which fed gasoline downward into the intake headers. No air scoops were used, air being drawn directly into the venturi out of the "Vee" between the cylinders.

4- The standard cylinders extend into the crankcase and create a shallow sump when the engine was in the inverted position. No special provisions were made to keep oil out of the pistons.

The engine was tested on the dynamometer on February 5th and 7th of 1919 and complete data recorded relative to power, torque, B.M.E.P., fuel and oil consumption, and friction horsepower.

The engine ran well with the exception of trouble experienced with overheating of exhaust valves and spark plugs. A number of plugs had to be changed during the test and exhaust valve clearance had to be adjusted on several occasions to compensate for gradually reduced clearances caused by the elongation of the valve stems from heating. No trouble was experienced with plug fouling from over oiling even though that had been an expected problem.

The only problem other than that experienced with overheating valves and plugs was excessive oil consumption, which was approximately five times normal.

The overall conclusions drawn were that the Liberty was well adapted to running in the inverted position and performed well but, to be practicable, methods would have to be developed to overcome the valve and plug heating problem and to reduce oil consumption. Although these tests were relatively short in duration, they felt they proved that inverted operation would not present very many problems.

NEW INTEREST IN INVERTING THE LIBERTY COMES IN 1923

Active interest in this project ceased at this point. It would be almost exactly four years before the inverted Liberty was taken up again. As we will see, when more extensive testing was carried out a number of additional problems needed resolution before such operation would be relatively trouble free.

The encouraging results of the February 1919 tests were duly recorded and without doubt noted by a number of Engineering and flying personnel in the Air Service. The project, however, was shelved awaiting perhaps future interest and funds. Funds would be found very short for a number of years but renewed interest in the project was expressed in January of 1923.

Just what initiated this interest is not recorded, but at that date the Chief of the Air Service, Major General Mason M. Patrick, sent to the Engineering Division a formal order to rework a Liberty for inverted operation and install it in an aircraft for flight testing.

On February 10, 1923 work was begun on converting a Packard engine for operation in the inverted position. At about the same time modifications were started on DH-4B AS 22-528 P-256 to receive the engine.

Experience gained from the 1919 conversion led to additional modifications to this second engine. 1- A gear-type fuel pump, installed to act as an auxiliary oil pump,

Here are three views of the modified Packard engine of 1923. Most of the external modifications mentioned in the text can be seen in one or more of these views. (National Archives photos courtesy of Kim McCutcheon)

P-256 at McCook Field with inverted Packard Liberty installed. The plane has been made ready for its first flight with Lt. James Doolittle at the controls. Only the cowlings remain to be installed. (Courtesy of Peter Bowers)

was used to scavenge oil from the camshaft housings, they now being on the bottom rather than top of the engine. 2- Adapters were fitted between the distributor heads and the camshaft housings to prevent oil from entering the distributor heads. 3- Three fourth-inch holes were drilled through the main bearing webs at a position just between the cylinder barrels to allow oil to flow from one end of the crankcase to the other. In the inverted position the former upper crankcase half now formed the sump with the cylinder barrel extensions into the crankcase keeping oil from draining into the cylinders. 4- The oil scavenging pump line (normally on the bottom) was tapped at the front end and a half-inch line carried to the bottom of the vee to pick up oil. A second scavenge line was connected from the gear compartment to the scavenging pump housing. 5- An auxiliary Hispano scavenging pump was mounted on the rear of the crankcase and driven directly by the crankshaft. This pump picked oil up from the crankcase and delivered it to the gear housing. 6- A pair of baffles were installed to prevent oil from working into the generator. 7- The intake manifolds were fitted with adapters to allow correct mounting of standard Zenith carburetors. 8- Water flow through the water jackets was reversed. 9- The standard Liberty oil breathers were removed and the holes plugged, and a Packard filler was placed on top of the crankcase where the drain and bail cap is normally found.

Although these changes permitted reasonably good operation, it was soon found that a number of other modifications would eventually be found advisable.

On February 23 the engine was sent to the dynamometer lab for tests. Testing and minor modifications continued

C-6 scavange pump for the distributor heads as installed in late 1923. (National Archives photo courtesy of Kim McCutcheon)

XCO-6 P-360 AS 23-1236 shown here with an inverted water-cooled Liberty as power. (Courtesy of Peter Bowers)

for the next few months. The engine was delivered to the final assembly department on June 14, 1923 for installation in DH-4B P-256 in preparation for flight testing. The first flight took place on September 5, 1923 with Lt. James Doolittle as pilot with a flight duration of 14 minutes. Several other short flights were made in the next few days with other pilots

The first cross country flight was made on September 9 by Lt. F. O. Carroll from McCook Field to Toledo. A forced landing was required due to an oil leak in the pipe connection between the oil pump and the oil tank. The McCook Field civilian engineer in charge of the development of the inverted Liberty was Mr. F. W. Heckert. He not only directed the project but often flew as observer and filed most of the official reports.

Flight testing showed further work would be required to handle the continued seepage of oil from the camshaft housing by the slinger which would eventually fill the distributor heads. The first modification to correct this problem was to mount a C-6 fuel pump below the distributor heads in the vee driven by a flexible shaft, this pump acting as a scavenge pump for the two heads. This change corrected the problem sufficiently well to allow the accumulation of about 100 hours of flight-testing in P-256.

Other changes were made, including the removal of the top-mounted Packard breather and in its place two standard Liberty breathers were mounted on extensions (to clear the engine bearers) and inverted.

P-256 was flown to the Pulitzer races in St. Louis which were held on October 4, 5 and 6, 1923.

In the mean while, work was progressing in the building of the Engineering Division XCO-6 and Boeing XCO-7, which would both be used in testing inverted water-cooled Liberty engines and also the inverted air-cooled Liberty. In addition, Grover Loening had been contracted to build the COA-1 amphibian for the inverted water-cooled engine.

A second engine was ordered built up in inverted configuration (Buick AS 37873 GY). On January 28, 1924 the work was reported as going slowly because "practically the entire mechanical personnel has been engaged on overhauling and reconditioning engines for the World Cruise." It was completed and given a full power run on February 26.

This engine was installed in P-256 and its engine transferred to Loening for installation in the COA-1. The airplane was ground-tested on May 29 and 31, 1924. The plane was put on water on June 7 in preparation for flight testing. While taxiing for takeoff it hit an obstacle in the water and was wrecked. As a result, no further flight tests were attempted until January 18, 1925, at which time the plane was flown from Mitchel Field to Bolling Field, D.C. On February 2 it was flown from there to Langley Field and then on to McCook Field on February 13. It was given an official performance test on March 16, 1925 (Reported in Air Service Circular No. 540).

An additional four engines were ordered converted and by the end of August of 1924 they had been finished and tested, giving a total of six since the project was restarted in 1923. Nothing more seems to be heard of the original 1919 engine. Those six engines are listed below in the order I believe they were built and one of the airplanes it powered is also listed. Some powered several different planes over their life.

XCO-7B shown above with inverted L-12 installed. At right is seen the engine as modified for this airplane. Note the generator and distributors mounted on top of the engine. Split oil pumps occupy the housings on the ends of the camshafts and the water pump is located where the generator once was. (National Archives photos courtesy of Kim McCutcheon)

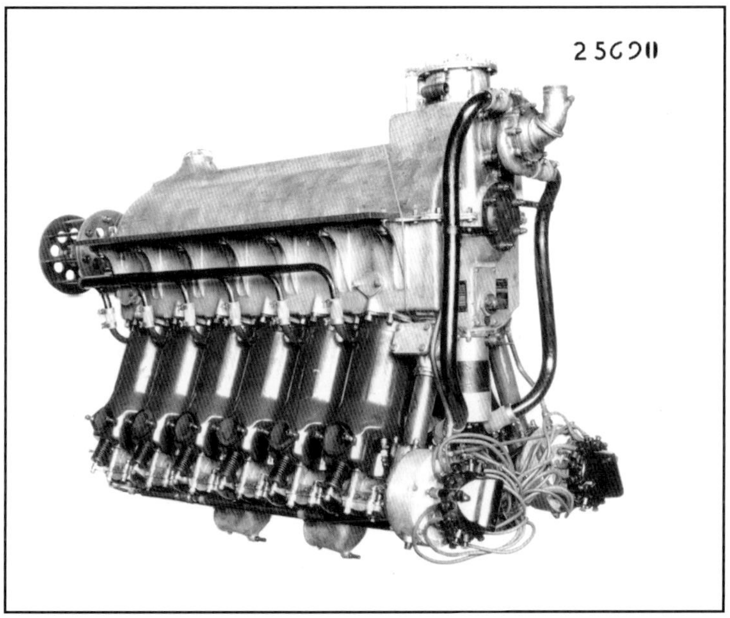

The two pictures above show (left) the first change from the top-mounted generator and distributor which Loening objected to. The water pump, back at its standard position, resulted in steam pockets so it was moved back to its former position in the vee and the generator was mounted in the old water pump location (right). Note that both engines are Allison production engines carrying their data plates. (National Archives photos courtesy of Kim McCutcheon)

Packard	AS 21996 Y	Mfg. 4567	XCO-6
Buick	AS 37873		DH-4B
Lincoln	AS 26281 Y	Mfg. 2835	XCO-7B
Ford	AS 29533	Mfg. 40	COA-1
Packard	AS 71614 Y	Mfg. 6369	
Packard	AS 21930	Mfg. 4560	COA-1

AS 21930 was shipped to Loening for installation in their second plane.

By June of 1924 an order had been placed with Allison for the production of 20 engines of standardized design. The Loening COA-1 prototype had been ordered in 1923 (23-1234) and three additional in 1924 (24-6,

24-7 & 24-8). In 1925 an additional nine production models were ordered (25-226-234).

The engine to be installed in the XCO-7 was modified in several respects in an attempt to both fix the camshaft housing oil problem and make the engine a better fit to the airplane. The oil pump was removed from the top of the engine and replaced by a split pump with the scavenging half mounted on the end of the left camshaft, and driven by it, and the pump section on the right camshaft. The distributors were moved to the top of the engine in the prior location of the oil pump. The generator was moved to the same location. The generator was replaced by a drive extension on the bottom of which was mounted the water pump. The pump's high location had been found to sometimes result in steam pockets; the lower location eliminated this problem as well as allowing better pump efficiency.

Loening, however, objected to this arrangement on his airplanes as the distributor/generator location reduced pilot visibility. This modification also increased the engine weight to 960 pounds.

Design standardization had surely not yet arrived. Changes were made in the Allison engines both during and after production. The next modification was to put the generator back in its original location, put the original oil pump back in its original location, put the water pump back in its original location, build extension housings for the ends of the camshaft housings in which were mounted small sump pumps for clearing the leakage from the housings. Drive extensions passed through these extension housings to drive the distributors.

Once again the steam pocket problem arose and thus the water pump was once again moved to the old location of the generator. This time the generator was moved to the water pump location.

One additional variation is known which must have been generated at about this time. It was probably tried as an alternate to the generator and distributors on top of the engine. In this case the same split oil pump mounted on the ends of the camshaft housings was used along with the water pump in place of the generator. The distributors were moved to auxillary drives in the center of the camshaft drive shafts. The generator was driven by a takeoff at the base of the extension for the water pump. With this setup the oil pump is not on the top of the engine which allows a lower engine cowl. The only picture of this variation the author has seen is used in an Allison ad of September 1, 1928 and it has no contract plates and thus may represent an Allison design which did not become a standard. (See the reproduction of the ad at right for an illustration.).

THE STOUT METAL AIRPLANE CO.
AND FORD REQUEST INFORMATION

In the Engineering Division history report on the inverted Liberty is the note that Mr. W. B. Stout of the Stout Metal Airplane Co. and Mr. H. A. Hicks of the Experimental Department of the Ford Motor Co. came to McCook Field on April 3, 1925 to obtain as much information as possible on the Inverted Liberty Twelve

September 1, 1928

ALLISON
Engineering Company
INDIANAPOLIS

We Design and Build
Airplane Engines
Reduction Gears
Superchargers

Eleven years' experience in the development of Aircraft Power Plants and Power Transmissions.

Allison Steel Back Bearings

ALLISON V1650
An interesting development of the Liberty Engine for inverted installation.

Loening CO-A1, the first of a series of this type for which the inverted water-cooled Liberty was built. Early in the development of this engine Engineering Division officers were expressing the opinion that it was so promising that it seemed likely that all future Air Service observation and pursuit aircraft might be powered by inverted engines. They also expressed the opinion that the engine had great commercial potential. Neither prediction, of course, materialized. (Courtesy of Peter Bowers)

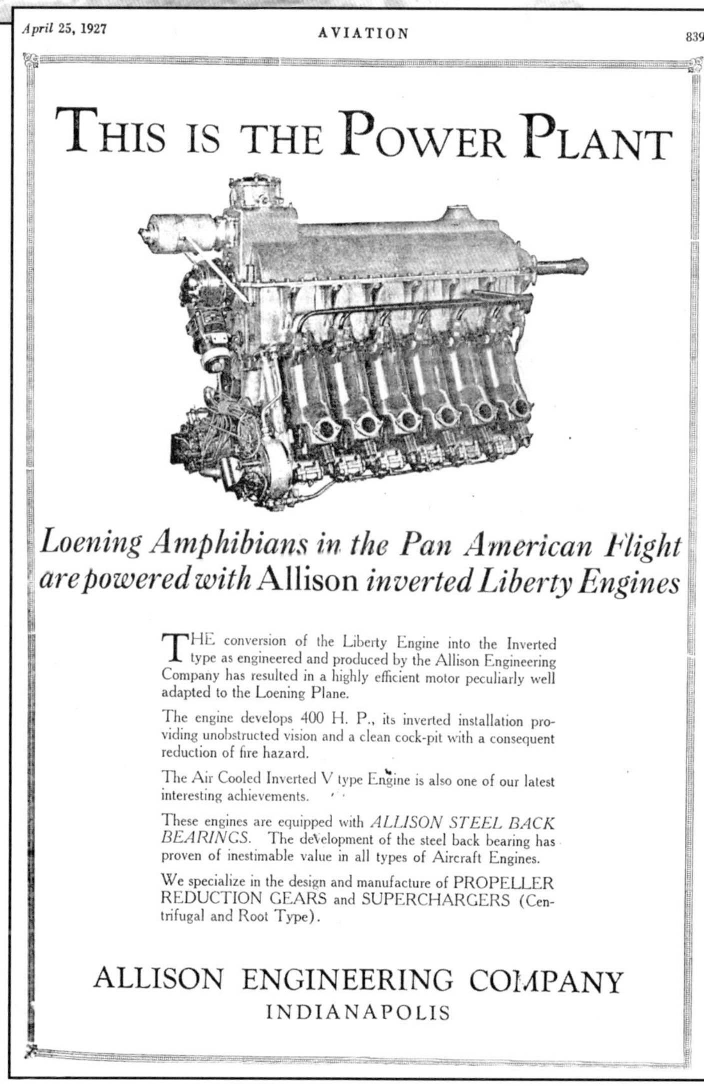
April 25, 1927 AVIATION 839

THIS IS THE POWER PLANT

Loening Amphibians in the Pan American Flight are powered with Allison *inverted Liberty Engines*

THE conversion of the Liberty Engine into the Inverted type as engineered and produced by the Allison Engineering Company has resulted in a highly efficient motor peculiarly well adapted to the Loening Plane.

The engine develops 400 H. P., its inverted installation providing unobstructed vision and a clean cock-pit with a consequent reduction of fire hazard.

The Air Cooled Inverted V type Engine is also one of our latest interesting achievements.

These engines are equipped with *ALLISON STEEL BACK BEARINGS*. The development of the steel back bearing has proven of inestimable value in all types of Aircraft Engines.

We specialize in the design and manufacture of PROPELLER REDUCTION GEARS and SUPERCHARGERS (Centrifugal and Root Type).

ALLISON ENGINEERING COMPANY
INDIANAPOLIS

development. They requested complete sets of blueprints and parts lists and the use of any patterns needed to cast special parts needed for the engine.

Permission was given and orders issued to produce and ship to Dearborn, Michigan the requested information. Stout and Ford thus had all the information they needed to enable them to convert standard Liberty engines to inverted configuration.

At this time Stout Metal Airplane was just putting into production the Liberty-powered 2AT with probably three having been finished. This airplane used a standard upright Liberty. All 2AT airplanes produced used standard Liberty engines. If Stout had plans for the use of an inverted Liberty it has not surfaced and the author has no information regarding any attempt to build such an engine by Ford or Stout.

BLUEPRINTS AND 6 ENGINES TO THE NAVY

On April 24, 1925 the data requested by Stout was assembled and made ready to ship. One set was assembled for Stout, two sets for the Navy and five sets for the Engineering Division Shop and files.

On April 28 the Factory Planning Branch issued Experimental Order 730-192 covering the building of six sets of parts required to build inverted Liberty engines by the Engineering Division to be forwarded to the Navy as soon as possible.

It is about this time that the Engineering Division history ends. The engine has been developed to a production state and blueprints and specifications have been sent to the Navy, Stout Metal Airplane, Allison and the Engineering Division. This is not to say that no further changes were made, but all major ones had been done.

In 1925 the Navy contracted Loening to build five OL-2 aircraft for them. These were the Navy equivalent to the Army COA-1. The six inverted engines being built by the Engineering Division were for these airplanes. Just why these were built by the Army instead of Allison is not known. In 1926 the Navy contracted Loening to build six OL-2 models. There is no record of Allison contracting engines for these either and even though I currently have no direct evidence they were built by the Engineering Division, logic says they probably did build them. The information I currently have makes no further mention of these engines. However, it would seem quite likely the six inverted engines to be built for the Navy would be built around Navy stock engines sent to the Engineering Division for that purpose. Why should the Navy use (and pay for) Army engines when they had a large surplus of their own they would like to either use or sell on the surplus market (which was not very active)?

Since the Navy ordered a total of eleven Loenings, which were the users of this engine, they probably had an additional six engines built later. If Navy engines were used for these twelve then the engine's government serials would be in the Navy series and start with a 0 (zero).

ALLISON BUILDS THE INVERTED LIBERTY 12

As I have already noted, Allison was contracted to build 20 inverted Liberty engines in about June of 1924 (order 2159-24). Some time during Fiscal Year 1926 they were contracted to build another 15 engines. Contract price of these engines was $1,472.00 each. So far as is currently known these 35 engines were all that were contracted. By July 1, 1925 (start of Fiscal Year 1926) probably no more than 10 of the 20 engines ordered from Allison in June of 1924 had been delivered. The 15 ordered in 1926 no doubt were produced as a continuation of the earlier order. Allison contract production therefore would have run from late 1924 to late 1926.

Engines built to the standards of June 1924 weighed 898 pounds without the starter. All of this late series of engines, including those built by the Engineering Division, were equipped with Allison Steel-Backed rod and main bearings in addition to all changes incorporated in the standard "1921 Model." Allison-built engines all incorporated the air scoops and fog-crankcase ventilation pipes seen on the engine pictured in the September 1, 1928 Allison ad on page 310. All previously built Engineering Division engines would have eventually been retrofitted to the latest standards.

Allison initiated several changes during production which improved performance and reliability. A large portion of these changes involved improvements in various methods of shaft packing and provisions for draining accumulated oil from various unwanted locations. In addition they initiated the installation of sil-chrome tulip exhaust valves and Randall Graphite exhaust valve guides on all engines.

During 1927 the Air Service initiated a project of replacing the US-52 carburetors used on these engines with Stromberg NA-Y5AS models. This carburetor had about the same performance as the US-52 but also had an idle adjustment which the former lacked. What instigated the change was the availability of NA-Y5 and NA-Y5A carburetors removed from Curtiss D-12 engines under a project to upgrade these engines to NA-Y5D carburetors. These carburetors were campaigned to modify them to new NA-Y5AS standards to adapt them to the Liberty.

EXTENSIVE CIVILIAN AND MILITARY USE PREDICTED

By the time it seemed that inverting the Liberty was going to meet with success the Air Service began to express high hopes for the inverted Liberty as a service engine as well as more-modern inverted engines which might follow. In an Air Service News Letter written by F. W. Heckert on May 31, 1924 for public release he made the observation that the Liberty was particularly adapted to inversion because of the extension of the cylinder barrels into the crankcase which provided an effective sump in the inverted position. In addition, operation in this position seemed to be bound for success from testing carried out up to that time. He noted as well that the Curtiss D-12 would also be amenable to the inverted position and experiments with its operation in that position would shortly be made. Although I feel sure it was tried, it evidently did not prove to have promise as it resulted in no extensive following project as it surely would have since the D-12 was a service engine which saw extensive and very successful use.

In addition Heckert noted "In view of the success of this equipment, plans have been made to build four Liberty engines, especially designed for the inverted position and should these prove successful, it is possible that

KNOWN SERIAL NUMBERS OF INVERTED WATER-COOLED LIBERTY ENGINES
(Of the 59 engines built, serials of 39 are known)

AS Serial	Original builder
21996	Packard
37873	Buick
26281	Lincoln
29533	Ford
71614	Packard
21930	Packard
(The above six are the original 1923-25 conversions)	
31209	Ford
31606	Ford
23865	Lincoln
25289	Lincoln
25424	Lincoln
5169	Nordyke & Marmon
(The above six are early production engines by Allison)	
26222	Lincoln
26228	Lincoln
26385	Lincoln
26821	Lincoln
27093	Lincoln
27108	Lincoln
27277	Lincoln
27655	Lincoln
28156	Lincoln
28158	Lincoln
28385	Lincoln
28438	Lincoln
28582	Lincoln
28645	Lincoln
28735	Lincoln
29018	Lincoln
29217	Lincoln
29387	Lincoln
77324	Lincoln
77347	Lincoln
77385	Lincoln
77386	Lincoln
77409	Lincoln
77420	Lincoln
77445	Lincoln
77477	Lincoln
77599	Lincoln

The final 27 engines listed above were all in storage at Little Rock but none of the first 12 listed came from there. Why they picked Lincoln engines for most of the production conversions is not known. These numbers would indicate that, for most production they picked new Lincoln engines from storage.

the inverted engine will be adopted as standard for all corps observation and pursuit type airplanes."

In a history of the inverted Liberty compiled on May 9, 1925, for Major J. F. Curry, head of the Engineering Division, First Lt. C. H. Ridenour, Acting Chief of the Power Plant Section, also noted the advantages of the inverted engine and expressed the opinion that it would "find a very decided favor" as a commercial aviation engine.

The inverted Liberty was successful and there were other following inverted engines, notably the inverted air-cooled Liberty, the inverted Packard 1500 models and the inverted air-cooled Wright 1460. None, however, were built in quantity and a complete switchover to inverted types for the observation and pursuit types never occurred.

Allison took note of the observations stated above and evidently had similar feelings. They began to use the inverted Liberty in their advertisements as early as 1927. The larger aero magazines carried periodic listings of available American commercial aircraft and engines and their basic specifications from 1919 until the advent of WW II. They solicited this material from the builders. Allison noted the availability on special order (no price stated) of the inverted water-cooled Liberty, which they called the Allison V-1650, from 1927 through about 1932.

We have been fortunate that we have found an Allison internal document which appears to have been generated in 1929-30 just after the death of James Allison when the company was on the market which contains an itemized list of "Work Designed, Built & Sold From 1919 to 1929" as well as design projects and work in progress in 1929-30. Names of the items and numbers built are noted. It listed 40 inverted water-cooled Liberties as built. Since we know 35 were built on Air Service contracts, then 5 must have been built for commercial sale.

TOTAL INVERTED WATER-COOLED LIBERTY PRODUCTION

Therefore, to recap, total known production of this engines was:

Air Service:	1919	1
	1923-1926	18 (12 Navy engines)
Allison:	1924-1929	40
	Total	59

This engine was known by a number of different names and this can sometimes be confusing. Allison called it the Allison V1650. It is also sometimes referred to as the IV-1650. In this case I believe the name has been applied "after the fact," so to speak, perhaps occasionally in government references but mostly by later historians who applied the name as the logical one for the engine to distinguish it from the standard engine. In documents of the period Air Service material invariably refer to it as the Inverted Water-Cooled Liberty. And lastly, it was assigned the designation of V-1650-13 in the Air Force publication *Model Designations of USAF Aircraft Engines* which I discuss on page 288. In that publication it was also noted that prior to the 1933 system the engine was called the IV-1650.

Although considered as distinct models of the Liberty, these engines were not assigned any distinctive serial range of their own. All were conversions of standard L-12A engines. In all official documents they were referred to by their originally assigned numbers. They retained the data plates given them when originally manufactured. Using information gathered from a number of surviving documents we can assemble a list of most of their numbers, which I have done on the prior page. These represent known Army engines, I have no reports of Navy ones. If the Navy engines were converted from Navy engines, as I suspect, then their serials would start with zero as all Navy Liberty engines (except a few very early ones) carried Navy numbers.

THE AIR-COOLED LIBERTY

In January of 1924 the Engineering Division assigned to S. D. Heron the project of design, construction and tests of air-cooled cylinders for the Liberty engine. The intention was primarily to prove whether or not it was feasible to operate a modern in-line engine with air cooled cylinders.

The cylinder design was based on the type "J" air-cooled cylinder. New camshafts with longer duration and higher lift were designed as were new camshaft housings, rocker system, pistons, piston pins, bushings for the small end of the connecting rod and air-cooled cylinder. Standard Liberty crankcase, rods, vertical cam drive shafts and ignition system were used. A gear-driven centrifugal

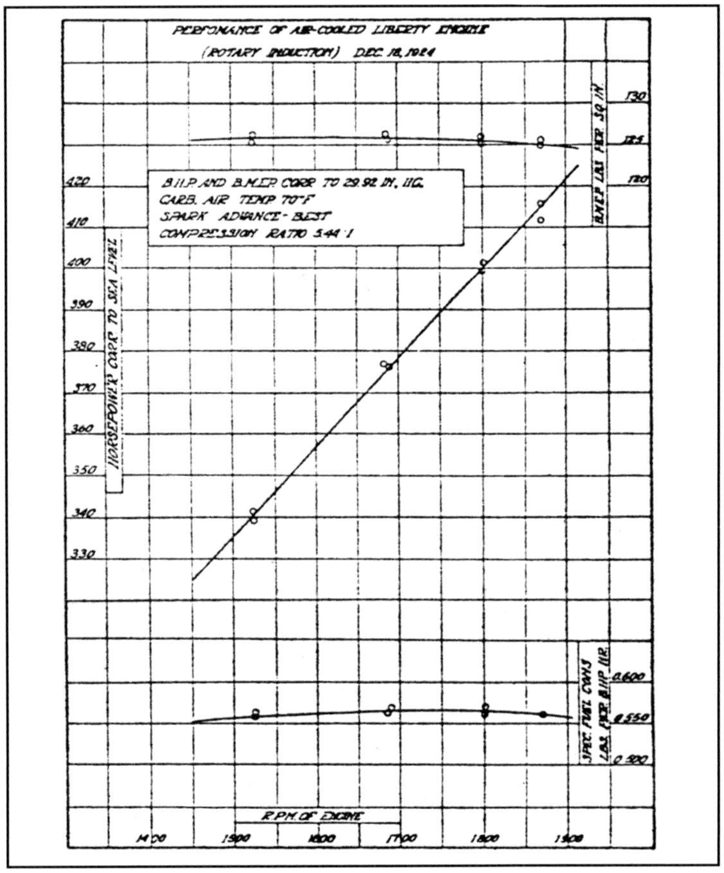

December 18, 1924 power graph on the first air-cooled Liberty with rotary induction. Although the supercharger used had clearances too great to enable it to produce any boost, it gave good fuel distribution and prevented power drop off at higher RPM. BMEF remained relatively constant all through the RPM range. (Author's collection)

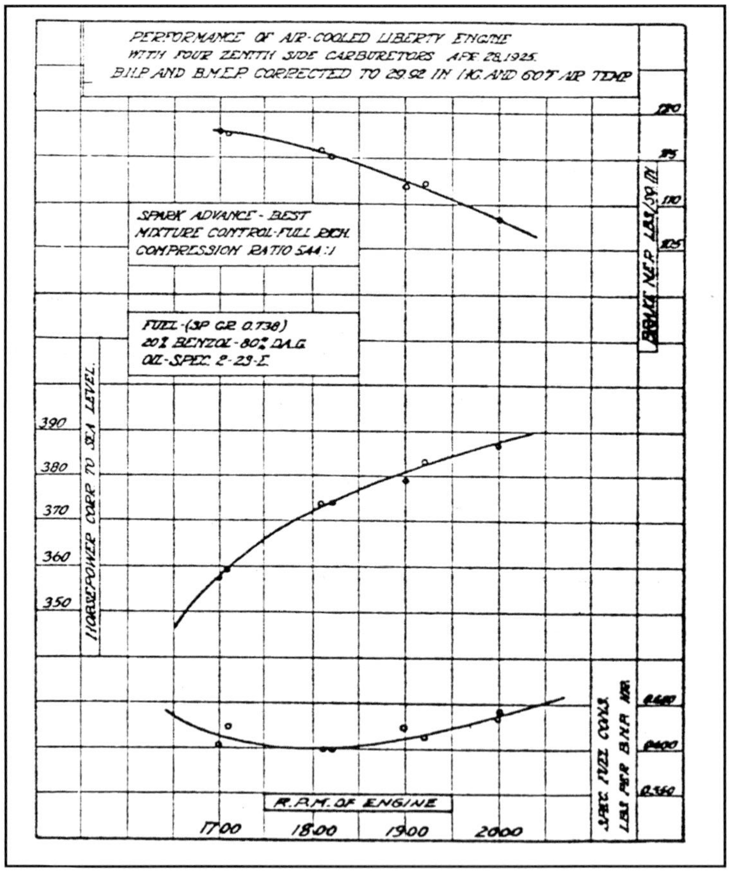

April 28, 1925 power graph of the first air-cooled Liberty when equipped with four outside-mounted Zenith carburetors in place of rotary induction with a single carburetor. The graph looks more like one would expect from a 1,410 cubic inch long-stroke engine running out of breathing ability at the higher RPMs. (Author's collection)

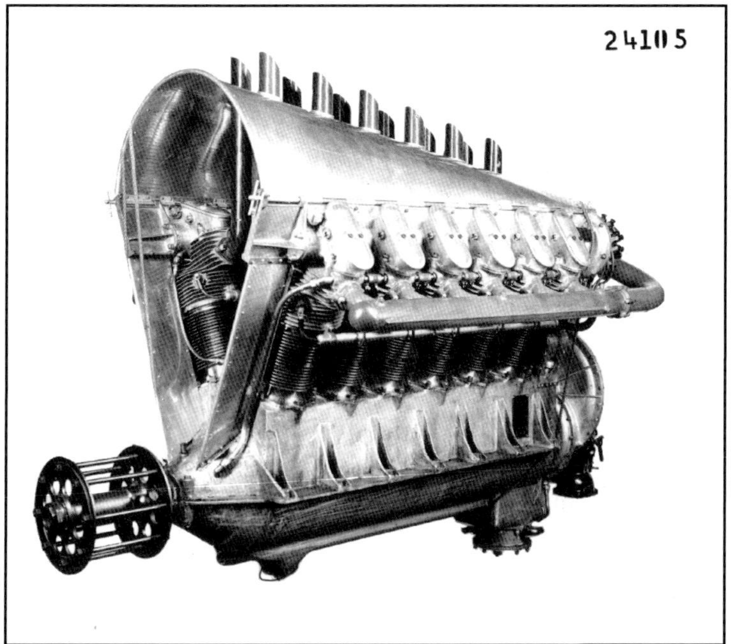

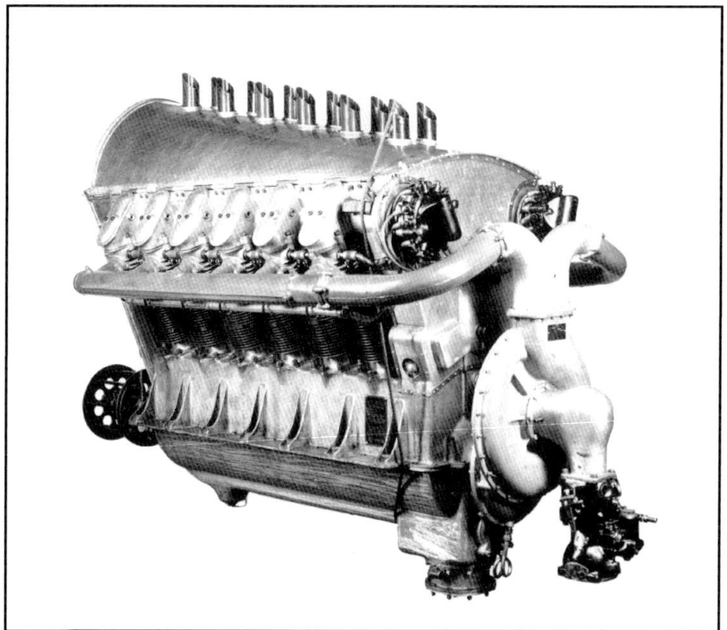

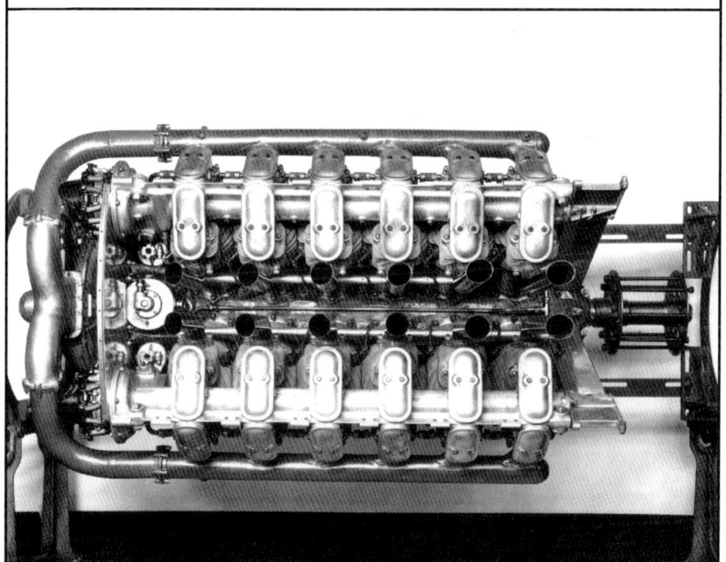

All illustrations on this page are of the first air-cooled Liberty built. This engine was assembled by the Engineering Division with the necessary conversion parts contracted to Allison. This engine was designed to be operated in the upright position. The top two and right-center photographs show it with rotary induction and as originally assembled. After repeated failures of the supercharger drive the induction system was replaced with four outside-mounted carburetors, as shown in the left-center and right-bottom photographs, to allow continued testing while the drive was redesigned.
(National Archives photographs courtesy of Kim McCutcheon)

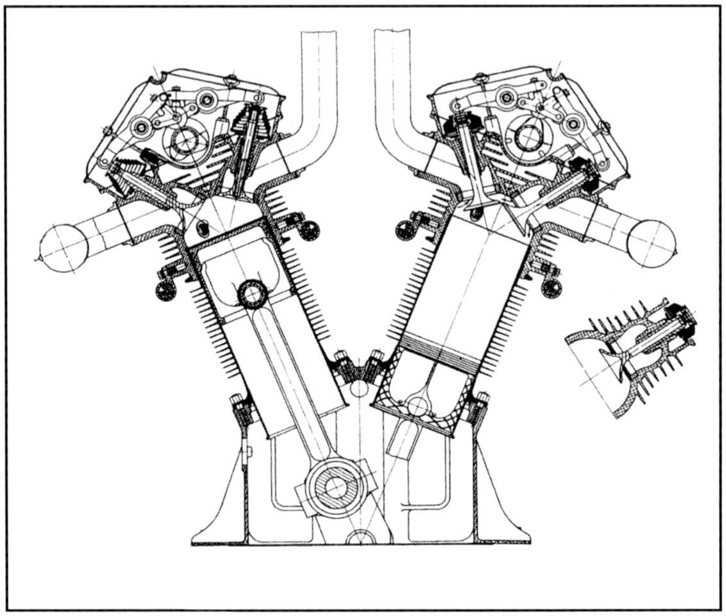

Above is a sectional drawing of the air-cooled Liberty as originally concieved and built by the Engineering Division in 1924. The original drawing carried the approval signature of S. D. Heron dated 2/18/24. Note it is in the upright position. At right is a sectional drawing of the engine in inverted position as produced by Allison in 1926 and after. It was not only inverted but the valve actuation system was redesigned both to correct early problems and to better accommodate inverted operation. (Author's collection)

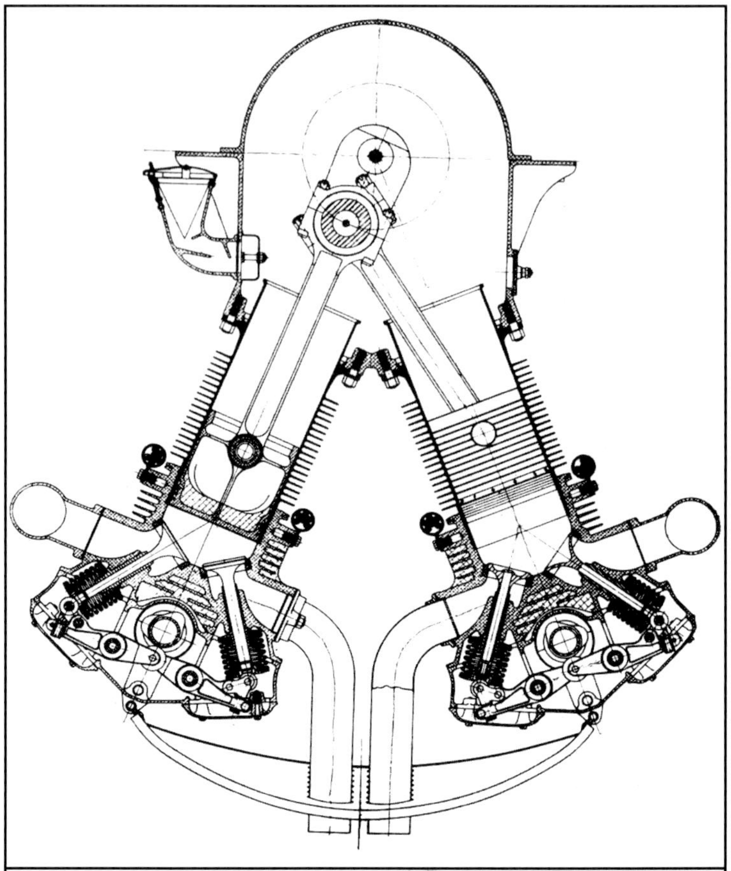

supercharger was used with a 7.3 to 1 drive ratio, resulting in an impeller speed of 12,500 RPM. The supercharger used on this first engine was an old one with excessive clearances and one impeller blade broken off. The opposite blade was removed to balance the impeller. Although crude, this supercharger was used primarily as a rotary induction system so as to avoid the necessity of lengthy design of an induction system and allowing the use of a single carburetor, the Stromberg NA-S-8J. In addition, salt-cooled exhaust valves, Allison steel-backed main and rod bearings and stub-tooth gears were fitted.

Technical Bulletin No. 38 of the first quarter of 1924 noted that drawings for all non-standard parts required had been completed and bids for their construction had been requested from various airplane engine manufacturers. Allison was awarded a contract to produce sufficient parts for two engines. Parts delivery was somewhat late and assembly of the first engine by the Engineering Division was not started until September of 1924.

This engine was of upright design and assembled with air cowling, exhaust stacks and propeller hub weighed 945 pounds.

Assembly of the first engine was completed and it was mounted on the air-cooled testing stand on November 19,

Engineering Division XCO-6C P-360. This airplane was used earlier for testing the inverted water-cooled L-12 also built by Allison. Here it has the first Allison-built air-cooled V-1410 installed. (Author's collection)

1924. It was first jacked in (3 hours at speeds up to 800 RPM) and then run under light load for another 4 hours on November 21. ("Jacked in" is a term commonly used during the era. It meant to turn a new or rebuilt engine over with an external drive source to lap in the moving parts before running it under its own power.) After a number of tests resulting in some minor failures and their corrections, which were not unexpected, a full-power test was run on December 18, 1924.

Maximum corrected horsepower at 1,900 RPM was just over 420 HP. Maximum corrected BMEP was 126 lb. The bore and stroke were 4 5/8" x 7" giving a displacement of 1,410 cu. in. Compression ratio was 5.44:1. Manifold pressure, because of the missing rotor blades and loose clearances, was maintained at about 1 atmosphere throughout the operating range of the engine.

From these initial tests it was concluded that air-cooling of modern high-power V-type aircraft engines was perfectly feasible, thus opening a large field of applications to air-cooled engines of any size.

Testing continued for the next two months with good results but showing up various weaknesses in the newly designed parts. One area showing a number of problems was the supercharger drive, which suffered numerous failures. On January 30, 1925, after yet another supercharger drive failure it was decided to rebuild the engine without it and instead supply fuel with four outside-mounted Zenith LS-52 carburetors. In this manner testing of basic components could continue until the supercharger drive problem was resolved. When so modified, the total engine weight was reduced to 916 pounds.

Allison planetery gear and outer drive ring gear as used on the VG-1410. The spring coupling is hidden within the aluminum front gear case. (Courtesy of Rolls-Royce / Allison)

Engineering Division XCO-6C P-360. The same airplane pictured on the prior page but now powered by the first Allison-built VG-1410. Note the slight cowling changes, different propeller and longer nose because of the longer engine. (Author's collection)

Overall performance of the engine suffered significantly from the loss of the rotary induction benefits as can be seen from the power graph generated with this system on April 28, 1925 (see graph on page 314).

A 50-hour standard test was begun on this engine on February 15, 1925 and completed on March 17. No major failures were encountered. A number of items were noted as needing redesign for improved durability.

Thus far no record has surfaced giving the serial number of this first engine. However, photographs show it to have a Lincoln data plate.

After numerous modifications, a second 50-hour test was initiated on April 4, 1925. Various changes were made to correct problems encountered and testing continued using this engine until June 7, 1926. At that date it was decided to discontinue testing of this first engine and "use the parts from it in the other air-cooled engines." The Engineering Division history of the development of

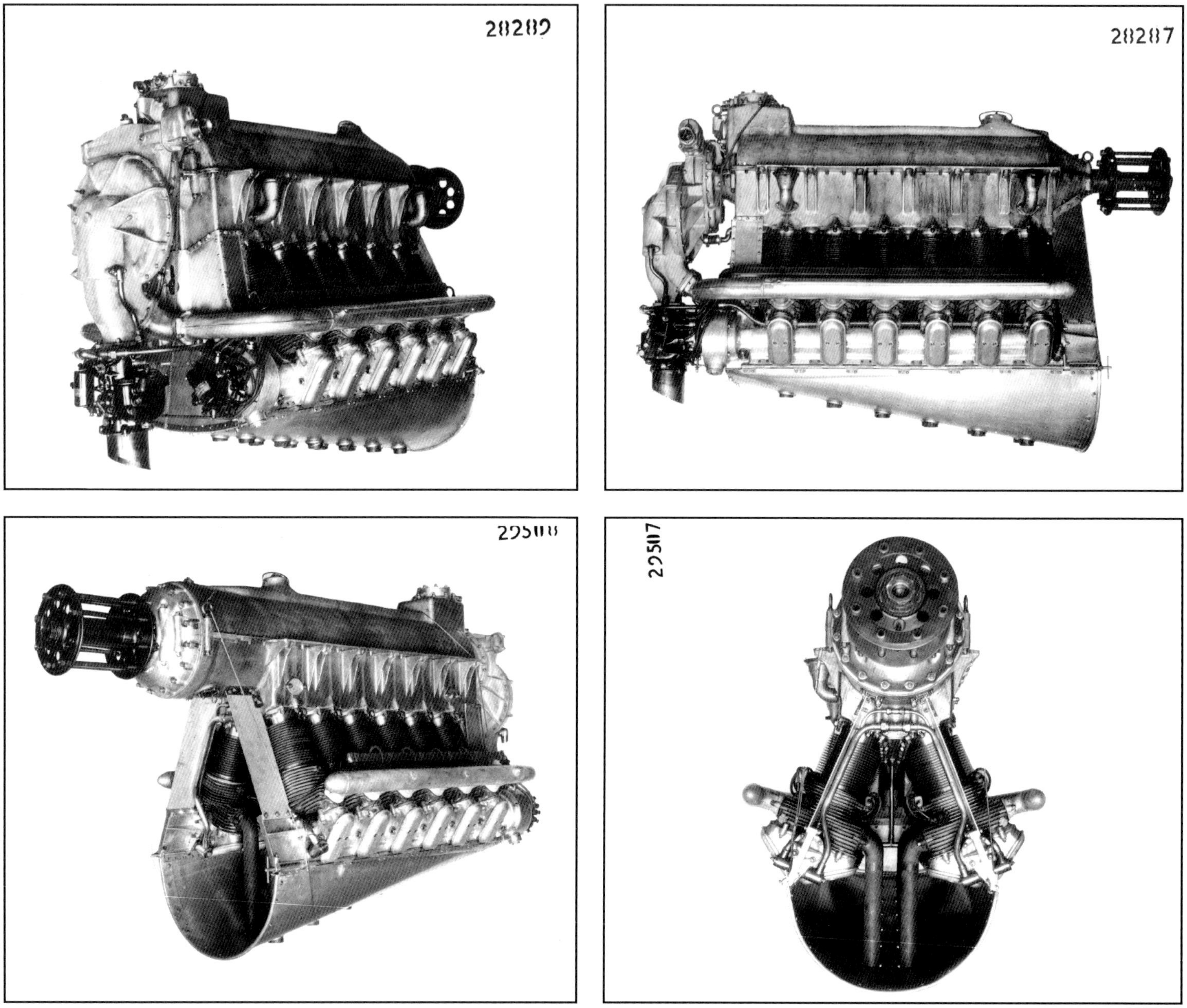

Above at the top-left and right is the first Allison V-1410 direct-drive air-cooled engine. The camshaft assemblies were of the same design as that of the Engineering Division prototype, with slight modifications in oiling because of the inverted position. Below-left and right is the first Allison VG-1410 geared air-cooled engine. This engine also used the early design camshaft assemblies. Both engines had provision for hand-cranking at a takeoff on the supercharger drive and no provision for electric starter. Both engines were photographed here with the larger air hoods used on the test stand. The flying hood was smaller. One Stromberg NA-S8J carburetor was used on the intake side of the 12,500 ft-GE supercharger driven at 9.6 times crankshaft speed. With this supercharger setup a boost of 2.3 in. Hg. was provided at 1,930 RPM. Power at that speed was 449 HP. (National Archives photographs courtesy of Kim McCutcheon)

the air-cooled Liberty indicates only one engine had been built even though parts to convert two engines had been received from Allison. Thus we see that the first engine, an upright one, had been built, tested from November 21, 1924 to June 7, 1926 and then scrapped.

INVERTED AIR-COOLED ENGINES ORDERED BUILT BY ALLISON

By the later part of 1925 the Engineering Division was confident that the air-cooled Liberty was going to be successful and placed an order with Allison for three examples, one direct and two epicyclic-geared versions. The direct-drive version was a conversion of the standard Liberty 12 engine and was given the model designation V-1410. The geared engines were conversions from the lot of 100 epicyclic-geared Liberty 12B engines the Army had requested to be transferred from the Navy in February. These were designated the VG-1410 model.

The first engine received, the direct-drive, was tested briefly and then installed in XCO-6C P-360 and flown to the air races at Mitchel Field in October of 1925. P-360 had been fitted with an inverted water-cooled engine for testing and was converted for installation of the air-cooled engine and fitted with nicely streamlined cowling.

These first three engines carried early design camshaft assemblies and a revised supercharger drive which incorporated a provision for hand-cranking the engine. The first Engineering Division example had no starting provision. It was never test-flown so starting was always done on the dynamometer.

The first air-cooled engine was an upright design and carried the two engine breathers in the same location as was standard on the Liberty, the right side. Inverted versions had the crankcase modified by drilling two additional breather ports on the opposite side, moving the breathers to the new location and capping the old locations. This way the breathers remained on the right side of the aircraft with the engine in the inverted position.

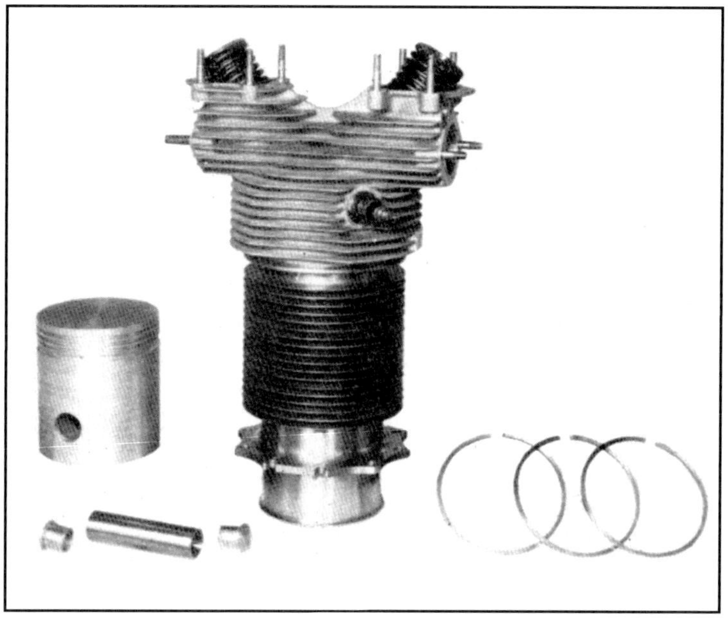

Air-cooled Liberty cylinder assembly, piston, rings and grunion pin. The cylinder barrel was steel with fins machined integral. The head was aluminum screwed on the cylinder barrel with a shrink fit. In addition, a steel band was shrunk over the aluminum around the joint. The aluminum pistons were of 4 5/8" bore and a 7" stroke was used. (Author's collection)

KNOWN SERIAL NUMBERS OF AIR-COOLED LIBERTY ENGINES
(Of the 14 engines built 6 serials are known)

AS Serial	Manufacturer's No.	Comment
26-94	1	direct
26-21	4076	
26-23	4038A	
26-26	3750	
26-27	5239	direct
02425	3572A	geared

The Engineering Division prototype was a Lincoln engine but its serial is unknown.

The recorded AS serial of 02425 is a mystery and could be a typo in records but is probably the Navy number of the original engine as the engine was of Navy origin.

If some logic prevails and the two orders (first 3 and then 10) carried consecutive numbers within each order, then the AS numbers may have been 26-94 through 26-96 and 26-21 through 26-30. 26-94 MNFG NO. 1 was the first Allison contract engine.

Other changes were made in the inverted models to allow proper oiling in this position. This was effected by so arranging the system that the oil from the crankcase drains into the camshaft housings from which it is returned to the oil tank by double scavenging pumps on each housing.

The two geared engines of this order were delivered somewhat later than the direct model, probably because of the extra time required to design and build an Allison spring coupling and incorporate it into the gear system. In addition, the geared engines were conversions of water-cooled 12B engines which had to be transferred from Navy inventory.

Testing of these first three Allison engines continued both on the block and in flight while negotiations progressed with Allison regarding further orders and the modifications which would accompany them. Initial plans were to order an additional eight direct and four geared engines but by the time they were actually ordered the numbers were changed to six direct and four geared. Thus a total of thirteen engines were ordered from Allison, seven V-1410 and six VG-1410.

Both orders were placed during fiscal year 1926 so all engines should have carried 26-xx serials. Unfortunately

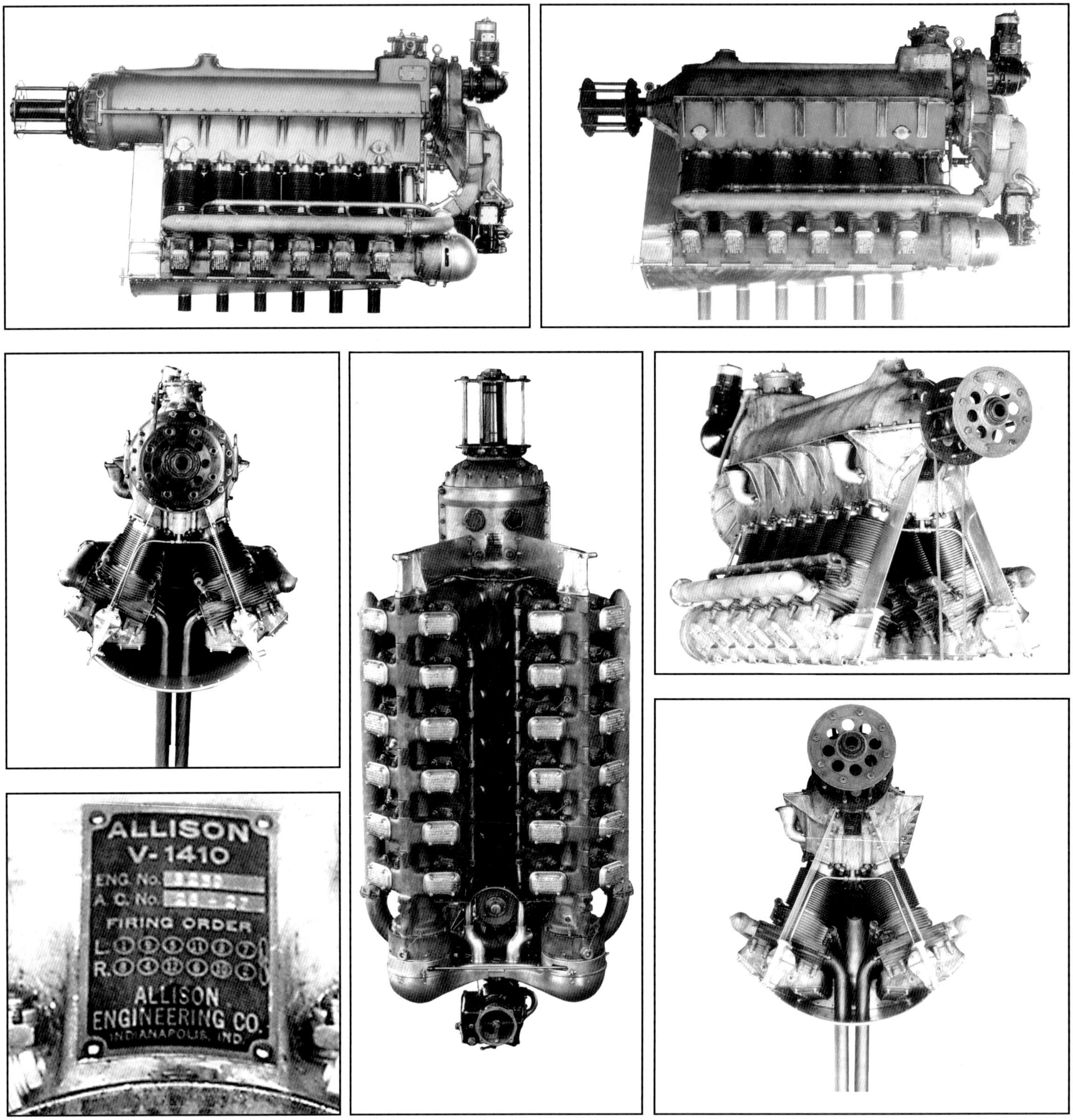

The above illustrations are of the geared and direct production versions of the Allison-produced air-cooled Liberty VG-1410 and V-1410. Top-left, center-left and bottom-center are VG-1410. Top-right, center-right and bottom-right are V-1410. Bottom-left is the data plate of V-1410 A.C. NO. 26-27 Eng. NO. (MNFG. NO.) 5239. The top view is characteristic of both models. In the bottom-center and right-center view the exhaust stacks and air cowl have been removed. (Courtesy of Rolls-Royce / Allison)

I have not found records indicating the specific serials assigned to these contracts. I have found various test records and other references showing serials of individual engines but assembling all known serials does not, unfortunately, generate a very logical pattern of assigned numbers. The first direct-drive Allison test noted in the Engineering Division history is of engine 26-94 whose Manufacturing No. is noted as 1. Although numbering of the first engine would seem to indicate an intention to start manufacturer's numbers at 1 and progress upward, this was not followed. All currently known records of other serials give manufacturer's numbers as four digits, and none consecutive or even close to it. It would thus appear possible that the manufacturer's number of the original engine from which the conversion was made was used, at least in the case of the direct-drive engines. The serials of engines produced is of no great consequence but the author has always had a deep interest in such things so I do hope this information surfaces eventually.

In contrast to the treatment of the inverted water-cooled engines, these engines, with the exception of the Engineering Division prototype, were given their own data plates carrying their assigned model names, V-1410 and VG-1410, and serial numbers.

THE ALLISON PRODUCTION ENGINES

The second Allison contract engines had camshaft drive assemblies of revised design as shown on the sectional drawing on page 316. These were considered the production design.

Although Allison issued information to any interested publications indicating either of these models could be ordered from Allison for civilian use, there is no indication that they sold any beyond the Army contract for thirteen. They were, however, Department of Commerce licensed.

TECHNICAL DESCRIPTION OF THE PRODUCTION DESIGN

The engines use a standard Liberty crankcase, crankshaft and rods, either L 12A or L 12B, depending upon whether geared or not. Cylinder bore is reduced from 5 to 4 5/8" in order to allow more space between cylinder banks for cooling. The use of smaller pistons allows higher operating speeds.

Valve lift is higher than the Liberty and timing is extended. A centrifugal-type rotary induction fan (supercharger) pulls fuel-air mixture from a Stromberg NA-S8E carburetor and serves to both pulverize the fuel droplets and deliver a uniform fuel-air mixture and overcome frictional resistance of the intake manifolds and valves at high speed. Coupled with the higher speed of the engine and the slight boost provided by the supercharger, power is slightly higher than the standard Liberty.

Above are two views of what is believed to be the final V-1410 produced by Allison. The illustrations show the engine to carry most of the characteristics of the production engines except that it was built in upright configuration with its oiling system modified to accommodate that change. (Courtesy of Rolls-Royce / Allison)

Cylinder barrels are steel with cooling fins machined integral. Heads are aluminum, screwed and shrunk on the cylinder barrels. They are additionally secured by a steel band shrunk over the aluminum around the joint. Valve seats are heat-treated aluminum bronze; guides are alloy steel; both inserted with shrink fit.

Valves are cobalt-chrome, salt-cooled. The springs and outer ends of the stems operate in an oil bath which carries away heat conducted through the stems. Valves are 2 1/2" in diameter, the same as the standard Liberty, but have greater capacity because of the increased lift, longer duration and improved port design.

Oil is supplied at 100 pounds pressure. The cylinder barrels project into the crankcase, and the space around them forms a sump which is scavenged from both ends. The camshaft housing is quite large, as it carries a flood of oil to cool the valves; it is scavenged from both ends. The oil collected by the several scavenging pumps is circulated through jackets around the carburetor and carburetor elbow, where its heat is transmitted to the mixture.

The open space between the cylinder banks is closed by a sheet-metal arch across the open part of the vee, and a back plate; thus a deep wedge-shaped box is formed, open at the front. Air from the propeller slip stream enters the front under slight positive pressure and flows out between the cylinders. Cooling proved to be very simple and efficient. As noted on the illustrations, the opening size of this hood is smaller on flight engines than on engines being subjected to block testing.

Main and connecting rod bearings are Allison steel-backed-type. A steel shell replaces the ordinary bronze; it is lined with babbitt, and a layer of bronze is fused to the outside to form a bearing for the forked rod.

Ignition is, as was the case with all prior engines, by standard Liberty Delco battery-type. On the production engines the ignition system was shielded from radio interference by encasing with sheet metal covers.

On geared engines a 5:3 Allison epicyclic reduction gear is used, with an Allison spring coupling added to protect the teeth from shock. The original Liberty 12B gears did not have this device.

Department of Commerce rating for both engines was 410 HP @ 1,800 RPM, and 430 HP @ 1,900 RPM. Guaranteed fuel rate was .53 pounds per HPH and oil consumption .03 pounds per HPH. The VG-1410 weighed 1,125 pounds and the V-1410 1,000 pounds.

The last of the Army order was delivered in 1927 and civilian listings were carried by aero magazines from 1928 to about 1932.

A SECOND UPRIGHT AIR-COOLED

The Allison archives yielded photographs of what appears to be a final variation of the V-1410 of which we currently have no history. All of the Air Service-contracted air-cooled engines were built in inverted configuration, the first three with most of the early features of the single Engineering Division-built engine and the last ten with revised valve drive and various other features to better accommodate inverted operation.

This last engine carried the late design valve drive, ignition shielding and low flying-type air hood. It had a unique supercharger to allow the carburetor to be mounted for upright operation but offset to allow clearance for mounting an electric starter on the supercharger gear drive. Crankcase breathers are on the right side as is normal for an upright Liberty and no alternate location holes have been cut in the opposite side for inverted operation breathers.

This would be the eighth V-1410 built by Allison. It may have been sold on the commercial market but I currently have no knowledge of its ultimate use.

THE AIR-COOLED 1410 IN SUMMARY

The first engine designed by the Engineering Division was built at McCook Field by the Engineering Division in 1925. This engine was of upright design and direct-drive.

Allison was contracted to build three additional engines of a similar design but in inverted configuration.

Above and to the right are two views of the second type NACA Allison Roots supercharger. (The above picture is courtesy of Rolls-Royce / Allison). ***The right view is from a 1929 SKF Bearing ad.*** (Author's collection)

One was direct-drive and two with epicyclic-gearing. The epicyclic-engines were conversions of Navy Liberty 12B models. These three were delivered in late 1925 and early 1926.

Allison was then contracted to build an additional four direct-drive and six gear-drive inverted engines with improved valve drive, shielded ignition and improved inverted lubrication design. These were delivered in 1926 and 1927.

Allison built a final upright engine with the late design features.

Total: Two upright direct-drive, one Engineering Division and one Allison. Five inverted direct-drive and eight inverted gear-drive by Allison. Grand total: fifteen engines of all types.

The air-cooled Liberty was considered a successful engine. The initial object had been to use a proven, even though obsolete, engine base to prove it was feasible to air-cool an inline engine. This had been accomplished and thus the Air Service instigated the design of a modern engine built around the air-cooled cylinders. This resulted in the Wright inverted air-cooled V-1460. See the side-bar, on page 324, for more details on this engine and its uses.

AIRPLANES WHICH FLEW WITH THE AIR-COOLED LIBERTY

The inverted air-cooled Liberty is known to have been flown in only a few aircraft, all of which were on an experimental basis. It powered the Engineering Division XCO-6C P-360 in both direct and geared version. It was later flown in the XCO-7B, DH-4B, Douglas XA-2, the first of the Attack Bomber series, XCOA-1 Loening Amphibian, the Curtiss P-1 Hawk P-400 and the twin-engine XLB-3 Keystone bomber.

THE ALLISON-BUILT ROOTS SUPERCHARGERS

Allison built a number of superchargers during the early years of the company but the earliest Liberty associated one currently known is a Roots-type which was built for NACA in 1926. NACA continued to experiment with Roots superchargers for the Liberty until at least 1931 and used several slightly different designs. Although the initial design was laid out by NACA, later versions may have been designed and built by Allison.

According to a 1930 list generated by Allison to document their activities from 1919 to that date, they built 3 #2 and 34 #3 Roots-type superchargers during the period. Some were for the NACA Liberty project and the remainder for other aircraft projects, many of which went overseas.

WORK DESIGNED, BUILT & SOLD FROM 1919 to 1929

3	Race Cars (Won 1 & 3 Indianapolis 1919 Race, $50,000. Sheepshead Match Race. $10,000.)	
200	Planetary Geared Liberty	
20	Planetary Liberty Gear Units	
1	Direct-Drive	
2	Reduction Gears	U.S.S.
2	Reduction Reverse Gears	*Shenandoah*
5	Clutch	
8	Miami 12 Motors	
5	Miami Light Plants	
1	Miami Pump	
1	Miami Air Comp.	
1	W-1-A Reduction Gear	
1	20,000 ft. Supercharger	
1	7,500 ft. Supercharger	
1	Clutch Drive - M. O. Reeves	
2	Wright T-2 Gears	
1	Wright T-3 Gear	
1	Supercharger - Duesenberg Race Car	
1	2 Speed Liberty Motor & Gear	
1	DeBothozat Helicopter	
1	Ice Machine	
1	Curtiss 2 Motor Gear	
44	Packard 1500 Gears	
42	Packard 2500 Gears	
1	Manley Drive G. M. C.	
1	4 Motor Gear	
34	#3 Roots-Type Superchargers	
1	Road Shock Measuring Device	
6	#2 Roots-Type Superchargers	
3	Universal Test Engines	
25,000	Steel-Back Liberty Conn Rod Bearings	
3,000	Steel-Back Liberty Main Bearings	
30,000	Steel-Back Bronze Lined Bearings	
6	Tachometer Testing Machines	
7	Packard Roots-type Superchargers	
13	Inverted Air-Cooled Liberty Motors	
1	Caminez Cam Motor	
1	Water Brake	
40	Inverted Water-Cooled Liberty Motors	
1	X-4520 Motor	
3	16 ft. Reversable Propeller Hubs	
1	Marine Reduction Gear	
1	Lasley Turbine Drive	
2	50 HP Reduction Gears	
1	Packard 2775 Gear	
2	Prop. Shaft Assembly - Power Plant Cooling - 4 ft.	
3	Prop. Shaft Assembly - Power Plant Cooling - 8 ft.	
3	Single Mesh Liberty Reduction Gears	

DESIGNS AND STUDIES MADE AND SOLD FROM 1919 TO 1929

Design and study of W-1-A Cylinder Head (Air Corp)
6-Cylinder 100 HP Marine Diesel Engine (Navy)
400 HP Radial Engine (Air Corps)
Angular Drive. Two Propellers from One Engine (Navy)

WORK DESIGNED BUT NOT SOLD FROM 1919 TO 1929

Packard Reduction Gear for 1A-1237 Engine
Packard Reduction Gear for 1A-2025 Engine
2 Speed Reduction Gear - Goodyear Balloon
600 HP Navy - Wright Planetary Reduction Gear
Goodyear Balloon Drive Gearing
ANW-2 700 HP Reduction Gear

WORK DESIGNED, BUILT OR IN PROCESS 1929-30

U.S. Navy Aircraft Diesel Engine
Goodyear - Allison Propeller-Drive
400 I. G. Aircraft Motor
Steel-Back Bronze-Lined Bearings

The above Allison internal document appears to have been generated after the death of James A. Allison, either preceeding the sale of the company to Eddie Rickenbacker or after his purchase and pending its sale to Fisher Brothers Investment Trust. The 1929-1930 addendum was added later. The list was not generated in chronological order. Although very comprehensive, this list was not absolutely complete. Some well documented projects noted in this chapter do not appear. In addition, the Liberty rebuilding program was not mentioned even though it was a major income source.

THE WRIGHT AIR-COOLED V-1460/V-1560

In 1928 the Air Service contracted Wright to build a modern version of an inverted air-cooled V-12 engine patterned roughly after the V-1410. Although the engine resembled the V-1410 it was a very different design, retaining little of the Allison V-1410 features beyond the design of the cylinder. The bore and stroke were 4 7/8" x 6 1/2" and displacement was 1,460 cubic inches. It was built in an inverted vee formation with an included angle of 60°. Weight was 925 pounds and output was 525 HP @ 2,300 RPM. It used a one-piece crankcase.

After testing, the design was revised and a second version produced which used a two-piece crankcase split at the crankshaft center. Some other changes were also made but ratings and displacement remained the same. It is believed that 14 V-1460 models were built from 1928-1930.

The engine was then given some minor modifications and its bore and stroke increased to 5" x 6 5/8" which increased its displacement to 1560 cubic inches and its power rating to 600 HP @ 2,400 RPM. Weight increased to 960 pounds. This version became the V-1560. Two were built in 1931.

The first V-1460 was test flown in the same Curtiss P-1 used for testing the V-1410. After converted to this configuration the airplane was assigned the designation XP-17. Another early engine was test-flown in the same XCOA-1 Loening in which the V-1410 was tested. With the V-1460 it was called the XCOA-1A. The Loening success resulted in the use of the V-1460 in the special XO-10 model and 8 OA-2 airplanes.

Two fighter designs were initiated to use the more powerful V-1560, the Curtiss XP-18 biplane and the XP-19 low-wing monoplane. Both projects were eventually canceled before completion. Ultimately the air-cooled radial won out over the inline air-cooled as the future air-cooled type of preference.

The top three illustrations are of the first Wright V-1460. In this illustration it is equipped with the larger-type air intake as used on the test stand. The engine had built-in rotary induction and a single-piece crankcase. (All photos are Wright factory courtesy of Kim McCutcheon)

The bottom four illustrations are of the improved version represented by the last 13 engines built. This engine is equipped with a flight-type air intake. The split crankcase easily identifies the production version. The V-1560 probably looks the same.

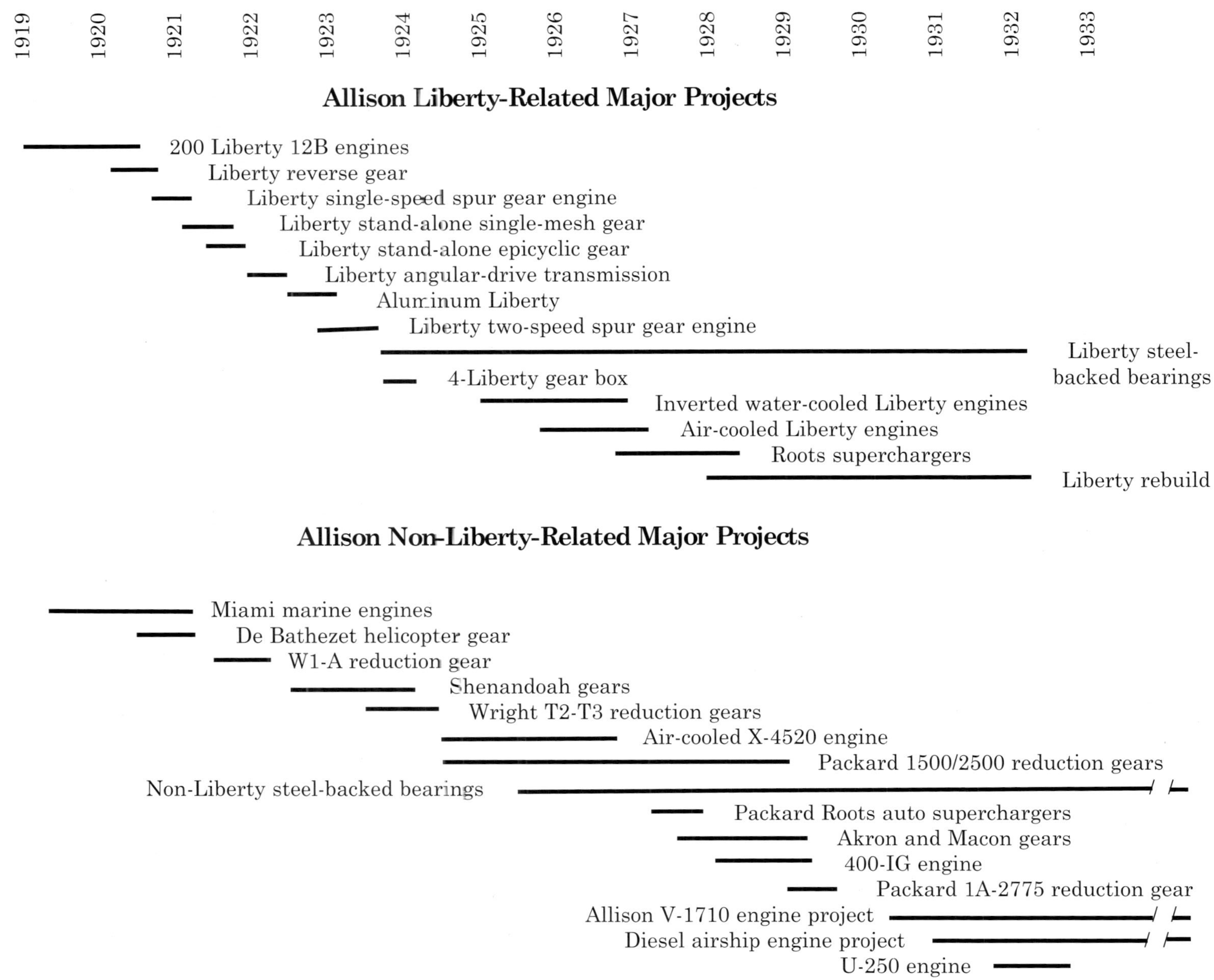

Notes: The above chart is intended to give the reader an idea of what work Allison was involved in during the period they were engaged in Liberty projects. Other than the Liberty, the only full engine projects undertaken by Allison from 1919 to 1933 were the Miami marine engine, the air-cooled X-4520, the 400-IG, the Diesel airship engine and the U-250 engine. The V-1710, Diesel airship engine and steel-backed bearing production continued beyond the period of this chart. Dates represented by the chart are approximate only. A number of the less significant non-Liberty projects are not shown.

Martin MB-2 Bomber P-162 equipped with Form B superchargers. This was the only multi-engined plane this type was ever installed on. It set a number of altitude records and encouraged the Air Service to later install Form D superchargers on an additional 20 Martin Bombers. (Courtesy of Peter Bowers)

Chapter 9

Supercharging the Liberty

Not long after the Allies began using airplanes in combat it became evident that an advantage in altitude was a fighting advantage. The maximum altitude an airplane can obtain is dependent upon a number of things, of course, but principle among them is the amount of power its engine can produce. Air density decreases as altitude increases. The less dense the air the less combustible mixture can be brought into the engine's cylinders and thus less power can be generated.

As an example, the power of the average airplane engine is reduced to half its sea level output at 15,000 feet. Thus if you consider a 400 HP engine, such as the Liberty, it will produce only 200 HP at that altitude. One must either increase the size of the engine or find a way to keep the input air to the engine at sea level pressure. Increasing the engine size increases its weight and thus to a large degree counters the advantage of the power gained. An example of a potential replacement engine contemporary to the Liberty would be the Engineering Division W-1 of 1920. That was an engine rated for 700 HP at sea level. It developed only 351 HP at 15,000 feet, or 49 HP less than a Liberty at sea level. The weight of the Liberty was about 845 pounds. The W-1 weighed 1,815 pounds.

The alternate method is to keep the input air to the engine at sea level pressure at that altitude (or beyond). This is called supercharging. There are a number of ways to accomplish this and if one can do so with a device which does not add a great deal of weight, then it would be better than using a larger engine. As we shall see, the later method was chosen and by about 1921 it had been accomplished with a turbosupercharger which only increased the weight of the Liberty to 1,010 pounds and kept its output at 400 HP at 15,000 feet (and beyond).

EARLY ALLIED SUPERCHARGER DESIGN HISTORY

Serious experimental work with supercharging devices was begun by the Allies early in 1915. The British effort was initially directed toward gear-driven types of blowers. These included reciprocating, rotary and centrifugal compressor designs. The reciprocating type was abandoned early because of its inherent low efficiency. The rotary and high-speed centrifugal compressors both presented problems with breaking the highly loaded drive gears during changes in engine speed. Because of this difficulty the British switched their interest to the exhaust gas-driven turbine driving a centrifugal compressor (the turbosupercharger) because it eliminated the gear-drive and its difficulties.

The French seemed to have bypassed the gear failure difficulties as their interests were in the exhaust-driven type almost from the start. Their leading expert in the compressor field was Professor Auguste Rateau, an inventor and industrialist who had founded a company which specialized in compressors of all types as well as steam turbines and industrial pumps and fans.

Rateau developed a turbosupercharger which was tested with great promise on Spad, Breguet and A.L.D. aircraft. As a result of these successes the British had an example sent to them. They fitted it to a RAF-4D 100 HP air-cooled 12-cylinder engine and powered an RE-8 airplane for flight-testing. This was the British impetus to switch to the turbosupercharger over gear-driven types.

The Italians were also experimenting with superchargers but were confining their efforts to gear-driven centrifugal types. Their success, up to the time America entered the war, had been limited.

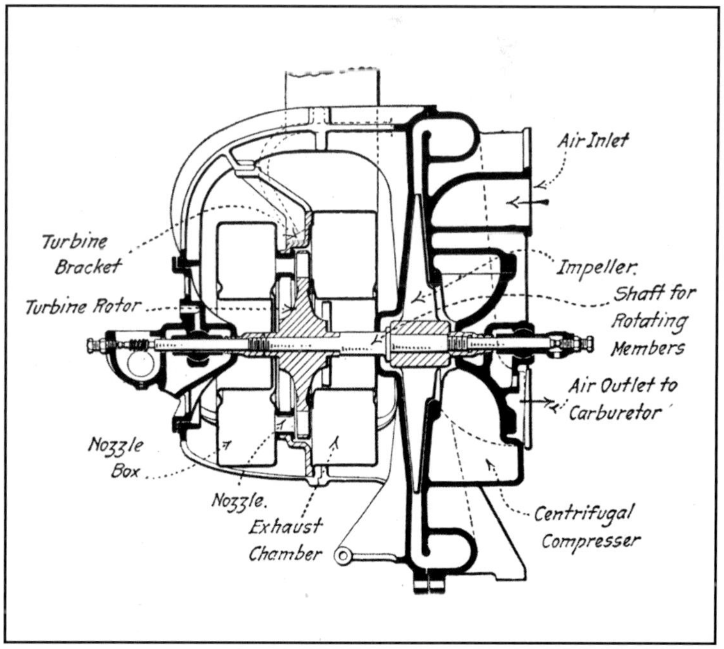

Longitudinal section of the Rateau turbosupercharger as shown in the January 1919 issue of the Bulletin of the Experimental Department Airplane Engineering Division. It would thus represent the design as of late 1918.

Although both the British and French had flight-tested several supercharger designs, neither had any design near ready for production.

AMERICA INITIATES A PLAN OF SUPERCHARGER DESIGN

This was the Allied state of the supercharger art when America entered the war in April of 1917. By June of 1917 the Bureau of Aircraft Production had studied all information available from Europe, decided to initiate its own design efforts and assigned Mr. E. H. Sherbondy the task of investigating the matter of supercharging aircraft engines from all angles. By the time Sherbondy had reached the point of actually beginning to lay out a design to be followed, the Liberty engine series had been designed and the Liberty 12 ordered into production. Thus Sherbondy's turbosupercharger was designed for use on the L-12A.

A second study, aimed specifically at turbosupercharging, was initiated at the behest of Dr. William F. Durand, chairman of the National Advisory Committee for Aeronautics. Dr. Durand was formally a professor at Cornell University and had been there when Sanford A. Moss was a student there working on his Doctor's thesis on gas turbines. After graduating, Dr. Moss was hired by General Electric to do experimental work in the field of turbine design.

Thus when Dr. Durand received a report of Allied progress in supercharger development in mid-1917 he thought of Dr. Moss and contacted him asking if he would undertake for the government a study of the problem of supercharging aircraft engines, more specifically the Liberty 12. He specifically referred to a turbo compressor of the type developed in France by Rateau as the type they were interested in. He did so in a letter dated November 19, 1917, and on the same date sent a letter to Mr. E. W. Rice, director of the turbine department at General Electric's River Works at West Lynn, Massachusetts where Dr. Moss was employed. In it he requested the company authorize Dr. Moss to "give such attention to this problem as may be necessary to give to us the results of his experience in the study of such matters."

It appears, from a letter written by Dr. Moss some years later (1942) in reference to the subject, that initially there was some reluctance by General Electric officials at the Lynn River Works to "take up the project." However, Dr. Durand sent Mr. Jesse Coats, a former G.E. engineer at Lynn and then Dr. Durand's first assistant, to Lynn to help "put the proposition across," which he did.

Thus late in November of 1917 G.E. and Dr. Moss began the turbosupercharger project which eventually resulted in the production by G.E. of thousands of turbosuperchargers used on aircraft engines from 1918 until the end of the large piston aircraft engine era.

THE SHERBONDY GEAR-DRIVE SUPERCHARGER

As part of the requirement to "investigate the matter from all angles" between June of 1917 and May of 1918, Sherbondy laid out the design of both a gear-driven centrifugal supercharger and a turbosupercharger. The gear-drive type, however, was not pursued further and no machine was ever built for testing.

Even so, he explained its design in some detail in his 1920 book, *Textbook of Aero Engines*. Its centrifugal compressor was driven by spines directly from the rear of the crankshaft at 16-times engine speed. The gear-drive theory is shown in the illustration below and a sectional drawing of the actual design is shown on page 329. The result was a very compact unit containing the complete gear-drive and compressor in one unit.

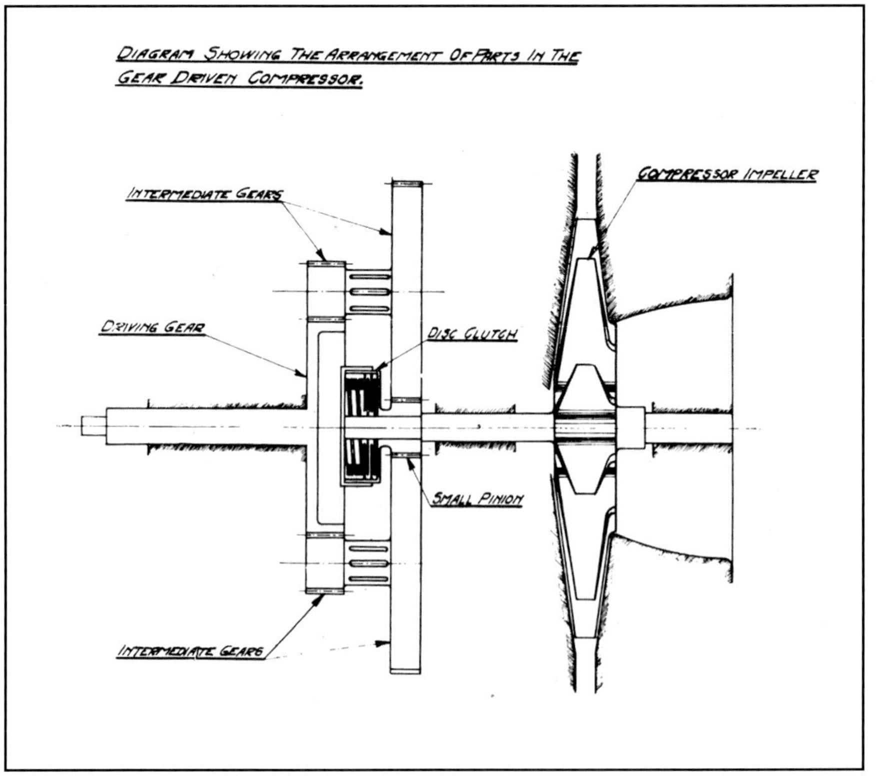

Using the theory diagram above as a reference, the reader can more easily understand the sectional drawing shown on the next page. The design resulted in a very compact unit. (Author's collection)

The gear train consists of a driving gear, two intermediate gears and a small pinion gear. A disc clutch is interposed between the impeller shaft and the small pinion gear which delivers power to it and the shaft connecting the two intermediate gears was slotted to give a certain flexibility at that point. The intention of the disc clutch was probably to dampen torsional vibration, or to protect the impeller from sudden changes in engine speed.

THE SHERBONDY TURBOSUPERCHARGERS TYPE NO. 1 AND NO. 2

He instead had concentrated on the turbo type and built the first of an eventual three designs. On May 7, 8 and 9 of 1918 design No. 1 was submitted to steam trials at the plant of the DeLaval Steam Turbine Co. of Trenton, NJ. An alternate, or second design was also constructed with the principal difference between the two being the number of turbine buckets (72 in the No. 1 machine and 53 in No. 2) and the design of the nozzle box. This No. 2 design was given steam trials at the same location on June 23 and 24.

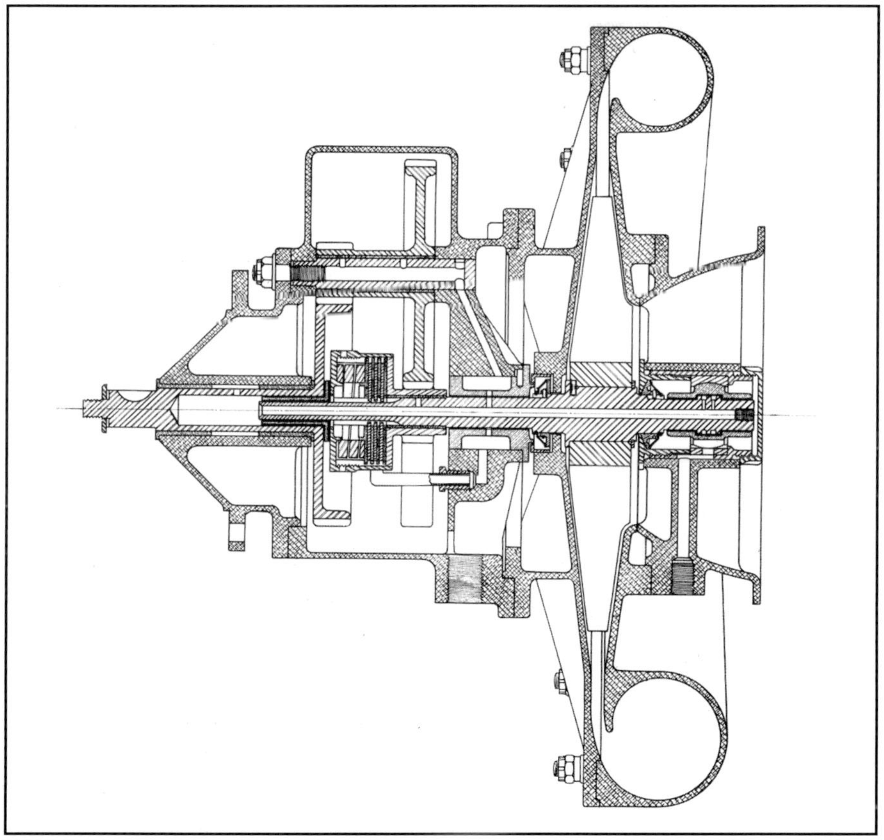

Sectional drawing of the Sherbondy gear-driven centrifugal supercharger designed for the Liberty 12. This machine was never actually built but does represent the first gear-driven unit designed for the Liberty. (Engineering Division photo from author's collection)

SHERBONDY AUTOMATIC BOOST CONTROL

The Sherbondy turbosuperchargers all used an automatic control device designed to keep engine power constant, relative to throttle position, at all altitudes. It was accomplished in the following manner:

A constant pressure chamber is provided which is sealed so that the pressure will remain constantly at 14.7 lbs. per sq. in. On one side of this chamber a diaphragm is placed which is free to move in or out, depending upon the difference of pressure within the chamber and without. This diaphragm is connected to a piston rod which operates a piston valve in the control casting. This piston, by its movement, opens or closes a set of ports which lead oil under pressure from a reservoir to the valve pistons. The latter, in turn, open or close the valves, allowing the exhaust gas from the motor to escape without passing through the turbine.

At ground level, these by-pass valves are open, allowing the gas to escape into the atmosphere, while the turbine runs just fast enough to furnish the power to overcome the losses of the air in passing through the blower. As the airplane equipped with this device ascends, the by-pass valves slowly close, causing more gas to pass through the turbine and increasing its speed, maintaining the blower discharge pressure at sea level value. In this manner the power is kept at a constant value. The device can be seen in the sectional drawings of both the No. 2 and No. 3 units. (Features of the device were covered by patents 1,346,563 and 1,346,564 which were applied for on March 28 and April 8, 1918.)

TESTING MOVES TO MCCOOK FIELD

After steam tests were completed the two Sherbondy superchargers were moved to McCook Field and installed on a Liberty engine and given tests on a dynamometer.

The first unit was shipped to McCook Field and had begun tests by the first week in June. By July 13 the No. 2 unit was being fitted to a Liberty engine so it could also undergo tests. The weekly reports of the Engineering Division noted the first two Sherbondy machines as being under test continually from mid-June. During much of this period parts were being swapped between the units and only a single unit was undergoing tests at any given time. The report of October 12-19 noted that Packard Exp. No. 1 engine was being used for the purpose. Reports of the remainder of October and into November continued to note the engine in use as Exp. No. 1.

Considerable trouble was encountered with warping of the gas casing which resulted in deflection of the nozzles to the point that they would contact the turbine wheel. In addition the diaphragm which divided the inlet gas from the exhaust gas warped out of shape, changing the nozzle shape and allowing gas to flow from nozzle to nozzle.

Additional problems encountered were seizing of the turbine rotor because of the failure of the rear bearing of the turbine. This occurred several times and it was decided that air cooling of that bearing was not adequate and water cooling would be needed. The high-heat of operation also caused about half the buckets in the turbine wheel to become loose. Numerous parts became distorted from the heat and either cracked or caused contact between moving parts and the various housings. As is the case in many designs, the theory seemed sound but in practice they did not work out. In this case much of the trouble was due to the lack of construction materials which were able to withstand the 1500°F exhaust heat of the engine.

THE SHERBONDY NO. 3 TURBOSUPERCHARGER

Machine No. 3 was designed to eliminate the problems found in No. 1 and No. 2. It was tested with steam in the same manner as No. 1 and No. 2 and then shipped to McCook Field and installed on Packard MNFG. NO. 1 U.S. NO. 17435 for tests.

As can be seen from the sectional drawing of this No. 3 design, the layout was considerably altered even though the basic design remained much the same. Water cooling of the rear bearing was added but it was later determined that the design did not allow enough space to supply sufficient cooling water.

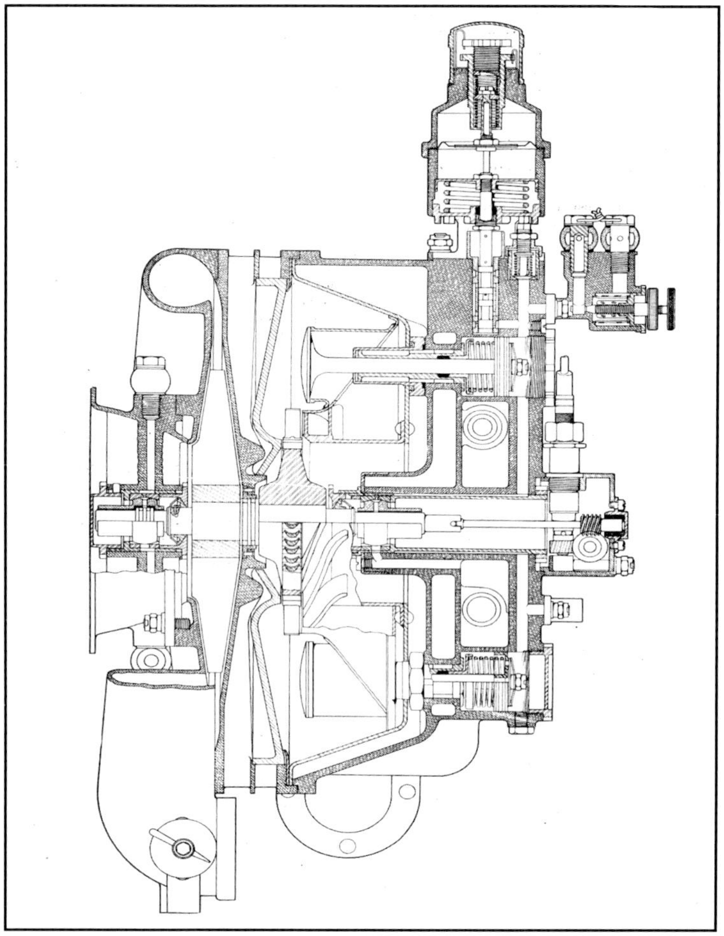

Sectional drawing of the Sherbondy turbosupercharger No. 2. Note the atmospheric boost control on top of the unit and the two exhaust bypass valves. The No. 1 design was very much like it. (Author's collection)

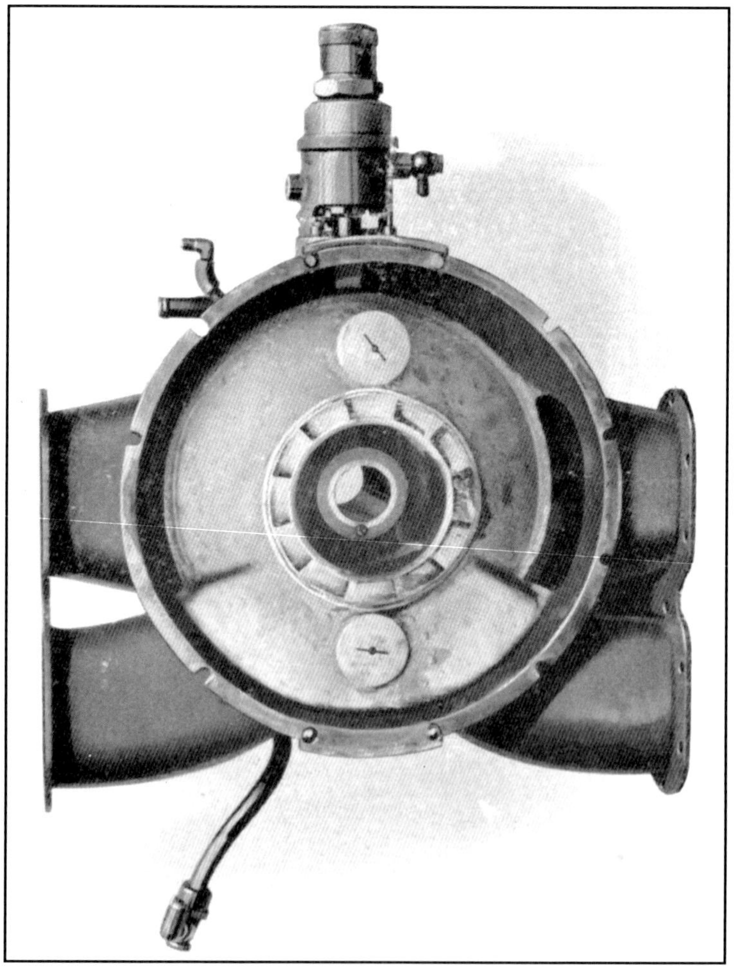

Sherbondy turbosupercharger No. 2 nozzle casing showing blistering after tests. Note the exhaust bypass valves in the top and bottom of the casing. (Author's collection)

Sherbondy No. 2 turbosupercharger mounted on a Liberty engine. (Author's collection)

Rotor shaft of the No. 3 Sherbondy turbosupercharger carrying the compressor wheel and turbine wheel, the only moving part of the supercharger. (Author's collection)

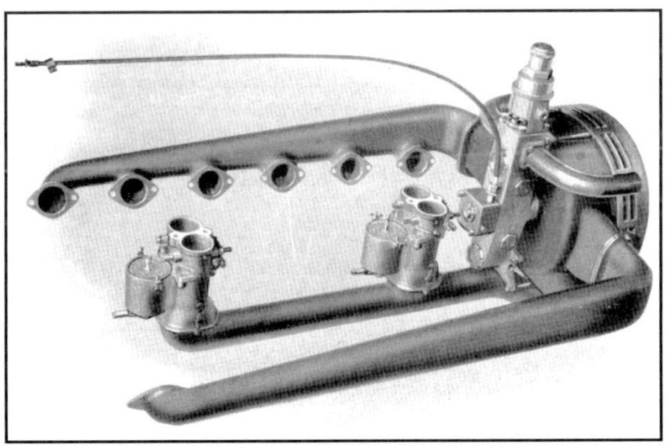

Sherbondy No. 2 turbosupercharger from the rear unmounted. In this view the exhaust manifolds, carburetor manifold and the two standard Zenith carburetors are visible. (Author's collection)

Sherbondy No. 2 turbosupercharger mounted on a Liberty engine. The No. 1 machine differed very slightly in design and was close to the same in exterior appearance. (Author's collection)

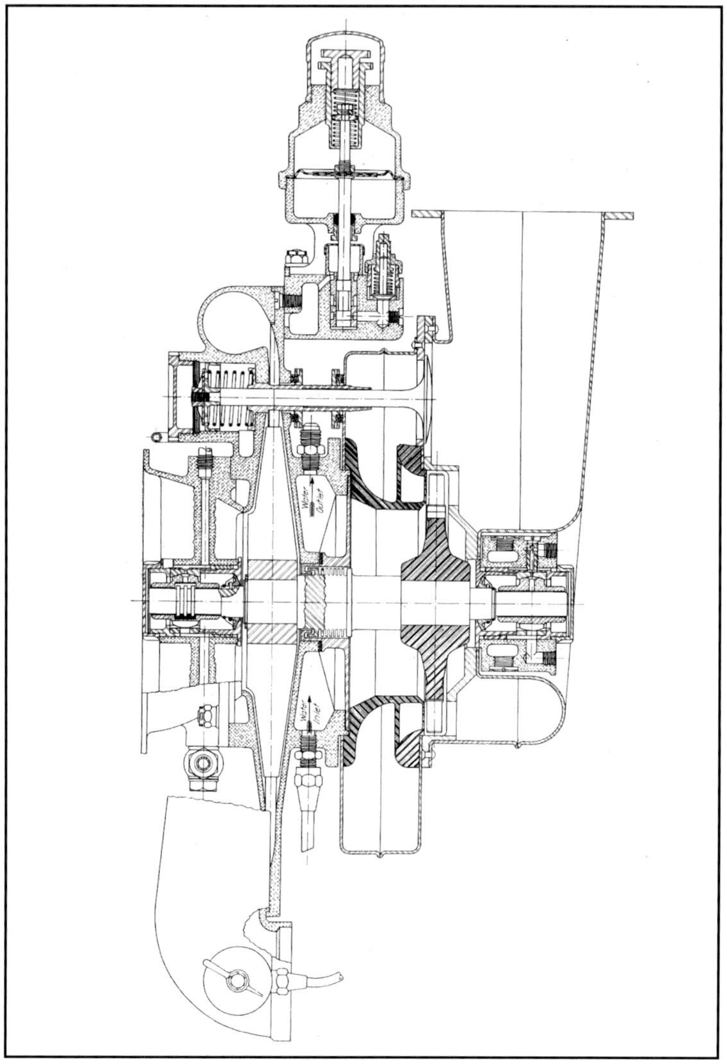

Sectional drawing of Sherbondy turbosupercharger No. 3. Note that both exhaust bypass valves are above the turbine wheel and the machine is much shorter front to back than No. 2. (Author's collection)

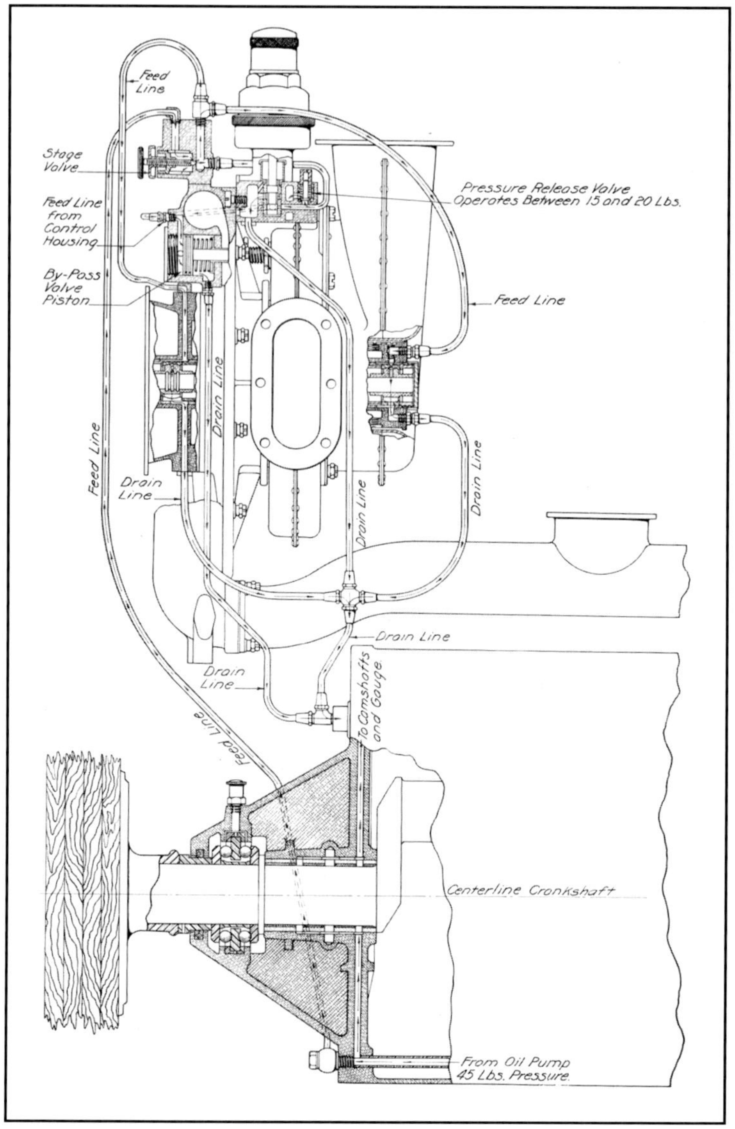

Oiling system of the No. 3 Sherbondy turbosupercharger showing the method of circulation and the connections to the standard Liberty lubrication system. (Author's collection)

A 4-hour endurance test was begun during the week of December 7-14. It resulted in numerous problems. Bearings were burned-out in spite of the water cooling feature. The layout changes reduced the problems caused by heat expansion but by no means eliminated them. After a little under 3 hours into the 4-hour endurance run the motor was stopped to correct problems caused by loosened oil and water connections.

After sitting some 15 minutes and cooling, it was found that the turbine wheel could not be moved because the rear bearing had melted due to excessive heat. In addition, the bypass valves were stuck in the closed position. In addition, the outside ring had moved out of alignment with the deflector fins by 3/32". The results were carried in a report dated December 16, 1918.

Although the Sherbondy turbosupercharger had a number of novel features and certainly appeared to be a well-made and quality finished unit, its design did not overcome the problems presented by the materials then available. The Moss design was, at this point in time, much more advanced and less troublesome. The November 11, 1918 Armistice had been rapidly followed by the cancelation of any aeronautical project which could not be seen as fulfilling an immediate need. By late December the Sherbondy project had been canceled. I will take up the history of the Moss supercharger design next and we will see that work on it was also halted for a brief period. However, it was very shortly taken up again as it was showing much more promise.

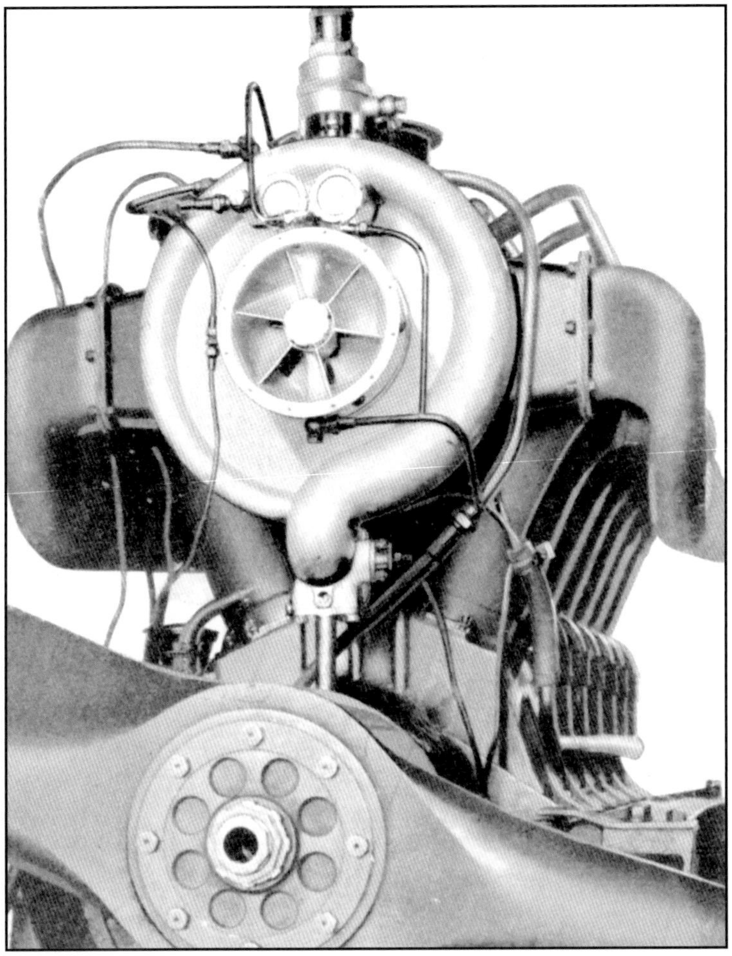

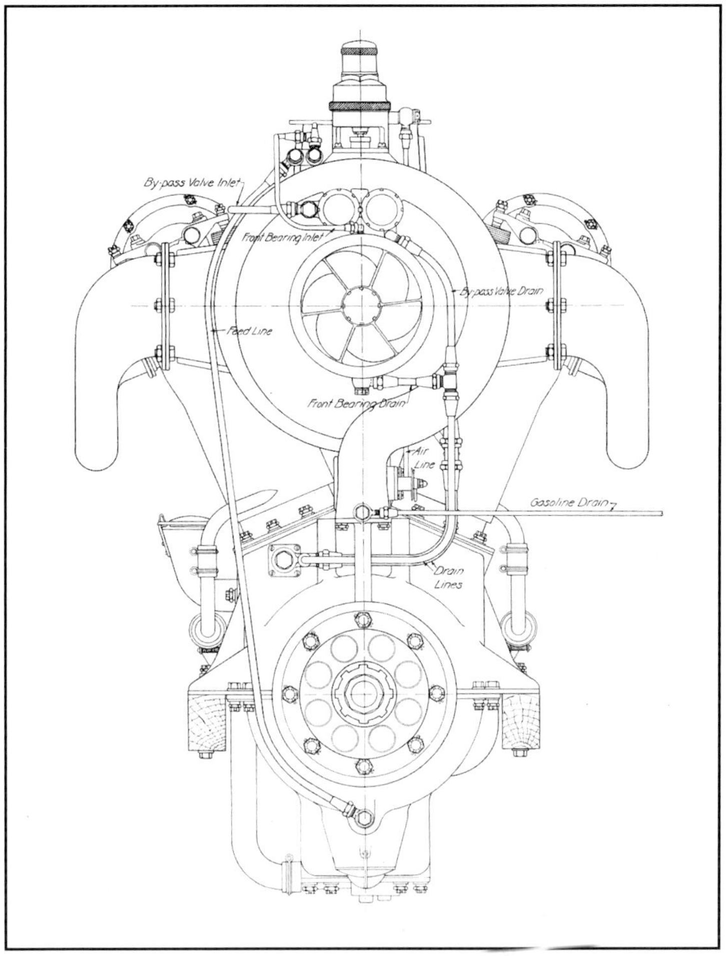

Left: The No. 3 Sherbondy turbosupercharger mounted on a Liberty engine and on the test stand. Right: Front elevation of the No. 3 Sherbondy turbosupercharger mounted on a Liberty engine. None of the three Sherbondy machines built passed any endurance test. As a result, development was halted shortly after the Armistice and none of the three was ever subjected to flight tests. (Author's collection)

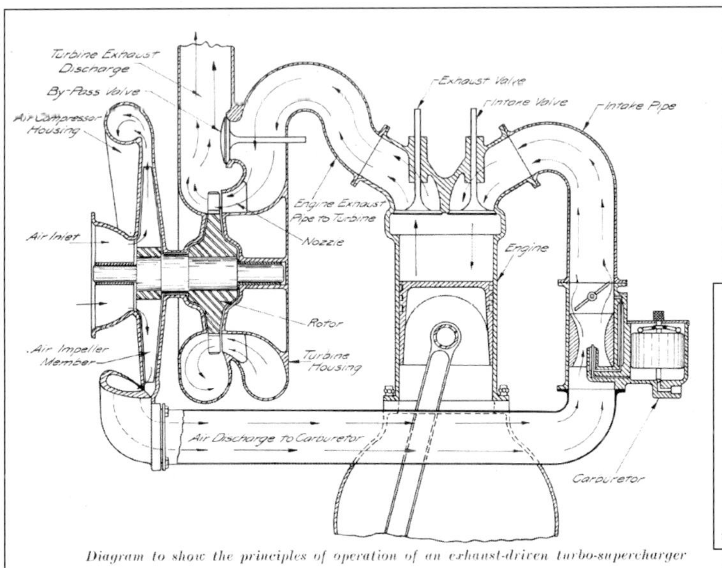

Diagram to show the principles of operation of an exhaust-driven turbo-supercharger

At left is a diagram showing the general principles of operation of the exhaust-driven turbosupercharger. Below is a diagram showing the distribution of combustion energy of an internal combustion four-cycle engine of the period. The reader will note that only about 28-30% of the energy is actually transformed into useful work. 40-50% of the energy is lost as exhaust. The turbosupercharger taps this lost energy and uses it to maintain sea level pressure input air to the engine. In doing so it can maintain sea level power output from an engine to 25,000 ft. and above, an altitude at which engine power would normally be reduced to about 26%.

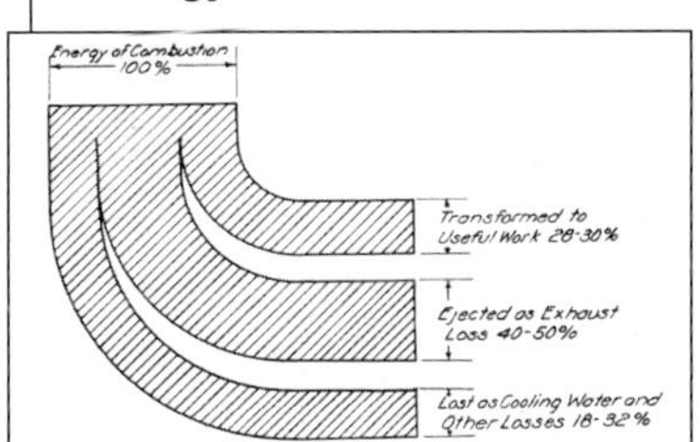

Diagram showing combustion energy distribution of turbo-supercharger and engine cycle

Sherbondy took out a number of patents on various features of these three supercharger models. All were applied for between May and August of 1918 and issued on various dates between 1920 and 1925. There numbers were: 1,346,565, 1,400,323, 1.436,976 and 1,547,695.

Before I leave the Sherbondy I will note that some historical documentation referencing this design call it the Form A-1. This material was generated in 1926 purely for historical purposes and the author feels that the Form number "A-1" was certainly generated "after the fact," and that the Sherbondy was never known by that name when it was an active project. Logically the Sherbondy would have been called the "A" and the Moss would have been the "A-1" since the Sherbondy actually preceded the Moss.

Form A Moss turbosupercharger photographed at McCook Field. This is the only picture the author has seen with a data plate on the unit. At left is an extreme enlargement of it. Most of the stamped numbers cannot be read but it can be seen it is a 20,000 ft. unit and also that the plate was not attached but just sitting on the top strengthening web for the photograph. (Courtesy of the Ludvigsen Library)

The term "Form" was first used by GE and can be seen on the Form A data plate pictured below-left. The Sherbondy was actually built in three different models, all of which seem to have been included in the later designation of Form A-1.

DEVELOPMENT OF THE MOSS TURBOSUPERCHARGER

As already noted, Dr. Moss began work on his supercharger design late in November of 1917 (Order CS 78), some five-months behind the start of the Sherbondy effort. In spite of the later start, the Moss machine had been designed, constructed, steam-tested at GE's Lynn River Works and was delivered to McCook Field on June 4, 1918, about the same time the first Sherbondy arrived. After preliminary tests, the machine was fitted to Lincoln Liberty engine MNFG NO. 59 U.S. NO. 23486 and operational test begun on June 19. This engine was assigned to this project and continued to be used with the No. 1 Moss supercharger through the end of 1918. This design became known as the "Form A."

Testing continued from June 17 to about July 7 and showed minimal problems. The rear bearing was water-cooled and proved to be reasonably trouble free. The supercharger was run under two controlled conditions, with the waste gate almost totally open and the compressor output connected to the carburetor intakes, or with the waste gates closed and the output of the compressor routed to the open air. The first setup would turn the compressor just fast enough to supply moderate supercharging to the engine. This moderation was necessary to prevent severe pre-ignition and engine destruction which would occur with high boost at McCook Field's altitude of 750 feet with the very low octane fuels in use. Even so, with the 470 HP developed during one of these tests spark plugs failed and additional problems developed as well. The second setup would turn the turbine and compressor at maximum RPM and subject them to maximum stresses and also subject the turbine to maximum heat exposure and accomplish both without destroying the engine.

The supercharger complete with all mounting brackets and hardware and the exhaust manifolds weighed 170 pounds. Since the exhaust manifolds weighed 53 pounds the additional weight above that usually quoted for the Liberty engine alone would be about 117 pounds.

At that point it was decided that the design was worthy of continued tests at altitude. A much refined and improved version of the truck-mounted test stand used on Pikes Peak to test U.S.A. 8 No. 2 engine a year earlier (see page 55) was ordered built. The weekly report of August 24-31 noted its completion. The complete outfit left McCook Field on September 1, 1918 headed for Colorado

Form A Moss turbosupercharger photographed at McCook Field in July of 1918, prior to Pikes Peak testing. An open-air test shed can be seen to the rear of the engine. The front air intake is in the front center of the compressor housing and the rear one can be seen on the lower-left. (Courtesy of the Ludvigsen Library)

Form A Moss turbosupercharger showing blower and turbine wheel. The top of the rear compressor housing and the nozzle box have been removed. The rear water-cooled bearing housing and air entry tube are visible in this view. (Courtesy of the Ludvigsen Library)

Form A Moss turbosupercharger photographed on Pikes Peak in late 1918. The complete unit is shown removed from the engine and includes the standard Zenith carburetors and linkage. As compared to the earlier photograph at right, an air duct has been added to provide additional cooling to the rear bearing. (Courtesy of the Ludvigsen Library)

Form A Moss turbosupercharger photographed at McCook Field. Note "LEFT" written with chalk on the left exhaust manifold to avoid inadvertently mounting the left and right in reverse. The 3/4 left, left and front views seen on this and the prior page are the full views most often seen of this first Moss unit. In most cases, however, the versions seen have been airbrushed by the Engineering Division to remove the background, a common practice done to official release photographs to remove detracting background material. (Courtesy of the Ludvigsen Library)

The Pikes Peak test truck at McCook Field just after completion in August 1918. The truck was a Packard and appears to be a Government "3-E 13 Special No. 2" model of 1918. The Liberty engine is mounted on a roller-bearing cradle torque stand (shown with the supercharger removed) which was equipped with a 63.025 in. brake arm which extended through the instrument board to bear on the platform of a very accurate scale. In this manner torque could be measured and horsepower accurately calculated. Power was absorbed by the propeller which could be of a design to absorb full power at any desired speed. Two radiators were mounted, one to the left of the engine just behind the instrument board and one to the right. The propeller slipstream was sufficient to maintain normal temperatures. Oil and fuel containers were mounted on the truck platform. All required operating and test instruments and controls were mounted on the instrument panel. The truck thus became a complete portable dynamometer laboratory. (Courtesy of the Ludvigsen Library)

Above is shown the test rig under canvas cover still mounted on the railroad flat car on which it was transported from McCook Field to Colorado Springs, Colorado. At left is the test rig set up on Pikes Peak at 14,109 feet above sea level. The engine is running and an operator is sitting on an elevated platform above the fuel barrel. Another is standing by the truck. Notice the chains on the rear wheels. (Courtesy of the Ludvigsen Library)

Springs, Colorado and then to the top of Pikes Peak. Initial engine testing was done in Colorado Springs from September 6-13, it was then driven up the mountain where testing at 14,109 ft. continued from September 14 through October 7, 1918.

Although some problems were encountered, they were overcome on-site. The lessons learned would result in refinements in later models. No serious overheating was encountered and it was found that the supercharger (designed as a 20,000-foot unit) could maintain the engine's full sea level power at 14,109 feet. Output of the engine under full power at McCook Field was 354 HP @ 1,800 RPM. Output at that speed on Pikes Peak was 355 HP. Engine output at 1,800 RPM without any supercharging was 230 HP.

Upon return to McCook Field the decision was made to continue testing in actual flight. The October 19-26 weekly report noted that the Moss-equipped Liberty engine was being installed in a LePere plane for tests under the supervision of Dr. Moss. This was P-53, the aircraft which went on to considerable fame as a record-making high altitude craft.

Shortly thereafter orders were issued to build an additional supercharger, test it on a Liberty engine at McCook Field and then take it to Pikes Peak for altitude testing (Orders CS-212, 213 and 214). All three orders were canceled on November 13, two days after the Armistice.

OTHER WAR TIME SUPERCHARGER DESIGNS

Apparently the Engineering Division let it be known that they were interested in considering any supercharger designs outside engineers might wish to submit as several were examined. None, however, showed sufficient promise to hold the Division's interest for long. Most did not show enough merit to leave much of a trail of testing or engineering analysis.

THE FERGUS TURBO COMPRESSOR

The earliest superchargers the author has noted, other than the Sherbondy and Moss designs, seem to have been initiated very early in 1918. In late 1917 or very early 1918 a "turbo compressor" design was submitted to the Engineering Division by Fergus Motors of America, Inc., of Newark, New Jersey. Division engineers deemed the design worthy of in-depth testing and placed an order for three examples under order CS-84. Since preliminary testing was done at the DeLaval Steam Turbine Co. this was probably an exhaust driven type. The author has as yet found no test reports of the device, however.

The history of this order can be followed in the weekly reports which note the first unit was delivered during the first week of June of 1918. The remaining two were reported one half complete at this time. A few tests were made and during the third week of June this unit was returned to the factory to have several "minor" changes made. In the meantime work was halted on the remaining two.

During the last week of June Fergus reported they were testing unit No. 1 and had started work on No. 2. They reported that the No. 1 unit would be shipped to McCook Field on July 15 and that No. 2 had finished tests at the DeLaval Steam Turbine Co. which revealed a defect in the rear blower casing which had to be corrected, following which No. 2 would be shipped to McCook Field.

Nothing more is heard of the Fergus machines until the second week of August when the first two units were reported as under tests and work was started on the No. 3 unit during the week.

The third machine was reported ready for assembly on the first of November. It was sent to DeLaval for testing during the first week of November and was expected to ship to McCook Field about November 19. During the following week it was reported that No. 3 machine had been run for 4 hours at 28,500 RPM with the maximum being 31,500 RPM and that it would be shipped to McCook Field the following week. It was reported as received at McCook Field about the first of January in 1919.

Unfortunately no tests or photographs of any Fergus turbosupercharger has surfaced. Although the design evidently had merit and there are no indications of any catastrophic failures during McCook Field testing, the design was soon dropped and all supercharger work directed to the Moss design.

LE BLANC/KERR GEAR DRIVE COMPRESSOR

Some development work on a gear-driven centrifugal supercharger was started in France under the direction of the AEF in late 1917 or early 1918. The design was initiated by Monsieur Le Blanc with the aid of Captain Kerr. The project was brought to McCook Field and actual construction was apparently done by Allison Engineering. The design, however, was not a success and was dropped, probably after the Armistice.

THE GATES TURBO-BOOSTER BLOWER

The above name was applied to this device by the Engineering Division in their test report serial No. 15 of April 1918. Generally the term "turbo" is applied to exhaust-driven compressors but in this case it was belt-driven. It was manufactured by M. E. Gates and W. P. Gates of Chicago, Illinois and is described below.

"The general construction of the blower may be said to be somewhat like a turbine, it having a rotor about 10 inches in diameter with both impulse and reaction blades. The driven pulley on the rotor is very small, about 1 1/2 inches in diameter, while the driving pulley is very much larger, about 10 inches in diameter. The drive is by means of four pure rubber belts which are stretched very tightly over the two pulleys. The driving pulley is to be connected directly to the engine and run at engine speed, thereby allowing the driven pulley and rotor to run much faster, due to the belt drive reduction upwards of 14,000 RPM. The air outlet of the blower is to be connected directly to a special carburetor, the entire arrangement being expected to give greater power at high altitude."

The test was executed by driving the blower with an electric motor at various speeds and with two different kinds of belts and measuring the air driving capacity under several pressure loads.

It was found that the blower would deliver only 311 cubic feet of air per minute against a pressure of .49 pounds whereas the requirements to supply a Liberty engine is about 800 cubic feet per minute at 1600 RPM against at least 2 pounds of pressure.

Thus its capacity was found to be entirely unsatisfactory. In addition, the belt slip was found to be very high and the belts would heat and begin to go to pieces in a "very few minutes." The device was thus found to be entirely unsatisfactory both in capacity and method of drive and was rejected.

WHY THE LEPERE WAS PICKED AS THE FLIGHT TEST PLANE

The LUSAC-11 (P-53) project was also briefly suspended. As noted in the report of the first flight tests given in *Technical Orders No. 10* of November 1919, "--the suspension of hostilities was followed by a period of uncertainty as to the future, during which all further work on superchargers was abandoned. As soon as the Engineering Division was reorganized, plans for continuing work on the supercharger were resumed and arrangements were made with the General Electric Company to

LUSAC-11 P-53 in its early configuration with a Moss Form A supercharger. A small auxiliary radiator was added below the wing to boost cooling at high altitudes. This was found necessary because of the reduced cooling ability of the less-dense air at high altitudes. Although not usually a problem with engines which were not supercharged because their power output and thus heat generation reduced as altitude was gained, it is a problem when the supercharger maintains sea level power, and thus heat generation, at high altitude. (Author's collection)

Top (McCook 5830) and bottom (McCook 3603) showing the Form A as modified for its first flight tests. The top view shows more clearly the square air input at the top for cooling the turbine. An additional air input is provided at the lower-right. (Courtesy of the Ludvigsen Library)

put the supercharger in condition for flight test. This included a general overhaul of the machine with the design and construction of new thrust washers, new exhaust headers and a new compressor casing, the necessity for this having been demonstrated by the tests at Pike's Peak."

The plane and engine were ready and testing was commenced on July 12, 1919. The LePere LUSAC-11 had been chosen over all other available planes because it appeared best suited for the purpose. "The structural strength, especially around the engine mounting, was excellent. The seating arrangement was favorable to close co-operation between pilot and observer. A nose radiator was out of the question as the supercharger occupies most of the space between the front cylinders and the propeller. The excellent performance of this type was favorable to a good performance when equipped with a supercharger. Airplane P-53 was selected for the installation, as this airplane was slightly tail heavy and for that reason the additional weight of the supercharger on the nose was not objectionable." The supercharger added about 110 pounds of weight. In addition, the LePere used a top wing radiator which relieved the need for a conventional nose radiator.

SUPERCHARGER CHANGES MADE BEFORE FLIGHT TESTS

As soon as it was determined that the supercharger project was to be continued, the unit was returned to General Electric for rebuild and installation of those modifications which had been determined as advisable following the Pikes Peak tests.

These consisted of the installation of redesigned thrust washers, new exhaust headers and a new compressor casing. The external differences in the exhaust manifold and compressor casing can be seen by comparing the illustrations on pages 330-331 with those on page 335.

It should be noted here that the basic design of the Form A, and the following Form B, C and D, were that of Dr. Moss and he continued to be the chief design engineer controlling the project at GE. However, once this became an Engineering Division project, ultimate control of design suggestions and acceptance became that of Engineering Division engineers. From this point in time a number of design improvements and even basically new designs were at the initiation of the Engineering Division, the design of the Form F being probably the best example. General Electric was the contractor that refined and built the designs and even participated directly in their flight testing.

P-53 with Moss Form A supercharger with modified compressor casing and exhaust manifolds. The illustration on the prior page shows the installation from the other side but after installation of the wings. (Author's collection)

Engine failure of the supercharged P-53 on July 12, 1919. The test report stated "On landing it was found that a connecting rod had broken and the crankcase was practically cut in two." Obviously a safe landing was made. Whether or not the engine was still running was not stated. (Courtesy of the Ludvigsen Library)

FIRST FLIGHT TESTS IN P-53

The engine was then installed on P-53 and given a ground run test. A couple of trials were made during which supercharging was allowed while on the ground. The resulting preignition caused some damage to the engine and required in an engine change.

The new engine was installed and the first flight was made on July 12, 1919 with Major Schroeder as pilot and Lieutenant Elsey as observer. This flight proved to be disappointing as engine trouble developed shortly after the supercharger was cut in. Upon landing it was found that a connecting rod had broken and cut a large hole in the crankcase. The problem was determined to have been caused by engine overheating.

P-53 in flight with early configuration Form A supercharger and the additional above wing radiator fitted to overcome cooling problems. This radiator is only seen in late 1919 photographs. This image comes from an in-flight photograph taken over McCook Field which showed much of the field in the background. The version most often seen has had most of the background view of the field air-brushed away in order to present the airplane without distraction. Unfortunately, in that version the above wing radiator was removed along with the background. (Courtesy of the Ludvigsen Library)

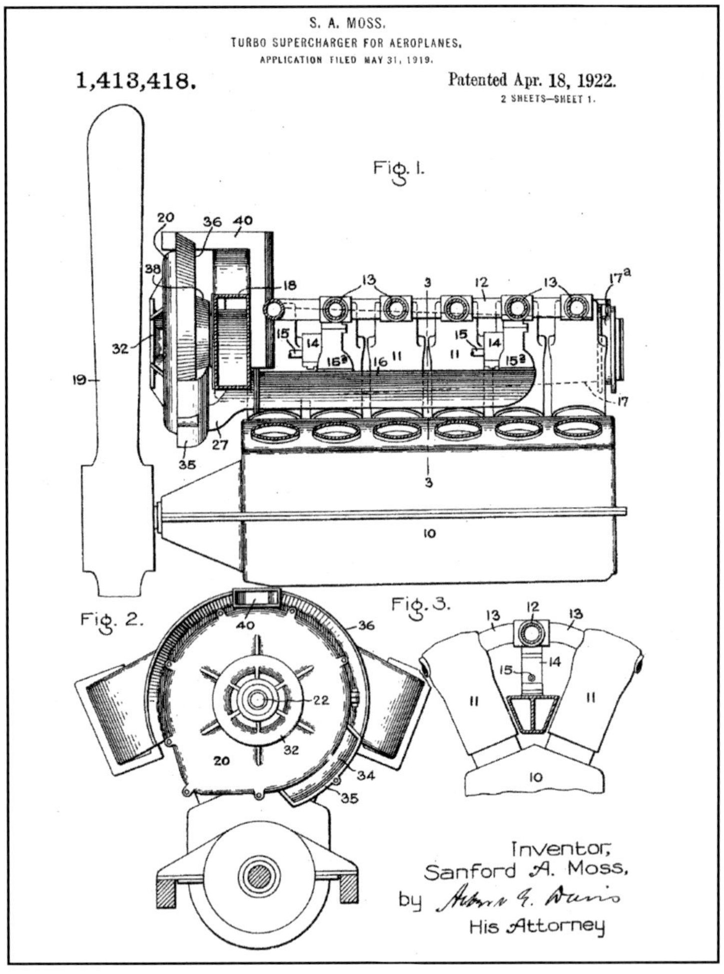

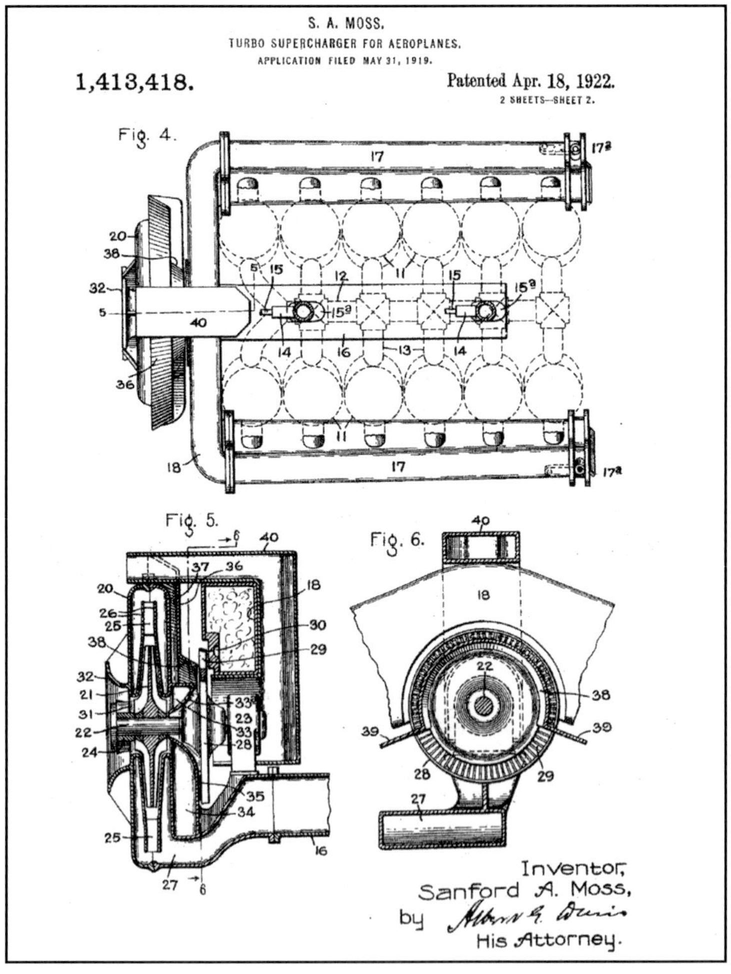

Here are the two sheets of drawings of Moss patent 1,413,418 covering the Form A supercharger in its confuguration as modified before flight testing. The newly installed air duct for cooling the turbine and rear bearing is clearly shown, as are the waste gates and various other design features. (Author's collection)

A new engine was installed (Nordyke & Marmon US No. 4672) and a thermometer mounted in front of the wing radiator and flight tests continued. A flight was made on July 23rd which showed that heating problems were being caused by hot exhaust gasses flowing back over the engine and into the wing radiator. Exhaust was also causing much discomfort to the pilot and observer. As a result an exhaust pipe was added to the waste gate exits and a deflector was added to the left-side of the turbine wheel to deflect exhaust out of the right-side where the prop blast would carry it clear of the radiator and crew.

A flight was made on August 2 which gave good results. At an altitude of 18,400 feet a level flight speed of 130 MPH was obtained. Some engine overheating was still noted so upon return it was decided to install an additional auxiliary radiator above the wing.

After a number of additional flights it was decided to make a flight to the maximum altitude obtainable. This resulted in many flights being made, several of which resulted in the setting of new two-man world altitude records as noted below, the last of which stood until early 1921.

September 6, 1919	28,500 feet
September 24, 1919	30,900 feet
October 4, 1919	31,821 feet

The reason for so many flights was that the supercharger was designed to operate at an altitude of about 20,000 feet and when operated at speeds required to obtain enough power to exceed about 22,000 feet it began to have mechanical failures. Most of these were displacement of rotor buckets which resulted in rotor imbalance and excessive vibration. This in-turn resulted in broken oil or water lines. Turbine repairs and improvements were made between flights until they were halted temporarily after the October 4 flight.

After a short time testing was resumed and on February 27, 1920 Major Schroeder set a one-man world altitude record of 33,113 feet. The flight was certainly the most eventful of Schroeder's life as it almost ended it. Schroeder had left McCook Field in P-53 and climbed for one hour and forty-seven minutes when his oxygen supply failed. He rapidly passed out and the plane then went into an out-of-control dive. The recording barograph indicated an altitude of 36,130 feet at the start of his decent. (With various corrections factored in, the official altitude was later set at 33,113 feet.) This barograph plotted time and altitude and showed that during this dive the airplane lost 30,600 feet in altitude in 2 minutes and 3 seconds giving an average rate of decent of 190 MPH. During the most rapid part of the dive a speed of almost 300 MPH was reached. The plane miraculously did not shed its wings during the decent and Schroeder regained consciousness at about 3,000 feet and managed to regain control and glide to a landing at McCook Field. He suffered from the effects of oxygen starvation, rapid pressurization and his eyes were frozen open when he was found in his plane.

THE FORM B

At the conclusion of this series of tests a number of changes were suggested. "Improvements in future designs may be looked for mainly along the following lines: greater simplicity in supports, bearings and gas-tight joints, better means of preventing the large rise in temperature of the compressed air, modification of the shape of the exhaust manifold and size and location of the waste gates to prevent back pressure on the engine and alteration in the nozzle area to better handle the volume of gas leaving the engine." Their incorporation resulted in the next design which was named the "Form B."

It was also noted that a controllable pitch prop was very much needed so as to take advantage of the engine's power at both low and high altitude. This last item was not to be perfected for several years, however.

Incorporation of the above changes would alter the appearance of the exhaust manifold, the shape of the nozzle box and add an aftercooler. It would also cause other more subtle changes which may not be evident in external appearance, among them was a change from water-cooling of the bearings to air-cooling. This feature carried through to the Forms C, D and 35,000' model as well.

Work was immediately started on the Form B. This version did not remain in use very long and probably only a few units were built. For this reason its history has not been easy to trace. Current evidence indicates that one was used for testing on P-53, two were mounted on a Martin MB-2 bomber, P-162, and one was flight-tested on DH-4B P-131 and one on DH-4B P-190.

This new design differed considerably in outward appearance from the Form A, the principal reason being the addition of an air cooler (aftercooler) to reduce the temperature of the compressed air fed to the carburetors. This cooler had to be exposed to the slip stream in order to do this. This required the air output point on the compressor to be moved to a new location. On the Form A this output came from the bottom of the compressor and fed an air plenum in the engine vee which directly fed into the two Zenith carburetors. In order to allow the aftercooler to be in the slip stream it was placed above the supercharger and engine. This required the compressor air output to be taken from the top and the aftercooler to be attached at that point. The supercharger appearance was thus much changed.

Form B supercharger mounted on a Martin MB-2 bomber P-162. McCook Field photo 9822. This photo was Air Service captioned as "Form B Supercharger, a later development which enabled Lt. Macready to break Major Schroeder's record." In this case it was equipped with the GE-designed tubular aftercooler. (Courtesy of Peter Bowers)

Form B supercharger (with honeycomb aftercooler) on P-53 (McCook Field photo 10646) taken just after Lt. Macready set a new one-man world altitude record of 34,509 feet on September 28, 1921. All Form B, C and D superchargers used aftercoolers (any one of several designs) and had front compressor cases so similar in design they are almost impossible to tell apart. The externally distinguishing item seems to be the flat design exhaust manifold and nozzle box entry. (Author's collection)

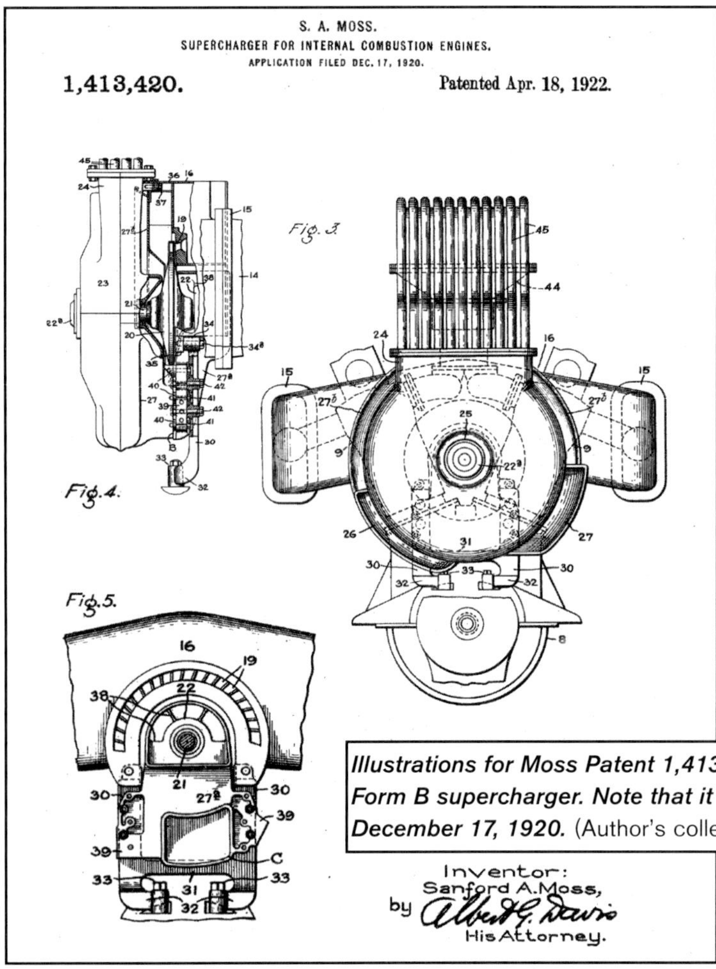

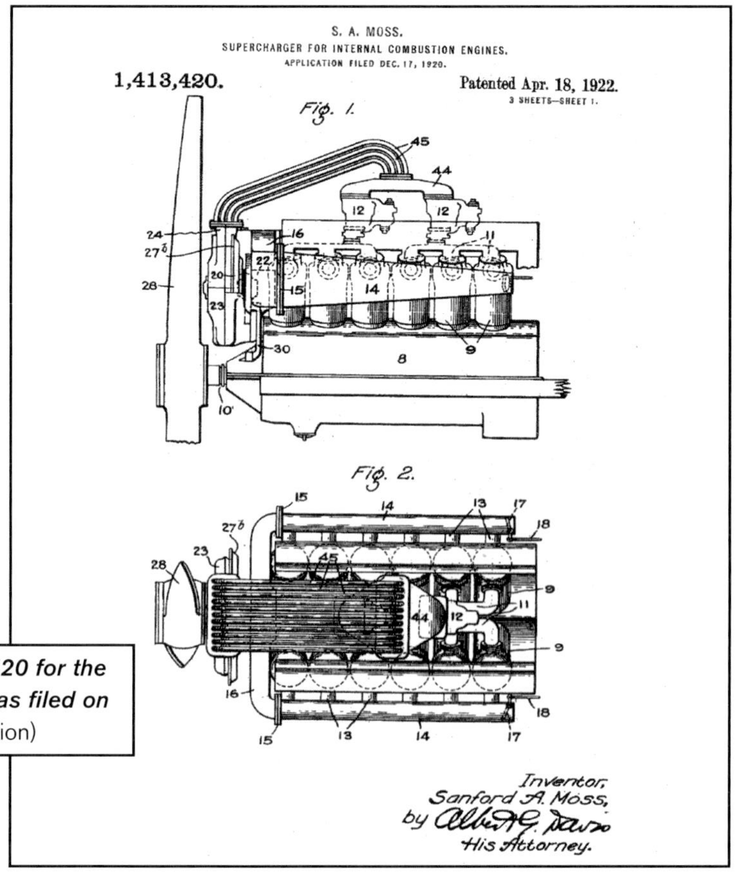

Illustrations for Moss Patent 1,413,420 for the Form B supercharger. Note that it was filed on December 17, 1920. (Author's collection)

This new air entry point for the engine required the use of carburetors with air intake on the top rather than on the bottom. The Engineering Division had been working on an "inverted" carburetor since late 1918 and had by 1920 decided on a design by Stromberg designated as the NA-L5 (see page 274). Although it was originally intended that the inverted carburetors be adopted for use on all Liberty engines, it was eventually found that their performance was not sufficiently improved over the Zenith to warrant a blanket change. They did prove to be just what was needed for the new design superchargers to feed air input to the top of the carburetors. By July of 1922 the carburetor was being noted as designed for use with superchargers and by October of 1922 it was noted that it was the standard carburetor for use on Form B and Form D superchargers. 150 examples had been ordered in 1922 in anticipation of the arrival of the fairly large quantity of Form D superchargers then on order.

It is evident by the application date of December 17, 1920, for the Moss patent 1,413,420 issued for the design of the Form B, that the first unit had probably been built by January of 1921 or slightly before.

THE FORM C

In the Spring of 1920 the Wright Model H engine was favorably considered for pursuit planes and it was decided to provide a turbosupercharger for this engine. On August 30, 1920 Contract 305 was let to GE for the construction of two turbosuperchargers for that engine of the same general design as those being built for the Liberty 12 but sized to accommodate the Wright Model H operated at 2,000 RPM at 20,000 feet with a turbine speed of 22,000 RPM.

The first unit was delivered to McCook Field in late 1921, installed on DH-4B P-188 and flight-tested on January 22, 1922. Some minor changes were made to overcome bearing oiling difficulties, exhaust gas distribution and aircraft balance followed by satisfactory flight tests by Lt. Macready with Mr. David Gregg as observer on January 32, 1922. The only negative observation of note was the lack of effectiveness of the GE tubular aftercooler.

After several other flight tests by Lt. Nelson, a flight was made to 20,000 feet with full readings taken which indicated sea level manifold pressure could easily be maintained to that altitude. As this installation was solely to test the operation of the supercharger, and that had proved satisfactory, the supercharger was removed and the plane released for other work.

Meanwhile the second unit had arrived from GE and was installed on Orenco D-1 P-161. Lt. Macready made the first flight test with this plane on March 23, 1922 and reported its operation was quite satisfactory. On March 28 he made a climb test to 20,000 feet which took 33 1/2 minutes. In a later test that time was improved to 25 minutes.

During the later flight testing trouble was experienced due to cracking of the nozzle box around the nozzle ring at the point where the exhaust pipes enter. This was due to the lack of expansion joints to accommodate the unequal expansion caused by hot exhaust gasses. They felt the problem could easily be overcome if need be.

Flight-testing continued for some time but by the time the formal report, serial 2027 of August 3, 1922, was published, it had been decided not to proceed further with this model. The reasons were as follows: "At the present time the trend of pursuit engine development is more toward a 12-cylinder engine on the lines of the Liberty or Packard. A side supercharger has been designed and constructed for (the) Packard 1237 engine and seems to offer considerable promise. The resistance is less than with the present front type of supercharger and the visibility and ease of mounting arrangement is improved.

"In view of the foregoing, and until tests have been completed with the side type supercharger, it is suggested that no further development work be done with the General Electric Form "C" supercharger."

The side-type supercharger, as we shall see, proved quite satisfactory and became the Form F series. It was the replacement for all the front-type GE superchargers. Thus only two Form C units were built. None are known to have survived.

Current evidence indicates the first batch of Form D superchargers for the Liberty were probably ordered from General Electric at about the same time as the Form C units. Their designs were the same except for capacity and mechanical differences required for the different engines for which they were built. The earliest photo of a Form D is McCook Field 12049 showing it on a DH-4B. This is an earlier numbered photo than the earliest known Form C photo (No. 12246). In addition, the only known surviving Form D supercharger (in the Smithsonian collection at the NASM) carries a data plate dated April 18, 1922.

THE FORM D

The first batch of Form D superchargers had been received at McCook Field at least by April of 1922 and possibly a little earlier. The actual number is currently unknown but was likely at least six as it was expected to equip a number of Liberty 12 engines for testing. Initially, at least, the P-53, an MB-2, and one or more DH-4B airplanes. These units were rated to produce 860 cu. ft. per minute of sea level pressure air at 20,000 feet with a

The top two illustrations show Form C supercharger No. 1 installed on DH-4BW P-188 and are McCook Field photos 12246 and 12247. The center two and bottom right illustrations show Form C supercharger No. 2 installed on Orenco D-1 P-161 and are McCook Field photos 12560, 12561 and 12556. These are believed to be the only two Form C superchargers built. All photos were taken in early 1922. (National Archives photos courtesy of Kim McCutcheon)

turbine speed of 22,000 RPM. In actual practice they were capable of even better performance.

The first instruction manual for this model was dated June 8, 1922 and used McCook Field photo 12049 seen on the following page to show an installed unit. By the time this manual was published there were already two types of thrust washer arrangements in use on the impeller/turbine shaft.

The shaft was supported on two air-cooled babbitt bearings lubricated by a 1/4" oil line connected to the front oil feed to the camshafts. This feature was also used on the Form B, C and 35,000' model. The principal changes from the Form B were a new exhaust manifold and nozzle box.

Since testing of MB-2 P-162 with the Form B superchargers had been very successful it had been decided to equip the last 20 of the 50 on order from Curtiss with Form D superchargers. (AS 68508-68527) Technical Bulletin 34 of the second quarter of 1923 noted that in addition to the 20 supercharged Curtiss NBS-1 aircraft then being tested at Curtiss, 12 Form D superchargers were being assembled at Fairfield Air Intermediate Depot for installation on DH-4B airplanes for service tests.

NASM Form D Moss supercharger mounted on Packard L-12 MNFG NO. 4293 U.S. NO. 21925. The supercharger is General Electric No. 5936 Form D 20,000-foot 22,000 RPM manufactured April 18, 1922. It is the only known surviving Form D supercharger. (Author's photo)

Two examples are pictured on page 261, both of which were on engines under test in DH-4B aircraft on which propeller failures occurred.

Although we have found no definitive production figures for the Form D supercharger, known usage would indicate at least 60 and possibly 70 units must have been built. One is known to have survived in the NASM collection, it is GE Number 5936 with a build date of April 18, 1922, which would make it one of the first production batch. It is mounted on Packard Liberty U.S. NO. 21925 MNFG. NO. 4293.

Form D supercharger on a DH-4B McCook Field photo 12049 and probably the first Form D supercharger delivered in early 1922. (Courtesy of the Air Force Museum)

Above photos 14235 and 14234 show P-53 with its first Form D installed early in 1922. (Courtesy of the Air Force Museum)

THE EXTREME ALTITUDE TURBOSUPERCHARGER

The success of Macready's September 28, 1921 flight with a Form B supercharger appears to have not only insured continued development of the 20,000-foot type, which resulted in the Form C and D, but inspired a new "extreme altitude" type.

Air Service Bulletin 141 published March 1922 described what was referred to as the 35,000 ft. supercharger, as follows:

"Resume: A turbine supercharger has been designed and built specifically for extreme altitude operation. The recent altitude record climb of 40,800 ft. altitude (Indicated altitude of the Lieut. J. A. Macready flight of Sept. 28, 1921) has accentuated the feasibility of an airplane to operate at these high altitudes. The new supercharger is rated to give sea level pressure at 35,000 ft. altitude, but judging from the performance of the 20,000 ft. machines, the absolute ceiling of a plane equipped with this new supercharger will probably depend chiefly on propeller limitations and limits of human endurance.

"General description: In its general appearance the 35,000 ft. supercharger is like the Form "B" 20,000 ft. machine. It is a turbine compressor set driven by the red hot exhaust gases of the Liberty 12 engine. The compressor casing is of aluminum alloy and stands only about 20" high (exclusive of the air cooler). The entire outfit as shipped from the factory (not counting the exhaust manifold of the engine) weighs about 140 lbs. *(Author's note: The Form D weighed about 110 pounds.)* This additional weight carried by the plane will increase the power of the Liberty 12 at 35,000 ft. by at least 280 HP (taking the brake power proportional to atmospheric density), an increase of 2 HP for each extra pound carried."

Above is the latest photo of P-53 known to the author. The McCook Field photo number of 19859 dates it in the era of November-December of 1923. It is well past the time at which testing of the Form D began on P-53 (see McCook Field photo 14235 which dates about January of 1922) and is within a year of the retirement of P-53 to the museum on November 11, 1924. The supercharger seen in this photo is not, of course, the Form D but fits the description of the 35,000' supercharger found in Air Service Bulletin 141 dated March 1922. (Courtesy of Peter Bowers)

The paper goes on to describe some of the methods used to produce a turbine and single-stage compressor which would compress the required 2,200 cu. ft. of air per minute. That ability required a compressor speed of 33,000 RPM which gave an impeller tip speed of 1,500 ft. per second. The shaft power necessary to compress that quantity of atmospheric air was calculated as about 110 HP. The supercharger was given an overspeed test to 41,000 RPM without failure.

The author has found no formal test papers or photographs of this model but the described 20" height of the compressor case is almost exactly the same as the Forms B, C and D machines. With its described general appearance being like the Form B, I would be inclined to think it would very closely resemble the Form B machine.

As a matter of fact, the latest photograph of P-53 the author has is McCook Field photo 19859. This photo shows the airplane in its latest known color and marking configuration, the same as photo 14235 showing it with a Form D supercharger. However, in 19859 it is fitted with a supercharger which looks very much like a Form B but with a slightly larger aftercooler. The author believes that this is the 35,000' supercharger. The McCook Field photo number of 19859 dates it in the time period of November-December 1923.

Technical Bulletin 37 of November-December, 1923 records the failed attempt by Lt. J. A. Macready, in P-53 equipped with this supercharger, to set a new record. Upon reaching an indicated 36,000' (later calibrated to approximately 32,000') the Kingsbury bearing around the turbine shaft melted and froze the shaft.

It was noted that further attempts would be made, but reports of them have not been found. Although logic might seem to say this would have become the Form E, had it been successful, that was not the case. During this general period of time GE also designed and built a gear-driven centrifugal supercharger for use on the Liberty and Curtiss D-12 and it was this supercharger to which the designation Form E was applied.

The side-type supercharger was in early development at this same time and had just been installed and flight tested on a Liberty powered DH-4B. Those tests rapidly proved so encouraging that all future development was dropped from the front type unit and centered on the new side-type, soon to become the Form F.

By early first quarter of 1924 a side-type had been installed on Engineering Division TP-1 in place of its Form D unit and testing begun in earnest. Macready's September 28, 1921 record was not broken by an American until he did so on January 29, 1926 using a side-type supercharger on the XCO-5A. Thus the record-breaking capabilities expected of the 35,000' model never materialized. The Form D was the end of the successful front-type.

THE FORM F

I noted during the discussion of the Form C that the Engineering Division had designed and built a new design turbosupercharger called the side-type. Its design is credited to Lt. E. T. Jones, who became acting chief of the Power Plant Section replacing Major George E. A. Hallett in 1922.

Although the same basic principals were used, this was physically a major redesign done to eliminate as many of the drawbacks found in the front-mounted layout as possible. To explain these problems and the methods used to obviate them with the new design, I will quote extensively from report serial 2365 of June 6, 1924, on the new "side-type turbosupercharger."

This is the first Form F supercharger built. It was installed and tested on a Packard 1A-1237 engine and the engine installed in DH-4B P-188, the same plane previously used for testing Form C supercharger No. 1 in January of 1922. Testing of the Form F began in August of 1922 and after very successful bench and flight tests, the unit was returned to GE to be converted for installation on a Liberty 12. Note that the aftercooler is below the engine as that is the location of the carburetor on the 1A-1237. (Author's collection)

DRAWBACKS OF THE NOSE-TYPE

"In order to appreciate the advantages of the side-type supercharger, it might be well to consider briefly the defects of its predecessor, the nose-type. This supercharger has been used successfully for the past three years and has been responsible for a number of World's records and for a marked increase in the performance of the airplanes on which it has been installed. However, it has a number of faults and serious defects which the side-type supercharger was designed to eliminate.

"The turbine wheel rotates in a chamber formed by the nozzle box and the back of the compressor casing. The exhaust gasses after expanding and driving the turbine wheel must leave through a somewhat restricted opening between these two parts. As a result, the turbine wheel is continually surrounded by hot exhaust gas with little effective cooling, and, under these conditions, the turbine buckets frequently stretch and break. Failure is due to the fact that the stress set up in the rotating buckets exceeds the elastic limit of the material at the high operating temperature. It is necessary to provide a steel baffle plate back of the compressor casing to insulate it from the hot gases which discharge directly onto it. The nozzle box is only 1/8" from the two front cylinders and heat is transmitted from the nozzle box to the water jackets. It is necessary to wrap the front water hose connections and the front spark plug wires with asbestos tape in order to keep the rubber from disintegrating on account of the heat.

"To provide sufficient room for the supercharger between the front of the engine and the propeller, it was necessary to recess the propeller hub about 1", making a special hub necessary.

"The air cooler, located on top of the supercharger outlet, interferes with the pilot's vision and, together with the location of the supercharger and nozzle box, makes it impossible to cowl the engine in. Inverted carburetors are necessary and a rich mixture is used to keep the exhaust gas temperature below the point at which the turbine buckets stretch.

"Due in part to its location and in part to its general design, the nose-type supercharger is difficult to adjust and requires constant attention to keep it in running condition."

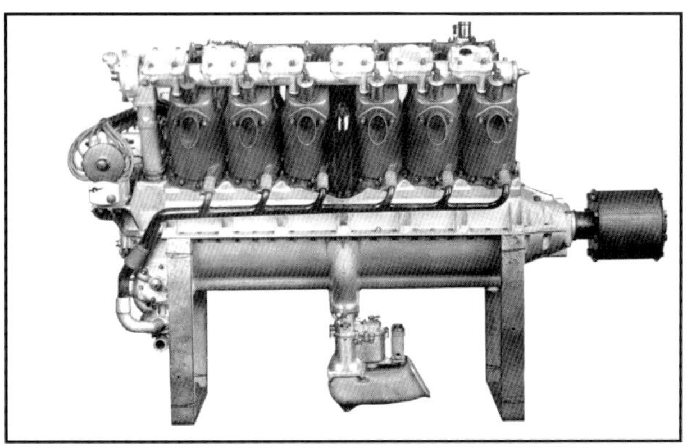

Above left is DH-4B P-188 (McCook Field photo 14656) showing the first Form F supercharger with its bottom-mounted aftercooler clearly visible. At right are two views of the airplane in this configuration. Flight-testing began in August of 1922 and was very successful with a maximum indicated altitude of 32,160' being reached. At left is illustrated a standard Packard 1A-1237 engine showing its normal bottom-mounted carburetor. (Author's collection)

THE FORM F DESIGN ADVANTAGES

"The first problem was to keep the turbine wheel cool and to provide a free discharge passage for the exhaust gas. This was done by mounting the supercharger parallel with the side of the engine — and having the turbine wheel overhung and exposed to the propeller slip stream. The nozzle box is of circular section which resists warping. It is located between the compressor casing and the turbine wheel so that the exhaust gases pass through the turbine buckets, directly into the open air. There is, therefore, a minimum of back pressure on the exhaust, and the turbine wheel and buckets are kept cool and do not stretch. As the compressor unit is next to the engine, the cowling can pass between the compressor and the nozzle box, thus all hot parts, that is, the manifolds, nozzle box, and turbine wheel are exposed to the propeller slip stream, while the other parts are inside the cowling and offer no additional resistance.

The first Form F after being modified for use on a Liberty 12. It was returned to McCook Field in September of 1923 and installed on DH-4B P-289. (McCook Field photo 19564) The aftercooler is now located in front of the supercharger with its output air plenum feeding an air manifold in the engine vee which feeds two standard Zenith updraft carburetors. The new side location of the Form F resulted in the abandoning of the Stromberg inverted carburetors and their special manifolds and a return to the original Zeniths. (National Archives photo courtesy of Kim McCutcheon)

"The air cooler is located in front of the compressor casing — and the air after passing through the core of the cooler, flows over the nozzle box and turbine wheel. The head resistance of the supercharger is thus kept to a minimum yet the air cooler has all the advantages of free air flow. It also offers no obstruction to vision or to the mounting of machine guns or armament on the airplane."

The two bearings were of plain-type lubricated from the camshaft oil feeds just as the front-type had been. Total weight of the units as built for the Liberty was 20 pounds less than the lightest nose-type.

"One particular advantage of the side-type supercharger over former types is, that it may be assembled and adjusted on the work bench and all clearances checked before it is mounted on the engine. The supercharger and engine may be removed as one unit from the airplane without disassembly."

It sounds so simple when explained that one might wonder why it was not done that way in the first place. However, engineering at its very best is rarely able to get to the best design on the first try.

THE FIRST FORM F IS BUILT FOR THE 1A-1237 PACKARD AND THEN CONVERTED TO A LIBERTY 12 UNIT

The above description, taken from the report which was written after successful tests on the Packard 1A-1237 engine, led to the supercharger being converted for the Liberty. Originally the aftercooler was mounted on the bottom of the engine just in front of the Packard's bottom-mounted carburetor (see illustrations on the prior pages), in which case the air flow to the turbine wheel was totally unobstructed.

When this design was first conceived the Packard 1A-1237 was being seriously considered as a potential standard pursuit engine. As a result it was decided to build the first sample to fit that engine. Its construction was contracted to GE and the unit was delivered to McCook Field on August 5, 1922. Judging from known contracts versus delivery dates of other supercharger models, it would be reasonable to estimate the date of its order at perhaps a year earlier, or about September of 1921.

After a short period of bench tests the engine with supercharger was installed in DH-4B US 63757 P-188, the same airplane in which the first Form C supercharger was tested in January-February of 1922. Testing was so successful that it was decided to remove the supercharger and send it back to GE to have it rebuilt for use on a Liberty 12. It would appear that any idea of adapting a quantity of Packard 1A-1237 engines to supercharging had been rapidly dropped even though testing of the combination had been quite successful. P-188 continued in use for some time afterward and was surveyed on August 18, 1926.

The first Form F supercharger installed on P-289 showing the very neat installation after the cowling is installed. The data plate is shown at left. (National Archives photo courtesy of Kim McCutcheon)

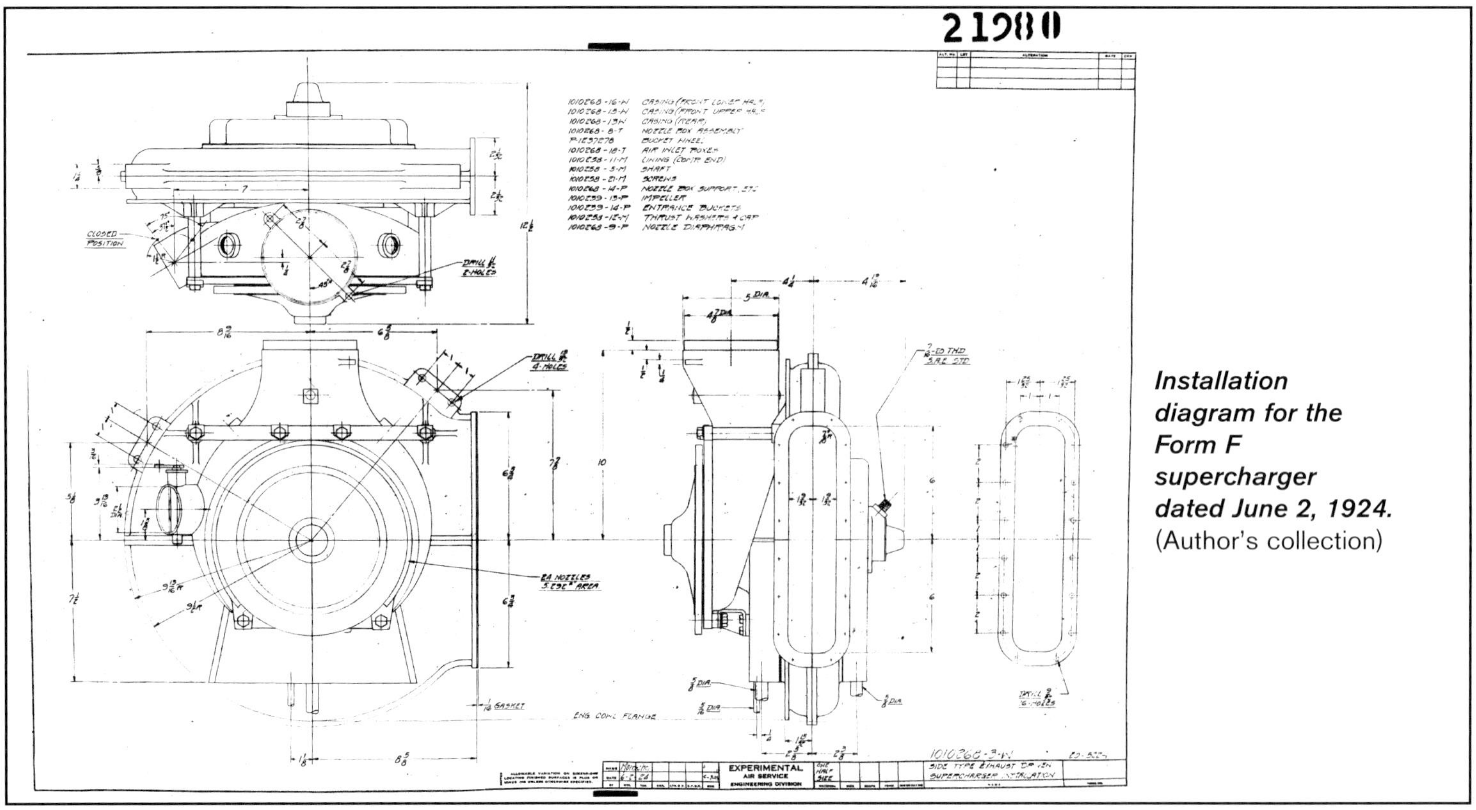

Installation diagram for the Form F supercharger dated June 2, 1924. (Author's collection)

The rebuilt supercharger was returned to McCook Field on September 18, 1923 and after bench tests on a Liberty 12 was installed on DH-4B US 22-1110 P-289. Flight tests proved the supercharger was trouble-free and would maintain sea level pressure up to about 30,000 ft. even though its altitude rating was 20,000 ft. at 24,000 RPM.

The Form F supercharger was known simply as the "Side-Type Supercharger" from the time of its conception until the revision of the original unit to suit the Liberty had been completed. The data plate of the revised machine was stamped as a "Form F" as it was delivered on September 18, 1923. It would thus appear that the designation "Form E" had been assigned some time between March of 1922 and September of 1923 (which was for a double-sided gear-driven centrifugal-type designed by GE for use on the Liberty and Curtiss D-12), leaving "Form F" for the new side-type.

By the first quarter of 1924 several additional Form F superchargers had been ordered with the proposal they be tested on both Curtiss D-12 and Liberty 12 engines. I do not know if any were actually tested on a D-12 at this early date, but Engineering Division TP-1, which had been operated for some time with a Form D supercharger and side radiators, was converted to use a Form F side-type and wing radiator. It was reported that at an altitude of 20,300 ft. it gave a high speed of 149.8 MPH and better maneuverability than a single-seat pursuit at that altitude.

By the third quarter of 1924, 15 additional turbos had been ordered for new DH-4M2 airplanes under construction by the Atlantic Aircraft Corp. When so equipped these were designated DH-4M2S. Delivery was probably accomplished early in 1925.

THE FORM F-1

By the time report serial 2365 of June 6, 1924 was issued there was already in the plans a revision of the Form F which would replace the engine oil-fed plain bearings with ball and roller-type and redesign the impeller and turbine along with some other changes.

Technical Bulletin 47 of the second quarter of 1926 reported that 15 Form F-1 superchargers had been ordered from GE for service installation use on both Liberty 12 and Curtiss D-12 engines. They were described as follows: "These superchargers are being purchased for service installation, and in them previous difficulties experienced with this type of supercharger have been eliminated by redesign. The principal changes involve an improved air inlet flow, the use of ball and roller bearings to eliminate oiling difficulties, and the use of a new iron base alloy bucket material which is expected to overcome necking of the turbine buckets." This change also included the use of a demountable turbine wheel.

Form F-1 supercharger installed on XCO- 5A P-305. This airplane set a number of altitude records piloted by Lt. Macready. (Courtesy of the Air Force Museum)

NACA DH-4M2 test airplane with a Form F-1A supercharger installed. NACA's installation design was quite different from that of the Air Service. They used the inverted Stromberg carburetors used with the Form B and D superchargers which required different intake manifolds from the usual side-type installation. In addition, the use of a standard DH-4 nose radiator required the opposite bank exhaust line to be routed over the engine rather than in front of it. NACA conducted a number of tests comparing Liberty engine performance with turbosupercharger and Roots-type, using the same airplane. (Courtesy of Peter Bowers)

The above photograph was captioned by Air Corps as "Side-type Supercharger - Form F-1A designed to give sea level pressure up to 30,000 ft. - on special, altitude airplane XCO-5A." All standard F through F-1A units are designed for 20,000 ft. sea level pressure ability so this is no doubt a special unit. It may, however, be what became known as the Form F-2 experimental unit. (National Archives photo courtesy of Kim McCutcheon)

The Form F-1 was assigned for use on engines up to the 400 HP class, which included the Curtiss D-12. Additional machines may have been ordered as it was changed from experimental to limited standard classification on July 27, 1927. It continued in that status until April 1, 1932 when it was reclassified to obsolete. Stocks were not exhausted until April 20, 1935.

THE FORM F-1A

The same Technical Bulletin 47 mentioned above carried the first description of a new design which would shortly become known as the Form F-1A. It too was a 20,000 ft. capacity unit but incorporated the new features of a variable nozzle area, demountable turbine and impeller, and a split turbine wheel which permitted disassembly for bucket replacement.

The Form F-1A thus was first used under test conditions in mid-1926. It was designated as standard on May 14, 1928. On January 16, 1932 its classification was reduced to limited standard. All stocks were exhausted by July 14, 1937.

THE CHRONOLOGY OF SUPERCHARGER DEVELOPMENT AS APPLIED TO THE LIBERTY

1918	April - Gates supercharger tested and abandoned. June - July - Sherbondy No.1 & No. 2 start tests. June - Moss Form A begins test. First Fergus unit starts test. September - October - Moss Form A tested on Pikes Peak December - Sherbondy No. 3 starts 4-hour endurance test.
1919	January - Sherbondy project canceled. Fergus project canceled. May 31 - Patent for modified Form A applied for. July - Modified Moss Form A testing in Le Pere P-53 starts.
1920	February 27 - Schroeder sets 33,113' altitude record following which plans for a new Form B start. December 17 - Patent for Form B applied for.
1921	January - First Form B finished.
1922	September 28 - Macready sets 34,509' altitude record with Form B.
1923	January - First Form C delivered. March - 35,000' turbo-type built. April - First Form D probably delivered. August - First side-type (to become Form F) built. August - Form C abandoned. August - L-6 with rotary induction tested.
1924	Summer - during this period GE built first Form E centrifugal type. September - First side-type (now named Form F) for Liberty tested.
1925	February - First Form F-1. November-December - 1st Liberty Roots built for NACA & Engineering Division December - Eng. Div. 7,500' & 20,000' centrifugal received from builder.
1926	January 29 - Macready sets American altitude record of 35,900' with Form F-1 on XCO-A5 May - June - Eng. Div. purchases 3 Roots superchargers for tests. June - First Form F-1A. August - First flight test of NACA Roots September - British test ground boosted L-6
1927	
1928	September - First Form F-2 October - McCook Field flight-tests Roots with 2:1 gear drive.
1929	
1930	February - NACA compares Roots to Form F-1A in flight test.
1931	
1932	The last Army use of the Liberty with a supercharger applied, probably dated no later than some time in 1931.

THE FORM F-2

The Air Corps "Type Designation Sheet" for the F-2-series of superchargers (which lists types F-2 through F-2H) lists the Form F-2 as an experimental type for use on the Liberty 12, Curtiss D-12, Packard 1550, Curtiss 1550 and Packard 2500 engines, and describes it as "including all details incorporated in A, B, C, D & F & F1 and to include 360° nozzle diaphragm, removable buckets, ball and roller bearings, variable nozzle area control, demountable impeller and turbine." The entry is not dated but probably its first use would have been in late 1928.

The October 1928 issue of the Material Division Activity report carried a note that "A new development of this type (the Form F) is the Form F-2 for application to 600 h.p. engines. This unit incorporates a 360-degree nozzle ring divided into nine sections and controlled by three blast gates. The turbine buckets are clamped between two plates of the turbine wheel in such a way that they can be removed for replacement without destroying the wheel."

The author has found a close-up photograph of the Macready high-altitude XCO-5A P-305 fitted with a turbosupercharger which fits the above description in that it has a 360° nozzle box. However, the photo is captioned as a Form F-1A designed to give sea level pressure up to 30,000 feet. This would be a special unit in that all prior types were designed as 20,000 ft. units. Also it was called a "Form F-1A" and not a "Form F-2." In addition, only a single blast gate is visible. The photo is dated June 16, 1928 and the adjustable pitch prop carried the date 4-17-28. Its external appearance conforms to the description of the Form F-2 but probably this is a one-of-a-kind modification of a Form F-1A built to make another attempt at the world altitude record.

The Type Designation Sheet gives no dating for the F-2, just lists it as experimental and obsolete. Its replacement, the F-2B was also listed as experimental with a first-use date of January 23, 1930. This and other information leads the author to estimate the dating of the Form F-2, at least for Liberty use, as between September of 1928 through the early 1930s.

The Form F-2 was the last of the turbosuperchargers to be used on the Liberty. The last Liberty was flown by the Air Service on September 13, 1935. It was not supercharged but the removal of the Liberty from service use at this time marked the end of any possibility of service supercharged engines.

The GE turbosupercharger continued to be developed for later engine designs and continued as advancements of the Form F, reaching the F-8 designation for the new Allison V-1710 by the late 1930s.

Dr. Sanford A. Moss
Courtesy of the Ludvigsen Library.
1872-1946

S. A. Moss attended the University of California and attained his bachelor's and master's degrees in Engineering. He then went to Cornell University and attained his doctorate with principal studies in gas turbines and centrifugal air compressors.

He joined General Electric in 1903 and carried out research in the areas in which he majored while attaining his doctorate.

In 1917 he was requested by the National Advisory Committee for Aeronautics and the Bureau of Aircraft Production to design a turbosupercharger for the Liberty aircraft engine.

His efforts resulted in the General Electric Forms A, B, C, D and F turbosuperchargers and, ultimately, the WW II turbosuperchargers used on thousands of America's high-altitude bombers and pursuit planes.

For perfecting this supercharger Dr. Moss and the Army Air Corps were awarded the Collier Trophy in 1941.

He retired in 1938 but when war broke out he went back to GE to help perfect the turbosupercharger for the extensive war use it was to see.

DEVELOPMENT OF THE GEAR-DRIVEN CENTRIFUGAL SUPERCHARGER

The earlier 1917/1918 gear-drives by Sherbondy and Le Blanc/Kerr have already been discussed. No further serious work appears to have been attempted in this area until mid-1922.

STURTEVANT

In about August of 1922 the Engineering Division accepted for testing a gear-driven centrifugal supercharger from the B. F. Sturtevant Co. of Hyde Park, Boston, MA. It was designed by H. E. Morton. It was briefly described late in 1926 by David Gregg, an Engineering Division engineer who was in charge of supercharger development up until about this time.

"This supercharger was a very interesting design that incorporated an oil clutch to reduce the acceleration load on the impeller. The design was rather too complicated for the Air Service to develop, although it had many interesting features."

We hear no more regarding this design and it would appear that it was abandoned as impractical, probably before the end of 1922.

THE ENGINEERING DIVISION GEAR-DRIVE

The Engineering Division designed and built their own single reduction-type gear-drive supercharger in the summer of 1922. It completed its first serious running test in November and was reported in Technical Order No. 30 of that month. The machine had been given a 3 1/2-hour run on a dynamometer with one hour at full load and 15 minutes at 15 percent overload. All moving parts were reported in excellent conditon. The gearing was of herring bone-type.

This initial design was of 10,000 ft. rating and was later developed into a 20,000 ft. unit as well. By 1926 these two sizes were reliable enough to be used in limited quantities on Liberty and Curtiss D-12 engines in flight testing.

THE GE FORM E

Some time during the summer of 1923 General Electric introduced a new gear-driven centrifugal supercharger which they named the "Form E." This double-geared model was designed as a 35,000 ft. unit for 400 - 500 HP engines and was thus suited for the Liberty and Curtiss D-12.

The author has only seen images of it as applied to the Liberty and in this case it was applied to the input side of standard Zenith carburetors just as was the Form F turbosupercharger. Thus the same method of fuel pressure balance must have been used. A box-type honeycomb aftercooled was inserted between the supercharger and carburetor input.

The supercharger was supplied with two valves: a butterfly-type on its input and a barrel valve on its output.

The Form E GE gear-drive centrifugal supercharger seen applied to the Liberty at left. Note the input throttle butterfly valve on the left-side and the output barrel valve on the top. The output feeds an air plenum which inputs a honeycomb-type aftercooler which then feeds the input air plenum to the two carburetors. At right is a view of a bare centrifugal unit without gear-drive attached and with the housing cast to input air from the opposite side as the Liberty-mounted unit. (Author's collection)

In early 1927 David Gregg, former supercharger engineer for the Engineering Division, referred to this and the (soon-to-be-introduced) GE Form G as the standard centrifugal superchargers of the Air Service. Although testing continued for some time, the Form E never saw extensive use.

The reader will note by referring to the chronology chart on page 352 that during the period from January 1922 to early 1924 rapid advances in supercharger design were being made. The Forms C, D, E, F and G were all introduced during this period. It even appears that the first Form F was built (as a 1A-1237 supercharger) and under test when the Form E was introduced. It was, however, during this period still being referred to simply as the "side-type" supercharger. By the time it was decided that it should definitely be developed, Form E had been assigned to the new GE gear-drive supercharger and it became the Form F. Shortly thereafter GE introduced the Form G.

THE GE FORM G

Sometime in early 1924 General Electric introduced the gear-driven Form G supercharger. This was a single inlet unit designed to have the carburetor on its input side and have the capability to provide sea level air at up to 20,000 ft. depending on the ratio of the gear-drive.

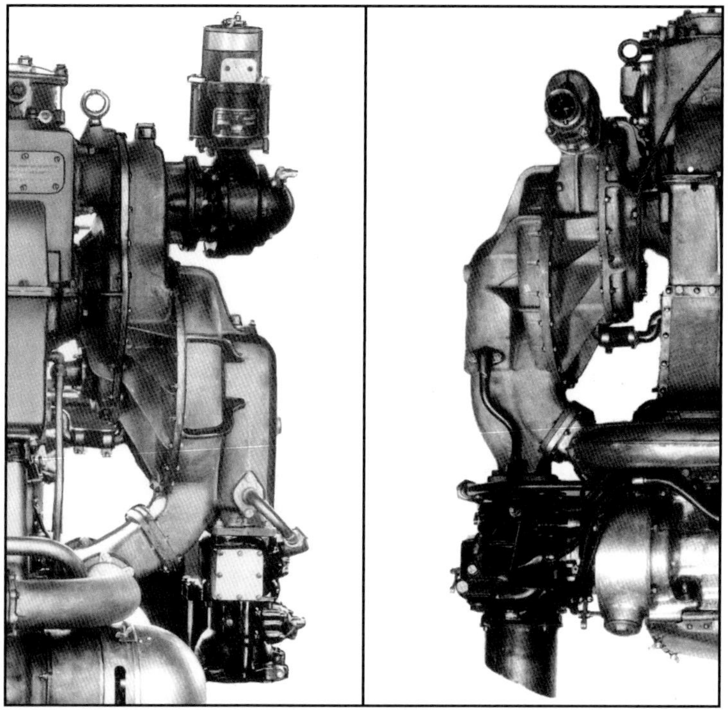

Form G superchargers applied to inverted V-1410 Liberty engines with provision for Eclipse electric starter on left and hand-crank starter on right. (National Archives photographs courtesy of Kim McCutcheon)

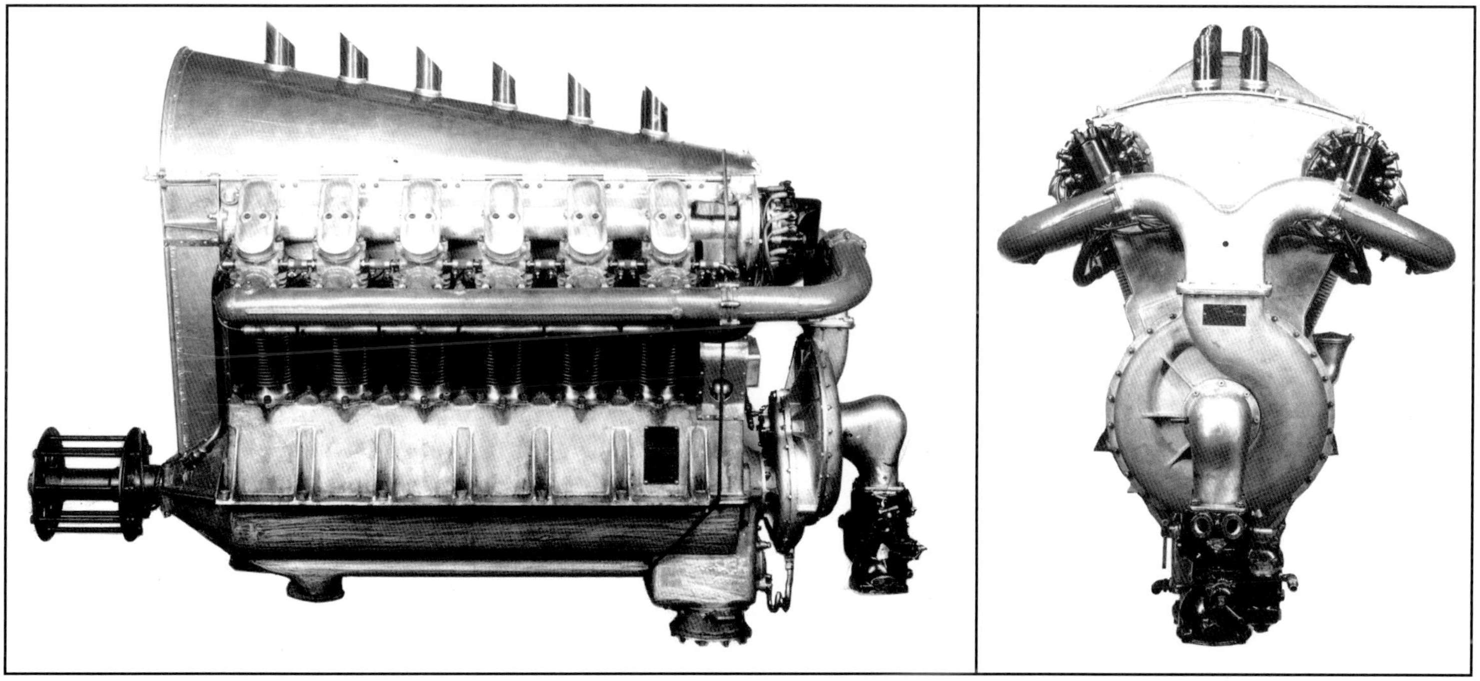

The first Engineering Division V-1410 air-cooled Liberty fitted with an early Form G supercharger. These two photographs are McCook numbers 24107 and 24109 and were taken in November of 1924. The air-cooled Liberty was designed to have the exhaust on the inside of the vee and intake on the outside. The standard water-cooled Liberty had two carburetors inside the vee and the exhaust outside. This supercharger as applied to the standard Liberty would have fed a special set of intake manifolds inside the vee. (National Archives photographs courtesy of Kim McCutcheon)

This model was used on the new air-cooled Liberty V-1410 introduced in September of 1924. As noted in the Allison chapter, the gear-drive was still giving problems well into 1925. As used on the V-1410 it provided rotary induction and was used to obtain good mixture distribution from a single carburetor rather than provide boost. Under such conditions no aftercooler was needed since input air temperature was not significantly raised.

It was also used on both the standard water-cooled Liberty and the Curtiss D-12. In both of these cases the supercharger output fed special inside-the-vee intake manifolds and the regular carburetors were replaced by a single one mounted on the supercharger intake.

Fifteen Form G superchargers were used on the V-1410 air-cooled Liberty. An additional fifteen may have been built for use on the D-12 and standard Liberty engines. The author knows of no surviving units.

THE ENGINEERING DIVISION 7,500 FT. AND 20,000 FT. GEAR-DRIVE CENTRIFUGAL

In late 1925 or very early 1926 the Engineering Division designed two new gear-drive centrifugal superchargers, a 7,500 ft. and a 20,000 ft. unit. Both were contracted to be built by Allison. By June of 1926 they had been delivered and were under test.

Engineering Division 20,000' gear-drive centrifugal supercharger applied to a Curtiss D-12. This supercharger could also be fitted to the Liberty. Note the aftercooler mounted above the front of the engine. The photograph number dates it as mid 1926. (National Archives photograph courtesy of Kim McCutcheon)

They were sized for use on the Liberty and Curtiss D-12. Although no technical information has thus far surfaced, a photograph (McCook Field 30781) of the 20,000 ft. model fitted to a D-12 has been found. In this installation it is mounted and driven from the rear of the engine. Its output feeds two air plenums which feed an aftercooler mounted above the front of the engine in the slipstream. The output of the aftercooler feeds an air plenum which feeds input air to carburetors in the vee. Installation on a Liberty could be expected to be similar.

THE LIBERTY 6 WITH ROTARY INDUCTION

In about August of 1922 the Engineering Division decided to test a rotary induction system on a Liberty 6. The design chosen was one which they had tried in 1920 on an air-cooled radial and reported on earlier.

The test was brief and not encouraging and was reported in Test Serial 1926. The device was of rather simple and basic design and was described in the resume given in TO 28 of August-September 1922 as follows: "A modification of the Smith rotary induction system, a detailed description of which is given in Serial No. 1457 entitled, *Report on Test of Rotary Induction System for Radial Air-cooled Engines,* was applied to the Liberty 6 engine. In form, this system is a centrifugal pump which distributes the air-fuel mixture from the carburetor to the induction pipes leading to the cylinders. It consists essentially of a cast iron housing in which revolves an aluminum rotor or impeller."

The result of the test and its conclusions were given in summary: "The apparatus showed no improvement in operation over the standard carburetion and manifolding on the Liberty 6. However, with the rotary induction system attached it was plainly evident that the distribution and performance were better with the system in operation than when not in operation and that the best performance was obtained with the highest speed of the rotor. For these reasons, it appears that the use of the apparatus might be beneficial when an engine has poor distribution, due to improper manifolding."

It is not clear whether the test was conducted in an attempt to improve the performance of the Liberty 6, but that is not likely. By this time the Liberty 6 had been redesigned, a batch of ten ordered built by Wright and probably all had been delivered. It would seem that the Liberty 6 was available and made a good test case. No further work was done relative to using rotary induction or supercharging on the Liberty 6, at least not in this country.

THE GROUND-BOOSTED BRITISH LIBERTY 6 AND OTHER BRITISH LIBERTY SUPERCHARGING EFFORTS

The Royal Aircraft Establishment (R.A.E.) put forward the suggestion in July of 1925 that further research into ground-boosting of aircraft engines be carried out preliminary to initiating the design of a new engine specifically for that purpose, and that it be carried out on a six-cylinder Liberty.

There was good reason for choosing this engine - and it was not because they wished to use the design as the basis for a new purpose-built supercharged aero engine. After all, the Liberty was getting to be a rather elderly engine so far as aero engine designs were concerned. The reasoning was actually quite straightforward.

Initial tests were to run under boost sufficiently high to increase engine output to approaching twice normal. The object was to: 1- Study the effects on the bearings of the increased loading; 2- Investigate the problem of boosting a multi-cylinder engine up to such power output as the standard crankshaft would stand; 3- Applying thereto the principle of the intermittent flow exhaust-driven turbine.

The Liberty lent itself well to such tests because first, it was a well-tested engine whose performance was known. Second, ample engines and parts were available. Third, they were going to test a six-cylinder engine at very high power and the Liberty twelve with six cylinders removed fit the requirements nicely.

The project was approved and work began in September of 1926. The cylinders were modified in that they were fitted with bosses for indicator valve connections for checking cylinder pressure, and four stud flanges fitted to the exhaust and inlet ports. 4.5:1 pistons were used and "Puma" connecting rods fitted to withstand the increased stresses. Standard intake valves and springs and special "silichrome" steel exhaust valves were used. The engine was operated on its six starboard cylinders, the port bank and all associated hardware having been removed.

An externally driven and controllable supercharger with a special horizontal Zenith duplex carburetor on its pressure side was used. The carburetor had been specially built for use on previous flight tests by the R.A.E. of a supercharged Liberty 12 in the DH-9A. A special intake manifold constructed to withstand 80 PSI in pressure in case of backfire was used. Summer grade fuel with 20% Benzol was used. The intake manifold was water jacketed and its water temperature was controlled as was cooling water temperature.

Input air temperature to the carburetor was maintained at 54°. This was noted as mainly due to the very high moisture content of the source air supply. The result was that even with a boost pressure of 5.88 PSI (about 11.97 in. hg) and ignition advance of 42°, preignition was not experienced. Although the report did not note that there was a specific purpose in maintaining such high moisture and low temperature, the result was an early example of water injection controlling preignition and allowing a supercharger boost at sea level without destructive results.

After 5 1/2 hours run-time, the engine was disassembled and examined. It was found that it had one broken valve guide, stuck top rings in four pistons, all exhaust valves stretched out of tolerance and a fracture in the crankshaft at no. 6 driving web. Boost pressures used had produced outputs as high as 280 BHP @ 1,650 RPM.

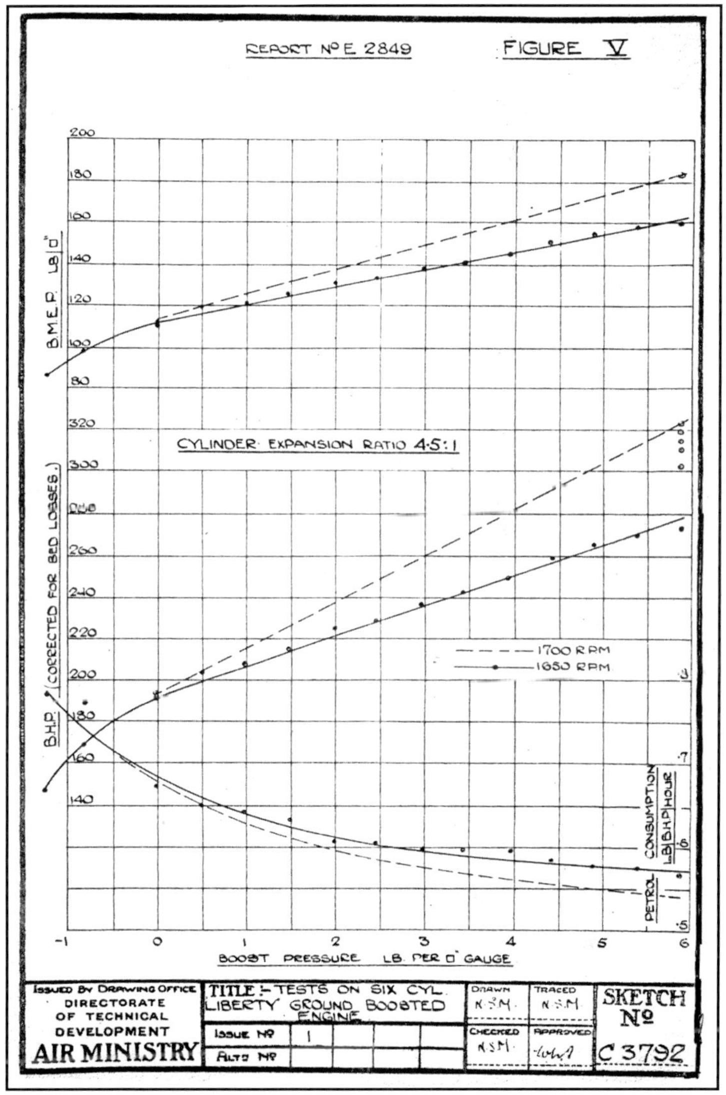

R.A.E power graph of the supercharged Liberty 6 test conducted between February and July of 1927. Outputs are plotted at both 1,650 and 1,700 RPM with boost pressures from -1 to 6 PSI. (Courtesy of the British public records office)

After replacement of the crankshaft and correction of any other problems noted, the engine was set up and run an additional 5 1/2 hours, 2 of which were at 323 BHP @ 1,700 RPM with 183 BMEP. Work on the project was considered complete at this point. Total run time, including initial calibration and testing, was 47 hours.

Initial unboosted test of the engine produced 194 BHP @ 1,700 RPM. The 5.88 PSI boost had produced an addition 129 BHP in output. As a point of interest, the test showed that power output increased linearly in relation to increase in boost pressure within the range applied.

Although the author has not had the opportunity to research it in detail, British records and references so far noted indicate they carried out a number of tests with supercharged Liberty engines, both in-flight and on the bench. Flight-testing seems to have been conducted primarily with the DH-9A. Most references are to the use of a turbosupercharger but it is not clear whether it was British or American.

THE ROOTS SUPERCHARGED LIBERTY

In early 1926 both the Engineering Division and NACA developed an interest in investigating the use of Roots-type superchargers on aircraft engines. Both chose the Liberty 12 as the engine on which testing would be done, probably because it had been the engine on which most gear-drive centrifugal and turbosupercharger development work had so far been carried out. The object was to find how practical and beneficial the use of this type was as compared to prior types tried.

TB 47 of April-May-June 1926 reported that the ED had purchased three Roots units for development tests, two 14,000 ft. types for use on bomber type aircraft and one 20,000 ft. unit for high-altitude observation types.

By October of 1928 it was reported that the 20,000 ft. Roots was undergoing flight tests in a Liberty-engined observation plane with 2:1 driving gears. Higher ratio driving gears would be tested as soon as they could be obtained. The specific type was not noted but would likely have been an aircraft type with which performance with prior supercharger types was well documented, perhaps the DH-4B or TP-1.

No data on the design of either type noted above has so far been found. Considering this was within the same time frame as the NACA Roots designs discussed next, it is quite possible the Engineering Division was testing NACA designed and Allison-built units as a separate venture.

First design NACA Roots supercharger applied to a Liberty 12A. This unit was designed at NACA request by Clark Thompson Research in late 1919 and is believed to have been built by Allison Engineering. The large tank mounted on top of the unit is an air receiver which acts to reduce the pulsating air characteristics inherent to the Roots-type supercharger. (Picture from NACA Report No. 230)

Unfortunately the author has found no further references to Air Service testing of the Roots supercharger. It is evident, however, that overall results were not sufficiently good to warrant continued development as the Roots supercharger never saw significant use on aircraft engines.

THE NACA MODEL I ROOTS

NACA also took up an interest in the Roots supercharger and test reports on their efforts indicate they began at about the same time as those of the Air Service. As you will see, theirs continued at least until February of 1930.

The first official report was contained in NACA Report No. 230. NACA interest in the Roots-type of supercharger began in 1918 and, at their request, Clark Thompson Research made a complete design-analysis and in October of 1919 submitted drawings for a Liberty 12 size unit. NACA contracted the unit's construction and it was delivered to Langley Field for ground-testing. It is currently believed that Allison Engineering was the builder as it is known they were the builder of the later design units. Exactly when the first unit was actually delivered to NACA is currently unknown but the first report was not issued until about August of 1926. This

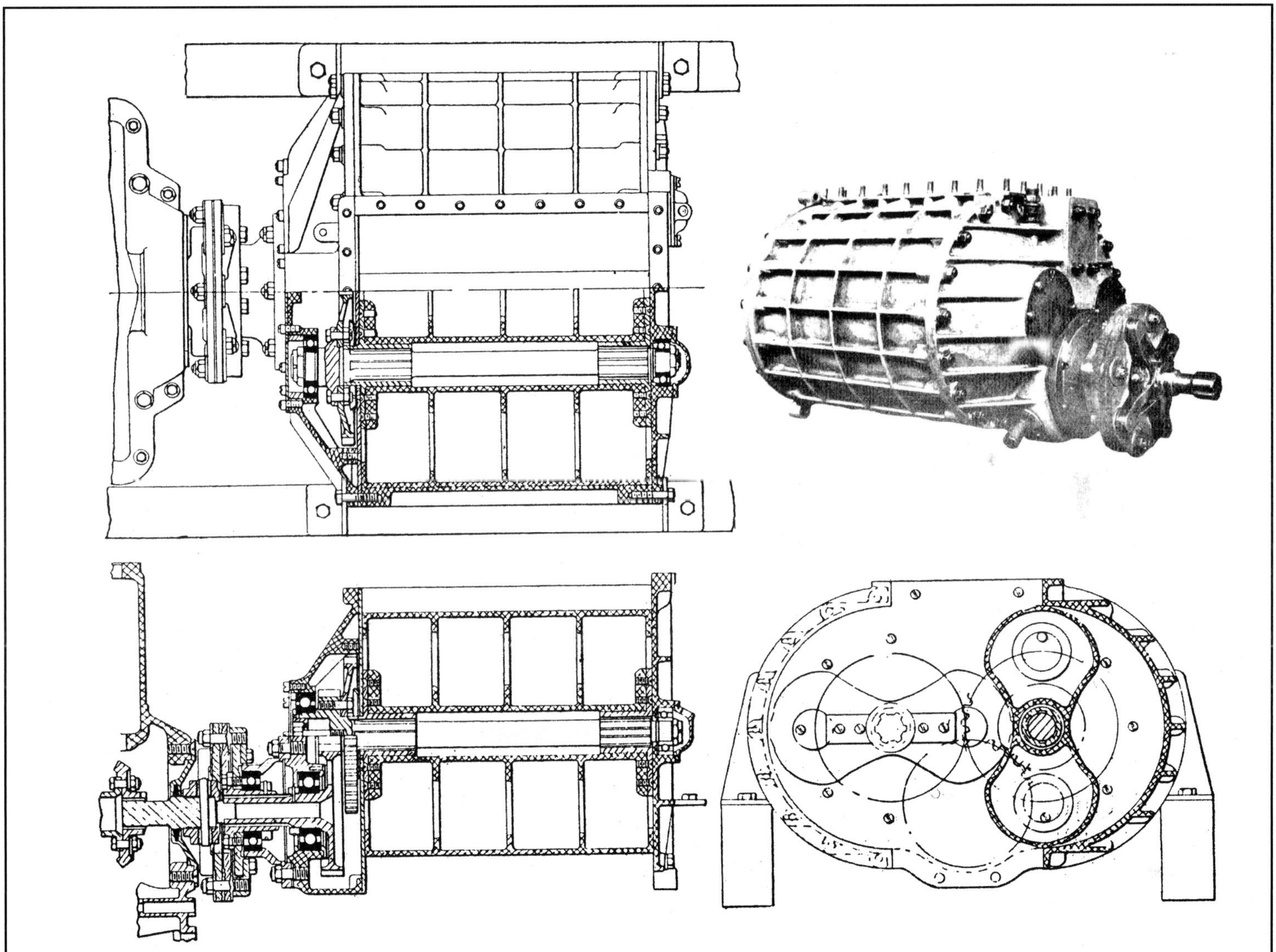

Sectional drawings of the first design NACA Roots supercharger along with an illustration of the completed unit at upper-right. Note the flexible coupling and spline drive for connection to the rear of the engine crankshaft. The input shaft is connected to the gear drive with a shear pin designed to give way upon excessive torque. In this manner the supercharger will be disconnected from the engine in case of its mechanical failure. Mounting brackets were bolted onto the front and rear of the main case. (Illustrations from NACA test reports)

leads the author to believe it was probably not until late 1925 that work began.

Following successful ground tests the unit was adapted to a Liberty engine. Initial tests used standard Zenith carburetors but eventually inverted Stromberg NA-L5A carburetors were settled on. The unit size was determined by a combination of delivery requirements to supply sea level pressure to 15,000 feet and space limitations imposed by the necessity of mounting it between the Liberty's engine rails.

These considerations resulted in establishing the following proportions: diameter of rotors, 9 1/2 in.; length of rotors 11 in.; distance between rotor axes, 5.833 in.; rotor speed, 1.5 crankshaft speed. Displacement per revolution was 0.509 cu. ft.

Mounting brackets were separate steel fabrications which were bolted to the front and rear of the main case casting and can be seen in the sectional drawings as well as in the illustration on page 359. As originally designed, the unit was to be mounted directly to, and be supported by, the engine crankcase and have no flexible coupling. The flexible coupling was added to isolate the gear-drive from crankshaft torsional vibrations and compensate for any misalignment.

By December of 1926 ground-testing had been completed and the engine fitted first to a DH-4M2 and later to a Douglas DT-2. The DT-2 was subsequently modified to a float plane and tested. In both cases the engine was equipped with a standard supercharged Martin bomber propeller. Two sets of drive gears were used: 1.5:1 to maintain sea level pressure to 12,000 ft. and a 1.94:1 to maintain sea level pressure to 20,000 ft.. The DT-2 was only tested with the higher ratio drive.

The DH-4M2 was flown to an altitude of about 31,000 ft. with 1.94:1 drive gears. The much heavier DT-2 (land version) only obtained an altitude of about 23,000 ft. The weights with supercharger were 3,960 and 5,211 pounds respectively.

Performance with the supercharger was found to be very good in both cases. One of the most satisfying characteristics found was that the supercharger did not produce much heating of the intake air to the carburetors and therefore no aftercooler was required. Suitable controls were found to be simple in both design and operation. Supercharger operation is controlled by a bypass valve on its delivery side which is mechanically linked to the engine throttle control. As the throttle is advanced the bypass valve is closed. The pilot must refer to a sealed altimeter connected to the carburetor air intake to be sure he does not advance the throttle sufficiently to cause a below sea level pressure at that point, which would result in preignition and engine damage.

Added weight of the aircraft charged to the supercharger installation was 185 pounds for the DH-4M2 and 166 pounds for the DT-2. An auxiliary cooling radiator (which is not included in the 185 pounds) was required on the DH-4M2 (the same as was required on turbosupercharger installations on this airplane), but the DT-2 did not require one.

NACA DH-4M2 equipped with the first type NACA Roots supercharger. This is the same airplane in which NACA tested the Form F-1A turbo supercharger. (Image from the NACA flight report No. 263)

THE NACA MODEL II ROOTS

Beginning in 1927 NACA began to experiment with different capacity Roots superchargers. Comparisons were made between the operation of units with 11" (the size of the original or Model I), 8 1/4" and 4" rotors. These were driven at different speeds to see if they maintained good efficiency and durability at higher speeds. The object was to determine if unit size could be reduced and drive speed increased and capacity thus be retained. The tests proved this to be practical and a new design was produced for a 8 1/4" unit of 0.382 cu. ft. capacity. This was called the "NACA Model II" and Allison Engineering was contracted to build their test unit.

This new model supercharger was given extensive tests during early 1929 using gear-drives of 1.616, 1.957, 2.4 and 3 to 1 ratios. These corresponded to altitude ratings of 7,000, 11,500, 17,000 and 22,000 feet respectively.

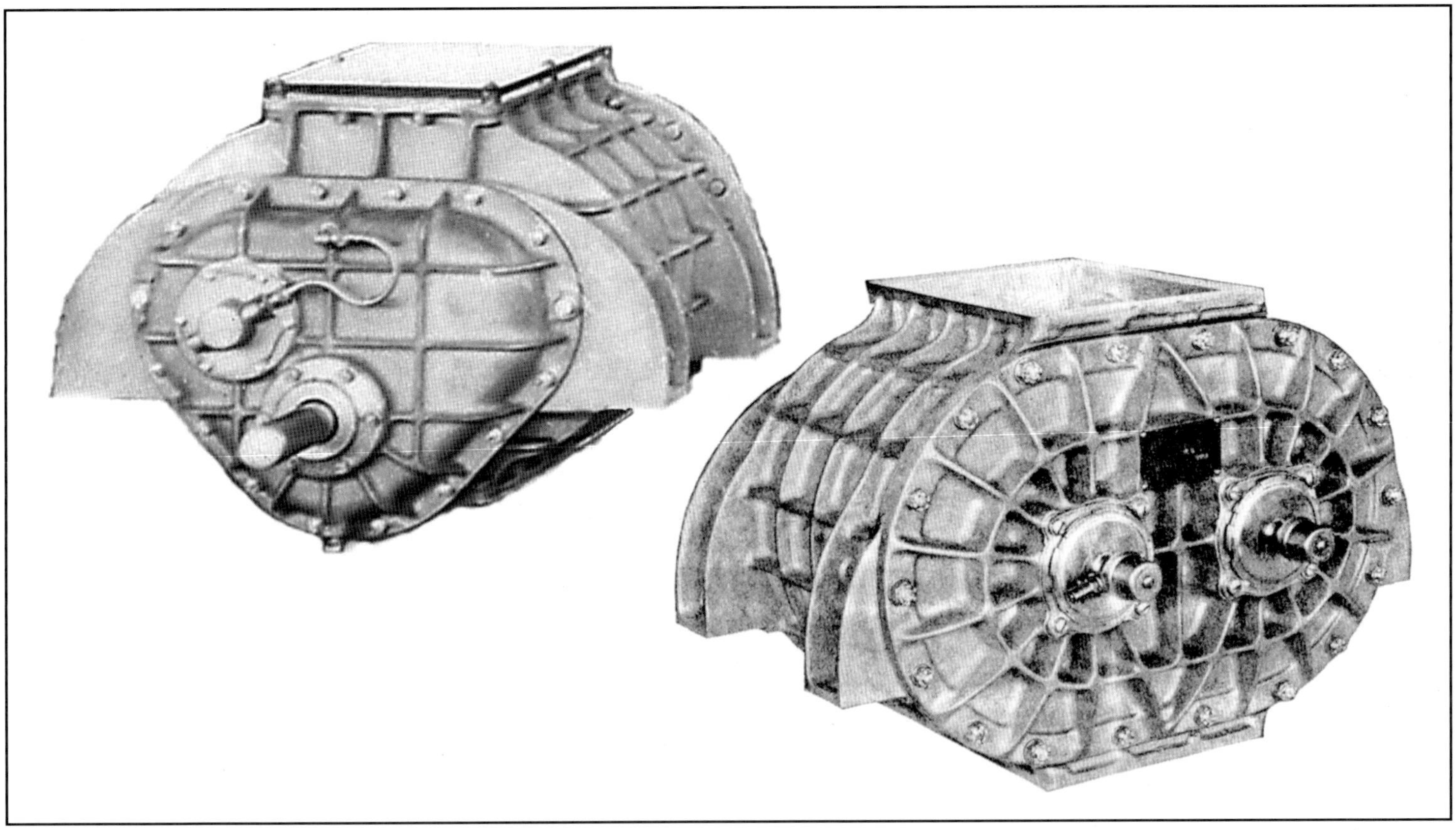

Allison-built NACA Model II Roots supercharger. Above are front and rear views of the unit with no gear-drive attached. The separate gear-drive was designed to provide a flexible coupling and which ever gear ratio was desired. (Author's collection)

NACA Report No. 327 of March 13, 1929 reported the results of flight testing with this new model. Capabilities were found to be very close to the theoretical calculations. For example, the 3:1 drive 22,000 ft. version would maintain sea level carburetor pressure to 22,200 ft.

The superchargers worked well and proved mechanically reliable. On the negative side was the fact that the 22,000 ft. unit required 24% of the engine output to drive it at rated altitude whereas the type F-1A turbosupercharger only cost about 14% of engine power at 22,000 ft.

The last reference the author has found on the NACA Roots supercharger is Report No. 355 of February 25, 1930 in which they reported the results of comparative flight test of their DH-4M2 equipped alternately with a Form F-1A turbosupercharger and a Model II NACA Roots with 22,000 ft. rated gear-drive.

The end results showed that the rate of climb for each installation was practically the same and maximum altitude was close to the same, being 26,400 ft. for the Roots and 27,175 ft. for the turbo. Top speed at high altitude, however, showed a significant difference. The turbo actually has a maximum speed advantage at all altitudes but it did not become significant until about 10,000 ft. By that altitude the turbo speed was 132 MPH compared to 122 MPH. At 22,000 ft. the turbo had reached its maximum speed of 143 MPH whereas the Roots had reached its maximum speed of 122 MPH at 16,000 ft. and was down to 121 MPH at 22,000 ft. The speed difference continued at about 20 MPH up the maximum obtainable altitude by both supercharger versions.

PRODUCTION TOTALS

Probably only one example of the Model I NACA Roots was built. It is also believed that only one Model II was built for NACA but Allison records indicate that they built a total of six of these units. Possibly one or more of these went to the Engineering Division.

Allison records also note that they built a total of 34 Model III (#3) Roots superchargers. The Model III unit used 9 1/2" x 4" rotors and displaced half the volume per revolution of the Model II. Theoretically it could produce the same output as the Model II if driven at twice the speed.

The number built indicates it saw considerable application, but on which engines we currently do not know.

DH-4B P-190 AS 63181 at McCook Field with a Form B supercharger installed. In the background behind the tail is another DH-4B with a Form B supercharger installed, probably P-131. Almost totally hidden behind the tail is a Martin bomber, quite possibly P-162, also with Form B superchargers. (Courtesy of Peter Bowers)

Chapter 10

Post-War Aero Use of the Liberty Worldwide

At the close of the war the United States, of course, had a large stock of new Liberty engines. Great Britain also had a large stock, as did France. The later two, however, were not in the same league as the United States. By the time engine production was completely halted, which was not until January of 1919, a total of 21,138 Liberty engines of all types had been delivered. The British received 2,252 and the French 3,575. The Italians had proposed ordering 3,000 engines for Caproni Ca-5 bombers but the actual order never materialized, thus they only received 5 engines for testing.

As would be expected, the United States, Great Britain and France were the largest users of the engine for aero purposes. In addition, however, following the war Russia became a producer of Liberty engines and a significant user. That story is told in detail in the following chapter but their aero use will also be included here as part of the overall subject of study.

Added to these four countries were a number of others worldwide. The reason is, of course, easy to see. There was a vast surplus of Liberty engines. They were state-of-the-art at the end of the war and thus remained relatively up-to-date for a few more years. They were also much cheaper to purchase than any equivalent current production aero engine. For some years the four countries mentioned above kept the engine in active military use and this tended to also promote civilian use.

designs produced in the future. Likewise, the Navy had about 2,000 surplus engines which were usable for the same purposes.

This situation, in combination with drastically reduced funding for a number of years following the war, led to extensive use of the Liberty engine in new aircraft designs.

The large surplus of Liberty engines began to be available to civilian purchasers in mid-1919 and continued so until the last government supplies were exhausted some fifteen years later. Initial prices were about $2,775 each but had dropped to about $350 by 1931.

Although the incentive to use the Liberty in new civil aircraft was great there does not seem to be any evidence that they powered large quantities of this type aircraft. The writer has made a detailed study of the uses of the Liberty engine in aircraft worldwide and the evidence does not substantiate a large civil usage. Appendix F documents what I have found in this regard. Although the reader will see from its study that use of the Liberty for aircraft power was very extensive and probably spanned a greater number of years and was applied to a larger variety of aircraft models and types than can be claimed by any other piston aircraft engine ever built, the civil airplanes which used the engine actually account for a very small number in comparison to the number of surplus engines eventually disposed of.

UNITED STATES USE

During the war both the Army and Navy made the decision that the Liberty 12 would be their principal high-power engine. By the end of the war the Army had been assigned 5,198 engines which had gone directly to aircraft manufacturers. Most were used producing 4,701 DH-4 airplanes which were distributed both here and abroad. The Navy received 4,231 Liberty 12 engines (128 L-12A and 4,103 L-12N) of which 2,208 were used for the production of 1,681 Navy aircraft of all types.

Thus, at the end of the war there were about 5,000 surplus engines available to the Army for use maintaining existing aircraft and for any new

Curtiss Carrier Pigeon I. 11 airplanes of this Liberty-powered model were built for National Air Transport in 1925 for use on contract air mail routes. (Courtesy of Peter Bowers)

U.S. CIVIL USE

Fifty-two different models of U.S. civil airplanes used the Liberty for power. Between them they account for about 175 engines. Of the numerous Liberty powered airplanes which operated in South America and Europe in the years following the war, almost all were surplus U.S. military airplanes and did not use surplus Liberty engines. As I will note later, a fairly large number of surplus Liberty engines went to Russia as well as a few to Norway and Holland.

The first Liberty-powered U.S. civil aircraft appeared immediately following the war and the last appears to have been built in about 1927. Some no doubt remained in operation into the early 1930s.

THE AIR MAIL SERVICE

One of the most significant civil users was the U.S. Air Mail Service. The Air Mail Service was initiated under Army operation on May 15, 1918, starting with six specially modified Hispano-Suiza-powered Curtiss JN-6H airplanes, US 38262 being the first airplane.

Within a few months several Liberty-powered Curtiss R-4L airplanes were put into service because of their higher speed, longer range and increased load capacity. The Air Mail Service was officially transferred from the Army to the Post Office Department on August 10, 1918. Shortly thereafter several Hispano-Suiza-powered Standard JR-1B airplanes were added to the fleet. Also 100 DH-4 airplanes were ordered for the Air Mail Service through the Bureau of Aircraft Production in late 1918. At this point in time the war was still in progress and airplanes and engines still in production.

By the time the first anniversary of the Air Mail Service arrived, transcontinental routes were being established and, because of its availability, durability and carrying capacity, the DH-4 had been chosen as the principal mail plane.

During the early years of the Air Mail Service a number of different airplanes were tried in service, including the Italian Caproni CA-33, Handley Page, Curtiss HA Mail, Twin DH with two Hall-Scott engines, Martin MB-1 Mail and others.

By the end of 1920 transcontinental routes had reached San Francisco and the plane used was the Liberty-powered DH-4. On July 1, 1921, the DH-4 was made standard equipment for the Air Mail Service.

In July of 1924 transcontinental "through" air mail flights were inaugurated, that is day and night flights carried the mail coast-to-coast in under 30 hours elapsed time. In anticipation of the event the Post Office issued, in August of 1923, a series of three air mail stamps covering rates across one, two or three zones (the country had been divided into three east-west postal zones) with values of 8, 16 and 24 cents. Pictured were two views of the DH-4 and an air mail pilot's wings, a tribute to the venerable DH-4 and its Liberty engine.

Air mail service made rapid progress and was given even more momentum by the passage of the Kelly Bill on February 2, 1925. This legislations was "An act to encourage commercial aviation and to authorize the Postmaster General to contract for air mail service." This act resulted in the issuing of contracts to private carriers. Contracts were awarded to bidders found acceptable to the Post Office Department. The bidding was for specified routes and their contracts, and thus routes, were assigned numbers in the chronological order awarded.

By mid-1926 ten contracted routes were in operation. (CAM-1 through CAM-10. The CAM stood for "Contract Air Mail.") Transcontinental routes were still being handled by the Post Office with DH-4s but mail volume increased so dramatically that DH-4 load capacity was continually exceeded. By late 1925 it was time for a replacement airplane. The Post Office Department initiated a competition among the various manufacturers to design a replacement. Although there were several submissions, ultimately it was the Douglas Mail, a modification of the Douglas O-2 observation plane currently in production for the Army, which was chosen.

Douglas Cloudster as modified to airliner configuration after initial production as a cargo carrier in 1921. Only one of these Liberty-powered airplanes was built. (Courtesy of Peter Bowers)

Douglas submitted the M-1 model (it was originally called the DAM-1 for "Douglas Air Mail-One" but later shortened to M-1) which was tested in the spring of 1925 and judged successful. No immediate government order resulted, however.

In early 1926 the newly formed Western Air Express, a company which had been awarded one of the CAM contracts, ordered 6 Douglas M-2 models. These were slight improvements of the M-1.

In March of 1926 the Post Office ordered 50 M-3 models. These differed from the M-2 only in minor details. Before the entire order was delivered Douglas redesigned the M-3 to increase its load-carrying capacity, primarily by increasing its wing size. The Post Office liked the modification and changed the last 40 of their order to M-4 models. Western Air Express later purchased an M-4 from Douglas which was designated M-4A.

National Air Transport, another CAM contractor, initially used the Liberty-powered Curtiss Carrier Pigeon, but in July of 1926, when the Post Office relinquished all routes

At left are the three long-haul air mail stamps issued in 1923 to cover the three east-west mail zones. Above is one of the three denominations (10, 15 and 20 cents, all of the same design) issued in 1926-27 to cover the rate-by-weight system which replaced the zone system in 1927. In both types the Liberty-powered DH-4 was pictured. (Courtesy of the Smithsonian)

DH-4B modified for mail use by LWF and in service to the Post Office Department. The front cockpit has been replaced by a mail storage locker, landing lights installed on the ends of the upper wings and navigation lights installed above them and on the rear edge of the rudder. The Liberty engines were updated with all the latest improvements. (Courtesy of Peter Bowers)

to private carriers, purchased all 10 of the Post Office M-3s and 8 of their M-4s. They eventually added other M-4 models and operated as many as 24 Douglas Mail planes.

The above two operators kept the Douglas Mail planes in service until the very early 1930s and then replaced them with larger airplanes.

There were a number of other Liberty-powered mail planes used by both the Post Office and private carriers. The reader is directed to Appendix F for a list and illustrations of any not mentioned here.

U.S. ARMY USE

The Army adopted the Liberty as the standard high-power aero engine for combat use early in World War I and developed a number of airplanes around it during that time. During what is usually referred to as the pre-designation period of 1917-1920 (this period actually began with the first Army airplane, the Wright Model A of 1908, but only 1917 and after concern us in this study) the Army either designed itself, requested private designs or accepted private designs of 30 different airplane models with Liberty power. These airplanes were intended for many different duties, including observation, combat fighter planes, bombers, mail planes and others. During this period airplanes were known by either the names given by their builders or by those assigned by the Army in the cases where it was the designer/builder.

Douglas O-2K AC 29-179. This airplane, XP-529, was the first of 59 of this model contracted in fiscal year 1929 and represents the last significant production of a new design, Liberty-powered airplane by the Army. The O-2 Douglas line was an Army stalwart which saw production through an improved line of models from the O-2 of 1924 to the O-2K of 1929 with a total of 246 of all models being produced. (Courtesy of Peter Bowers)

In 1921 the Army switched to a designation system in which airplanes were assigned designations based on their intended use such as CO for Corps Observation, LB for Light Bomber, O for Observation, etc. Most airplanes designed and built before this time, but remaining in active use, retained their original designations, the DH-4 and its later variations being a prime example.

The Army had so many new Liberty engines in storage as well as thousands in use in airplanes built during the war, the DH-4 again being a prime example, that they continued to design new airplane models around the Liberty until officially ordered to no longer do so in 1929. Thus they were designing new airplanes around the Liberty from 1917 to 1929. That by no means ended its useful life in Army service as it continued to power airplanes (and lighter-than-air craft) until late 1935. From 1917 to the end of the use of Liberty power by the Army, some 250 aircraft models totaling about 5,800 airplanes and three lighter-than-air ships were put in service.

The last airplane to fly for the Army with Liberty power was a Douglas Ambulance Airplane ZC-1C Air Corps 27-204 (built in 1927) which was powered by a Lincoln-built Liberty 12A which had been built in December of 1918.

U.S. NAVY USE

Although the Navy initially showed no interest in the Liberty in 1917, they rapidly changed their mind and by 1918 had adopted it as their standard for flying boats. During the war the Liberty was to be their primary power for combat-type aircraft. The Navy air arm was never intended to be as large as that of the Army and thus their allotment of Liberty engines was much smaller. They did receive 4,231 engines, either as the power of constructed and delivered airplanes or as new engines delivered into storage.

The Navy was quicker than the Army to switch from water-cooled to air-cooled engines. In addition, of course, it had less stocks of stored Liberty engines to find a use for or sell as surplus. As a result of these and other factors, the Navy retired the Liberty from use sooner than did the Army.

Counting the NC boat series (NC-1 through NC-10) as a single model, the Navy powered about 33 models of airplanes with the Liberty engine. These accounted for a total of 2,152

F-5L in flight. One of 225 built by the Naval Aircraft Factory in 1918. The Navy continued this model airplane in active service until 1931. (Courtesy of Peter Bowers)

airplanes, about 223 of which were various models of the DH-4 transferred from the Army. The last Navy plane with Liberty power of which the author has found good records appears to be the F-5L, the last of which were removed from service in January of 1931. The Loening OL-2 and OL-4 may possibly have lasted slightly longer.

Using a little arithmetic we come up with a total of 7,729 U.S. military aircraft powered by the Liberty over the years 1917 to 1935. Add to that about 147 civil aircraft with Liberty power and we come up with a total of 7,876 U.S. airplanes with Liberty power.

BRITISH USE

The British ordered a total of 8,000 Liberty 12 engines before the war ended but in the end settled for a delivery of 2,252 engines. British historians report that a total of 867 DH-9a airplanes were delivered and accepted for RAF service by December 31, 1918. In addition, about 200 DH-10/10a airplanes were powered by the engine. As heretofore related, deliveries of Liberty engines to the British fell short of expectations, both as to quantities and delivery dates. The first 10 Liberties arrived in England in March of 1918 and the first DH-9a (C6122) was tested with Liberty power on April 19, 1918. This aircraft served the RAF from that date to 1931 and was known as the "Ninak."

The first British squadron to begin receiving DH-9a airplanes for operational service was the No. 110 Squadron. The squadron moved to Kenley on June 15, 1918 and began receiving airplanes. It rapidly came to full 18 aircraft strength and on August 31, 1918 flew to Bettoncourt in France to join the Independent Force, RAF and flew its first mission on September 14.

Other squadrons began receiving Ninaks shortly after the 110 Squadron and by October 31, 1918, the RAF had a total of 405 DH-9as on charge, of which 149 were in France and the remainder in various units and in storage in the UK. At least 53 are reported to have been used by the U.S. Marine Corps Northern Bombing Group, beginning in September of 1918.

The DH-10 was not ready for production as early as the DH-9a and that fact combined with slower than anticipated deliveries of the Liberty to the British resulted in a total production of only 223 airplanes, with about 200 being Liberty powered.

The difference between the DH-10 and DH-10a was the location of the engine nacelle and not the type of engine used. The DH-10 nacelle was located mid-way between the wings and the DH-10a nacelle was located directly on top of the lower wing.

The DH-10 was named the "Amiens" by the British and saw service in England, Germany, France, India and Egypt. It was used in many of these regions for mail service and was withdrawn from service in 1923.

The above two aircraft accounted for at least 1,267 engines for original equipment. Add to this normal spare engine requirement and military usage must have amounted to at least 1,500 engines. British military surpluses would have been a relatively modest 750 engines. Commercial usage was, however, very limited.

Two commercial aircraft with Liberty power were built shortly following the war by Bristol. The first was the Bristol Braemar Mark II (type 25) built originally in response to an Air Board request, this four-engine triplane was first flown on February 18, 1919. After some testing the plane was wrecked while taking off at Martlesham in late 1921 or early 1922.

A second airplane was authorized in April of 1919 which was intended for commercial passenger service. This was the Pullman (type 26) which was first exhibited at the Aero Exhibition at Olympia in West London from July 9-20, 1920. Another four-engine triplane, it was designed to carry 14 passengers or a combination of passengers and freight. The airplane saw limited testing and was dismantled in late 1920.

Bristol Pullman. This triplane was designed to carry 14 passengers and a crew of two. Wing span was 81' 8". Two Liberty 12 engines were carried in each nacelle, one tractor and one pusher. It carried the RAF service serial C4298 and was Bristol c/n 3753. For a short while (April-May 1920) it carried the British civil registration of G-EASP. It was test-flown for a short while, providing data for future designs and then was dismantled in late 1920. (Courtesy of Peter Bowers)

The Director of Military Intelligence in Washington submitted an information request to the Military Attache in London regarding types of British planes using Liberty engines and on June 17, 1920, received a reply stating that the only types in military service so equipped were the DH-9a and DH-10. The report further stated that the following commercial airplanes were also powered by the Liberty. No numbers were given for any airplanes mentioned but probably the listed commercial planes involved single examples being tested as no records indicate any production aircraft were ever built.

Listed were:

Airco	DH-4a
	DH-16 (a passenger conversion of a DH-9a)
Bristol	Pullman
Handley Page	0/400
	V-1500
Vickers	Vimy - commercial

FRENCH USE

The French took delivery of 3,575 Liberty engines and 3,310 sets of spares. What information the author has found indicates they installed these engines in somewhat more than 400 Breguet aircraft, perhaps as many as 425, and one Farman 31 C2, one Morane-Saulnier type A. N. L. and one Voison Bn2 on an experimental basis.

This accounts for only about 428 engines plus a few spares. That leaves a sizeable surplus of 3,147 engines. Brigadier General William Mitchell headed a European aeronautical inspection trip dispatched by the Air Service in the winter of 1921-1922. This mission made a detailed study of aeronautical progress since the war in France, Italy, Germany, Holland and England. The report submitted by this mission was quite detailed and noted most aircraft under development by the various countries as well as those in service. The French report was very detailed as to types under development and mentioned none intended for use with the Liberty.

The report noted that progress in motor development had been retarded to a certain extent by the mass of wartime motors still on-hand and went on to supply a table showing all French motors on-hand (in storage) and available for aircraft use. 1,000 Liberty engines were listed. If this list can be considered accurate then we must assume that a surplus of perhaps 3,000 Liberty engines at the end of 1919 had been reduced to 1,000 by the end of 1921. Although I have no specific information as to just where these engines went, it seems likely that a substantial number probably were sold to Russia. The reader will find a detailed discussion of the Russian use of the Liberty in the chapter following. Russia used the Liberty 12 in aircraft and tanks totaling 5,526. They put the engine into production in their own factories as the M-5 and built a total of 2,909, the remaining 3,091 Liberty 12 engines needed were imported from the United States, England and France.

French Breguet 14 A2 9094. The French built in excess of 400 Liberty-powered Breguet 14 airplanes and kept them in service until about 1928. (Courtesy of Peter Bowers)

ITALIAN USE

The Italians had procured only five Liberty engines during the war but evidently purchased a number of additional engines in the immediate years following as Caproni is known to have built several large airplanes utilizing the Liberty for power. Indications are that in most cases only single examples were built. The one exception is the Caproni Ca-42 of which some historians say 12 were built and some say 6. There is confusion even here in that two of this number are reported to have been sent to the U.S. for evaluation during the war whereas reliable American Air Service records indicate Italian-built Caproni airplanes operating in the U.S. consisted of one Ca-44 (Ca-5) powered by three Fiat A-12 engines, one Ca-33 (Ca-3) powered by three Isotta-Fraschini type V-4-b engines (both of which were purchased by the U.S.), one Ca-41 (Ca-4) with three Isotta-Fraschini type V-5 engines and one Ca-33 (Ca-3) with three Isotta-Fraschini type V-4-b engines. The last two airplanes were owned by the Caproni Company and both were lost in crashes.

The other Caproni airplanes using the Liberty for power were the Ca-43, Ca-46 and Ca-60. The Ca-42 was a variation of the Ca-4. The Ca-43 is reported to be a single example fitted with floats. The Ca-46 was a Ca-5 variation.

The Ca-60 was built by Caproni after the war and was the most unique of the lot. This was Caproni's bid for a 100-passenger airliner which some said was intended for transatlantic flight. It was a seaplane with three sets of triple wings and was powered by eight Liberty 12 engines. Built in 1920, this airplane made its maiden flight in Italy from Lake Maggiore on March 4, 1921 and after lifting some 60 feet off the water the plane stalled and dropped with so much force the hull was severely damaged. The pilot then attempted to taxi it back to the dock and in so doing forced water into the hull and sank it.

This airplane was the last Caproni with Liberty power and the last Italian airplane the author knows of which used a Liberty engine.

RUSSIAN USE

Other than the United States, Russia was the world's largest user of the Liberty engine for aircraft. They powered at least 3,022 airplanes with either a Liberty 12A or the Russian-built M-5 replacement. They had Liberty-powered airplanes in active service from mid-1922 when they first acquired a few Liberty-powered DH-9a airplanes from the British, until mid-1932 when they took their last Russian-built equivalent, the R-1, out of service.

The ill-fated Caproni Ca-60 which was built in 1920 and crashed on its maiden flight on March 4, 1921 from lake Maggiore in Italy. This nine-wing airplane was powered by eight 400 HP Liberty 12A engines and was designed to carry 100 passengers and a crew of 8. It was 77 ft. long and had a wing span of 98 ft. 5 in. Weight fully loaded was 55,100 lbs. Estimated cruising speed was 80 MPH and range 410 miles. (Courtesy of Peter Bowers)

They built 2,641 Liberty-powered R-1 (Reconnaissance type 1) airplanes and, in addition, built 124 MR-1 types (Float-equipped R-1). A short production run of 30 reconnaissance types of improved design known as the R-3 were built with Liberty power between 1925 and 1928.

In addition to the "R" types mentioned above, two fighter types ("I" for Istrebitel, the Russian word for fighter) were built. The first model, the I-1, was built from 1923 to 1927 and the second, the I-2, from 1924 to 1929. Total production was 14 and 213 respectively.

Also, between 1923 and 1928 a half-dozen single prototype airplanes were built with Liberty power and in 1924-1925 110 Liberty-powered Fokker C-IV airplanes were imported from Holland for service use.

Once the Russians decided the Liberty was to become a service-type engine, they began the process of putting it into series production in their own factories. In the meantime, however, they supplied their needs by importing surplus U.S.-built engines from whatever sources they could find. These included all three countries in possession of large quantities, i.e., the United States, Great Britain and France.

Shortly before the last R-1 airplanes were retired as obsolete, the Russians initiated the production of a new series of tanks designed around the American Christie, the BT series. This series was powered by the Liberty/M-5 engine and added some 2,504 vehicles to the engine requirement. Thus the Russians powered 3,022 airplanes and 2,504 tanks with the Liberty/M-5 engine for a total requirement of 5,526 engines not including spares. M-5 production in their own factories is known to have been 2,909 engines. If we round out the requirements to about 6,000 engines including spares this leaves a requirement of about 3,091 imported engines. I do not know the exact quantities imported from each country but what is currently known I have related in the following chapter.

Ski equipped Russian I-2 shown in typical winter environment. This airplane greatly resembled the Curtiss R-2/R-3 of 1923-1925. (Author's collection)

The story of the Russian-built and Liberty-powered airplanes is both complex and interesting, as is the story of their implementation of Liberty/M-5 engine production. Liberty engine production outside the United States was done in both Russia and Great Britain. The reasons behind it are interwoven, very complex and very interesting. I have chosen to devote a chapter to the subject and it follows this one.

For the purposes of this chapter we can sum up the use of the Liberty in Russian airplanes by noting that they used a total of about 3,132 airplanes so equipped between the years of 1923 and 1934.

DUTCH USE

The Fokker Company was the only airplane builder in Holland and they built no Liberty-powered planes for use within the country that I know of. They did, however, build a number for other countries. The U.S. Navy purchased three model T.III torpedo planes which they designated FT-1 (Bu A numbers A6008-6010) which were delivered in 1923. Also in 1922-23 the Army purchased seven airplanes derived from the Fokker C-IV assigned the designation CO-4. In 1922 the Army also purchased two Fokker F-IV airplanes which they designated as T-2s. One (AS 64234) was converted to a two-litter ambulance and designated A-2. The second (AS 64233) was fitted out for long-range flight and became the record-breaking coast-to-coast non-stop machine.

During this same 1923-1924 period the Russian Government ordered 110 Liberty-powered Fokker C-IV airplanes.

PORTUGUESE USE

Portugal purchased one DH-9a from the British government in 1924 and maintained it in service for several years.

SPANISH USE

In 1922-23 the Spanish government also purchased eight British DH-9a airplanes. They continued in use until 1936.

GERMAN USE

The author has only found references to one German airplane with Liberty power, the Heinkel HD-24L. This was a trainer-type aircraft. A number were built with most going to the Swedish Air Force. Normally powered with BMW engines of various sizes, only a single example reported was built in 1926 with Liberty power.

Heinkel HD-24L. The only German-built airplane constructed after WW I the author knows of which utilized Liberty power. Designed in 1926 as a training airplane, it was normally powered by BMW engines of lesser power than the Liberty 12. (Courtesy of Peter Bowers)

NORWEGIAN USE

Norway showed an interest in the Douglas DT-2 airplanes which the Navy had first ordered in 1921 as the DT-1. This model was refined and the last two (of the initial order for three) were delivered as DT-2 models. The Navy ordered an additional 38 of this model from Douglas, the last being delivered in 1924.

In the meanwhile, the Norwegian Navy had negotiated the purchase of a DT-2 which they wanted to be supercharged. Douglas replied that they could only furnish it with a standard Liberty. The airplane was delivered in 1925 along with a spare engine. Both engines were supplied with late-type intake manifolds, heavy stub-tooth timing gears and jump-gap Delco ignition. This model was known as the DT-2b and had some minor differences from the U.S. Navy DT-2 and actually was much like the Douglas World Cruiser as the reader will note from the illustrations.

The Norwegian Navy assigned the airplane number F.80 and obtained a license from Douglas to build additional DT-2b airplanes in the Naval Workshops Marinens Flyfabrik, Horten. From 1927 to 1929 an additional four were built there, all with Liberty power. The required Liberty engines were imported from France through surplus dealer C. Marabini. Engine requirements were set at one service and one spare for each plane. Thus a total of 10 Liberties were required, eight were French surplus. These last four airplanes were assigned the numbers F.82, F.84, F.86 and F.88.

The Naval Air Arm asked the Norwegian Embassy in Washington to assist in getting the U.S. Navy to supply Liberty maintenance information in 1926.

The airplanes remained in service to the early 1930s, at which point they began to re-engine the Norwegian-built ones with A. S. Panther IIA engines which had been selected as standard for both Army and Navy air arms. When so altered they became DT-2c models. It appears that the conversions were made between 1932 and 1934 and involved all but the original F.80. The original Douglas airplane remained Liberty powered at least until 1940 when the Germans invaded the country.

The Navy equipment inventories of both 1928 and 1930 listed 10 Liberties and noted that time between over hauls was estimated at 300 hours. The 1938 inventory

Douglas DT-2b sold to the Norwegian Navy in 1925 and remained in service use at least until the fall of Norway to the Germans in 1940. This DT-2b model greatly resembled the Douglas World Cruiser, which was built during this same period. Four more were built in Norway under license. (Courtesy of Knut Erik Hagen)

listed eight engines, all in need of overhaul before being used. In 1940 F.80 was in need of an engine but none were available. The Liberty thus served the Norwegian Navy from 1925 to 1940.

Records have been found indicating that F.84, which had been operating with Panther power since 1934, was captured by the Germans in 1940 and taken to Horten to be fitted out for use in fisheries patrol. F.86 was destroyed by a German vessel on April 10, 1940. It had been converted to a DT-2c by installing a Panther engine in 1932.

AUSTRALIAN USE

As might be expected, immediately following WW I the British supplied as part of the "Imperial Gift" a number of aircraft with which the Australia Air Force was to be established. The decision to do this was finalized in 1919 and the aircraft received in Australia in 1920. This initial group consisted of 28 DH-9 and 30 DH-9a aircraft. With these airplanes the Royal Australian Air Force (RAAF) was officially established in 1921.

Of these two initial airplanes only the DH-9a was Liberty-powered, of course. The DH-9a airplanes were assigned serials A1-1 through A1-30 and the DH-9 airplanes A6-1 through A6-28. One of the 30 DH-9a airplanes was lost in 1920 immediately after receipt and was replaced by a DH-9a purchased by the Australian government.

These 30 airplanes served with the RAAF in various capacities from 1920 until April of 1930 when the last eight survivors were struck off charge and either scrapped or given to flying clubs.

The remaining eight had been approved for write off on April 2, 1930. A1-14 appears to have been the last to actually fly in service, having logged its last flight on May 3, 1930.

CANADIAN USE

British DH-9a airplanes were provided to the Canadian government after WW I in the same manner as to the Australians. In 1920 twelve were shipped to Canada where they became the initial airplanes of the soon-to-be-formed Royal Canadian Air Force (RCAF).

These airplanes served from 1920 until 1929 and were used in a number of various capacities. One, G-CYAJ, was involved in Canada's first trans-continental flight whose purpose was to draw public attention to the newly formed Air Board and to the RCAF and the future possibilities of aircraft in passenger and air mail transportation. The flight was accomplished using a number of DH-9a and Fairey seaplanes and pilots, G-CYAG and G-CYBF being two of three DH-9a airplanes involved.

One of the 30 British DH-9a airplanes given to the Australian Air Force in 1920. They were serial numbered A1-1 through A1-30 when the Royal Australian Air Force was formed in March of 1921. (RAAF official photograph)

At least one additional DH-9a is known to have served in Canada. It was a surplus British airplane purchased in 1920 by Major W. T. Blake with the intention of making a round-the-world trip in 1922. It was the first civil registered Canadian DH-9a (G-EBDF). The flight was terminated after a crash in Calcutta.

A number of Curtiss HS-2L seaplanes were put in service in Canada. Seaplanes were found to be more useful than land planes during the early years because of the vast areas which had no air fields on which to land. Seaplanes could, however, operate from rivers and lakes virtually anywhere within the country. Twelve were used by Navy units beginning in 1928 and at least 30 are known to have been in RCAF service from 1920 to 1929. In addition, two Curtiss H-16 seaplanes were in RCAF service from 1922 to 1924.

Other DH-9a and Curtiss seaplanes were very likely in civil service in the country but the author has not attempted to research this area.

ARGENTINE USE

Argentina followed the trend of a number of other South American countries following WW I and showed a markedly increased interest in aviation. Since South America had almost no aviation industry of its own the countries there looked to Europe and the United States as a source of aircraft. Surplus WW I airplanes supplied the majority of their newly found needs.

The Argentine Navy purchased eight Curtiss HS-2L airplanes in 1921. These were assigned the numbers T-1 through T-8, after modification to their needs, and put in service in 1922. Three were lost in a hanger fire in 1923 and apparently another to a crash-landing not long thereafter and four replacements were purchased in 1925. The HS-2L series remained in Argentine service until 1931.

In 1921 eight of the somewhat larger Curtiss F-5L airplanes were also purchased and likewise modified, assigned the designations B-1 through B-8 and put in service in 1922. These airplanes were formed into a Patrol Squadron and were used for fleet patrol service until 1930 when they were replaced by British Southamptons. They were then moved into various other roles and continued in service until 1935.

In 1922 the Argentine Navy also purchased four Italian-built Dornier Do J Wal flying boats for inclusion in their Patrol Squadron, assigning them the numbers B-9 through B-12. These were powered by a pair of Liberty engines mounted push - pull in a single nacelle on top of the wing above the fuselage. These remained in service until 1932.

BRAZILIAN USE

The Brazilian Navy purchased six Curtiss HS-2L flying boats and operated them from late 1918 to 1923. Assigned numbers were 10 through 15. In 1923 they apparently replaced them with Curtiss F-5L airplanes and added to their fleet at the same time, as they purchased 14 of the F-5Ls in 1923 and kept them in operation until 1930. Numbers 311-316 and 321-326 were assigned and the F-5Ls remained in service until 1930.

Brazilian Naval Aviation lays claim to the first Air Mail service in South America with service by Curtiss HS-2Ls 10 and 11 between the Capitol and the Great Island during naval maneuvers in August of 1919.

Argentine Curtiss HS-2L in Navy service carrying the number T-3. A total of 12 airplanes of this model were purchased between 1921 and 1923 and served until 1931. (Courtesy of the Argentine Department of Naval History)

Brazilian Curtiss HS-2L No. 13. One of six purchased by the Brazilian Navy and operated from 1918 to 1923. (Courtesy of the Brazilian Department of Naval History)

CHILEAN USE

Chile followed the same pattern as the other large countries of South America and, following WW I, purchased up-to-date airplanes for their military services. In the case of Chile they purchased 20 DH-9a airplanes from the British for their Air Force in 1919 and kept them in active service until 1928.

No Liberty-powered Navy airplanes were acquired until 1926 when, after noting the experience of Argentina with Liberty-powered Dornier Do J Wal flying boats, purchased eight of them in 1926. Evidently they gave good service as they remained in active service status until 1944.

PERUVIAN USE

The only Peruvian user of Liberty-powered airplanes was the Navy. Although they acquired no great quantity they appeared to like variety in their airplanes. In 1920 they purchased five Curtiss HS-2L and two F-5L flying boats, followed in 1921 with a single H-16.

The H-16 only remained in service for a single year, probably being lost in an accident. The last F-5L left service in 1924 and the last HS-2L in 1926.

One of the eight Liberty-powered Dornier Do J Wal flying boats purchased by Chile in 1926. These airplanes continued in Chilean service until 1944. (Courtesy of the Argentine Department of Naval History)

The F-5L was evidently replaced by an order for 4 new Douglas DT2b airplanes in 1926. The four were probably a continuation of the single one of this model ordered by Norway in 1925. These remained in use until 1933.

MEXICAN USE

Mexico acquired its first Liberty-powered airplane in 1919, a Curtiss HS-2L. Although it remained in active use until 1921, no additional Curtiss flying boats of any type are known to have been acquired.

In 1924 the first of a total of 19 DH-4B airplanes were purchased from U.S. Army stocks. These were kept in service until at least 1928. During this period the Mexican National Workshops of Aeronautical Construction (Talleres Nacionales de Construcciones Aeronauticas) built a series of three special biplanes called Quetzalcoatl Biplane "6-E-136" which were briefly powered by the Liberty engine.

Use of the DH-4B airplanes evidently led to further interest in U.S. Air Service Liberty-powered airplanes as Mexico purchased eight O-2C airplanes from Douglas in 1926. Later versions were also purchased which carried the designation O-2M ("M" for export military) but these were air-cooled radial-powered. The O-2 series saw service into the middle 1930s.

CHINESE USE

The Chinese purchased nine Curtiss HS-2L flying boats in 1919-1920. Russia sold eleven M-5 powered R-1 airplanes to the Chinese in the early 1920's and some historians note that ten Liberty-powered Douglas O-2C airplanes were purchased in 1926. As in the case of Mexico, the Douglas airplanes were followed by the purchase of a number of radial engine models in later years.

Peruvian Douglas DT-2b. Four of these airplanes were purchased by the Peruvian Navy in 1926 and remained in service until 1933. (Courtesy of Sergio de la Puente)

SIGNIFICANT ACHIEVEMENTS WITH LIBERTY POWER

Although America's armed services had other engines with which they powered their airplanes, the Liberty was their principal high-powered one and was soon developed into one of good reliability. During the years from the end of WW I until the late 1920s airplanes powered by this engine were chosen to attempt a number of very significant flights. Almost all were successfully completed.

AIR SERVICE "FIRST ROUND TRIP" ACROSS THE UNITED STATES - 1919

The Air Service was just beginning to gain confidence in the abilities of their airplanes for long distance flight by the end of the war. Little cross-country flying had been felt necessary or attempted, either in this country or in Europe, up to this time.

The first transcontinental Air Service flight, actually "trip" would be the better term, was made by a group of Curtiss JN-4 airplanes. Five planes left San Diego and headed east on December 4, 1918. One soon became lost in fog and landed at El Centro, California. Its pilot wrecked the plane attempting to take off and ended up shipping it back to San Diego by train.

The remaining four planes continued the flight. They arrived in El Paso on December 7th, Houston on the 11th, New Orleans on the 14th, Montgomery on the 16th, Americus, Georgia on the 17th and Jacksonville on the 18th. The flight was made in relatively short increments, stopping for fuel, minor repairs, and overnight rest along the way. The trip had required some 14 days but no major difficulties had been encountered.

Successes such as this soon led to other such flights. Major Theodore C. Macauley, Commander of Taliaferro Field (near Fort Worth) received permission to attempt a coast-to-coast flight early in 1919. He chose a DH-4 and equipped it with tanks for an additional 57 gallons of gas

Five Douglas O-2C airplanes ready for delivery to Mexico in 1926. An additional three were later purchased for a total of eight. (Author's collection)

and 10 gallons of oil. He and his mechanic, Private Staley, departed Fort Worth headed for Rockwell Field at San Diego. They would begin their flight from there on January 21, 1919. Although a few difficulties would be encountered along the way, this trip progressed much more rapidly than did the earlier JN-4 effort. Failure of the generator followed by the battery (which was then the only power source for the ignition) forced them down near Hot Wells, Texas. Here they connected the lighting battery as a temporary power source and flew on to Pecos, where they made the necessary repairs. From here they flew on, arriving at Baton Rouge on the 23rd and flew on to Americus, Georgia and then south to Arcadia, Florida (about midway down the state) on January 24, 1919. They had made the trip of a somewhat longer distance than the prior JN-4 flight in about 4 days rather than 14.

The return flight was not so uneventful. Heading back up-state toward Jacksonville on the 25th, the major had engine trouble (type not stated) and force-landed near the north end of Lake Okeechobee. After obtaining help he pulled the airplane out of the weeds and put it on a barge for shipment to Fort Lauderdale and from thence to the Marine base in Miami for repairs.

They resumed their trip on February 1, 1919, and headed north to Americus and then westward. Shortly after leaving Americus they started losing water and overheated the engine, causing it to stop in-flight. In landing the major hit a wire fence and crashed the plane. The wreckage was picked up by Souther Field personnel and sent to the repair depot at Montgomery. The Souther Field commander gave the major a replacement DH-4 with which to continue the trip.

Between Montgomery and West Point, Mississippi heavy rain damaged the propeller, requiring a landing at Payne Field near that city for a replacement. The heavy rain had softened the field too much for a takeoff so the major and Staley took a train to Fort Worth and returned two weeks later to resume the flight. On February 17, 1919, they took off for Taliaferro Field and arrived later that day. This completed their round trip, the westward portion not being so uneventful as had been the eastward.

THE FIRST TRANS-ATLANTIC CROSSING 1919

On April 1, 1913 the London *Daily Mail* announced it was offering £10,000 (about $50,000 at the then-current rates of exchange) to the first person to fly the Atlantic. The terms of the flight were reasonable for the time. The flight did not have to be non-stop but only a single aircraft could be used. All intermediate landings had to be at sea. A pilot could land at sea and board a ship to eat or sleep. Landing at sea for fuel or repairs was allowed but in any case of landing for any reason the journey had to be taken up from the same point. The starting and finishing points could be anywhere in Newfoundland or the British Isles. Total elapsed time for the flight could not exceed 72 hours.

A number of aviators from all over the world announced their intention of competing for the prize but all efforts were put on hold by the outbreak of war in Europe in the summer of 1914.

By the summer of 1918 the German tide of battle was beginning to turn. Although the end of the war was not then thought to be imminent, the *Daily Mail* renewed its trans-Atlantic offer in July of 1918. It was not long (November 11) before the war did end. This was immediately followed by a number of announcements of preparations for cross-Atlantic flights.

The U.S. Navy already had in process the building of the NC series of flying boats, the result of a wartime need for transatlantic aircraft conceived by Admiral David W. Taylor in the summer of 1917. Not long thereafter the newly-designed Liberty 12 was successfully tested (August 25) Admiral Taylor noted in a memo to his staff that he thought the new United States motor gave good promise of success and would provide the required power for a transatlantic flying boat.

Obviously a great amount of detail will be omitted in the telling of the story which follows (several books have been written on it), but shortly thereafter Glenn Curtiss, America's most eminent designer and builder of flying boats, was called to Washington to discuss the design of such an aircraft.

The ultimate outcome was a biplane design utilizing a large seaplane hull with its tail group suspended on outriggers. The object became a balance between the ability to lift enough payload to include the aircraft structure, sufficient fuel to carry it the required distance, and enough power to carry it out. The distance goal being to either fly non-stop from Newfoundland to Ireland, or to fly from Newfoundland to the Azores and then to Portugal and then to England.

Although initially called the TH-1 model by Curtiss, the eventual designation chosen was NC for Navy-Curtiss. On January 8, 1918, the Curtiss Engineering Company was given a construction contract to build four airplanes. They would be built in sections subcontracted to various builders and assembled by Curtiss. The largest and most difficult sections were, of course, the hulls. Of these, Curtiss would build one, Lawley & Sons (boat-builders of Boston) would build two and Herreschoff Company (boat-builders of Bristol, Rhode Island) would build the last.

After the usual seemingly unnecessary delays in progress, NC-1's parts (the first of the four which were assigned NC-1 through NC-4 designations) were finally

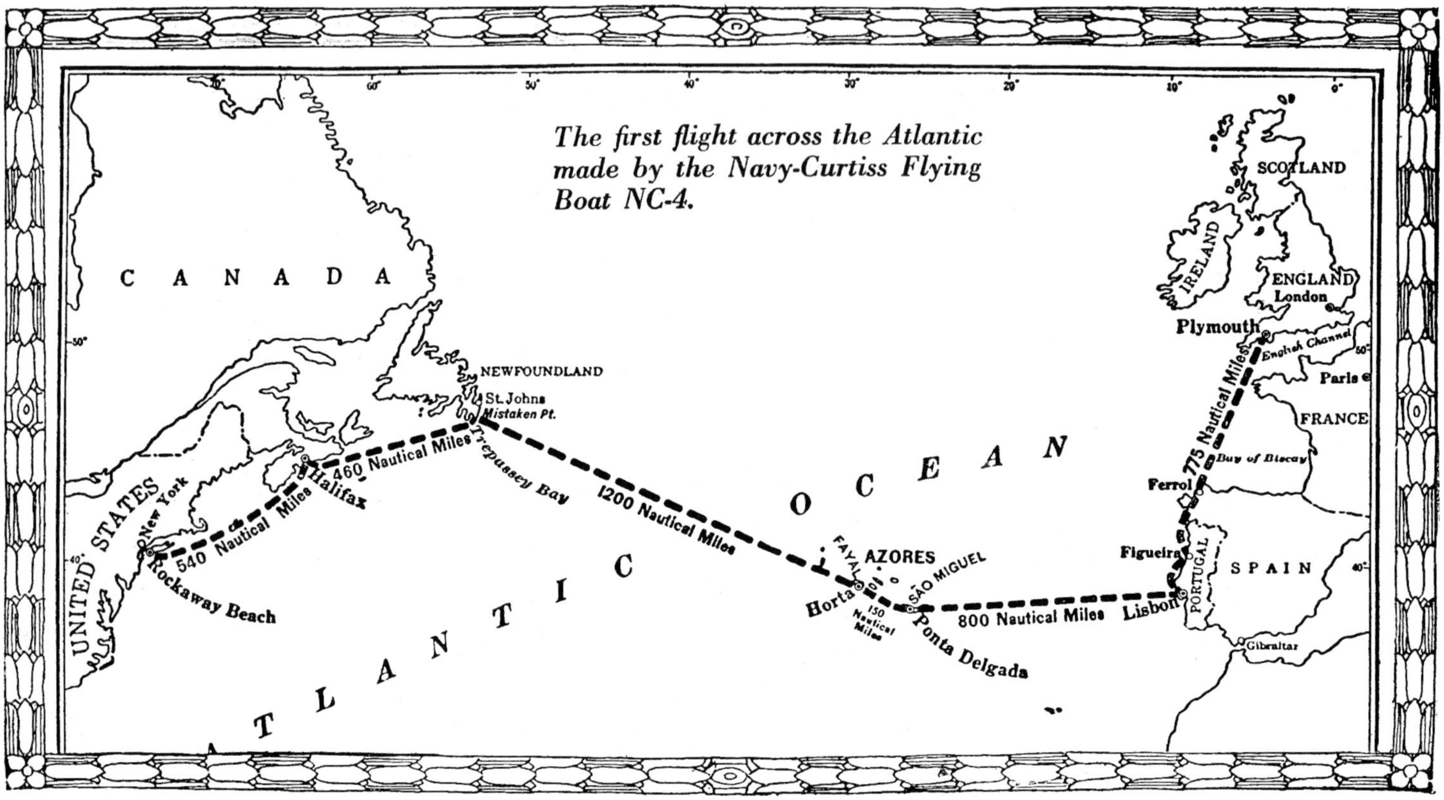

Map of the flight of the NC flying boats. NC-1, NC-3 and NC-4 left Newfoundland on May 16, 1919 and arrived in the Azores at about 1 PM on the 17th. NC-1 and NC-3 both landed at sea just short of their destination after becoming lost in fog. Neither were able to finish the journey. NC-4 made it to Horta and then Sao Miguel. On May 27, after a long wait for good weather, NC-4 left for Lisbon, Portugal. She arrived at about 10 PM. The NC-4 had become the first airplane to fly the Atlantic. From Lisbon the final destination of Plymouth, England was reached on May 31, 1919. (Map from an advertising publication of the Curtiss Airplane Co. of 1919)

delivered and the airplane assembled in the Navy hanger built for it at the U.S. Naval Air Station at Rockaway Beach, NY. It was first towed out of the hanger on October 4, 1918, about a month before the war ended. Later that day its three Liberty engines were started and after a warm-up she was taxied out for a brief flight. This was followed by another five brief takeoff and landings for a total of a little under five minutes flying time that first day.

From the initial flight until the end of November the plane logged about 28 hours in the air, several of which were used flying from Rockaway to Washington, D.C. and then to Hampton Roads and back to Rockaway. This last trip had taken place during the last few days before the signing of the Armistice when the great airplane's debut was overshadowed by building news of the termination of war.

Within the week a Handley Page V-1500 took off from Crickle-wood, England for a flight over London with 41 persons aboard and set a world record for passengers carried on an airplane. The Navy's pride in their new flying boat was too great to let this challenge go by. With a design gross takeoff weight of 22,000 pounds the Navy was sure it could exceed that record and so on November 26 they put 46 additional Navy personnel aboard and with pilot and copilot took off at a gross weight of 23,578 pounds with 48 on board. The next day it was decided to try for an even 50 and fuel load was reduced by 91 gallons and 48 put aboard. A flight of 1,000 yards at 50 feet was made and NC-1 returned to the dock with a new world record. Passengers were counted as they went ashore and it was discovered that one extra had stowed away. The new record was 51 persons carried.

As one might very well expect, a very significant factor in this project was how much fuel these airplanes could carry and would that amount get them to their proposed destination. It seemed that all factors were being pushed to the limit. The gross weight that could be carried was determined by a combination of aircraft design and power. During initial stages of the project the Liberty was available only in its 330 HP version but later the high-compression 400 HP version became standard. This allowed a higher gross weight. In the end it was even decided to change from three to four engines to gain even more lifting ability. Even then, the theoretical range was so close to the required distance that both the Navy's Bureau of Steam Engineering and the BAP's Liberty Experimental Engineering section were assigned the task of improving Liberty fuel economy. This was C. H. Vincent's last assigned project before he left BAP service at the end of the war. The Army and Navy working together succeeded in accomplishing this with improved carburetor design.

The Armistice put a brief hold on the remaining three airplanes but it was soon decided that it was worthwhile to complete them all. Commander John Towers of the Chief of Naval Operations Section had proposed on October 31, 1918, that the NC flying boats be used to mount a Navy effort to be the first to fly across the Atlantic. The proposal had resulted in the appointment of a board to study it and report their findings. A positive report was made on February 1, 1919, and Secretary of the Navy Josephus Daniels gave his approval on February 4.

NC-4 on tour after completion of the first cross-Atlantic flight. She carried four high compression Liberty engines with two in the center nacelle mounted one tractor and one pusher. (Courtesy of Peter Bowers)

Once this decision had been made, work began on preparations for the flight. These included committing and preparing a large number of ships which were to serve as tenders and ocean station vessels along the chosen route, which was Newfoundland - Azores - Portugal - England. In addition, radio and navigation testing was carried out off Hampton Roads using mostly F-5L flying boats. One of them was equipped with geared Liberty 12B engines. These showed good performance and better fuel economy than standard engines. It was hoped that this type would be available to power at least one of the airplanes on the flight but the bugs were not considered ironed out sufficiently by flight time. The engines under test in the F-5L were two of the four initial experimental engines built by Packard and transferred to the Navy in the last part of 1918. No production engines of this type had yet been delivered as they had just been ordered in April or May of 1919.

Testing was still being carried out to determine the best engine configuration, both from a performance standpoint and to determine whether the greatest range could be obtained with three engines or four. Three engines weighed less but could lift less weight and therefore less fuel. Four engines weighed more but could lift a lot more fuel. It was determined that the added fuel more than balanced the added weight and fuel consumption of the added engine and the four-engine configuration was chosen. The configuration was to be a tractor engine on each side between the wings and a tandem pair in an above body nacelle with one tractor and one pusher.

During the production of the four boats NC-1 had been the first to fly. NC-2 was used as a test plane to determine the best configuration and NC-3 and 4 finished to that configuration. NC-4 first flew on May 1.

The construction and testing of the four boats, was carried out under the stress of a very short schedule. In the midst of that, on March 27, the NC-1 was caught in an unexpected gail while anchored outdoors and was badly damaged, requiring wing replacement. Since no replacement wing was available it was decided to cannibalize NC-2 for parts and use only three airplanes on the attempt.

The three airplanes had a 1,000 mile flight from Rockaway to Newfoundland before they could actually start the venture. Even before that flight they endured a seemingly endless series of mishaps which included everything from a hanger fire to a near-fatal encounter with a moving propeller by a machinist mate.

They overcame them all and the three NC boats left Trepassey, Newfoundland at about 6:00 PM on May 16, 1919, bound for the Azores. The three flew through the night without problems but ran into heavy fog early the next day. The three became separated during the early morning hours. NC-1 lost direction in the fog and found it necessary to land at sea to allow use of her radio direction finder, a very primitive radio set by today's standards. It could not be reliably operated while the engines were operating because of the ignition interference.

NC-3 also became lost and after realizing she was badly off course and getting low on fuel decided to land and await conditions clear enough to get reliable sextant navigation and to stop the engines and allow better RDF.

NC-1 had drifted off-course to the north and had landed at 1:10 PM. NC-3 had drifted off-course to the south and landed at 1:30 PM. Both were within the Azores island chain but had never sighted land.

NC-4 had managed to stay on-course and broke out of the cloud cover and sighted the island of Fayal. From there she flew on to Horta where she landed at 1:23 PM.

Both NC-1 and NC-3 had sustained enough damage during the landing in heavy seas to prevent them from taking off again even if conditions improved. NC-1 drifted north and was sighted by the Greek freighter *Ionia* and her crew rescued. NC-1 eventually sank in the heavy seas.

NC-3 managed the additional 205 miles to Sao Migeul in the next 53 hours by drifting and taxiing. She was now in too bad a condition to be able to repair and fly so the trip would be completed only by NC-4.

After waiting several days for favorable weather, the NC-4 left for Lisbon, Portugal at about 11 AM on May 27 and landed about 10 PM. The Atlantic had been flown.

After appropriate celebrations, preparations were completed and NC-4 flew on to her final destination of Plymouth, England, where she arrived on May 31, 1919.

The object of the Navy's efforts with the NC boats had been largely scientific - that is that they were out to prove it was feasible to fly the Atlantic. However, being the first to do so also had great meaning in the effort. Several countries had aspired to have their flyers be the first across the Atlantic but in the end it became a race between America and England with the British attempting to accomplish it in a non-stop flight. Doing so would not only bring them the glory of being first across but also the prize offered by the London *Daily Mail.*

Although Commander John Towers and all the crew of NC-4 received much congratulation and wining and dining first in Europe and then in the United States, it was to be ten years before they received awards of a more concrete nature. In February of 1929 Congress approved the awarding of a Congressional Gold Medal by the President to Commander John H. Towers and the crew of NC-4. As approved by the House Committee on Naval Affairs the citation read: "In recognition of Commander John H. Towers *extraordinary achievement* in conceiving, organizing, and commanding the first trans-Atlantic flight; and Lieutenant Commander Albert C. Read, United States Navy, commanding officer; Lieutenant Elmer F. Stone, United States Coast Guard, pilot; Lieutenant Walter Hinton, United States Navy, pilot; Lieutenant H. C. Rodd, United States Navy, radio operator; Lieutenant J. L. Breese, United States Naval Reserve Force, engineer; and Machinist's Mate Eugene Rhodes, United States Navy, engineer *in making the first successful trans-Atlantic flight, in the United States naval flying boat NC-4, in May 1919.*"

The British flyers would also fly west-to-east from Newfoundland but fly directly to Ireland, the shortest non-stop route across. Such a flight would cover about 1,800 miles from takeoff to landing. Flying via the Azores required a non-stop flight of some 1,400 miles. At this time in aviation development engine reliability had reached the point that operation for the number of hours required to travel such distances was not considered so much of an obstacle as carrying enough fuel to run them that long. The NC boats could carry enough fuel to make the Azores with a safe margin, but not Ireland.

Two British planes and their crews had arrived in Newfoundland and had been assembled and tested by the time the NC boats were just beginning their trip from Rockaway to Newfoundland. Only bad weather conditions in the north Atlantic was preventing them from starting their flights. Neither was a seaplane so their flight had to be non-stop, not only because the rules required it but because they could not land at sea.

The first was a Sopwith piloted by Harry Hawker with Kenneth MacKenzie-Grieve as navigator and the second a Martinsyde with Frederick Raynham as pilot and C. W. F. Morgan as navigator.

The Britishers were still waiting for good weather when the NC boats took off for the Azores on May 16. On May 18 the word came to the Britishers that the weather was favorable. The NC boats were by this time in the Azores also awaiting weather before NC-4 could complete the final leg of their trip across. The two British teams prepared to take off. There was still a possibility they could be the first across.

Hawker took off in the Sopwith at 3:18 PM and was shortly over open water headed for Ireland. About an hour later Raynham started down the runway in the heavily loaded Martinsyde, was at the point of lift off and was suddenly hit by a strong side wind which turned the plane to the side and pushed it down slightly. The landing gear hit the ground and collapsed and the plane skidded to a crashing stop.

Although Hawker did not know it at the time he was now the only British plane in the race. The American team was sitting in the Azores awaiting good weather before leaving for Lisbon. They received word by ship's radio that Hawker was away and Raynham had crashed. Hawker had only 22 hours of fuel so by the afternoon of the 19th word should be received of his arrival in Ireland or he would be down at sea.

CREW OF U. S. MARTIN "ROUND THE RIM FLYER"—Left to right: Colonel Hartz, Lieuts. L. A. Smith and E. E. Harmon, Sergeants John Harding, Jr., and Jeremiah Tobias

Martin Bomber No. 2 has been in constant use for a year at Bolling Field, Washington, D. C. With more than 10,400 miles of cross country work already to its credit, this plane was started off on an 8,000 mile flight around the rim of the United States, in command of Col. R. S. Hartz. The colonel and his crew of four have just completed their unparalleled pioneering trip.

Martin airplanes have demonstrated beyond a question the greatest ton-mile efficiency of any aircraft yet produced, and are rapidly proving that the important centres of the United States can be connected by aerial transportation with but a few hours' flying.

THE FOLLOWING IS A COMPARISON OF A FEW ACTUAL FLIGHTS BY MARTIN BOMBERS WITH THE BEST RAILROAD TIME:

FROM	TO	PILOT	DISTANCE	MARTIN AIRPLANE TIME	RAILROAD TIME
Cleveland	Dayton	Eric Springer	215 Miles	1 hour, 50 min.	6 hours
Washington	New York	Capt. Francis	225 Miles	2 hours	5 hours, 15 min.
Cleveland	Washington	Eric Springer	350 Miles	2 hours, 58 min.	14 hours
Cleveland	New York	Eric Springer	450 Miles	3 hours, 20 min.	14 hours
Dayton	New York	Capt. Francis	635 Miles	6 hours, 40 min.	18 hours
Macon	Washington	Capt. Francis	630 Miles	6 hours, 15 min.	21 hours

The superiority of performance which characterizes Martin Airplanes is the result of ten years of experience in building aircraft of a superior quality

The Glenn L. Martin Co.
CLEVELAND
Contractors to the U. S. Army, Navy and Post Office Departments

A Glenn Martin ad of 1919 taking advantage of the recent "Around-the-Rim" flight by the Air Service using a Martin MB-1 bomber. Here the crew is featured with the airplane in the background. Two of the crew, Lowell Smith and John Harding, also participated in the 1924 "Round-the-World" flight. (Author's collection)

Nothing was heard regarding Hawker until a radio message from the American freighter *Lake Charlottville* was intercepted on the 23rd. They had found the wreckage of Hawker's plane at 49°40' north, 29°08' west (about halfway between Newfoundland and Ireland) with no one on board. On the afternoon of the 25th a radio message was intercepted that Hawker and Grieve had been found and safely landed in Scotland.

A few hours out of Newfoundland Hawker's radio had gone dead. His radio range had only been 60 miles but now it was gone altogether. They flew on but began to encounter cumulus clouds and much turbulence. Throughout the night they alternated between dodging clouds and flying through them. Somewhat after daylight they encountered a cloud barrier they could not fly around or over so went through into turbulence so great they decided to dive for the surface and hope to get out of it. Part way through the engine quit.

They glided through and broke-out about 100 feet above the water, all the while trying to restart the engine. It suddenly came back-to-life but checking the gauges revealed overheating. With no way to replace lost cooling water (they could not land at sea) there was no chance of reaching Ireland, which was some 500 miles away.

A check of navigation indicated they had drifted about 130 miles off-course south and were now in the shipping lanes. The decision was to reduce power, nurse the engine along as long as possible and criss-cross the shipping lane and ditch along-side a ship if possible.

Luck was with them and they sighted a ship and ditched close to it. It was the Danish ship *Mary* en route from Norfolk, Virginia, to Aarhus, Denmark. The ship, however, had no radio. Thus no one knew of the rescue until the 25th when the ship crossed close to the Lloyd's station Butt of Lewis on the north tip of Scotland and passed the news by blinker signal.

The Admiralty was notified and dispatched the destroyer *Woolston* to intercept the *Mary* and take the flyers on-board. A few hours later they were landed at Thjurso, Scotland, and put on a train to London.

The British did shortly claim the title of being the first to fly non-stop across the Atlantic but none of those involved in the first attempt were among the flyers. On June 14-15, 1919, Captain John Alcock, pilot, and Lieutenant A. Whitten Brown, navigator, flew a two-engine Vickers Vimy from Newfoundland to Clifden, Ireland, in 16 hours and 12 minutes (1,880 miles).

U.S. POTENTIAL NON-STOP ATLANTIC FLIGHT

The Air Service had planned to fly a Martin bomber across the Atlantic and even picked a plane and pilot (Martin MB-1 and Captain Roy Francis) and proceeded to fit it out. However, before they could complete the testing they felt was required, the British Vickers Vimy accomplished the feat. A non-stop cross-continent flight was then contemplated and on July 25 the plane was sent to Hazelhurst Field in New York in preparation. Before anything further could be accomplished an unexpected strong wind severely damaged the Martin MB-1 at Hazelhurst. Evidently there were those who had serious misgivings about the proposed flight and this setback provided an excuse to cancel the project. The cross-Atlantic flight from Newfoundland to Ireland was one of about 1,800 miles. A cross-continent flight would be about 2,400 miles. The Air Service did not again mount an attempt until General Patrick approved 1st Lts. Oakley Kelly and Muir Fairchild's Fokker T-2 proposal in 1922. More about that later.

AROUND-THE-RIM FLIGHT 1919

In the meantime, another Martin MB-1 (SC 39056 P-88) took off from Bolling Field on July 24, 1919, on the first leg of a trip around the continental United States. Piloted by Lt. Colonel Rutherford Hartz, this was to be a combination of an endurance test of the Martin bomber and its Liberty engines and an Air Service "good will" tour to boost interest in aviation and Air Service enlistments and, at the same time, chart routes and locate landing fields.

The trip was not an attempt at long-distance flight but was made in relatively short hops from field-to-field. It began on July 24 and ended at the same field on November 19, 1919. From Bolling they flew to Augusta, Maine, then to Camp Lewis, Washington, San Diego, California, Houston, Texas, Pinehurst, North Carolina and back to Bolling, with numerous stops in between.

During the trip seven forced landings were made: two were because of instrument failures, one because of a forest fire in Montana and four because of weather.

NEW YORK CITY TO NOME, ALASKA AND BACK - 1920

Late in 1919 Brigadier General Mitchel set in motion a plan to send a flight of four DH-4B airplanes on a round-trip from New York City to Nome, Alaska, and return. It was stated that the object was to lay out an air route so that if required, air units could be moved to Asia by direct flight and, in addition, to collect data for improvement of the DH-4B and photograph unmapped portions of Alaska.

By the spring of 1920 four planes and crew and a route had been selected. The Air Service called it "The year's greatest aerial event." Captain St. Clair Streett was flight commander and pilot of plane No. 1. Sergeant Edmund Henriques was his mechanic. Plane No. 2 was piloted by 1st Lieutenant Clifford C. Nutt, 2nd in command, with navigator and engineering officer 1st. Lieutenant Erik H. Nelson as passenger. Plane No. 3 was piloted by 2nd Lieutenant Clarence E. Crumrine (also photographic officer) with mechanic Sergeant James D. Long as passenger. Plane No. 4 was piloted by 2nd Lieutenant Ross C. Kirkpatrick, who also acted as information officer, with mechanic Master Sergeant Joseph E. English as passenger.

The flight left Mitchel Field on July 25, 1920 bound for Erie, Pennsylvania. Progressing steadily north-west they crossed into Canada just north of Portal, North Dakota. Long endurance flying was not the object of this endeavor so stops were made in Erie, Grand Rapids, Minneapolis and Fargo before landing at Portal. The flight continued across Canada with several more stops before crossing into Alaska just north of Dawson. From there they went to Fairbanks, Ruby and then on to Nome to land after a 4,502-mile trip. The trip had been completed in 50 hours flying time but had occupied 40 days as they had had various delays caused by weather, repairs after bad landings and other assorted reasons, along with time spent being wined and dined along the way.

They left Nome on August 31, 1920, and returned via the same route to arrive at Mitchel Field on October 20, 1920. The next day they flew on to Washington, D.C. to accept the congratulations of Generals Menoher, Mitchell and Pershing. Although we would not think of this as a particularly great feat today, it was hailed at the time as surpassing the recent British non-stop flight of the Atlantic and ranking high against the NC-4 cross-Atlantic flight. The flight was referred to as "one of the most hazardous and stupendous aerial events yet attempted in any country."

The flyers received the Mackay Trophy for 1920 for their efforts.

COAST-TO-COAST IN UNDER 24 HOURS 1922

Although it had been attempted without success in early 1921, in 1922 1st Lieutenant James H. Doolittle, after several months of planning and preparation, on September 2 flew a DH-4B from San Diego to Jacksonville with a single 85-minute stop at Kelly Field, Texas, for gas and service, in 22 hours 35 minutes. The coast-to-coast in under 24 hours barrier had been broken.

COAST-TO-COAST NON STOP 1923

The idea of making a coast-to-coast non-stop flight was conceived by Lieutenants Oakley G. Kelly and Muir

Fokker T-2 AS 64233 McCook Field P-253, photographed at McCook Field (photo 14967) in 1922. This was the first airplane to fly coast-to-coast non-stop, which it did on May 2-3, 1923. Prior to this time it was also used to set a new world duration record of 36 hours 4 minutes and 31 seconds as well as six speed records over various long distances. (Courtesy of Peter Bowers)

S. Fairchild in the winter of 1921-22 while stationed at McCook Field. The obstacles to be overcome were first to find an airplane capable of making the flight, and second to get permission to attempt it.

The two men decided to place a well-marked map on Kelly's office wall to kind of announce the plan and wait until Col. Thurman Bane, field commander, happened to see it in passing and become interested in the project. They felt sure he would then back the project and do what he could to get the required permission from Washington.

The ploy worked and Col. Bane indicated he would back the attempt if an airplane could be found to try it with. The two men scanned the specifications of potential airplanes but found nothing currently suitable.

On December 4, 1920 the Air Service had contracted Anthony H. G. Fokker of the Netherlands to build them two single-engine F-IV monoplanes. These were large Liberty-powered airplanes designed to have an open pilot's cockpit forward of the wing and carry eight passengers in an enclosed cabin. Given the Air Service designation of T-2 (transport of the second design) they were completed and shipped to the United States aboard the transport *Cambria* from Antwerp, Belgium on March 19, 1922.

After arrival at Brooklyn they were shipped to McCook Field to be assembled and tested. Lieutenant Kelly was assigned as project officer to fly the acceptance tests. The first plane was assembled and after the first flight test on June 1, 1922 by Kelly he immediately decided the airplane had the potential to be modified for the coast-to-coast attempt. After the other plane passed load-testing and all results were forwarded to General Patrick, approval to make the flight in a T-2 was given on August 10, 1922.

From this point it only remained to suitably modify the airplane to have dual controls and carry enough fuel and oil to make the flight. In the end total fuel capacity was increased to 725 gallons and maximum takeoff weight to 11,000 pounds. A flight path and direction (that is, east-to-west or west-to-east) now had to be picked.

Although this may sound relatively simple it actually was not as these requirements would tax the aircraft and pilots to their limits. In order to carry enough fuel to make the trip the airplane would be so heavy it would be barely able to take off and would not be able to gain much altitude until it had burned off considerable weight in fuel. Thus the course and direction became critical in that mountain ranges had to be crossed possibly at a point in the flight that the airplane was not yet light enough to fly at the required altitude.

Timing of the actual departure was also critical because weather was a very serious factor. Having to fly over or around weather disturbances could mean the difference between success and failure.

During the process of making the necessary preparations for the flight Lieutenant Fairchild was injured in an accident while testing a controllable pitch propeller and had to be replaced as the second pilot. In his place Lieutenant John A. Macready was chosen.

In the end it took three attempts before success was obtained. On the first try a west-to-east course was chosen. The plane was shipped to Rockwell Field, San Diego to await good weather.

On October 5, 1922 the first attempt began. After taking off at a gross weight of 10,695 pounds it took several minutes to reach an altitude of 200 feet. Less than 50 miles out they ran into fog and had to circle for an hour waiting for it to clear so they could proceed. It did not clear and they thus had wasted too much fuel to feel confident they would reach New York.

They decided to alter the plan and stay aloft as long as possible and try to break the existing endurance record of 26 hours 19 minutes and 35 seconds set by Stinson and Bertaud on December 29, 1921. They returned to Rockwell Field and relayed the request to the ground and received permission. They succeeded in staying aloft for 35 hours 18 minutes and 30 seconds but because the required sealed barograph was not on board to verify the flight, it was not accepted as an official record. It had, however, provided valuable information about fuel consumption and instilled added confidence in the airplane and its engine.

Favorable weather was again reported on November 2 and they prepared and took-off at daylight November 3, 1922 again headed for New York, some 2,780 miles as the course was plotted.

All went well as they progressed across country. A small water jacket leak was noted about 450 miles out of San Diego but if it grew no worse it would not be a problem so they flew on through the day. By the time they passed Tucumcari, New Mexico it was 5:30 PM and dark. They were 1,020 miles from San Diego.

Just beyond Terre Haute, Indiana Kelly noted that the water supply was being rapidly depleted and on inspection found the engine was squirting small streams of water from both banks of cylinders. Just past Indianapolis, at 7:10 AM on November 4, the engine began to overheat and it was decided to turn back to and land at Indianapolis, which they did at 7:15 AM. Upon getting on the ground the throttle was pulled back to idle and the engine promptly seized. They had come 2,060 miles in the effort but failed to complete the trip.

The engine for this and the first attempt had been Nordyke & Marmon AS NO. 5142 MNFG. NO. 745.

T-2 RECORD FLIGHTS

A new engine was installed and the plane flown to McCook Field. Here it was prepared for a series of attempts at several world records. The purpose was to be dual - to gain further experience for another attempt at the coast-to-coast flight and, of course, to break the records.

The flight was made on April 16-17, 1923 with Ford engine AS NO. 30393. The world duration record was again taken, but this time it was official. Time was 36 hours 4 minutes and 31 seconds with a distance of 2,516.5 miles. In this same flight speed records were set for distances of 1,500, 2,000, 2,500, 3,000, 3,500 and 4,000 km. They ranged from 70.79 to 73.00 MPH and although not impressive speeds were nevertheless records for the distances. In addition the world's weight-lifting record of 10,800 pounds with a single Liberty engine.

Kelly and Macready received the 1922 Mackay Trophy for breaking the world's duration record.

Records sometimes did not stand long and a couple of these are extreme cases. The speed over the distances of 1,500 and 2,000 km. were taken by Lieutenant Harold R. Harris in a DH-4L with an extra gas tank mounted in the rear cockpit on April 17, 1923, almost immediately after the prior records had been set by the T-2. He had flown over the same course as Kelly and Macready using the same timers and observers since they were already in place. His speeds for the two distances were 114.35 and 114.22 MPH.

COAST-TO-COAST FINALLY

Preparations were then made for a third attempt at the coast-to-coast flight. This time they decided to make the attempt east-to-west. The plotted course in this direction was somewhat shorter at 2,445 miles and the higher western mountain ranges would be easier to cross with the lighter load of fuel that would be on board at that point in the flight.

The plane was flown to Roosevelt Field, Long Island and there awaited report of good weather. Records are not clear as to whether the same engine used on the record flights of April was still powering the plane or whether Buick engine AS 38062 MNFG. NO. B607, which is currently in the airplane, was used.

Weather reports were good on May 2, 1923 and the plane was made ready and rolled out on the field. After purposely delaying two hours to make sure they would reach Tucumcari, New Mexico at dawn to be able to check navigation prior to entering the mountains to the west, they completed the ground run-up and opened the throttle for takeoff. Only after being aided by the pushing of the ground crew were they able to start moving on the dirt field.

T-2 took off at 11:36 AM central standard time at a loaded gross weight of 10,850 pounds and barely got off the ground. For the first few miles the plane barely cleared power and phone lines.

As there yet were no aeronautical charts in existence, Rand & McNally road maps were used for navigation across the continent. Fortunately these showed not only cities and towns and roads but also railroads and rivers.

The flight was relatively uneventful but, of course, very tiring. The flyers alternated piloting and reached Tucumcari at dawn as scheduled with 1,670 miles behind them. They crossed the Continental Divide just beyond Show Low, Arizona at 10:00 AM at 10,200 feet altitude. Distance, 2,080 miles.

They flew on to San Diego and arrived at 2:26 PM, landing 26 hours 50 minutes and 38 3/5 seconds after takeoff. Total distance was 2,470 miles and average speed 92.05 MPH.

Both pilots were later awarded the Distinguished Flying Cross as well as the Mackay Trophy for 1923 for the accomplishment. This was the third time Lieutenant Macready had received the trophy, having been first awarded it in 1921 for breaking the world's altitude record in the Liberty-powered LePere.

The airplane was flown back to McCook Field and then to Washington where it was placed in the Smithsonian collections, where it still resides at the NASM.

FIRST IN-FLIGHT REFUELING 1923

It did not take long before flying officers at Rockwell Field, California and their commander, Major H. H. Arnold, came up with a plan to break the endurance record very recently set by the T-2. They designed an in-air refueling system whereby a DH-4B with an additional 50-gallon tank with a wide filler opening was installed behind the rear cockpit. A second DH-4B fitted with extra tanks could then be flown above and ahead of it and drop a hose which would be inserted in the tank. Fuel would be transferred down the hose to the tank. From there it would be transferred to the main tank.

DH-4B over Rockwell Field, California during in-flight refueling and a world flight duration record attempt. (Author's collection)

The flyers of the record attempt plane, AS 23-462, were 1st Lieutenants Lowel H. Smith and John P. Richter. The first attempt was made on June 27, 1923, but had to be aborted after a little over six hours when the plane burned out its generator. The next day a second attempt was made and this time they flew continually for almost 24 hours before fog kept them from being able to refuel and they had to land.

Not ones to give up, the two lieutenants took to the air again on August 27, 1923 and landed 37 hours and 15 minutes later on the 28th to not only set a new endurance record but also a new distance record of 5,300 kilometers and six new speed records for distances from 2,500 to 5,000 kilometers.

This time they had used two refueling planes. The only problem they had with the record plane this time was having the engine cut out several times the first day. They rapidly diagnosed to problem as loss of fuel to the engine and tried anything they could think of to resolve it. Being rather limited in that regard, they used a wrench to bang on a fuel valve which also had a strainer and found it apparently loosened the stoppage as the engine picked up again.

They continued to have the problem and relieved it in the same manner for another 24 hours or so until they finally came down. The original intention of these flights had been not only to break the endurance record but to continue to fly the plane until the engine wore out and refused to continue running. They figured this would take at least four days and four nights, or in excess of 100 hours. As it turned out, with the fuel problems being a constant threat, they satisfied themselves with breaking the old record.

They did, however, on October 25, 1923 execute the first cross-country refueling endurance flight with the same airplane by flying from the Canadian border in British Columbia to the Mexican border south of San Diego by refueling in-flight at Eugene, Oregon and Sacramento, California, covering a distance of 1,300 miles in 12 hours 13 minutes.

FIRST FLIGHT AROUND THE WORLD 1924

In the spring of 1923 the Air Service, heavily influenced by efforts in that direction by Brigadier General Billy Mitchell, became very interested in mounting a

"Flight Around the World" attempt. Much like the first flight across the Atlantic of 1919, there was a growing world-wide interest in accomplishing the feat. Almost every nation with a major interest in aeronautics was talking of having a "native son" be the first to accomplish it. In the case of the Air Service it was hoped its accomplishment would spur new national and congressional interest in aviation which might result in increased financial support of the service, something that was sorely needed.

Certainly the recent aeronautical accomplishments of the Air Service had proved such a flight was well within the realm of possibility. A group of officers was appointed to initiate the planning of a route and evaluate available aircraft. The Air Service had no airplane capable of the intended flight but the Navy had purchased several Douglas DT-2 airplanes which had good potential as they were convertible from wheels to floats, a requirement of this flight, and had good range. The DT-1 - DT-2 airplanes of the Navy had followed Douglas' first aircraft, the "Cloudster." The "Cloudster" had been built for Douglas' partner David R. Davis who intended to use it to make the first non-stop coast-to-coast flight. It was the first airplane in history able to lift its own weight in useful load.

It had left California on June 27, 1921 with Eric Springer as pilot headed for the East Coast but made a forced landing at Fort Bliss, Texas because of engine failure. Although it may well have accomplished the task with another try, Davis lost interest and sold his interest in the company and the venture died.

After recommendations were reviewed, the War Department instructed the Air Service to gather data on the Douglas Cloudster and the Fokker F-V airliner, a plane with similar characteristics, and purchase an example of each.

When Douglas was approached with the request for the Cloudster data he instead proposed a modified version of the DT-2, the prototype of which he said he could deliver within 45 days of receipt of a contract.

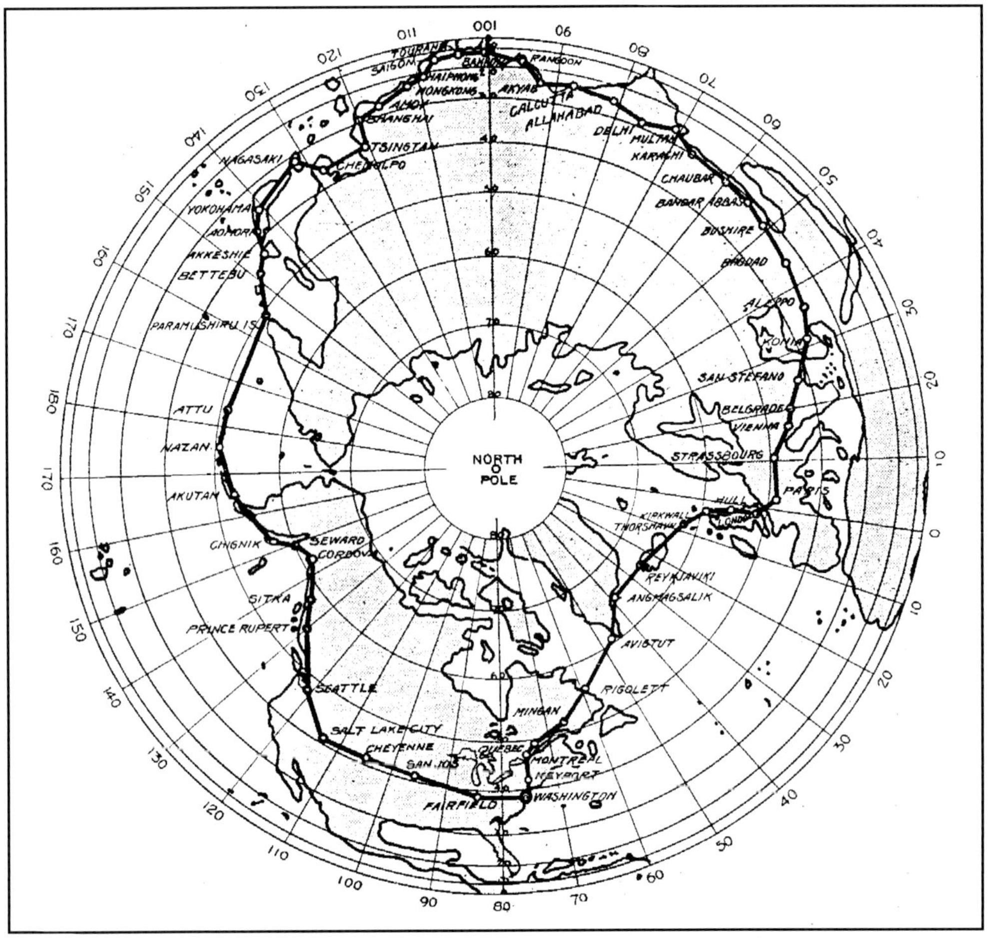

Map of the "Round-the-World" flight as originally planned. With the exception of the return trip across the United States it was executed as planned. They actually flew across the country to Santa Monica, California and then up the coast to Seattle. (Author's collection)

The proposal was accepted, a contract signed on August 1, 1923, and the airplane delivered on schedule. Thus the prototype Douglas World Cruiser (DWC) was born. After testing and acceptance, four additional standardized versions were ordered. The last of the four was delivered on March 11, 1924.

With their 582-gallon fuel capacity the airplanes had a range of about 2,200 miles. They were capable of easy conversion from wheels to floats and back, a necessary feature for their route was a combination of over land and over sea.

The route was picked and an elaborate support system set up along the way. Just as the Navy had made all possible preparations to ensure success of the first Atlantic crossing in 1919, the Army did all they could to insure the success of this venture. Fuel, fifteen spare engines, fourteen extra sets of floats and a very large assortment of air frame replacement parts were placed at critical points around the world as well as aboard the several ships that were to provide support along the way.

The flight of four aircraft was scheduled to leave Seattle on April 2, 1924. They actually began the trip from Santa Monica, California on March 17 and, after a delay caused by fog, left Seattle on April 4, 1924, headed around the world in a westerly direction.

The four planes were:

AS NO.	Side NO.	Name	Crew
23-1229	1	Seattle	Maj. Frederick L. Martin & Staff Sgt. Alva L. Harvey
23-1230	2	Chicago	Lt. Lowell Smith and Lt. Leslie Arnold
23-1231	3	Boston	Lt. Leigh Wade & Staff Sgt. Henry H. Ogden
23-1232	4	New Orleans	Lt. Erik Nelson & Lt. John Harding

The four DWCs took off from Sand Point Field at Seattle on April 4, 1924. They headed across Canada to Alaska (the complete route can be followed on the accompanying map). The *Seattle* crashed into a mountain on the Alaska Peninsula while flying in fog. The plane was destroyed but the crew walked out after a ten-day ordeal.

Flight command then moved to Lt. Lowell Smith in the *Chicago*. After some delay in the Aleutian Islands from weather, the remaining three craft continued and reached Japan on May 23. This was not intended to be a race for speed, other than the desire to complete it before another country's efforts might be successful, so there were many delays along the way for weather, to effect repairs and planned maintenance, or even to be wined-and-dined by local dignitaries.

Floats were replaced with wheels at Calcutta. Prior to that time they had flown along the Asia-Pacific coast to avail them of water landing.

From Calcutta they flew westward across India, the Middle East and Southern and Central Europe. They arrived in Paris on July 14, 1924, and departed for London on July 16.

They left London on July 17 and headed West for the Orkney Islands, Faroe Islands, Iceland, Greenland and Canada. West of the Orkney Islands the *Boston's* oil pump failed and they landed at sea. The *Chicago* managed to pass a message that they were down and their location to the destroyer *USS Billingsby*.

The USS *Billingsby* and the Cruiser USS *Richmond* arrived and the USS *Richmond* attempted to haul the *Boston* on board but the tackle broke and dropped the plane back in the sea. Damage to the plane was heavy and it was decided that towing it was the only option. They headed for the Faroe Islands but gail-force winds came up and before the islands were reached the plane capsized and sank.

Only the *Chicago* and *New Orleans* were left. They had flown on to Iceland. From there they would go to Greenland and then on to Labrador. From Labrador they went to Nova Scotia and upon landing sighted the prototype DWC, now christened *Boston II* at anchor in the harbor. General Patrick had ordered the plane flown from Langley Field so that Lt. Wade and Sgt. Ogden, by now transported by ship to that location, could finish the trip and participate in the celebrations.

***Douglas World Cruiser* New Orleans.** (Courtesy of Peter Bowers)

Back to three, the flight left Nova Scotia bound for Boston. Here they changed from pontoons to wheels for the last time. From Boston they went to New York and then to Washington, to be greeted by the President and his First Lady, and the full cabinet.

After much celebration the flight continued across the United States bound for Seattle to complete the circumnavigation of the globe. They arrived on September 28, 1924.

They had flown a total of about 24,345 miles in 371 hours flying time over a period of 175 days. Obviously a great deal of time was spent on the ground either awaiting good weather, the arrival of various repair parts, performing planned maintenance or just for rest and recreation.

ENGINE PERFORMANCE

How had the Liberty engine performed? Overall, about as expected but occasionally giving unwanted grief. A brief resumé of that performance is in order.

First, it should be understood that at that time, 1924, average TBO figures for the Liberty and most other contemporary engines (see page 258) was 100-150 hours. Planners knew about how many hours would be required to make this flight over the route planned and they placed new engines at appropriate points along the way so that the engines could be replaced (overhauled) before they had a high potential for failure.

All engines were replaced at the starting point, Seattle. Thus all started with fresh engines. The engines were all rebuilt to then-current standards with the exception that all cylinders were modified to reinforce the heads in such a manner as to help prevent cracked water jackets, probably the most prevalent problem encountered with the Liberty and other contemporary engines with this type of cylinder cooling. This was a McCook Field developed modification which was applied to all the DWC Liberties, but not all Liberties.

Fuel was supplied by an engine-driven gear pump which was backed up by a wind-driven pump as well as a hand-powered wabble pump for dire emergencies. All engines had Eclipse electric starters with back-up hand cranks.

The first mechanical problem was experienced by *Seattle* when it had to land between Seward and Chignik, Alaska. A hole had developed in the engine crankcase and all oil had leaked out. This was certainly an unusual problem to encounter but nevertheless an engine problem. Flight time on this engine was about 25 hours.

The engine was changed and *Seattle* flew on to join the rest but encountered fog and, while trying to avoid what they thought was a mountain, flew into the ground and destroyed the plane. They survived with minor injuries and walked out but that was the end of the flight for *Seattle*.

The three remaining airplanes flew on to Tokyo and stayed there for six days performing planned maintenance and repairing anything that showed a need. Engines were changed as planned. The old engines had about 90 hours operating time.

They then resumed the flight but only about 10 hours flying later *Chicago*'s radiator sprung a leak and the engine overheated and self-destructed. They landed in the Tonkin Gulf. The others left supplies and flew on to arrange for a new engine, which was delivered and installed in a period of just 71 hours. It might sound easy but be assured, it wasn't.

The flight was resumed and at Calcutta all planes changed engines and installed wheels and new wings. Flight time from the last scheduled engine change was about 81 hours.

Just before reaching Ambala, India the *New Orleans* cracked a water jacket on one cylinder. At their scheduled stop at an RAF base at Ambala the cylinder was replaced with a British part as they had a major repair base at nearby Lahore with a stock of Liberty spares.

The next day the flight took off for Multan. The route passed along the edge of the Sind Desert and along the way they flew through a sand storm which cut their visibility to a few yards. After reaching the British base at Multan they rested and cleaned the sand out of everything. They took off the next day for Karachi and all went well until the *New Orleans* suffered engine trouble. An exhaust valve broke and fell into the piston with the end result of a broken rod which went through the block. The pilot throttled back to 1,100 RPM and by some good piloting and good luck managed to continue flying 75 miles to Karachi. After landing the engine was changed. As a precaution, the engines of the other two planes were changed at the same time although they only had about 24 hours time on them. The sand storm had aparently taken its toll and they wanted to avoid further engine failures.

The flight resumed and they made their way across Europe and not far from London stopped at Brough for a scheduled engine change before crossing the Atlantic. The engines had 65 hours flight time (and would have had about 90 hours had they not had the unscheduled change at Karachi). They remained at Brough for thirteen days.

After leaving England and heading across the Atlantic, *Boston* broke an oil pump and was forced to land at sea. Although no one was hurt and the plane was not damaged, as already related, the plane was accidentally destroyed while trying to haul it aboard the Cruiser USS *Richmond*.

The remaining two planes continued to Canada and then to the United States. Just outside of Boston the *New Orleans* lost its timing gears and had to force-land. The plane was repaired the next day and continued.

The two remaining planes had been joined by *Boston II* in Canada and the three finished the trip across the United States with only one further incident, between Santa Monica and San Francisco *Boston II* had to land and replace its battery before continuing on to Seattle.

Thus we see that the total flight was scheduled to take about 375 hours. They started with new engines and were scheduled to have three engine changes. This would keep TBO to a safe, under 100 hours. Spare engines and repair parts were distributed along the route to take care of any eventualities.

In the end, there were five engine failures: 1- An unexplained hole in the crankcase; 2- A cracked cylinder; 3- A broken rod; 4- A broken oil pump; 5- Broken timing gears.

Problem number 1 was one of those "one in a million" type. Problem 2 was not uncommon but usually was not a catastrophe and was not in this case. 3 was more than likely a result of the sand storm. The carburetors of these engines had no intake air filters so the sand in the intake air went right through the engine. Problem 4 involved an accessory failure which was not at all common. All engines were equipped with stub-tooth timing gears so problem number 5 would not be expected either. Fortunately no airplane was lost directly as the result of engine failure. The *Seattle* was lost due to weather and the *Boston* was lost by a secondary accident during retrieval after a forced landing.

Much to its credit, the Liberty carried the *New Orleans* 75 miles to its destination after breaking a rod and punching a hole in its crankcase.

Among the numerous honors bestowed on the flyers was the Mackay Trophy for 1924.

During the same year in which the Unites States was mounting the effort to circumnavigate the world by air and succeeding, several other nations were also attempting the same feat. All met with failure. A British plane went down in the North Pacific, two Frenchmen crashed their plane at Shanghai, two Portuguese flyers abandoned their effort at Nacao, two Argentines crashed at Hanoi and an Italian plane crashed in the North Atlantic.

The three OL-2 Loening amphibians at the Boston Navy Yard being disassembled in preparation for loading them aboard the* Peary *for the voyage north. (Courtesy of the Naval Museum)

THE NAVY-MACMILLAN ARCTIC EXPEDITION OF 1925

The idea for this expedition was originated in the fall of 1924 by Eugene F. McDonald, Jr., an explorer and the man responsible for Zenith Radio Corporation's prominence in the radio industry, and Donald F. MacMillan, an explorer of considerable note. Both held the rank of Lt. Commander in the Navy.

The idea was to explore the last unexplored region in the Northern Hemisphere, a 3-million square mile area between the north coast of Alaska and the North Pole. In addition they would use Zenith's newly developed short wave radio transmitter-receivers to prove their worth in long- and short-distance communications between ship-and-base and ship-to-ship. They would also gather further knowledge of arctic fish and birds as well as magnetic and meteorological data for future navigation in arctic regions.

By mid-February these two men had discussed the project with Navy Secretary Curtiss Wilbur and had decided that fixed-wing rather than lighter-than-air craft would be the better choice. The most logical choice of airplanes was the Loening amphibian currently being developed by the Army. The first of this series, the COA-1 was first tested in May of 1924. Three production airplanes were ordered in 1924 (24-6, 24-7 and 24-8). In 1925 an additional nine were ordered (25-226-234). After the Navy became interested in the airplane for use in this expedition they placed an order with Loening for five of the type which they designated as the OL-2.

Because of the urgency placed on this project by the Navy and also through President Coolidge, who was a personal friend of McDonald, the Navy order for OL-2 airplanes received priority over the Army order and was delivered first. They were Navy numbers A6980-6983 and A7030. The first three airplanes were assigned to the expedition.

Lt. Commander Richard E. Bird was, during this period, trying to develop an arctic expedition of his own. It appears that upon being briefed on March 28, 1925 of the MacMillan plans he modified his so as to coincide with theirs and, as a result the two were combined and Bird was placed in command. MacMillan was second in command.

Two ships would carry the expedition to northern waters, the Arctic schooner *Bowdoin,* owned by MacMillan, and the steam-powered yacht *Peary*, owned by McDonald. The *Peary* would be modified to carry the three Loenings on board.

The three Loenings were delivered in late May and early June of 1925 and modified as required for the project and flown to the Boston Navy Yard where they were loaded aboard the *Peary*. The two ships sailed north on June 17, 1925 and arrived at Godhavn, Greenland in late July.

From there they sailed up the west coast of Greenland to Etah, a small village on Smith Sound. Here they assembled the three planes and this became their base. On August 3, 1925 Lt. Schur took NA-2 airplane up for its first test flight and became the first man to fly over the Greenland ice cap. The water from which they had to fly would only remain unfrozen for about three weeks so they had to make the best use of what flyable weather came to them, and this was probably less than half of that three weeks.

Operating in terribly cold temperatures on the ground and flying open cockpit airplanes, the crew suffered greatly. In spite of this they managed to fly more than six thousand miles and view from the air some 30,000 square miles of terrain, a large part of which had never been seen by man before.

Most of the expedition goals were accomplished and the new Loenings and their inverted Liberty engines performed well. The engines revealed as yet unresolved problems with oil leaking from the camshaft housings. This was still the early stages of development for this engine. More knowledge was gained but no major problems occurred.

THE PAN-AMERICAN GOODWILL FLIGHT OF 1926

In 1926 the Army planned and organized a five-plane flight intended as a "Goodwill" tour of South America. Picked for the flight was the inverted Liberty-powered OA-1A Loening Amphibians. The Navy OL-2 airplanes used in the Navy-MacMillan Arctic Expedition had already proved the airplane and its engine were reliable and by this time the engine had been further perfected.

The flight would originate at San Antonio and cross most of Mexico over land. It would then move to the coast of Central America and from that point travel mostly along the coast of the South American continent, taking care to visit every country on the continent.

The flight route of the Pan-American Goodwill Flight of 1926-1927. (Author's collection)

Five airplanes were sent on the trip. All were equipped with extra fuel tanks allowing 10 hours of flight. Like the Around-the-World flight, the airplanes were named for cities. The flagship was the *New York* and the remaining four named *San Antonio*, *San Francisco*, *Detroit* and *St. Louis*.

The flight was scheduled to cover 18,524 miles and end in Washington, DC. It started on December 21, 1926, and ended at Bolling Field on May 2, 1927.

The route was set and advance officers traveled it and arranged for storing fuel, spare parts and spare engines along the way. 56 flying and 77 delay days were scheduled. The delay days were to accommodate scheduled maintenance, diplomatic meetings and ceremonies. A total of 133 days. The flight actually took 59 flying days and 74 delay days and was completed on schedule, even though the division of time was slightly different than planned.

As would be expected during this early period in aviation, there were a few problems. There were one or two bad landings which were not disastrous but did require a repair or two to the aircraft.

As with the Around-the-World flight, installation of previously overhauled engines placed along the way was planned. In this case, however, the pilots performed their own maintenance. Considering the shorter distance and a cruising speed of about 100 MPH, probably a single engine change was scheduled.

Although the flight in general went well, it was marred by one serious accident. On February 26 the *Detroit* and the *New York* collided in mid-air while approaching the Argentine Air Service Field at Palomar, Buenos Aires. The crew of the *Detroit* were both killed and both planes destroyed.

The remaining airplanes continued and arrived at Bolling Field, Washington on May 2, 1927, to be greeted by President Coolidge and other dignitaries. The flyers were all awarded the Distinguished Flying Cross and the Mackay Trophy for 1926, the crew of the *Detroit* posthumously.

In December of 1927 the *San Francisco* was transferred to the Smithsonian Institution. It was restored in 1964-1965 and is currently displayed in the NASM in Washington.

This was the last year the Mackay Trophy was awarded to pilots for some type of performance in a Liberty-powered airplane. In 1919 it had been awarded to seven pilots of DH-4 airplanes and the pilots of two SE-5s and one Fokker for performance in the 1919 Transcontinental Reliability and Endurance Test in which they flew from the Atlantic to the Pacific and return. A large number of airplanes and their pilots participated in this event and it received much publicity, a good deal of which turned out to be bad and before it was all over a number of planes crashed resulting in a total of seven fatalities. Many thought the price too high.

In 1920 the trophy went to the Nome, Alaska and return flyers. In 1921 it went to John Macready for his world altitude record in the LePere P-53. In 1922 it went to Macready and Kelly for their World's Duration Record. In 1923 it went to the same two men for their Non-Stop Coast-to-Coast flight. In 1924 it went to the pilots of the Around-the-World airplanes. In 1925 it did not go to a Liberty flyer but instead to the winners of the Pulitzer and Schneider Cup races. In 1926, of course, it went to the Pan-American flyers.

Although 1926 did not end the significant usefulness of the Liberty as an aircraft engine, it did mark the last year in which an undertaking as notable as those just related was completed.

KNOWN SIGNIFICANT ACCOMPLISHMENTS AND RECORDS SET BY AIRPLANES POWERED BY LIBERTY ENGINES

	Date	Airplane	Accomplishment or Record
1918			
	11/28/18	NC-1	World's record number of passengers taken aloft, 51
1919			
	4/25/19	F5-L	World flying boat endurance record of 20 hours 10 min. closed circuit
	5/8-27/19	NC-4	First aerial crossing of the Atlantic
	9/6/19	LUSAC-11 P-53	Two-man world altitude record 28,500', Maj. Schroeder & Lt. Elfrey
	9/24/19	LUSAC-11 P-53	Two-man world altitude record 30,900', Maj. Schroeder & Lt. Elfrey
	10/4/19	LUSAC-11 P-53	Two-Man world altitude record 31,821', Maj. Schroeder & Lt. Elfrey
1920			
	2/27/20	LUSAC-11 P-53	World altitude record 33,143', Maj. Schroeder
	7/25 - 10/20/20	DH-4	Four DH-4 airplanes fly New York to Nome and back.
	12/20/20-3/10/21	NC-5 & 6 & 12 F-5Ls	Fly from San Diego to Panama and return.
1921			
	2/21-24/21	DH-4	Transcontinental speed record 22 hours 30 min. San Diego to Jacksonville, FL. with one stop for fuel.

	Date	Airplane	Accomplishment or Record
	3/15/21	LUSAC-11 P-53	Two-man world altitude record 32,000', Maj. Schroeder & Lt. Elfrey
	9/28/21	LUSAC-11 P-53	World altitude record 34,509', Lt. Macready
1922			
	-/-/22	LUSAC-11	American record number of consecutive loops, 454 executed in 1 hour, 54 min. and 10 sec. by Lt. Ralph J. Johnson at Carlstrom Field.
	3/15/22	Martin MB-1 P-162	World altitude records (supercharged)
	8/1/22	Martin MB-1 P-162	World altitude record (supercharged) 24,830'
	9/3/22-2/9/23	H-16	Plane named *Sampaio Correia II* flys from Pensacola, FL to Rio de Janeiro, Brazil
	10/13/22	LUSAC-11 P-53	Wins Liberty Engine Builders Trophy race at 128.8 MPH.
1923			
	3/29/23 & 4/17/23	DH-4L	World records - speed over 1000 KM 127.2 MPH speed over 1500 KM 114.4 MPH speed over 2000 KM 114.2 MPH
	4/16-17/23	Fokker T-2	World records - speed over 2500 KM 71.7 MPH speed over 3000 KM 71.5 MPH speed over 3500 KM 70.4 MPH Maximum distance - 2541 miles Duration 36 hours 4 min. 34 sec.
	6/6/23	F5-L	World duration for seaplanes with 500 KG payload 7 hours 35 min. 54 sec. with distance of 466 miles.
	6/7/23	F5-L	World seaplane altitude record with 1000 KG payload, 7,979' World seaplane greatest payload carried to 2000 meters, 1000 KG
	8/27-28/23	DH-4B	World record duration (with refueling in flight) 37 hours 15 min. 14.8 sec. Distance 3,293.26 miles.
	10/3/23	NBL-1	Six-engine plane sets American records with 2000 KG payload: Duration - 1 hour 47 min. 10.5 sec.
	10/25/23	NBL-1	Greatest payload carried to 2000 meters - 4,409 pounds.
	10/25/23	DH-4B	First cross-country refueling endurance flight Canada to Mexico, 1300 miles in 12 hours 13 min.
1924			
	3/17/24 to 10/10/24	DWC	Two Douglas World Cruisers are the first planes to circumnavigate the world.
	3/27/24	TP-1	World altitude record with 250 KG payload, 29,462'
	5/17/24	DWC	First airplane(s) to fly across the Pacific.
	5/21/24	TP-1	World altitude record with 500 KG payload, 28,143'
	5/22/24	TP-1	American altitude record with payload of 1000 KG, 16,732'
	6/28/24	DT-2	American records with 500 KG payload: Duration - 9 hours, 11 min. 53.4 sec. Distance: 590.3 miles.
	6/28/24	CO-5	American speed record for 100 KM with 500 KG payload, 112.346 MPH.
	6/28/24	MB-2	American record for bomber speed for 500 KM with 500 KG payload, 74.90 MPH.
	10/1/24	MB-2	World altitude record with supercharged bomber. 17,000' with 3,500 lb payload.
	10/2/24	NBS-1	American record for duration carrying 1000 KG pay load, 2 hours, 13 min. 49.6 sec.
	11/7/24	Loening Air Yacht	World record for speed for 1000 KM class C-2 seaplanes, 101.642 MPH.
1925			
	4/13/25	Stout Air Pullman	Ford Airlines starts operating route between Dearborn, Chicago and Cleveland.
1926			
	1/29/26	CO-5	American altitude record of 38,704'.
	6/6/26	3 Loening OA-1A	Start Alaska Aerial Survey Expedition.

Chapter 11

The Russian- and British-Built Liberties

In this chapter I will discuss, as the title indicates, the Russian- and British-built Liberty. The actual chronology of the subject can be a little confusing, so I will try to explain the events leading up to production of the engine in these two countries in such a way as to make it as clear as possible.

In Chapter 3 I discussed the British contracts for Liberty 12 engines to be used by their forces. In Chapter 5 I discussed the delivery to the British of their contracted engines and how they used them. In Chapter 6 I discussed use by the United States of the Liberty 12 in tanks and related vehicles.

Thus the reader has followed the Liberty's U. S. history up through the Christie tank of 1928-31 and has seen that the Liberty-powered Christie was also ordered first by Russia and then by Great Britain. I will try to explain in this chapter just why these two countries chose to power later developments of these two Christie tanks produced in their own countries with a variation of the Liberty each put into production in their own country. In doing so, I will also show that Russia's first production of the Liberty was for aero use.

EARLY BRITISH LIBERTY HISTORY 1918-1930

The British, of course, took delivery of a number of Liberty engines and began using them in aircraft during 1918. They continued their active use in aircraft until about 1930. They intended to use them in Mark VIII tanks but the war ended before they ever did so. British active military use of the U. S.-built Liberty thus ran from 1918 to 1930.

RUSSIAN USE OF THE LIBERTY

"There was submitted a request of the Russian Government Purchasing Commission in Washington that Lieutenant de Veyher, a Russian naval architect, be permitted to familiarize himself with the U. S. 12, and the board granted permission."

The above is a quote from the Aircraft Production Board minutes of October 31, 1917 and represents the earliest documented interest of the Russian government in the Liberty. On this same day Ford signed its first production contract. The first Liberty 12 was flown on October 21 in a Navy HS-1 and the Army first flew it in a DH-4 on October 29. By October 31 no production engines had been completed, only experimental or prototypes.

Although one can see that Russian interest was early, we find no references to the results of this familiarization. Early Russian aeronautical development relied almost entirely on foreign aero engines, some produced there and some imported. It is therefore quite logical that they had an interest in the Liberty and its development, as did England, France and Italy.

RUSSIAN MILITARY HISTORY 1914-1920

In order to understand just why nothing resulted from the above Russian inspection of the Liberty, we must have a brief review of Russian history of this period.

On June 28, 1914 Archduke Franz Ferdinand was assassinated in Sarajevo. Austria-Hungry declared war on Russia on July 28, 1914. On August 3, 1914 Germany declared war on France. On August 4, 1914 Britain declared war on Germany and Austria-Hungary, the Central Powers.

Russia, Great Britain, Italy and France formed the Allied Powers. Russia was ill-fitted in either leadership or military power to stand against the enemy and suffered heavy losses and military reverses at the hands of the German Army. These losses plus great economic hardships of the time contributed to a pending Russian revolution.

On March 12, 1917 a Provisional Russian Government was formed. This became known as the *March Revolution.* Tsar Nicholas abdicated on March 15, 1917 and this was followed by the collapse of the tsarist government and much of the economy of Russia. Aleksander Kerensky was appointed Prime Minister of the Provisional Government on August 6, 1917.

It is under this weak Provisional Government that the above mentioned request for information regarding the then new Liberty engine was made on October 31, 1917. Even while this official request was being honored, the Provisional Government was being overthrown by the Bolsheviks.

Vladimir Lenin was elected Chairman of the Bolshevik Soviet Council of People's Commissars which now controlled the Russian Government.

The Provisional Government was arrested by the Bolsheviks on November 7, 1917. This became known as the *October Revolution.* One of the early edicts of this new government was the nationalization of most manufacturing plants and property within the country.

Aleksander Kerensky, who managed to escape arrest, gathered loyal troops and resisted the Bolsheviks. This resulted in civil war within Russia with the new People's Government being known as the "Red Army" and the opposition as the "White Army."

The White Army was joined by many who opposed the new Lenin regime, including land owners and factory owners who had lost their property to nationalization, devout members of the Russian Orthodox Church who objected to the new government's atheism and royalists who wanted to restore the monarchy, among others.

On December 22, 1917 the People's Government, who were now in control of Russia, and Germany began peace talks and on March 3, 1918 the Treaty of Brest-Litovsk was signed. This peace treaty removed Russia from the war and greatly relieved military pressures on Germany as she no longer had to fight Russia on the eastern front.

Although the withdrawal of Russia from the Allies was a military setback, it was not Germany's salvation and some eight months later, on November 11, 1918, Germany signed the Armistice and the war ended.

The Russian civil war raged for some three years, ending in late 1920. It is estimated that about 800,000 soldiers from both sides died during the war as well as another 8 million who died from starvation and disease.

RUSSIAN AVIATION INDUSTRY

The Russian aviation industry was in its infancy at the outbreak of WW I. Although some progress had been made with designs of their own, virtually all aircraft and aero engines produced within the country were foreign designs. In addition, they imported foreign-built aircraft and engines for use by the military.

The largest of several aircraft factories in use during the period was the Duks factory in Moscow. It had built aircraft since 1910 and by 1917 had built a total of 1,733 airplanes and employed 2,500 workers. Its production ability was 2 airplanes per day.

In the autumn of 1917 the Russians received an incomplete set of plans for the de Havilland DH-4 and passed them on to the Duks factory with the intention of building the plane under license. This was a plan started by the Provisional (White) Government with an agreement with Airco of Hendon, England.

The object was to replenish the supply of older Voisen, Farman and Morane-Saulnier airplanes which made up a large part of the Russian military aircraft and which were rapidly dwindling. Motive power was to have been the Italian Fiat engine. The timing of the October 31 request to examine the Liberty might have indicated they were considering it as a possible power plant since the Americans were already doing so.

The designer who was to adapt the airplane to Russian production was Nikolai N. Polikarpov. In his archives were found notes regarding plans to build the DH-9 as well. The Duks, Lebedev and Anatra factories were to build 650 examples. These were Provisional (White) Government plans and were, of course, never to be fulfilled.

The October Revolution and takeover by the Lenin (Red) government resulted in the nationalization of all of the aircraft industry. The civil war caused the disintegration of the national economy and, for example, the number of workers at the Duks factory was reduced to 540. Among them, however, was Nikolai Polikarpov.

Although all work on the DH-4 and DH-9 project was officially halted, Polikarpov, as a personal venture, continued work converting the British DH-4 plans to metric with the hope that at some future date the project might again become officially sanctioned. The aircraft industry was now under Red control.

The Allied powers attempted to aid the White army in the revolution because doing so might bring Russia back into supporting the Allies fight against Germany. Ultimately, this attempt was not successful, but it did bring a few DH-4, DH-9 and DH-9a aircraft to Russia in the form of both airplanes provided to the White army and a few planes left by the British when they eventually abandoned the effort after the end of the war. A number of these planes then fell into the hands of the Red army, being either captured from the White army or recovered after abandonment by the White Army during various retreats. It is believed that about 20 airplanes were so acquired.

Several of these airplanes were tested at the Nauchno-Opytny Aerodrome (NOA-the Scientific Testing Field) at Khodynka in Moscow. The results were so good that they decided that a plane of similar design would be very useful as a general purpose aircraft for the new Red Air Fleet. However, in the interest of obtaining such aircraft as rapidly as possible, it was decided to copy the DH-4 rather than design a new, but similar airplane.

RUSSIAN DH-4 PRODUCTION

The work of assembling the required plans was well under way because of the continued interest and efforts of Polikarpov and the availability of the actual airplane made the completion of the work fairly rapid. The plane was then put into production in the Duks factory, which had been nationalized and renamed Gosudarstvenny Aviayzionny Zavod 1 (State Aviation Factory No. 1, or GAZ-1) on December 30, 1918.

A total of 20 airplanes of this design were built and referred to simply as "Aircraft DH-4 with 240 HP Fiat A 12 engine." Performance was disappointing because of the relatively low power and poor construction of the airframe, thus the short 1920-1921 production run.

In addition, in 1922 some 20 DH-9 airframes (minus engines) were purchased from Great Britain and most equipped with purchased 260 HP Mercedes engines. With these, they had about 60 DH-4/DH-9/DH-9a-type aircraft in service, including a small number of Liberty-powered DH-9as.

THE R-1

It was decided to produce in place of the DH-4 a copy of the DH-9 and plans were rapidly drawn up to facilitate it in the same GAZ-1 factory. It was designated as the R-1, the R stood for "Razvedchik" or reconnaissance. The end of 1920 had brought an end to the revolution, more stability and increased interest and funding for the newly nationalized aircraft industry. By 1922 it also brought a more standardized system of designating aircraft and their engines. R-1 represented the first of the reconnaissance series of airplanes.

Both the DH-4 and the following variations of the DH-9 were built without the benefit of a license (as would be the M-5 version of the Liberty engine which followed close behind). As one Russian historian put it, "It would be perhaps unfair to accuse the Soviets of illegality in view of the extraordinarily difficult circumstances prevailing at the time and the devastation and chaos in the aftermath of the civil war, apart from the international isolation imposed on the USSR."

This airplane was built in several variations with the differences mainly caused by the use of different engines for power. Production began in 1922. Early airplanes were powered by the 260 HP Mercedes D IV engine. It is estimated that 200 of the Mercedes-powered version were built in 1922-23.

At the same time the decision was made to build the R-1, a decision was also made to build a copy of the Liberty 12. The resulting engine would become known as the M-5.

RUSSIAN AERONAUTICAL ENGINE PRODUCTION

During World War I Russia had several factories which built aero engines, almost all of which were foreign engines either built under license or assembled from imported parts. From the outbreak of the civil war, due to

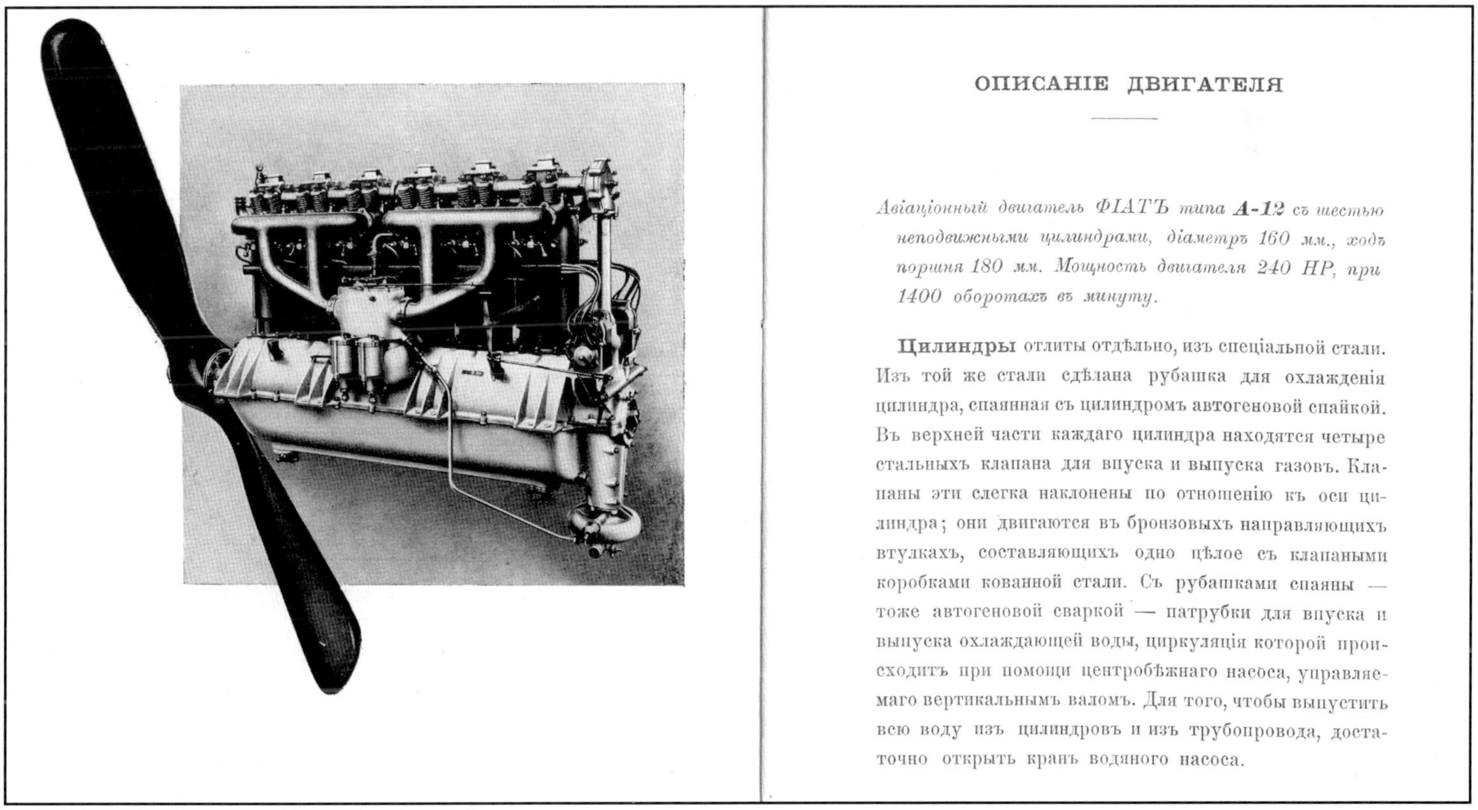

ОПИСАНІЕ ДВИГАТЕЛЯ

Авіаціонный двигатель ФІАТЪ типа ***А-12*** *съ шестью неподвижными цилиндрами, діаметръ 160 мм., ходъ поршня 180 мм. Мощность двигателя 240 HP, при 1400 оборотахъ въ минуту.*

Цилиндры отлиты отдѣльно, изъ спеціальной стали. Изъ той же стали сдѣлана рубашка для охлажденія цилиндра, спаянная съ цилиндромъ автогеновой спайкой. Въ верхней части каждаго цилиндра находятся четыре стальныхъ клапана для впуска и выпуска газовъ. Клапаны эти слегка наклонены по отношенію къ оси цилиндра; они двигаются въ бронзовыхъ направляющихъ втулкахъ, составляющихъ одно цѣлое съ клапаными коробками кованной стали. Съ рубашками спаяны — тоже автогеновой сваркой — патрубки для впуска и выпуска охлаждающей воды, циркуляція которой происходитъ при помощи центробѣжнаго насоса, управляемаго вертикальнымъ валомъ. Для того, чтобы выпустить всю воду изъ цилиндровъ и изъ трубопровода, достаточно открыть кранъ водяного насоса.

Descriptive page from the Russian manual for the Fiat A-12 engine fitted to the GAZ 1 - built DH-4. (Author's collection)

the general chaos in transportation, foreign supply and the Russian economy, all these factories ceased production. Early in 1918, production was restarted at the Gnome & Rhone (originally French-owned) and Motor factories in Moscow.

The December 30, 1918 nationalization which had renamed the Duks factory GAZ-1 effected all aeronautical factories. Gnome-Rhone became GAZ-2 and Motor became GAZ-4. The former French Renault (Rybinsk) plant became GAZ-6. After the Ukraine was liberated from the Whites, the Dyuflon & Konstantinovich factory (DEKA) in Alexandrovsk was added to the engine factories as GAZ-9.

Total engine production during the civil war was very small and it took until 1926 for significant stability and production improvement to be realized. Beyond 1926, production rose rapidly and significantly. By 1934 it exceeded 10,000 annually.

Considerable additional financial resources were poured into the aeronautical industry beginning in 1921. Even so, it would be some years before any but foreign-designed engines would be built for Russian aero use.

RUSSIAN
AERO ENGINE PRODUCTION 1918 - 1926

Fiscal year	Engine production - all types
1918-1919	79
1919-1920	77
1920-1921	81
1921-1922	12
1922-1923	50
1923-1924	70
1924-1925	157
1925-1926	342

Beginning in 1919, increased engineering and production effort was put into the utilization of licenses obtained before the revolution as well as copying additional foreign designs. In addition, native designs were encouraged.

Initially, there was no set system of engine designation and each developing factory or engineering staff named their engine whatever they liked. Generally they were given the same model name the copied engine carried. In 1922, however, a standardized system was introduced whereby each engine was assigned a number prefixed by the letter M, which stood for the Russian "motop." All engines which had been in production and adopted by the Air Force Service were assigned numbers. From this point, numbers were assigned to the various organizations which were carrying out engine development.

Numbers were assigned to each development project as it was approved. In the long term, this resulted in some numbers being assigned to a projected engine that was never to reach production. In some cases, the project itself may have been cancelled before even a prototype was completed. There were therefore numbers for which no engine was ever built and a number for which only prototypes were built.

M-1 was assigned to the in-line 150 HP 6-cylinder RBVZ-6 which was built in the Motorniy factory in Petrograd. A total of 45 were built in that factory in 1917-1918 and, reportedly, another 20 were completed later in the Fili plant.

M-2 was assigned to the 110 and 120 HP versions of the same Le Rhone rotary built at the same GAZ-2 plant starting in July-August of 1919. A state of mass-production of the model was not reached until 1925. By the time production ceased in 1927 about 2,000 had been built. The 1918-1926 production chart shows that a total of 868 aero engines of all types were built between those dates, thus 1927 production must have exceeded 1000.

M-3 (also known as the Reno-220) was assigned to the V-12 French 285 HP Renault WC which was assembled from imported parts in the Russkiy Reno factory in Petrograd. A total of about 98 were built in 1916-1917.

M-4 was assigned to the license-built 200 HP Hispano-Suiza V-8. A license was obtained to build this engine in 1916 but nothing resulted until an order was given to build 100 at the GAZ-4 plant in December of 1918. The year 1919 was spent setting up tooling and mastering the required technology. The first engine was built in July of 1920 and the model passed its type tests on March 24, 1921. Production was halted in 1922 after 36 had been built. The engine was now considered obsolete.

M-5 Liberty

This brings us the the Russian-built Liberty. It was recognized that the R-1 needed a 350 to 400 HP engine in order to give acceptable performance. Preparations were begun in May of 1922 to put the Liberty into serial production in the GAZ-2 plant in Moscow. They had no engineering drawings to work from so they took an American engine, disassembled it and generated drawings in metric from the actual engine. Just why it was felt important to record what company built the copied engine we do not know. But they did and it was a Ford-built Liberty.

Initial production setup began in September of 1922. One can imagine the difficulties encountered. Except for the short production run of the M-4 Hispano-Suiza copy built in GAZ-2, the Russians had very little experience with aluminum castings of the size and complexity of those required for the Liberty.

Before I progress further in the history of the M-5, I should note that a few of these early "M" numbers were eventually assigned to other engines years later, and

The M-5 Liberty. This engine is in the Russian Air Force Museum (Monino). (Courtesy of Harry Woodman)

SPECIFICATIONS OF THE M-5

V-12 in-line 45° water-cooled

Power	400/440 HP
Bore/stroke	127 mm x 178 mm
Displacement	27.01 liters
Compression ratio	
early	5.3:1
late	5.4:1
Direct drive	
Weight	
early	420 kg (926 lb)
late	410 kg (904 lb)

this was one of them. In this case the designation was assigned as M5, without the usual hyphen. This is a 5-cylinder 150 HP radial brought under development in the 1990s by the Voronezh Engine-Building Design Bureau in Voronezh. The engine is intended for general aviation and light aircraft.

Another Liberty M-5 problem which took considerable time and effort to resolve was the production of crankshafts. Best results are obtained by forging these, but no drop forges of sufficient size were in existence in Russia at the time and initial runs of crankshafts were manufactured by turning them from solid billets of steel. The resulting parts were far more expensive to manufacture as well as being inferior in strength.

As one would expect, early M-5 engines were more costly to produce than their American counterparts and had a much shorter life.

The prototype passed its 50-hour type test in December of 1923 and the model was placed in serial production in 1924. As would be expected, production rose slowly and by the end of 1925 it is estimated that about 200 had been built.

Development and production was headed by A. A. Bessonov and M. P. Makarouk. The engine was as close to an exact copy of the original as metric conversion would allow.

At the same time the M-5 started serial production, so did the M-5-powered R-1. Production rate of the airplane followed closely that of the M-5 engine and it is estimated that the same 200 were built between 1924 and 1926 as were engines.

Although we do not currently know just when this occurred (but probably 1924), production of the M-5 was also set up at the old Bolshevik factory in Leningrad. Available records indicate that a total of 480 engines were built at the Leningrad plant from its start of production until it was halted in 1929. 2,349 engines were built at the GAZ 2 plant from 1923 until production was stopped in 1930. Production was resumed at that plant for a short period in 1932 and an additional 80 engines were produced. **Total production of M-5 engines was thus 2,909.**

A few changes are known to have been made during production. First, the ability to forge crankshafts was gained. In the meantime, however, a considerable number of crankshafts were imported from the Atlas and Packard companies in the U. S. The GAZ 2 plant produced a complete engine series using American crankshafts. Known as the M-5 Zh series, these engines were thus mixed Imperial/metric. Fuel pumps, some fuel system components (probably including carburetors) and complete ignition systems were imported from abroad. Next, a change to tulip-shaped valves was made. The design of the gun synchronizer was changed and the driving gears of the valve train were strengthened. The latter change was applied to the American Liberty during the 1920s and it is reasonable the Russians would adapt the change as well. It has also been noted that the number of light alloy parts used was increased. Just what they were is unknown but they resulted in a weight reduction of some 22 pounds. In addition, later production engines had their compression ratio raised from 5.3:1 to 5.4:1.

RUSSIA PURCHASES SURPLUS AMERICAN-BUILT LIBERTIES

In 1924 the Soviet government contracted to purchase various models of aero engines from Great Britain, Germany, France and the U. S. In the summer of 1924 the People's Commissariat for Foreign Trade was ordered to import 250 Liberty engines. About 100 had been delivered by mid-September. An additional 400 were ordered from the U. S. in January of 1925. During 1925 deliveries of U.S. engines averaged about 20 per month.

According to a report by Onufriev, Director of GAZ-1, it was noted on March 11, 1926 that it was planned to purchase 500 engines from the United States. Gar Wood, the industrialist and builder of high-class boats and marine-converted Liberty engines, once said that he had sold some 1,500 Liberty engines, which he had bought as surplus, to the Soviet Union to help them build up their air force.

Beginning in 1924 surplus Liberties were also purchased in Great Britain. Although no specific information has surfaced in this regard, it is highly likely that Liberty engines were imported from France as well as they had a known surplus stock of 1,000 in 1922 and were actively selling various models of aero engines to Russia in the 1920s.

A British DH-9a shown in service in the near east. Note the auxiliary radiator under the fuselage for hot climate service. It was such an aircraft as this that the Russians copied it as the R-1. (Courtesy of Peter Bowers)

A production R-1 with M-5 power. A second plane of the same model can be seen behind the tail. The "ELECTROSILIA" on the fuselage translates as Electric Power. (Courtesy of Harry Woodman)

FURTHER DEVELOPMENT OF THE M-5

The M-5 had barely reached a state of production when in late 1923 the fledgling Russian Air Force set in place a plan for future engine development. This included both the design and production of new indigenous engines and the updating and improvement of foreign engines in current manufacture.

This included the M-5, even though at this point it had yet to achieve series production. Work was begun on designing modernized versions of the Liberty. Their main directions were: 1- modernize the design to better suit their production capabilities; 2- to increase power ratings.

Support for such efforts was not wholehearted. Some thought it would be better to instead build more modern and powerful foreign designs such as the Napier Lion, Wright T4 or Lorraine-Dietrich 12E. However, by late 1924 (things moved slowly in the bureaucracy) the Liberty was in full production and it had become the primary service engine. The end result was an official decision to develop an engine family based on the Liberty.

Several projects were initiated with the aims being both increasing power and improving the basic engine while keeping the same power ratings.

Efforts to increase power to the 500-600 HP range were directed in two basic directions: 1 - Increase the number of standard cylinders; 2 - Improve performance by increasing compression and RPM.

The M-9 Liberty X-16

Work on number one was started first. On November 10, 1923 the proposal of an X-16 engine was approved. On April 1, 1924 the prelimary project was approved and design work was completed by September of 1925 under the direction of G. M. Gorokhov. The engine was designated as the M-9. Although work continued for some time, it was ultimately decided the engine would be too difficult to build. The project was terminated and a prototype was never completed. The design called for 127 mm x 178 mm bore and stroke, 16 cylinders yielding 2,197 cu. in. and between 500 and 530 HP output with standard 5.4 to 1 compression.

The M-10

The M-10 was a proposal to improve the M-5 by giving it magneto ignition and changing the included angle to 60°. This project was approved in late 1924 to be developed in the Moscow factory. The prototype was supposed to be ready by mid-1925. When January of 1926 arrived the engine was only half complete. By this time production of the standard battery-type ignition components had been successfully initiated in the USSR. This relieved the pressure of building the M-10. The prototype was completed in August of 1926 and passed its required tests during the period September-December 1926. Because the ignition production problem of the M-5 had been resolved, no serious consideration of producing the M-10 was undertaken and the project died.

The M-14

Another approach to power increase was taken by the engineers at GAZ 4 headed by A. D. Shvetsov and V. A. Dobrynin. Their design became the M-14. It retained the basic design of the M-5, changed the included angle to 60°, increased the bore to 130 mm (5.12") and decreased the stroke to 170 mm (6.69"). The resulting displacement was 27.1 liters (1,654 cu. in.). Compression ratio was raised to 6.0 to 1, magneto ignition was used and RPM increased. Output was 450-480 HP. Weight was 415 kg (915 lbs.)

The design work was carried out between May and September of 1927 and the first prototype completed in October of 1927. Factory testing was carried out in March of 1928 and between September and November extensive testing was carried out at NII VVS (Scientific Testing Institute of the Air Force) which resulted in numerious failures and the return of the engine to the factory for improvement.

A second prototype was submitted for tests and passed. Factory No. 24 was then instructed to build a batch of 24 engines. At this point it was intended that the M-14 become the basis of a new family of engines. In December of 1928 the decision was altered and all work on the engine was cancelled.

The Inverted M-14P

Factory No. 24 had been carrying out a related design project in which the M-14 would be inverted and designated the M-14P. The design was supposed to have been completed in 1926 but due to lack of funding it was not finished until March of 1928. A prototype was to be completed

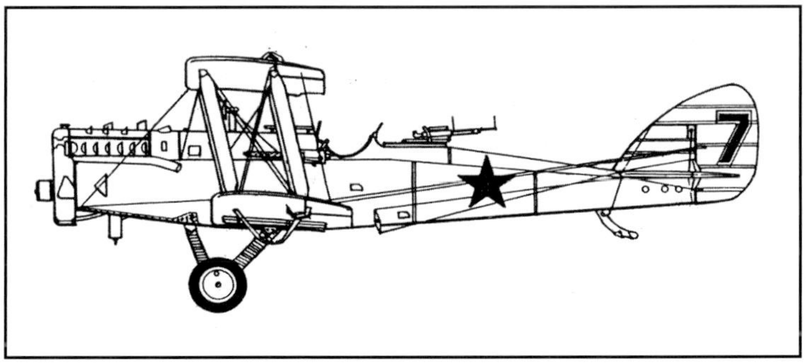

Standard R-1 with full armament and under wing bomb carriers and extra hot weather radiator under the nose. The markings shown are typical of the 1920s and 1930s. (Courtesy of Harry Woodman)

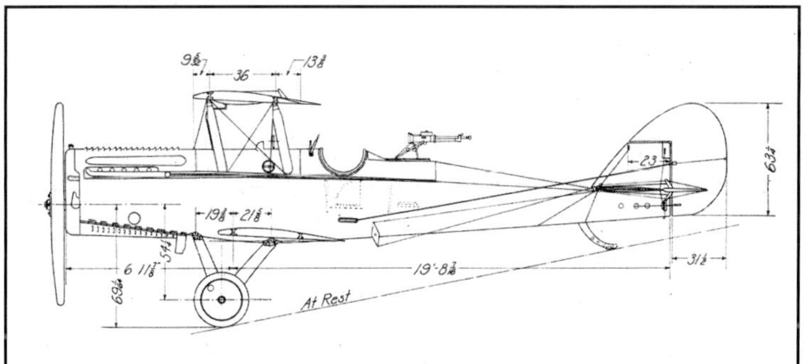

American DH-9a drawing from the BAP Bulletin of the Experimental Department, Airplane Engineering Division, November 1918. (Author's collection)

by November 1, 1928 but the same series of decisions which resulted in halting the M-14 project finished this one as well. No prototype was ever built.

THE 6-CYLINDER M-16

As part of the "new family" of engines to be based on the M-14 an in-line six-cylinder design was also started at Factory No. 24. This engine consisted of a single bank of cylinders from the M-14 and was intended for training aircraft. The design work was completed in early 1928. Specifications were for 824 cu. in. displacement, 250 kg (551 lbs.) weight and 200-250 HP output. (It is not believed that a prototype was built and that the project was halted with the M-14P project.)

THE NAMI W-18 PROJECT

Engineers at the Scientific Automotor Institute undertook some parallel development of the M-5 in an effort to increase its power. The first attempt was taken up by engineers Aleksandrov and N. R. Brilling. Their aim was to enable a reliable increase in RPM and thus raise rated power to 450 HP. Their approach was to use roller bearings for the mains, increase the width of the crank journals and replace the double carburetors with three.

The project was dropped in 1925 while still in the design stage but some of the new features were utilized in a later and more ambitious project at the same location to design a "W" configuration of the M-5.

In May of 1925 N. R. Brilling and V. M. Yakovlev received approval to design a "W" configuration version of the M-6 whose power ratings would be in the 600-700 HP range. It would use the standard (127 mm x 178 mm) 5" x 7" bore and stroke of that engine resulting in a displacement of 2,472 cu. in.

Brilling brought with him several of the features designed for the now abandoned high-speed M-5 just described and applied them to the W-18 design. The project was approved in October of 1925 and design development continued from December 1925 to May 1926.

The resulting design used roller main crankshaft, articulated rods of circular cross section, magneto ignition, shortened pistons and, of course, a new crankcase. By the time the design was completed little of the original design of the M-5 remained and thus one of the objects, to utilize as many existing M-5 parts as possible, had been almost totally lost. Lack of funding caused the project to be dropped after 1926 but the NAMI engineers kept it alive on their own and attempted in September of 1926, without success, to get GAZ-4 to agree to build a prototype.

Given the designation of V-18-I and W-18 at various periods in its life, the Russian W-18 Liberty died on the drawing boards in late 1926.

MAGNETO M-5

The idea of designing a magneto for the M-5 was revived briefly in late 1927 at Factory No. 24. The project was active from September of 1927 until February of 1928. An engine was so equipped and put under test between December of 1928 and March of 1929.

Russian historians refer to the engine as a prototype and note that it was "under construction" from March to November of 1928. Whether this modification made significant changes to the engine itself is not clear. The Engineering Division tested several types of magnetos on the Liberty but in this case it involved only adapting a magneto and its drive to existing engines and thus was only an accessory change. Russian references seem to indicate extensive changes may have been made in their case.

THE TURBO M-5

In 1935 a GE turbo-supercharger was installed on a M-5-powered R-1 and tested from July to October of that year. Published images seem to indicate it was an F-1 type. This was the first use of a turbo-supercharger in the Soviet Union. Russian historians note, however, that the interest was in the turbocharger and the M-5 was used as a test bed only because of convenience and not from any desire to put the M-5 into service with a turbocharger. It seems an interesting historical parallel that the first use of an aircraft turbocharger in the U.S. was also on a Liberty engine some 17 years earlier.

CONTINUED AIRCRAFT USE AND THE MR-1

R-1 production in the GAZ-1 plant got into full swing in 1926 and continued until early 1930. Between 1924 and 1930 they built 1,695 of the aircraft. GAZ-31 (Taganrog) took up production as well in 1929 and continued it until 1932. They built a total of 946 to make a grand total of 2,651 aircraft of the type with M-5 power.

The R-1 was so well liked that there was soon proposed a variant of it with floats. Wooden floats were designed by Polikarpov and installed on an R-1 and trials took place on the Moscow River in the autumn of 1925. After testing had proved that the float struts required strengthening, and this was accomplished, the seaplane was put into serial production at the GAZ-31 plant in 1927. It was given the designation of MR-1 (Maritime Reconnaissance Seaplane type 1).

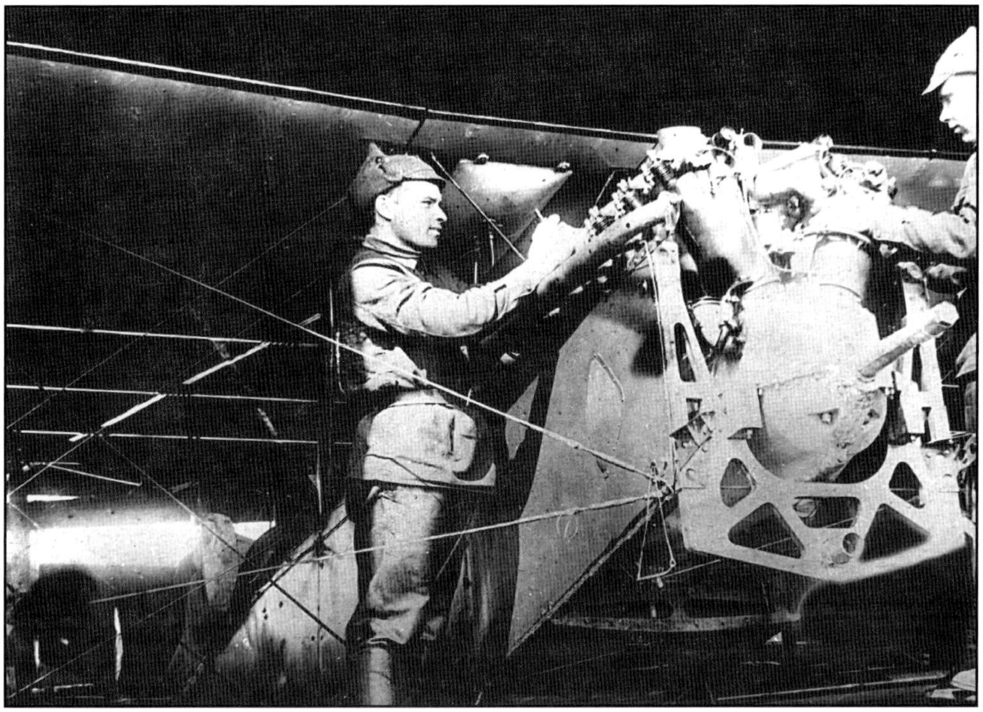

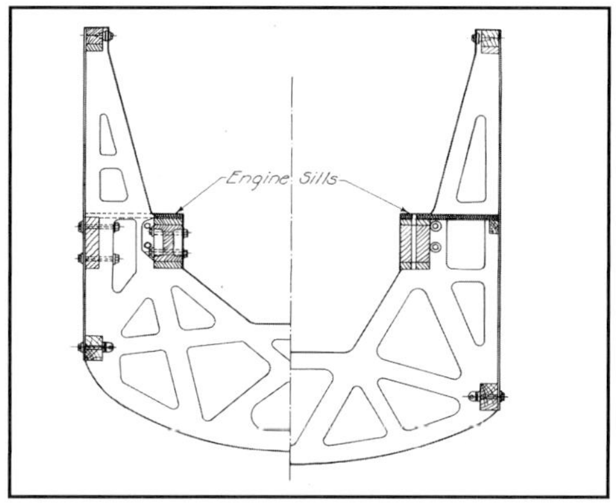

Left: Russian R-1 and M-5 engine receiving service in a hanger facility. Above: half sections of the U. S. DH-9a fuselage at the first two engine bearers. Note that the Russian R-1 bearer design is very similar, but not an exact duplicate of the American design. (Photo courtesy of Harry Woodman) (Section drawings from the author's collection)

Production continued through 1929 with a total of 124 being built. It was in that year that GAZ-31 took up full-scale production of the R-1 and GAZ-1 began to phase out its building.

THE PM-2

There was also a variation of the MR-1 built in the fall of 1927 which was fitted with aluminum skinned floats rather than wooden ones. The bottoms were skinned with smooth sheet and all other surfaces with corrugated sheet. Very few of this variation were built because of the difficulty in obtaining the aluminum sheet. In this configuration, the plane was known as the PM-2. (Polikarpov and Muintzel) The designer of the floats was a man named Muintzel, who is thought to have been a German who was working at the Russian Junkers plant at Fili. This would seem logical as Junkers was building corrugated aluminum-skinned aircraft during that period.

MILITARY HISTORY OF THE R-1 AND MR-1

These aircraft provided a substantial part of the Soviet air fleet into the early 1930s as is evidenced by a January 1, 1931 Russian Navy report on the air strength of that service. On that date the Navy had 2,929 aircraft (total serviceable and unserviceable). 1,687 of these were R-1-based aircraft and of that number 1,588 were M-5-powered. This represented in excess of 54% of their fleet.

We have no similar figures for the Russian Army, but total M-5-powered R-1/MR-1 production by January 1, 1931 was about 2,460. This would leave 872 for Army service and 305 yet to be built before production ceased in 1932.

Russian MR-1 (PM-2) equipped with Muintzel metal floats. (Courtesy of Harry Woodman)

This airplane saw extensive service throughout the country and a number were exported to friendly countries. In May of 1924 several were flown to Tehran, Iran (just south of the Caspian Sea) and were purchased by the Iranian government. Later in 1924, six R-1s were delivered to Afghanistan as a result of the Soviet-Afghanistan Treaty of February 28, 1921 and another group was delivered in 1925.

The last leg of the delivery flight of this last group involved an 807-mile flight from Tashkent (in southern Russia) to Kabul which carried them over the 14,494 ft. Hindu Kush mountain range. The Soviets had begun to help Afghanistan establish an air force in 1924 and had sent a mission with pilots and instructors to help set up schools and flying fields.

THE R-1/MR-1 IN CHINA

The Chinese-Soviet Treaty of May 31, 1924 brought military cooperation between China and Russia. China was a prime target for conversion to communism in the early 1920s and military aid was a good way to help influence them. Beginning in 1925 and continuing until 1927, the Russians sent aviation missions to China with staffs of pilots, instructors and ground crews. They were equipped with R-1 airplanes.

The purpose of the aviation missions was to aid the Chinese Kuomintang government in suppressing the various rebellious northern War Lords. In June of 1926 the Kuomintang launched the "Northern Expeditions" against these War Lords with the aid of their Russian allies' numerous R-1 airplanes. Both reconnaissance and bombing missions were carried out and during the 35-day siege of Wuchang, the capital of Hopei province, the R-1s dropped 7,054 lbs. (3,200 kg) of bombs.

Soviet-Chinese relations deteriorated in 1927 over Soviet interests in the Chinese Eastern Railway in Manchuria. The Chinese Nationalists attempted to regain the railway by force in 1929. This resulted in clashes between the two military forces. During this campaign the Soviets used 60 aircraft, including a number of river-based MR-1 floatplanes. In the end, the Chinese were overwhelmed and abandoned the railway to the Russians.

R-1s USED TO HELP SUBDUE NATIVE POPULATION

As the various Central Asian areas were taken over by the Red regime, severe resistance was met from the indigenous population. A long series of uprisings occurred in Bukhara, Bashkiria, Turkestan and Ferghana. The principal resistance was from a guerrilla organization known as the Basmachi, with the struggle lasting until the early 1930s. Records indicate that at least 30 R-1s, in addition to several other types, were used during these "pacification" programs.

The R-1 and MR-1 continued in Russian military service until the late 1930s and gained the reputation of a rugged and reliable aircraft and engine combination. In many ways it served the same purposes as did the American DH-4. It was used for reconnaissance, bombing and training and, much as did the American DH-4, served as the backbone of the Russian air fleet while it was in service.

By the time the Russian M-5 engine was put in production, the original Liberty had already set numerous distance and duration records. The Russians immediately put the engine to work. These early accomplishments were recorded and preserved in the March/April 1998 issue of *Air Enthusiast* magazine from which I extracted them with appreciation.

NOTABLE EARLY LONG-RANGE R-1 FLIGHTS

February 12 - March 19, 1925
V. B. Inshakov and Tausen; Flight from Moscow-Lipetzk-Kharkov-Gomel-Smolensk-Moscow. This was a test of the first M-5 engine built by the GAZ-2 factory. Total distance flown was 775 miles (1,250 km).

March 21 - April 1, 1925
F. S. Mastagaev and N. N. Kurbantov; Moscow-Smolensk-Leningrad-Moscow. This was a test of the first M-5 engine built by the GAZ-4 factory. Total distance flown was 1,054 miles (1,700 km).

June 10 - July 13, 1925
This was a group flight of six airplanes and the object was to flight-test Soviet planes and engines in a first flight across the Gobi Desert. Two R-1s were included in the flight. Pilots were M. M. Gromov and M. M. Volkovoinov and technicians were E. V. Radzevich and V. P. Kutzentsov. Moscow-Ulan to Bator-Peking. The 52-hour flight covered 4,015 miles (6,476 km).

August 30 - September 2, 1925
The above flight was continued by the R-1s; Peking-Seoul-Tokyo. This trip was 1,736 miles (2,800 km) and was the first crossing of the Yellow Sea by a landplane. Total distance of the combined flights was 5,751 miles (9,276 km). (Note of interest: The Douglas World Cruisers had visited Japan between May 22 and June 2, 1924, and China between June 5 and June 16, 1924. Thus the Liberty engine was not new to these countries.)

M-5 Liberty-Powered R-1 and MR-1 Production 1924 - 1932

Plant Aircraft	1924-25	1926	1927	1928	1929	1930	1931	1932	Totals
GAZ-1 R-1	200	382	387	445	225	56			1,695
GAZ-31 R-1					312	329	303	2	946
GAZ-31 MR-1		49	43	32					124
								Grand total	2,765

June 4 - June 12, 1926
M. Snegiriov; Moscow-Karkhov-Sevastopol-Rostov-on-Don-Borisoglebsk-Lipetzk-Gomel-Smolensk-Kiev-Vitebsk-Leningrad-Moscow. This was a 56-hour flight and the total distance was 4,030 miles (6,500 km).

July 14 - July 16, 1926
Y. A. Moisev and P. V. Morosov; Moscow-Khrakov-Rostov-on-Don-Mineraly Vody-Baku-Tehran. The flight duration was 17 hours, 27 minutes. The R-1 returned on July 24, 1926. The total distance flown was 1,922 miles (3,100 km).

July 16 - July 20, 1926
P. Kh Mezheraup and M. P. Golovanov plus journalist M. E. Kolstsov; Moscow-Kharkov-Sevastopol-Ankara. This was a publicity flight in an R-1 named *Krasnaya Zvesda* (Red star). Upon arrival at Ankara, Turkey the airplane was presented to the Turkish government. The flight took 11 hours and 16 minutes and covered 1,202.8 miles (1,940 km).

THE R-III

In 1925 a decision was made to improve upon the crew vision of the R-1 airplane by reconfiguring the fuselage. In this new design the top of the fuselage was at the level of the top wing and the bottom spaced above the lower wing. This allowed the crew to have their eye level above the upper wing. The designer was M. M. Shishmarev. One might think that any gains in upward vision might be countered by possible losses in downward vision.

One sample airplane was built and tested. Its performance was reported as good but the ultimate decision was to not interrupt R-1 production with a new design and only one R-III was built.

DEVELOPMENT OF THE ANT-3/R-3

The designer of this airplane was A. N. Tupolev, the aeronautical engineer who had designed the ANT-2, the first Russian all-metal fuselage aircraft. The ANT-2 was a deep fuselage airplane of much smaller dimensions which used a 100 HP Bristol-Lucifer engine. The connections between it and the ANT-3 were its designer (ANT for A. N. Tupolev), its metal fuselage, and its deep fuselage design. The gunner stood upright while manning his position.

The design was ordered in January of 1924 and by April the process had been set in motion. The Liberty-powered prototype was completed in July of 1925 and it was subjected to trials until April of 1926 with M. M. Gromov as pilot and V. S. Vakhmistrov as observer.

Testing went well and it was decided to build several airplanes for testing and propaganda purposes. Following the recommendation of M. M. Gromov, these were powered by the larger 450 HP Napier Lion. Performance was better than with the 400 HP Liberty (the prototype was powered by an American-built Liberty) but when the ultimate decision was made as to the engine of choice, the Russian-built M-5 Liberty was chosen. It had been decided that the Napier Lion would be too costly to put into production and the M-5 had, by that time, already entered series production.

By this time the designation of the airplane had been set as the R-3 (third reconnaissance design). It was put into production in the GAZ-5 plant with work starting in November of 1925. The production run was only about 30 aircraft and was completed in late 1928.

One of these planes was sent on an extensive propaganda trip which lasted from August 20, 1927 to September 1, 1927. The pilot was S. A. Shestakov and the mechanic D. V. Fufaev. The flight left Moscow and arrived at Tokyo via Sarapul - Omsk - Novosibirsk - Krasnoyarsk - Irkutsk - Chita - Blagoveschensk - Nanjang (Korea) - and

Drawing of the R-III aircraft design. (Courtesy of Alexandre Savine)

ANT-3/R-3 with Gromov and Radavich in Koinigsberg. (Courtesy of Peter Bowers)

Okayama. The return took 11 days and the total trip covered 20,000 km in 153 flights.

The R-3 was replaced by the R-3LD. This was the same airplane with a 450 HP Lorraine Dietrich engine. 79 of that model were built using a batch of 100 purchased engines. (Probably 12 Eb models.)

The R-4

The R-1 was still in full production when Polikarpov decided it might be beneficial to modernize the design somewhat in 1927. He modified the R-1 airframe by lowering the engine slightly and moving it forward 5 1/2". In conjunction with this, the radiator was reduced in frontal area and a small retractable radiator was added under the fuselage. This allowed fitting a smoother cowl and spinner. An all-metal undercarriage replaced the standard wooden one. Small changes were also made in the tail surfaces.

The prototype was officially designated the R-4 and given trials from June 13 to July 3, 1928. Although its performance was somewhat better than the R-1, it was not considered enough improvement to justify putting the design into serial production. By this time Polikarpov's R-5 design was also under development with the prototype about ready to fly. It was seen as the better choice for the future and was eventually put into production and took over the duties of the R-1. Only a single R-4 was built.

DEVELOPMENT OF THE LIBERTY M-5-POWERED FIGHTERS

By mid-1922 the earlier economic woes of the 1919 - 1921 period were starting to ease. As Russian historian Mikhail A. Maslove noted in relation to GAZ-1, the primary aircraft factory, "Between 1919 and 1921, it lacked orders so it was short of necessary resources. This naturally led to a collapse of the work force and the more skilled staff was leaving. The technical department of this factory was lacking the necessary material support; for example, the paper was of newspaper quality and the pencils were of carpentry type. Amongst the administrative staff there was a notion that the technical documentation and aircraft drawings were the best material for packing their food rations."

I can attest to the paper quality of some of their aircraft publications by noting the quality of the several Russian aero engine manuals of the period in my collection. I have three dated 1917 which covered the various foreign engines in use in Russia at the time. All three are printed on newsprint-quality paper.

As noted prior, some 20 DH-4 copies were built in GAZ-1 in 1920-1921 and about 200 R-1/DH-9 with Mercedes power were built in 1922-1923. R-1 production continued and beginning in 1924 they were M-5-powered.

During the above mentioned period GAZ-1 underwent numerous changes. Toward the end of 1922 a directive

An Il-400b at Khodynka Airfield in Moscow in the summer of 1925.
(Courtesy of Harry Woodman)

from the Main Directorate for Arms Industries (GUPV) of the Supreme Council of National Economy (VSNKh) had changed the GAZ-1 technical department into a design department. Nikolai Polikarpov, formally lead technician in that department and the man who prepared the necessary drawings to build the DH-4 copies and the DH-9/R-1 copies, now headed this design department.

By February of 1923 Polikarpov had been transferred to the design department of the Main Directorate of United Aircraft Works. Here he continued his work as an aircraft designer. His post as technical director of GAZ-1 was assumed by Dmitri Grigorovich.

The first order of business for the new Workers' and Peasants' Red Air Fleet was development of a new fighter-type aircraft which would be equal to those of the Western Powers. Polikarpov had been working on a biplane for this purpose since mid-1922. Known as the GUPV-23, the design had met with approval and was even recommended for construction but, for various reasons, was never advanced beyond that stage. Probably the next design by Polikarpov was seen as more promising before any physical progress had been made regarding production.

THE IL-400
(IL-400a)
(IL-1)

Work began on this new design in January of 1923 and it received official sanction and funding by GUPV on May 23, 1923. Polikarpov decided on a monoplane design of rather conventional construction. That is, it was wood-framed with fabric covered wings and tail surfaces and plywood-skinned fuselage. The feature of being of monoplane rather than biplane design was rather new for the time, however.

The Russian prefix for fighter-type aircraft was "I" (Istrebitel) and was used in the early part of their history in a number of variations which I will try and explain as they are used.

An early decision to be made in this design was with what engine it was to be powered. Importing foreign-made engines was very expensive so a domestic product was sought. At this time only two engines with the potential power required were in the offering, the M-5 and the M-6. At this time neither of these engines were in production but both were scheduled. The M-5 was, of course, a copy of the 400 HP Liberty and by the summer of 1923 production preparations were well advanced at GAZ-2. The M-6 was a copy of the French 300 HP Hispano-Suiza to be built at the GAZ-9 plant (Alexandrovsk) and by this time preparations for its production were also well advanced, but not as far as the M-5 it would appear. The M-6 did not pass type tests until May of 1925 whereas the M-5 did so in December of 1923. Ultimately the production run of the M-6 lasted only from 1925 to 1926 with a total of only 329 engines built.

The choice of engine was the M-5 Liberty and the first prototype aircraft was designated as the IL-400, which stood for Fighter-Liberty-400 HP. It was completed in the summer of 1923 and was no doubt powered by an American-built Liberty since the M-5 was not yet available.

Its first flight test was attempted on August 2, 1923 and the airplane was found to be exceedingly tail heavy and even after a number of high-speed taxi runs they were unable to raise the tail. Several changes were made in an attempt to correct the situation. The undercarriage was moved toward the rear but this afforded little improvement. After that, the tailplane incidence angle was set to maximum negative.

Following this change, a flight attempt was made by test pilot Kostantin Artseulov on August 15. He was able to get the airplane off the ground but was unable to control its ascent. It climbed at a very steep angle and to stop this, the pilot cut the engine. The resulting crash totaled the airplane and nearly did so to the pilot as well. The flight had lasted 21 seconds.

It was discovered that the cause of the problem was the extreme rearward center of gravity (-52% MAC) of the plane. Work was immediately begun to design a replacement without this problem.

This new design became known as the IL-400b to distinguish it from the original design. Thus, at this point, the original became, retroactively, the IL-400a. During this same period some officials at the producing factory, GAZ-1, had made a decision to call these the IL-1 (IL-400a) and IL-2 (IL-400b), meaning Fighter-Liberty-type 1 and type 2. Later, after the IL-400b became a second prototype rather than a production design, the production aircraft was called the IL-3. That decision was later overruled by higher authorities and the resulting production airplane was designated simply as the I-1, or Fighter-type 1.

THE IL-400b
(IL-2)

Initiation of an improved design was approved on October 4, 1923 and by February 19, 1924 construction was begun on the new airplane. It was completed on June 20. Although a number of men were involved in the design of the various parts of the airplane, overall design was credited to Nikolai Polikarpov.

This new design resulted in several changes. The cockpit was moved forward about a foot, the center of gravity moved forward, the header-type radiator was replaced by a Lamblin-type mounted beneath the fuselage, allowing the streamlining of the nose section. In addition, the area of the wing and tail sections was increased and their coverings changed from fabric to corrugated sheet aluminum.

Beginning in July, 1924, the plane was subjected to testing by several pilots, including Artseulov, the pilot who had crashed and suffered so severely in the IL-400a. They tested it until February of 1925 and reported that this design flew and handled well. After several recommended modifications were made it was turned over to the NOA for testing and evaluation. Reported performance was: maximum speed- 170.3 MPH; landing speed- 62.1 MPH; climb to 16,404 ft.- 14 - 16 minutes; ceiling- 24,278 ft.

They also reported positively and recommended production of 5 to 10 airplanes for unit evaluation. It was demonstrated before P. I. Baranov, Commander in Chief of the USSR Air Fleet and he granted permission and allocated finances to produce the unit level batch of airplanes.

THE I-1
(IL-3)

As explained above, the first production batch of airplanes were known as the IL-3 model and were delivered in February of 1926. By June of 1926 the designation had been changed to I-1.

The first I-1 was c/n 2885. A number of design changes were incorporated with the resultant airplane now having wings of wooden construction with fabric skin and a honeycomb-type radiator which was retractable into the fuselage. Flight test reports were favorable and compared with some characteristics of other Russian aircraft currently under tests or in use, such as the I-2 fighter by Grigorovich (about which we will learn more shortly).

Much of the testing of this aircraft was being done by one of their best pilots, Filippov. Unfortunately, Filippov and his observer were both killed in the crash of a two-seat fighter they were testing. The crash was attributed to the failure of the wing structure of the airplane. The wings of the I-1 were built with the same system and this resulted in the grounding of the I-1 while its wing was redesigned.

Russian M-5 powered I-1 at Khodynka aerodrome in the winter. (Courtesy of Harry Woodman)

Upon accomplishment of this task, series production was once again started. Initial plans had been for the production of 80 of these aircraft but ultimately this would not be. Incorrect center of gravity continued to cause flight problems, principally spin recovery. Several cases of very difficult spin recovery led to altering the center of gravity followed by tests. In one such case one of their best pilots, Mikhayl Gromov, was to test the latest alteration and was instructed to wear an Irvin parachute as a safety precaution. The test commenced on April 25, 1927 and after failing to recover the I-1 from a spin, he bailed out at 120 meters and landed safely. This was the first recorded safe decent by parachute by a Russian pilot. This marked the end of the I-1 production program.

Russian material conflicts in the noted total production numbers. Some reports say 12 I-1 aircraft were built in addition to the two prototypes, making a total of 14. *(Some historians say a total of 33 were built.)*

The use of this model was not quite at an end, however. Starting in the summer of 1928 a series of tests were carried out to test handling characteristics with the center of gravity moved to various locations by using a special movable ballast. One of the tests carried out was high-speed dives to test the structural strength of the wing. Dives were carried out to speeds of 400 km/h (248.5 mph).

By 1929 the fighter was considered obsolete. Extension of its useful life by replacing the Liberty with a 450 HP Lorraine-Dietrich engine were considered but abandoned. The last of the I-1 aircraft could be seen parked in storage areas at Khodynka aerodrome in 1933.

THE I-2 SERIES

—

I-2 PROTOTYPE

Fighter design was being carried out on a competition basis. At the same time Polikarpov was designing and building the prototype IL-400, Dmitri Grigorovich was developing, also at GAZ-1, a rival fighter which was initially known as the I-1, but for clarity we will refer to as the I-2 prototype This design was a Liberty-powered biplane with wooden construction with the forward section plywood covered and the remainder fabric covered. The forward section of the fuselage was well streamlined and the radiator was carried between the landing gear struts.

This aircraft was completed and first tested in early 1924. Performance was found to be 143 mph maximum speed, with a ceiling of 19,685 ft. Although performance was in many respects good, it was found to have the same excessive rear center of gravity problem as did Polikarpov's IL-400, as well as poor climb and stability in addition to cooling problems.

I-2

In spite of this, the aircraft was considered to have good potential and Grigorovich was directed to redesign it for further evaluation. He did so and the result was significantly changed from the prototype. The slabsided fuselage was replaced with one which was much more rounded and streamlined. This second plane was completed in the autumn of 1924 and submitted to testing. The problems encountered with the first prototype had been resolved and, in addition, a top speed of 155 mph was now possible.

By this time it would seem that the designation decision referred to above had been made and the IL-400 series of designs were to become the I-1 and the Grigorovich fighter would be the I-2. At this point the

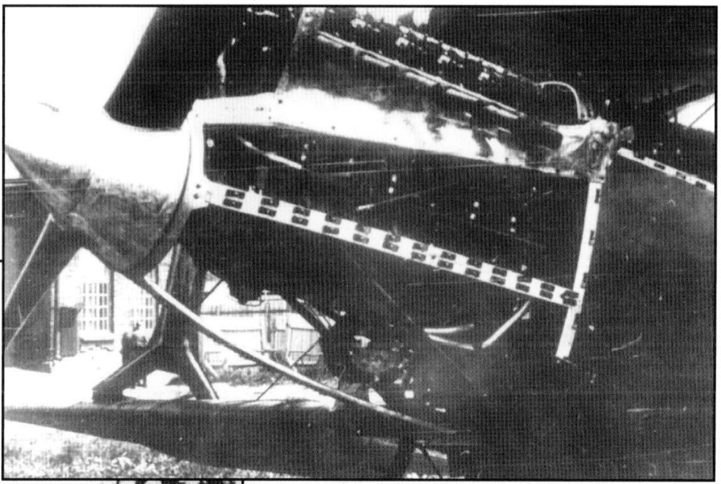

Left: An I-2bis resting in an embarrassing position following a nose over. Above: A close view of the M-5 installation in a I-2bis. (Courtesy of Harry Woodman)

decision was made to put the design into series production. Although general performance was now considered acceptable, there were several additional items of design that needed resolution. They included relocation of some controls, enlarging of the cockpit and additional changes to the cowl area.

I-2bis

This became the production design and was known as the I-2bis. The engine cowling and upper fuselage section back to the cockpit was built of frame sections with duraluminum skin. The engine support framework was made of welded steel tubing. The remainder of the fuselage panel was plywood. Overall, the airplane was quite streamlined and resembled, in many respects, the Curtiss R-6 and PW-8 airplanes of 1922-24.

Production commenced at GAZ-1 and later at GAZ-23. Between 1926 and 1929 164 were built at GAZ-1 and 47 at GAZ-23 for a total production of 211 I-2bis. This resulted in 213 aircraft in the series. All were M-5 powered.

The FOKKER C-IV

During the early 1920s Russia was building her Air Force as rapidly as money and technology would allow. As we have seen, one area of immediate concern was reconnaissance aircraft. Their new R-1 did not get into produc-tion until 1924. In the meanwhile, they placed an order with Fokker of the Netherlands for a total of 110 Liberty-powered C-IV airplanes to be used for this purpose.

Sixty were delivered between September of 1923 and early 1924 and the remaining fifty before the end of 1924. The first batch were C-IV models and the second batch C-IV-B. Later Soviet documents referred to them as FS-IV and FS-IVbis models. Fokker acquired the Liberty engines (with which they powered them) from the Aircraft Disposal Company in England. They were, of course, surplus British military Liberty engines which were built in the United States. Any airplanes which at a later time required engine replacement received Russian-built M-5 engines.

These aircraft were issued to reconnaissance units throughout the country. A number were replaced by R-1s beginning in 1925 and continued to be for some time, the C-IVs being transferred to other duties such as staff transport, military training schools, etc.

By the end of 1931 seventeen C-IVs remained in military service with the last one being withdrawn from service in 1934. During 1928 five were loaned to Dobrolet for use in airmail service between Moscow and Irkutsk. The test was found to be successful and in 1929 twenty three were transferred to the mail service and assigned civil registrations CCCP-150 to CCCP-172.

A number of additional C-IVs were transferred to civil registration and used for photographic survey and other experimental work. The last civil C-IVs were withdrawn from service in 1934.

Considering the fact that a large number of Liberty engines were imported to Russia and that almost 3,000 were produced in Russian factories, it is certainly likely that at least a few may have been used to re-engine either military or civil aircraft of which we have no current knowledge.

THE LIBERTY-POWERED RUSSIAN TANK

Russian tank development following the revolution was a slow process. Their initial tanks were captured British and French examples. They set up the Krasnoye Sormovo factory in 1919 to produce copies of the 1917 Renault light tank. Progress was very slow and it was mid-1920 before the first example left the factory.

The Renault copy continued in production in small quantities. During 1920 and 1922 competitions were initiated for new designs and a handful of experimental tanks were built from accepted designs.

One of the Fokker C-IV airplanes transferred to the Dobrolet (Russian Volunteer Air Fleet Co.) for mail service. (Courtesy of Peter Bowers)

1927 saw the development of the MS-1 light tank, which was based on the Renault light tank. Eventually, after improvements, called the T-18, these tanks were produced in the Leningrad Weapon Factory beginning in 1928. By the end of 1928 Russia had only 92 tanks.

Russia started its first 5-year plan in 1929 and tank development was high on the list of goals. A commission was set up to go to Britain and the United States to purchase tanks. In Britain they purchased 15 Mark II medium tanks, 15 Vickers-Armstrong 6-ton tanks, 26 Carden-Lloyd Mark VI tankettes and 8 Carden-Lloyd amphibian tanks. In the United States they were interested in purchasing T-1 light tanks but claimed they were unable to obtain acceptable price or conditions of sale and technical support from the manufacturer. Russia was not officially recognized by the U. S., however, and that raises the possibility that the sale would not have been sanctioned anyway. They did purchase 2 Christie M 1930 wheel-track tanks. These were delivered without turrets and classified as tractors to avoid having the U. S. government prohibit the sale, which it appears they intended to do. I will add more detail to that story shortly.

THE RUSSIAN CHRISTIE

The above mentioned Russian commission went abroad on December 30, 1929 and first visited Germany, Czechoslovakia, France and Great Britain. They then went to the United States and made the unsuccessful attempt to purchase T-1 tanks at J. Cunningham and Sons Co. They then visited J. Walter Christie at his U. S. Wheel Track Layer Corp. in Rahway, New Jersey to examine his revolutionary new M 1928 high-speed tank. This was early 1930. Christie's tank had been demonstrated to the U. S. Ordnance at Fort Myer, Virginia on October 28 of 1928, and then subjected to tests before the Tank Board in November of that year.

These tests resulted in some rather spectacular performances for a tank, having obtained speeds of 70 MPH on wheels and 42 MPH on tracks. Such performances received wide publicity which had not escaped the Russian military. At the time of the discussions with the Russian commission, discussions were probably still under way between Christie and the U. S. Ordnance Depot for an order for a single tank for testing by them. The U. S. tank, an improved version of the M 1928 with a turret, designated as the M 1930, was delivered on January 19, 1931 (some four months late).

The Russian order was consummated with a written agreement with U. S. Wheel Track Layer Corp. signed by Christie and A. V. Petrov, president of AMTORG (the trade organization representing the USSR in the U. S.) on April 29, 1930. In it they agreed to purchase two Christie M 1930 tanks, to be delivered no later than four months from the signing of the agreement, spare parts, the rights of production, sale and use of the design within the boarders of the USSR for ten years, the transfer of patent rights and a guarantee of technical assistance in setting up production and for the tanks, but not for the engine since the Russians already had the Liberty in production as the M-5. The agreed-to price was $60,000 for the delivered tanks, $4,000 for delivered spare parts, and $100,000 for patent and production rights and technical assistance.

As noted above, the Russian order specified the tanks as M 1930. They actually arrived in Russia early in 1931 and thus must have been completed at about the same time as the first test tank delivered to the Ordnance Depot in January of that year. (See Chapter 6.)

In July of 1930 a Russian representative, N. M. Toskin, arrived in New York and visited the Christie factory and took delivery of 127 pages of blueprints of the tank. These were forwarded to Moscow where they arrived on August 9, 1930. They were sent to appropriate authorities for study and on November 21 the basic Christie design was approved for production.

It should be noted, however, that the original design was recognized as needing revisions before being totally acceptable as a military vehicle and the resulting production tanks were altered in a number of ways from the delivered examples.

The Russian Christie tanks were supposed to have been finished for delivery by September, but were not completed and delivered until late in December. Obviously, this was at almost the same time the U. S.-ordered tank was delivered.

In preparation for production of their version of these new tanks, the Russian government had directed AMTORG to purchase Liberty engines and their inspector, Captain L. Tharp, a U.S. test officer for the Tank Board, noted that they had purchased "several hundred."

The two Russian Christies were delivered to Pier 6, Staten Island, New York on December 23, 1930. In order to avoid problems related to the sale of military goods to Russia, a country with which the U. S. had no official diplomatic relations, they were delivered through the Russian trade organization AMTORG as tractors, without turret or armament. Christie had advised the State Department that they were commercial tractors and were on the pier ready for shipment. The State Department approved the shipment. Several days later, on December 27, officials of the State Department were informed by Lt. Col. J. K. Crain of the War Department that the Christie "tractors" were actually tank chassis. Suddenly the State Department was not so anxious to approve the shipment. By then it was too late to interceed as the shipment had left New York on December 24.

The American Christie as delivered to Russia in 1931. (Author's collection)

They arrived in Russia in early 1931 and were immediately subjected to testing. An early decision which had to be made was what type of military designation would be applied to these new tanks. In a letter regarding the Christie the head of the UMM noted that the "American Christie tank does not meet the requirements of a tank-tractor-armored combat vehicle, and has not been accepted into the inventory, it will not receive the comprehensive designation "T." The suggestion to give it the designation "ST" or "BT" (skorostnoy tank or bystrochodya tank, both of which translate to "fast tank" in English) is more rational." The decision was to use "BT."

Testing continued through May and, although a number of problems were experienced, the overall impression of the tanks remained very favorable. A high-level meeting of various members of the tank corps and production managers resulted in a tank production program for 1931 which included the building of 100 BT tanks. These would, of course, be of a design slightly revised by the Russians and more suited to their needs.

Russian Christie BT-1 under test, having just broken the right idler wheel mount. These prototypes were furnished without a turret. (Author's collection)

THE CHRISTIE BT-1

The prototypes assumed the designation of BT-1, or Christie BT-1. The first production model was to be called the BT-2.

THE BT-2

Production was started in September of 1931 in the Kharkov Locomotive Works (KhPZ) Factory No. 183. The first three BT-2 prototypes were completed in October without armament and participated in the November 7 Moscow parade. Production quantities were extended and production continued into 1933. Total production quantity was 620. Power was, of course, by the Russian-built M-5 Liberty or imported U.S. Liberty.

The convertible feature of the Christie was retained in all the BT series of tanks produced by the Russians. The tracks could be removed and the tank run on the bogie wheels as road wheels with the front wheels being steerable. For this reason, the inward sloping shape of the front hull section can be seen in all the BT models. This was necessary for clearance of the steered wheels.

The BT-2 was made in several variations, with the differences being in armament. The principal two types were those having a single 37mm cannon and a single machine-gun ball-mounted beside it, and those having machine-guns only. The latter having three turret mounted types.

The machine gun only model was not well received and consequently production reverted to those models mounting the 37mm cannon in combination with either one or two machine guns.

THE BT-3 and BT-4

These two models were experimental types that never went into production with less than a dozen built. The BT-3 had solid wheels, a feature carried into the BT-5 series, and a 45mm main gun. In 1937 one of these was converted into a bridge-laying tank, but without success. Also during this period, a second one was converted into a flame-throwing tank. Again the experiment went no farther.

The BT-4 carried two machine gun turrets and was intended for infantry support. It was decided that there was really no valid requirement for this type and it was dropped.

A BT-2 under test in what must be a rather famous picture and an event long remembered by the crew. (Author's collection)

Russian BT-2. Note the great similarity to the Christie BT-1 prototype on page 408. (Author's collection)

The Liberty as used in the American Christie tank. Note the very similar design of the fan and flywheel to that used on the Russian BT series; also the use of the hand crank on the rear of the engine. (Courtesy of R. P. Hunnicutt)

THE BT-5

In 1932 instructions were given to adapt the new 45mm Model 1932 tank gun to both the T-26 and BT series of tanks and to use a common turret. The Kharkov Factory brought out the BT-5 incorporating these features. This resulted in a larger cylindrical turret carrying the new gun with a single machinegun beside it in a co-axial mount replacing the ball-type.

This turret design did not last long before an improved type was introduced which allowed better access (twin hatches in place of a single one) and better storage provided by a new larger bustle in the rear of the turret.

In addition to the gun and turret changes, the BT-5 also used bogie wheels with a dished design rather than spokes and, in addition, the suspension system was beefed up.

Production continued until 1935 with a total of 1,884 being built. As was the case with the BT-2, these tanks were powered by both Russian-built M-5 and U.S. Liberty engines. Between 1924 and about 1934 approximately 3,000 U.S. Liberty engines were imported.

A number of experimental variations were made over the BT-5s life, including a snorkel version (BT-5PKh), one armed with a 76.2mm gun for use as mobile light artillery (BT-5A) and a flame-thrower version, none of which were built in quantity. It was also fitted as a command model (BT-5V) with radio equipment.

The BT-5 was replaced in 1935 with the BT-7. (The BT-6 existed on paper only and was never built.) This new tank resembled the BT-5 but was of all welded construction rather than riveted. It also used a new more powerful engine (the M 17T which was based on the German BMW aero engine) and had a redesigned power train.

BT-2 General Data
(620 built - see text)

Length	18'0"
Width	7' 4"
Height	7' 3"
Combat weight	24,300 lbs
Engine	M-5 45° V-12
Power	400 HP @ 2,000 RPM
Power/weight ratio	36.4 HP/ton
Gearbox	4 forward, 1 reverse
Speed on road	52 km/h track 72km/h wheels

Armament

37mm M1930 gun (also see text)
1 or 2 7.62mm MG
Main armor thickness 13mm

The Russian-built M-5 with cooling fan and flywheel as used in the BT series of tanks. (Courtesy of Harry Woodman)

TG-1 (Tank Grotte-1) also designated T-22. Three of this model were built as prototypes with M-5 power. (Author's collection)

BT-5 tanks on maneuvers with a BT-5V command tank in the lead. (Author's collection)

USEFUL LIFE OF THE BT-2 THROUGH BT-5

Both of these models were in active service for a number of years with, of course, the BT-5 serving latest. A number of BT-5s (about 50) went to Spain to serve with the Russo-Spanish Republican tank units in 1938. A large number were among the tanks which were met by the German advance in 1941 and did not fair well. Their 1935 armor and firepower were no match for the new German tanks.

A fast, modern and useful tank when it was designed, the BT-2 and BT-5 tanks served well from 1932 to the late 1930s, but the development of the more modern WW II tanks left them at an untenable disadvantage. The remaining BT-2s and BT-5s were relegated to training duties only after the German invasion in 1941.

THE T-22 PROTOTYPES

During the early 1930s the Soviets were developing their own designs of medium tanks while at the same time utilizing a foreign design for their BT series. Work was initiated on two different designs: the TG-1 (Tank Grotte-1) designed by a German engineer named Grotte, which was given the Army designation T-22, and a more conventional design designated the T-28 by a team headed by N. Tsiets. The T-22 was completed in 1932 in three variations, all of which were M-5-powered. One had a 37mm cannon and four machineguns, one a 76mm cannon and four machine-guns and one a 37mm cannon, a 76mm cannon and one machinegun.

This tank was quite advanced for the time and was very streamlined in appearance. It had pneumatic steering and suspension as well as being a wheeled-track vehicle.

BT-5 General Data
(1,884 built)

Length	**18'0"**
Width	**7' 4"**
Height	**7' 3"**
Combat weight	**25,400 lbs**
Engine	**M-5 45° V-12**
Power	**400 HP @ 2,000 RPM**
Power/weight ratio	**34.8 HP/ton**
Gearbox	**4 forward, 1 reverse**
Speed on road	**52 km/h track**
	72km/h wheels

Armament

45mm M1932 gun (also see text)
1 or 2 7.62mm MG
Main armor thickness 13mm

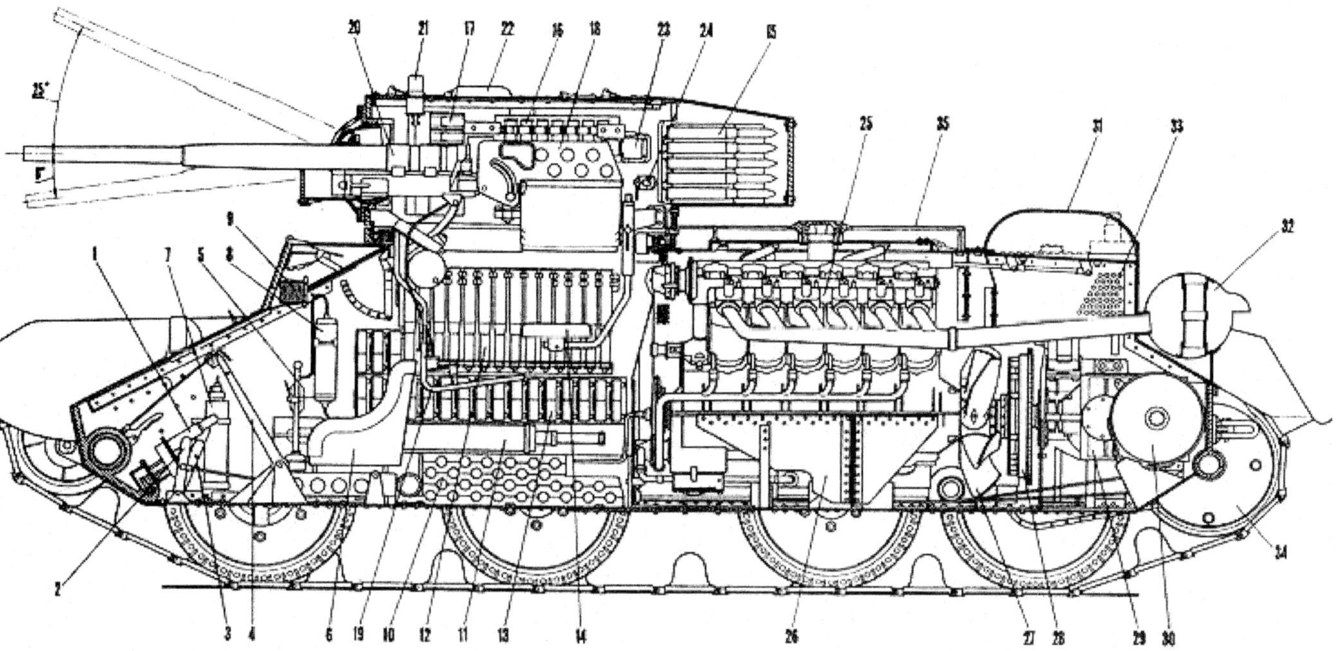

Sectional drawing of the late model BT-5. Note the turret bustle and the fan and flywheel mounted on the rear of the M-5 engine. (Author's collection)

It was rejected because of its complexity and expense of production. It weighed 50,000 lbs, was 24' 6" long, 9' 8" wide and 9' 3" high and had a road speed of 35 kph. The more conventional T-28 was chosen in its place.

THE BRITISH CRUISER TANK

In September of 1936 a British Military Mission attended the maneuvers of the Russian Army. On the mission was Lt. Col. G. Le Q. Martel, a pioneer of British tank development from the 1920s, during which time (1926) he had designed the Morris-Martel tankette. Martel, then a Major, had gained considerable tank experience while serving as GSO-2 on the staff of Tank H. Q. in France during World War I. He had been appointed Assistant Director of Mechanization at the British War Office earlier in 1936. (Martel would eventually be Lieutenant-General Sir Gifford Le Quesne Martel.)

Among the Russian tanks participating in the maneuvers were their BT fast tanks. British notes do not say which models were observed, but they probably included only BT-2 and BT-5 models as his report specifically noted their use of a Russian copy of the American Liberty. Martel was very favorably impressed by the speed and maneuverability of these tanks and specifically noted the origin thereof as being the suspension design of the American Christie and the high power and light weight Russian copy of the American Liberty engine.

Upon his return to England he immediately pressed for the purchase from America of two Christie tanks for tests. He said that in his opinion, the nearest British rival, the 13 ton A9 Cruiser Mark I powered by a 150 HP engine, had no chance of ever coming up to the performance of the BT. He further stated: "It is thought that we may be able to produce a tank as good as or better than the Russian tank by using the existing design of the Christie suspension and producing an engine similar to the Liberty engine in this country, while fitting our own pattern of fighting cab. By taking an existing well-tried design of suspension and transmission long delay may be avoided and production may be possible within the limited time available." An order was placed in October, £16,000 having been allocated for the purpose.

The order specified immediate delivery of a "1932 Model," followed later by a second tank of updated design. Although the agreement specified this would be Christie's

The British Christie A13.E1 after delivery in November of 1936. Although delivered to England almost six years after the Russian Christies were delivered, the only noticeable exterior difference was the mufflers mounted above the engine compartment, and these were probably added after delivery. The assigned number T2086 has been applied which indicates the picture was probably taken in early 1937. (Courtesy of the Tank Museum at Bovington)

"1932 Model" the tank actually delivered was the M 1930 which had been rejected by U. S. Ordnance. If it had been modified from its original configuration it resulted in no noticeable exterior changes. From Christie's past history it would seem more likely that he had just given the 1930 Model an updated name to make the British think it had improved features from the original model. (Unfortunately, the model name 1932 picked conflicted with that assigned to another tank built by Christie, his M 1932 Flying Tank, which was quite different from the one delivered.) The contract also called for future delivery of a second later design which Christie referred to as his "1937 Model." The first tank arrived in London in November of 1936, accompanied by Christie, and was delivered to MEE on November 17. The second arrived some time later and will be discussed below.

It was decided that Morris Commercial Cars Ltd. would act as purchasing agent for the War Department. The vehicles were described as "1 commercial and convertible Christie truck tractor of Series I and 1 of Series II."

THE FIRST CHRISTIE

Records indicate the first Christie arrived fitted with Navy Liberty engine No. 371 which had an eight-blade cooling fan of 30" diameter and 35º pitch at the tips. Although it had a 41 ½" turret ring, no turret was provided with it. The tank was almost immediately assigned the designation of Model A.13.E1. Tank number T2086 was assigned to it some time later.

The hull was 5/8" plate on the front and 1/2" elsewhere. The gross weight was 9 tons. Height without turret was 5' 7 1/2"; length 18' 1 1/4"; width 7' 3 1/8"; distance between track centers 6' 5". The Liberty engine as tested developed 411 HP @ 2,000 RPM.

The following description of this tank is, for the most part, extracted directly from various entries found in the Mechanization Board reports of the period.

The tank was taken to the testing area at Farnborough and loaded to a gross weight of 10 3/4 tons to simulate the carriage of a turret and full War Department fighting equipment. It was then subjected to testing aimed at finding any weak points of design and correcting them in the proposed designs to follow.

"The hull which is rectangular in cross-section with a pointed nose and square rear portion is divided into three main portions. The forward part forming the pointed nose is occupied by the driver and the controls. Immediately behind the driver is the fighting compartment provided with a turret ring of 41 1/2" diameter. No turret was received with the vehicle. The rigidity of this hull has been ensured with a minimum weight by an ingenious construction of inner and outer walls connected by a lattice girder arrangement. This is a very necessary feature because the side plates are long and their strength has been curtailed by the slot provided for the upward movement of the bogie wheels. The power unit is a Liberty aeroplane engine. The cooling system includes two large vertical radiators installed longitudinally in the space between the engine and the inner walls of the hull. There is no oil cooling system. The rear end of the crankshaft is connected to a light flywheel and clutch which forms an assembly with a ring gear and electric starter. The output of the clutch is connected to a special form of gearbox. The cross shaft comprises steering brakes and clutches, and terminates at both extremities in spur pinions which mesh with the final drive spur wheels attached to the sprockets. Only one set of brakes is provided for steering, stopping and parking. The controls which operate the clutches and steering brakes are connected to long hand levers in the driver's compartment and the pedal control for the clutch and those of the engine are of comparatively conventional design. The suspension comprises two small diameter idler wheels mounted on a heavy cross tubular member at the front of the vehicle, in four large diameter twin wheels with rubber tires attached to either side of the hull by a system of levers connected to long coil springs which provide an exceptionally large absorption. The three rear bogies have vertical spring cases and the front bogie arrangement embodies a horizontal coil spring. The sprockets comprise two discs carrying four rollers which engage with tongues on the tracks and by this means the drive is effected. The track shoes consist of flat pieces of steel, the pin joints forming a type of bar tread."

The tank also had the standard convertible feature common to this series of Christie tanks: the tracks could be removed (about 30 minutes were required to do so) and the tank run on its bogie wheels and steered with the front set of bogies.

THE SECOND CHRISTIE

Nuffield was expecting the arrival of the second tank and Christie in January of 1938. It did indeed arrive by mid-month and was delivered to the Mechanization Experimental Establishment (MEE) and demonstrated before members of the War Office and the MEE board. It was delivered, as was the first Christie, without a turret. The hull was of 5/8" plate and the gross weight was 6 tons. A turret would have added approximately 2 tons. Height without turret was 4' 6"; length 17' 3"; width 6' 9 1/2"; distance between track centers 5' 11 1/2". The tank was

The A13.E2 which was the first pilot model of the new Christie-based A13 Cruiser Mark I. This tank would be assigned the number T2085. The similarities to the A.13.E1 can be seen, as can some of the differences. (Courtesy of the Tank Museum at Bovington)

powered by a Curtiss D 12 engine rated at 470 HP @ 2,600 RPM. With this power-to-weight ratio the tank was able to attain a speed of 64.3 MPH. The suspension design was similar to that of the first tank. Suspension was found to be harsh at speeds above 30 MPH. Tracks had a pitch of 5" and showed no tendency to jump the driving rollers. Although some brake overheating and track mal-alignment was reported, no extensive testing was attempted. By this time, considerable commitment had already been made to designs derived from the first Christie tank and this second one presented nothing new they wanted to pursue. (As we will see, by this time the first experimental A13.E2 had been delivered and was undergoing tests and the second, the A13.E3, was near completion.) After having completed only 48 miles the vehicle was returned to the designer, as it was decided not to proceed further with a vehicle of this type.

The A13.E3 which was the second pilot model of the Christie-based A13 Cruiser Mark I. This tank was not completed and submitted to testing until after the first contract for 65 A13 Mark I tanks had already been signed. It was assigned T3642. Note the shorter pitch tracks than on the A13.E2. (Courtesy of the Tank Museum at Bovington)

Christie called this tank as his "1937 Model" in his British-related documents but its description matches that of his M 1936 flying tank which he says he later modified into a M 1938 and it would certainly appear that they were one and the same.

THE A13 CRUISER PILOT MODELS AND SELECTION OF THE LIBERTY

By the end of 1936 it had been decided that the building of a series of tanks based upon the Christie design was due serious consideration. Financial approval to build two prototypes (pilot models) at Morris Commercial Cars was sought and obtained and design was started in early 1937.

Detail drawings were nearing completion by March 18, 1937. Construction probably did not commence on the first Pilot, A13.E2, until mid-summer. By this time Lord Nuffield had set up a new organization in Birmingham specifically to manufacture tanks and munitions, Nuffield Mechanizations Ltd. It was in this facility that the pilot models were completed, although most initial design work was done at Morris Commercial Cars.

The first pilot, A13.E2, was completed and given initial trials on October 7, 1937 and delivered to MEE on October 13 without gun or shell racks. Until this shortage was corrected, a mock up gun and weights were substituted.

Trials of the original A.13.E1 Christie continued along with those of the new A13.E2. A tentative order for 50 production A13

The A13 Mk I, also known as the Cruiser Mk III. This was the first production tank of the A13 series designed and built by Nuffield Mechanizations & Aero, Ltd.
(Courtesy of the Tank Museum at Bovington)

tanks was contemplated but in a January 20, 1938 meeting held at Nuffields, it was noted that the order was expected to be changed to 65. The order, T5114, was indeed placed on January 22 for 65 A.13 Mk I tanks.

A13.E3 was under construction and was delivered in the middle of February of 1938, almost a month after the placement of the first production order. At a February 16 meeting held in the War Office it was decided, from various failures which had occurred in both the A13.E1 and A13.E2 which were felt were caused by the strains resulting from excessive speed, that the engines would be governed to 30 MPH. Doing so limited engine speed to 1,500 RPM and thus output to about 340 HP.

For those who wonder why the Nuffield Liberty was rated so much lower than was the U. S. Liberty when, for all practical purposes, it was the same engine with some improvements, this was the reason. Engine speed, and thus also power, was governed to keep tank speed down to 30 MPH and thus reduce wear and tear on the tank as well as the engine. With no governor, the engine was capable of driving the tank in excess of 40 MPH. At such speeds on any but level ground both the tank and crew were likely to sustain damage.

ENGINE SELECTION

In the meantime, there was some investigation regarding potential engine selection. Col. Martel had been very favorably impressed with the performance of the Liberty with which the Russian BT was powered and had so indicated in his report. In addition, the Christie which was purchased and tested was Liberty-powered.

Tests were carried out on the Liberty, the Napier Lion and the Thornycroft RY/12 engines. Any one of the three would need some amount of redesign to enable its use as a tank engine if its basic design was found to be useable. The object was, of course, to come up with an engine which would produce the required amount of power and, hopefully, be relatively light in weight. One of the requirements was that it be able to operate well on Grade III (65 octane) fuel.

The Thornycroft RY/12 was a 60° V-12 marine engine of 5 3/4" bore and 7" stroke, giving a displacement of 2,181 cu in (35.75 liters). With a compression ratio of 5.1 to 1 and a maximum speed of 2,000 RPM it produced 537 HP. Torque was 1,594 ft. lb. @ 1,300 RPM. Its primary shortcoming was a weight of 2,900 lbs. without flywheel. It was accepted for use in the A14 pilot model. Later a V-8 version of this engine, which was to develop in the area of 400 HP, was requested of Thornycroft as an alternate to the Liberty for A13 tanks. Nothing ever came of the 400 HP version and the V-12 engine only saw use in the single A 14 pilot model.

The Napier Lion Series XIA engine was rated at 500 HP and weighed about 1,000 lbs without flywheel. It was a 12 cylinder W engine with a bore of 5 1/2" and stroke of 5 1/8" giving a displacement of 1,462 cu in, or 29.36 liters. The compression ratio was reduced from 6 to 1 to 4.3 to 1 to restrict its output for tank use and reduce its tendency to detonation. Although it would operate satisfactorily on aviation-grade fuel, on 65 octane Grade III fuel it suffered from severe detonation at low speeds under full throttle. It was determined that the power available for tank use was 465 HP @ 2,500 RPM. Maximum torque was 1,155 ft. lbs. @ 1,300 RPM. Because of its detonation tendency, it was removed from consideration.

The Liberty was a V-12 with a bore of 5" and a stroke of 7" and was rated at 400 HP. With its 5.2 to 1 compression ratio it produced 411 HP @ 2,000 RPM and 1,274 ft. lbs. of torque @ 1,100 RPM. Displacement was 1,649 cu. in. or 27 liters.

A13 Mk I - Cruiser Mk III General Data
(65 built - see Appendix A)

Length	**19' 9"**
Width	**8' 4"**
Height	**8' 6"**
Combat weight	**31,360 lbs.**
Engine	**Nuffield Liberty Mk I or II**
Power	**340 HP @ 1,500 RPM**
Power/weight ratio	**21.68 HP/ton**
Gear box	**4 forward, 1 reverse**
Speed on road	**30 MPH**

Armament

2 - pdr. QFSA
.303 Vickers MG
Main armor thickness 14 mm

THE LIBERTY IS CHOSEN
THE NUFFIELD LIBERTY Mk I

Since performance of the Liberty was acceptable and there were no other production engines available in the power class, the Liberty was settled on as the logical choice and officially adopted on June 28, 1937.

Performance in the A.13.E1 Christie led to various alterations to the original Liberty engine to better adapt it to tank use. Solex, Ltd. was given the job of designing a new induction system to make use of more modern carburetors which would be anti-spill at accute operating angles and to place the carburetors in a more accessible position for maintenance and adjustment. These were governed-type carburetors and would restrict maximum RPM to 1,500 plus or minus 100 RPM.

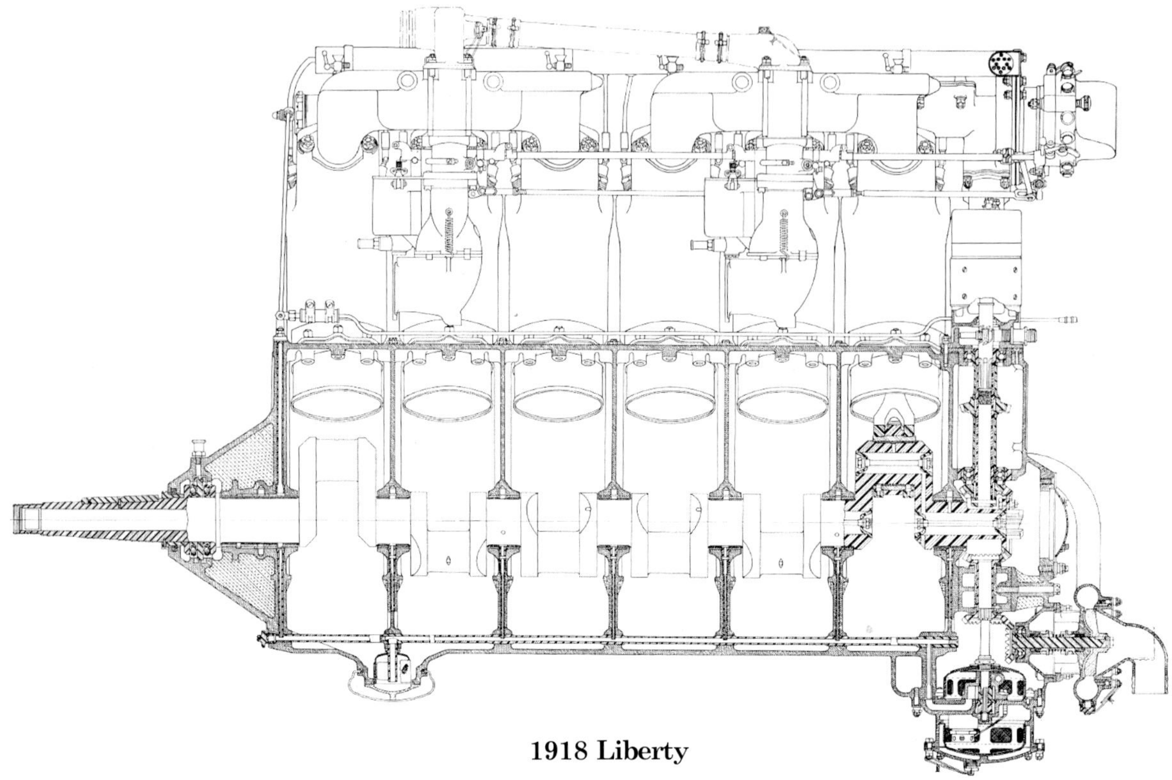

1918 Liberty

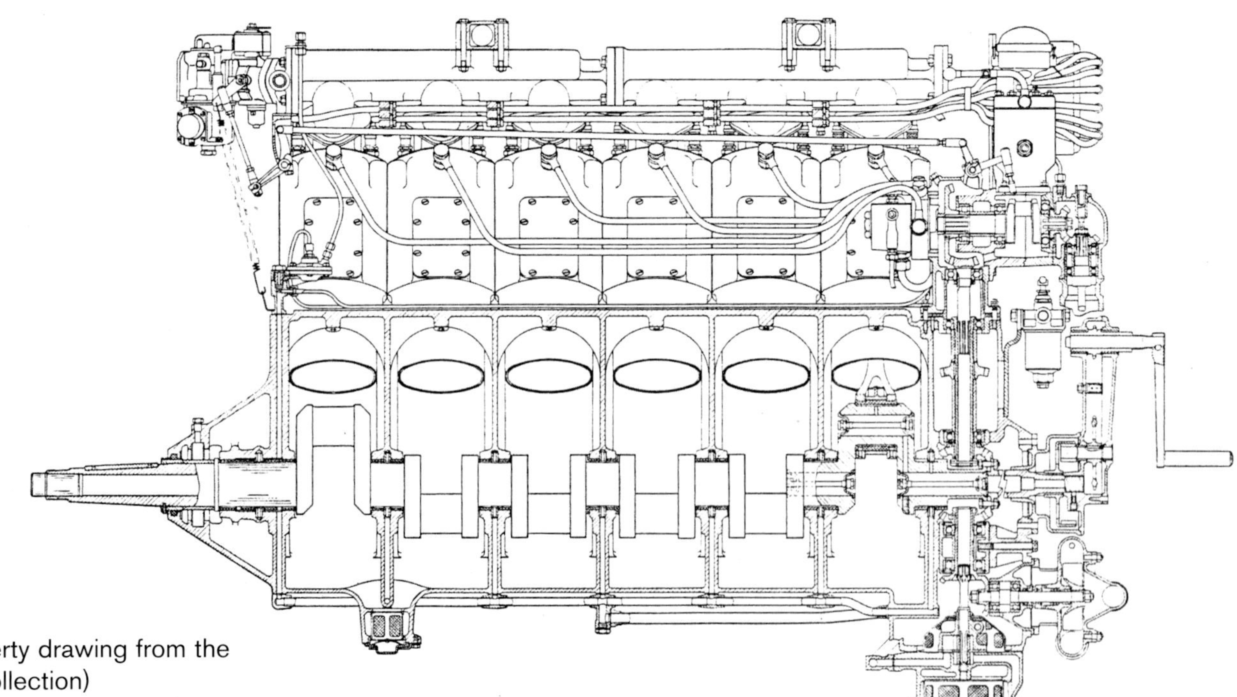

(1918 Liberty drawing from the author's collection)
(Nuffield Liberty drawing courtesy of the Tank Museum at Bovington)

Nuffield Liberty Mk II

1918 Liberty

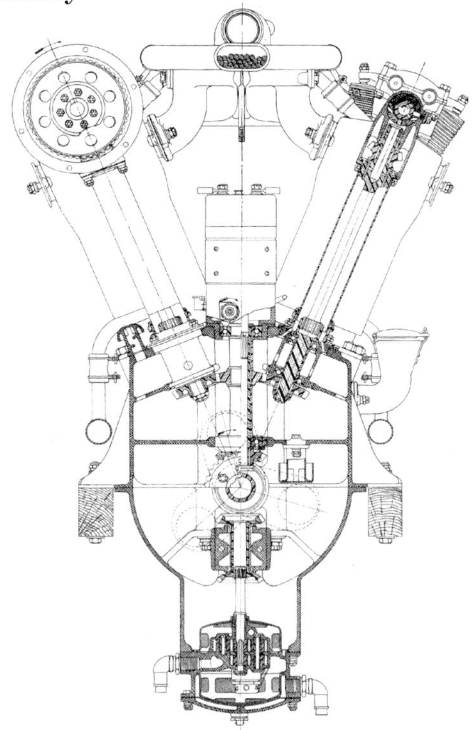

1918 Liberty Tank design cylinders

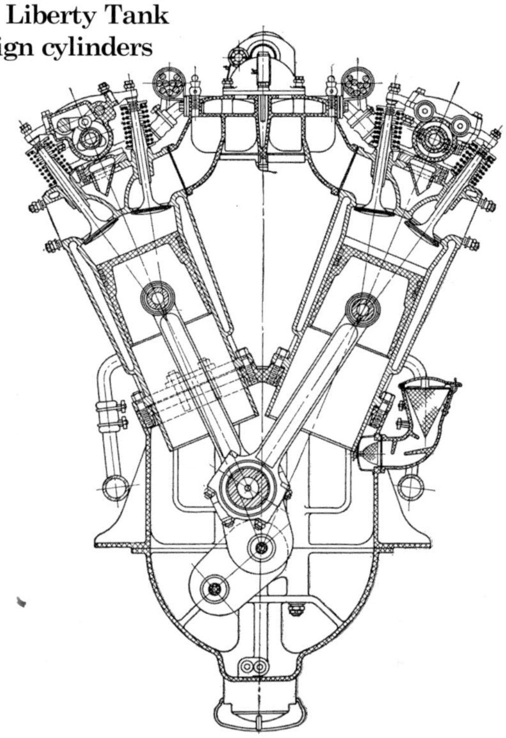

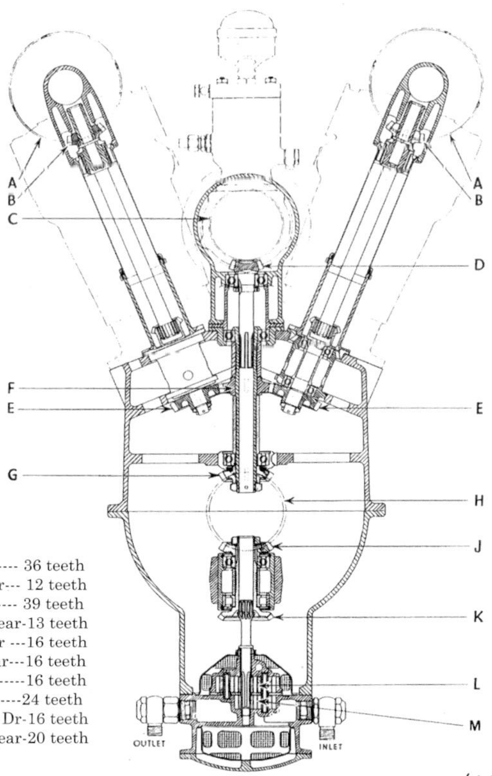

A-Camshaft gear-------- 36 teeth
B-Camshaft Dr. gear--- 12 teeth
C-Compressor gear ---- 39 teeth
D-Compressor Dr. gear-13 teeth
E-Splayed shaft gear ---16 teeth
F-Intermed. Dr. Gear---16 teeth
G-Main Dr. gear----------16 teeth
H-Crankshaft gear------24 teeth
J-Water & oil pump Dr-16 teeth
K-Water pump Dr gear-20 teeth

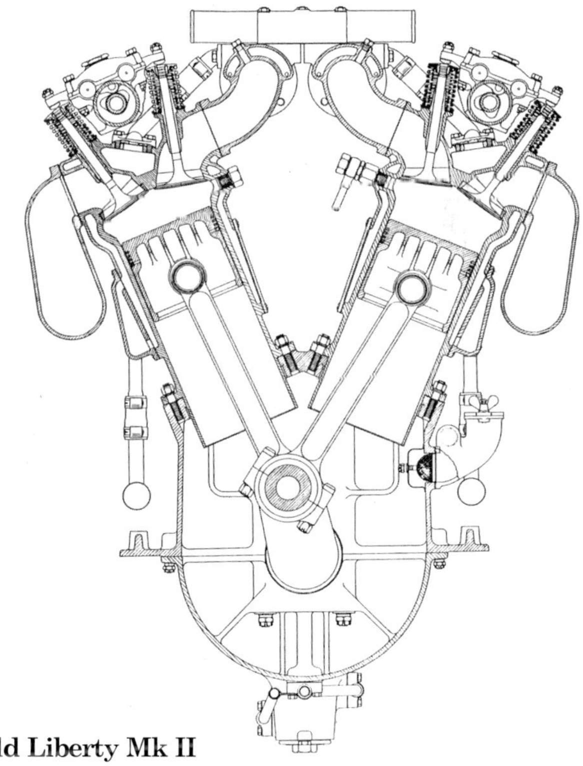

Nuffield Liberty Mk II

(1918 Liberty drawing from the author's collection)
(Nuffield Liberty drawing courtesy of the Tank Museum at Bovington)

Work was also undertaken to design a better method of crankcase breathing, strengthen the timing gear to reduce wear, and provide a better method of locating the crankshaft against end thrust.

As soon as the A.13.E1 had been submitted to extensive trials, it was evident that spare engines were needed. The placing of the order for the two pilot models also necessitated the acquisition of engines to power them. A half dozen or so were found in England. As a matter of fact, those involved were so sure that a tank production commitment was imminent, and that the official engine of choice would be the Liberty, that we find the following mid-December, 1936 note in the Munitions Board correspondence file from Nuffield Mechanization and Aero:

"As for the engine, if manufactured in U. K. there might be some delay, but Mr. Boden (Nuffields) thought that if some 30 of these were brought from the U. S. A. then it would be ahead of other components for which dies would have to be made, unless of course, they, too, could purchased from the same source. The two pilot models should, therefore, take about six months to supply and production tanks were visualized at the rate of some four per week provided the bullet proof plate became available in time."

It would appear that these initial engines were acquired through Walter Christie. Edward Christie's biography of Walter Christie, *Steel Steeds Christie*, carries a reproduction of a letter from Morris Commercial Cars dated January 18, 1937, ordering 25 Liberty engines and a complete set of drawings to be delivered "on the same basis as the three already ordered." The price was $1,800 each. Thus Christie furnished 28 engines in addition to the one furnished with the A.13.E1.

In January of 1937 Nuffield Mechanizations Ltd. was formed to manufacture tanks and munitions in Birmingham and later Coventry. On September 24, 1937 it was combined with Wolseley Aero Engines Ltd. (a company which was purchased by Lord Nuffield, then W. R. Morris, in February of 1927) to become Nuffield Mechanizations and Aero Ltd.

It would be this organization which would be the parent company for the production of both the A13 tank and the Liberty engine chosen to power it.

It would appear that the initial decision was to acquire existing Liberty engines in sufficient quantity to supply needs and modify them for use in the A13 tank. Overhaul orders have been found for some of these engines and from them it appears that their practice was to either renumber these U. S.-built engines with a number series beginning with K-1 and progressing upward, or to add this series of numbers to their already existing serials for tracking purposes.

How many such U. S.-built engines were acquired, we do not know. The first official order for British-built Liberties, to be known as Nuffield Liberties, was not placed until January of 1939. By that time the first production order for 65 A13 Mark I Cruisers, also produced by Mechanizations and Aero, had all been delivered. In addition to this production order, of course, the two pilot models had been built. This would account for 67 engines plus spares which must have been supplied with U. S.-built engines. **About 75 such engines must thus have been acquired before Nuffield production commenced.**

The first order for the Nuffield Liberty was apparently placed at about the same time as was order T6551 for 30 (originally for 65 but reduced to 30 partway through the production run) A13 Mk II tanks from London Midland & Scottish Railway. T6551 was dated January 26, 1939 and on the order card was the note that engine order E6621 was for Liberty engines for these tanks.

THE FIRST BRITISH PRODUCED LIBERTY IS ORDERED — THE NUFFIELD LIBERTY Mk II

Just how they were going about the job of getting these engines into production is not very clear. The decision to use the Liberty had been made back on June 28, 1937 yet we find nothing regarding an official order until 18 months later. At this point we must note, however, that a number of rather odd things went on in Great Britain regarding tank production during this period of time. Lord Nuffield had a great interest in British military preparations and did many things at his own expense to prepare for production of munitions. Perhaps setting up for the production of the Liberty was one of them.

It would appear therefore that Lord Nuffield had set up the former Wolseley aero-engine factory at Common Lane, Birmingham for tank and Liberty engine manufacture commencing probably in early 1938. In the mean time, as noted above, World War I U. S. Liberty engines were acquired either in England or the U. S. and modified as required.

Scattered throughout the Munitions Board notes of 1937 and 1938 are comments regarding various changes that were made to the Liberty to improve it in some way or to better suit it for tank use. The carburetor and intake manifold redesign was done by Solex and the Munitions Board (M.B.). In addition, M.B. pattern concertina air cleaners were fitted and the original ignition system was replaced by one designed by Messrs. Joseph Lucas, Ltd.

Most of these changes were applied originally to U.S.-produced Liberty engines, now known as the Liberty Mk I, as used in the prototype and first production batch of A13 Mk I tanks. An additional change which was incorporated in the design of the new production engines, which were known as the Nuffield Liberty Mk II, was a compressed air starter by Heywood. This starter was in

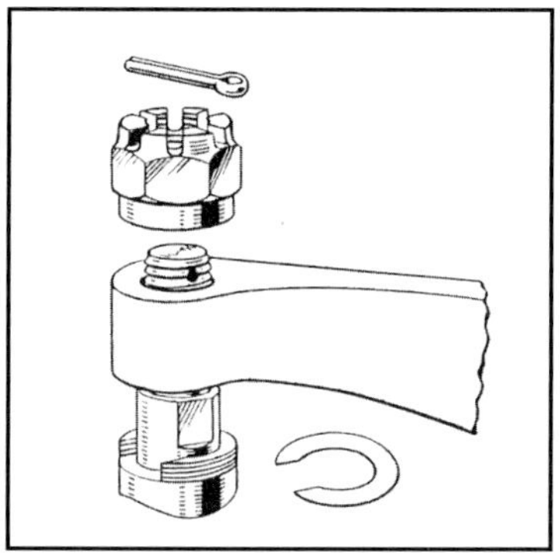

Valve adjustment design was the same on the Nuffield Mk II as on the original 1918 Liberty. Above the NL Mk II is shown on the left and the 1918 Liberty on the right.

Heywood Air Pump

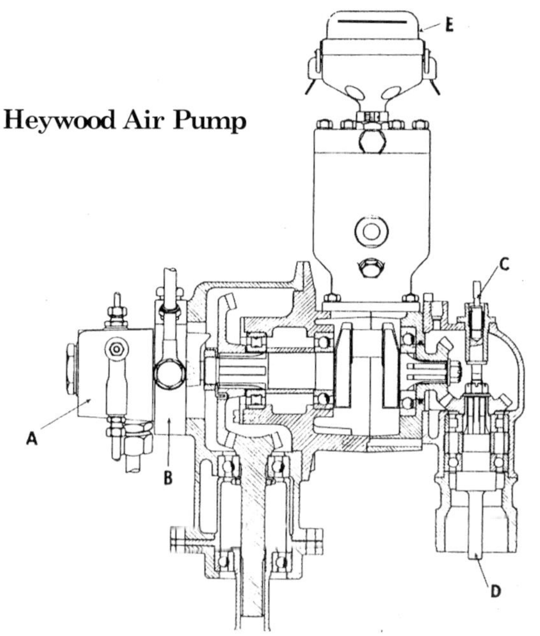

FIGURE 16.

A. Atomiser.
B. Distributor.
C. Revolution counter drive.
D. Drive to power traverse pump.
E. Air filter.

Shown above is the Heywood single-cylinder air-pump used initially on NL Mk II engines. The cylinder was water-cooled by the engine system and the bevel gearing and crank pin were lubricated by the engine oil system with oil overflow returning to the engine sump through the compressor drive housing. Above right is the starting system layout. (Drawings courtesy of the Tank Museum at Bovington)

Diagram of Heywood Air Starter Installation

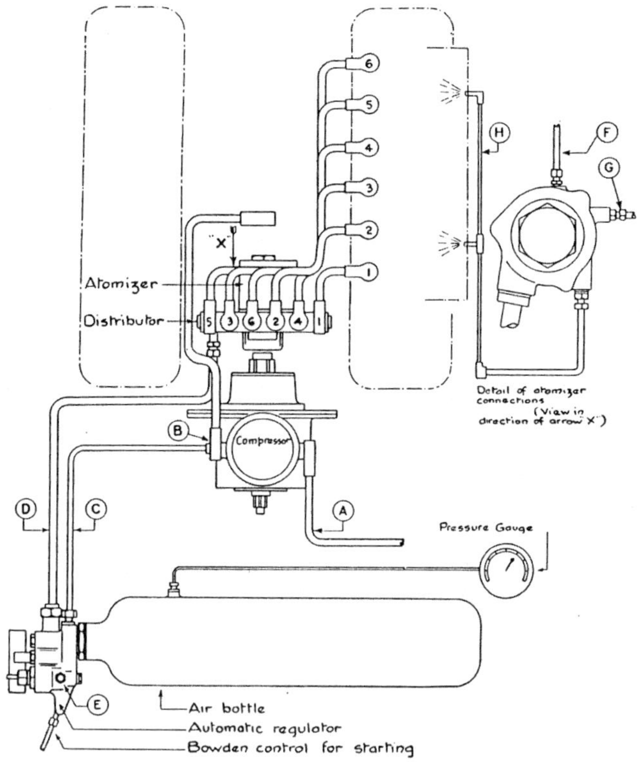

FIGURE 19.

A. Cooling water to compressor.
B. Water delivery, compressor to manifold jacket.
C. Compressor to air bottle.
D. Air bottle to distributor-atomiser.
E. Exhaust from automatic regulator.
F. " Ki-gass " pump to atomiser.
G. Overflow from atomiser.
H. Atomiser to induction manifold.

By early 1941 the Heywood starter had been abandoned and removed from most A13 tanks and their engines. The pump was replaced by the assembly shown to the right. The power traverse pump drive and the tachometer drive was retained as shown.

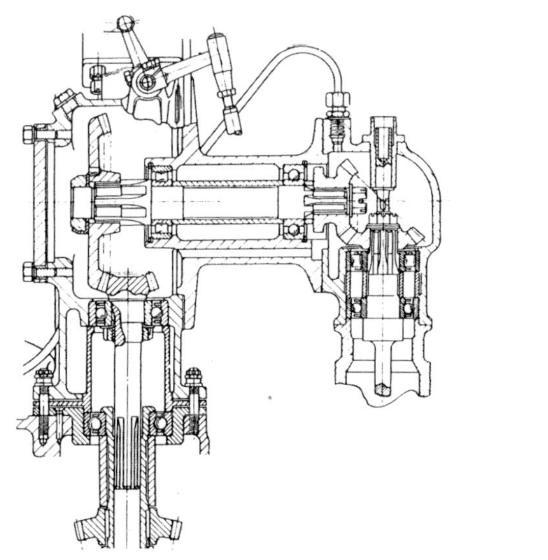

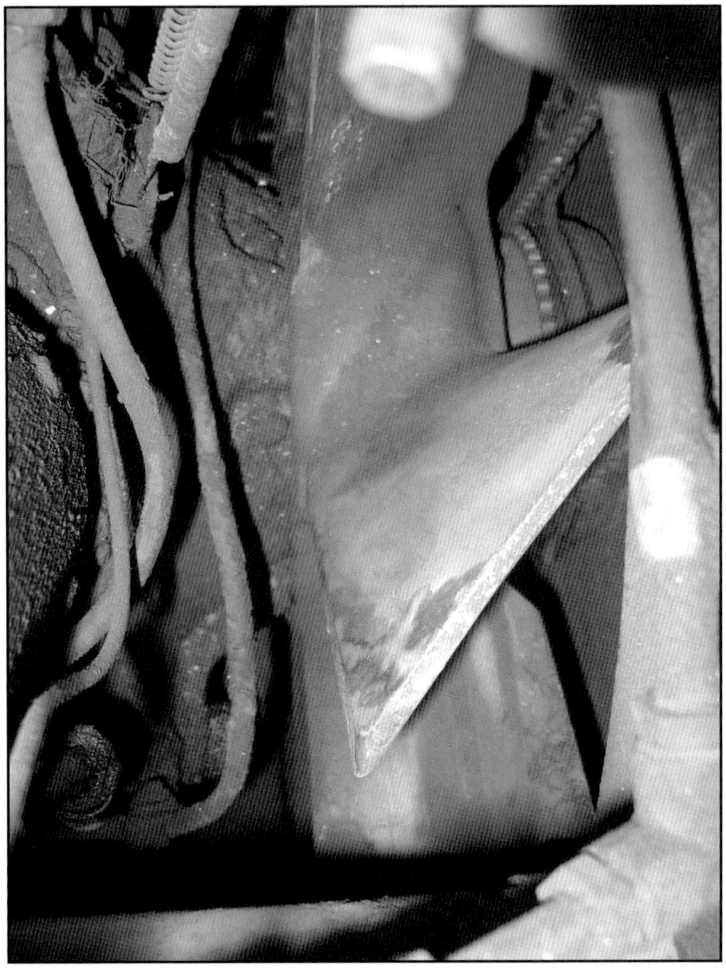

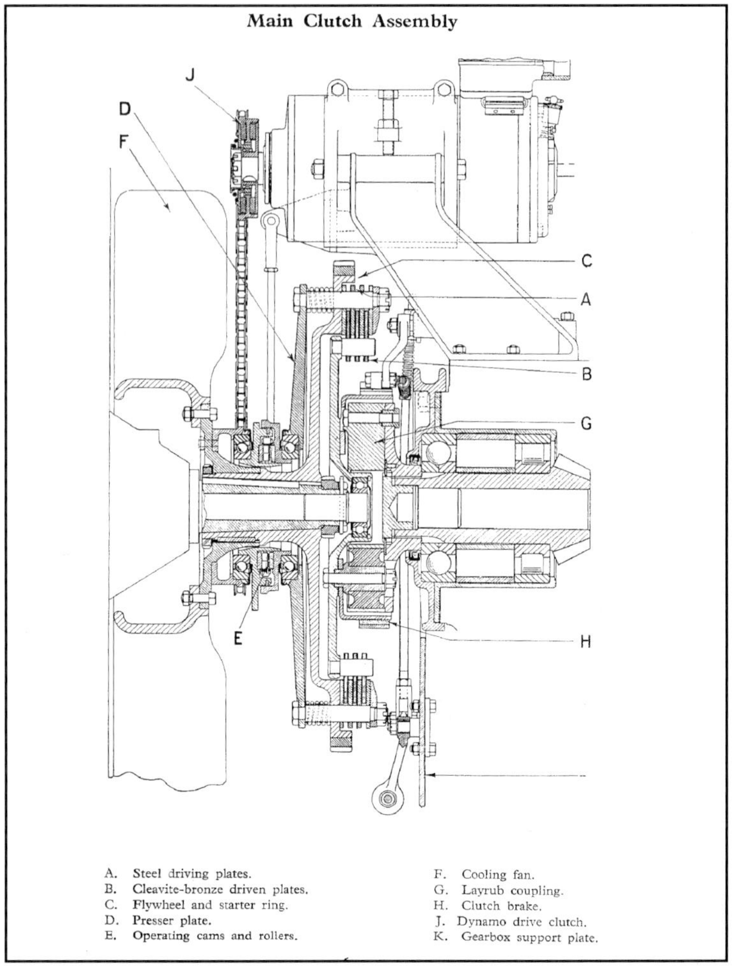

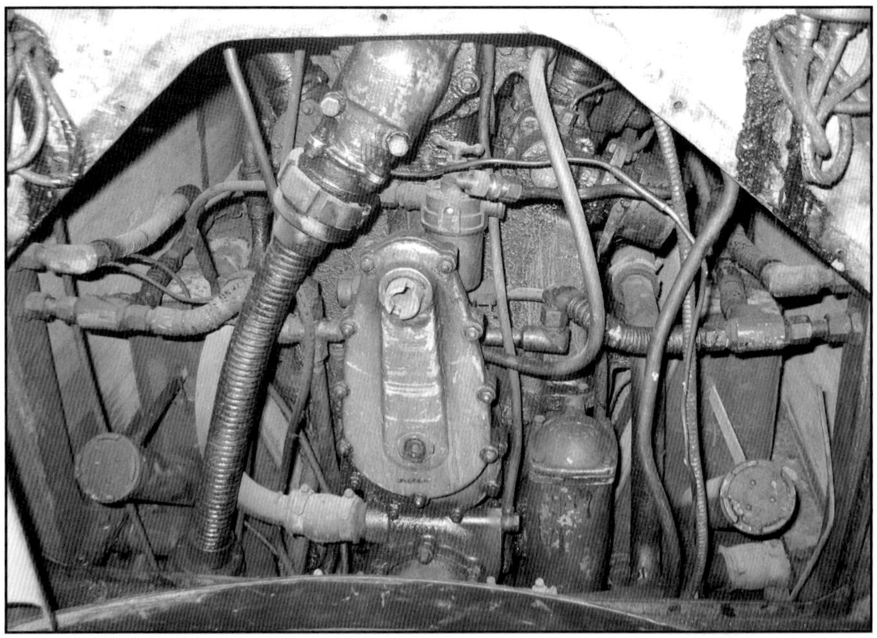

The above-right sectional drawing is from the February 1941 edition of the A13 Mk II instruction manual and shows the cooling fan, clutch drive and generator of the NL Mk II engine. The engine nose is to the left. The top-left picture shows a NL Mk II engine and fan as mounted in a A13 Mk II tank from about the same angle but slightly above. At bottom-right is a view of the same NL Mk II engine as seen through the open operators access hatch. A number of the accessories may be seen, including the hand starter which was provided only on NL Mk II engines. Below-left is a picture of the ID stamp of the engine pictured. The engine is serial N174 Nuffield Liberty (NL) Mark IIA. The broad arrow over T 233 is the inspector's stamp. (Photos and sectional drawing courtesy of Jamie Fairchild) (The tank photographed is in the collection of the Tank Museum at Bovington)

addition to the 12-volt electric one which engaged a ring gear on the flywheel. This was an air injection starter which utilized high pressure air injected into the right bank of cylinders to force the pistons down and rotate the engine to start it. This required injection ports in the cylinders that were not present in original Liberty cylinders. It was recorded in the MB minutes that the engine of A13.E3, which was a modified U.S. Liberty, was fitted with one of these starters on August 19, 1938, but after only a few miles of running the compressor casing fractured. It therefore seems possible that at least part of the later Mark I Liberty engines were also fitted with these starters when they were refined. This would, however, require replacement of the cylinders to which the air injectors were applied (the right bank of cylinders).

A13 Mk IIA - Cruiser Mk IVA. (Courtesy of the Tank Museum at Bovington)

Thus far, no picture or sectional drawing of a British Liberty Mk I have surfaced, although there are many references describing the various changes applied to the original U. S. Liberty engines which converted them to Liberty Mk I engines. From these descriptions, the resulting engine would be almost identical to the Nuffield Liberty Mk II as seen in the drawings reproduced on prior pages.

THE NUFFIELD LIBERTY Mk II

On pages 416 and 417 are sectional drawings of the 1918 U. S. Liberty and the Nuffield Liberty Mk II. You can easily see that the Mk II is little different from the original Liberty in basic design. Almost all changes involve accessories. The exception to that is the use of cast iron cylinders which, you will note, have air-starter injection ports on the inside of the combustion chamber. Also note that only the left bank of cylinders are connected to this starter system. All cylinders have the port, since only a single design of cylinder was made, but only the right bank is connected.

In the interest of showing the similarity, I have shown the cylinders that were designed in 1918 for tank engine use during WW I (but never produced). In addition to the use of cast iron cylinders, the Nuffield engine used Stellite-faced exhaust valve seats.

Intake manifolds were redesigned to move the location of the carburetors to the output shaft end of the engine for accessibility but remained of cast aluminum. Carburetors were twin Solex RZNHP-type with self starter, governor

A13 Mk II - Cruiser Mk IV. (Courtesy of the Tank Museum at Bovington)

A13 Mk II-IIA - Cruiser Mk IV-IVA General Data
(270 built - see Appendix A)

Length	**19' 9"**
Width	**8' 4"**
Height	**8' 6"**
Combat weight	**33,040 lbs.**
Engine	**Nuffield Liberty Mk II**
Power	**340 HP @ 1,500 RPM**
Power/weight ratio	**20.58 HP/ton**
Gear Box	**4 forward, 1 reverse**
Speed on road	**30 MPH**

Armament

2 - pdr. OQF
or 3.7" howitzer
.303 Vickers MG
or Basa MG
Main armor thickness 30 mm

and accelerator pump. The engine was governed to 340 HP @ 1,500 RPM. It was capable of about 410 HP @ 1,700 RPM.

In place of the generator of the original Liberty was a right angle drive for the air compressor which was mounted between the cylinder banks. This compressor maintained 620 lb. per sq. in. in a 750 cu. in. storage bottle for engine starting. The compressor crankshaft is driven on the engine end and its anti-engine end supplies a power takeoff to drive the tachometer and the power traverse pump.

The original Delco distributors were replaced with Lucas distributors with remote-mounted ignition coils. Ignition remained 12V battery type dual with two plugs per cylinder, K.L.G. RML/30 or RV7/2 types being used.

Exhaust manifolds were two-piece cast iron with sliding socket joints.

The engine had three available methods of starting: 1- Heywood high-pressure air injection starter type SH8.1/1; 2- 12V C. A. V. Bosch-type S583 mounted to the gearbox casing (this starter had a Bendix-type gear and starter pinion which engaged the flywheel ring gear when activated by remote controlled solenoid switch); 3- a hand-crank on the rear of the engine of similar design to the Navy hand starter supplied on the Christie A.13.E1 prototype. The Nuffield Mk II was the only British Liberty supplied with a hand-crank starter.

Cooling was accomplished by two large capacity gilled-tube radiators housed almost vertically, one on each side of the engine compartment. The cooling system held 22 gallons. An eight-bladed fan was bolted to the engine flywheel hub drawing air in through gratings over the radiators, and then through the radiator gills, and exhausted it through the transmission compartment and out the rear of the hull. This system was very similar to the design of the Christie tank, including the eight-blade fan mounted on the flywheel and the use of two side-mounted radiators. **Approximately 325 Nuffield Liberty Mk II engines were built.**

HISTORY OF THE A13 TANK SERIES
A13 Mk I - CRUISER Mk III

The first order for 65 A13 Mk I - Cruiser Mk III tanks was given to Mechanization and Aero on January 22, 1938. The first pilot model, A13.E2, had been completed in very early October and the second pilot, A13.E3, was under construction. The first tanks from this new order were completed in April and the last in December. Due to numerous problems which had to be resolved with the design of brakes, steering, transmission and cooling, deliveries were held up. By summer of 1938, Col. Martel suggested that production be halted until July of 1939 to allow all pending problems to be soundly resolved. His request was denied by the Director General of Munitions Production and production proceeded. The first production tank, T4385, was delivered on December 30, 1938 powered by Liberty Mk I engine No. K8. The remainder were delivered before summer of 1939.

Thus, from the start of design in December of 1936 until delivery of the first production model was a remarkably short two years. This model saw combat service in the 1st Armored Division in France in 1940 and in the 7th Armored Division in Libya in 1940-41.

The A16.E1 prototype near completion. This proposed Battle Cruiser used 30 mm armor, the suspension and propelling machinery of the A13 and the fighting arrangements of the A14. The hull was lengthened by about 1 ft. and an additional pair of bogie wheels added. (Courtesy of the Tank Museum at Bovington)

A13 Mk II - CRUISER Mk IV

In early 1939 the requirement was laid out for 30 mm armor and one of the A13 pilot models was tried with 20-30 mm armor to see what effect the 1,000 lb. increase in weight would have on performance. The result was found to be minimal and an up-armored version of the A13 was ordered into production as the A13 Mk II - Cruiser Mk IV.

Even though Nuffields was the parent company and designer of the A13 tank, the first contract for this new version was given to London Midland & Scottish Railway (LMS). It was awarded on January 26, 1939 for 65 tanks but later reduced to 30. At the same time this order was given, so was one for 65 Nuffield Liberty Mk II engines to Nuffield Mechanization and Aero (M&A).

In February of 1939 M&A was also contracted to build the A13 Mk II - Cruiser Mk IV, their first contract being for 95. Part way through this contract the A13 Mk IIA - Cruiser Mk IVA was introduced, the primary difference between it and its predecessor being the use of a Besa MG in place of the Vickers and a new main gun mantlet. 31 of the 95 tanks of this contract were completed as A models.

Three more Nuffield contracts were let for A13 Mk II - Cruiser Mk IV tanks for a total of 145. (See Appendix A for contract details.) Of these, 30 were A models. Resulting production was therefore 209 A13 Mk II - Cruiser Mk IV and 61 A13 Mk IIA - Cruiser Mk IVA. The last tank of this model was completed on January 10, 1940. Cruiser Mk IVs were used in France in 1940 by the 1st Armored Division and by the 7th Armored Division in the Western Desert in 1940-41.

THE BATTLE CRUISER

Col. Martel and Nuffields had no more than gotten started on the A13 series designs when Martel also expressed the need for a fast heavy cruiser tank as well. Specifications and scale models were made up for two different plans. The first was designated A14, designed by the Chief Superintendent of Tank Design and a single pilot model contracted to LMS. This tank was powered by a V 12 Thornycroft engine. The pilot, A14.E1, was delivered and put on trials in early 1939.

A16.E1 Experimental Battle Cruiser General Data (One built)

Length	20' 6"
Width	9' 2"
Height	8' 11'
Combat weight	45,000 lbs.
Engine	Nuffield Liberty Mk II
Power	340 HP @ 1,500 RPM
Power/weight ratio	15.11 HP/ton
Gear Box	4 forward, 1 reverse
Speed on road	30 MPH

Armament

2 - pdr. & 7.92 mm MG in main turret
2 aux turrets with 7.92 mm MG
Main armor thickness 30 mm

THE A16 BATTLE CRUISER

The second was designated A16 and assigned to Nuffield. Both of these tanks had 30 mm armor. A16 was in fact, a heavy version of the A13. Design work had been started on the A16.E1 (pilot) on November 27, 1937, just after A13.E2 had started trials. The pilot was ordered from Nuffield on February 27, 1938. (A13.E3 had been delivered and was in trials. Nuffield had started production of the A13 Mk I but not yet completed a tank.) Progress on the A16.E1 was slow and on June 23, 1938 Col. Martel reported that Nuffield had informed him that Nuffield had promised completion by the end of 1938, but owing to other commitments, any production could not start before the end of 1940.

On July 13, 1938 the Director of Mechanization requested a second A16 pilot be ordered, noting the increased importance of this type in the future. Financial sanction was asked and received and A16.E2 was ordered on July 27, 1938.

A16.E1 was delivered early in 1939 (by which time they were close to completing their first A13 Mk II - Cruiser Mk IV). This tank was powered by a Nuffield Liberty Mk II. As originally conceived, its Liberty engine was still rated at 414 HP @ 1,900 RPM and the proposed final weight was 21 tons. This would have given a power-to-weight ratio of 19.7 HP/ton. As finally delivered, weight was up 1 1/2 tons and the engine had been down-rated to 340 HP @ 1,500 RPM. The resulting power-to-weigh ratio was 15.1 HP/ton.

Testing was carried out for at least a year and a half following delivery even though the design was abandoned much earlier in favor of the A15. The second pilot was never completed.

A13 Mk III - CRUISER Mk V

In the meantime the A14.E1 had been delivered by LMS and put on trials in early 1939. It was found to be excessively noisy and complicated and slower than the A13.E3 pilot equipped with the same 30 mm armor. It was therefore decided to abandon that design and have LMS instead work on an improved version of the A13 which was to use as many A13 parts as possible, have 30 mm armor, and have a lower profile. A wooden mock-up was approved at the end of April of 1939. LMS was at this point just getting into A13 Mk II production on their contract for 65 tanks.

The lower profile required a lower engine and an enlarged version of the Meadows Flat-12 used in the A17 Tetrarch was used. This new tank became the A13 Mk III - Cruiser Mk V, was put into production in mid-1940, and was officially named the "Convenanter." You can see the

A15 Mk I - Crusader Mk I - Cruiser Mk VI. Note the auxiliary MG turret. (Courtesy of the Tank Museum at Bovington)

effective switchover by observing the dates of completion by LMS of tanks of contract T6551 in Appendix A, the last of which were in January of 1940.

Total production of the A13 Mk III was 1,771 but it was continually plagued with mechanical and cooling problems which were never totally resolved. As a result, this model tank was never used in combat but was relegated to training rolls for the remainder of its useful life.

THE A15

The designation A15 had been assigned as a "1938 Class Medium" tank but its design had been delayed in the planning stage due to repeated changes in requirements. The completion of the A14.E1 and the A16.E1 along with the progress made in testing and production of the A13 series of tanks brought with it the need to review alternate designs for a heavy cruiser.

Nuffield had been approached as the builder of the A13 Mk III but had instead suggested, as an alternative, redesigning the A13 Mk II as a medium cruiser. They felt that such a design would be much better suited to their production methods and easier and faster to get into production. The Liberty engine and most drive gear of the A13 Mk II was suggested rather than the proposed Meadows of the A13 Mk III since the Liberty was already in production. It was also proposed that the overall height of the Liberty could be reduced in order to be able to lower the profile of the tank and achieve an overall height close to that proposed for the A13 Mk III.

This proposal was accepted and the design was assigned the A15 designation. On June 27, 1939 a pilot and 100 A15 Mk I tanks were contracted with Nuffield's munitions branch, Mechanization & Aero. The order was then changed to 200 plus the pilot. The pilot was delivered on April 9,1940. Production deliveries began very early in 1941 and were completed by fall.

Nuffield became the parent company for a group of nine companies engaged in A15 production. Before the end of 1939 orders had been placed with two of these companies for an additional 200 tanks. By the end of 1941 total orders reached 5,000 for the various models produced. (See Appendix A for contract statistics.)

In mid-1940 the British Army began to give names to the various tanks. The A13 Mk III was named the "Covenanter" and the A15 series became the "Crusader."

A15 Mk I-II-III - Cruiser Mk VI General Data
(Named the Crusader in mid 1940)
A15 Mk I = Crusader Mk I = Cruiser Mk VI
A15 Mk II = Crusader Mk II = Cruiser Mk VIA
A15 Mk III = Crusader Mk III = no Cruiser Mark was assigned, logically it would be the Mk VIB

Length		**19' 8"**
Width		**8' 8"**
Height		**7' 4"**
Combat weight		
	Mk I&II	**42,560 lbs.**
	Mk III	**44,240 lbs.**
Engine		
	Mk I& II	**Nuffield Liberty Mk III**
	Mk III	**Nuffield Liberty Mk III, IIIA, IIIB or IV**
Power		**340 HP @ 1,500 RPM**
Power/weight ratio		
	Mk I&II	**16.00 HP/ton**
	Mk III	**15.37 HP/ton**
Gear box		**4 forward, 1 reverse**
Speed on road		**27 MPH**

Armament

Mk I	**2 pdr. plus 7.92 mm Besa MG Aux. turret 7.92 mm Besa MG 40 mm main armor**
Mk II	**2 pdr. plus 7.92 mm Besa MG 49 mm main armor**
Mk III	**6 pdr. plus 7.92 mm Besa MG 51 mm main armor**

THE NUFFIELD LIBERTY Mk III

As mentioned above, one of the requirements of the initial A15 design was a height reduction of the Nuffield Liberty engine in order to be able to keep the overall height of the tank within criteria. This was accomplished by a redesign of the oil pump which reduced its height and removed the filter mounted below it, relocation of the water pump and its bevel drive from below the crankshaft to above it, and raising the bottom level of the lower crankcase. The latter was aided by changing the location of the crankshaft main bearing oil feed lines from inside the bottom of the lower crankcase to external lines feeding from the side of the lower crankcase. The result was an engine 6 3/8" lower than the Mk II.

Basic specifications of bore, stroke and output remained the same as the NL Mk II. Power continued to be limited to 340 HP @ 1,500 RPM by use of a carburetor governor.

The only other significant known change was the use of a new design thrust bearing which required no periodic lubrication. For this reason the thrust bearing oil cup located above the output shaft was replaced with a plug.

The air compressor had been removed from the NL Mk II engine midway through its use because the air starting feature was abandoned. This pump was brought back into use on the NL Mk III, not for air starting but to supply high-pressure air for braking and turning control. It was mounted in the same position as before but minus the air start control and distributor on the engine output shaft end.

You will note that the side view sectional drawings show cylinders with no air injection ports yet the end view shows ports with plugs and no connections. I suspect that old cylinders with plugged ports were used until supplies were exhausted.

The A15 also used a new cooling fan system which substituted a pair of chain-driven 19" six-bladed fans for the eight-bladed fan on the flywheel of the A13. The chain was impregnated with grease before assembly and was supposed to require no maintenance, other than maintaining tension, for 1,000 miles. It was driven by a sprocket mounted on the engine output shaft.

This new system was the source of much trouble and would see several redesigns over the life of the A15 series tank. Cooling problems and breakdowns of the fan drive were some of the most common complaints leveled at this tank series.

At least three different chain-drive designs were used before it was abandoned and a shaft drive system was substituted. The changes required for this new shaft-drive system created the new Mark IV Nuffield Liberty. In the meantime, the A15 had progressed through another Mark. The NL Mk IV would first see use in the Crusader III. **Approximately 3,123 Nuffield Liberty Mk III engines were built.**

SECRET

MARK	ENGINE	TRACKS	HULL	TURRET	ARMAMENT	REMARKS.
I	LIBERTY MK. III	10·7"	RIVETED SKIN OUTER ARMOUR BOLTED	WELDED SKIN OUTER ARMOUR BOLTED.	2 PDR. & 1-7·92 M/M. BESA M.G.	WHEN FITTED WITH AUX. TURRET CRUSADERS I & II
II	LIBERTY MK. III.	10·7"	RIVETED SKIN OUTER ARMOUR BOLTED.	WELDED SKIN OUTER ARMOUR BOLTED.	2 PDR. & 1-7·92 M/M BESA. M.G.	CARRY ONE EXTRA 7·92 BESA M.G.
III	LIBERTY MK. III. IIIA, IIIB OR IV	10·7"	RIVETED SKIN OUTER ARMOUR BOLTED.	WELDED SKIN OUTER ARMOUR BOLTED.	6 PDR & 1-7·92 M/M BESA M.G.	
III A.A. I	LIBERTY MK. III. IIIA, IIIB OR IV.	10·7"	RIVETED SKIN OUTER ARMOUR BOLTED	SINGLE SKIN WELDED	40 M/M. BOFORS.	
III A.A. II	LIBERTY MK. III IIIA, IIIB OR IV.	10·7"	RIVETED SKIN OUTER ARMOUR BOLTED.	SINGLE SKIN WELDED.	2-20 M/M OËRLIKON & 1-VICKERS G.O. ·303.	
III A.A. III	LIBERTY MK. III IIIA, IIIB OR IV.	10·7"	RIVETED SKIN OUTER ARMOUR BOLTED.	SINGLE SKIN WELDED	2-20 M/M OËRLIKON.	EXTENDED TURRET, RE-ARRANGED STOWAGE & NEW SIGHTING ARRGTS.
III. GUNTRACTOR I.	LIBERTY MK. III, IIIA, IIIB OR IV.	10·7"	RIVETED SKIN. OUTER ARMOUR BOLTED WELDED SUPER STRUCTURE	NONE.	NONE	

DRAWN BY TRACED BY. J.H. CHECKED BY E.L.W. FOR D.T.D.	A.F.V. NOMENCLATURE. CRUSADER SERIES	MK III GUN TRACTOR I." ADDED. 9-2-44	T.D. 14745. 1 SHEET SHEET 1. ORIGINAL DATE 20-9-43.

(Courtesy of the Tank Museum at Bovington)

MORRIS MOTORS ENGINES BRANCH BECOMES THE SECOND NUFFIELD LIBERTY PRODUCER

It soon became evident that Liberty production capabilities would have to be increased beyond M&A capabilities. To that end Nuffield committed Morris Motors Engines Branch to Liberty production. Just when Nuffield began to set Morris up for production, we do not know. Since it was one of the Nuffield Group of companies, that could have happened at most any time. The first contract specifically assigned to Morris was dated August 15, 1941 and was T 5179 for 1,250 Mark III engines.

The Morris plant at Courthouse Green, Coventry was assigned Nuffield Liberty production (and, as we will see later, eventually Meteor tank engine). Before the adoption of the Meteor as the principal tank engine, Morris had

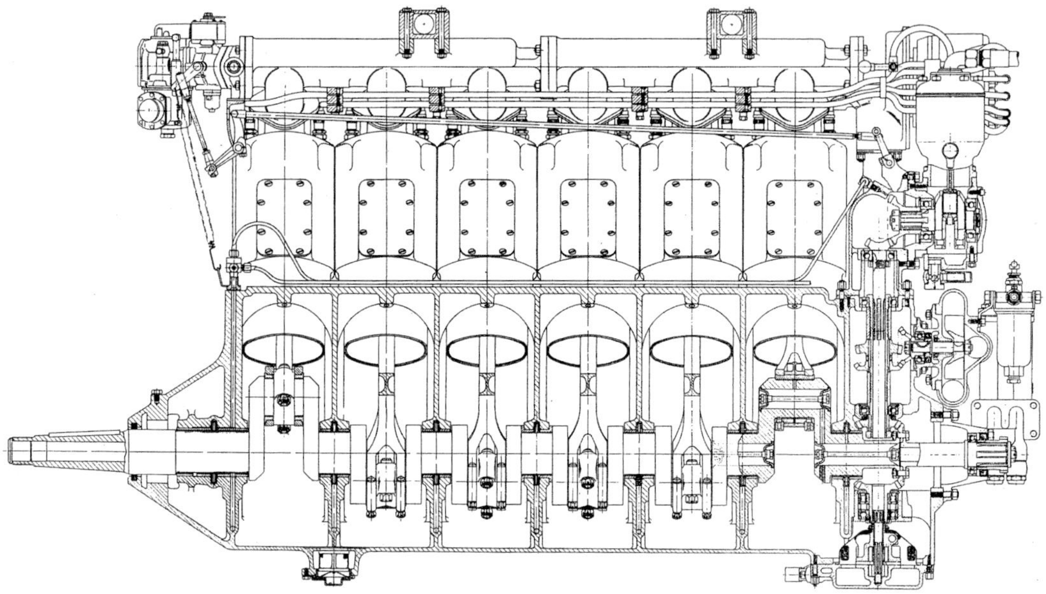

Nuffield Liberty Mk III

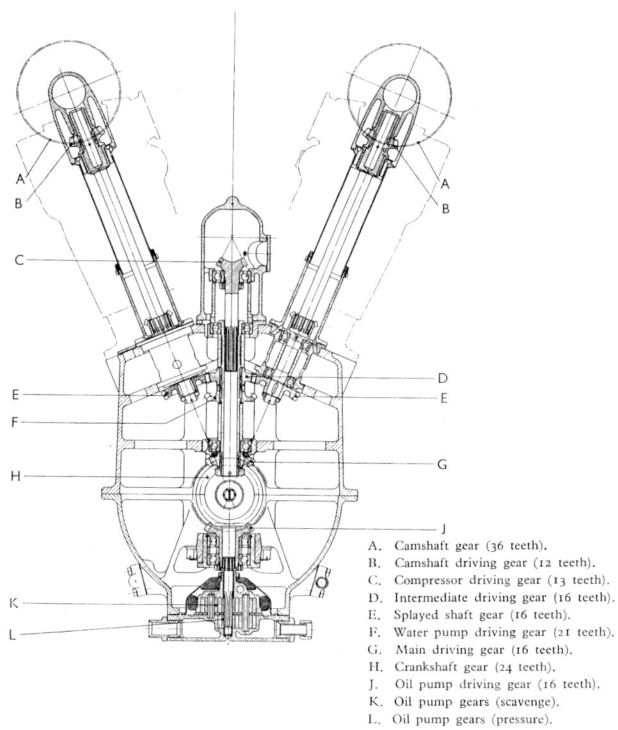

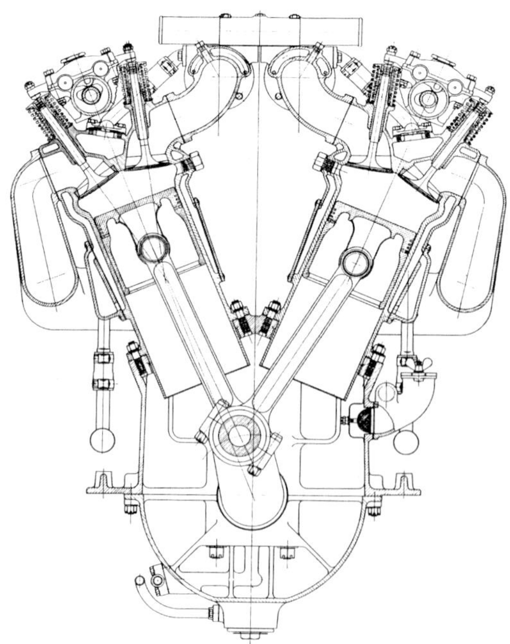

(Drawings courtesy of the Tank Museum at Bovington)

received contracts for a total of 6,227 Nuffield Liberty engines, all of which were for Mark III models.

The Nuffield Liberty would evolve through two more model designs (Marks) before production ceased and contracts which originally specified Mark III engines were altered to later marks. Morris thus, no doubt, built Mark III, IV and V models.

A15 CRUSADER MODELS

The A15 was built in three basic models and became the principal British tank from the spring of 1941. In May of 1941 the 6th Royal Tank Regiment in the Middle East was supplied with Crusader Mk I tanks and they saw combat service in Operation Battleaxe in the relief of Tobruk in June of 1941.

This and the following Crusader II and III were used in North Africa campaigns through 1943, the Crusader II coming into field service early in 1942 and the Crusader III by autumn of that year.

The three basic models were modified for use in a number of roles which extended the life of the tank when it became obsolete as a combat tank late in the war. By this method and use as a training vehicle, the Crusader continued active service until the end of WW II.

Crusader I - Main turret with 2 pdr. gun and MG. aux. turret with MG. 40 mm armor.
Crusader I CS - Close support model with 3" howitzer replacing 2 pdr.
Crusader II - Same main guns as Crusader I but aux. turret eliminated and armor up to 49 mm.

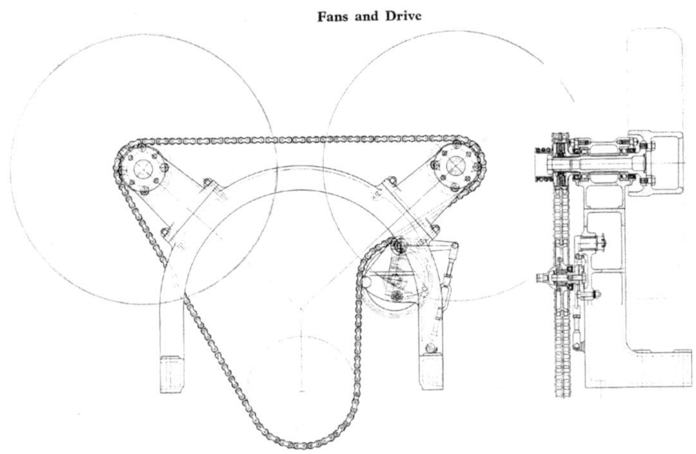

The first design chain drive cooling fan in which the chain was pre-lubricated. This system proved unreliable.

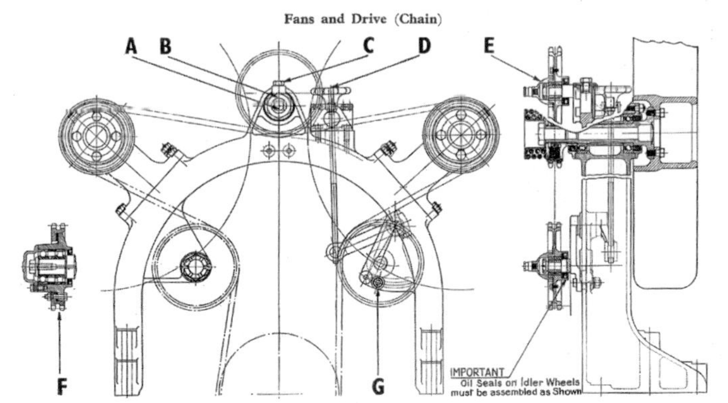

Late type chain drive cooling fan in which a second adjustable sprocket is added at the top. The idler sprocket on the left was added still later to give the same chain contact area to both fan sprockets.

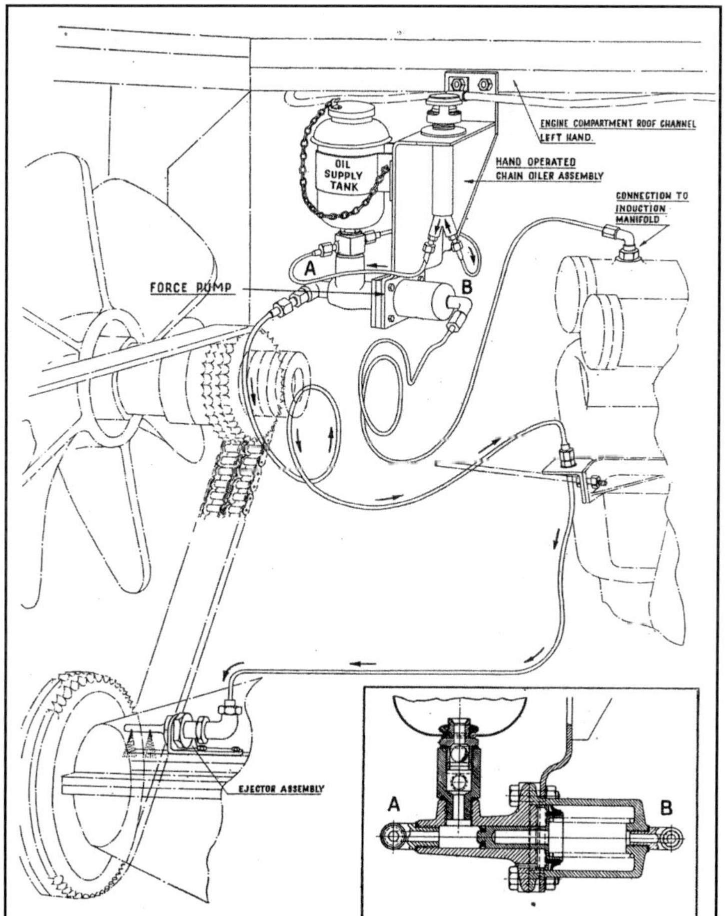

Chain lubrication system which was added when it was found that the prelubricated chains were not giving satisfactory service. The oil spray pump was operated by changes in engine intake manifold vacuum.

Above illustrations show the various designs of the chain-driven cooling system used on the Crusader tank series. (Courtesy of the Tank Museum at Bovington)

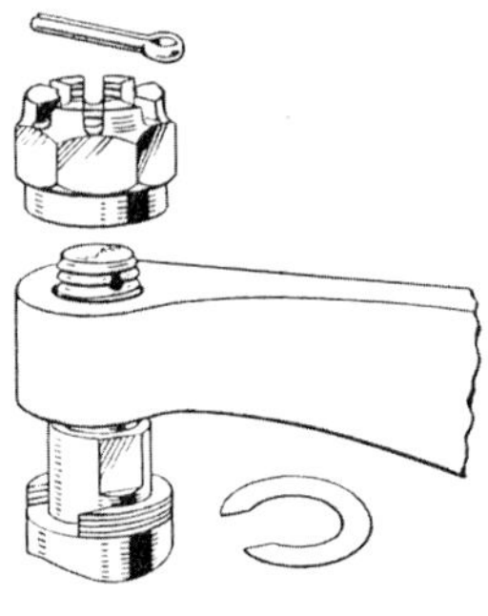

Original Nuffield Liberty valve adjustment design.

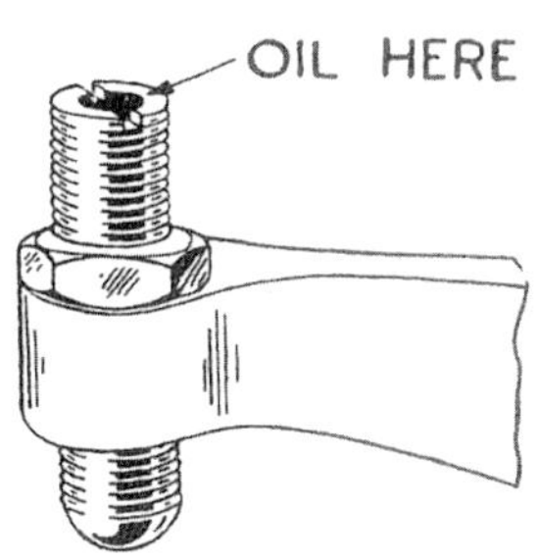

Improved adjustment design implemented during Mk III production.

Nuffield Liberty Mk III as viewed from the right-front. Insets at left show valve adjustment design changes. (Courtesy of Jamie Fairchild)

Removal of a Nuffield Mk III engine from a Crusader tank using a jib and block and tackle rigged on the turret. (Courtesy of Mike Kendall)

A crusader tank with its engine access panels removed for major maintenance revealing the output end of the Nuffield Mk II engine. Most of the chain-driven fan system has been removed but the main drive chain can be seen still looped over one of the intake headers. (Courtesy of Mike Kendall)

Crusader II CS - Same as Crusader I CS except for use of 49 mm armor.
Crusader III - Main turret gun replaced with 6 pdr and armor upgraded to 51 mm. Produced from May of 1942.
Crusader III AA Mk I, AA Mk II, AA Mk III - Crusader III tanks with the turret removed and replaced by a single Bofors 40 mm AA gun, a twin 20 mm Oerlikon AA gun and an AA Mk II with radio equipment relocated respectively. The AA versions were produced for the invasion of NW Europe in 1944 to provide defense against aircraft but were withdrawn from use shortly after the invasion at Normandy in June of 1944 when Allied air superiority made enemy air attacks a rarity.
Crusader II Gun Tractor - Crusader II converted as a fast tractor for the 17 pdr anti-tank gun and crew. It was widely used during the NW Europe campaign during 1944-45.
Crusader ARV - Crusader of any model with turret removed and fitted as an armored recovery vehicle.
Crusader Dozer - Crusader of any model with turret removed and fitted with dozer blade for use by the Royal Engineers for construction work.
Crusader AMRA - Crusader equipped with anti-mine roller attachment.

Shaft type cooling fan drive used by the Crusader Mk III and on the Cavalier Mk I.

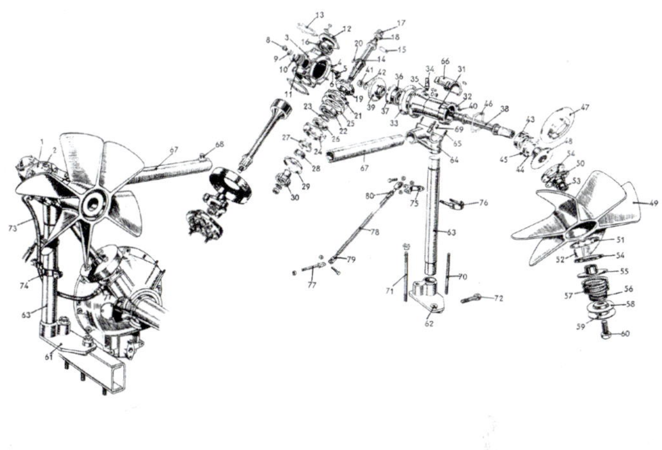

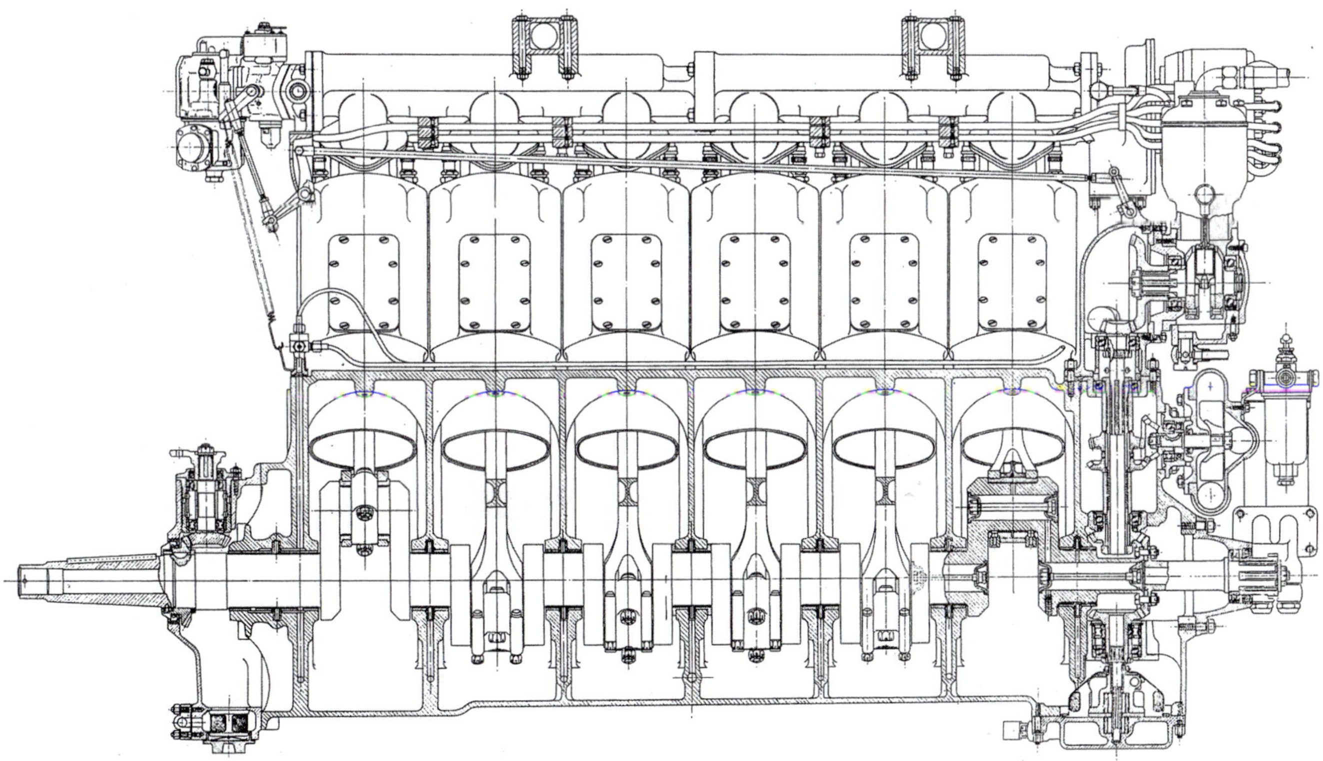

Nuffield Liberty Mk IV

(Drawings on this page courtesy of the Tank Museum at Bovington)

Production by basic model was as follows:
Crusader Mk I - 291
Crusader Mk II - 2,312
Crusader Mk III - 2,362
Crusader Bofors gun carrier - 35

NUFFIELD Mk III MODIFICATIONS (Mk IIIA and Mk IIIB)

The NL Mk III was recorded to have been built in three types, the Mk III, Mk IIIA and Mk IIIB. These changes must have been minor ones as the tank manuals do not make any specific mention of them. The fact that they existed has only been found in one place: a document titled A.F.V. NOMENCLATURE CRUSADER SERIES and dated February 9, 1944. Because it is the only source, and also contains other information about the Crusader tanks which may be of interest, it is reproduced on page 425.

At least five changes are currently known to have taken place during the production of the NL Mk III. One was the use of an improved valve adjustment mechanism. Until this change took place, the adjustment was exactly as it was on the original 1917 design.

In addition to the above, at some point during production the compression ratio was raised from 5.2:1 to 5.3:1. Rated power remained 340 HP @ 1,500 RPM with RPM controlled by a carburetor governor.

The third change was to enlarge the pressure oil cross-feed passage under the center main bearing which supplied oil from the filter to the main bearing oil gallery pipe on the opposite side of the lower crankcase. The increase in the size of this passage can be seen in the side sectional

U.S. Liberty

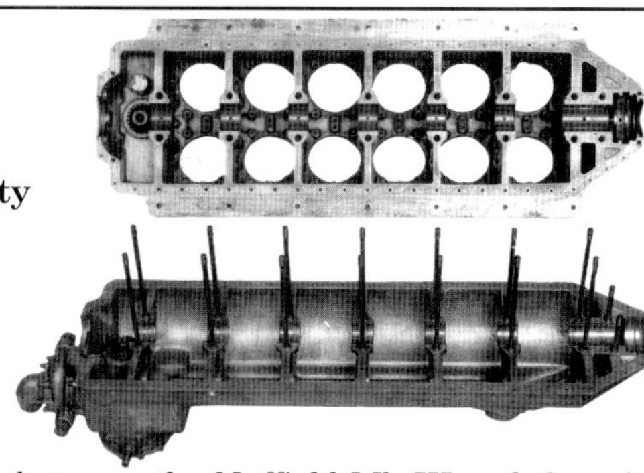

The similarity between the Nuffield Mk III and the original 1917 Liberty crankcase is easily seen. They were not, however, direct copies.

(Courtesy of the Tank Museum at Bovington)

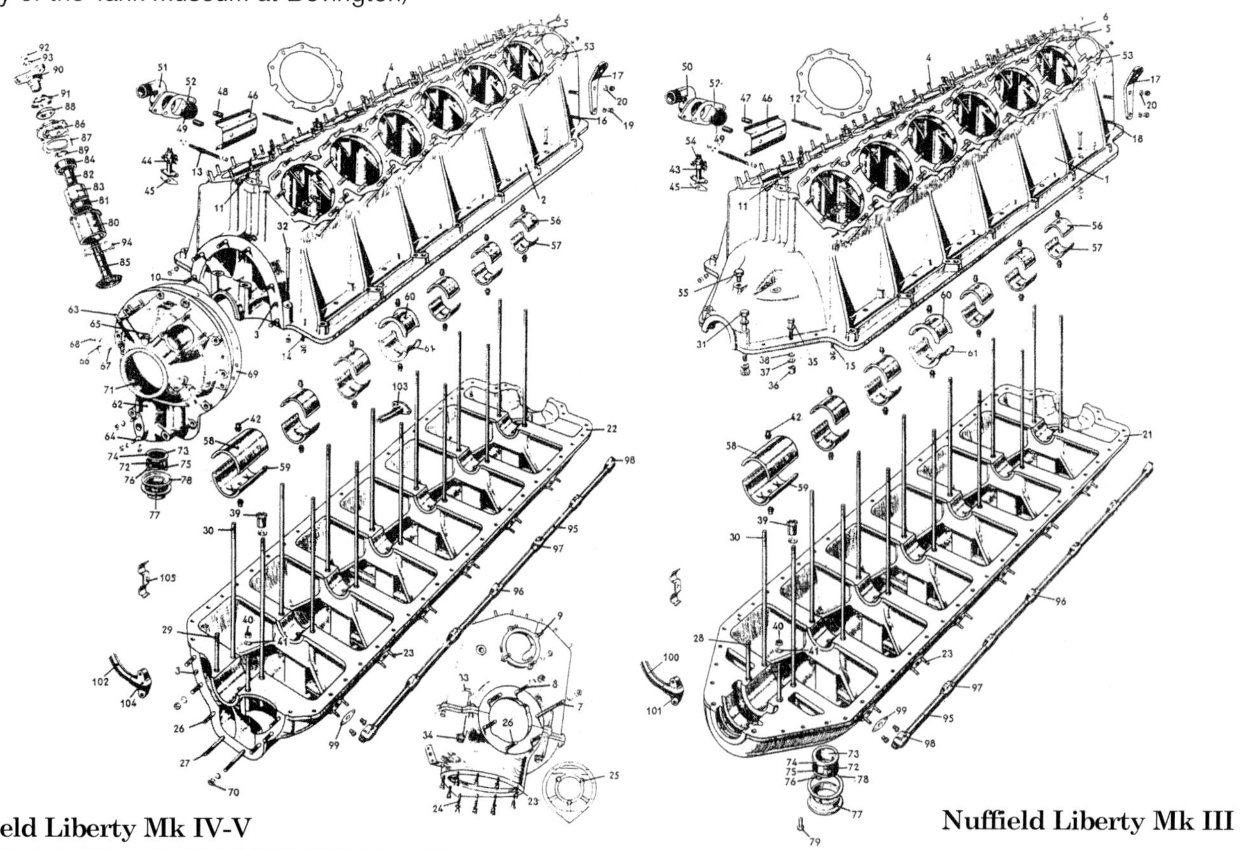

Nuffield Liberty Mk IV-V **Nuffield Liberty Mk III**

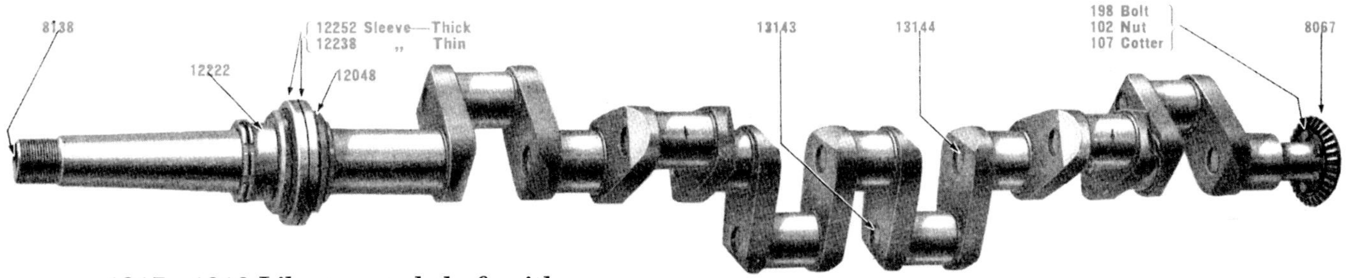

1917 - 1918 Liberty crankshaft with thrust bearing attached.

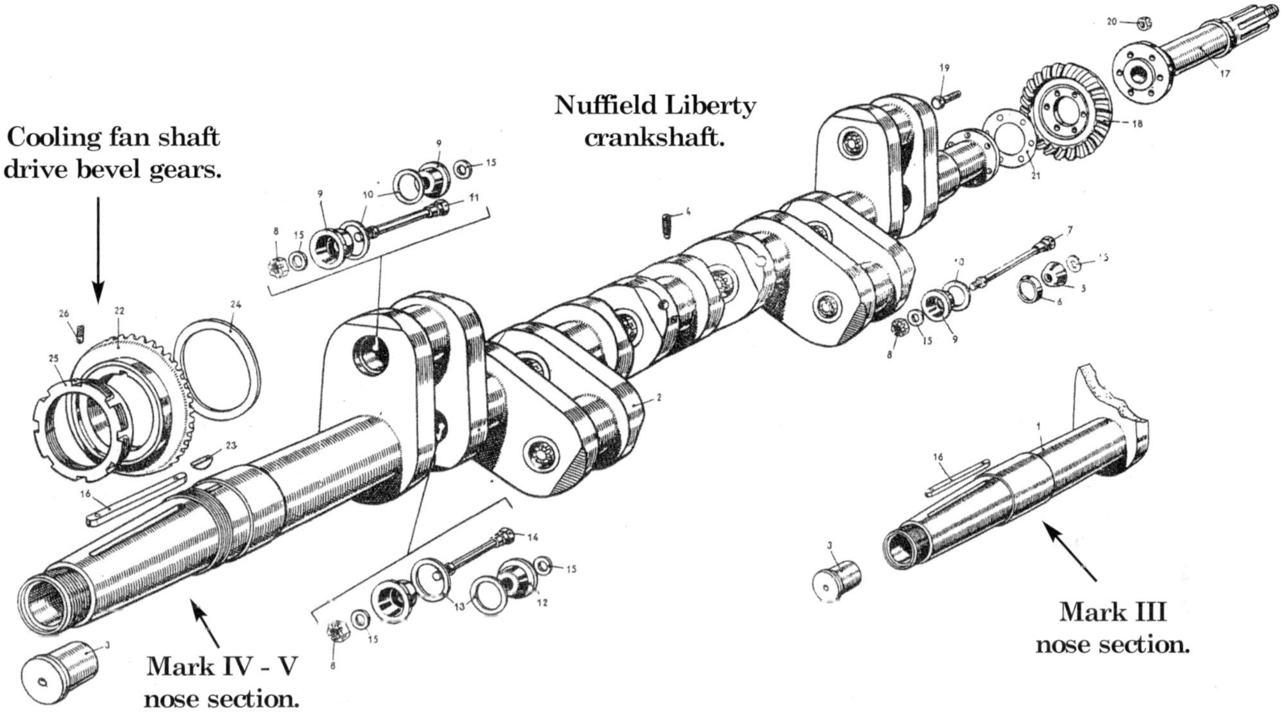

Nuffield Liberty crankshaft.

Cooling fan shaft drive bevel gears.

Mark IV - V nose section.

Mark III nose section.

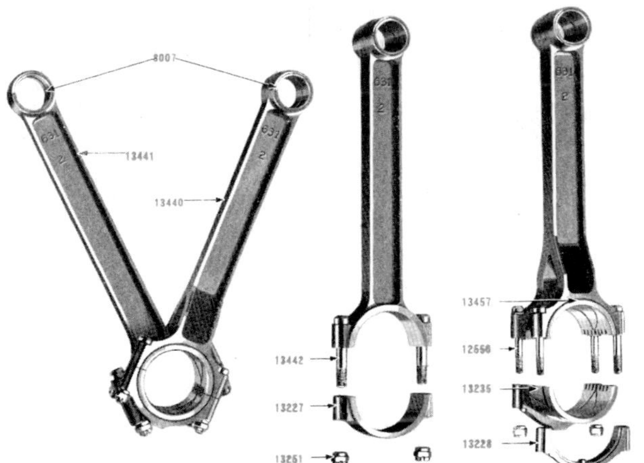

1917 - 1918 Liberty connecting rods.

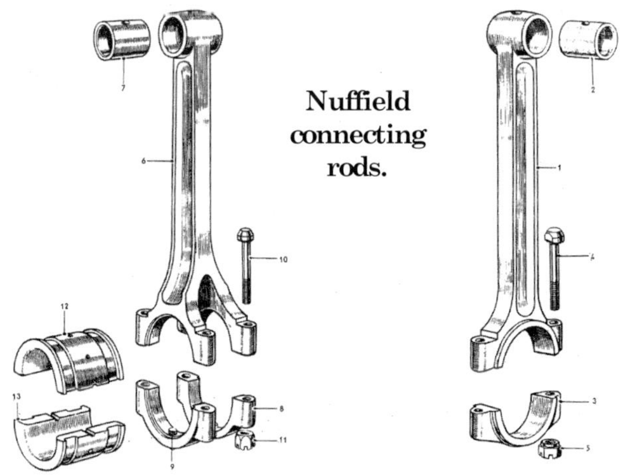

Nuffield connecting rods.

(From the author's collection and the Tank Museum at Bovington)

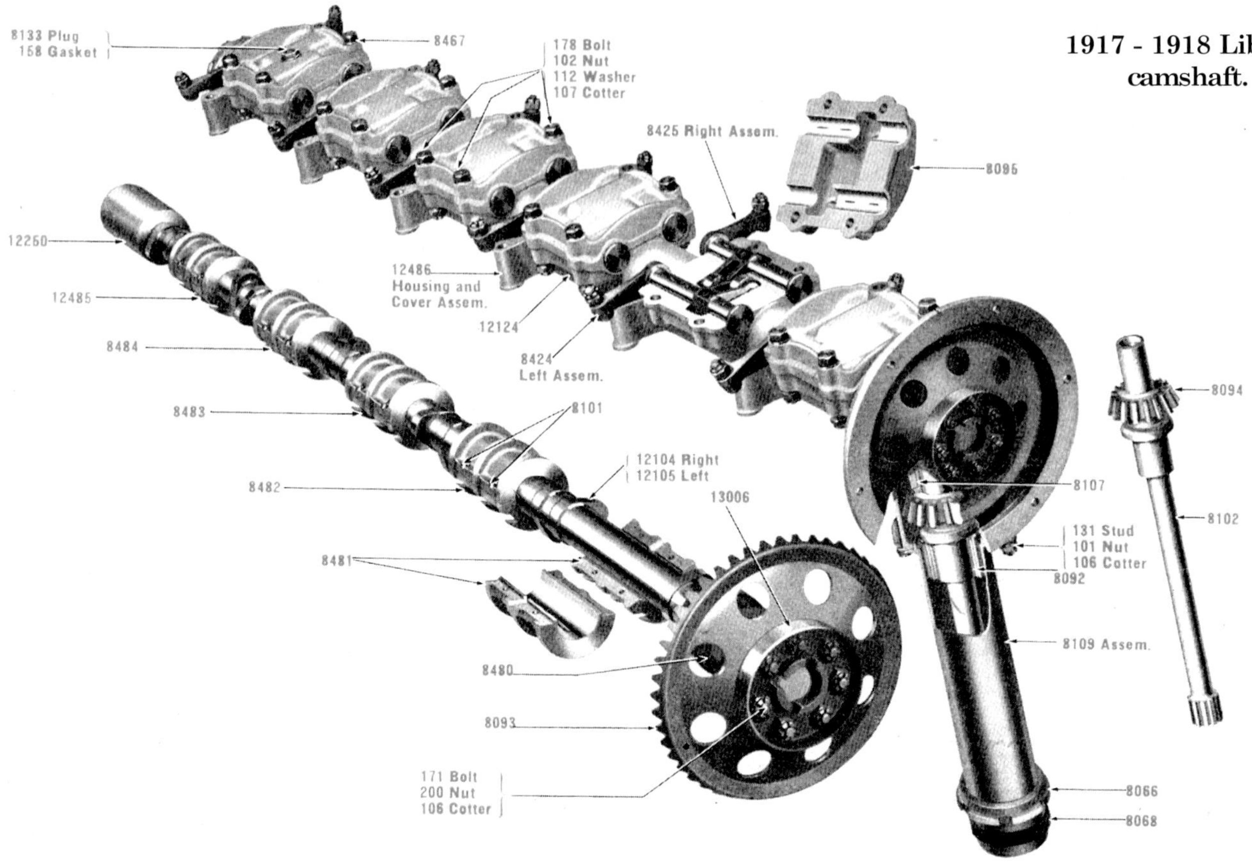

1917 - 1918 Liberty camshaft.

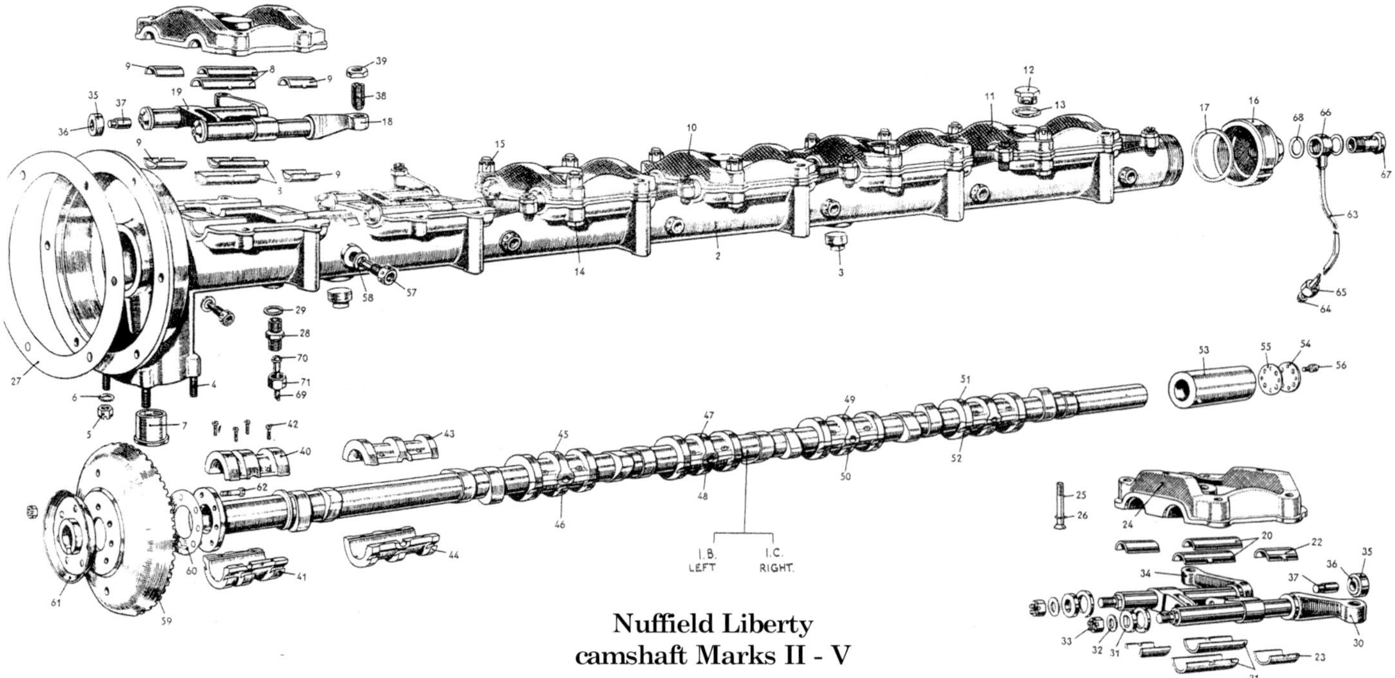

Nuffield Liberty camshaft Marks II - V

(From the author's collection and the Tank Museum at Bovington) drawings of the Mk IV as compared to the Mk III. The feature was carried over into the Mk IV.

Additionally, the design of the water pump was changed twice after repeated reports from North Africa of failures of this accessory.

I do not know whether these changes were the only ones applied after introduction of the Mk III, nor which change or combination of changes resulted in changes of Mark designations to IIIA or IIIB. All of these changes appear to have been applied to the Mk IV on introduction.

THE NUFFIELD LIBERTY Mk IV

The principal reason for this new engine design was to accommodate a new cooling fan drive system in an attempt to eliminate troublesome fan drive failures of the chain drive system. This system consisted of two angle drives driven from the output end of the crankshaft by bevel gears. This system required changes in the crankshaft and both crankcase halves. At the same time, the crankcase sump was moved to the front of the crankcase.

A slightly different air compressor pump was used on the Mk IV engine in that it was driven at half engine speed rather than at three-quarters engine speed as was that of the Mk III. The differences required by the change in bevel gear ratio can easily be seen in the sectional drawings of the two Marks.

Although most A. F. V. historians note that all Crusader III tanks were powered by Nuffield Mk IV engines we see, from the document reproduced on page 425, that this is not entirely accurate. An unknown number were powered by Nuffield Mk III, IIIA and IIIB model engines. I believe the number involved was relatively small and since we currently have no statistics in this regard, I will make no attempt to guess at that number.

Thus I will calculate the number of Nuffield Mk IV engines based on an assumption that all Crusader III tanks were powered by that model. There were 2,362 tanks of this model built and assuming the normal complement of spares of 20% (472 engines), then **approximately 2,834 Nuffield Liberty Mk IV engines were built.**

THE CRUISER Mk VII - A24 CAVALIER

Very early experience with the Crusader and Covenanter prototypes, along with experience gained in Libya and France, led the Ministry of Supply to request a heavier cruiser design. The object was to increase armor, provide a heavier gun, the 6 pounder (which was also eventually applied to the Crusader), more power and better reliability.

Nuffield Mechanizations and Aero proposed a design based on the Crusader which they already had in production. Front hull armor was 65 mm, turret armor was 75 mm. The turret ring diameter was increased to 60 inches to better handle the new 6 pdr. Battle weight was 59,360 pounds.

Three pilot models were ordered on January 31, 1941 to be delivered in the autumn. The designation A24 was assigned and the name Cromwell I.

At the same time Nuffield submitted this proposal, Leyland Motors submitted a plan to use a chassis similar to the Crusader but with a naturally aspirated version of the Rolls-Royce Merlin as power along with the new Merritt-Brown gearbox being used in the Churchill tank. Although this was well accepted, it was realized that the current very high demand for any Merlin production capacity would not immediately allow building a tank version of this engine (eventually to be known as the Meteor). It was decided, therefore, to build the Leyland tank initially with Liberty power with the ability to repower it with the Meteor when that engine became available.

The Leyland design was assigned the designation A27L and the name Cromwell II. The eventual Meteor-powered version would be the A27M and Cromwell III. The use of the Cromwell name would shortly change, but this was the initial plan.

With the pilot models of the A24 not yet delivered, an initial order for 500 production Cruiser Mk VII tanks, by now known as the Cavalier, were ordered from M&A on June 20, 1941. A block of 820 numbers was assigned to this order, T129620-T130119, but no additional tanks were ordered to use up the block.

THE NUFFIELD LIBERTY Mk IVA

The Nuffield Liberty was up-rated to 410 HP @ 1,700 RPM governed speed. Governing was still accomplished with a carburetor-type governor.

The need to increase the reliability of the Liberty engine as well as its power was recognized by all who were involved in the tank program. The only immediate accomplishment was to increase the power of the engine. Again, I point out that the Liberty had always been capable of producing in excess of 400 HP if allowed to operate at high enough RPM. Ratings were always kept moderate in earlier tank use in order to increase engine life and, in the lighter tanks, to keep vehicle speeds down and thus prevent excessive damage to both the vehicle and its occupants.

The overriding factor in this case was the need for more power. The A24 Cavalier was some 15,000 lbs. heavier than its predecessor, the A 15 Mk III, which was in turn 11,000 lbs. heavier than the A 13 Mk III. To be even more to the point, the first tank the British powered with the Liberty rated at 340 HP @ 1,500 RPM was the A 13 Mk I which weighed 31,360 lbs. The A 24 had a combat

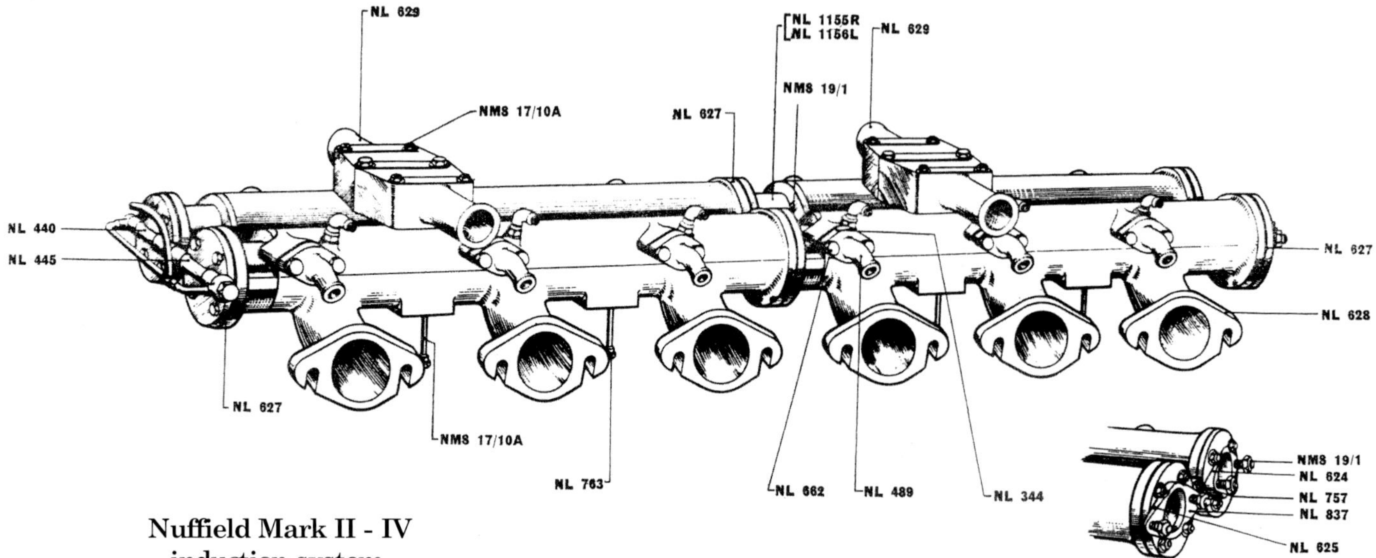

Nuffield Mark II - IV induction system.

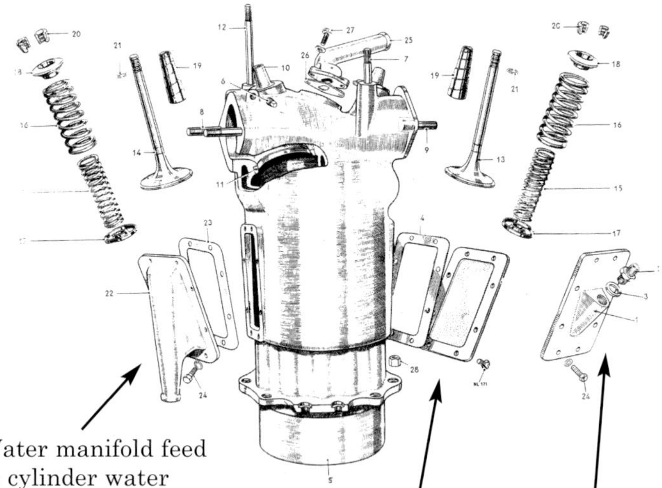

Water manifold feed to cylinder water jacket as used on Mk IV and V. Mark II and III used a slightly different design. (See engine illustration.)

Solid plate used to close the intake side water jacket access port on all marks.

Intake side water jacket access plate with hot water line to intake manifold found on cylinders 1 and 6 of both cylinder banks on Mk IVA and V only.

Nuffield cylinder as used on Mark III through V engines. Mk II was the same except for the addition of air-start ports.

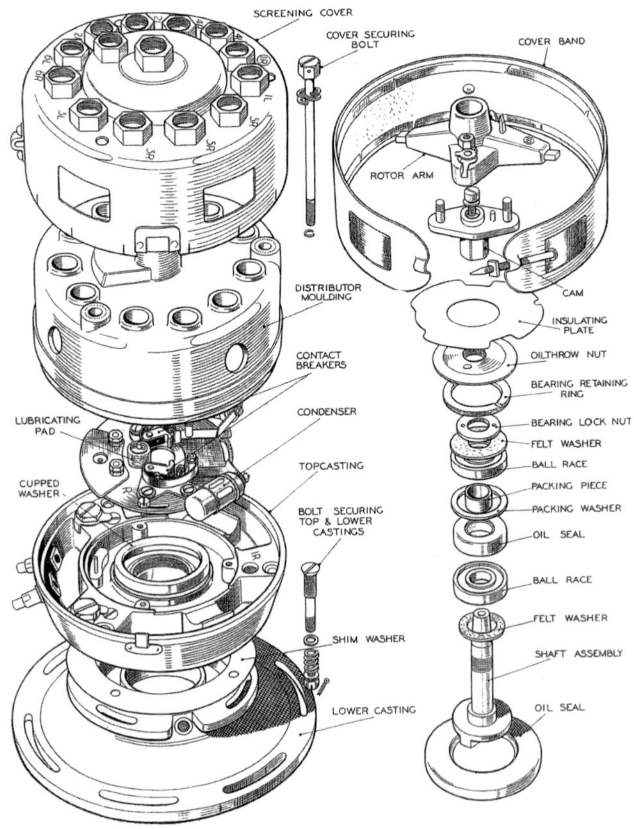

Nuffield Liberty distributor.

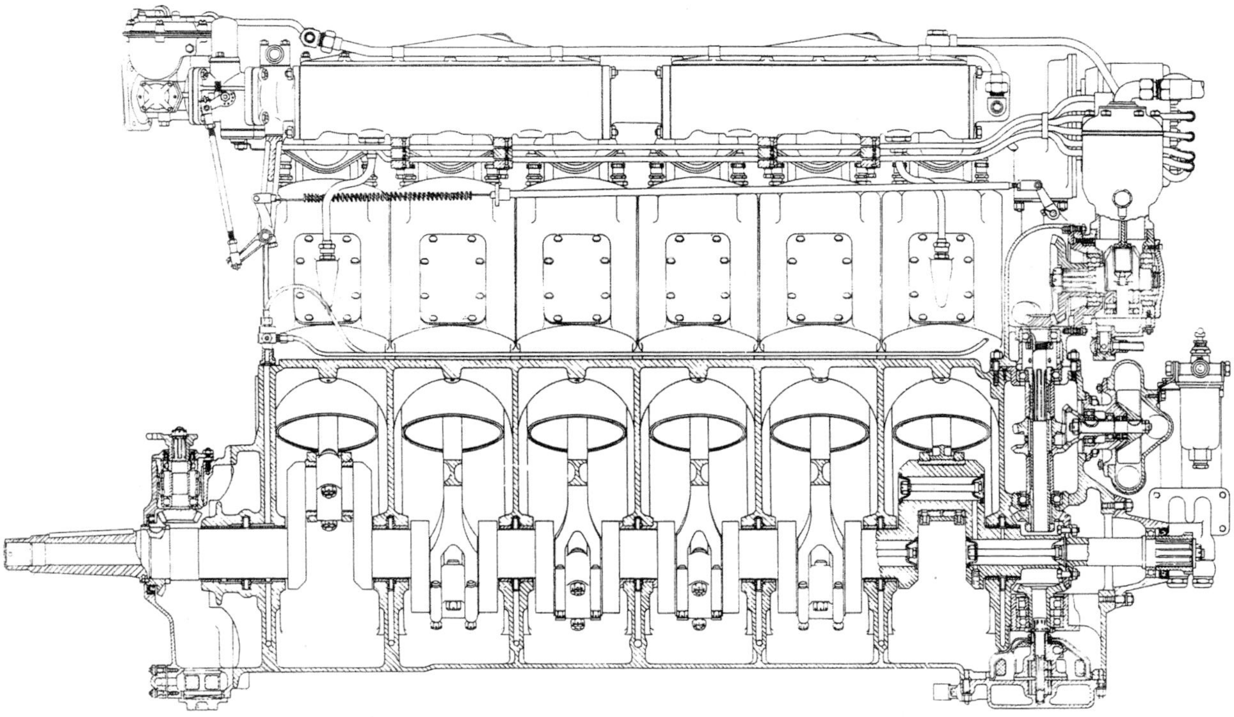

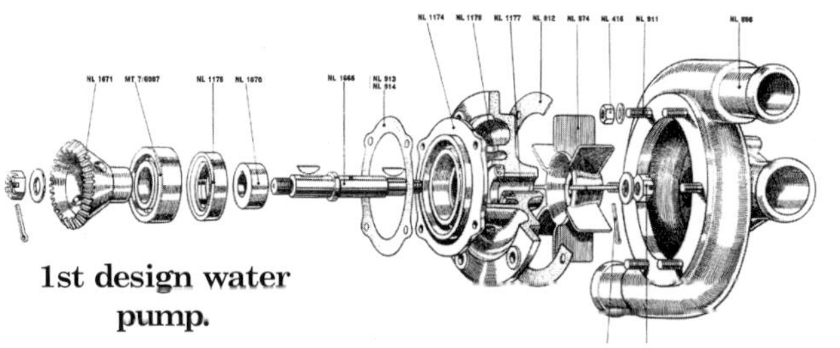

1st design water pump.

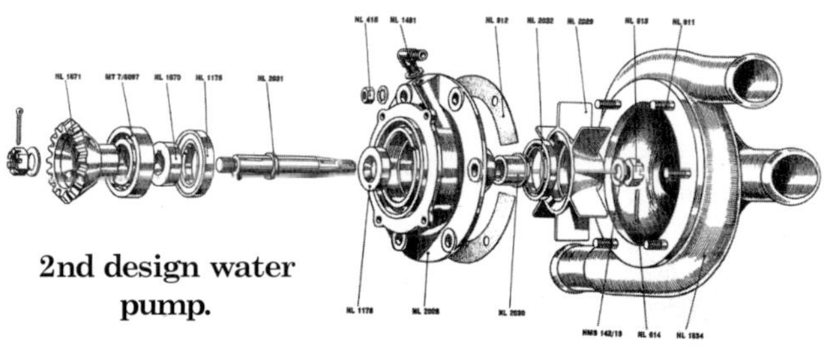

2nd design water pump.

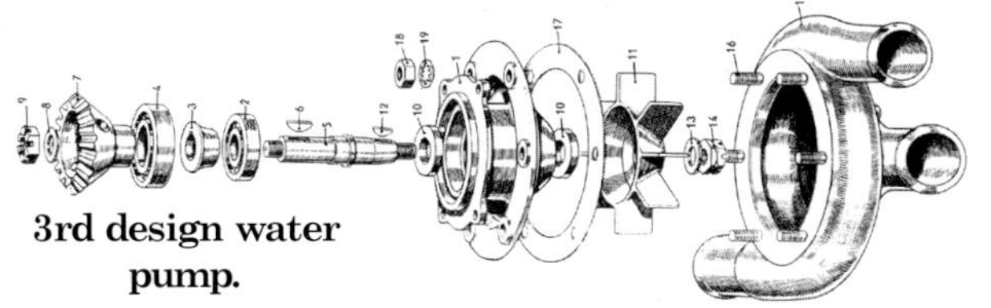

3rd design water pump.

The three designs of water pumps used in the Nuffield Liberty. All three are believed to have been initiated during Mark II production.

Nuffield Liberty Mark IVA

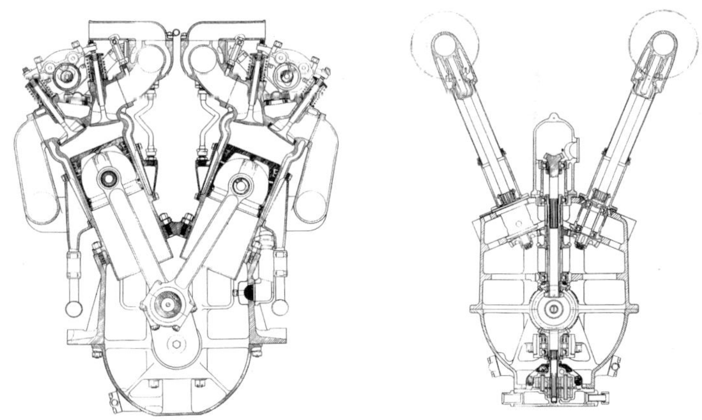

The primary difference between the Nuffield Mk IV and IVA was the use of a different design intake manifold and carburetor dictated by the requirements of the new tank in which it was used, the A24 Cruiser Mk VII, known as the Cavalier. In conjunction with the introduction of the Cavalier, the engine was up-rated to 410HP @ 1,700 RPM. Although a different model carburetor was used, the increase in power was obtained entirely by governing at the higher speed of 1,700 RPM. (All drawings on this and the prior page are courtesy of the Tank Museum at Bovington)

Nuffield Liberty Mark IVA. This model was the same as the Mark IV except for the use of a different intake manifold and carburetors and some minor differences in accessories. (Courtesy of the British Motor Industry Heritage Trust)

Nuffield Liberty Mark IVA. (Courtesy of the British Motor Industry Heritage Trust)

weight of 59,360 lbs. which was approaching double the weight of the first tanks.

The first approach to the problem was to adapt the Nuffield Liberty Mk IV to the new tank by modifying any accessories as required for the new environment and changing the governed speed of the engine upward in order to increase power output. This resulted in an interim model known as the Mk IVA which carried a rating of 410 HP @ 1,700 RPM. Initial production of the A24 Cavalier was fitted with this engine. Just how many tanks and engines this involved was not recorded.

THE NUFFIELD LIBERTY Mk V

Meanwhile the engine was subjected to somewhat of a redesign which resulted in the Mk V engine. Unfortunately not many details of the changes instigated have come to light. Other than the various changes to the accessories which can be seen by comparing the illustrations of the different models, the only obvious change was to the oil feed system to the lower-half of the crankcase and thus to the main bearing galleries.

In the Mk III and IV engines oil was fed from the pressure filter on the front of the engine to an external pipe

A24 Mk I - Cruiser Mk VII General Data
(503 built - see Appendix A)

Length	**20' 10"**
Width	**9' 5 1/2"**
Height	**8'**
Combat weight	**59,360 lbs.**
Engine	**Nuffield Liberty Mk IVA or V**
Power	**410 HP @ 1,700 RPM**
Power/weight ratio	**13.81 HP/ton**
Gear box	**4 forward, 1 reverse**
Speed on road	**24 MPH**

Armament

6 pdr. OQF
1 or 2 Besa 7.92 MG

Armor thickness
65 mm front hull
76 mm front turret
hull minimum 20 mm

An early Cavalier A 24 pictured at the Nuffield Mechanizations and Aero plant. (Courtesy of the Tank Museum at Bovington)

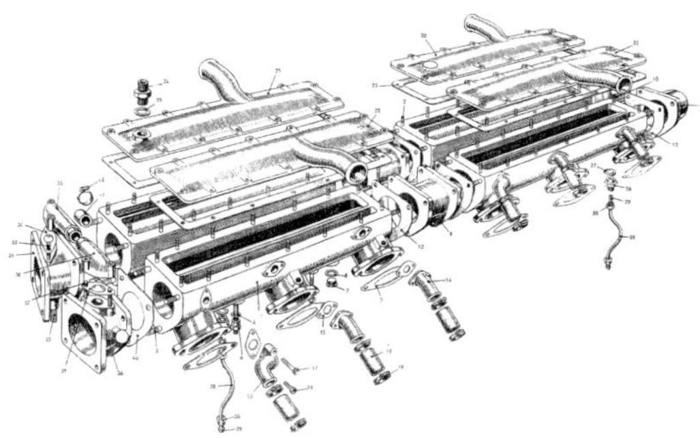

LEFT: Induction system as used on the Mk IVA and V engines powering the Cavalier tank.

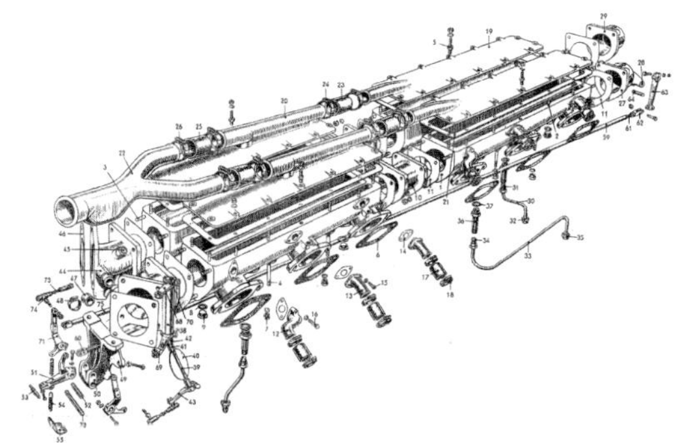

RIGHT: Induction system used on the MK V engines powering Centaur tanks.

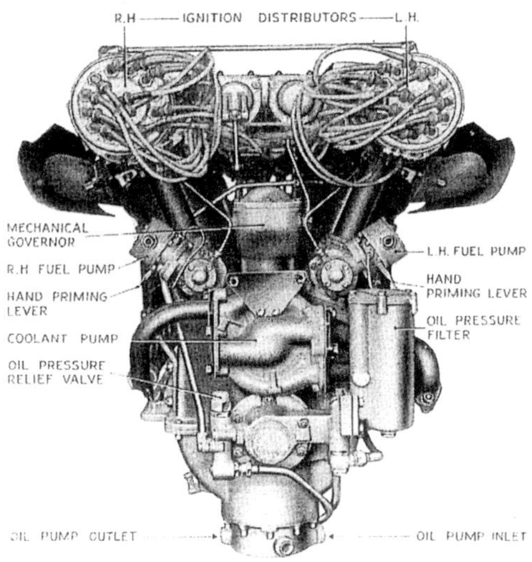

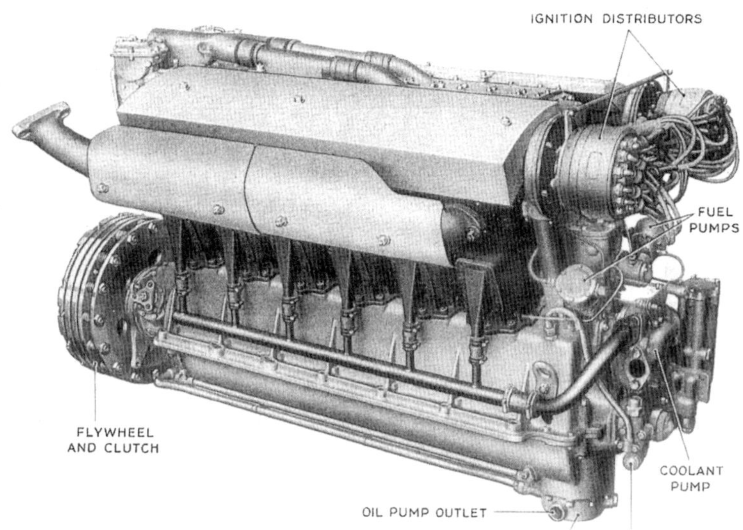

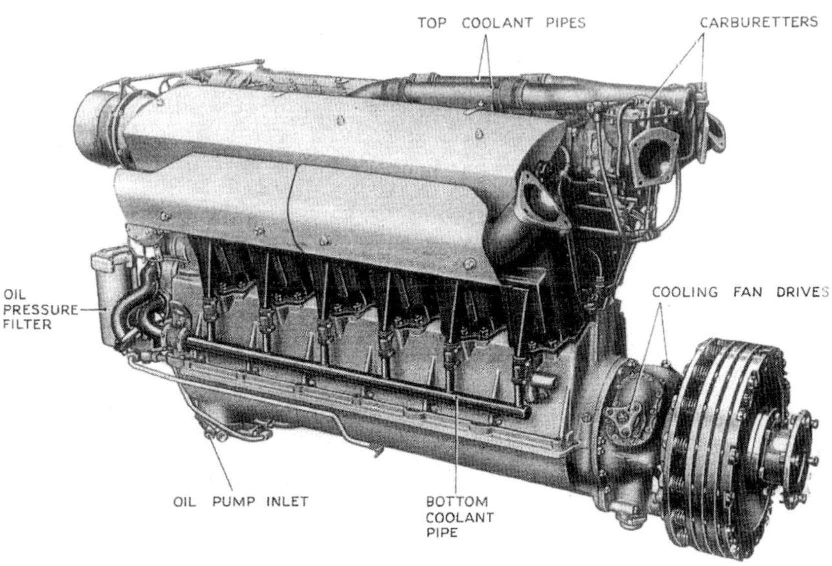

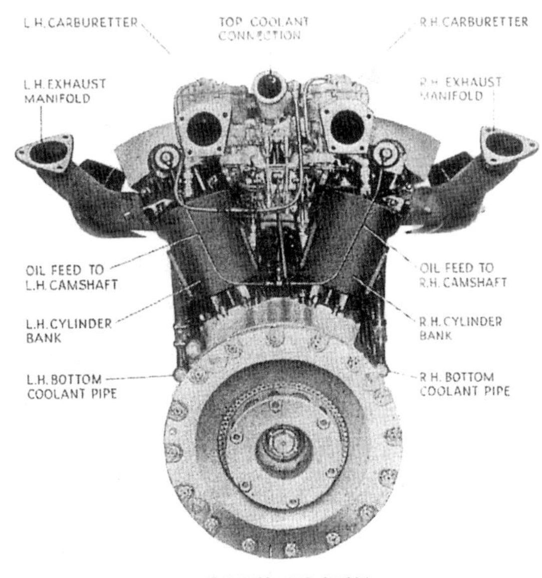

Above are pictured the early version of the Nuffield Liberty Mk V configured as used in the Centaur Mk I tank. Notice the external oil gallery pipe on the right side of the lower crankcase just above the feed line from the rear sump back to the oil pump (top-right illustration). (Courtesy of the Tank Museum at Bovington)

and thence down the left side of the lower crankcase to a horizontal oil passage drilled in the center crankcase web. This passage communicates the oil to an external gallery on the other side of the crankcase which in turn feeds oil to horizontal passages in the remaining crankcase webs. The external feed pipe and oil gallery are easily seen on the illustrations of the Mk III, IV and IVA engines.

On the Mk V engine the feed pipe from the pressure filter remained but the cross feed must terminate in an internal oil gallery similar to that used in the original Liberty as the external gallery on the opposite side now connects only the end main bearing journals, the center five being fed internally. Evidently the external gallery between the two largest main bearings was a precaution to insure adequate flow of oil. The last version of the Mk V engine also eliminated this final external oil gallery. Evidently they either increased flow capacity in the internal gallery or found the safety factor afforded by the external connection between the end mains was unnecessary.

The Mark V engine was the last version of the Nuffield Liberty that the British used in their tanks. There was yet another model tank beyond the A24 Cavalier which would be powered by the venerable Liberty and I will take up its history shortly.

A24 CAVALIER HISTORY

The Cavalier was considered an interim type while awaiting getting the Meteor into production. It was used heavily for training as gun tanks and in 1943 about half were converted to OP (artillery observation) tanks for use as observation for artillery officers. In this version it saw considerable use in NW Europe during the 1944-45 campaigns by artillery regiments of armored divisions.

A few were converted to ARV (armored recovery vehicles) configuration. First produced in 1942, the A24 saw use in one form or another until the end of the war.

THE HISTORY OF THE CHANGE FROM THE NUFFIELD LIBERTY TO THE ROLLS-ROYCE METEOR AS THE PRIME TANK ENGINE

Even though it will be necessarily brief, I believe it is in the interest of this history to include the story behind the switch from the Nuffield Liberty to the Rolls-Royce Meteor as prime tank power in the later part of World War II.

Rolls-Royce was, by 1940, intensely involved in the production of the Merlin aircraft engine for war use. About as much of the Rolls-Royce factory facilities as could be applied to that task were at work on it. Mr. W. A. Robotham, head of their chassis design and development division at Belper, was considering just what his facilities and people might be doing toward the war effort, since they were not suited to work on aero engines and might otherwise be disbanded.

The fact that an adaptation of the 1917 Liberty aircraft engine was in use as the principal tank engine, that it had some problems which needed solving and that before long tank development would find it wanting in power was no secret. Robotham felt that this was an area in which he, and perhaps Rolls-Royce, might help find some solutions. He then contacted H. Spurrier, a director of the Leyland Company. Spurrier was quite familiar with the tank situation because Leyland had been requested to contract the production of the A15 Cruiser Mk VI tank (which they did not).

The solution brought to mind was the adaptation of one of the Rolls-Royce aero engines to tank use, thereby obtaining a much more modern engine design with more reliability as well as much higher power potential. Adaptation of both the Kestrel and Merlin were considered. The Kestrel was felt to be wanting in ultimate power potential, although certainly more powerful than the Liberty. As to size, it would fit as a replacement.

The Merlin was felt to be better suited. It was more powerful and its size would also allow it to fit as a Liberty substitute. The fact that it was in heavy production and there might be a possibility of using substandard aero engine parts for tank engines probably was also considered.

This was in very late 1940. The Crusader was in production and use. The devastating evacuation of the British forces at Dunkirk had occurred and France had fallen to the Germans. The need for a better tank engine was already obvious. The German armored divisions had proved greatly superior to the British and heavier and more powerful British tanks would be required.

Specifications of a tank version of the Merlin were worked out and the name Meteor given it. Those specifications and their comparison to the Nuffield Liberty of the period (NL Mk III) are listed below. The Liberty would eventually be up-rated to 410 HP in later versions simply because the power was absolutely necessary. Doing so was at some sacrifice in reliability, however, as such output was pushing the engine to its full capabilities.

	Meteor	Liberty
Weight	1,450 lb	1,350 lb
Horsepower	550	340
Lbs/HP	2.64	3.98
Height	31"	36.4"
Length	49.5"	61.25"
Width	30.25"	28.0"

E. W. Hives of Rolls-Royce was in agreement with the proposition of developing the Meteor for tank use and intensive work was begun in this direction. Rolls-Royce was not interested in producing the engine themselves as they had all they could do to keep up with required Merlin production. Leyland was considered the potential producer. They had already been proposed by the Directorate of Tank Design as a possible producer of Liberty engines. They felt, however, that they could put the Meteor into production as rapidly as they could the Liberty.

In March of 1941 a meeting was held with the Directorate of Tank Design in which a number of objections to the proposition were raised. Robotham felt none of them were justified and managed to persuade the Director to allow a trial of a Crusader tank equipped with a Meteor engine.

Such a trial was undertaken in April and proved successful. On the 27th Robotham received a verbal intimation that the Meteor was to be put into production within some ten months and that the target output was 150 to 200 engines per month by the spring of 1942.

On May 16, 1941 Leland was given a firm contract for 1,200 Meteor engines. Design specifications called for the engine to be suitable for the Cruisers Mk VI (Crusader) and Mk VII (Cavalier), adaptable to three different transmissions and be installationally interchangeable with the Liberty. Leland was to proceed with tooling and expressed the expectation of being in production by February of 1942.

Little had transpired in that direction in the next month and Robotham expressed concern to Hives. It appears that Leyland felt, at least this is what Robotham thought, after a study of the drawings of the Meteor (provided by Rolls-Royce) that the production task might be far more difficult than they had first thought. In addition, tests of the prototype Meteor in the Mk VI tank had shown up cooling problems when using the radiators and cooling system of the Liberty.

Leyland then proposed the possibility of producing a 350 HP Diesel of their own design, an engine that had already been proposed and rejected. And later the possibility of producing both types was proposed.

By July 4, 1941 little had been done at Leyland toward the fulfillment of their Meteor contract. Rolls-Royce had made much progress in the design of the engine and had, by this time, completed three prototypes. Tests of these engines in tanks had continued to reveal cooling difficulties. Leyland seemed to consider them almost insurmountable whereas Rolls-Royce had no such convictions. By mid-July it had become clear to many that Leyland now had no intentions of producing the Meteor.

On July 21 Sidgraves suggested to the Chairman of the Supply Council that the Ministry of Supply should order its engines through the Ministry of Aircraft Production (MAP). It was evident by this time that if there was to be any Meteor production Rolls-Royce would have to take responsibility for production as well as development. On July 23 Robotham discussed the problem with General Crawford, Deputy Director-General of Tank Supply, who suggested that the Meadows Company, which was producing its own flat twelve engine for the tank program and had expressed an interest in producing the Meteor, should be contacted.

On July 29 at a Ministry of Supply conference it was decided to cancel the Leyland Meteor contract and to order the engines through the MAP. Rolls-Royce was to now have full responsibility for design and development and also be responsible for production, but not as the producer. At about this time the MAP also agreed to release some types of Merlin parts for the Meteor project but stipulated that a separate factory would have to be provided to produce special parts and to provide final assembly.

Eighty-five percent of the Meteor parts were common to both the Meteor and Merlin. It was decided that Meadows and Austin would produce those parts special to the Meteor, and Rolls-Royce would be able to provide the remainder from "unbalanced" Merlin production, that is, able to produce parts surplus to Merlin needs.

The Meteor tank engine which was destined to replace the Nuffield Liberty as the principal British tank engine. (Courtesy of Rolls-Royce)

The plan was to set up Meadows and Austin as special parts producers, along with several subcontractors, and Meadows as an assembling and testing facility. Rolls-Royce would use their Belper facility to carry on all design work as well as building 25 additional test engines. All of this was to be carried out without affecting the production of Rolls-Royce aero engines. The original Leyland order for 1,200 Meteor engines had been reissued as an order for 1,000.

On September 26, 1941 Rolls-Royce was informed that 1,900 engines would be required for the 1942 program and a further 2,100 by June of 1943.

The Meteor was by now considered a success and had been chosen as the primary power plant for the Mk VII Cavalier tank. (This was, of course, to change when the Meteor was not available and the Liberty was used instead.)

Work was also progressing on the Mk VIII tank which was to be able to use either the Liberty or Meteor engine. Birmingham Railway Carriage and Wagon Company was initially the "parent" of this design but that responsibility was later switched to Leyland.

By the end of October 1941 there were six Meteor engines in existence. Rolls-Royce was making good progress on Meteor design and prototype construction but work at Meadows was badly lagging as they had meanwhile received an order for an additional 1,000 of their own engines.

U. S. PRODUCTION OF TANK ENGINES CONSIDERED

In October 1941 the Ministry of Supply instructed the British Purchasing Commission in Washington to explore the possibility of obtaining 4,000 tank engines in the United States. Three different options were looked at. The first was to find a builder of Liberty engines. Quite naturally the Packard Company was approached with this proposition. They declined but instead proposed an adaptation of their 1500 aero engine of the 1927-30 era. This engine easily produced the 600 HP required. This proposal was rejected as the British felt they did not want yet another engine design to adapt to their tanks, nor another one adapted from a now obsolete design.

At this point J. M. Reid, the Rolls-Royce representative at Packard, handed over to the Packard engineers a descriptive handbook of the Meteor. Packard was at this time in full production of the Merlin engine for both British and American use. Maurice Olley suggested that Packard should produce a cast-iron version of the Meteor which could be produced in their auto plant.

A test engine was rapidly built and submitted to Mr. Knudsen, who headed American munitions production and would have to approve any Lend-Lease engine requisition for the British.

Packard had not kept it a duplicate of the Meteor. They had made a number of design changes, adapting some features of their current marine PT engine and otherwise modifying the design as they felt might improve it or aid in manufacture. The cylinder block was two-piece as the Packard Merlin. The block, head and crankcase were all cast-iron. Bore and stroke were 5.4" x 5 1/2" resulting in a displacement of 1,512 cu. in. This was a 1/2" reduction from the Meteor/Merlin stroke of 6" and reduced the 1,649 cu. in. displacement by 137 cu. in. The reason for this change is not known. With its 5.1:1 compression ratio the engine produced 425 HP @ 1,800 RPM and 525 HP @ 2,500 RPM on 75 octane fuel. Outputs of 600 HP and above could only be obtained with supercharging. Overall weight with accessories would be approximately 1,700 lbs.

Packard estimated they could produce 1,425 by the end of 1942 without affecting Merlin production. They would not proceed, however, unless they had total design control over the engine as well as government A1A priority for the acquisition of tools and materials.

This proposition was rejected because of the changes in the Merlin design and because Knudsen felt the engine was not well adapted to automotive factory production. He then proposed instead that Packard should build the Ford V-8 tank engine. During all these proposals the British expressed great concerns that the contracting of Packard to build any type of tank engine might affect Merlin production.

The Packard proposition seems to have been the only serious attempt at American production of a British tank engine. It came to an immediate conclusion when Packard received a contract for an additional 14,084 Merlin aero engines on January 31, 1942. With this new requirement in-hand, Packard could no longer consider taking on the additional engine project.

MEADOWS PRODUCTION DIFFICULTIES

Meadows was not making much progress in the production of special Meteor parts nor setting up facilities for assembly and testing. Part of this was, of course, due to lack of effort by Meadows but much was due to lack of force in the policy of the Ministry of Supply. That lack of force may have been due to the hope held by the Ministry for the production of Meteor engines by Packard. That hope fell by the wayside with the new Merlin order at Packard and about 30 days later, on March 4, 1942, Meadows production of the Meteor was finally given A1 priority.

The first production Meteor was put on test in the Meadows plant on March 28, 1942. Even so, little hope was held that Meadows would ever produce in large quantities. It was felt that not more than 200 engines could be expected from them by the end of 1942.

This, and the abandonment of the search for U. S. producers, prompted active efforts to find other producers of Meteor parts as well as assembly and test plants. Arrangements were made with U. S. producers to supply both machine tools and parts, in particular crankshafts. In the latter case, arrangements were made to have factories which were then making Liberty crankshafts switch over to Meteor crankshafts as soon as possible.

The producers ultimately chosen were Rolls-Royce-Meadows (the original production group), Rover and Morris Motors Engines Branch. Morris was currently the second producer of the Nuffield Liberty, the first being Mechanization & Aero Ltd. (M&A) (both were part of the Nuffield group of companies). Morris had received its first Liberty contract in the summer of 1941 to supplement M&A production and shortly was in possession of contracts for 6,227 engines.

Rolls-Royce had, of necessity, become the parent company. Meadows had delivered about 200 engines by the end of 1942. Rover and Morris began to switch over to Meteor production during 1943 although most engine output was still accounted for by Meadows during that year. By the end of February, 1943 some 1,490 Meteor engines had been delivered by the combined three producers.

Morris was scheduled to complete their Liberty contracts by the end of 1943 and swing to full Meteor production. In January of 1944 Meteor parentage was switched to Rover, thus relieving Rolls-Royce of the responsibility. By October 12, 1944 3,000 Meteors had been delivered and enough tanks fitted with them to have a significant number participate in the 1944 invasion of Europe.

THE CROMWELL TANK SERIES

As already noted, the A24 was initially to be called the Cromwell I, the Leyland A27L (Liberty-powered) the Cromwell II and the Leyland A27M (Meteor-powered) the Cromwell III. This did not last long, however. The A24 was renamed the Cavalier, the A27L was renamed the Centaur and the A27M took over the Cromwell designation.

THE CRUISER
Mk VIII - A27L
CENTAUR

While waiting for the Meteor engine to get into production, the Leyland design was produced as the A27L Centaur and powered by the Nuffield Liberty Mark V engine. This and the A27M Cromwell were designed to have the same engine compartment and drive system so that when the Meteor could be brought into full production, Centaur tanks could be converted to Cromwell configuration (and thus become Cromwell tanks) by replacing the Liberty engine with a Meteor. This was the ultimate fate of a large number of Centaur tanks by the end of the war.

A27L Centaur Mk I at the factory with hull machine gun still not mounted. (Courtesy of the Tank Museum at Bovington)

A27L CENTAUR MODELS

In the meantime, the Centaur was put into heavy production to try and fill the need for heavy cruiser tanks. By the time its production ceased 3,434 had been built. They were built in a number of variations to fulfill many roles:

Centaur I: Original production version with 6 pdr.
Centaur III: Main armament of 75 mm gun.
Centaur IV: Close support model with 95 mm howitzer.
Centaur OP: Conversion for artillery observation officers with dummy gun and extra radio equipment.
Centaur AA Mk I: Turret replaced to take Polsten AA cannon.
Centaur AA Mk II: Turret replaced to take Polsten AA cannon with modified radio location.
Centaur ARV: Any model with turret removed and fitted as an armored recovery vehicle.
Centaur Kangaroo: Any model with turret removed and fitted for use as an armored personnel carrier.
Centaur Dozer: Any model with turret removed and fitted with dozer blade for use by the Royal Engineers for construction work.

A27L - Cruiser Mk VIII General Data
(3,434 built - see Appendix A)

Length	**20' 10"**
Width	**9' 6"**
Height	**8' 2"**
Combat weight	**63,600 lbs.**
Engine	**Nuffield Liberty Mk V**
Power	**410 HP @ 1,700 RPM**
Power/weight ratio	**12.89 HP/ton**
Gear box	**4 forward, 1 reverse**
Speed on road	**27 MPH**

Armament

6 pdr. QFC (Mk I) or
75 mm (Mk III) or
95 mm howitzer (Mk IV)
1 or 2 Besa 7.92 MG

Armor thickness
65 mm front hull
76 mm front turret
hull minimum 20 mm

CENTAUR HISTORY

Below are listed a few of the more significant events participated in by Centaur tanks or uses to which they were applied. Obviously the list is not all inclusive.

A number of the close support models with 95 mm howitzers were used by the Royal Marines on D-Day providing support fire while on board LCTs off shore. They afterward landed on the beaches on June 6, 1944.

Although certainly a few battle tanks participated in various operations during the 1943-1945 period, probably the majority of service as fighting vehicles was provided by the Anti-Aircraft, OP and ARV types. Records seem to indicate that most of those in battle-tank configuration were eventually used for training purposes after the Meteor

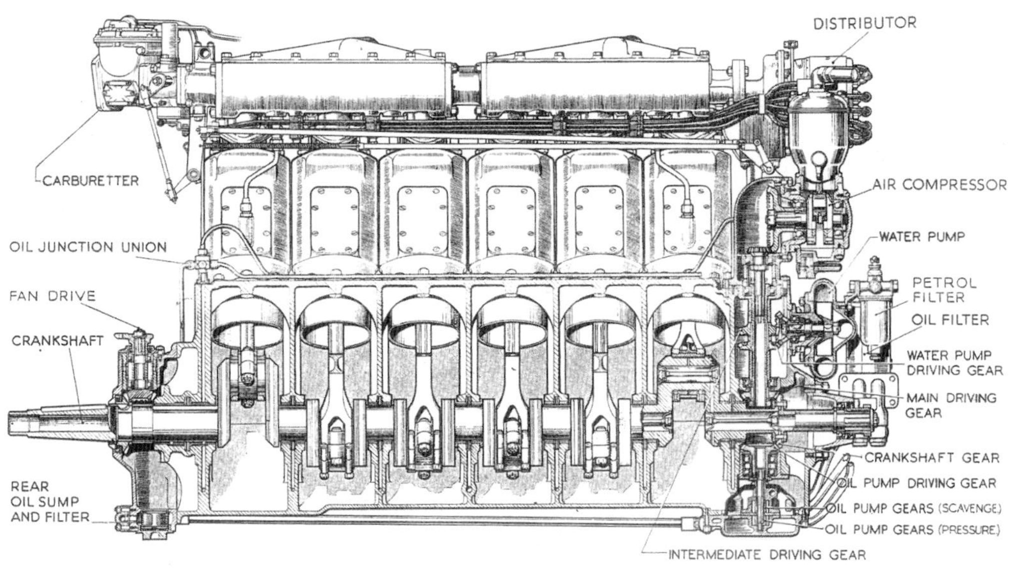

Nuffield Liberty Mark V with A24 Cavalier-type induction system. (Courtesy of the Tank Museum at Bovington)

engine became available for use in the Cromwell version, or for converting a Centaur to a Cromwell.

Probably its most far-reaching use was as armored dozers used for needed construction in war zones and for clearing rubble to allow passage of supply vehicles to the front.

THE ARGOSY AMPHIBIAN

Although some British interest in an amphibious tank was shown as early as late-1940 by Michael Dewar, head of the British tank mission to Washington, there seemed to have been no follow up to his report of inspecting an American prototype in Florida around Christmas of that year. By 1942, after a lot of water had passed under the bridge, so to speak, interest perked up. As a result of War Office inquiries in December of 1942, the British Truck Mission in the U.S. was asked to inquire regarding obtaining a supply of the American Duck (GMC DUKW-353) wheeled amphibian.

After being told that all American supply was needed for our own forces, the matter was directed back to the Ministry of Supply who replied that although a wheeled type amphibian might be designed and built in Great Britain, all capacity for design and production of a tracked type was tied up in tank production.

Apparently unknown to the ministry was the fact that the Nuffield Organization was currently developing such a vehicle on their own at their Morris Commercial Cars division. That vehicle was known as the Argosy. The first model, the Argosy 1, was privately tested in the spring of 1943. Following some lessons learned, it was modified and again tested in July of 1943, this time under the auspices of the Fighting Vehicle Proving Establishment.

Testing of the first Argosy continued and at the same time a revised model, the Argosy 2, was built. In October of 1943 the Argosy 2 went on trials. Trial reports indicate it performed very well in the role for which it was designed, a vehicle to transport supplies from ship-to-shore. It, however, had no ramp to allow rapid loading and unloading of supplies and personnel. It could not perform the duties of the American Landing Vehicle Tracked (LVT) and, evidently, that was the ability that was being sought. Thus the Argosy was not accepted for production and there were only two built.

Both vehicles were powered by the Nuffield Liberty Mk IVA. Propulsion was by tracks on land with their design being based on that of the Valentine tank. Propulsion on water was by two propellers mounted in tunnels at the rear of the vehicle and steering by two rudders mounted above them.

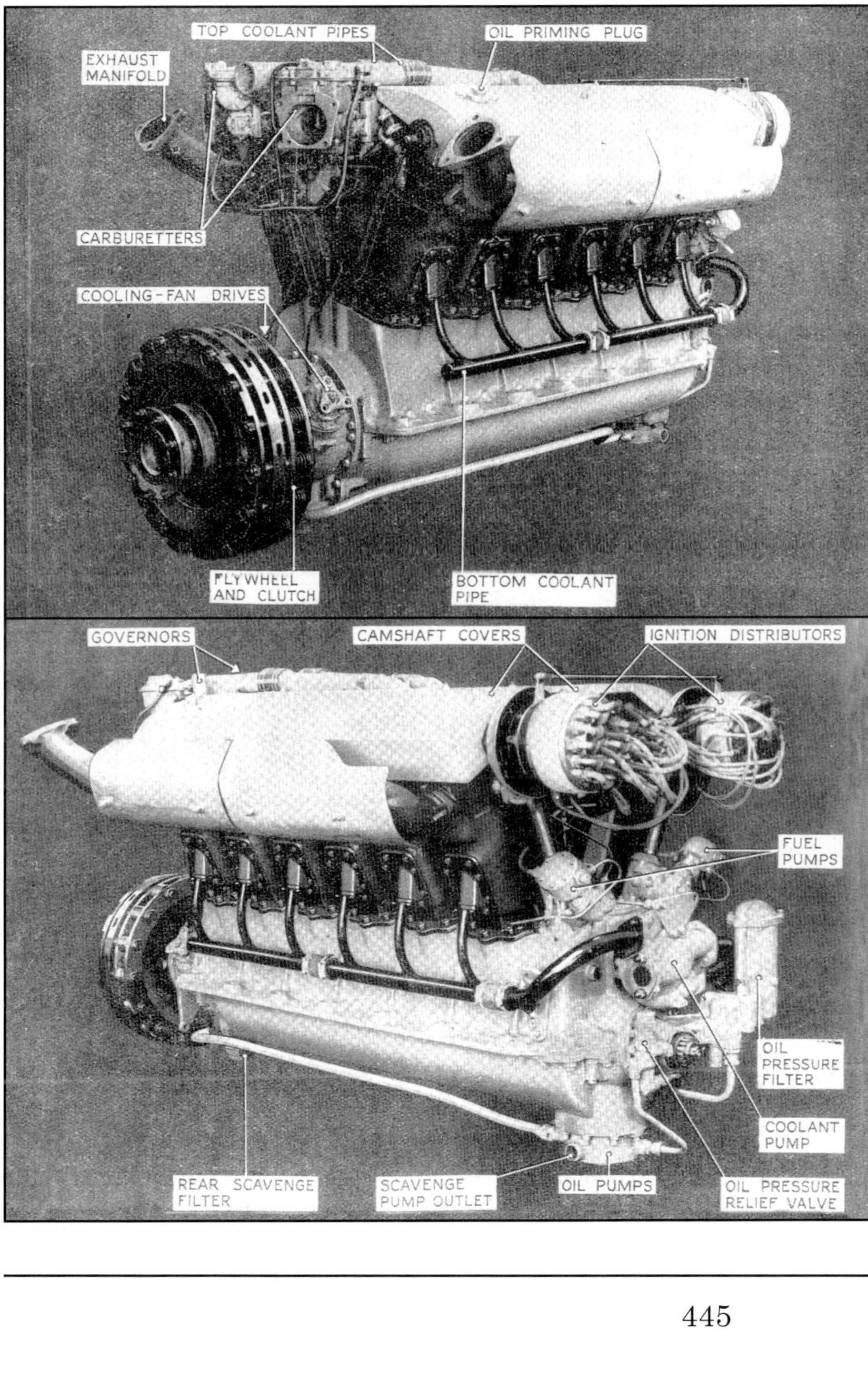

At left are two views of the Nuffield Liberty Mk V engine in its last version. The illustrations are from the Centaur Mk III - IV manual. Notice the oil return line from the rear sump to the scavenge pump remains but the outside oil gallery pipe is no longer present.
(Courtesy of Jamie Fairchild)

THE LAST LIBERTY-POWERED MILITARY VEHICLE

To the Centaur no doubt belongs the title of the last Liberty-powered military vehicle in active use. It continued to be used in Great Britain as a training vehicle for some years after the end of World War II, probably into the very early 1950s. The author has observed a Mark V engine with a rebuild plate dated in 1952 indicating the British must have used that particular engine beyond that date.

The most positive evidence of use, however, is documented Greek use. Greece received its first tanks in May of 1946 as part of a British military aid program when they received 52 Centaur tanks along with Marmon Herrington Mk IV carriers. These tanks saw action before the end of that year in the Greek civil war with the Communist People's National Army of Liberation.

This Communist-led civil war finally ended with the capture of the last of the Greek Communists on August 28, 1949. On October 16 a formal surrender was announced.

Just how much the Centaur tank had to do with the ending of this civil war I do not know. However, the Centaur tanks continued in use by the Greek Army through the 1950s and perhaps as late as 1960. According to the official Greek Armored Corps history, the last of the Centaurs were sold as scrap in 1962. One remains on display at the Armored Corps School at Avlona.

We can thus say that the Liberty engine was in continuous military use from 1917 to at least 1960.

General statistics of the Argosy

Length	25' 7"
Width	9' 9"
Height	11' 4"
Weight unloaded	28,000 lbs.
Weight loaded	46,000 lbs.
Engine	Nuffield Liberty Mk IVA
Power	governed to 340 HP @ 1,500 RPM
Speed on road	23 MPH
Speed on water	4 knots

Argosy 2 British Amphibian built by Morris Commercial Cars in 1943. (Courtesy of The Tank Museum at Bovington)

A Greek Centaur tank during exercises in Western Macedonia, Greece in the summer of 1957. The underside view is one not often seen. (Courtesy of Christos Kalfopoulo)

Greek Armored Corps exercises during the summer of 1957 near the city of Amyndaion in Western Macedonia, Greece. 10 Centaurs line up with Kolfopoulo standing on the front of his tank. (Photo courtesy of Christos Kalfopoulo, son of the soldier posing in the picture)

Chapter 12

Marine Conversions of the Liberty

The earliest commercially built marine Liberties the author has noted were those done by the C. C. Smith Boat & Engine Company of Algonac, Michigan in early 1919.

Industrialist Garfield (Gar) A. Wood had purchased the Chris Smith-built boat *Miss Detroit* in 1916 and had Smith build *Miss Detroit II* in 1917. Both of these boats were powered by eight cylinder 250 HP Sterling marine engines. He won the 1917 Gold Cup race with *Miss Detroit II.* In 1918 he asked Smith to convert an experimental Curtiss V-4 aircraft engine for marine use and build a racing hydroplane named *Miss Detroit III* to use it. This was accomplished and the boat was raced in the 1918 Gold Cup competition and won.

THE SMITH-LIBERTY

In 1919 Wood had several Liberty engines converted for marine use by Chris Smith. He replaced *Miss Detroit II* and *III* engines with one. He then entered both in the 1919 Gold Cup race. *Miss Detroit III* won and *Miss Detroit II* placed second. Also entered in the 1919 Gold Cup was a third Liberty-powered boat built by C. C. Smith, the *Eleventh Hour*. At about the same time Wood had another Liberty converted and installed in a Smith-built cruiser named *Gar Jr. Gar Jr.*'s engine was credited to Grant in a 1920 C. C. Smith Boat & Engine Co. ad. These appear to have been the first uses of the Liberty as a marine engine.

The conversion was a very basic one and consisted of coupling the engine output (propeller shaft) to a Snow & Petrelli marine clutch and reduction gear which in turn coupled to a Cross v-drive whose output drove the propeller shaft. In this setup the engine sat in the rear half of the boat with the driver and mechanic seated behind. The engine output shaft faced forward and drove the v-drive through the clutch and gearbox. The v-drive reversed the direction of the output shaft. The propeller shaft thus passed under the engine and to the rear of the boat. Cooling water was input to the water pump from overboard and after passing through the engine was exhausted overboard. Engine exhausts were dry and directly to the atmosphere in the case of racing boats, and through a water-cooled exhaust manifold in pleasure boats.

There were a number of builders of this type clutch and gearbox as well as v-drives. Thus it was only necessary to adapt available equipment to the engine and propeller shaft.

Miss America ***with its pair of Liberty marine engines. These conversions were done in the C. C. Smith Boat & Engine Co. plant in Algonac, MI in early 1920, probably under the direction of Howard Grant. In some Smith magazine ads of the period they were called "Smith Twin-Six" engines. When the boat was still under construction it was displayed in the National Motor Boat Show in February of 1920 with a sign calling the motor the "Grant Motor."*** (Courtesy of Mariner's Museum)

THE GRANT-LIBERTY

After this time Chris Smith evidently decided he did not want to make a business of converting Liberty engines and into the gap stepped Howard M. Grant. Grant was a business man and former government inspector of Liberty production. Some early marine racing historians claim his first Liberty marine conversions were accomplished by assembling a few Liberty engines from reject parts he had acquired as an insider when they were disposed of as scrap.

However, as I noted in Chapter 7, the Army put 1,000 engines on sale on April 17, 1919. This is early enough to have been the source of the first Gar Wood conversion by Chris Smith as well, of course, as the earliest known Grant-Liberty conversions. Grant could have built his first Liberty conversions from reject parts but that would not seem necessary or likely.

It would thus appear that after from four to perhaps six at most Smith-Liberty conversions had been built in 1919, the C. C. Smith Boat & Engine Company got out of the Liberty conversion business and was replaced by the Grant Marine Motor Company, whose conversions were known as the Grant-Liberty conversions.

Grant set up business at 1251 Beaufait Ave., Detroit, MI. In his ads and brochure he listed his factory address as Bellevue and Confield Avenues in Detroit

He had the Liberty conversion business to himself for only a short while. It did not take long for others to enter.

Grant appears to have put a minimum of expense and engineering into his conversions. He used existing standard components as much as possible. Engine modifications would have been limited to add-on components such as the starter, exhaust manifolds, marine water pump, fuel pump, etc.

The clutch and reverse gear and v-drive, if required, would be stand-alone units manufactured by others.

A marine engine builder who had been in the business in St. Paul, MN since 1905 was Auto Engine Works. This company specialized in the design and building of gears of all types, marine transmissions, clutches and reverse gears, and during some years even its own marine engines. During the period following WW I they had built and marketed two models based on the same engine, a

Grant-Liberty conversion as pictured in an April, 1920 Power Boating ad. The engine is shown with water-cooled exhaust manifolds and a Bijur front-mounted starter, the same type used on the Navy NC boats for the cross-Atlantic flight. (Courtesy of Keith Brayer)

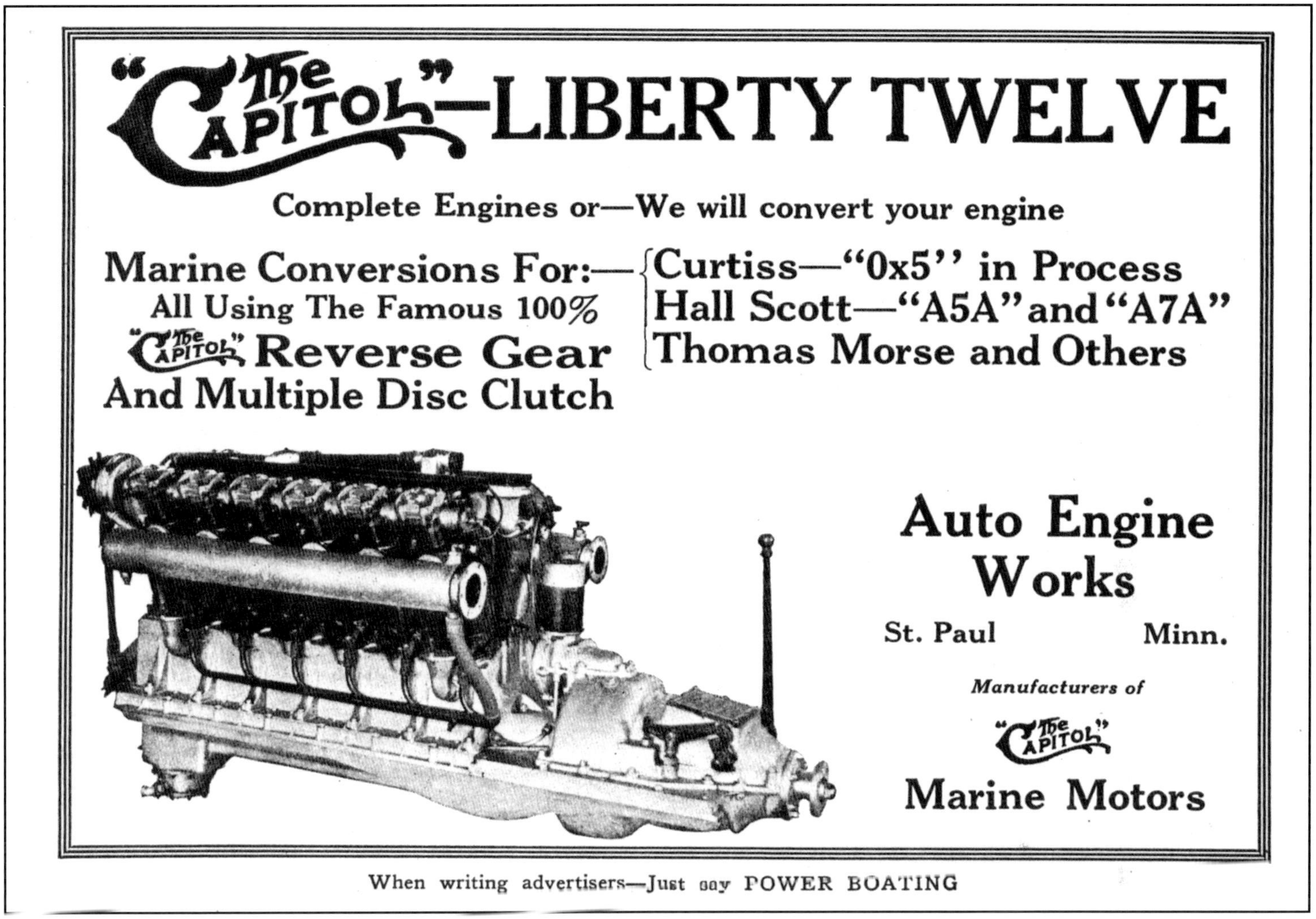

Capitol ad of January, 1924 which served to announce the availability of their marine-converted Liberty engine. Capitol was one of the few builders of marine conversions who designed and built their own clutch and reverse gear. (Author's collection)

four-cylinder 4 1/2 x 5 1/2 unit but rated at either 34 HP @ 1,000 RPM or 65 HP @ 2,000 RPM. An eight-cylinder model of the same bore and stroke was added in 1924 and rated at 135 HP @ 2,000 RPM. All their engines had built-in clutch and reverse gears designed by themselves. They marketed the engines under the trade name "The Capitol" marine motor.

The end of the war had brought back the civilian market. It had also put large numbers of surplus aero engines on the market and it did not take long for boat buyers and boat builders to realize these high-power, light-weight engines could be adapted to marine use at overall coasts much lower than comparable marine engines. A number of companies went into the business of converting various models of these engines for marine use. Auto Engine Works was one of the earliest and most successful of these companies.

The first conversion they did was of a four-cylinder Hall-Scott A-7a. They introduced this conversion in January of 1923 and advertised it heavily in most of the boating magazines for the remainder of the year. It was a good seller and they immediately began designing similar conversions for other WW I aero engines.

The next conversion announced was for the Liberty 12 in January of 1924. The half-page announcement ad noted several important items: 1- they would either supply complete converted engines or they would convert your engine; 2- they were in the process of developing conversions for the Curtiss OX-5; 3- they also had conversions for the Hall-Scott A-5a and A-7a, Thomas Morse and others.

The Capitol Curtiss OX-5 (OXX-6) conversion was announced in an ad of June 1924. In the same magazine Grant Marine Motor Co. announced a "special offer to boat builders", the "Grant-Curtiss" (OX-5), and noted that

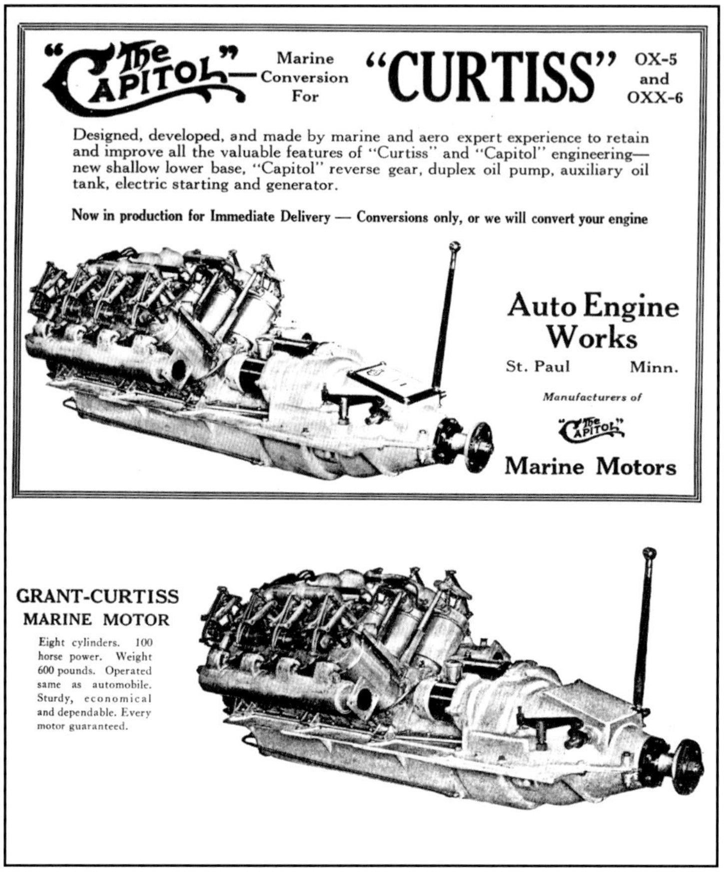

Above is the Capitol ad of June 1924 announcing their new Curtiss conversion. Reproduced below it is the illustration portion of the Grant Marine Motor Co. announcement of their Grant-Curtiss, the same illustration as used in the Capitol ad but with the normal identifying Capitol script blanked from the gear box cover plate. Obviously Grant had chosen to market the Capitol conversion rather than build his own. Capitol had similar arrangements with other marketers as the years went by. (Author's collection)

Above are the cover plates from the gear boxes of a Grant-Liberty and a Capitol Liberty of the same era. They are illustrated at the same scale, i.e., they are of identical size. These plates came from sample stocks of an aluminum foundry which supplied Capitol and a number of other marine engine builders. The illustration brings out a number of things regarding the Capitol conversions. First, it certainly appears Grant had the same arrangement with Capitol regarding Liberty conversions as with Curtiss conversions, that is, that Grant marketed the Capitol-Liberty conversion under their own name. Second, other marketers also sold Capitol conversions; in this case Detroit Marine Garage and they made no attempt to pass themselves off as the builder as did Grant. Another marketer that followed the practice of Detroit Marine Garage was Johnson Motor Products of New York. (Author's collection)

Chris-Craft offered a standardized 37 MPH runabout equipped with the Grant-Curtiss motor. Also specifically noted (but not illustrated) was the availability of the Grant-Liberty. The interesting thing about these simultaneous announcements is that they use the exact same illustration of the Curtiss OX-5 conversion. The Capitol illustration shows the usual cover plate on the gear box with the script "Capitol" logo clearly visible. On the Grant ad the cover plate is blanked out.

The importance of pointing out this arrangement is because it appears Grant had the same one with regard to the Grant-Liberty. Grant does not appear to have stayed in the business much beyond 1925 as that is the last year the company is listed in boating magazines in their regular listings of marine engine builders. His listed engines were the Grant-Liberty and Grant-Curtiss.

THE LEE-LIBERTY

C. C. Smith Boat & Engine and Grant Marine Motor Co. were not the only ones to enter the Liberty conversion business in 1919. In that year the Lee Motor Co. of Norristown, PA brought out a much more refined unit which in many regards set the standard for most builders who followed.

The engine was offered in either a low or high-compression model with ratings being 400 and 450 HP respectively. Cylinders were either original aero-type specially treated for salt water or cast-iron at a slight extra cost. The original water pump was replaced with a double-gear-type which, along with a DeJon 12-volt generator and a hand-crank starter, was mounted on a separate gear-drive bolted on the front of the engine and driven off the crankshaft. The generator and water pump were driven at 1/3 engine speed. The hand-starter appears to have been a Navy type.

On the output drive shaft was connected a combination DeJon electric starter with Bendix drive and flywheel with ring gear, all mounted in an aluminum housing. Behind this housing was another containing the clutch and reverse gear and output shaft. This unit was made special for Lee by Joe's (Snow & Petrelli Mnfg. Co. of New Haven). Water-cooled exhaust manifolds were included.

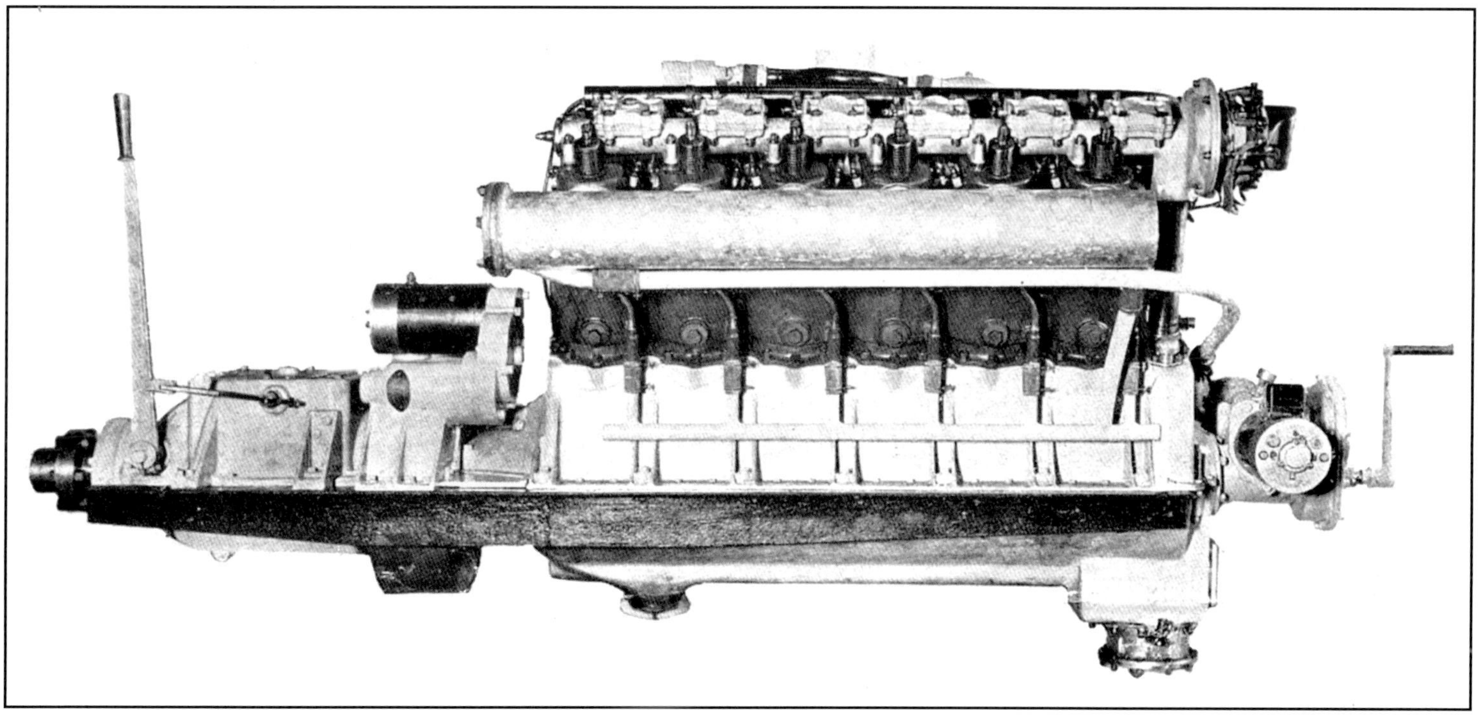

Above right is a Lee Motor Co. Liberty marine conversion brochure from 1920-21. Note the trademark at the top of the page indicating the company dealt in both marine and aircraft motors and was apparently operated by the Templeton Brothers. Bottom illustration shows the engine with optional cast-iron cylinders. Similarities to the Gar Wood conversions which followed can be noted. (Author's collection)

The complete unit was mounted on a pair of 3/8" x 4" side rails and included an oil cooler of their design. It was offered in either left or right rotation for dual installations and was priced at $4,500 with standard cylinders and included the oil cooler, ammeter and oil gauge.

Although brochures noted that they had been built since 1919 and "hundreds are now in use," the brochures are not dated and the author has thus far never seen a magazine ad nor boating magazine engine builder listing for this builder. Although the conversion appears to have been a well-designed and built one, there is no evidence they stayed on the market beyond, perhaps, the early 1920s.

THE GAR WOOD-LIBERTY

Gar Wood used Liberty marine conversions beginning in 1919, built first by C. C. Smith Boat & Engine Co. and then, by early 1920, by Grant Marine Motor Co. It appears the *Miss Detroit II* and *Miss Detroit III's* Liberties were Smith-Liberties and all Liberty-powered Gar Wood boats built by C. C. Smith Boat & Engine during 1920-1921 used Grant-Liberty engines. These would include *Gar Jr.*, *Miss America*, *Miss America II*, *Baby Gar*, *Miss Detroit IV*, *Miss Detroit V* and *Gar Jr. II*.

In late 1916 Gar Wood bought controlling interest in the C. C. Smith Boat & Engine Co. Smith continued operation of the boat business, but Wood could now get his racing and pleasure boats built on demand, so to speak. As the end of 1921 approached, Gar Wood came to the decision that he would go into the boat-building business on his own. He purchased the remaining interest in C. C. Smith Boat & Engine Co. and took over their Algonac, Michigan, boat yard at Pointe du Chene. Wood would organize his new boat company, Gar Wood Inc., with its plant at that location and Smith would build a new plant at another Algonac location.

Smith incorporated his new boat-building company as Chris Smith & Sons Boat Co. on February 15, 1922. Chris

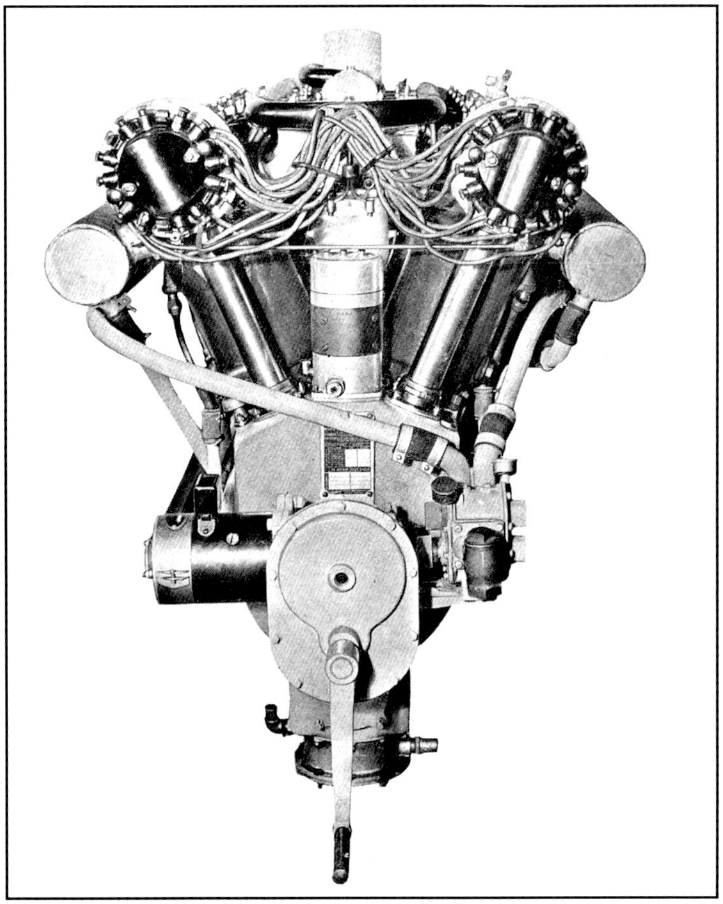

Above and right are end views of the Lee-Liberty marine conversion. It can be seen that this particular engine was converted from a Packard Navy Liberty. The top cover plate of the reverse gear is marked Lee Motor Co. Norristown, PA. (Author's collection)

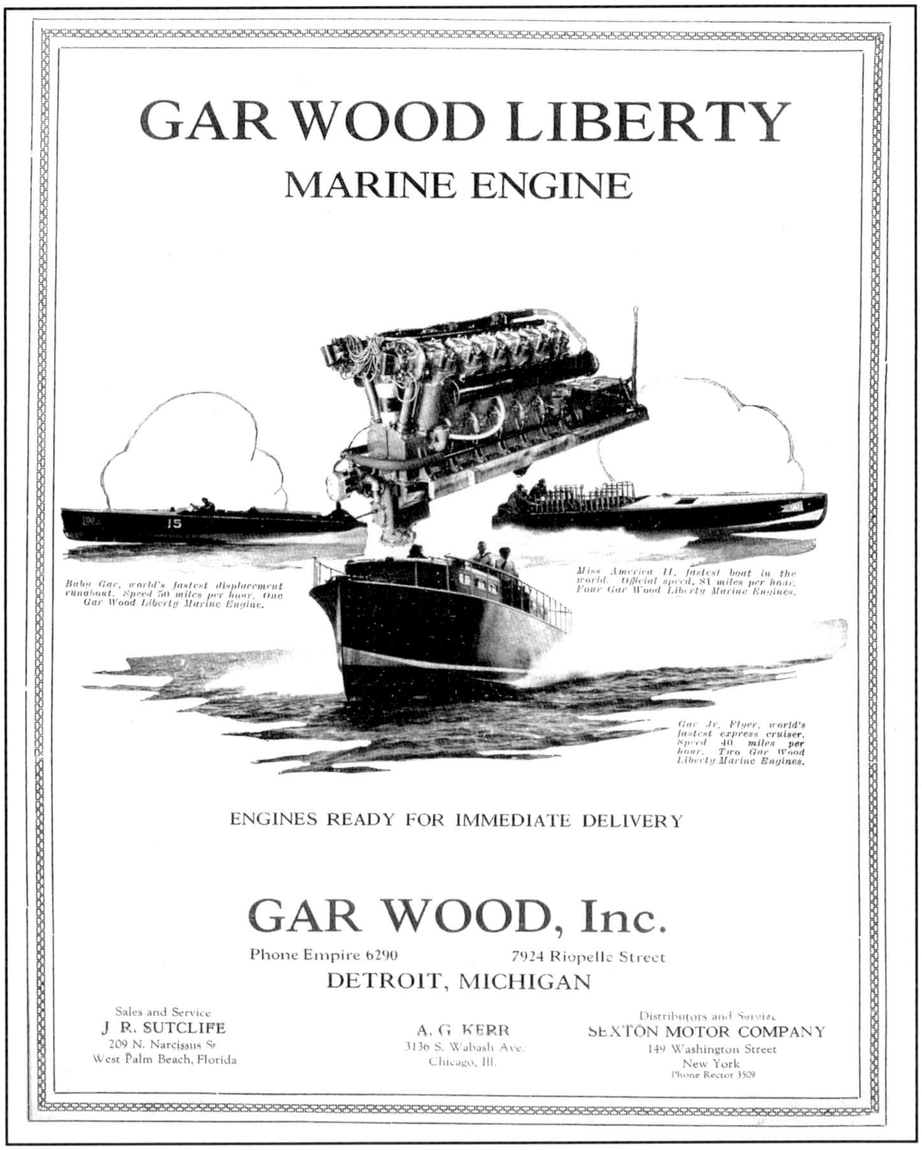

1924 Gar Wood Liberty Marine Engine brochure. Although built at the Detroit Marine-Aero Engine plant, Wood always chose to refer to the Liberty engines as Gar Wood engines and the Fiat marines as Detroit Marine-Aero engines. (Author's collection)

Smith & Sons built the first 22 Baby Gar runabouts under contract at their new plant while Wood was getting his newly acquired plant into production.

Wood realized that he would need a source of engines for his new boat-building enterprise and set up a special company which would be both a supplier of needed engines and a separate business in itself. This was the Detroit Marine-Aero Engine Co. It was incorporated on December 30, 1921 and was initially set up at the location of Gar Wood's Wood Hydraulic Hoist Co. at 4196-98 Bellevue Ave. in Detroit. Before the end of 1922 Wood had moved its plant location to 409 Connecticut Ave. and its office location into his Wood Hydraulic Hoist/Gar Wood Inc. office at 7924-60 Riopelle St., Detroit. By October of 1922 it was advertising the availability of marine-converted Liberty and Fiat engines to boat builders and individuals.

Although the engine-building company had actually become operational first, it was shortly followed by the Gar Wood Inc. boat-building company. When Wood took over the old Smith boat yard he incorporated Gar Wood Inc. with its business offices located with his hoist business offices which had very recently been moved to 7960 Riopelle St. in Detroit.

Wood immediately began advertising a new commercial version of his Baby Gar boat design powered by either a 450 HP marine Liberty or 300 HP marine Fiat, both were built by Detroit Marine-Aero. He had contracted Joe's Gears to build the clutch and reverse gear and they would continue to build them for both the Fiat and Liberty engines as long as Gar Wood offered the engines for sale. The only known changes in the gear box was to add a forward set of mounting flanges in 1926.

The starter and flywheel/ring gear were designed and built by Wood. The design concept was the same as used on the Lee-Liberty already discussed but instead of being a single assembly, was built as two individual units. The flywheel/ring gear was a unit and was mounted on the output shaft between the engine and reverse gear. The starter and Bendix drive were built as a unit and attached to the front of the crankcase by the front four crankcase bolts. The Gar Wood Liberty was always offered in both left and right rotation. Cylinder water jackets were treated to reduce corrosion but cast-iron versions were never offered by Wood.

From 1922 - 1924 the Liberty was rated at 450 HP @ 1,850 RPM and called the 450 HP Gar Wood Marine Liberty. During the latter part of 1924 they began using the name T-24 In 1925 the rating of the standard version was reduced to 400 HP @ 1,700 RPM and a new 500 HP @ 2,150 RPM high-power version introduced. They were named the T-25 and TS-25 respectively. "T" indicated twelve, or twin six, and "S" indicated super-carbureted. The TS model swapped camshafts between cylinder banks which resulted in having the intake on the outside of the cylinder banks and the exhaust on the inside. Two duplex

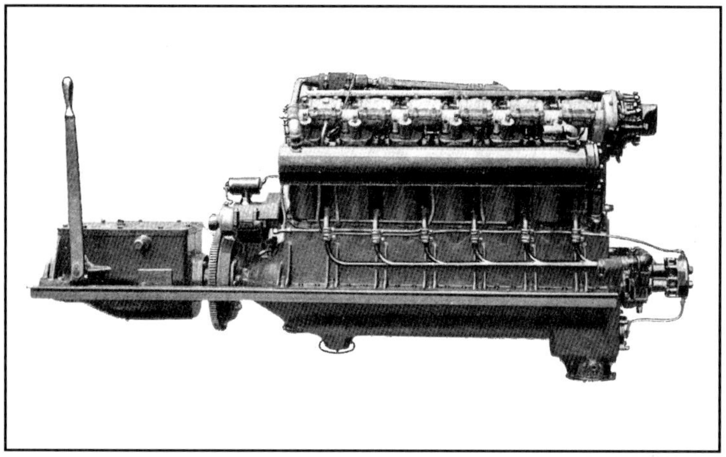

Gar Wood T-24 Liberty of 1924. The engine and reverse gear weighed 1,250 pounds and was priced at $4,500. The flywheel and ring gear was an open unit mounted on the output shaft between the engine and reverse gear. The starter and Bendix were built as a separate unit and mounted on the engine nose. Note the reverse gear had mounting flanges on the rear only. The engine was rated at 450 HP @ 1,850 RPM. (Author's collection)

1926 French brochure published by Societe Francaise de Fabrication Aeronautique which was located at 79, Route de Saint-Cloud, Rueil, just west of Paris. Inside was illustrated the T-26. (Courtesy of Kim McCutcheon)

Zenith carburetors were used on each cylinder bank. Better breathing and higher RPM resulted in the 100 HP increase in power. The number 25 indicated model year.

The only foreign material the author has found which was aimed specifically at the Gar Wood Liberty is the French one of about 1926 shown at left. Also in 1925 Howard W. Lyon of New York became Wood's exclusive boat distributor and made them available in England and Europe through Claude Grahame-White of 12 Regent St., Pall Mall, London. S.W.1. The engines would have been available as well. (See page 488.)

United States distributors for Gar Wood engines, and in some cases boats as well, were Sexton Motor Co., 149 Washington St., New York City; Hubbard H. Erickson & Co., 3037 Western Ave., Chicago, IL.; A. G. Kerr, 1200 4th Ave South, Seattle, WA; J. R. Sutcliff, 209 N. Narcissus St., West Palm Beach, FL ; and Atlantic Boat Yard Co., 201 Southwest 6th St., Miami, FL. In Cuba the distributor was A. M. Puente & Co., Egido 10, Havana, Cuba.

These two models continued to be listed each year with the same ratings and the model names progressively changed each year through the models T-30 and TS-30. In 1931 - 1934 they were listed as models T and TS with no year designation following. The model TS was dropped and only the T was listed in the final years Gar Wood offered the engine, 1935 - 1937.

During most of the years Gar Wood offered the Liberty marine for sale to individuals and boat builders he was offering various Gar Wood boat models powered by it as well. Initially these were the Baby Gar models and between 1922 and 1934 he sold about 96 of them. During that period express cruisers were also popular and Wood built at least 14 powered by from one to as many as five Gar Wood Liberties. The combination of the Depression

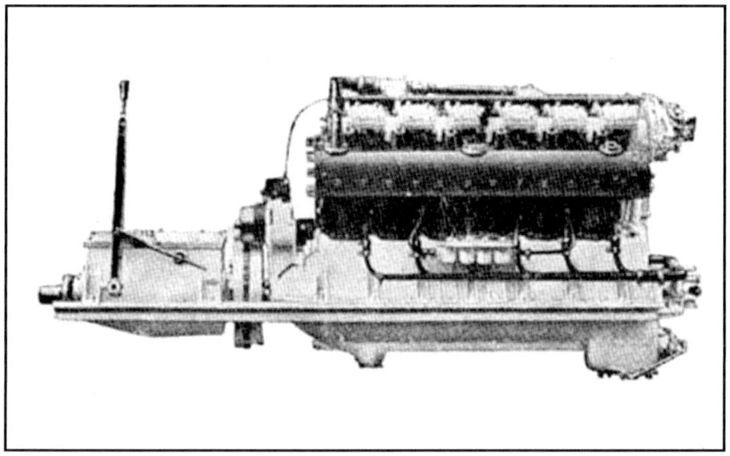

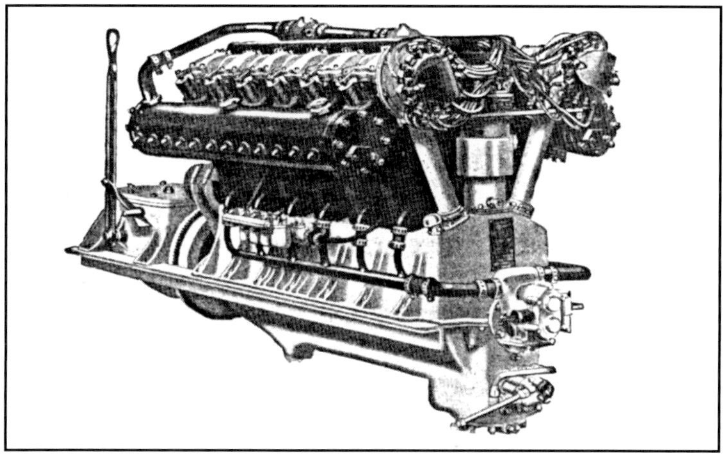

Above are two views of the T-26 engine. Only minor changes were made in either the T or TS models through the end of production, most of which would be slight changes in accessory design or routing of water lines, etc. Note the reverse gear had mounting flanges front and rear starting with this model. (Author's collection)

Above are left and right views of a TS-26 engine. The left image is from a Gar Wood brochure and the right is of a surviving engine. The data plate at lower left is from T-25 engine No. 219, indicating 219 Gar Wood Liberty engines had been built up to that time. (Author's collection)

and the advent of more modern marine engines in the same power range served to drastically reduce the sales of Gar Wood Liberty engines and the boats he powered with them. We have no specifics regarding individual engine sales, but Wood's Liberty-powered boat sales were 17 in 1930, 6 in 1931, 4 in 1932, none in 1933, 2 in 1934 and a single boat in 1935, the last year Wood listed a production boat with Liberty power.

Detroit Marine-Aero Engine Co. initially intended to do a large business in both airplane engines and marine conversions of them. One of their earliest ads announced in the March 27, 1922 issue of *Aerial Age Weekly* magazine that they had "purchased the stock of airplane motors received by the Government as its share in the division of spoils by the Allies" and they named several of the makes purchased. There must have been some degree of truth in the statement as a large number were listed in a 1925 liquidation ad which appeared in the November 2, 1925 edition of *Aviation* magazine.

Among their popular marine conversions was the 300 HP Fiat. This model was the only marine conversion besides the Liberty which was continually advertised from 1922 through 1937. The Air Service inventory of that engine in

1919-21 was 389. By April 1922 they had 10 remaining. Wood was no doubt the purchaser of the 379 sold.

THE GAR WOOD LIBERTY IN EUROPE

A French distributor was located at 79, Route de Saint-Cloud in Rueil in 1926 and probably later. They issued a brochure illustrating and describing the T-26 model but making no mention of the TS.

Detroit Marine-Aero Engine Co. continued operating as an individual business through 1929. At that time it was dissolved and Gar Wood Inc. took over its facilities and business.

We have no record of how many Gar Wood Liberty engines were sold but surviving Gar Wood Inc. boat production records allow us to closely estimate the number of Liberty engines used in Gar Wood boats. Including the various boats built for Gar Wood himself, about 175 Liberty engines were installed in Gar Wood boats. To these can be added perhaps another 225 sold to individuals and boat builders as well as those sold in Europe. Grand total production from 1922 to 1937 was about 400 engines.

List prices of Gar Wood Liberty Engines (left or right rotation)

Year	Model	Weight	Price
1922	GW Liberty		
1923	GW Liberty	1,250 lb	
1924	T-24	1,250 lb	$4,500
1925	T-25	1,350 lb	$4,500
1926	T-26	1,350 lb	$5,000
	TS-26	1,500 lb	$6,500
1927	T-27	1,450 lb	$5,000
	TS-27	1,475 lb	$6,500
1928	T-28	1,500 lb	$5,000
	TS-28	1,550 lb	$6,500
1929	T-29	1,500 lb	$5,000
	TS-29	1,550 lb	$6,500
1930	T-30	1,500 lb	$5,000
	TS-30	1,550 lb	$6,500
1931	T	1,450 lb	$4,000
	TS	1,550 lb	$5,500
1932	T	1,450 lb	$4,000
1933	T	1,450 lb	$4,000
1934	T	1,450 lb	$4,000
1935	T	1,450 lb	$4,000
1936	T	1,450 lb	$4,000
1937	T	1,450 lb	$4,000

SPECIAL GAR WOOD LIBERTIES

Gar Wood built a number of special racing Liberty engines over the years that are worthy of mention and I will describe them in the order in which they were built.

Wood instigated the building (by Grant Marine Motor) of the first reverse rotation marine Liberties for use in the *Miss America, Miss Detroit IV, Miss Detroit V* and *Gar Jr. II.* When he started building marine Liberty engine conversion on his own in 1922 he offered them in left or right rotation at the same price.

The Gold Cup engine rules changed from no limits to a maximum of 625 cubic inches displacement in 1922. Wood liked the Liberty and used one from which he had removed one bank of cylinders and destroked the other to 5.25". This resulted in a six-cylinder engine displacing 618.5 cu. in. which finished 7th (*Baby Gar Jr.*). Also two Capitol-converted Liberty 8 engines whose normal displacement was 1,100 cu in were destroked to 625 cu in and entered. They finished 3rd (*Redbird*) and 10th (*Bluebird*).

Another popular boat race of the 1920s was the International Sweepstakes. Engines used in this race were limited to 1,350 cu in. A standard 1,650 cu in Liberty would therefore have to be modified in order to run. For the 1923 race Wood destroked three to 1,349 cu in and ran them in *Teddy* (finished 1st), *Baby Gar II* (finished 6th) and *Bruin* (finished last).

For the 1924 Gold Cup Wood ran his 618.5 cu in Liberty in *Baby America* and finished 5th. Rules for the 1924 Sweepstakes were modified to allow full-size Liberty engines if the boat hull length was 32' or longer. Several boats used standard Liberties but a number of reduced displacement Liberty engines were also entered, including two 1,349s, a 1,249 and a 1,060, the latter being a Gar Wood engine built to race in the 1060 APBA class. *Miss Detroit VII* with a 1,349 Gar Wood Liberty was the winner.

Wood's last shot at the Gold Cup race was taken in 1927. Again he used Liberty engines but in this case they were not just reduced in displacement by destroking, they were totally redesigned with about the only Liberty component still in use being the crankcase and cam drive. It was a double overhead cam four valve design using an aluminum head assembly for each bank of cylinders. Bore was 3.828" and stroke 4.500" giving a displacement of 625 cu. in. Power was about 375 HP @ 3,500 RPM.

Twelve separate water-jacketed aluminum cylinder barrels, using steel sleeves, were bolted to the crankcase. Cooling water was fed to them from a manifold for each cylinder bank in the same manner as the standard Liberty. The aluminum head on each bank was fed water from the cylinder barrels. Two Miller carburetors fed each bank from the outside.

The 1940 Gold Cup Tinker Toy built by Gar Wood Jr. at his father's Algonac boat yard and powered by one of the two Liberty Gold Cup engines he built for the 1927 event. The data plate for one of them is shown on the right. The NO. 23701 is the original U.S. number of the engine used for the conversion, a Lincoln-built engine. Note the vertical dry stack exhausts on the engine. Gar Wood Jr. is at the extreme right standing on the dock with a life jacket. The first Liberty-powered boats to run in the Gold Cup were Miss Detroit II & III *in 1919. Tinker Toy was the last in 1940.* (Author's collection)

One of the two Gar Wood Liberty Gold Cup engines built in 1927 for the Gold Cup races. With a bore and stroke of 3.828" x 4.5" the engine displaced 625 cu in, the maximum allowed for the 1927 Gold Cup class. Double overhead cams drove four valves per cylinder. When this photo was taken by the author, both engines were in the collection of Craig Magnusson.

Capitol Liberty marine conversion shown with the early "K"-type reverse gear. Factory photo probably taken in late 1926 or early 1927. This engine is shown equipped with the optional cast-iron cylinders type 25 CA (first type). This reverse gear model was supplied from introduction in 1924 until the LA-12 model was introduced in 1931. The cover plate shown is the style in use from mid-1927 to about 1931. (Courtesy of Jim Aamodt & Author's collection)

He built two engines and raced them in *Baby Gar VII* and *Baby Gar VIII*. Both boats were built very light in an attempt to get more speed. As a result both boats suffered structural damage to their bottom planking and withdrew in the sixth and seventh laps.

A month later Wood entered *Baby Gar VIII* in the Detroit Sweepstakes race with the same engine. This was an endurance race comprised of a single heat of 150 miles and was thus a severe test of both boat and engine. This time the water was relatively calm and the boats suffered little damage. Of the seven entrants only three finished the race. *Baby Gar VIII* finished third.

Nothing more is heard of these engines until Gar Wood's son, Gar Jr., installed one of them in the 18' 9" step hydroplane *Tinker Toy*, built specially for the 1940 Gold Cup race. Her speed was very disappointing, being almost 20 MPH under the top runner's 70 MPH. Wood dropped out after eight laps.

Both engines have survived in a private collection and will some day be restored to running condition.

THE CAPITOL-LIBERTY

I noted the Capitol marine conversion of the Liberty when covering the Grant-Liberty but will now go into the entire history of the Auto Engine Works' Liberty marine engines.

The first commercial Capitol-Liberty conversions were advertised in January of 1924 and used their "Model K" reverse gear. With the Capitol gear the flywheel-ring gear and starter were built into the transmission rather than being separate units as they were in the Lee, and later Gar Wood conversions.

They had gained some earlier experience when they first converted two Liberty 8 engines for use in the 1922 Gold Cup, and two Liberty 12s for use in the 1923 Sweepstakes race.

The Model K gear was designed for the Liberty but was later used on any conversion of similar size that Capitol did, the Packard 1A-1237 being one example. The transmission provided 1:1 ratio in both forward and

Capitol LA-12 introduced in 1931 and the standard model until the end of Capitol Liberty sales in 1938. It was available in left or right rotation, high or standard compression and standard or cast-iron cylinders. Weight with standard cylinders was 1,400 pounds and with cast-iron was 1,700 pounds. At lower-left is the reverse gear cover plate and available Capitol angle drive which had a step-up ratio of 1 to 1.368. (Author's collection)

reverse, a feature they claimed as exclusive for a number of years. Lubrication was accomplished with engine oil fed through the hollow crankshaft. The standard front crankshaft plug was replaced by a special Capitol one which allowed oil to exit the front of the crankshaft and feed the gear box. An outside oil line then allowed its return to the engine sump. The gearbox was covered under patent No. 1,571,210 of February 2, 1926. Its application was filed on February 7, 1924.

Engines were supplied with clutch and reverse gear, propeller shaft coupling, special triple-gear water pump, new design oil pump, water-jacketed exhaust manifolds, a set of tools and a dash panel fitted with a Delco double ignition switch and ammeter, oil pressure gauge, water radimeter, oil radimeter and dash light. The panel was designed to receive a tachometer but it was not furnished.

Prices changed over the years but in 1928 they were as follows:

Std. rotation high or low comp. & steel cylinders	$3,175
Reverse rotation	3,410
Std. rotation low comp. only with cast-iron cylinders	3,565
Reverse rotation	3,800

This clutch and reverse gear continued in use until about 1931 at which time it was redesigned and became known as the LA-12 model. The case design was changed but parts manuals indicate that few interior parts were altered.

At this same time they introduced the LA-12-GR model. This was a LA-12 model with a special gear case which allowed the attachment of a herringbone reduction gear of either 2 to 1, 2 1/2 to 1 or 3 to 1 ratio. The reduction gear was provided with a separate disengaging clutch and was designed to drive propellers of from 36" to 45"

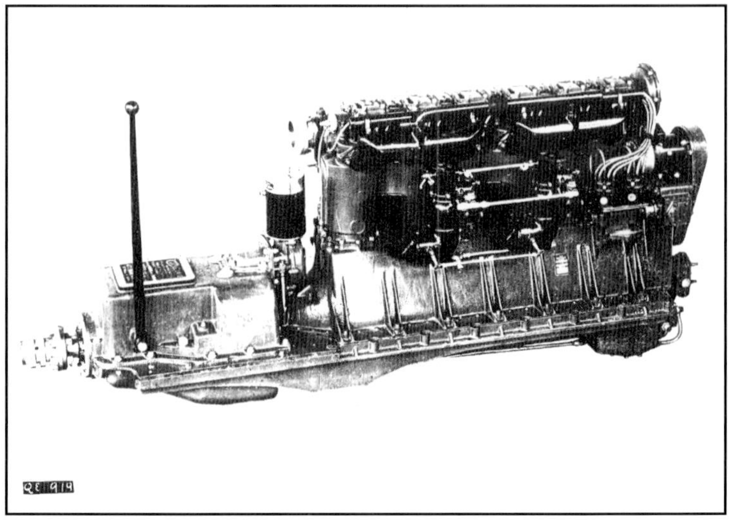

Two Capitol conversions of Wright-built Liberty 6B engines. The left one has an early-type "K" reverse gear and the right has the late LA-12-type introduced in 1931. Both had their original Dixie magneto ignition replaced with Delco battery-type. The generator is mounted on the ignition gear drive housing and is driven from the front accessory drive, either by a belt or a series of spur gears. The two examples which have survived are equipped with late-type reverse gears. (Courtesy of Bob Visser)

The Aeromarine N.A.L. Liberty. Capitol must have purchased the engine surplus. A number of rare Capitol Liberty conversions have survived (two Liberty 6B conversions and the conversion of the only Trego-built Liberty accepted by the Air Service). Possibly this engine did as well. Unfortunately the dark color the engine was painted prevents this photograph from yielding much detail. (Courtesy of Bob Visser)

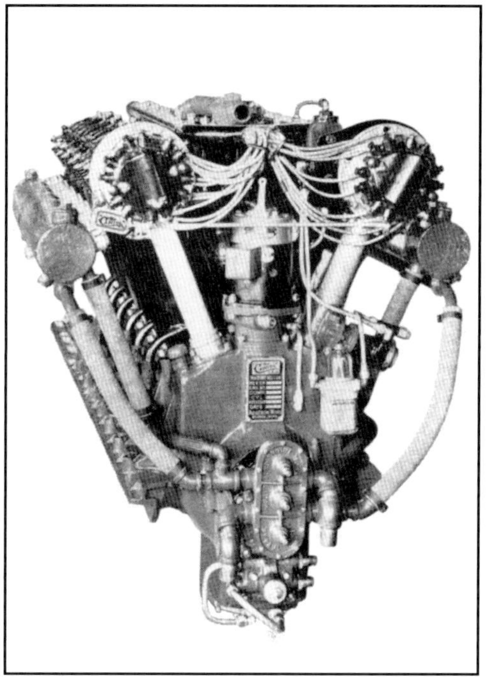

Capitol LA-12-GR introduced in 1931. The extra transmission on the rear is a reduction gear and was available in three different ratios, 2 to 1, 2 1/2 to 1 or 3 to 1. Mounted on special heavy side rails, the complete unit weighed 2,750 pounds equipped with cast-iron cylinders and 2,450 pounds with standard cylinders. At right is an end-view of the LA-12 but is also representative of the pre-1931 model as well. (Author's collection)

diameter. Both the LA-12 and LA-12-GR were offered in left or right rotation, with standard or cast-iron cylinders and in high (5.42 to 1) compression or standard (4.89 to 1) compression. Rated power was 420 HP and 380 HP respectively.

The LA-12 and LA-12-GR continued to be offered until at least 1938 and possibly later. Although the L-12 was their standard Liberty model, they purchased surplus Liberty engines of any type available and then marine-converted and sold them. They are known to have converted at least seven and probably eight Liberty 6 engines. Three of these are known to have been Wright-built engines. Considering the fact that a total of only 16 Liberty 6 engines were built, this is a very high percentage. Two conversions of Wright-built engines are known to have survived.

Two Liberty 8s were Capitol-converted in 1922 and raced in the Gold Cup of that year, *Redbird* (G 8) and *Bluebird* (G 11). Both were reduced to 625 cu in by reducing their stroke.

A Capitol photograph has survived which shows one of the rarest Liberties built, the Aeromarine N.A.L., was purchased and converted by Capitol prior to 1931. If the converted engine survived it may some day turn up in a boat.

Their U.S. representatives were Detroit Marine Garage, 351 Motor Boat Lane, Detroit, MI; J. Frank Knorr, 528 NW Seventh Ave., Miami, FL; Frank L. Morse, 316 Wilmington Blvd., Wilmington, CA; Pacific Marine Supply Co., 1223 Western Ave., Seattle, WA, Liberty Marine Aero Motors Co., 479-481 1st Ave., New York City; Johnson Motor Products Inc., 518-522 West 57th St., New York City and Vancouver Shipyards Ltd., 1877 West Georgia St., Vancouver, B.C. Capitol marked both reverse gear cover plates (example on page 455) and dash panels with representative's names on request.

Capitol Liberty dash panel. (Author's collection)

Two photographs have been found which show a pair of early-type Capitol conversions in a British shop. They are equipped with remote control chain sprockets on the reverse gear lever shafts as if they may be a pair of opposite rotation engines intended for dual installation. This could be either the shop of a British representative or a customer preparing them for boat installation. Currently the author has found no direct reference to a British or other European Capitol representative.

The total number of Liberties converted by Capitol is unknown. Capitol was probably the largest builder, however. They actively sold them from 1924 to 1938 or later and more surviving Capitol conversions have been noted than any other builder.

THE JOHNSON MOTOR PRODUCTS JOHNSON MARINE LIBERTY and GLOBE MARINE ENGINE

Johnson Motor Products purchased large quantities of surplus Liberty engines (see pages 280-281) and sold them as aero engines. Beginning in late 1924 or early 1925 they also widely advertised marine conversions with illustrations being of Capitol engines. They did not mention the Capitol name and indicated the engines were their own. Ad illustrations show control panels and gear cover plates marked "Johnson Motor Products" and "Johnson Marine Liberty." Johnson must have either purchased the conversions from Capitol and installed them, or sent engines to Capitol for conversion.

Although the company may have sold a few examples as standard Capitol conversions, it was not long before they were marking them as the "Johnson Marine Liberty"

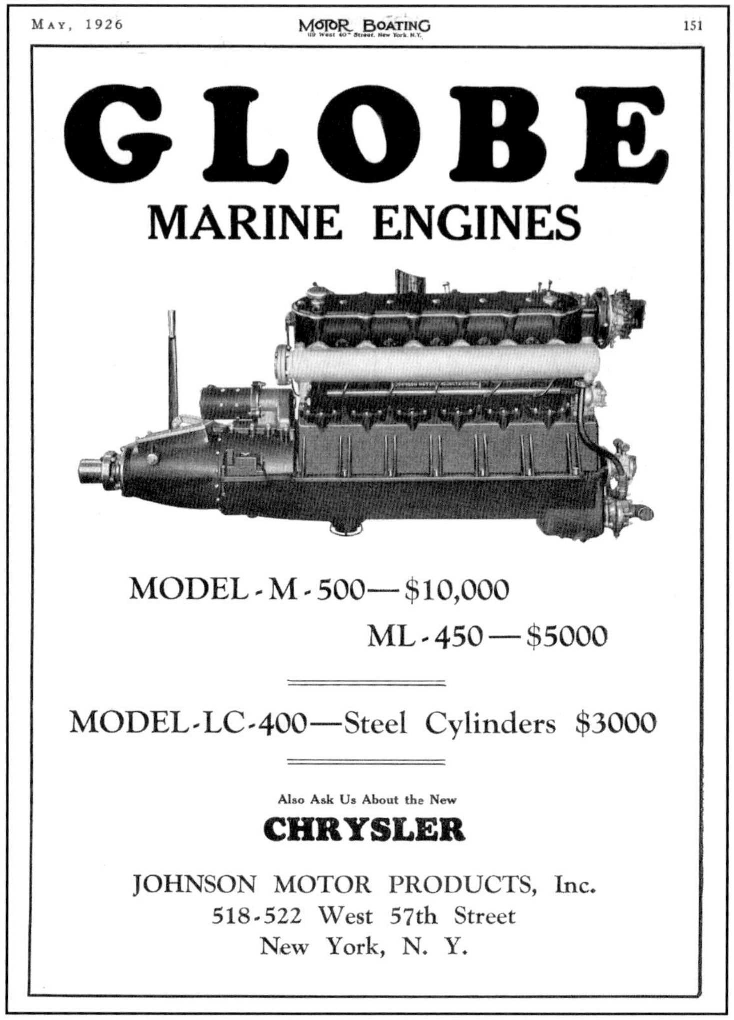

(Above and below courtesy of Keith Brayer)

Johnson Motor Products M-500 Globe

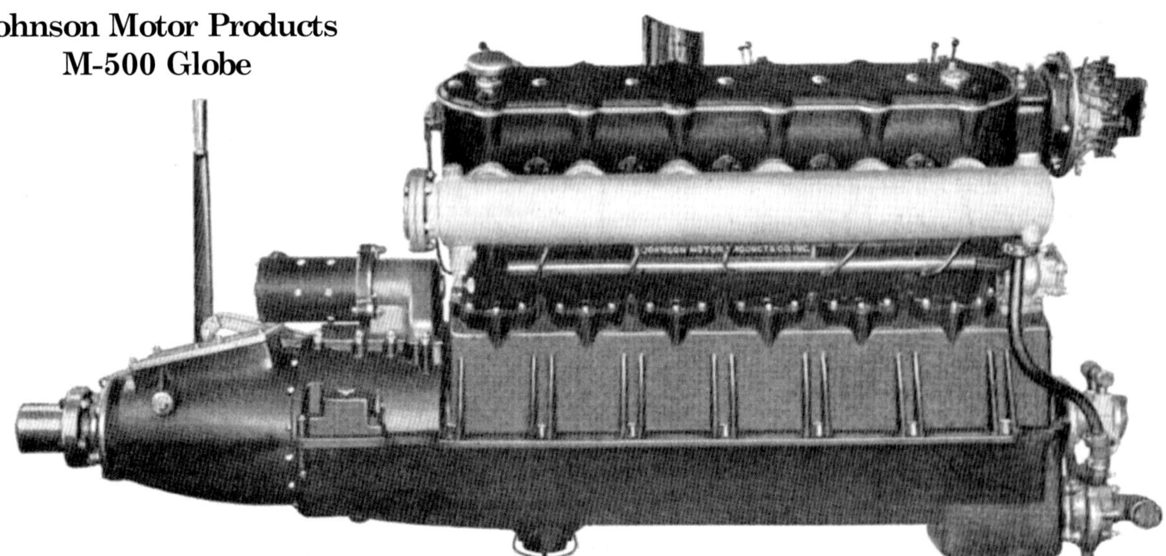

March 1929 MoToR BoatinG

425 H. P. Lyon-Tuttle Liberty

Service
- Maintenance
- Repairs
- Installations
- Overhauling
- Alterations
- Storage

Construction

Sea-Lyon Runabouts
- Sea-Lyon Commuter
- Sea-Lyon 65
- Sea-Lyon 60
- Sea-Lyon 45
- Sea-Lyon 40
- Sea-Lyon 35

Used Runabouts
of all makes, from $1,000 to $10,000.

See page 223

THE Lyon-Tuttle Liberty is the fastest and finest marine development of this famous motor. For the past decade various engineers have been improving and refining marine models of the twelve cylinder Liberty. All the best of their experience and discoveries are embodied in this Lyon-Tuttle motor.

Fast enough for the speediest runabouts and racing hulls—strong and sturdy enough for express cruisers and commuters—this motor offers abundant speed and power at a cost per horsepower considerably lower than any other high grade marine engine. Priced at $5,000 complete, f. o. b. New York. Right and left hand for single and twin screw installations.

You are invited to come up to our new plant and service station at City Island and see our boats and motors under construction. You will find this a different kind of shipyard from any you have ever known. With perfect facilities, machinery and equipment for every operation of boat and motor construction, maintenance and repair, we are prepared to render a quality of service that exceeds every standard of the past.

Stop at our Hotel Barclay Showroom, see the Sea-Lyons on display, and make an appointment for a demonstration.

THE LYON-TUTTLE CORPORATION
KING STREET, CITY ISLAND, NEW YORK CITY
Telephones: City Island 1645-1646

Salesrooms: **HOWARD W. LYON, INC., HOTEL BARCLAY**
532 LEXINGTON AVE. (at 49th STREET) NEW YORK
Telephones: Vanderbilt 4445-4446

The only runabout builders with complete plant and service facilities in New York

16

(Above and below illustrations from the author's collection)

engine and selling them under their own label. These were in fact Capitol conversions and one could not tell the difference, only the name changed. The 450 HP Capitol version was rated at 1,850 RPM and weighed 1,250 lbs.

Evidently sales were so good that they felt justified in investing a fairly large sum in the design and manufacture of a modernized version of the Liberty. They had Paragon build a reverse gear and an engineering staff design and build a set of new cylinder banks with enclosed valve train and cast-in-block cylinders. The standard Liberty carburetors and ignition system were retained as was the Liberty bore and stroke. Thus standard Liberty pistons and rods were probably used. Rated power was 500 HP @ 2,000 RPM and weight 1,500 lbs. The new engine was introduced about January of 1926, called the Globe Marine Engine and priced at a hefty $10,000. At $10,000 this was the most expensive Liberty conversion on the market at more than double the average price and $3,500 more than Gar Wood's 500 HP TS engines.

Even with this modernized design, I am sure sales were dismal. The engine did not stay on the market long and it appears to have been gone by 1929 and perhaps earlier. The standard Johnson Marine Liberty disappeared with it.

THE LYON-TUTTLE LIBERTY

Howard W. Lyon had first become a distributor of Baby Gar Runabouts in late 1924 and had set up offices and a display area at 725 East 135th St. in New York City. By July of 1925 he was advertising in all the popular boating magazines that he was the exclusive distributor of Baby Gars with offices and show room in the

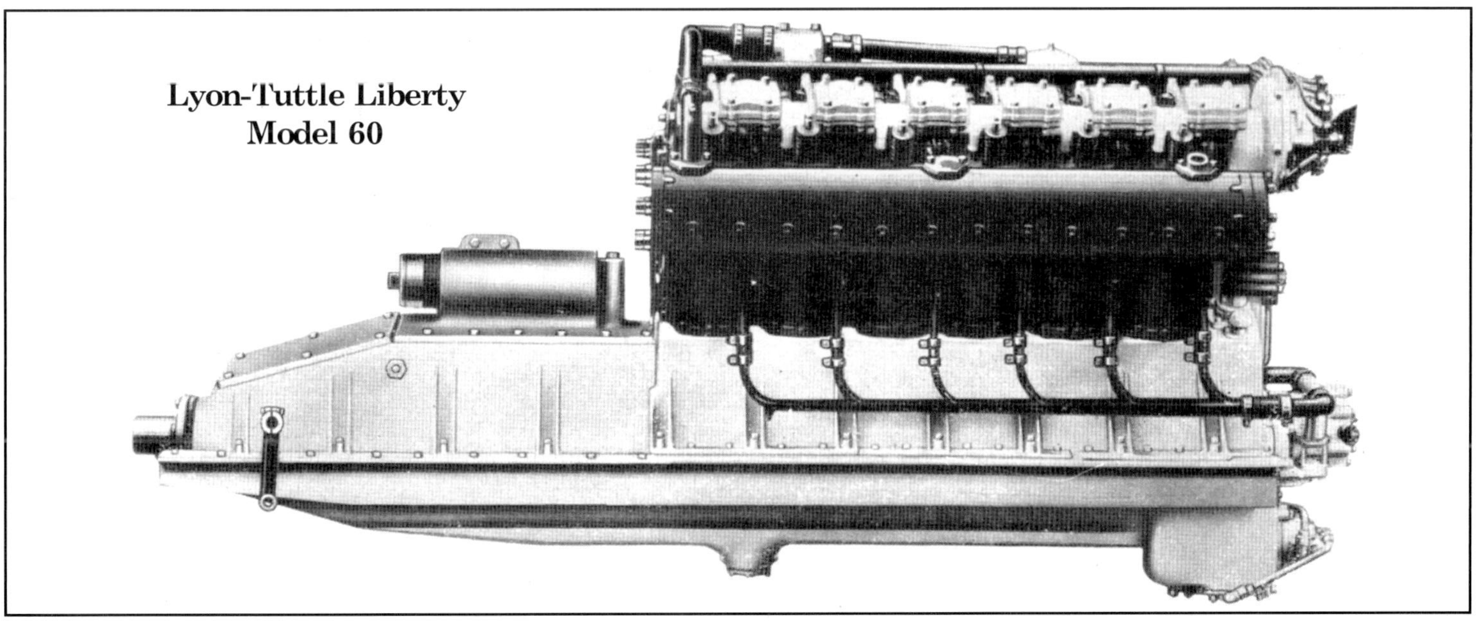

Lyon-Tuttle Liberty
Model 60

Gotham National Bank Building, 1819 Broadway, New York.

His New York sales offices were moved to the Hotel Barkley, 530 Lexington Ave. in December of 1926. During a period of a little over a year during 1926-27 he sold 14 33' Baby Gars and 41 26' Baby Gar Jrs., about 40% of Wood's total production of the period.

In December of 1926 it was announced that Howard Lyon would take over the distributorship of the entire line of Gar Wood boats and establish dealerships in several cities across the country.

Howard Lyon was a salesman of extraordinary talent but he and Wood began to have severe differences of opinion. By mid-1928 they had parted company. Wood took over his own distributorships and Howard W. Lyon set up his own boat building company, took over a boat yard on King Street, City Island, New York and converted his Hotel Barkley showrooms to display his own line of Sea-Lyon boats.

A 1930 Sea-Lyon ad showing their extensive line of high class boats. The model "60" was powered by their Lyon-Tuttle Model "60" Liberty Marine. The "60" referred to the speed obtainable by the 36' runabout. (Author's collection)

Gar Wood's top-of-the-line cruisers and Baby Gar boats were powered by high-power Liberty marine engines and Lyon recognized the drawing power of having these high-class engines offered. He decided to offer his own Liberty marine engines and joined with Mr. Tuttle to set up the Lyon-Tuttle Corp. and build conversions at his Island City boat yard. The new engine was announced in a full-page ad in March of 1929.

The engine was listed in various boating magazine lists of leading marine engines in 1929 and 1930 but not thereafter. It was born just prior to the crash and died shortly thereafter. The market would not support the 36' Sea Lyon 60 with his Liberty at $12,500. The engine alone sold for $5,000.

The engine was rated at 425 HP @ 2,000 RPM, weighed 1,350 pounds and used Joe's gear and Lyon-Tuttle starter. The reverse gear was built into a well matched gear case and resulted in one of the better looking Liberty marine conversions built. It was referred to as the Lyon-Tuttle Model 60 Liberty.

A Rosenfeld photo has survived which shows the Lyon-Tuttle engine assembly room with five of these engines under construction. That probably represented at least a third of the total production.

THE VIMALERT LIBERTY

The Vimalert Co. Ltd. of 807-835 Garfield Ave, Jersey City, NJ was the last company to build Liberty marine conversions. The first reference the author has found of this activity is February of 1930. If indeed this was the date of their entry then they started just when the Depression was really getting bad. An interesting choice since Lyon-Tuttle had entered the market the year prior and by now was already on the way out. Both Capitol and Gar Wood were, by this time, well-entrenched in the market.

In 1933 Vimalert purchased a number of surplus Packard 2500 aero engines and converted them for marine use as well. The Liberty had been listed simply as the Vimalert Marine engine up to this time. It now became necessary to differentiate between the two and the Liberty became the Vimalert M-12 and the Packard the Vimalert Packard and eventually the P-2500.

Initially the M-12 was rated at 400 HP @ 1,850 RPM, weighed 1,800 pounds and used a special Joe's gear, regular service generator and De Jon starter. By late 1930 they were already refining the conversion. Both high-compression MH-12 and low-compression ML-12 were now offered. The MH-12 carried the 400 HP @ 1,850 RPM rating and the ML-12 a 360 HP @ 1,800 RPM rating. They installed flame arrestors on the carburetor intakes and a special Robert Bosch generator on the gear case behind the starter. It was available in both left and right rotation.

In 1932 they introduced the Vimalert Duplex, a system whereby two M-12 Liberty engines were mounted on heavy rails connected to a central gear unit so either one or both engines could drive a single propeller shaft and be controlled from a central location. Total weight of the duplex unit was 5,500 pounds. Propeller shaft output ratio was available at 1.58 or 1.34 to 1 or direct-drive and left or right rotation was available. They advertised the advantages of installing two of these units in larger boats such as Coast Guard cutters and noted there were a number in current use. Such use resulted in 1,680 HP being installed in an engine room of 12'x16'. They also noted that their 700 HP Packard P-2500 engines could be used with Vimalert Duplex gears giving an output of 1,400 HP

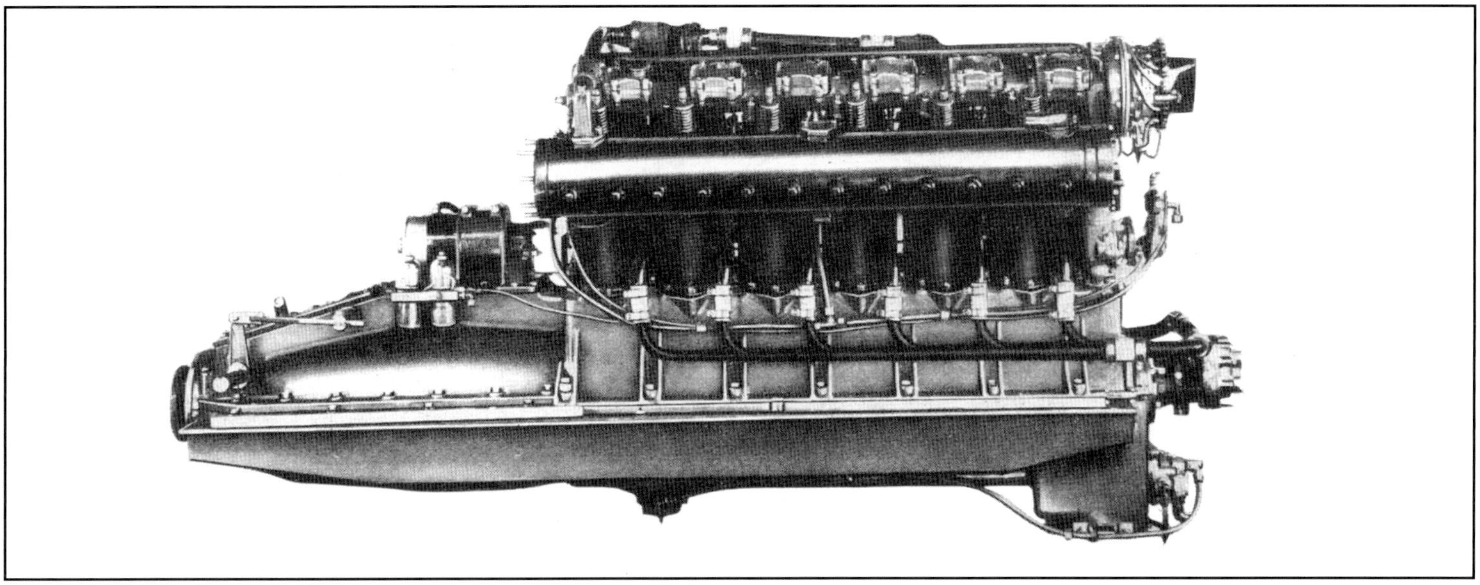

These three views are of the M-12 Vimalert as it was initially offered in early 1930. Carburetor air intakes were through standard Air Service manifold. Standard 12V Air Service generator was used. Water-jacketed exhaust manifolds were round cross section.
(Author's collection)

per duplex. Although it is doubtful that the Packard was ever sold in that configuration, it is known that a number of Coast Guard cutters were equipped with Vimalert Liberty Duplex units from 1932 to past the end of WW II.

Unfortunately the author has never encountered a Vimalert ad nor a marine engine listing which gave a price for these engines.

Vimalert actively marketed their Packard marine conversions through 1938 and their Liberty conversions into WW II. They also designed a new 1200 HP marine engine which they licensed the Sterling Engine Co. to build as the Sterling Admiral. The first of these were delivered in late 1940 for use on some of the initial experimental PT boats.

Although there seem to have been no major changes made in the M-12 engine after 1931, a number of refinements were made well into WW II, many of which were the result of extensive use by the Coast Guard, both as single units and duplex versions. Engines were provided with heat exchangers for fresh water cooling, cast-iron cylinders were offered, radio shielded ignition, original ignition coils built into the distributor caps were bypassed and replaced by individual external coils. The author has observed supplements to the Duplex/M-12 maintenance and operation manual dated as late as February of 1944.

Vimalert was the last company to offer marine-converted Liberty engines and, as a result of their sales efforts to the Coast Guard, the Liberty remained in active U. S. Military use until at least 1945 and probably a year or two later.

Although marine-converted aero engine sales represented a significant portion of Vimalert's business, and

The Vimalert M-12 as it was offered from late 1930. Carburetors now have flame arrestors with Vimalert logo; the generator has been moved to the top of the gear box; new exhaust manifolds with square top section and Vimalert name have replaced the original design. (Author's collection)

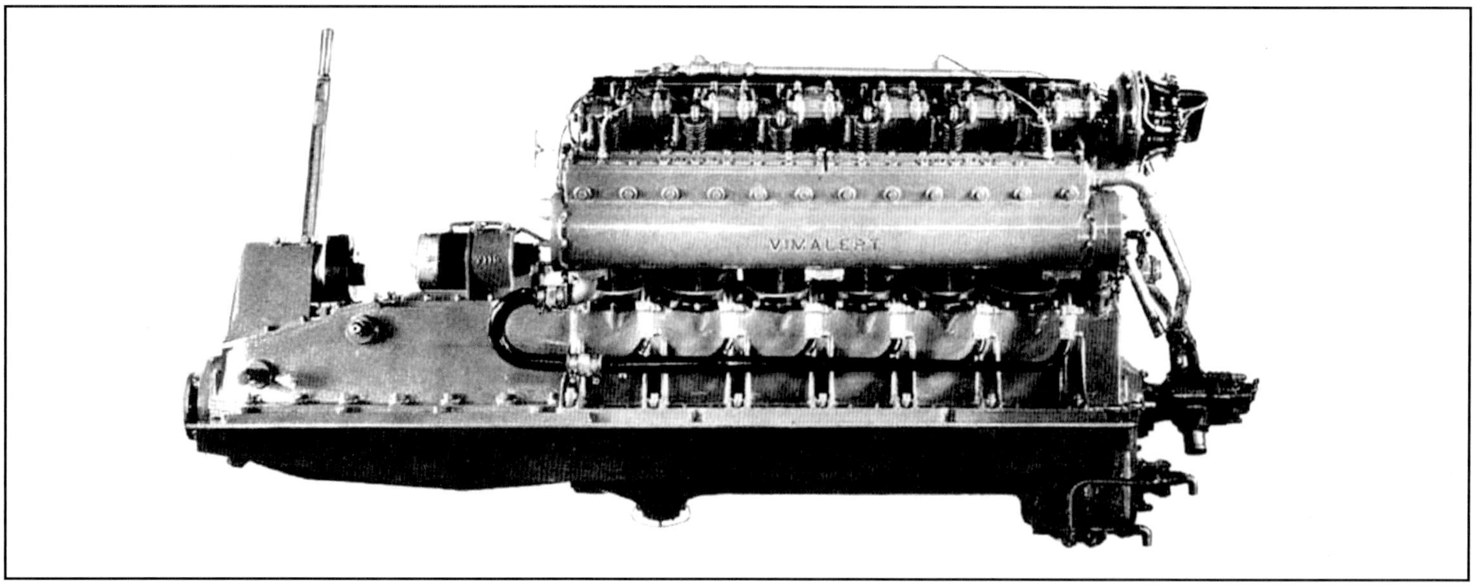

was no doubt the initial purpose of forming the company, it was not long after its 1930 start that Vimalert also became a buyer and seller of large quantities of surplus civilian and military aero engines and aircraft.

A political article appearing in the January 11, 1937 issue of *Time Magazine* had a great deal to say regarding Vimalert's request to the State Department for permits to sell and ship various aeronautical items to the Spanish Loyalist Government for use in the Spanish Civil War and the fact that the company's owner was Robert Cuse, a naturalized Latvian of Jersey City, NJ. The permit had been granted and he was given permission to sell some $2,777,000-worth of aviation material, including 18 obsolete airplanes consisting of Laura Ingall's Lockheed Orion, Powell Crosley's Northrop, seven discarded American Airlines Vultees and Harry Richman's Lady Pace and a few others. Most of the rest was composed of 411 airplane motors and enough parts to make 150 more. The motors and parts had come from Army surplus stocks at Dayton.

It was noted that "envious airplane dealers observed that Mr. Cuse had picked up his consignment at junk prices and should make between $1,500,000 and $2,000,000 on the deal."

I have no information regarding the 411 engines he was selling to Spain but two months later (April 1937) Vimalert placed a full-page ad in *Aviation Magazine* offering for sale two new and one used Vultee V-1A airplanes and a total of 514 surplus government aircraft engines.

There was also an inference in the *Time* article that Vimalert had, in the past, sold surplus aircraft equipment to Russia through Amtorg. This could well have included some Liberty engines for aircraft use since Russia was still an active user of the Liberty in the early 1930s. The author has an undated copy of "Specifications and Description of the Vimalert-Liberty Type V-12-A Aviation Engine." It describes the engine in detail and notes it is completely tested and equipped with all the latest improvements before sale and shipment. It also notes the careful packing it gets for export.

Although they certainly could have existed, the author knows of no other "production" marine conversions of the Liberty. Certainly a few were converted by individuals using stock "bolt-on" reverse gears and perhaps V-drives as there was an ample choice available. This method was undoubtedly used by a number of early race boat enthusiasts to produce one-of-a-kind craft.

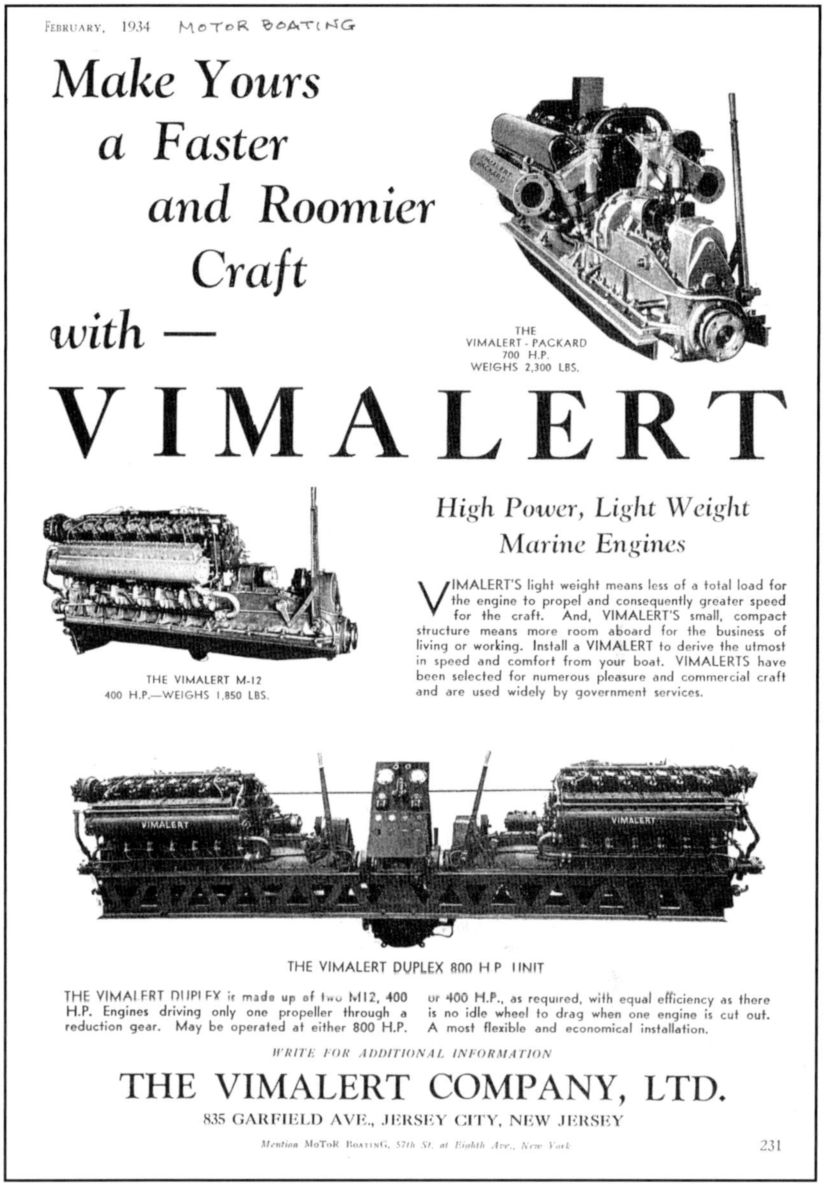

February 1933 Vimalert ad describing their line of marine engines. The M-12 was first offered in 1930, the Duplex M-12 in 1932 and the P-2500 Packard in 1933. (Courtesy of Keith Brayer)

UNIQUE LIBERTY MARINE ENGINES

An early and very interesting marine conversion was done on Liberty 8 No. BX-15. This was the last Liberty 8 built and probably one of the first sold as surplus. The engine made its way into the ownership of Harry N. Snavely of Pittsburgh, Pennsylvania and by the summer of 1922 he had given it a very basic marine conversion, the type usually made on aero engines destined for use in race boats. A Stromberg carburetor was adapted to modified intake manifolds to which long intake pipes were attached. A combination flywheel and clutch was mounted on the

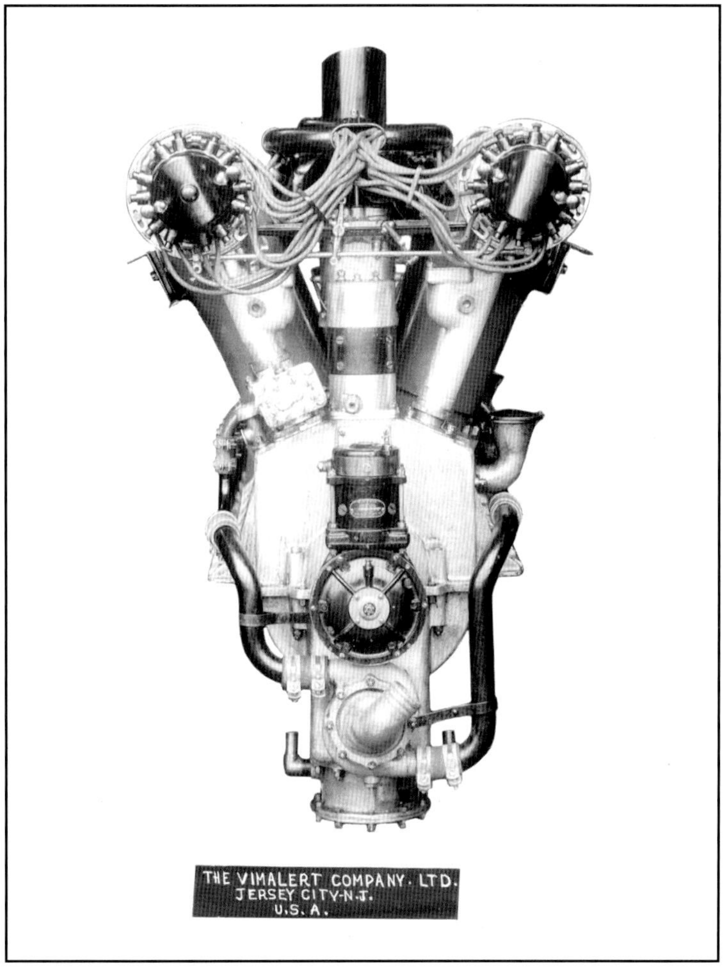

VIMALERT DUPLEX UNIT

800 H.P.—Weight: 5500 Lbs.

Comprises two M-12 Engines, end to end, with a Reduction Gear between, both Engines operating one Propeller Shaft. Either Engine can be used alone or both be operated together.

One of the new fleet of U. S. C. G. Patrol Boats. Powered with Vimalert Duplex Units. Designed by Eldredge-McInnis, Inc.

From a January 1938 Vimalert ad which listed the M-12, P-2500 Packard with added supercharger for 1,000 HP output if desired, and the Duplex Liberty unit. (Author's collection)

The Vimalert Liberty as pictured in their specifications and description of the Vimalert-Liberty Type V-12-A Aviation Engine which offered the engine for sale and export. Equipped with Bijur starter and machine gun synchronizer if specified. (Courtesy of Kim McCutcheon)

propeller shaft with an output shaft driving forward to a V-drive. Starting was accomplished with a manual starting bar connected to an adapter on the rear of the crankshaft.

The engine was installed in a small, custom-built single-step hydroplane named *Liberty The Second* and taken north to Pennsylvania's Conneaut Lake in preparation for participating in a number of racing events to be held the last week end of August and the first week end of September. After a number of test runs it participated in several events on the first weekend, winning the 15-mile event.

The boat supposedly was capable of great speed and rumor was that she had unofficially broken the "world speedboat record" with a speed of 78 MPH and was set to make it official the next weekend. Such was not to be as she capsized during testing on September 2, 1922 and sank rapidly. Neither the driver/owner nor mechanic were injured and, in spite of numerous efforts over the next two years, divers were never able to locate and retrieve the boat.

In 1980 two local boating enthusiasts with an interest in local boating history, William Houghton and Brian Simpson, began an effort to locate and raise the boat. They spent the last two years in actual diving efforts and finally on July 13, 1985 found and raised the boat. It was in remarkable condition after resting in 50' of water and a few feet of silt for 63 years.

Above: **Liberty The Second** ***just after refloating on July 13, 1985. The boat and motor had been under water for 63 years but were little damaged. Restoration required a little over a year. Right: The boat and engine as it appeared after restoration. Reinstallation is almost complete with the clutch-to v-drive shaft remaining. The engine is Buick-built L-8 BX-15.*** (Courtesy of Dan Styduhard)

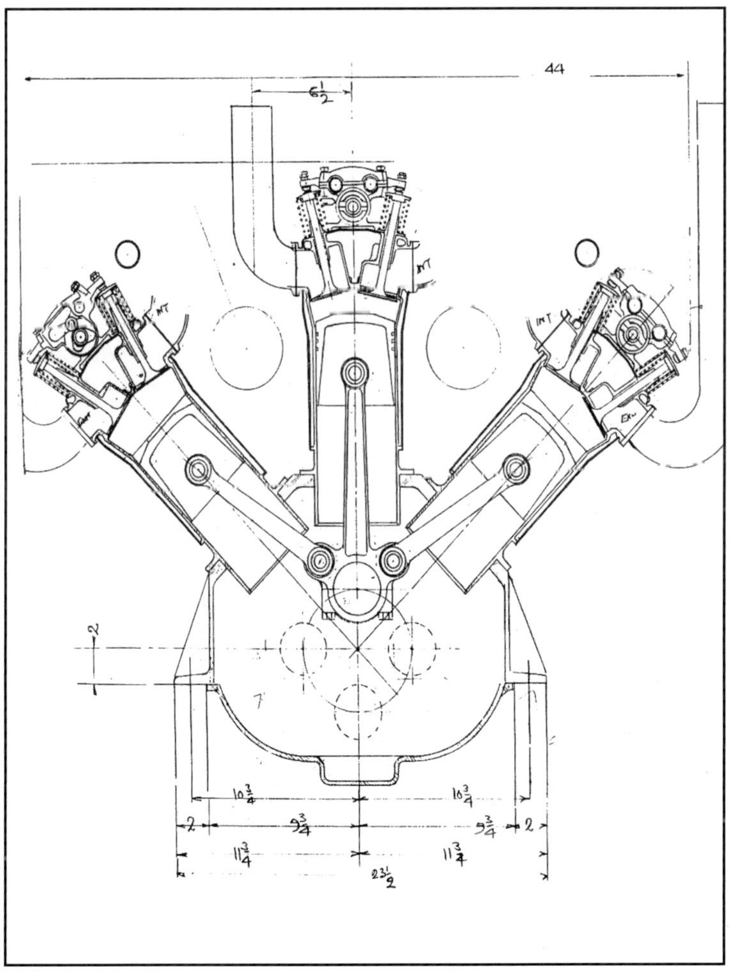

Left: the layout drawing for the Miller-Liberty W-3300. Although the bearing caps split in the center of the crankshaft the crankcase itself is extended 2" farther down. Special Miller-designed master and articulated rods were necessary. Above: one of the two engines built during testing in the Miller shop. (Courtesy of Mark Dees)

At left is Miss Los Angeles II during the first heat of the 1929 Harmsworth. The photo at right gives the reader an idea of how much space is required to mount two engines the size of the W-3300. The two gentlemen were visiting boating enthusiasts from Great Britain and not the actual driver and mechanic. (Courtesy of Mark Dees)

Both boat and engine were restored to museum standards and have been run a number of times on the lake in which it sank. Its permanent residence is in the Lakeside Museum on the waterfront in Conneaut Lake Park. BX-15 is one of only three surviving Liberty 8 engines. The first Liberty 8 built is in the NASM and BX-12 is in the Air Force Museum.

Whether or not the boat was ever capable of the claimed 78 MPH is highly debatable. The claim at the time was 450 HP for the engine. Upon restoration it appeared to be little modified from its original design. Considering the fact that official tests of serial BX-5 on April 18, 1919, gave its maximum power as 317 HP @ 2,000 RPM, the figure of 450 HP for BX-15 seems somewhat the figment of local imagination.

The world water speed record was then 80.567 MPH and held by Gar Wood with the four-Liberty-powered *Miss America II.*

THE 24-CYLINDER LIBERTY MARINE ENGINES

For size, at least, the 24-cylinder Liberty was the largest built. Two have already been mentioned, the Engineering Division "X" engine of 1918 and the Aeromarine AL-24 of 1922 which was a fan design using four rows of six cylinders.

Here the two Miller-Liberty W-3300 engines sit in the shop of the Creger Corporation in Los Angeles after being removed from Miss Los Angeles II and sold to Harman Fengler (at right). Looking over the sizable engine with him is C. W. Atwater, his shop manager. Their hopes were to raise enough money to build a speed car capable of 300 MPH with which Peter De Paolo could break the World Land Speed Record. Funds were not forthcoming and the project died without a car ever being built. The picture is dated 9/6/30. (Courtesy of William Lewis)

THE MILLER-LIBERTY W-3300

The engine next to be discussed came into being in 1929. It was designed and built in Harry Miller's shop with Leo Goossen and L. A. Orsatti doing most of the layout work. The work was done for California boat racer Jim Talbot. Talbot coveted the British International Trophy (The Harmsworth) and the World Water Speed record, both of which were currently held by Gar Wood with his Packard-powered *Miss America VII.*

Talbot had raced in the 1928 Harmsworth event with *Miss Los Angeles,* a Hacker Gold Cup size hydroplane powered by a Miller 620 cu in engine said to develop 746 HP @ 4,000 RPM. The race was won by Wood in *Miss America VII* powered by two 1,000 HP Packard engines. Wood currently held the W.S.R. at a speed of 92.838 MPH with the same boat.

Talbot had found that the choice of a light boat with a small but powerful engine had failed to be the successful combination. He was now ready to take Wood's lesson to heart and put a great deal of power in a larger boat. Either unwilling or unable to acquire Packards similar to those used by Wood, he went to Harry Miller for suggestions. Miller came up with the idea of building two W-24 engines based on Liberty cylinders, valving and ignition. With good cam design and breathing such a design offered about 1,100 HP @ 2,600 RPM from 24 Liberty cylinders with a total weight of 2,000 pounds. Two such engines of opposite rotation were to power his new *Miss Los Angeles II.*

The two Packards he was to compete against would by then have individual outputs of 1,100 HP, each at a weight of 1,200 pounds.

The new W-24 Liberties would be built on newly designed aluminum crankcases and use Miller-designed articulated rods and crankshaft. The angle between each bank of eight cylinders was maintained at the standard Liberty 45° in order to utilize Liberty timing. New eight-cylinder cams and cam housing were required. A bank of four two-throat carburetors fed the left and center cylinder banks and four single-throat carburetors fed the right bank. Displacement was 3,298 cu in and the engines were usually called the Miller-Liberty W-3300. The dual ignition was generated by six eight-cylinder Northeast distributors mounted on opposite ends of the three cam assemblies.

By the time the 1929 Harmsworth was run Wood had built an additional boat, *Miss America VIII,* and both *VII* and *VIII* were entered. The British entered *Estelle IV* owned by Barbara Carstairs. When the race ended *Miss America VIII* was first, *Miss America VII* second and *Miss Los Angeles II* third. *Estelle IV* did not finish.

The Depression finally caught up with Talbot as it did many others and his racing activities came to an end. In 1930 the engines were sold to Harlan Fengler of the Cregar Corporation. He and Peter De Paolo, a race driver of considerable note, laid out the design of a new land-speed record car they called *America I* with which they intended to bring back the land-speed record to America. The record was then held by Major Seagrave of England. That record had been held twice before by Liberty-powered cars (*Babs* in 1926 and the *White Triplex* in 1928) so it was not inconceivable that it could happen again. Such was not to be the case, however. They were never able to raise the finances and the car was never built.

Supposedly the engines were later cannibalized to keep Liberty engines running in the Bombard water taxi fleet operating between the California mainland and Santa Catalina Island.

THE FAN 24 LIBERTY MARINE

We currently know little of this engine's history. We do not know when it was built but believe it was done for Sol Zitto for use in one of his *Zitto Plane* boats. It has been sitting in storage in one collection or another for perhaps the last 50 years. It has as yet never been disassembled and inspected. Thus we only know what we can ascertain from external inspection.

The layout is a fan-type 24-cylinder composed of four banks of six cylinders utilizing standard Liberty cylinders, ignition system and cam assemblies. In order to be able to utilize as many standard Liberty parts as possible the designer made the included angle 45° between each bank of cylinders. In this manner he could use standard Liberty ignition with each pair of outside banks operating as a standard twelve-cylinder engine. With the angle between the center banks also being 45° a standard set of intake manifolds and carburetors could be used to feed those banks. The two outside banks used special manifolds set up to use a single standard US 52 carburetor. The end result was that virtually all ignition components were standard and, with the exception of two special intake manifolds, so was the fuel-feed system.

The crankcase, crankshaft and master/link rod system were, of course, unique for the engine. On the rear of the engine was a starter driving a double set of step-down gears through a chain and sprocket system. At that position is also mounted a control panel made by Auto Engine Works of St. Paul, the builder of Capitol Liberty marine conversions. It would have been convenient for the builder to use an available panel. Thus this does not prove or disprove Capitol as the possible builder.

The engine was built to serve as a marine unit and is equipped with a clutch, flywheel and reverse gear which appears to have been built specially for the engine. Exhausts fed water-cooled manifolds for each cylinder bank.

Considering the use of standard Liberty components it would be logical to assume the output of the engine would be about double that of a standard Liberty 12A, about 850 HP @ 1,850 RPM.

The author knows of no other production Liberty conversions and doubts that there were any since they left no trail of attempting any wide-spread sales through advertising. The Lee Motor Company conversions do seem to contradict that theory as a nice set of brochures has survived which indicate not only a well-designed and executed conversion but one for which they claimed "built since 1919 and hundreds are now in use." Builders of such products were not above exaggeration if it would help sales. Letting the public think the item was so excellent that they were already selling in large quantity might well have been looked at as acceptable. Or perhaps in this case the author just missed the ads.

Certainly there were other marine conversions done on a one or two item basis but not as a production program.

The Fan 24 Liberty Marine. This four-bank 24-cylinder engine was built using as many standard Liberty components as possible. This was in large part made possible by maintaining 45° between each bank of cylinders. In this manner standard ignition components were used as well as most intake components. (Courtesy of Warren Greatbatch)

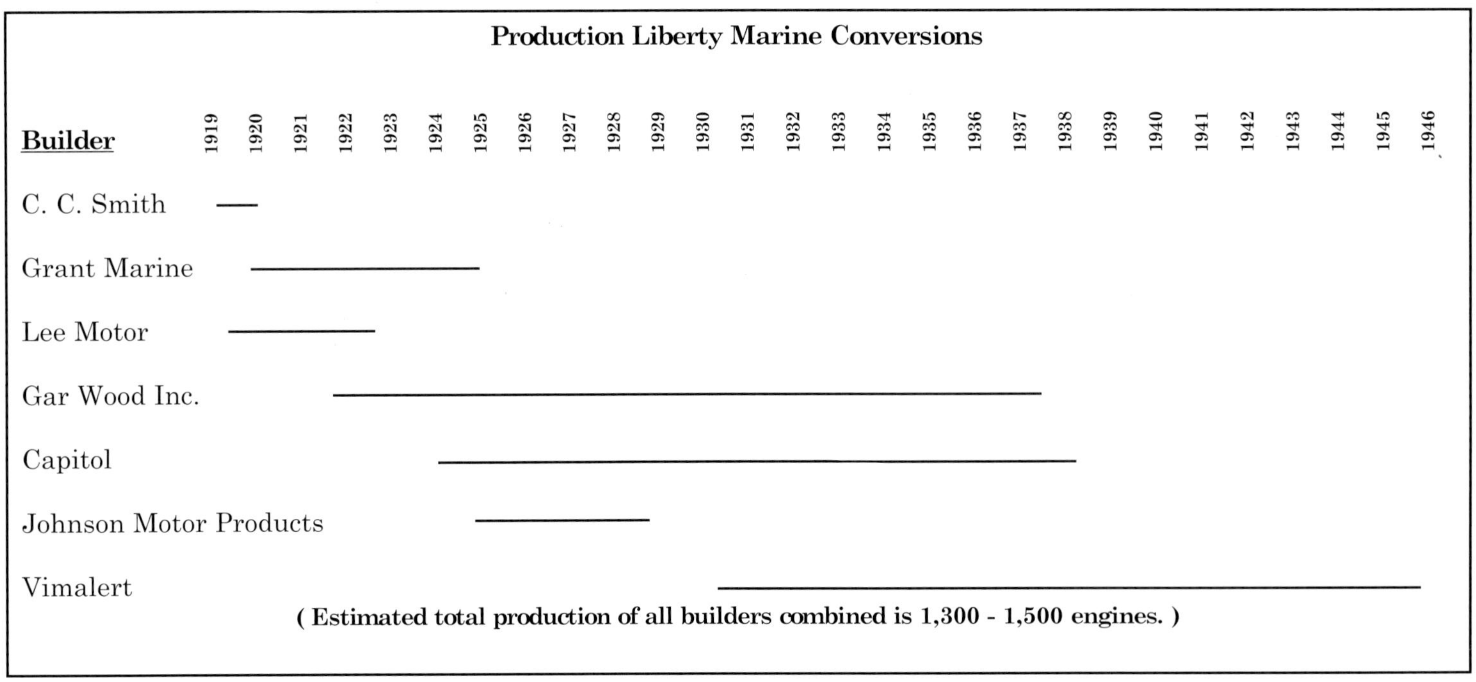

Chapter 13

Other Uses of the Liberty

This chapter will necessarily cover a variety of subjects and bring to light some unique uses to which the Liberty was applied. I will first go into the utility uses.

THE HOLLYWOOD WIND MACHINES

Beginning in the early 1920s and continuing into the late 1950s Hollywood movie producers utilized surplus Liberty engines as wind generators to give the effects of raging storms. By using multiple units, hurricane velocity winds could be generated over a fairly wide area.

Although their use for this purpose is well-known, little documentation seems to have survived as to the numbers of units involved nor what particular companies may have been involved in their manufacture. Nevertheless, the Liberty seems to have been the preferred engine for the purpose. It would have been the logical choice since it was probably the lightest and most powerful available and the price was reasonable.

Although there must have been some large studios that owned their own stock of these machines, the more common practice was to rent them when needed from companies in the business of supplying such resources. The units would be mounted on small flatbed trucks and transported to the location needed. Ray Sealock, of Hollywood, was a pioneer in that business.

Hollywood wind machines were made by hand using either surplus airplane radiators or special made ones. The above machine is on display in America's Packard Museum in Dayton, Ohio. The one at left was photographed by Al Shinnerer resting in a California boat yard and was probably destined to be converted to a marine unit.

Chances are that any hurricane scene you see in a 1920-50 motion picture was created using Liberty wind machines.

HOLLYWOOD ON LOCATION POWER GENERATORS

According to Ray Sealock, more Liberties were used to drive portable on-site power generators for movie studios when they were filming on location than were used as wind machines.

The units were trailer-mounted and towed on-site. Generator and motor were mounted in an enclosed and well sound-insulated cabinet and exhaust well muffled. Even though the unit might be located well away from the shooting location, ambient noise levels were kept to a minimum.

Liberty-powered units were in use at least to the early 1950s. The illustration at right is of one of the Sealock units in use in 1950.

Liberty power generator used for on-location filming by Hollywood studios. This particular unit was owned and leased by Ray Sealock as late as 1950. (Courtesy of Al Shinnerer)

LIBERTY LAND SPEED RACERS

Ralph DePalma set a new land-speed record of 149.875 MPH on February 12, 1919 in the *905 Racer*, a Packard race car powered by the 905-2 Packard which lent a number of its design features to the Liberty. On April 27, 1920 Tommy Milton drove his twin-engined *Duesenberg Special* to a record 155.046 MPH. Both runs were made at Daytona Beach. These records, although the fastest clocked in the world at that time, were not recognized as world records because Europe had not yet decided to recognize American records. This was because the American Automobile Association (AAA), which certified all racing and speed records in the United States, had no affiliation with the International Association of Recognized Automobile Clubs, which acted as the certifying body for European records. This situation changed in the late 1920s when the AAA became affiliated with the International Association of Recognized Automobile Clubs.

The European record (and thus recognized world record) was at that time 124.095 MPH and had been set on June 24, 1914 by L. G. Hornsted of Great Britain. The European record had been raised to 133.788 MPH on May 17, 1922 by Kenelm Lee Guinness of Great Britain; to 143.312 MPH on July 6, 1924 by a Frenchman; to 146.163 MPH on September 25, 1924 by Malcolm Campbell of Great Britain; to 150.766 MPH on July 21, 1925 by Malcomb Campbell (passing DePalma's 1919 record).

Parry-Thomas drives Babs down the ramp at Pendine Sands, Wales for the last time on March 3, 1927. Just after completing the first of two required runs for record qualification, Babs crashed and Parry-Thomas was killed. The car had just crossed the finish line and was traveling at about 179 MPH when the crash occurred. (Courtesy of the Ludvigsen Library)

Babs after being exhumed from its sandy grave in 1970 and application of considerable restoration efforts. In this photo a portion of the original body is retained to show its condition upon retrieval. The remainder of the car was well on the way to complete restoration. The job was carried to completion with the installation of a reconstructed duplication of the original body and the car is now on display at Pendine appearing just as it did in the picture on the prior page. (Courtesy of Neill Bruce)

In 1925 J. G. Parry-Thomas, a Britisher who had made something of a name for himself as a race driver, acquired the Liberty-engined *Higham Special* race car formally owned by Count Louis Zborowski, who had been killed during the 1924 Italian Grand Prix. Parry made some modifications to the car and took it to the seven-mile-long Pendine Sands in Wales in October 1925 to attempt to break Malcolm Campbell's record. The weather was so bad that no attempt could be made and he withdrew and returned for another try the following April.

In the meanwhile, Henry Segrave prepared his Sunbeam *Ladybird* and set a new record of 152.307 MPH at Southport Sands on March 16, 1926. This really put the pressure on Parry-Thomas and on April 27, 1926, he returned to Pendine Sands and drove the car, which he had renamed *Babs*, to a record speed of 169.298 MPH, breaking both Segrave's and Milton's marks. The next day he was out again and raised the record to 170.624 MPH.

In February of 1927 Captain Malcolm Campbell took his car *Bluebird* to Pendine Sands and established a new record of 174.20 MPH. This prompted Parry-Thomas to try to regain it. He took *Babs* back to Pendine Sands and, on March 3, 1927, made another run. The car was not running well, but after some carburetor work and a change of plugs, it went through the measured mile at 179.5 MPH. But when Parry-Thomas let off the throttle to begin to slow, the car swerved, flipped and landed again on its wheels and continued for some distance before sliding to a halt with a collapsed right rear-wheel. He, of course, failed to make the required return direction run for a record.

Babs was driven by chains mounted on both sides to drive sprockets on the rear wheels. Either before or during the crash, the right chain had snapped and lashed out, killing the driver.

Parry-Thomas was buried in the Byfleet Church graveyard near England's famous Brooklands race track. His friends decided that *Babs* should never run again and buried the car under the sand at Pendine. But the car refused to die. After more than 40 years a Welsh college teacher, Owen Wyn Owen, decided it would be fitting to exhume the car and exhibit it in Pendine Village. After

John Godfrey Parry-Thomas in Babs. The picture was taken some time in 1926 or early 1927 just before one of his record runs at Pendine Sands. (Author's collection)

At top-left is the White Triplex in early configuration. The driver sits with two engines behind and one forward. (AACA photo). Above-top is the car on April 22, 1928 with Ray Keech driving. (Phil Harms photo) Above-bottom shows the car on March 13, 1929 just after the crash which killed driver, Lee Bible. (Author's collection)

considerable government red tape was cut (the car was buried beneath a police station), *Babs* was dug up in 1969 and was, surprisingly, well preserved enough to be mechanically restored to running condition. A new body was constructed and the car returned to its original glory. It is now on display at Pendine Village.

Beginning in 1927 records set in the United States became also recognized World Records. This was a good step not only for the United States record holders but for the Europeans, because Daytona Beach provided the best high-speed running surface in the World.

Major Segrave brought his car, the *Sunbeam Special*, to Daytona Beach and raised Campbell's record to 203.790 MPH on March 29, 1927. In February of 1928 Campbell took his Napier-powered *Bluebird* to Daytona and regained the record with a speed of 206.956 MPH.

One last time prior to the Second World War, America took back the world land-speed record. Liberty power again prevailed, even if in the form of 10 year-old engines. Wealthy Philadelphia sportsman J. M. White had built the *White Triplex*, an 8000-pound car powered by three Liberty engines developing 500 horsepower each. Supposedly this car under tests had posted unofficial speeds of 253 miles per hour.

The *White Triplex* was shipped to Daytona Beach where it was driven by Ray Keech to a new record of 207.5526 mph on April 22, 1928. This was enough to take the record but it did not measure up to the 253 MPH claims made for the car.

Also being built and tested at this time was the Stutz *Blackhawk Special*. This 3000-pound car was powered by a modified Miller V-16 engine of 191 cubic inches displacement. It carried a very sophisticated and highly streamlined body. Unfortunately it crashed during trials on April 25, 1928, and Frank Lockhart, its driver, was killed. The vehicle was totalled. On this sad note rested the speed trials for 1928.

Major Segrave built a new car, the *Golden Arrow*, and brought it to Daytona Beach in March of 1929 in an attempt to take back the record. White had the *White Triplex* there, as well, with Lee Bible as the driver this year. The Napier Lion-powered *Golden Arrow* was put through the traps on March 11, 1929 for a two-way average of 231.362 miles per hour. On March 13, Lee Bible tried to drive the *White Triplex* to a record, but he lost control. The swerving car struck and killed Pathe News photographer Charles Traub before crashing and killing its driver as well.

Segrave would retain the title for another two years before once again relinquishing it to Campbell, who would not only regain it but periodically better it until by September 3, 1935, it was at 301.13 MPH.

If the Miller-Liberty speed car proposed by Harlan Fengler and Peter De Paolo (see page 474) had actually been built perhaps a Liberty-powered racer would have held the record one more time. It was not to be so and the *White Triplex* was the last with Liberty power to hold the World Land Speed record.

Chapter 14

Epilogue

The Liberty is a very unique aero engine, not so much because it was a unique design (because it was not), but because of its exceptionally long history of production and active military use. It was, for the time, an up-to-date design and would have been considered "state-of-the-art" when introduced. The only other design which was in current use which might be considered more modern was the aluminum monobloc. The only such engine in current successful production and use was the Hispano-Suiza, and it was a 200 HP engine and required a much smaller block than would have been required for a 400 HP Liberty.

Several other monobloc engines were in the process of design and production when design began on the Liberty, with none thus far successful. The Liberty's designers knew full well that successful design and production of a monobloc engine of this size had so far not been accomplished. These facts, however, were somewhat mute because a monobloc design did not lend itself to the requirements at-hand anyway. The design object was to produce a line of engines, four, six, eight and twelve cylinder, which shared as many common parts as possible in order to make manufacture as economical and rapid as possible. In addition, it had been specified that only proven designs be incorporated. Monobloc construction did not fit that requirement in that it would almost guarantee a long battle to come up with a successful product. In addition, at least two different designs would be required, a four-cylinder and a six-cylinder block. To allow total interchangeability required individual water-cooled steel cylinders.

The initial design for the Liberty was created between May 29, 1917 and June 4, 1917 and the decision to build eleven prototypes (six L-8 and five L-12) was made. After detailing engineering drawings, fabrication of parts began. The first prototype, an L-8, was delivered to the Bureau of Standards on July 3, 1917. This is somewhat of a record, from start of design to a delivered first engine in 36 days.

Work on the prototypes continued with L-8 No. 2 being ready for its first test-run on July 31, 1917. (L-8 No. 1 had been built as an engine to be used for demonstration to various committees and individuals and to verify the design could be easily assembled and disassembled for repair and parts replacement. It was never run.) L-12 No. 1 was completed and made its first run on August 13, 1917. After various modifications, mainly in the intake system, the engine was submitted to its 50-hour qualification test, which it completed successfully on August 25, 1917. Thus we have what was no doubt another first - beginning of design to successful completion of 50 hour qualification test between May 29 and August 25, 1917, a period of 92 days.

Initial production contracts were let during the first week of September 1917 and Packard completed the first production engine on November 19, 1917. Production was halted after the end of the war with the last production engine delivered on 3/15/19, number 20,479.

During this period of time several thousand were installed in American, British, Italian and French airplanes.

This is the history of the Liberty in an extremely condensed version from concept to end of production. Although Italian usage was very limited, the Americans, British and French kept the L-12A in active military and civil use into the mid-1930s. In addition, so did Australia, Canada and numerous countries of South America and Europe as well.

The U.S. Coast Guard first put the Liberty 12 in service for patrol craft in the early 1930s and continued such use until after the end of WW II, probably about 1950.

Russia even put the engine in production in their own factories from 1923 to 1932 and built 2,909 of them. (Their version was a metric copy of the WW I U.S. production engine.) In addition, they imported just over 3,000 U.S.-built engines. They used them initially in airplanes (from 1923 to 1933) and then in tanks from 1931 through the end of WW II.

The British found themselves caught short in the availability of a suitable tank engine when WW II broke out in Europe. They looked for a suitable size engine, which was already designed and proven, that they could rapidly put in production. The best they could find was the WW I Liberty 12. Initially, beginning in 1937, they used U.S.-built WW-I engines acquired both in the U.S. and in England. As soon as they could, they put the engine in production and eventually built some 11,000 of them in various configurations. They were used in a number of tank models by the British from 1937 to the early 1950s. After the war, they provided the Greek army with a number of surplus Liberty-powered Centaur tanks. The Greeks kept these tanks in active service from 1946 until at least 1960.

Thus we see that the Liberty engine was in continual military use somewhere in the world from 1917 to 1960. I know of no other engine design about which that can be said. The engine was obsolete by the mid 1920s so why was it continued in use?

Basically there were three reasons: 1- availability as a result of the excess production of WW I. Why was it produced in such quantities? Circumstances. Production was just coming into its own when the war unexpectedly ended. The Allies were begging for engines and had ordered several thousand. We had plans to use thousands and had aircraft in production and on order to use them. Liberty production was outrunning production of airplanes to use them but we expected those circumstances would shortly change. The war ended with a surplus to current needs already on-hand. Airplane production was almost immediately ended but contractual agreements with the Liberty producers allowed them to wind down production, if they wished, instead of cutting it off abruptly. This was allowed to prevent fast shutdown from causing them severe financial hardship. All but General Motors and Ford chose slow shutdown. These two auto makers would rather suffer some financial loss and get back into auto production as soon as possible as they were much more financially stable than the other builders. Thus we not only already had a surplus on-hand but added several thousand to it after the war ended.

The second reason was: 2 - In this country we cut financing of the Army and Navy to the bone after the end of the war. The armed services would have loved to supply their air arms with modern engines as more modern ones were developed, but they had no funds with which to do so. The Liberty was reasonably up-to-date and reliable at the end of the war and was refined considerably as to reliability by the early 1920s. It was available at almost no cost. Thus it and the DH-4, of which there were also a considerable stock on-hand, were used. And when the DH-4 became too obsolete they were updated and kept in use. A number of new aircraft designs were built around the Liberty as well. It was early July of 1929 before the Liberty became so obsolete that its use in any new aircraft was prohibited.

They realized that the continued use of the Liberty was severely inhibiting the development of more modern engines. Although that marked the beginning of the end for the Liberty, it was 1935 before the last Liberty-powered airplane was taken out of service. The list of aircraft in which the Liberty was used, both military and civil was indeed impressive.

The general reasons cited in item two held true to some degree with our WW I Allies as well.

The last reason: 3 - The Russian and British production and use of the Liberty in tanks and airplanes added about 14,619 to the total Liberty production. In the case of the Russians the choice of the Liberty was based on the early decision to copy it for aircraft use followed later by a need for an engine of its capability for tanks. Since it was already in production and the tank series they chose to build was based on the Liberty-powered American Christie, the Liberty was chosen.

In the case of the British they also based their tanks on the Liberty-powered American Christie and although they considered other engines for power, they rapidly found they had nothing of appropriate size available, and they decided on the Liberty because it was a proven engine even though no longer a modern one, was of appropriate size and could rapidly be put in production.

Thus we see a vicious circle of circumstances seems to have controlled the whole affair and led to probably the widest and most lengthy use of a single design of a piston aero engine that will ever be recorded.

Thus almost ends the story of the Liberty engine. It is still in active use in restored antique wooden boats around the world and will probably remain so for years to come. It also currently powers a small number of operating antique aircraft. Is there no end to it?

An excellent example of an antique wooden boat, Vogler Boy V, powered by a Capitol marine-converted Liberty. Restored several years ago by its current owner, Curtis Erickson. It sees heavy use in the boating season and is pictured at his home on Lake Tanwax, Eatonville, WA. (Author's photo)

Appendix A

Liberty Contracts

WW I U. S. Liberty production contracts by APB and BAP
Listed in assigned Air Service serial number order
(This list excludes the experimental engines assigned numbers in McCook Field's Engineering Division block 47828-47927.)

Manufacturer	A. S. serials	Model	Order #	Date	Contract #	Date	#Ordered	#Delivered
Packard	EXP 1-6	L-8	APB resolution of June 6, 1917			no contract	6	6
Packard	EXP 1-5	L-12	APB resolution of June 6, 1917			no contract	5	5
Nordyke & Marmon	4402-7401	L-12	10449	Sept. 5, 1917	1704	Sept. 7, 1917	3000	1000
Trego	7402-7901	L-12	10140		1648	Sept. 6, 1917	500	0
Trego	7402-7901	L-12	10450	Sept. 15, 1918	(7403-7901 canceled-transferred to Ord.)			1
Packard	17435-23434	L-12	10138	Oct. 11, 1917	1646	Sept. 4, 1917	6000	6000
Lincoln	23435-29434	L-12	10139	Oct. 11, 1917	1647	Aug. 3, 1917	6000	6000
Ford	29494-34493	L-12	30035	Oct. 31, 1917	2129	Nov. 22, 1917	5000	0
Ford	29494-34493	L-12	30035-A	Dec. 4, 1917	2129-A (price reduction)		5000	3950
Duesenberg	36998-37497	L-12	30087	Nov. 30, 1917	2318	Nov. 20, 1917	500	0
Duesenberg	Above order canceled, replaced by order 30087-B for Bugatti engines 40672-42671							
	Bugatti order was dated Jan. 11, 1918, for 2000 engines. Only 40 were delivered.							
General Motors	37503-38502	L-12	30071	Dec. 13, 1917	2266	Dec. 11, 1917	1000	1000
	Replaced by order		30071-5	Oct. 23, 1918	2266-5	Oct. 2, 1918		
Duesenberg	38510-39009	L-12	30126	Canceled			500	0
General Motors	43771-44770	L-12	30206	Jan. 21, 1918	none		1000	1000
	Replaced by order		30071-2	Mar. 29, 1918	2266-2	Mar. 4, 1918		
	Replaced by order		30071-5	Oct. 23, 1918	2266-5	Oct. 2, 1918		
Packard	47928-48136	L-12 school engines		Authorization 394 and 513			212	212
Packard	48137-48269	L-12 school engines		Order 710880			130	130
General Motors (Buick)	57209-66310	L-8	730155	Aug. 5, 1918	4433	Oct. 3, 1918	5000	0
	Production stopped Nov. 14, 1918. 15 already built transferred to order 730810 & numbered 94472-94486							
Willys-Overland	In above block	L-8	730179	Aug. 20, 1918	4456	July 29, 1918	3000	0
Packard	In above block	L-12B	730187	Aug. 12, 1918	4469	can. Dec. 4, 1918	100	0
Winton Motor	In above block	L-8	730223	Aug. 20, 1918	Canceled		1000	0
Packard	71272-77271	L-12	730369	Sept. 13, 1918	1646-A		6000	500
	Remaining production canceled after Armistice.							
Ford	Not issued	L-12	730407	Sept. 16, 1918			7000	0
	Production stopped Nov. 20, 1918							

Manufacturer	A.S.serials	Model	Order #	Date	Contract #	Date	#Ordered	#Delivered
Willys-Overland	Not issued	L-12	730520	Oct. 15, 1918	4914		5000	0
	Production stopped Nov. 13, 1918							
Lincoln	77272-88271	L-12	730584	Oct. 7, 1918	1647-6		11000	500
	Production stopped after Armistice. 77772-88271 not used							
Nordyke & Marmon	88272-90271	L-12	730585	Oct. 7, 1918	1704-A		2000	0
	Production stopped Dec. 7, 1918							
U. S. Navy *	90274-90441	L-12	720696	Oct. 14, 1918	none		168	0
	For Curtiss HS-2Ls on plane order 720696. Canceled after Armistice							
General Motors (Buick)	90444-93443	L-12	30071-4	Oct. 22, 1918	2266-4		3000	528
	90972-93443 remained unused, replaced by order 30071-5							
General Motors (Buick)	94472-94486	L-8	730810	Jan. 27, 1919	5463		15	15
	5000 ordered on 730155 under 57209-66310 which was canceled and 15 completed placed here.							
U. S. Navy *	94487-94492	L-12	721046	Jan. 29, 1919			6	6
	Possibly for 6 Curtiss HS-2Ls							
U. S. Navy *	94493-94498	L-12	721069	Mar 11, 1919			6	6
	For Curtiss HS-2Ls for Ford Island, Hawaii							

Note: The above information was provided by Robert B. Casari and is extracted with permission from his book *U. S. Army Aviation Serial Numbers and Orders 1908 - 1923.*

* Those contracts listed with the U. S. Navy as the manufacturer were most likely produced (or were intended to be produced) by Packard for delivery directly to the Navy and not actually built by the Navy.

Engineering Division at McCook Field orders & contracts for Experimental engines or associated items

Manufacturer	A.S. serials	Model	Order #	Date	Contract #	Date	# Ordered	#Delivered
Packard	EXP-1-2	L-12C	842	11/30/17	(replaced by C-5)		2	0
Hudson	EXP-1-2	L-4B	1373	12/31/17			2	2
Eng. Div.	?	L-1D	P.O. 31	2/-/18			3	3
Eng. Div.	?	L-1D	P.O. 70	3/-/18			3	3
Thomas-Morse	1-2 (47839-40)	L-6A	5257	3/5/18			2	0
Eng. Div.	T-1-T-2	L-2					2	2
Packard	EXP-1-2	L-12C			C-5	4/4/18	2	0
Packard	EXP-1-3	L-12C	4857	2/26/18			3	3
Packard	EXP-1-2	L-12E	4978	2/28/18	(replaced by CS-40)		2	0
Packard	EXP-1-3	L-12B			CS-24	4/29/18	3	0
Allison		L-12B gear sets			CS-34	4/29/18	3	3
Willys-Morrow		L-12B gear sets			?	5/-/18	3	3
Packard	EXP-1-6	L-12B	(replaced CS-24)		CS-43	5/8/18	6	6
Packard	EXP-1-3	L-12E	25		CS-40	5/7/18	3	1
Sherbondy		turbo	(research started June 1917)		?	?	3	3
GE Moss		turbo	(research started Oct. 1917)		CS-78	?	1	1
Packard	17821	L-12 500HP			CS-82	5/-/18	1	1
Thomas-Morse	1-2 (47839-40)	L-6A	(replaced 5257)		CS-85	5/31/18	2	2
Fergus Motors		Turbo			CS-84	6/6/18	3	3
Thomas-Morse	3-14 (47841-53)		L-6A		CS-95	6/6/18	12	0
Eng. Div.	EXP 7-12	L-8A	P.O. 737	6/4/18	(US 47854-57)		4	4

Manufacturer	A.S. serials	Model	Order #	Date	Contract #	Date	# Ordered	#Delivered
Eng. Div.	EXP 13-14	L-8A	add to P.O. 737	7/2/18	(US 47858-59)		2	2
Thomas-Morse	3-6 (47641-44)	L-6A	(replaced CS-95)		CS-125	6/25/18	4	4
Eng. Div.		L-1 six cycle		?	?	6?-/18	1	1
Packard	1-100	L-12B	730187		4469	8/12/18	100	0
Allison		L-12B gear sets			CS-203	9/10/18	3	3
Eng. Div.	EXP 15-16	L-8A	add to P.O. 737		10/1/18	(47865-66)	2	2
Eng. Div.	EXP-1	L-24	?	11/-/18			1	1

U. S. Post WW I Liberty Engine Contracts

In 1920 the Engineering Division contracted Wright to build 10 L-6B engines. These were updated versions of the original L-6A engines built by Thomas-Morse.

In 1921 Aeromarine submitted the design of, and test reports on, a 24-cylinder fan-type Liberty they called the AL-24 model. It is currently not known if the engine was built by Aeromarine on speculation or under contract. It was submitted but rejected as a suitable design.

Current information indicates the Navy contracted a single aluminum block experimental Liberty 12 engine from Aeromarine, Wright and Allison in 1923.

Allison built under Engineering Division contract 200 L-12B engines, probably one Aluminum L-12, one spur-geared two-speed, one spur-geared single-speed (PO 51411), thirty-five inverted water-cooled Project P-18A (plus five non-contract for civil sale) and 13 inverted air-cooled (plus one for civil sale) Project P-84 contract 927.

The Air Service Engineering Division built, in-house, eighteen inverted water-cooled engines and one upright air-cooled.

Russian Liberty Contracts

The Russians obtained, through various means, a number of British DH-4, DH-9 and DH-9a airplanes in the 1918-1920 era. They were impressed with the aircraft and produced their own copies. Most of the initial batch of Russian-produced airplanes were powered by purchased Italian 240 HP Fiat A12 or German 260 HP Maybach engines and the resulting performance was much inferior to those captured airplanes which were Rolls-Royce and Liberty powered. They therefore determined to manufacture their own copy of the Liberty which was to be known as the M-5.

The decision was made to proceed with M-5 production in May of 1922 and metric engineering drawings were begun at that time by disassembling a Ford-built U. S. Liberty and generating drawings from measuring the parts. Drawings were complete by September of 1922 and the engine was ordered into serial production. Production facilities were then set up at both GAZ-2 and GAZ-4.

The first prototype was produced in the GAZ-2 plant in Moscow and did not pass its 50-hour test until December of 1923. Actual serial production was begun in 1924. Total production had only reached 200 engines by the end of 1925.

The first production engine from GAZ-2 was sent on a long exhibition flight on February 12, 1925, and the first production engine from GAZ-4 was sent on a similar long-distance test flight on March 21, 1925. The M-5 was kept in production until 1932 with an estimated total of 2,909 being produced in the two plants.

In addition, about 3,091 1917-1918-built engines were purchased from the United States, England and France.

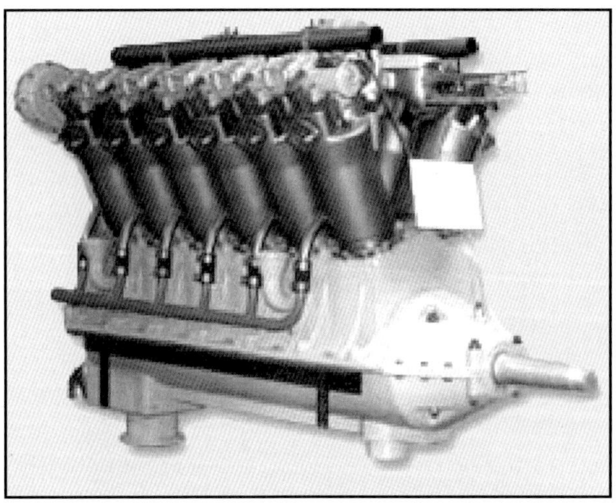

The Russian-built M-5 Liberty. This image is from the files of the Russian factory which originally produced the engines. A model of the R-1 airplane which it powered is mounted forward of the cylinder banks. (Courtesy of Samara JSC Motorostroitel)

British Liberty Engine Contracts
(These engines are generally referred to as the Nuffield Liberty)

The below list is of known British government contracts for Liberty engines to be built for use in tanks. This information came from surviving contract cards now in the collection of The Tank Museum at Bovington, England.

Manufacturer	Mark	Contract #	Total ordered	Noted as spares	Notes from order cards
M&A	NS (II)	E 6621	65	NS	For A13 Mk II tanks
M&A	NS (II)	E 6739	50	NS	For A13 Mk II tanks
M&A	II	T 7854	92	12	For Cruiser IV & IV A
M&A	II	T 9590	35	5	For Cruiser IV & IV A
M&A	III	T 7186	260	59	For Crusader tanks
M&A	III	T 414	130	30	For Crusader tanks
Morris Motors	III	T 5179	1,250	250	For Crusader tanks
Morris Motors	III	T/M 5859	3,296	659	For Crusader tanks
Morris Motors	III	T 847	1,681	219	For Crusader tanks
M&A	NS	T 2493	600?	NS	Number ordered not stated but was noted as for Cromwell I tanks numbers T129620 through T130119. These tanks were to become the Cavalier A 24 type. (This amounts to 500 tanks. These tanks were powered by Mark V-type engines. Where noted, it has been observed that on average 20% spares were ordered. An order for 500 tanks would thus result in an order for 600 engines.)
M&A	NS	T/M 6561	4,252	NS	For Crusader & Cromwell tanks. (These were probably Mark IV and V engines for Crusader III, Cavalier and Centaur tanks.)
		Total	**11,711**		

Notes:

NS = not stated
M&A = Mechanization & Aero Ltd.
Engines noted as Mark II = 242
Engines noted as Mark III = 6,617
Engines with Mark not stated = 4,852
Spares ordered = included in the total ordered
Although the Mark was not stated in contracts E 6621 and E 6739, these were for Mark II engines.

Actual Marks built were Mark II through Mark V. The designation of Mark I was assigned to the U. S.-built Liberty engines used in the original Christie from which the suspension was adapted for use in later Cruiser-type tanks and early prototype tanks built before the Nuffield Liberty went into production. In addition, there is excellent reason to believe that many of the early contracted A 13 series tanks were originally fitted with U. S.-built Liberty engines and later refitted with Nuffield-built Mark II engines.

The above contract card data probably represents accurately the number of Nuffield Liberty engines ordered. Without doubt, it does not correctly state the numbers of each Mark of the engine actually built. This is true because this engine was periodically redesigned and upgraded as its use progressed. By the time the last model was produced, it was a Mark V. Contracts were let for the number of engines needed to power the contracted number of tanks well ahead of requirements. As engine upgrades were made, contracts were revised to specify the newer models of engine. Although early tank contract cards were marked to show deliveries by day, engine contract cards carried no delivery information.

Also of interest is the fact that early engine contracts were done separately and had their own numbers. Later, beginning with contract T7854, engine contracts and tank contracts usually carried the same number. The final engine order of T/M 6561 is the only exception.

British tank contracts and production figures for Cruiser A 13 Series.
These were known as Cruiser Mark III, IV and IVA, corresponding to A 13 Mark I, II and IIA.

Manufacturer contracted	Date	Model	Contract #	Number ordered	Notes
Morris Commercial	2/15/37	A13.E1	T4127	1	Commercial and convertible Christie truck tractor. Delivered to MEE on 11/17/36 and assigned T2086.
M&A	1/22/38	Cruiser A 13 Mk I (Cruiser Mark III)	T5114	65	Noted as for tanks T4385-T4449. First completion shown as 4/6/38. From that time monthly completions were as follows: April - 5; May - 3; June - 11; July - 9; Aug -13; Sep - 11; Oct - 8; Nov - 3; Dec - 2.
LMS	1/26/39	Cruiser A 13 Mk II (Cruiser Mark IV)	T6551	30	This order was originally for 65 and then reduced to 30. First completion was on 8/14/39. From then completions were as follows: Aug - 2; Sep - 8; Oct - 5; Nov - 3; Dec- 8; Jan - 4. Noted as for tanks T7030-T7094. Not all numbers were used. It was noted that the engine contract for this order was E6621.
M&A	2/14/39	Cruiser A 13 Mk II (Cruiser Mark IV) Cruiser Mark IV - 64 Cruiser Mark IVA - 31	T6552	95	Noted as for tanks T9096-T9190. First completion shown as 1/1/39. The date on the contract card is later than the first completion because this was a continuation of contract T5114, the last delivery from which was made on 12/19/38. Monthly completions were as follows: Jan - 12; Feb - 15; Mar - 16; Apr - 17; May - 30; June - 5
M&A	12/1/39	Cruiser A 13 Mk II (Cruiser Mark IV)	T7854	80	Noted as for tanks T15215-T15294. First completion shown as 6/3/39, the same day as the last completion from contract T6552. Monthly completions were as follows: June - 25; July - 26; Aug - 22; Sep - 7.
M&A	6/4/40	Cruiser A 13 Mk IIA (Cruiser Mark IVA)	T9590	30	Noted as for tanks T18131-T18160. First completion shown as 9/10/39, the same day as the last shipment from contract T7854. Monthly completions were as follows: Sep - 13; Oct - 17.
M&A	6/3/40	Cruiser A 13 Mk II (Cruiser Mark IV)	T9714	35	Noted as for tanks T18096-T18130. First completion noted as 10/30/39, the same day as the last shipment from contract T9590. Monthly completions were as follows: Oct - 1; Nov - 16; Dec - 14; Jan - 4. Last completion was on 1/10/40.

Total by model = 65 A13 Mk I (Cruiser Mk III); 270 A13 Mk II&IIA (Cruiser Mk IV&IVA).
Total A13 series = 335

Notes: Although the data found on the contract cards can be assumed to be correct, the dates found on them (as shown in the date column) do not necessarily reflect the date of the actual contract. During the early period, these cards were kept in a central accounting location to keep track of completion status. It is not clear why some were dated some time after first completions had been made. It is obvious from the dates of completion that production was handled as a continuous operation. That can be seen from the above completion dates. It can also be seen in the M&A production switchover from the A13 to the A15 model. The first A15 was completed on 1/5/40 with only 3 others built in January. January completions of A 13 tanks was 4.

Contract cards for A13 and A15 tanks are hand-marked with daily completion numbers whereas cards for later model tanks are only marked with total completion. Completions do not necessarily equate with deliveries, particularily in the case of the A13 Mk I where there are documents indicating the first actual deliveries were made in December of 1938, yet the contract card clearly indicates completions started in April and the final tank was completed on December 19, 1938. I have read in one history a reference stating A13 Mk I tanks were stored for some time after completion, but nothing was said as to why this was done.

M&A= Mechanization & Aero Ltd. LMS= London Midland & Scottish Railway. MEE= Mechanization Experimental Establishment.

British tank contracts and production figures for Cruiser A 15 Series. These were known as Cruiser Mark VI & VIA, corresponding to Crusader I and II and III

The below data come from the original order cards for these models of tanks plus comments noted on the cards. One final document which allowed reasonably accurate distribution by Mark was a list of "Crusader WD Numbers" which listed all numbers by Mark and by constructor. This data represents initial delivery and cannot be used to determine final Marks or types as conversions were done on a large scale after delivery as needs changed.

Manufacturer contracted	Date	Model	Contract #	Number Ordered	Notes
M&A	6/27/39	Crusader I-II	T7186 Crusader I - 109 Crusader II - 92	201	Delivered as 1 pilot (T3646) Crusader I and 108 production Crusader I plus 92 Crusader II. WD numbers T15546-15645 and T16558-16657. The pilot delivered 4/9/40 and remainder completed by Sept. 1940. Notes on the order card indicate that 100 of this order was originally to be order T7855 for Cruiser A13 models assigned T15846-15945 and that order was cancelled.
Foden	?/?/39	Crusader I	T7710	100	Completed Nov. 1939 - Aug. 1940. WD numbers T15646-15745.
West's Gas Imp.	?/?/39	Crusader I-II	T7353 Crusader I- 82 Crusader II - 18	100	Completed as 82 Crusader I and 18 Crusader II from Jan. 1940 - Oct. 1940. WD numbers T15746-15845.
M&A	8/8/40	Crusader II	T414	100	Completed Sept. 1940 - Nov. 1940. WD numbers T43652-43751.
Fodens	8/8/40	Crusader II	T413	50	Completed Aug. 1940 - Dec. 1940. WD numbers T46212-46261.
West's Gas Imp.	8/8/40	Crusader II	T412	50	Completion dates unknown. WD numbers T46162-46211.
M&A	8/15/41	Crusader III	T5179	970	Completion dates unknown. WD numbers T126270-127269 (1,000) authorized. T126740-126749 and T127120-127139 not used. Delivered as 649 Crusader III, 68 Crusader AA II and 253 Crusader for rework to OP. T126270-126739 (470) were subcontracted to Fodens.
M&A	9/20/41	Crusader II-III	T5859 Crusader II - 540 Crusader III - 1392 Bofor gun carriers - 35	1967	Completion dates unknown. WD numbers T123633-126369 (2737 authorized). Delivered as 145 Crusader II, 825 Crusader III, 250 Crusader AA I, 332 Crusader AA II, 415 for reworking to OP and 35 as Bofor gun carriers. They were subcontracted as follows: T123633-123912 (280 Crusader III) to West's Gas, T123988-124312 (325 Crusader III) to Ruston-Bucyrus, T124478-124752 (275 Crusader II) to MG Car Co., T124788-125052 (265 Crusader II) to John Lysaght, T125118-125259 (142 Crusader III) to Minner's Safe T125261-125328 (68 Crudaser III) to Minner's Safe, T12547-125763 (290 Crusader III) and T125867 to various, T125984-126269 (286 Crusader III) M&A. 35 Bofors by M&A.
M&A	10/23/41	Crusader II	T847	1462	Completion dates unknown. WD numbers T43751-452213. Subcontracted as follows: T43752-43991 (240 Crusader II) to Morris Commercial Cars, T43992-44231 (240 Crusader II) to Ruston-Bucyrus, T444232-44401 (170 Crusader II) to MG Car Co., T44402-44521 (120 Crusader II) to John Lysaght, T44522-44715 (194 Crusader II) to Minner's Safe, T44716-44840 (125 Crusader II) to Fodens, T44841-44965 (125 Crusader II) to West's Gas, T44966-45213 (248 Crusader II) to M&A.

Total by Model = 291 Crusader I; 2312 Crusader II; 2362 Crusader III; 35 Bofor gun carriers. Grand Total=5000

British tank contracts and production for Cavalier - Cruiser Mark VII (A24)

M&A	1/31/41	Cavalier VII	T1782	3	Pilot models for Cromwell I, which was later changed to Cavalier Cruiser Mark VII (A24). T84618-84620. Completed in January 1942.
M&A	6/20/41	Cavalier VII	T2493	500	Tank Cruiser Mk VII. T129620-130119 production in 1943.

Total Cavalier 500 plus 3 pilot models = 503
The Cavalier used Mark IVA and V Nuffield Liberty engines.

British tank contracts and production figures for Cruiser Mark VIII - Centaur (A27L)

The below data comes from original order cards for this model tank and represents the models and configurations in which the tanks were originally delivered.

Manufacturer Contracted	Date	Models	Contract #	Number ordered	Notes
North British Loco.	?/?/?	Cromwell II	T11238	540	CANCELED see contract M2225 T132196-132735
English Electric	?/?/?	Centaur I	T3743	1	1 Pilot model. T171966 completed June 1942.
Leyland	?/?/?	Centaur I	T11234	865	863 production and 2 pilots.
Harland & Wolf	?/?/?	Centaur I	T11239	125	Originally marked Cromwell II and 390 ordered.. T132736-133125 and marked out.
John Fowler	?/?/?	Centaur I	T11240	715	Originally marked Cromwell II, then Centaur I. Noted as T133126-133840 and then marked out.
LMS	?/?/?	Centaur I	T11241	45	Original order was 350, then marked 75 and then 45.
English Electric	?/?/?	Centaur I	T11242	803	Originally marked Cromwell II. Original order was 1025, marked down to 975 and then to 803. Originally T133841-134865. Marked-out and changed to "First vehicle" T183800.
M&A	?/?/?	Centaur I-II Centaur I - 79 Centaur III - 58	T11243	138	Originally for 2350 and marked back to 138. Noted on back as delivered as 79 Mk I and 58 Mk III. Also marked T217772-220771 = 3000 and marked-out.
Leyland	4/29/42	Centaur I	T13136	2	Pilot models in mild steel. T171962-171963.
Leyland	8/17/42	Centaur I-II	M1300	290	Originally 380 and changed to 290. Marked "controlled by Leyland."
Leyland 132735	9/9/42	Centaur I	M2225	300	Marked Cromwell and changed to Centaur I. Originally for 540 and changed to 300. Also marked T132196- and then crossed-out.
Nuffield	?/?/?	Centaur AA I	M10748	150	Originally for 350 and changed to 150.

A total of 3434, including 5 pilot models, were ordered and delivered.

The following blocks of numbers were assigned to the program but not all, of course, were used.
Pilots - T171962-171966.
Leyland - Centaur I - T183800-188064 = 4265
T188064-190064 = 2000
T217772-220771 = 3000

Note: Most of the order cards for this model tank were not dated, thus the (?/?/?) in that location. It is thus difficult to be sure the orders are listed in proper order. For our purposes, that makes little difference. All models supposedly used Nuffield Mark V Liberty engines.

British tank names, production and engine models used.

Model Names	Number produced	Engines used
A13 E1 = Original British Christie	1	U. S. Liberty - British Mark I
A13 E2 = first pilot model	1	U. S. Liberty - British Mark I
A13 E3 = second pilot model	1	U. S. Liberty - British Mark I
A16 E1 = Heavy Cruiser pilot	1	Nuffield Mark II
A 13 Mark I = Cruiser Mark III	65	U. S. Liberty - British Mark I. Many were later re-engined with Nuffield Mark II engines.
A 13 Mark II = Cruiser Mark IV	209	Nuffield Mark II
A 13 Mark IIA = Cruiser Mark IVA	61	Nuffield Mark II
A 15 Mark I = Crusader I = Cruiser Mark VI	291	Nuffield Mark III
A 15 Mark II = Crusader II = Cruiser Mark VIA	2,312	Nuffield Mark III
A 15 Mark III = Crusader III	2,362	Nuffield Mark IV (a few early ones used III, IIIa or IIIb)
A 24 = Cavalier = Cruiser Mark VII	503	Nuffield Mark IVA or V
A 27L = Centaur = Cruiser Mark VIII At one time called Cromwell II	3,434	Nuffield Mark V
Total	9,241	

This brings total initial requirements to perhaps 75 U. S.-built Liberties equipped for tank use plus Nuffield models. The below calculations are based on initial production requirements plus an estimate of 20% spares. I have not attempted to guess how many early Crusader III tanks did not use Mark IV engines, but have assumed all used the Mark IV for calculating purposes. I have also made no attempt to guess how many Cavalier models used Mark IVA engines, but have assumed all used Mark V. The resulting figure comes reasonably close to the total contracted number of engines, which was 11,711.

	Initial	Spares	Total
Nuffield Mark II	271	54	325
Nuffield Mark III	2,603	520	3,123
Nuffield Mark IV	2,362	472	2,834
Nuffield Mark V	3,937	787	4,724
		Grand Total	11,006

JULY 17, 1925. THE MOTOR BOAT 5

THE GRAHAME-WHITE COMPANY

(C. GRAHAME-WHITE)

Beg to announce that they have been appointed

Sole Agents and Concessionnaires

for

THE BRITISH EMPIRE and EUROPE

for the World-famous

BABY-GAR

SPEED MOTOR BOATS

Guaranteed Speed 50 Miles per Hour.

Comfortable seating accommodation for ten persons.
Trial runs can now be arranged at Southampton.

For Delivery Dates and further particulars apply :

THE GRAHAME-WHITE COMPANY,

12, REGENT STREET, PALL MALL, S.W.1.

Telegrams: "Claudigram, Piccy, London." Telephone: Regent 2084.

B3

"Marine Motor Handbook." A simple explanation of all classes of internal-combustion engines for marine purposes. 2/6 net. [illegible] post free.

Advertising from the British publication The Motor Boat of July 17, 1925 in which the Grahame-White Company of London announced themselves as sole agent for Baby-Gar boats for the British Empire and Europe. As such, they would also sell Gar Wood Liberty marine engines. (Courtesy of Jamie Fairchild)

Appendix B

Liberty Production

U. S. WW I LIBERTY ENGINE ORDER - PRODUCTION

Manufacturer	Type	No. Contracted	No. Delivered	A.S. Serials	Mfg. Serials	Notes
Packard	L-8A	6	6	EXP	1 - 6	prototype
Packard	L-12A	5	5	EXP	1 - 5	prototype
Nordyke & Marmon	L-12A	3000	1000	4402-5401	1 - 1000	2000 canceled some were 12N
Trego	L-12A	500	1	7402	2 *	499 canceled - to Ordnance
Packard	L-12A	6000	6000	17435-23434	1-6000	4909N & 1091A
Lincoln	L-12A	6000	6000	23435-29434	1-6000	7 were 12N 5993 were 12A
Ford	L-12A	5000	3950	29494-33443	1-3950	1050 canceled
Duesenberg	L-12A	500	0	36998-37497	---------	canceled
GM (Buick)	L-12A	1000	1000	37503-38502	B-1-1000	
Duesenberg	L-12A	500	0	38510-39009	---------	canceled
GM (Cadillac)	L-12A	1000	1000	43771-44021	C-1-250	
(Buick)				44022-44770	B-1001-1750	

(The above order was with GM which split it between Cadillac & Buick. Cadillac built 250 and Buick 750.)

Engineering Division Experimental Series

On about February 1, 1918 a block of 100 serials starting with 47828 and ending with 47927 was assigned to the Engineering Division for use on experimental engines. They were used on Liberty and other types of engines. It is not known if all the numbers were assigned or used. Where assignment is known, it is listed. The Engineering Division did not assign an A. S. number to all experimental engines ordered. In some cases engines were marked EXP with no A. S. number. In some cases A. S. numbers would be assigned and EXP followed by a serial was used in the MNFG. NO. position of the data plate. Marking of EXP? in the A. S. column indicates it is not known if an A. S. number was assigned or not.

Manufacturer	Type	No. Contracted	No. Delivered	A.S. Serials	Mfg. Serials	Notes
Packard	L-12B	6	6	47833-47838	1 - 6	epicyclic
Thomas-Morse	L-6A	2	2	47839-47840	1 - 2	
Thomas-Morse	L-6A	4	4	47841-47844	3 - 6	
Eng. Div	L-8A	6	6	47854-47859	EXP 7-12	
Eng. Div	L-8A	2	2	47865-47866	EXP 13-14	
Hudson	L-4B	2	2	EXP?	1-2	experimental
Packard	L-12C	2	0	?	---------	geared-canceled
Packard	L-12C	3	3	-	EXP 1-3	exp. spur geared
Eng. Dept.	L1D	3	3	EXP?	1-3	test single cyl.
Packard	L-12E	3	1	EXP?	EXP 1	duplex - exp.
Eng. Div	X-24	1	1	-	EXP 1	experimental
Packard	L-12	1	1	17821	316	500 HP -see note

(Note: The 500 HP experiment was carried out on Packard engine A. S. 17821 MNFG. NO. 316 and was evidently an open-ended one. The contract was let in May of 1918, the objective was met by August but experiments continued until the end of the war.)

The following engines fall in the regular A. S. serial ranges.

Manufacturer	Type	No. Contracted	No. Delivered	A.S. Serials	Mfg. Serials	Notes
Packard	L-12A	212	212	47928-48136	A1 -212	school engines from reject parts
Packard	L-12A	130	130	48137-48267	A213-342	school engines from reject parts
GM (Buick)	L-8A	5000	0	57209- ? note	A ---------	canceled
Willys-Overland	L-8A	3000	0	57209- ? note	A ---------	canceled

Manufacturer	Type	No. Contracted	No. Delivered	A.S. Serials	Mfg. Serials	Notes
Winton Motor Car Co.	L-8A	1000	0	57209- ? *note*	*A* ---------	canceled
Packard	L-12B	100	0	57209- ? *note*	*A* ---------	epicylic geared-canceled.
Packard	L-12A	6000	500	71272-71771	6001-6500	5500 canceled. 415 were 12N 85 were 12A
Ford	L-12A	7000	0	none allocated	--------	canceled
Willys-Overland	L-12A	5000	0	none allocated	--------	canceled
Lincoln	L-12A	11000	500	77272-77771	6001-6500	10500 canceled
Nordyke & Marmon	L-12A	2000	0	88272-90271	---------	canceled
U.S. Navy	L-12N ?	168	0	90274-90441	--------	canceled
GM (Buick)	L-12A	3000	528	90444-90971	B-1751-2279	2472 canceled
GM (Buick)	L-8A	15	15	94472-94486	BX 1-15	replaced 1st order for 5000
U.S. Navy	L-12N ?	6	6	94487-94492	*note B*	ordered after Armistice
Wright	L-6B	10	10	94577-94586	*note C*	ordered after Armistice

Notes:

* - 499 of the contract was transferred to the Ordnance Dept. It is estimated that 250 of that number were actually completed.

A - **57209- ? :** Entries with this designation in the **A. S. serials** column indicate that the contracts noted were assigned a block of serials somewhere in the serial range 57209 - 66310, but exactly what the numbers were is not known. The contracts were all canceled and thus no numbers were actually used.

B - unknown

C - none

General: This chart shows the number ordered versus the number actually delivered of all regular and experimental contract engines. It also shows what manufacturer serials were assigned by each builder for each contract. As a general rule, each builder assigned his MNFG. NO. series beginning with “1” and continuing upward until his contract or series of contracts ended. In a few instances prefixes were added to the numbers. General Motors contracts were executed by both Buick and Cadillac. They used prefixes “B” and “C” to indicate the builder. The Buick-built Liberty 8 engine went through a slightly different sequence of events than was normal. No large production contracts were given for the eight until well into the war. Buick, Willys-Overland and Winton were set up for contracts. Of the three, only Buick had begun any production when the war ended. Buick had been able to do so because of already having been an L-12 producer. Only 15 engines had been built when the war ended and the decision was made to transfer those already built to a new order and classify them as experimental engines. Thus their Manufacturer serials were prefixed with “BX” for Buick experimental. Packard used no prefix on their regular production engines but used the prefix “A” on the serials of two contracts for “school” engines made up of reject parts to differentiate them from regular production engines.

Navy or 12N engines: Regarding the use of the term “12N”: Although there is no concrete proof as yet, I am reasonably sure the low-compression engines (12N) had standard data plates marked “12A" but stamped “LC” for low-compression. If this is not true, the most likely exception would be those six engines (94487-94492) actually built by the Navy. Data plates have been found for engines built for the Navy by Packard (brass plate of different design from the Air Service plates) which were marked “12A” and stamped “HC”. No plate of any description has yet been observed marked as “12N” nor marked “LC.” In other words, no low-compression engine plate of any type has yet been observed.

There were also several additional single-cylinder and two-cylinder test engines built that are not mentioned above. These were built “in-house” by the Engineering Division and not assigned A.S. numbers. Production of these test engines are covered in other charts.

WW I U. S. Liberty
ORDER - PRODUCTION TOTALS
1917 - 1919

Manufacturer	Single-Cyl.		Two-Cyl.		Four-Cyl.		Six-Cyl.		Eight-Cyl.		Twelve-Cyl.		24-Cyl.	
	Ord.	Del.	Ord.	Del.	Ord.	Del.	Ord.	Del.	Ord.	Del.	Ord.	Del.	Ord.	Del.
Packard									6	6	12462	6857		
Nordyke & Marmon											5000	1000		
Trego											500	1*		
Lincoln											17000	6500		
Ford											12000	3950		
Duesenberg											1000	0		
GM (Buick)											3000	528		
GM (Buick)									5015	15	1000	1000		
GM (250 Cadillac 750 Buick)											1000	1000		
Thomas-Morse							6	6						
Engineering Division	6	6	2	2					8	8			1	1
Hudson					2	2								
Willys-Overland									3000	0	5000	0		
Winton MCC									1000	0				
U.S. Navy											174	6		
Totals by type	**6**	**6**	**2**	**2**	**2**	**2**	**6**	**6**	**9029**	**29**	**58136**	**20842**	**1**	**1**

* - 499 of the Trego contract was transferred to the Ordnance Department. It is estimated that total production was 250.
(Grand total of all types 1917 - 1919. Ordered -- 67182 Delivered -- 21136)

Post-WW I U. S. Liberty
PRODUCTION TOTALS
1920 - 1930

Manufacturer	Six-Cyl. No. Built	Twelve-Cyl. No. Built	Twelve-Cyl. Inverted WC No. Built	Twelve-Cyl. Inverted AC No. Built	Twelve-Cyl. Geared No. Built	Comments
Packard		80				Commercial 1A-1650
Aeromarine		1				Experimental
Wright	10	1				The 12 was experimental
Allison		201	40	14	2	200 L-12B and 1 Aluminum cylinder block L-12
Engineering Div.			18	1		

Total Post War all types and makers = 368

WC = Water-cooled
AC = Air-cooled

TOTAL U. S. GOVERNMENT ORDERS - PRODUCTION BY MANUFACTURER 1917 - 1930

Manufacturer	Ordered	Delivered	Comments
Packard	12468	6863	Includes contract production plus experimental engines.
Nordyke & Marmon	5000	1000	
Trego	500	1*	Contract transferred to Ordnance after delivery of 1 engine. The number delivered to Ordnance is not currently known but is estimated at 250.
Lincoln	17000	6500	
Ford	12000	3950	
Duesenberg	1000	0	Contract replaced by one for 2000 Bugatti engines.
GM	10015	2543	Contract divided: 250 built by Cadillac and 2293 by Buick.
Thomas-Morse	6	6	
Hudson	2	2	
Willys-Overland	8000	0	
Winton Motor Car Co.	1000	0	
Engineering Division	17	17	
U.S. Navy	174	6	
Wright	11	11	1920 contract for L-6B engines plus probable aluminum block 12.
Allison	251	251	Post-war production of engines of all types.

Notes: The difference between the number ordered and delivered is, in all cases except Trego and Duesenberg, accounted for by the cancelation of the remainder of a contract because of the ending of the war.
The production and shipping record for wartime regular production of the twelve-cylinder model for all producers month by month can be found on page 100.

PRODUCTION LIBERTY 12 ENGINES SHIPPED LISTED BY MANUFACTURER AND AIR SERVICE SERIAL RANGE SHIPPED DURING EACH MONTH

	1917						**1918**								**1919**	
Manufacturer (number produced)	Dec.	Jan.	Feb.	Mar.	Apr.	May	June	July	Aug.	Sep.	Oct.	Nov.	Dec.	Jan.	Feb.	Mar.
Packard (6500)	17435 17457	17458 17496	17497 17566	17567 17694	17695 18003	18004 18486	18487 19061	19062 19742	19743 20392	20393 21020	21021 21923	21924 22557	22558 23129	23130 23434 71272 71382	71383 71750	71750 71771
Lincoln (6500)				23435 23450	23451 23544	23545 23671	23672 24015	24016 24499	24500 25260	25261 25947	25948 26997	26998 27815	27816 28924	28925 29934 77272 77771		
Ford (3950)							29494 29547	29548 29729	29730 30248	30249 31059	31060 32118	32119 33319	33320 33443			
Buick (2287)							37503 37583	37584 37715	37716 37918	37919 38111	38112 38502 44022 44189	44190 44597	44598 44764	44765 44770 90444 90843	90844 90970	90971
Cadillac (250)							43771 43821	43822 43915	43916 44021							
Nordyke & Marmon (1000)								4402 4418	4419 4460	4461 4519	4520 4827	4828 4970	4971 5211	5212 5401		
Trego Motors (1)										7402 (Remainder of the order for 500 went to Ordnance.)						

Packard delivered the first experimental L8 in July 1917 and the first experimental L12 in August 1917. Five additional L8 and four additional L12 experimental engines were delivered by Packard by about November 1917.

Russian Liberty Production

The decision to place the Liberty in serial production was made in May of 1922. The first factory in which it was to be manufactured was the GAZ-2 plant in Moscow. Metric engineering drawings were prepared by disassembling an American-built engine and measuring all parts.

Actual plant preparation began in September of 1922. It was December of 1923 before the prototype passed its 50-hour test. Serial production began in 1924. By the end of 1925 about 200 engines had been completed.

Production was also set up in the old Bolshevik factory in Leningrad, probably in late 1924 or 1925. Output from this factory was much slower than from GAZ-2 and production at it was ended in 1929 with 480 engines built.

Production was halted at GAZ-2 in 1930 after a total run of 2,349 engines. Production was restarted in 1932 for a short period with an additional 80 engines built.

Total production was thus 2,909 engines. To supply a total requirement of about 6,000 engines an additional 3,091 American engines built in 1917-19 were imported from the United States, Great Britain and France.

British Liberty Production

Initial production of the Nuffield Liberty was set up at Nuffield Mechanizations and Aero Ltd. plant at Common Lane, Birmingham, probably in early 1938. That engine was given a model designation of Nuffield Liberty Mark II.

As orders were placed with various builders for tanks of progressive designs which were to use the Nuffield Liberty engine, additional engines were ordered to power them. These orders were placed well ahead of actual need and were ordered at the then-current design Mark. Improvements in the engine were constantly in process and this resulted in progressive models through the Mark V.

Although most production contracts were originally issued for Mark II or III engines, actual production was switched to new designs as they came into being.

By the summer of 1941 it had become evident that the Common Lane plant would not be able to keep up with requirements and Nuffield committed his Morris Motors Engine Branch plant at Courthouse Green, Coventry as an additional producer.

The Common Lane plant built Mark II, III, IV and V engines, and the Courthouse Green plant Mark III, IV and V engines. Nuffield Liberty production was phased-out in 1943 as Rolls-Royce Meteor production came up to speed.

Nuffield Mechanizations and Aero was issued contracts for 5,484 engines and Morris Motors Engine Branch contracts for 6,227. This is a total of 11,711 engines and probably all were actually built. Nuffield engine production thus ran from 1938 through most of 1943.

See page 487 for estimates of individual model production.

Appendix C

Documented Tests July 1917 through June 1925

Record of known testing of Liberty engines from the first running in July of 1917 to August of 1918.

This first year could be considered the initial development period.
(It is strongly emphasized that this list absolutely cannot be taken as a complete list of tests done during this period, only those of which records have surfaced. Testing was carried out on five of the six prototype L-8 engines and all five L-12 prototype engines, plus a number of early production engines.)

Testing of the period July 25, 1917 to February 8, 1918

Engineering, development and testing during this initial phase was being done at the Packard factory as well as at the National Bureau of Standards in Washington, D.C.

Date	Description of tests
7/25/17 - 7/26/17	L-8 Exp. No. 2 tested for at least 20 hours, 15 hours at 9/10 power.
7/27/17	L-8 Exp. No. 2 tested on a truck platform for probably 4 hours.
8/13/17 - 8/17/17	L-12 Exp. No. 1 initial testing carried out through this 5-day period for about 42 hours of actual run time.
8/22/17 - 8/24/17	First 50-hour acceptance test of L-12 Exp. No. 1
8/24/17 - 9/2/17	L-8 Exp. No. 3 in the first Liberty to power a plane and accumulated 10.5 hours flying between these dates.
8/27/17	L-12 Exp. No. 1 runs 5-hour test for Navy.
8/28/17	L-12 Exp. No. 1 shipped to Washington for extensive use in redesigning intake headers.
9/2/17 - 9/26/17	L-8 Exp. No. 2 tested on Pikes Peak, CO.
9/10/17	L-8 Exp. No. 6 tested.
9/13/17 - 9/18/17	L-8 Exp. No. 6 run to do carburetor design for 4 to 8 hours per day
9/19/17 - 9/21/17	L-12 Exp. No. 2 tested and power graph plotted on 9/21/17.
10/16/17	L-12 Exp. No. 3 tested and power graph plotted during testing.
10/21/17	L-12 Exp. No. 3 powered Navy HS-1 on first L-12-powered flight.
10/29/17	L-12 Exp. No. 4 powered Army DH-4 on first L-12 Army flight.
11/17/17	L-12 Exp. No. 5 tested.
11/27/17	L-8 Exp. No. 3 tested and graph plotted.
12/4/17 - 2/8/18	Extensive break-down tests run on L-12 Exp. No. 5 and Packard production No. 12 engines. The number of running hours is not recorded but indications are they were in the hundreds. Reports indicate testing was carried out for perhaps 16 hours per day, seven days a week on two engines and lasted for about 60 days. In excess of 1,500 test hours must have been accumulated.

Testing of the Period January 1, 1918 to July 31, 1918

Beginning about January 1, 1918, the Engineering Division began assigning serials to their tests and listed them as Technical Reports. In June of 1920 they generated an index of them for reference. (Published in Air Service Circular No. 93 of June 20, 1920.) Test serials began at 1 and had reached 1300 by June of 1920. Those related to the Liberty engine are listed below with a brief description of the test. Most of the below testing was done at McCook Field on facilities set up there by Col. Vincent to make the facility the airplane and engine testing center for the Air Service. Construction of McCook Field was begun in October of 1917. Prior to this time, Liberty testing and development was done at the Packard factory and the National Bureau of Standards in Washington, D.C.

Test Serial	Date	Description of tests
1	1/7/18	Dixie magneto test on L-12.
2	1/9/18	Bijur starter test.
3	1/9/18	Gasoline tests on L-12.
6	2/9/18	1 cylinder Liberty test (L-1) ran 25 hours between 2/1-2/6.
7	2/23/18	Test of L-4 No. 1.
8	3/4/18	Test L-4 water pump.
9	3/12/18	Test LC L-12 Ordnance engine.
10	3/18/18	Special fuel test on L-1.
11	3/29/18	History of L-12 Prod. No 4.
12	4/2/18	L-1 muffler test.
14	5/14/18	Test of Dinshah engine tester on L-4.
15	4/12/18	Gates turbo booster test for L-12.
17	4/15/18	Power test of L-12.
18	4/18/18	Test of 1st design L-12 articulated-type connecting rod.
19	4/16/18	Test of special low-comp. intake and camshaft.
21	4/17/18	L-12 for DH-4 standardized.
22	4/18/18	L-12C No. 1 tested in U. S. C-1 plane.
23	4/18/18	Test of special quick-lift cam and valve springs.
24	4/19/18	Test of sliding vane gas pump.
25	4/19/18	Test of Navy L-12 with special intake headers for Ordnance.
26	4/22/18	Heat balance tests on Packard No. 5.
27	4/24/18	Standard engine test on Lincoln No. 59.
28	4/24/18	Standard Army test for DH-4 on Packard No. 320.
29	4/25/18	Oil test on special pistons.
30	4/25/18	Tests of special valve port water flow design.
31	4/27/18	Test of L-8 Exp. No. 2 on cradle stand.
32	4/29/18	Test of special pressed steel connecting rods run on L-1.
34	4/30/18	Tests of special fuels on L-1.
35	6/4/18	Test second design articulated rod on L-12 Packard No. 77.
36	6/24/18	Test of 4 ring Packard-designed piston on Packard No. 413.
38	6/29/18	Test of L-4.
39	6/24/18	Test of new design piston rings.
40	5/11/18	Oiling changes to prevent sooting and fouling.
41	5/1/18	Test of Packard L-12C with different camshafts.
42	5/11/18	Test of Stromberg NA-5-D on Packard No. 316.
43	5/11/18	Test of Packard L-12C No. 3.
44	5/8/18	Test of Delco Vibrator for start on compression on Packard L-12C No. 2.
47	5/10/18	Test of Bijur starter on L-12 Ordnance engine Packard 115.
49	6/24/18	Test of L-12 Packard No. 714 with special pistons.
50	6/16/18	Test of special pistons.
51	5/16/18	Test of Liberty pistons for effects of Energine & gasoline.
53	5/17/18	Test of special A.C. spark plugs in Liberty.
55	6/6/18	Test of L-1.

56	5/26/18	Test of L-12 AS 18212 for use on Curtiss R-4 P-20.
57	5/25/18	Standardization tests of 12 production L-12; 6 Packard & 6 Lincoln.
58	6/25/18	Report of L-12 AS 18212 after 60 hours flight in P-20.
59	6/6/18	Test of Westinghouse generator for L-4.
60	6/29/18	Test of Packard No. 5 before shipping to Glenn Martin for flight-testing.
61	6/29/18	Test of Packard No. 77 before shipping to Glenn Martin for flight-testing.
62	6/10/18	High-power experiment (500 HP) with Packard 316.
63	7/1/18	Report on Packard No. 719 after 75 hours in the air without overhaul.
66	7/3/18	Test the effects of water in the oil in the crankcase of L-12.
67-68-69	7/6/18	Tests of L-12 Packard No. 9, 77 and 299 before shipping to American Multiplane.
70	7/9/18	Test of Lincoln No. 25 received from Wilbur Wright Field.
71	7/11/18	Report on the test of L-8 Exp. No. 3 done on 11/27/17.
72	7/11/18	Test of L-6 engine.
73	7/11/18	Test of Zenith U.S. 52 carburetor on Packard No. 716 L-12.
**	7/19/18	Navy tested Packard No. 43 A.S. 17485/Navy No. 32. (**=No A.S. test serial assigned.)
102	7/2/18	Report on 60-hour endurance test on L-12.
121	7/9/18	Report on 75-hour endurance test of L-12.
133	7/15/18	Thrust meter for L-12.
137	7/16/18	Test of vacuum oil control on L-12.
140	7/16/18	Test of altitude control for Zenith 52 carburetor on L-12.
173	7/26/18	Test of L-8 Exp. No. 5.
177	7/9/18	Tests of latest changes on L-12 Zenith carburetor.
185	7/30/18	50-hour test on four ring L-12 piston.
192	7/31/18	Test on cast-iron Liberty piston.
195	7/1/18	Calculations of the Liberty 8 engine.
208	7/2/18	160-hour endurance test of L-12.
227	5/2/18	Gas velocities of intake system of Liberty 12.
229	6/3/18	Piston design study.
230	6/5/18	Comparison of Rolls-Royce reduction gear to plain reduction gear.
232	6/11/18	Report on L-4 starting motor.

Engineering Division Test Reports continued and reached test number 1300 by June 1920.

Notes:

The Engineering Division published a weekly report during this period which gave brief descriptions of projects they were working on, contracts they let and progress made. I have these reports for June 1, 1918 through and beyond August 1, 1918 but none before June 1, 1918. During June and July 1918 much Liberty-related work was carried out that does not appear in the above list. For example, much work was carried out on the Moss and Sherbondy turbo supercharger, epicyclic reduction gear design and manufacture, standardization of L-8 design and building of 8 test L-8 engines in-house (these were not all completed prior to August 1, 1918), building of 3 L-12C engines by Packard, building of an L-2 test engine, building of 6 L-6 engines by Thomas-Morse, building by Packard of a cannon Liberty 12, etc.

It should also be noted that I have included no tests of Liberty-powered airplanes in the above list, although there were numerous such tests, those done on the DH-4 and Bristol Fighter being two examples. The above list was intended to document testing related directly to the development of the Liberty engine itself during its first year, thus the lack of listing airplane related tests.

Official tests by the Engineering Division either of the Liberty and associated accessories or aircraft powered by the engine.
Period covered - August 1918 to June 1920.
(Except in special cases, propeller tests are not noted.)

Test serial	Date	Description of test
82	11/1/18	Experimental test of Liberty epicyclic reduction gear.
83	11/1/18	Strength test of LePere interplane struts and cross bracing system.
88	11/7/18	Water injection into intake headers of Liberty 12.
125	8/5/18	Design of epicyclic reduction gear for L-12.
197	8/2/18	Test of Claudel & Hobson carburetor on Liberty 12 Model A.
210	8/20/18	160-hour test on Packard engine No. 719.
218	8/1/18	Hartzell Walnut Propeller Co. propeller for U.S. Navy H-16 Flying Boat.
226	8/23/18	Liberty 12-A calculations.
231	8/17/18	Epicyclic propeller reduction gears.
235	8/30/18	Forked connecting rod test.
239	8/20/18	Engine test of vacuum oil control on Liberty 12.
241	8/26/18	Test of Liberty 12 (Packard) No. 719 equipped with 4-ring pistons.
242	8/20/18	Test of Liberty 12 (Packard) engine No. 801 with special intake headers.
243	8/1/18	Test on Liberty 12 No. 320 of heavy connecting rod and crankshaft.
244	8/22/18	Engine test on Liberty 12 No. 413.
245	8/24/18	Test of Liberty 12 epicyclic gear.
249	8/8/18	Flight test of flat tube DeH-4 radiator.
253	8/27/18	Test of special high-power Liberty 12-cylinder engine.
254	9/18/18	Comparative test of LePere and Liberty headers.
258	8/29/18	Comparative performance of Liberty with different types of gasoline.
294	8/23/18	Report on LePere U.S.A. - G.H.- 11.
296	8/21/18	Estimated balance of U.S.D. - 9A.
297	6/29/18	Report of the wreck of Lanzius plane. *(Noted here for its historical importance.)*
302	9/4/18	Report of wreck of DeH-4 at Indianapolis, Ind.
305	9/16/18	Exhaust header test on Liberty 12.
311	9/27/18	Test of Liberty 12 Model A engine No. C-250. *(The last Cadillac engine built.)*
346	10/2/18	Test of Liberty 12 after overhauling.
349	10/4/18	Test of performance of standard Liberty engine.
351	8/1/18	Intermittant oil feed Liberty 12.
378	10/14/18	U.S.D.-9 wooden wing ribs.
383	8/14/18	Handley-Page type rib of U.S.D.-9 chord.
388	8/26/18	Tail surfaces of Thomas-Morse MB-2.
393	9/21/18	Performance test of Caproni biplane with three Liberty 12 engines.
396	9/26/18	Wings of Thomas-Morse MB-2 fighter wings.
420	10/16/18	Test of Liberty 12 U.S. 18182 before installation in plane.
422	10/16/18	Test of Liberty 12 U.S. 18212 before installation in plane.
423	10/16/18	Test of Liberty 12 U.S. 18183 before installation in plane.
424	10/17/18	Test of La Char carburetor on Liberty 12.
425	10/17/18	Checking outlet water temperature during power test on standard Liberty 12.
426	9/30/18	Comparative table of performance of French 2-place machines with Liberty & others.
450	10/4/18	Investigation of external wire fittings in present use on DH-4 machines.
454	9/3/18	LePere C-11 tail surfaces.
463	10/17/18	Test Liberty 12 shipped to Curtiss Engine Corporation.
464	10/18/18	Resistance of Liberty 12.
465	10/18/18	Comparison of Simms and standard Delco ignition.
466	10/19/18	Comparison of special headers on Liberty 12 rear carburetor installation.
467	10/22/18	Test of syphon gasoline pump mounted on Liberty engine.

469	10/23/18	Test of articulated rods on Liberty.
470	10/23/18	Test Liberty 12 before installation in De9 A.
474	10/23/18	Test of Liberty 12 manufactured by Buick.
475	10/24/18	Test of Liberty 12 manufactured by Ford.
477	10/24/18	Test of Liberty 6 manufactured by Thomas-Morse.
478	10/25/18	Routine test of Liberty 12 equipped with Simms Magneto after installation.
486	10/5/18	Diagram on DH-4 prediction on vector.
487	10/11/18	Predictions for vectors for U.S.P. -2.
488	10/28/18	Gasoline consumption on Liberty 12 at maximum power at various R.P.M.
489	10/29/18	Comparative test of Liberty 8 equipped with different headers.
492	10/31/18	Test of Liberty 12 carburetor fitted with Brixton jets.
493	10/31/18	Comparison test of Liberty 12 with high and low compression.
498	9/19/18	Test of Turbo Supercharger made at the summit of Pikes Peak.
501	10/30/18	Test of Liberty 12 engine oil pump.
503	10/30/18	Routine test of Liberty 12 after overhaul.
504	10/30/18	Test of Liberty 12 before installing in DH-4B.
505	11/8/18	Test of Lincoln engine (206) U.S. 23630 before installing in plane.
509	11/21/18	Test of Cosmic connecting rod bearing for Liberty 12.
510	10/30/18	Changes in design of DH-4 wooden wing ribs.
515	11/7/18	Report on torque of 8-and 12-cylinder engines.
517	8/14/18	Determination of strength of Dayton-Wright DH-4 loading chassis.
520	11/14/18	Test of inverted Venus carburetor applied to Liberty engine.
534	11/21/18	Test of standard Liberty motor running on a mixture of 80 per cent Electra Aviation Gasoline and 20 per cent Benzol.
545	11/14/18	Weight schedule of the U.S.D. - 9A.
549	9/10/18	Performance of DH-4 plane.
550	11/21/18	Berkshire magneto of Liberty 12.
554	11/25/18	Routine test of Liberty 12 before installation in plane.
556	11/26/18	Comparitive test of Liberty 12 with double and single ignition.
560	12/2/18	Comparison of 4 Liberty engines built by Nordyke & Marmon, Ford, Cadillac & Buick.
564	12/4/18	Sand test of 2nd set of LePere wings.
568	12/5/18	Routine test of Epicyclic Liberty No. Exp 6 before shipment to U.S. Navy.
569	12/5/18	Test of 24-cylinder Liberty engine.
577	12/9/18	Test of Liberty 12 equipped with special inlet and exhaust elbows.
578	12/10/18	Description for plane design connecting rod for Liberty 12 engine.
588	12/12/18	Report of General Electric Co. Turbo Supercharger on Liberty 12 engine.
589	12/9/18	Radiator test on U.S.D. 9A cooling system.
590	12/17/18	Test Liberty 12 No. 732 U.S. 18181 after overhauling.
591	12/17/18	Test Liberty 12 No. 533 equipped with Keys piston rings.
592	12/17/18	Test of single-cycle and 6-cycle engine.
593	12/17/18	Double crankshaft on Liberty 12 Cannon engine.
597	12/9/18	Test of Simms 6-cylinder magneto.
603	12/20/18	Test of Sherbondy Turbo-Compressor on Liberty 12 Packard No. 1 U.S. 17435.
606	12/26/18	Standard design of engine control as applied to Liberty 12.
607	12/-/18	Routine test of Liberty 12 No. B-41 U.S. 37545.
608	12/27/18	Test of standard Liberty with and without water circulation around intake headers.
609	12/-/18	Test of twin-cylinder Experimental Liberty engine.
610	12/30/18	Routine test of Experimental Liberty Engine No. X-8 *(Probably this was Eng. Div. EXP-8 as it was delivered during the week of December 14-21.)*
618	1/3/19	10-hour test of herringbone planet gears in epicyclic Liberty engine No. 3.
621	1/7/19	Routine test of Liberty 8 BX-4 before shipment.
622	1/15/19	Liberty water pump.
623	1/15/19	Routine test of Liberty 12 No. B-74 U.S. 37573 before installation in plane.
624	1/8/19	Routine test of epicyclic Liberty 12 No. 3 U.S. 47835.

631	1/22/19	Current consumption of standard Liberty ignition system.
632	1/23/19	Standard Liberty generator outfit.
640	3/14/19	Test of McCook Field hand-starter.
662	1/27/19	Performance test of LePere U.S.A.C.-11.
677	1/23/19	Sand test of Polilio BVL-12 tail surfaces.
679	2/20/19	Fuel test of Liberty using a mixture of Benzol and gasoline.
680	2/20/19	Test of inverted 12-cylinder Liberty engine.
681	2/17/19	Test of Iron City Products Co. piston rings on Liberty cylinders.
682	2/20/19	Two-cylinder Liberty engine characteristics with two explosion phases.
688	2/25/19	Sand load test of Pomilio FVL-8 tail surfaces.
691	12/17/18	Hart adjustable pitch propeller S.C. No. 60528. *(Early report on this design.)*
692	12/19/18	Destructive Whirling Test of Hart adjustable pitch propeller S.C. No. 60529.
709	3/4/19	Center of gravity measurement of U.S.D. -9A reconnaissance plane.
725	3/13/19	Report on single-cylinder test engine, Test Order No. 23.
726	3/11/19	Report on sand test and stress analysis of wing structure of Pomilio FVL-8.
732	3/17/19	Sand test and stress analysis of Pomilio BLV-12.
733	3/11/19	Description of Moorehouse experimental oil pump for Liberty engines.
753	5/3/19	Test of water pump of 8-cylinder Liberty engine.
754	3/21/19	Standard engine report of 12-cylinder Liberty rated at 400 HP at 1,700 RPM.
764	3/27/19	Test of Porter oil pump as fitted to a Liberty engine.
797	5/6/19	Calculated performance of DH-4.
799	5/5/19	Sand load test of Pomilio FVL-8 landing chassis.
800	5/8/19	Wind tunnel test of USP-2.
809	5/19/19	Performance test of USD 9A showing drop in performance at high altitude.
810	5/16/19	Standard engine report of 8-cylinder Liberty rated at 290 HP @ 1,700 RPM.
829	5/15/19	Standard engine report of Thomas-Morse 6-cylinder Liberty rated @ 210 HP @ 1,700.
842	5/5/19	Report of analysis of USD 9A chassis and wheels.
863	6/10/19	Report of tests on standard Liberty 12 with various cooling water temperatures.
864	6/6/19	Test of standard Liberty 12 P-5098 sent to Bureau of Standards for altitude tests.
876	6/10/19	Test on single-cyl. Liberty of standard piston ring compression and mechanical losses.
886	6/19/19	Endurance test on USD 9A gasoline feed pump.
891	7/8/19	Sand load test on Pomilio BVL-12 chassis.
895	7/9/19	Test of wooden struts from DH-4 plane.
911	7/17/19	Report on vacuum gasoline system of LePere biplane P-53.
923	7/26/19	Cooling test of twin Martin bomber.
948	8/4/19	Test of Oildag in standard Liberty 12 engine.
955	8/15/19	Static test report of the LePere USAC-11 fuselage type X.
977	8/27/19	Test of standard Zenith, inverted Zenith and Stromberg carburetors on Liberty 12.
983	9/4/19	Test of standard Liberty 12 using 30-130-150 carburetor setting.
996	9/11/19	Report on DH-4 emergency floatation gear.
997	9/16/19	Report on static test of veneer-covered DH-4 wings.
1003	9/25/19	Report on J. V. Martin transmission.
1004	9/26/19	Report on investigation of wheel on DH-9 which turned turtle at McCook on 8/27/19.
1014	7/9/19	Test of A.E.D. propeller for LePere with Liberty 12 and Moss Supercharger.
1017	10/10/19	Static test of tail surfaces of LePere triplane USAO-11.
1032	10/24/19	Performance test of Pomilio BVL-12 bomber with Liberty 12 engine.
1043	11/3/19	Test of DH-4 with full radiator shutter.
1044	10/31/19	Ground test of Moss Supercharger.
1059	11/10/19	Test of vacuum oil control on Liberty 12.
1060	10/19/19	Test of Gallaudet transmission.
1064	9/5/19	Destructive whirl test of A.E.D. propeller for LePere P-53 with Moss Supercharger.
1071	11/26/19	Report on flight test with Moss Supercharger.
1086	11/25/19	Report on static test of fuselage of LePere Triplane USAO-11.
1088	11/18/19	Test of high-, low- & medium-compression on 6-cylinder Liberty engine.

1092	11/26/19	Investigation of used Liberty bearings.
1120	12/13/19	Report on static test and stress analysis of wing structure of LePere USAO-11.
1124	12/17/19	Wind tunnel test of LePere USAC-11.
1126	11/26/19	Test of standard Liberty 12 with Mercedes- and Benz-type manifolds.
1130	12/22/19	Official performance test of LePere Triplane with 2 Liberty engines.
1132	11/28/19	Official performance test of Glenn Martin Bomber with 2 Liberty engines.
1146	1/27/20	Test of various thicknesses of water jacket space on single-cylinder Liberty.
1148	1/30/20	Test of Monel Metal valves in Liberty single-cylinder engine.
1156	1/23/20	Report of preliminary test of Bijur ignition end starter motor for Liberty 12.
1160	2/10/20	Official test of Martin Transport showing comparison with Martin Bomber.
1172	2/13/20	Notes on comparison of Dixie Magneto and Delco ignition on Liberty 12.
1174	2/17/20	Report on static test of Dayton-Wright DH-4 tail surfaces.
1184	2/10/20	Report on A.E.D. propeller for Fokker plane with Liberty 6 engine.
1194	3/18/20	Report of static test of DH-4 P-34 wing cellule.
1196	3/16/20	Report of static test of landing chassis of DH-4 P-34.
1197	3/18/20	Report of static test of fuselage of Dayton-Wright DH-4.
1201	3/-/20	Report on Universal Test Engine.
1209	2/16/20	Report on 300 HP Hispano-Suiza (Wright Model M) with Liberty-type cylinders.
1212	3/19/20	Report of investigation of parts from LePere Triplane which was sand-tested.
1219	3/26/20	Test of 600 watt DC generator driven by silent chain from Liberty 12 on P-35.
1220	3/29/20	Performance test of Fokker D-VII with Liberty 6 engine.
1223	5/15/20	Report on wind tunnel test on a DH-4 model.
1225	5/5/20	Report on Dow Metal piston which failed in single-cylinder Liberty.
1232	3/19/20	Altitude chamber test of a standard Liberty 12.
1241	3/29/20	Performance test of Liberty 6 with Mercedes-type header with one and two carburetors.
1257	5/21/20	Report on centrifugal oil cleaner on Liberty 12.
1280	6/3/20	Comparative test of compact headers and standard headers on Liberty 6.
1283	6/7/20	Report on performance of five different design geared aviation engines.

Tests listed after number 1300 were found recorded in various official documents and did not always give an exact date for the test. Not taken from an official list of all tests, therefore not complete. Period covered is June 1920 to June 1925, which is not necessarily the last Liberty-related test.

Test Serial	Date if known	Description of tests	ASC=Air Service Circular
1311		Test of LUSAO 11 cooling system.	
1314		Test of auxiliary L-12 starting devices..	
1318		Test of Foster modifications to the L-12 Delco switch permitting running on battery alone.	
1319		Standard engine test of Hall-Scott L-6.	
1340		Test of inverted Stromberg NA-15 carburetor on L-12.	
1347		Test of magnesium alloy pistons on L-1.	
1348		Test of Canton Electrical Vaporizing Device on a Zenith carb.	
1357	8/31/20	Test of Martin Torpedo M-T with 2 L-12.	(ASC No. 139)
1383	9/17/20	50-Hr. endurance of Hall Scott L-6 No. 4519 .	(ASC No. 149)
1409	10/21/20	Test of GAX with 2 L-12.	(ASC No. 193)
1456	11/20/20	Performance estimate of LePere LUSAGH-11 with L-12.	(ASC No. 219)
1655	5/6/21	Test of Martin NBS-1 with 2 L-12 engines.	(ASC No. 290)
1694		Cooling test of LePere with L-12 and side radiators.	
1697	6/24/21	50-Hr. test of L-12 buzzer starter.	(ASC No. 302)
1711		Elimination of detonation in Liberty single-cylinder by use of fuel mixtures.	
1713		Valve opening diagrams for Eng. Div. W-1, L-12 and Hispano H and I.	
1719		Liberty cylinder operation on Recardo cycle.	
1724		Liberty cylinders with special oil grooves.	
1732	7/29/21	Carburetor settings for L-12 for dirigible	(ASC No. 311)

1739	7/28/21	Instructions for Stromberg NA-L5 inverted carb. For supercharged L-12 (ASC No.314)	
1744	9/-/21	Bearing loads on L-6	
1747		Effect of throttle opening on Liberty and Hispano engines	
1750		Elimination of detonation by fuel mixtures (follow-up of 1711)	
1751		Failure of Micarta prop on Supercharger L-12	
1777	9/-/21	Report on Aeromarine L-24 engine proposal	
1783		Variation of power output with spark advance - Wright H and L-12	
1800		Power output of L-12 supercharged at 40,000' from Oct. 5/21 flight	
1804		Fuel consumption tests with different carburetors on L-12, Hispano H, etc.	
1805		Oakley Monel Metal exhaust valve tested in 1921 model L-12	
1813		Breese single unit starter generator design for L-12	
1823		Lab test of Leece-Neville generator for L-12	
1838		Comparison of iron-skirt and standard oil grooved pistons.	
1844	1/24/22	50-hour test of 1921 Model L-12	
1850	1/25/22	Fuel consumption DH-4B with L-12	(ASC No. 346)
1860	4/-/22	Eng. Div. TW-1 P-200 with L-6 engine	
1863		Estimate performance of Orenco IL-1 with L-12	
1883		Bijur ignition end starter	
1907		Test of bellows fuel pump for L-12 and Wright. Supersedes Air Circular 281	
1918		Impact test of steel landing gear for Air Mail Service DH-4.	
1926		Test of Smith Rotary Induction System for Liberty 6	
1933		Test of Zenith 52 with different discharge nozzles for L-6 and L-12	
1942		Flight test of Hot Spot manifolds on L-12.	
1944		Performance test of Orenco D with supercharged Wright H.	
1951		Stress analysis of wings for DH-M2 Mail Plane.	
1953		Static test of Martin Bomber engine nacelle.	
1971		Static test of Eng. Div. CO-1 with L-12 power.	
1977		Performance test of DH-4M2 Mail Plane with L-12.	
1987		Description of Power Plant lab including 400 HP dyna. and single-cyl Liberty	
2010		Fuel consumption during climb of supercharged HD-4B	
2011		Test of three types of supercharger air coolers.	
2019		Test of cooling system of TW-1 with L-6	
2021		Change of exhaust temp. with mixture on SC L-12.	
2027	8/3/22	Development of GE Form C supercharger for the Wright H.	
2030	9/20/22	Failure of supercharger rotor buckets.	
2031		Report on structural strength of T-2 Transport	
2033		Comparison of high & low-comp. L-12 with GE Form D superchargers.	
2035		Cockpit heater for altitude flying in supercharged P-53	
2036		Oxygen system for pilot of supercharged P-53	
2041	11/13/22	Report on World's altitude record of 9/28/21 in LUSAC-11 with SC L-12	
2046	9/29/22	Performance test of Curtiss NBS-1 with 2 L-12	(ASC No. 435)
2075		Altitude test of Dixie Magnetos due to failures above 20,000 ft. on supercharged plane.	
2082		Test of Ball Manifold on L-6	
2092		Report on development of the inverted Stromberg NA-5L & NA-5LA	
2230	3/15/24	Performance test of Eng. Div. CO-5 with L-12	(ASC No. 452)
2365	6/6/24	Report on Eng. Div. Side Supercharger. Form F	
2456	1/12/25	Instructions for reinforcing Liberty cylinders	(ASC No. 504)
2459	2/9/25	Performance test of Douglas XO-2 with L-12	(ASC No. 509)
2489	3/19/25	Performance test of Curtiss XO-1 with L-12	(ASC No. 530)
2502	4/1/25	Performance test of Eng. Div. XCO-6 with inverted L-12	(ASC No. 521)
2513	4/1/25	Performance test of Eng. Div. CO-5 with L-12	(ASC No. 534)
2525	4/17/25	Performance test of Loening COA-1 with inverted L-12	(ASC No. 540)
2544	6/8/25	Performance test of Douglas C-1 transport	(ASC No. 546)

Appendix D

Specifications of the Liberty 12, 8 and 6 March - May 1919

Statistics reproduced below are extracted directly from BAP Airplane Engineering Division Standard Engine Report of the 12-Cylinder Liberty Motor.
McCook Field Power Plants Report #52 dated March 21, 1919.
The engine tested was Lincoln MNFG NO. 53 U.S. NO. 23481.
The engine pictured was Ford MNFG NO. 1635 U.S. NO. 31110.
Engineer in charge was George J. Mead.
Chief of Power Plants Dept. was Major Geroge E. A. Hallett.

Dimensions of Twelve-Cylinder Liberty Engine

Bore	5 in.
Stroke	7 in.
Number of cylinders	12
Stroke-bore Ratio	1.4:1
Piston displacement per cylinder	137.41 cu. in.
Total piston displacement	1648.92 cu. in.
Compression volume	31.12 cu. in.
Total volume of cylinder	168.53 cu. in.
Compression ratio	5.42:1
Percent compression	18.46
Compression pressure	110 lbs. gauge @ 120 RPM
	120 lbs. gauge @ 1500 RPM
Camshaft bearings	
Number on each shaft	7
Diameter	6 at .998"
	1 at 1.1235"
Material	6 of aluminum
	1 of bronze
Inlet valve	
Number per cylinder	1
Port diameter	2.5"
Lift	7/16"
Angle of seat	30°
Total area of opening	3.202 sq. in.
Tappet clearance	.013 - .016"
Exhaust valves	
Number per cylinder	1
Port diameter	2.5"
Lift	3/8"
Angle of seat	30°
Total area of opening	2.748 sq. in.
Tappet clearance	.019 - .021"
Valve springs	
Number per valve	2
Tension inlet (both springs)	
Valve open	62.2 lbs.
Valve closed	50 lbs.
Tension exhaust (both springs)	
Valve open	85.7 lbs.
Valve closed	71.5 lbs.
Internal spring tension	
Valve closed	26.5 lbs.
Valve open (inlet)	32.9 lbs.
Valve open (exhaust)	31.8 lbs.
Valve timing	
Inlet	
Opens	10° past top center
Closes	45° past bottom center
Exhaust	
Opens	48° before bottom center
Closes	8° past top center
Piston	
Type	Ribless
Area of head	19.63 sq in.
Material	Aluminum
Distance from ctr. of pin to top of piston	3 15/32"
Length overall	5 15/32"
Length of bearing in cylinder	3 3/4"
Clearance in cylinder	
Top	.0395 ± .0015"
Bottom	.020 ± .001"
Rings	
Number per piston	3
Tension	11-15 lbs.
Width	.24825"
Width of gap (ring in cylinder)	1/32 ± .010"
Pin	
Length	4 15/64"
Diameter	1.25"
Connecting rods	
Length of plain rod, c to c	12"
No. of bolts	2
Diameter	0.4365"
Thread	7/16-20 S.A.E.
Length of forked rod, c to c	12"
No. of bolts	4
Diameter	0.3115"

Thread	5/16-24 S.A.E.
Rod-Stroke ratio	1.714:1
Wrist pin bearing	
Length	2"
Diameter	1.25"
Material	Bronze
Big end bearing	
Length	2.25"
Diameter	2.375"
Material	Bronze shell, Babbitt lined
Crankshaft bearings	
Number	7
Material	Bronze shell, Babbitt lined
Diameter	2.625"
Length	#1 - 6 1.625" #7 4.25"
Carburetor	
Name	Zenith
Model	U.S. No. 52
Type	Double carburetor Annular nozzles
Size	
Diameter at the flange	52 mm.
Chokes, diameter	36 mm.
Jets	
Main	1.65 mm.
Comp.	1.70 mm.
Idling well	1.00 mm.
Ignition	
Type	Battery, generator
Manufacturer	Dayton Engineering Laboratories Co.
Model	D 5180 U.S.A. Liberty
Maximum spark advance	30° before top center
Maximum retard	10° after top center
Speed of ignition drives	
Generator	1 1/2 times C.S.
Distributors	1/2 times C.S.
Firing order	1L-6R-5L-2R-3L-4R -6L-1R-2L-5R-4L-3R
Rotation, direction of (facing propeller)	Counterclockwise
Spark plugs	
Number per cylinder	2
Location	Cylinder head
Size	18 mm/1.5 mm pitch (S.A.E. metric)
Name	A.C. or optional
Gap	.015 - .018"
Oil pump speed	1 1/2 times C.S.
Water pump speed	1 1/2 times C.S.

Reciprocating and centrifugal weights	
Pistons, complete with rings and pin	4.9 lbs.
Upper-end connecting rod	1.3lbs.
Lower-end forked connecting rod	4.4 lbs.
Lower-end plain connecting rod	1.9 lbs.
Total weights of connecting rods complete	
Forked	5.7 lbs.
Plain	3.2 lbs.
Total	8.9 lbs.
Valve (assembled)	0.8 lbs.
Loads from max. exp. pressure	
Assumed max. exp. pressure	510 lbs./sq. in.
Total load on piston head	10,011 lbs.
Load/sq. in. on wrist pin rod bearing	4,000 lbs.
Load/sq. in. on crank pin	1,872 lbs.
Load/sq. in. on main bearings	
No. 1 - 5 (incl.)	1,173 lbs.
No. 6	1,393 lbs.
No. 7	364 lbs.

Weights of 12-cylinder Liberty engine and parts

	Weight (lbs.)	Percent
12-cylinder assemblies, including valves	241.2	28.6%
1 crankshaft (heavy type), including gear, oil plugs, nuts & bolts, thrust bearing	105.8	12.5
1 upper crankcase complete, including nuts, studs, bearings, oil filters, flange nuts, & bolts	92.0	10.9
2 camshafts complete, including upper driveshafts & housings	78.6	9.3
1 lower crankcase complete, including studs, bearings, & pump bevel gear drive	70.0	8.3
12 pistons complete, including pins, pin retainers and rings	58.6	7.0
12 connecting rods complete, including bearings, bolts & nuts		
6 forked end 34.2		
6 plain end 19.2		
Total	53.4	6.3
1 propeller hub assembly complete	26.6	3.2
1 intake header assembly	26.4	3.1
2 Zenith US No. 52 carburetter assemblies complete including gondola scoop & vertical air intake	19.0	2.3
1 water pump assembly, including water manifolds, outlet elbows, hose connections & clamps	16.9	2.0
1 set of spark plugs and ignition header	13.8	1.6
1 set accessories' drives:		
2 camshaft driveshaft lower assemblies	6.4	

1 generator driveshaft 5.2		
total	11.6	1.4
1 generator	11.6	1.4
2 distributer head assemblies	10.4	1.2
1 oil pump & connections	7.5	0.9
Total engine weight	843.6	100.0

1 battery	10.5
1 switch & voltage regulator	2.4
weight of water in engine	45.0

Test Report Summary
Rated HP 400 @ 1700 RPM

Normal HP, full throttle	421 @ 1700 RPM
Fuel cons. @ normal HP	0.496 lbs./HPH @ 1700 RPM
Oil cons. @ normal HP	0.032 lbs./HPH @ 1700 RPM
Maximun HP	449 @ 1940 RPM
Maximum torque	1306 lbs. ft. @ 1540 RPM
Barometer	29.92 ins. Hg.
Air temperature	55° F
Maximum BMEP	119.4 lbs./sq.in. @ 1540 RPM
Compression pressure	110 lbs./sq.in. @120 RPM
	125 lbs./sq.in. @ 1500 RPM
Specific gravity of fuel	0.72

Weight dry, per HP at normal power		2.004 lb.
Material	4 of aluminum	

Friction Horsepower Run

RPM	Corrected engine HP	F.H.P.	Mech. Eff. %
1200	295	27.6	91.5%
1300	322	33.8	90.5
1400	348	38.3	90.1
1500	375	42.5	89.8
1600	398	49.1	89.0
1700	421	56.7	88.2
1800	442	65.4	87.2
1900	450	76.0	85.5
2000	433	88.0	82.2

Statistics reproduced below are extracted directly from BAP Airplane Engineering Division Standard Engine Report of the 8-cylinder Liberty Motor.
McCook Field Power Plants Report No. 67 dated April 18, 1919.
The engine tested and pictured was Buick MNFG NO. BX 5 U.S. NO. E (later assigned 94476)
Test engineer was Iskander Hourwich (Hourwich had until recently been a test engineer for the Experimental Engineering Section of the Detroit District Office under Charles H. Vincent.)
Engineer in charge was C. F. Taylor.
Chief of Power Plants Dept. was Major George E. A. Hallett.

Dimensions of 8-Cylinder Liberty Engine

Bore	5 in.
Stroke	7 in.
Number of cylinders	8
Stroke - bore ratio	1.4:1
Piston displacement per cylinder	137.41 cu. in.
Total piston displacement	1099.28 cu. in.
Compression volume	31.12 cu. in.
Total volume of cylinder	168.53 cu. in.
Compression ratio	5.42:1
Percent compression	18.46
Compression pressure	115.0 lbs. gauge @ 120 RPM
	130.5 lbs. gauge @ 1000 RPM
Camshaft bearings	
Number on each shaft	5
Diameter	4 at .998"
	1 at 1.1235"
	1 of bronze

Inlet valves	
Number per cylinder	1
Port diameter	2.5"
Lift	7/16"
Angle of seat	30°
Total area of opening	3.202 sq. in.
Tappet clearance	.013 - .016"
Exhaust valves	
Number per cylinder	1
Port diameter	2.5"
Lift	3/8"
Angle of seat	30°
Total area of opening	2.748 sq. in.
Tappet clearance	.019 - .021"
Valve springs	
Number per valve	2

Tension inlet (both springs)	
Valve open	62.2 lbs.
Valve closed	50 lbs.
Tension exhaust (both springs)	
Valve open	85.7 lbs.
Valve closed	71.5 lbs.
Internal spring tension	
Valve closed	26.5 lbs.
Valve open (inlet)	32.9 lbs.
Valve open (exhaust)	31.8 lbs.
Valve timing	
Inlet	
Opens	10° past top-center
Closes	45° past bottom-center
Exhaust	
Opens	48° before bottom-center
Closes	8° past top-center
Piston	
Type	ribless
Area of head	19.63 sq. in.
Material	Aluminum
Distance from ctr. of pin to top of piston	3 15/32"
Length overall	5 15/32"
Length of bearing in cylinder	3 3/4"
Clearance in cylinder	
Top	.0395 ± .0015"
Bottom	.020 ± .001"
Rings	
Number per piston	3
Tension	11 - 15 lbs.
Width	.24825"
Width of gap (ring in cylinder)	1/32 ± .010"
Pin	
Length	4 15/64"
Diameter	1.25"
Connecting rods	
Length of plain rod, c to c	12"
No. of bolts	2
Diameter	0.4365"
Thread	7/16-20 S.A.E.
Length of forked rod, c to c	12"
No. of bolts	4
Diameter	0.3115"
Thread	5/16-24 S.A.E.
Rod-stroke ratio	1.714:1
Wrist pin bearing	
Length	2"
Diameter	1.25"
Material	Bronze
Big end bearing	
Length	2.25"
Diameter	2.375"
Material	Bronze shell, Babbit lined
Crankshaft bearings	
Number	5
Material	Bronze shell, Babbit lined
Diameter	2.625"
Length	No. 1 - 4 1.625" No. 5 - 4.25"
Carburetor	
Name	Zenith
Model	G-D 55
Type	Double carburetor, Annular nozzles
Size	
Diameter at the flange	55 mm.
Chokes, diameter	40 mm.
Jets	
Main	2.40 mm.
Comp.	2.20 mm.
Idling well	1.00 mm.
Ignition	
Type	Battery, generator
Manufacturer	Dayton Engineering Laboratories Co.
Model	D 5177 U.S.A. Liberty
Maximum spark advance	30° before top-center
Maximum retard	10° after top-center
Speed ignition drives	
Generator	1 1/2 times C.S.
Distributor	1/2 times C.S.
Firing order	1L-4R-3L-2R-6L-1R-2L-3R
Rotation, direction of (facing propeller)	counterclockwise
Spark plugs	
Number per cylinder	2
Location	cylinder head
Size	18 mm./1.5 mm. pitch
Name	A.C. or optional
Gap	.015 - .018"
Oil pump speed	1 1/2 times C.S.
Water pump speed	1 1/2 times C.S.
Reciprocating and centrifugal weights	
Piston, complete with rings and pin	4.9 lbs.
Upper end connecting rod	1.3 lbs.
Lower end forked connecting rod	4.4 lbs.
Lower end plain connecting rod	1.9 lbs.
Total weights of connecting rods complete	
Forked	5.7 lbs.
Plain	3.2 lbs.
Total	8.9 lbs.
Valve (assembled)	0.8 lbs.

Loads from max. exp. pressure	
Assumed max. exp. pressure	510 lbs./sq. in.
Total load on piston head	10,011 lbs.
Load/sq. in. on wrist pin rod bearing	4,000 lbs.
Load/sq. in. on crank pin	1,872 lbs.
Load/sq. in. on main bearings	
No. 1 - 3 (incl.)	1,173 lbs.
No. 4	1,393 lbs.
No. 5	364 lbs.

Weights of 8-cylinder Liberty engine and parts

		Weight (lbs.)	Percent
8-cylinder assemblies including valves		163.1	25.5
1 crankshaft including gear, oil plugs, nuts, bolts, & hub key		72.0	11.3
1 upper crankcase complete including studs, bearings, & oil fillers		72.0	11.3
2 camshafts complete including upper drive shafts & housings		61.4	9.6
1 lower crankcase complete including studs & bearings		58.0	9.1
8 pistons complete including pins, pin retainers & rings		39.2	6.1
8 connecting rods complete including bearings, bolts & nuts			
4 forked end	22.6		
4 plain end	12.8		
Total		35.6	5.6
1 propeller hub assembly complete		26.0	4.1
1 intake header assembly		21.5	3.4
1 Zenith duplex G. D. 55 carburctor		9.1	1.4
1 water pump assembly including water manifolds, hose connections & clamps		12.7	2.0
1 set of spark plugs & ignition header		9.3	1.5
1 set accessories' drives:			
2 camshaft driveshaft lower assemblies	6.4		
1 generator driveshaft	5.2		
1 water pump bevel dr.	2.4		
Total		14.0	2.2
Stroke - bore ratio			1.4:1
1 generator		11.9	1.9
2 distributor head assemblies		10.2	1.6
1 oil pump (including pressure gauge & camshaft feed lines)		8.0	1.3
Nuts & bolts		7.2	1.1
Miscellaneous items		6.8	1.0
Total engine weight		638.0	100.0

1 battery	10.5
1 switch & voltage regulator	2.4
weight of water in engine	28.0

Test Report Summary
Rated HP 290 @ 1700 RPM

Normal HP full throttle	290 @ 1700 RPM
Fuel cons. @ normal HP	0.490 lbs./HPH @ 1700 RPM
Oil cons. @ normal HP	0.0211 lbs./HPH @ 1708.5 RPM
Maximum HP	317 @ 2000 RPM
Maximum torque	933 lbs. ft. @ 1200 RPM
Barometer	29.92 in. Hg.
Air temperature	55° F
Maximum BMEP	127 lbs./sq. in. @ 1200 RPM
Compression pressure	115.0 lbs./sq. in. @ 120 RPM
	130.5 lbs./sq. in. @ 1000 RPM
Specific gravity of fuel	0.72
Weight dry, per HP at normal power	2.2 lbs.

Friction Horsepower Run

RPM	Corrected engine HP	F.H.P.	Mech. Eff.
1200	212.4	16.4	92.6%
1300	228.3	19.9	92.0
1400	245.1	22.9	91.5
1500	260.9	26.0	91.0
1600	274.0	30.4	90.0
1700	290.3	34.6	89.4
1800	302.4	41.4	88.0
1900	309.6	44.3	87.5
2000	316.4	52.0	85.9

Statistics reproduced below are extracted directly from BAP Airplane Engineering Division Standard Engine Report of the 6-cylinder Liberty Motor.
McCook Field Power Plant Report NO.83 dated May 15, 1919.
The engine tested and pictured was Thomas-Morse MNFG NO. 5 U.S. NO. 47843.
Test engineer was Iskander Hourwich, engineer in charge was George J. Mead.
Chief of Power Plants Dept. was Major George E. A. Hallett.

Diminsions of 6-Cylinder Liberty Engine

Bore	5 in.
Stroke	7 in.
Number of cylinders	6
Piston displacement per cylinder	137.41 cu. in.
Total piston displacement	624.46 cu. in.
Compression volume	31.12 cu. in.
Total volume of cylinder	168.53 cu. in.
Compression ratio	5.42:1

- Percent compression: 18.46
- Compression pressure: 120.0 lbs. gauge @ 120 RPM; 129.0 lbs. gauge @ 1300 RPM
- Camshaft bearings
  - Number on shaft: 7
  - Diameter: 6 at .998"; 1 at 1.1235"
  - Material: 6 of aluminum; 1 of bronze
- Inlet valves
  - Number per cylinder: 1
  - Port diameter: 2.5"
  - Lift: 7/16"
  - Angle of seat: 30°
  - Total area: 3.202 sq. in.
  - Tappet clearance: .013 - .016"
- Exhaust valves
  - Number per cylinder: 1
  - Port diameter: 2.5"
  - Lift: 3/8"
  - Angle of seat: 30°
  - Total area: 2.748 sq. in.
  - Tapper clearance: .019 - .021"
- Valve springs
  - Number per valve: 2
  - Tension inlet (both springs)
    - Valve open: 62.2 lbs.
    - Valve closed: 50.0 lbs.
  - Tension exhaust (both springs)
    - Valve open: 85.7 lbs.
    - Valve closed: 71.5 lbs.
  - Internal spring tension
    - Valve closed: 26.5 lbs.
    - Valve open (inlet): 32.9 lbs.
    - Valve open (exhaust): 31.8 lbs.
- Valve timing
  - Inlet
    - Opens: 10° past top-center
    - Closes: 45° past bottom-center
  - Exhaust
    - Opens: 48° before bottom-center
    - Closes: 8° past top-center
- Piston
  - Type: Ribless
  - Area of head: 19.63 sq. in.
  - Material: Aluminum
  - Distance from ctr. of pin to top of piston: 3 15/32"
  - Length overall: 5 15/32"
  - Length of bearing in cylinder: 3 3/4"
  - Clearance in cylinder
    - Top: .0395 ± .0015"
    - Bottom: .020 ± .001"
  - Rings
    - Number per piston: 3
    - Tension: 11-15 lbs.
    - Width: .24825"
    - Width of gap (ring in cylinder): 1/32" ± .010"
  - Pin
    - Length: 4 15/64"
    - Diameter: 1.25"
- Connecting rod
  - Length c to c: 12"
    - No. of bolts: 4
      - Diameter: 0.3115"
      - Thread: 5/16-24 S.A.E.
  - Rod-Stroke ratio: 1.714:1
  - Wrist pin bearing
    - Length: 2"
    - Diameter: 1.25"
    - Material: Bronze
  - Big end bearing
    - Length: 2.25"
    - Diameter: 2.375"
    - Material: Bronze shell, Babbit lined
- Crankshaft bearings
  - Number: 7
  - Material: Bronze shell, Babbit lined
  - Diameter: 2.625"
  - Length: No. 1 - 6 1.625"; No. 7 4.25"
- Carburetor (2)
  - Name: Zenith
  - Model: 55 AS
  - Type: Single carburetor Venturi nozzles
  - Size
    - Diameter of flange: 55 mm.
    - Chokes: 40 mm.
  - Jets
    - Main: 1.80 mm.
    - Comp.: 1.55 mm.
  - Idling well: 1.00 mm.
- Ignition
  - Type: Battery, generator
  - Manufacturer: Dayton Engineering Laboratories Co.
  - Model: X 903 U.S.A. Liberty
  - Maximum spark advance: 30° before top center
  - Maximum retard: 10° after top center

Speed ignition drives	
Generator	at C.S.
Distributors	1/2 times C.S.
Firing order	1-5-3-6-2-4
Rotation, direction of (Facing propeller)	Counterclockwise
Spark plugs	
Number per cylinder	2
Location	Cylinder head
Size	18 mm./1.5 mm. pitch (S.A.E. Metric)
Name	A.C. or optional
Gap	.015 - .018"
Oil pump speed	1 1/2 times C.S.
Water pump speed	1 1/2 times C.S.
Reciprocating and centrifugal weights	
Piston, complete with rings & pin	4.7 lbs.
Upper end connecting rod	1.1 lbs.
Lower end connecting rod	3.7 lbs.
Total weight of connecting rod complete	4.8 lbs.
Valve (assembled)	0.8 lbs.
Loads from max. exp. pressure	
Assumed max. exp. pressure	510 lbs./sq. in.
Total load on piston head	10,011 lbs.
Load/sq. in. on wrist pin	4,000 lbs.
Load/sq. in. on crank pin	1,872 lbs.
Load/sq. in. on main bearings	
No. 1-5 (incl.)	1,173 lbs.
No. 6	1,393 lbs.
No. 7	364 lbs.

Weights of 6-cylinder Liberty engine and parts

	Weight (lbs.)	Percent
6-cylinder assemblies including valves	115.2	20.3%
1 crankshaft (heavy-type) including gear, oil plugs, nuts, bolts & thrust bearing	104.4	18.4
1 upper crankcase complete, including studs, bearings & oil fillers	77.2	13.6
1 camshaft complete, including upper driveshaft housing & lower driveshaft	44.6	7.8
1 lower crankcase complete, including studs, bearings & pump bevel drive gear	69.3	12.2
6 pistons complete including pins, pin retainers & rings	28.1	4.9
6 connecting rods complete including bearings, nuts & bolts	28.8	5.1
1 propeller hub assembly complete	22.1	3.9
1 intake header assembly	11.2	2.0
2 Zenith U.S. AS 55 carburetor assemblies, complete including scoops	11.7	2.1
1 water pump assembly, including water manifolds, outlet elbows, hose connections & clamps	11.5	2.0
1 set of spark plugs & ignition header	6.2	1.1
1 generator	12.5	2.2
1 distributor head assembly including external coils	9.5	1.7
1 oil pump & connections	9.0	1.6
Miscellaneous nuts & bolts	6.2	1.1
Total engine weight	567.5	100.0
1 battery	10.5	
1 switch and voltage regulator	2.4	
Weight of water in engine	21.0	

Test Report Summary
Rated HP 210 @ 1700 RPM

Normal HP, full throttle	221 @ 1700 RPM
Fuel cons. @ normal HP	0.526 lbs./HPH @ 1700 RPM
Oil cons. @ normal HP	0.0256 lbs./HPH @ 1700 RPM
Maximum HP	240 @ 1850 RPM
Maximum torque	729 lbs. ft. @ 1510 RPM
Barometer	29.92 in. Hg.
Air temperature	55°
Maximum BMEP	133.3 lbs./sq. in. @ 1510 RPM
Compression pressure	110 lbs./sq. in. @ 120 RPM
	129 lbs./sq. in. @ 1300 RPM
Specific gravity of fuel	0.72
Weight dry, per HP at normal power	2.45 lbs.

Friction Horsepower Run

RPM	Corrected engine HP	F.H.P.	Mech. Eff.
1200	163	9.6	94.4%
1300	178	12.1	93.6
1400	193	14.5	93.0
1500	207	17.5	92.2
1600	220	20.8	91.4
1700	231	25.5	90.1
1800	239	30.0	88.9
1900	238	34.2	87.5
2000	220	39.3	84.9

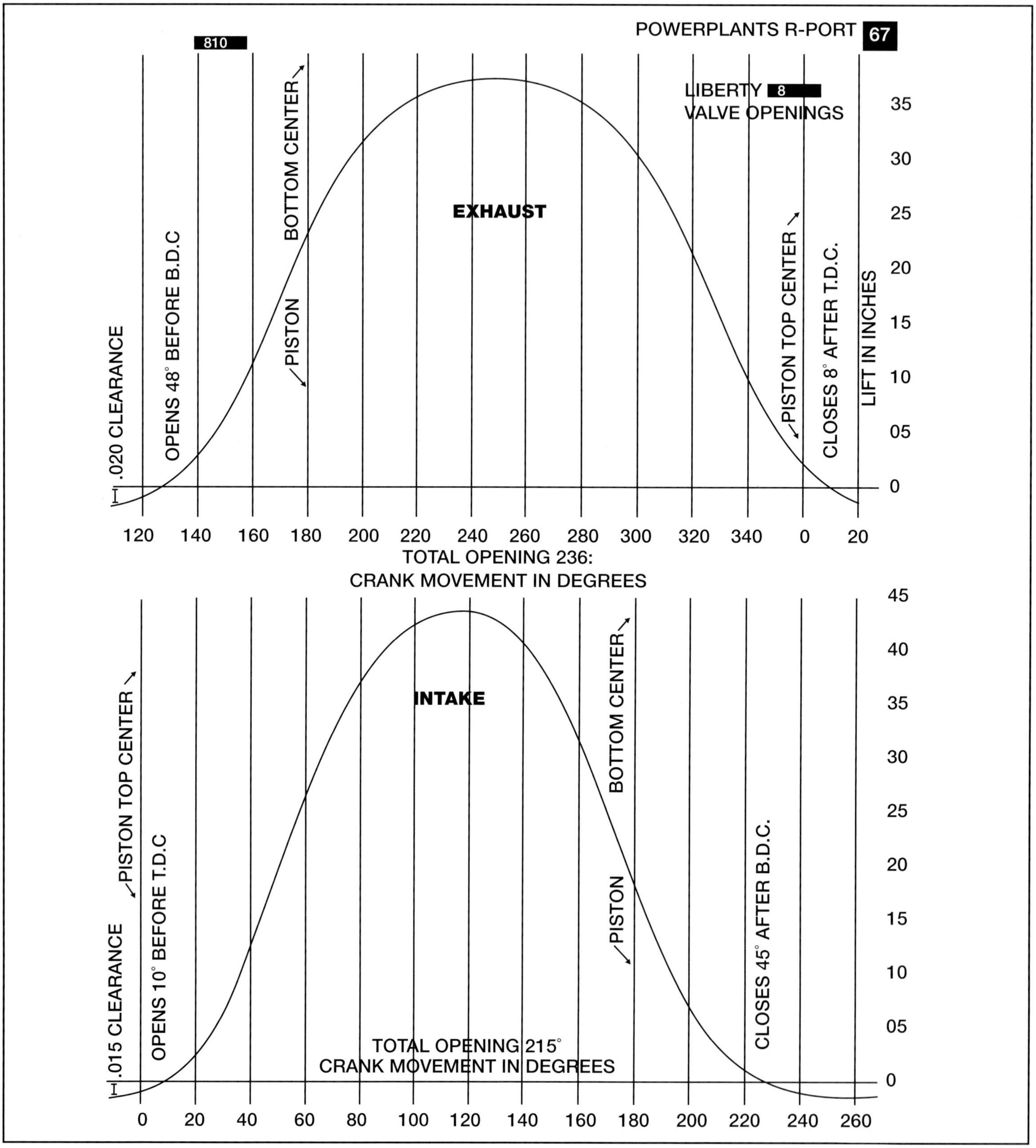

Valve opening profile for Liberty 6, 8 and 12 engines. This diagram is from the Liberty 8 test, but is identical for all three engines. (Author's collection)

Appendix E

The Military Life of the Liberty Worldwide

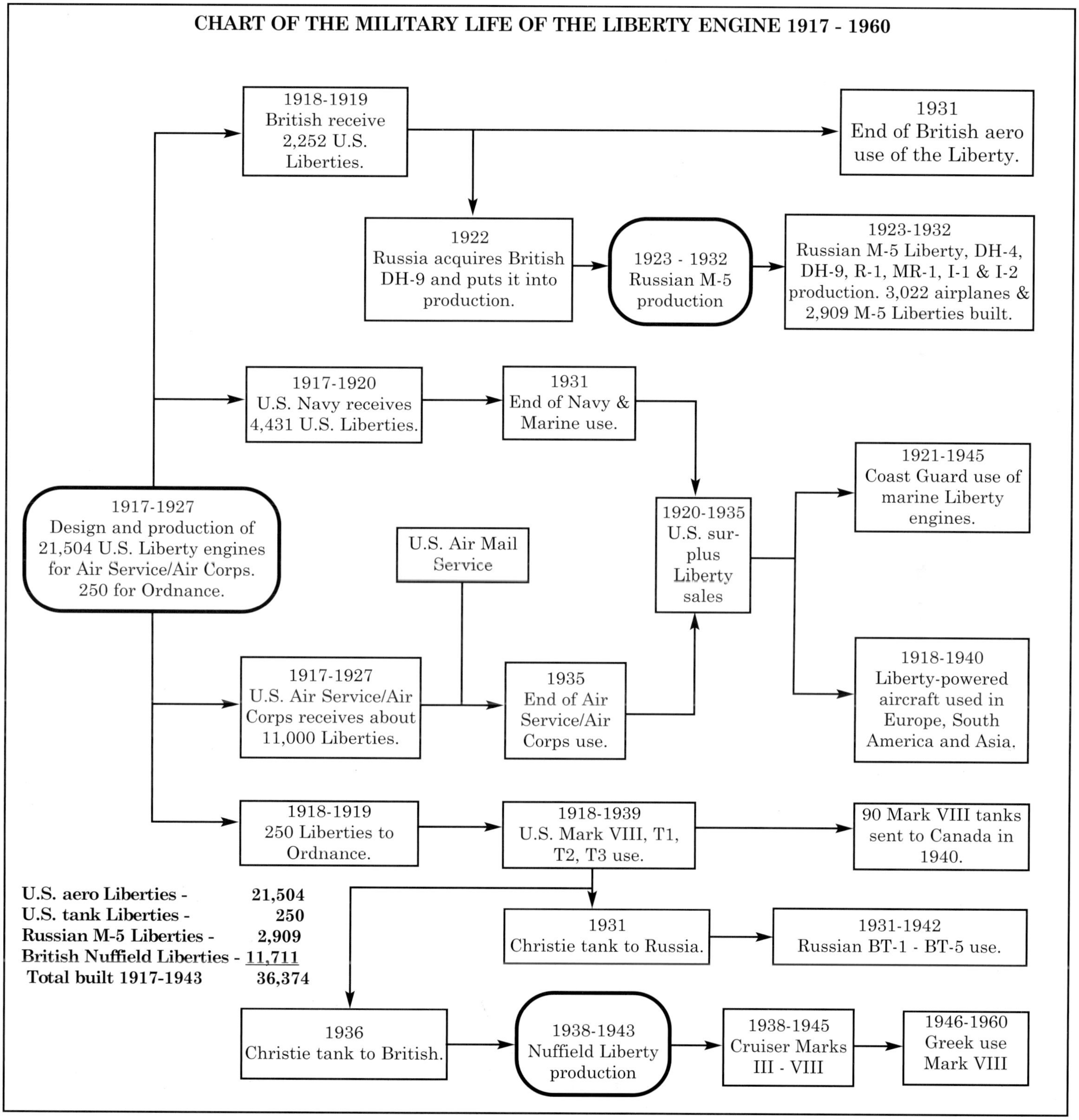

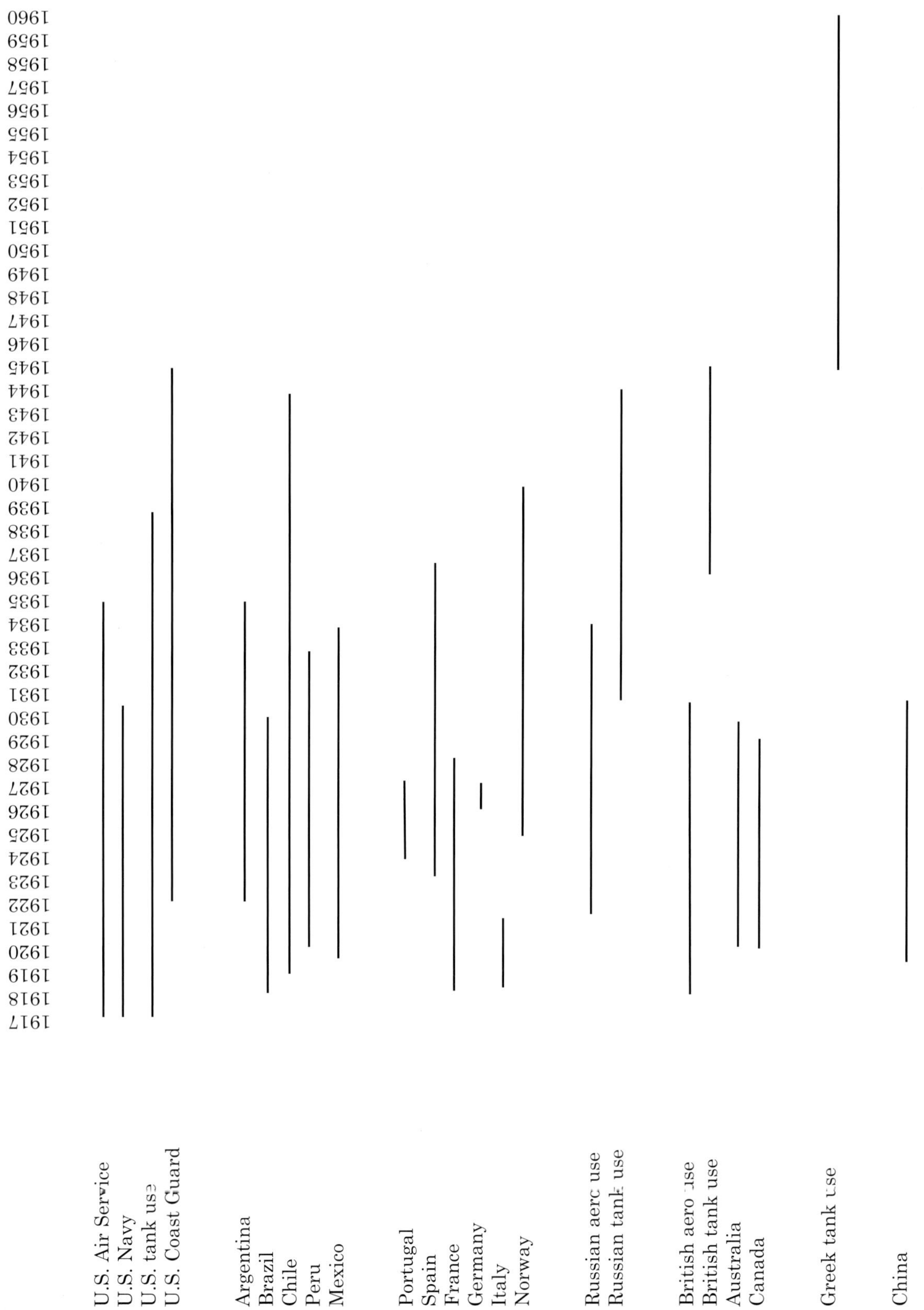

Liberty Military Use Expressed Chronologically including aeronautical, tank and marine power.

As can be seen, some version of the Liberty engine was in continuous active military service from 1917 to 1960.

Appendix F

Aircraft That Used Liberty Power

In this appendix are listed all aircraft I have found records of which were, at any point in their life, powered by a Liberty engine. I make no claim that this list is absolutely complete or absolutely accurate, even though I have made every attempt to make it so. The list contains aircraft originally powered with the Liberty as well as those reengined with a Liberty. Many are experimental or one-of-a-kind planes. Dates stated are, in most cases, dates of production. If the dates represent a period of use, it is so stated. Many of the military airplanes saw many years of use, some into the 1930s. In addition to several aero historians listed in the acknowledgements and references listed in the bibliography, I found Web sites such as Aerofiles and World Air Forces to be very helpful in compiling this list.

FOREIGN DESIGNS AND USES WITH LIBERTY POWER

	Number Built	Dates	Illustration page
Afganistan			
Russian R-1 - purchased	12	1924-1925	
Australia:			
DH-9A - supplied by the British	30	1920-1930 (use)	372
Argentina:			
HS-2L Curtiss - Argentine Navy purchased	12	1922-1931 (use)	373
F-5L Curtiss - Argentine Navy purchased	8	1921-1935 (use)	519
Dornier Do J Wal - Argentine Navy purchased	4	1922-1932 (use)	519
Brazil:			
HS-2L Curtiss - Brazilian Navy purchased	6	1918-1923 (use)	369
F-5L Curtiss - Brazilian Navy purchased	14+	1922-1930 (use)	
British:			
DH-4a	?	1920	
DH-9a	867	1918-1931 (use)	396
DH-10A	200+	1918-1919	520
Vickers FB-27 Vimy (some with Liberty)	?	1919	
Bristol Braemar II	1	1919	520
Bristol Pullman 26	1	1920	368
Handley Page O/400	?	1920	
Handley Page V-1500	?	1920	
Canada:			
DH-9a - supplied by the British	13+	1920-1929 (use)	198
Curtiss HS-2L	42+	1918-1928 (use)	
Curtiss H-16	2	1922-1924 (use)	
Chile:			
DH-9a - purchased by Chilean Air Force	20	1919-1928 (use)	
Dornier Do J Wal - Chilean Navy purchased	8	1926-1944 (use)	373
China:			
Curtiss HS-2L	9	1919-1920 (use)	
Russian R-1	11	1925-1928 (use)	
Douglas O-2C	10?	1926-1931 (use)	
Dutch:			
Fokker F-VIIA (at least two to U. S. with Liberty)	2	1926	
French:			
Breguet 14 A2, B2, Ap2, B2L	400+	1918-1928 (use)	208, 368
Breguet 16 Bu2	1?	1918	

Breguet 17 C2	10	1918-1923 (use)	
Farman 31 C2 (WW I one tested May 10, 1919)	1	1919	520
Morane-Saulnier type A.N.L.	1?	1919	521
Voison Bn2	1	1918	
Germany			
Heinkel HD-24L	1	1926	391
Iran			
Russian R-1 - purchased	6?	1924	
Italian:			
Caproni CA-42 (3 engine WW I triplane)	6?	?	521
Caproni CA-43 (3 engine WW I triplane with floats)	1	1918	521
Caproni CA 46 (3 engine WW I triplane)	?	1918	
Caproni CA 60 (8 engine triple-tandum)	1	1920	369
Mexico			
HS-2L Curtiss - Air Force purchase	1	1919-1921 (use)	
DH-4B - Air Force purchase	19	1924-1928 (use)	
Douglas O-2C - Air Force purchase	8	1926-1930 (use)	374
Quetzalcoatl Biplane 6-E-136 (Mexican built)	3	1925	
Norway:			
DT2b purchased from Douglas	1	1925-1940 (use)	371
DT2b built by Norwegian Naval workshops	6	1927-1940 (use)	522
Peru:			
HS-2L Curtiss - Navy purchased	5	1920-1926 (use)	
F-5L Curtiss - Navy purchased	2	1920-1924 (use)	
H-16 Curtiss - Navy purchased	1	1921 (use)	
DT2b - Navy purchased	4	1926-1933 (use)	374
Portugal			
DH-9a - purchased from British	1	1924	
Russia:			
DH-9a (captured British)	10?	1922	
Fokker (C-IV 60, C-IV-B 50)	110	1923-1924	404
R-1	2641	1926-1932	396, 522
MR1 (R-1 with wooden floats)	124	1927-1929	399
PM-2 (R-1 with metal floats)	1	1926	
IL-400a (IL-1)	1	1923	
IL-400b (IL-2)	1	1924	401
I-1 (IL-3)	14	1924-1926	402
I-2	213	1926-1929	370, 403
R-III	1	1925	399
ANT-3	1	1925	
R3	30	1925-1926	400
R4	1	1928	
Spain			
DH-9a - purchased from British	8	1922-1936 (use)	

U. S. CIVIL MODELS WITH LIBERTY ENGINES

	Number Built	Dates	Illustration page
Aerial Service Mercury	1	1925	523
Aeromarine BM-1	1	1924	
Aeromarine 75 converted F-5L	11	1920	523
Aeromarine 80 converted HS-2L	5?	1920	523
Aeromarine 85 converted HS-2L	1	1920	

Aeromarine A.M.C.	1	1924	
Aeromarine AM-1 mailplane	1	1923	524
Aeromarine AM-2 mailplane	1	1924	524
Aeromarine AM-3 mailplane	1	1924	524
Atlantic CO-4 mail (Fokker)	1	1925	525
Boeing B-1	1	1924	
Boeing 40 mailplane	1	1925	525
Burnelli RB-1 Lifting Fuselage Twin C/N 2	1	1923	525
Continental Motors Seaplane	1	1924	
CD Air Express	1	1925	526
Curtiss HA Mail	3	1919	526
Curtiss Eagle	3	1919	
Curtiss Carrier Pigeon I	11	1925	363
Curtiss Falcon Mailplane	13	1928	527
Curtiss R-4LM Mailplane	?	1918	527
Custom Biplane flying boat (Packard 1A-1650 engine)	1	1921	
Dare Variable Cambar Monoplane	1	1922	526
Dayton-Wright FP II	1	1922	528
Douglas M-1 Mail	1	1925	528
Douglas M-2 Mail	6	1926	528
Douglas M-3 Mail	10	1926	529
Douglas M-4 Mail	40	1926	
Douglas Cloudster	1	1921	364
Elias Mail M-1	1	1925	529
Elias Airmobile	3	1925-27	529
Fokker F-7-B-3M (Model 6)	1	1925	
Friesley Falcon	1	1921	530
Gallaudet Tourist C-3	1	1920	530
Huff-Daland Duster	2	1922	530, 531
Huff-Daland Petrel HD-31	1	1925	
Huff-Daland HD-22	1	1923	
Junkers-Larsen JL-12	1	1921	531
Laird Limousine	1	1922	531
Lawson 2 motor C-2	1	1919	532
Lawson 3 motor L-4	1	1920	532
Loening Air Yacht M 23	7	1922	532
Loening M-25	?	?	
Rinehart-Whelan Air Cruiser (Dayton-Wright KT)	1	1920	533
Sattco DH-4	1	1922	533
Sikorsky S-29A twin	1	1924	534, Cover
Sikorsky S-32	1	1926	534
Stout Air Pullman/air Transport AP-1/2AT	11	1924-25	534
Standard three-place (L-6-powered)	1	1924	
SVA two-place (L-6-powered)	1	1924	
Waterman 3L-400 (Packard LePere rebuilt)	1	1920	535
Yackey Transport (Breguet 14A2 rebuilt)	4	1925-26	535

PRE-DESIGNATION U. S. NAVY PLANES WITH LIBERTY POWER

	Number Built	Dates	Illustration page
Curtiss H-12L	19	1918	215, 536
Curtiss H-16 and variants	274	1918-1919	216, 536
Curtiss HA (2 versions HA-1, HA-2)	3	1918	536
Curtiss HS-1L (first to fly the Liberty 12)	912	1917-1918	55, 215
Curtiss HS-2L	213	1918-1919	537
Curtiss HS-3L (plus other manufacturers)	7	1918	537
Curtiss R6-L	30	1918	538
Curtiss R9	40	1918	538
DH-4 (transferred from Army)	51	1918	538
DH-4B & B1 (Transferred from Army)	172	1918-1919	539
Gallaudet D-4	2	1918	539
Martin MT (like Army MB-2)	8	1920	539
Martin MBT (like Army MB-1)	2	1920	540
Naval Aircraft Factory F-5L	225	1918	367, 540
Naval Aircraft Factory F-6L	2	1918	
Naval Aircraft Factory N-1	4	1918	540
Naval Aircraft Factory PT-1	15	1922	541
Naval Aircraft Factory PT-2	18	1922	541
Navy Curtiss NC-1	1	1918	541
Navy Curtiss NC-2	1	1919	542
Navy Curtiss NC-3	1	1919	542
Navy Curtiss NC-4	1	1919	377, 542
Navy Curtiss NC-5	1	1920	
Navy Curtiss NC-6	1	1920	543
Navy Curtiss NC-7	1	1920	
Navy Curtiss NC-8	1	1920	
Navy Curtiss NC-9	1	1920	543
Navy Curtiss NC-10	1	1920	543
NAF G-1 (nacelle completed but airplane canceled partially complete)		1921	548
NAF TG-1 (A6344) (L-6)	1	1922	543
NAF TG-2 (A6345) (L-6)	1	1922	544
NAF C-1	1	1918	544
Paul Schmitt float plane	1	1918	544
Sperry Triplane Amphibian	1	1918	545

(NAF = National Aircraft Factory.)

POST-DESIGNATION NAVY PLANES WITH LIBERTY POWER

	Number built	Dates	Illustration page
Boeing O2-B1 (DH-4M-1)	30	1925-1926	545
Douglas DT-1	1	1921	545
Douglas DT-2 (L.W.F.-NAF)	57	1922	546
Douglas OD-1	2	1929	547
Elias EM-2	6	1922	546
Fokker FT-1	3	1923	546
Loening OL-2 (V-1460 inverted water-cooled)	5	1925	387, 547
Loening OL-4 (V-1460 inverted water-cooled)	6	1926	547

PRE-DESIGNATION U. S. ARMY PLANES WITH LIBERTY POWER

		Number built	Dates	Illustration page
Breguet 14A P-148		1	1918	211
Caproni CA-5 Built by Standard & Fisher (three engines)		5	1918	212
Christmas Bullet (Civilian plane/Army engine L-6 US 47839)		1	1918	148
Curtiss R-4L - Liberty engine mail plane		6	1918	549
Curtiss-built Bristol USA O-1		27	1918	549
Cutriss CB "Battler"		1	1918	549
DH-4 variants (plus civil)		4846	1918-1934 (use)	55, 550
DH-9A (USD-9A, USD-9)		13	1918	550
Eng. Div. USAC-1	(L-12C)	1	1918	172
Eng. Div. USB-2	(L-8)	1	1918	154
Eng. Div Pomilio BVL-12		6	1919	550
Eng. Div Pomilio FVL-8	(L-8)	6	1919	155
Fokker D-VII	(L-6 US 47841)	1	1919	152
Fokker D-VII	(Hall-Scott L-6)	1	1919	551
Glenn Martin MB-1 and variations (GMB)		10	1918-1923	551, 552
Glenn Martin MB-1 Mail		6	1924	552
Glenn Martin MB-2 (became NBS-1)		110	1920-1927	326
Glenn Martin MB-2 (Curtiss built as NBS-1. 1 supercharged)		20	1923	552
Handley-Page O/400 built by Standard		107 (see note)	1918-19	553
John Multiplane		1	1919	553
J. V. Martin - central engines		1	1918	554
LePere LUSAC-11 and variants		31	1918	214, 346, 554
LePere LUSAO-11 twin-engine triplane		2	1919	555
LePere LUSAGH-11		2	1918	555
Lanzius		1	1918	555
L.W.F. -F (First Liberty engine to fly) (L-8)		1	1917	50
L.W.F. G-2		1	1917	556
L.W.F. Owl tri-motor twin fuselage		1	1918	556
Thomas Morse MB-1		2	1918	556
Thomas Morse MB-2 (two variations) (L-12C)		2	1918	172

Note: 7 complete planes were assembled and delivered to the Air Service in the U.S. 100 complete sets of parts were shipped to England but may never have been assembled.

LIGHTER-THAN-AIR

	Number built	Dates	Illustration page
Roma airship - six Liberties	1	1921-1922 (use)	267
RS-1 airship - four Liberties	1	1925-1927 (use)	270
RN-1 airship - two Liberties	1	1921-1924 (use)	267

POST-DESIGNATION ARMY PLANES WITH LIBERTY POWER

	Comments	Number built	Dates	Illustration page
AO-1 Atlantic		1	1924	557
XA-1 Cox-Klemin		2	1923	557
XA-2 Douglas	(V-1410 inverted air-cooled)	1	1926	557
A-2 Fokker	(converted T-2)	1	1923	558

BT-1 Douglas		40	1930	558
C-1 Douglas		9	1925	558
C-1A Douglas	(C1 with L-12B & Allison spur-gear)	1	1925	559
C-1C Douglas		17	1926	559
CO-1 Eng Div		2	1921	559
CO-1 Gallaudet		1	1922	560
CO-2 Eng Div		2	1922	560
CK-CO-2	Cox-Klemin	1	1924	560
XCO-4 Fokker	(C-IV)	3	1922	561
CO-4A Fokker	(C-IV)	4	1923	561
XCO-5 Eng. Div		1	1923	562
XCO-5A		1	1923	352, 562
XCO-5B		1	1923	563
CO-5		1	1923	562
XCO-6 Eng Div	(V-1460 inverted water-cooled)	2	1925	308
XCO-6B Eng Div	(L-12A)	1	1924	563
XCO-6C Eng Div	(V-1410 inverted air-cooled)	1	1927	316, 317, 563
XCO-7 Boeing		1	1924	
XCO-7A Boeing		1	1924	564
XCO-7B Boeing	(V-1460 inv. w/c & V-1410 inv. a/c)	1	1924	309, 564
XCO-8 Atlantic		1	1924	564
XCOA-1 Loening	(V-1410 inverted air-cooled)	1	1924	
COA-1 Loening	(V-1460 inverted water-cooled)	12	1925	311, 565
OA-1A Loening	(V-1460 inverted water-cooled)	15	1926	565
OA-1B Loening	(V-1460 inverted water-cooled)	9	1927	566
OA-1C Loening	(V-1460 inverted water-cooled)	10	1928	566
Curtiss Eagle		3	1920	566
DH-4B	Converted from the original DH-4 between 1919 and 1923. Some historians claim that as many as 1,500 DH-4 airplanes were converted to DH-4B configuration. Considering the fact that three different 1919 Army-wide inventories all show total DH-4 inventory at just under 2,500, the 1,500 figure seems high. Thus far, no complete records of these conversions have been found. Some conversions of specific interest to this study are listed below.			
DH-4B	(Form B supercharged)	2	1921	362
DH-4B	(L-6A engine)	1	1921	
DH-4B	(Allison spur-gear)	1	1921	
DH-4B	(Allison 2 speed spur-gear).	1	1924	302, 567
DH-4B	(V-1460 inverted water-cooled)	1	1923	307
DH-4B	(V-1410 inverted air-cooled)	1	1926	
DH-4M1 DH-4M2	DH-4 and DH-4B airplanes modernized by replacing the original wood-frame fuselage with a new steel-frame type in addition to other modifications. These were built in the 1923-25 era and various historians place the numbers thus converted at anywhere between 300 and 1,000. As with the DH-4B, no complete records of these conversions have yet surfaced.			
DH-4M2S	(supercharged)	9	1923	360
DWC Douglas	(Douglas World Cruiser)	5	1923	385
GAX Eng. Div.	(two engines)	2	1920	567
GA-1 Boeing	(two engines)	10	1921	567
IL-1 Orenco		2	1920-24	568
XLB-3 Keystone	(two V-1410 inverted air-cooled)	1	1928	568
XLB-5 Huff-Daland	(two engines)	1	1926	568
LB-5 Huff-Daland	(two engines)	10	1927	569
LB-5A Keystone	(two engines)	25	1928	569
XNBS-3 Elias	(two engines)	1	1922	569
XNBS-4 Curtiss	(two engines)	2	1922	570
XNBL-1 Barling/Witteman	(six engines)	1	1920	570

XO-1 Curtiss		1	1924	570
O-1A Curtiss		2	1925	571
XO-2 Douglas		1	1924	571
O-2 Douglas		46	1925	571
O-2A Douglas		18	1926	572
O-2B Douglas		6	1926	572
O-2C Douglas		36	1926	572
O-2D Douglas		2	1927	573
O-2E Douglas		1	1927	573
O-2H Douglas		73	1928	573
O-2J Douglas		4	1929	574
O-2K Douglas		59	1929	366, 574
O-5 Douglas	(DOS)	5	1925	574
XO-6 Thomas Morse		2	1925	575
O-6 Thomas Morse		4	1925	
XO-11 Curtiss Falcon		1	1927	575
O-11 Curtiss		66	1927	575, 576
O-11A Curtiss		1	1927	576
XP-17 Curtiss	(V-1410 inverted air-cooled)	1	1927	576
PN-1 Curtiss	(L-6 US 47841)	1	1921	577
S-1 Loening Air Yacht		9	1924	577
T-1 Martin	(former GMP)	1	1919	577, 578
T-2 Fokker	(Fokker F-IV)	2	1922	381
T-3 LWF	(L-12B, Allison spur & L-12A)	1	1924	297, 578
TP-1 Eng. Div	(supercharged)	2	1922	578, 579
TW-1 Eng. Div	(Wright L-6B)	2	1922	263, 579

Argentina

Argentine F-5L Curtiss. Pictured is number 8 of 8 F-5L airplanes purchased in 1921. After modification, they were put into service in 1922 and were used for fleet patrol until 1930. At that time they were moved to other service and kept in operation until 1935. (Courtesy of the Argentine Department of Naval History)

The Argentine Navy purchased four Liberty-powered Dornier Do J Wal aircraft in 1922 and added them to their fleet patrol squadron as B 9 through B 12. They remained in fleet service until 1932. (Courtesy of the Argentine Department of Naval History)

British - French

DH-10 E7851 of Egypt's No. 216 Squadron photographed probably in 1923. It is shown fitted with oversize tropical radiators. (Courtesy of Peter Bowers)

Bristol Braemar II powered by four Liberty 12A engines. One of two commercial airplanes built by Bristol shortly after the war. First flown in early 1919, it was wrecked during a takeoff in late 1921 or early 1922. (Courtesy of Peter Bowers)

French Farman 31 C2. A single unit was tested with Liberty 12A power on May 10, 1919. (Courtesy of Peter Bowers)

French - Italian

French Morane-Saulnier Type A.N.L. This single example was tested with Liberty 12A power in 1919. It was one of the few French aircraft, outside of the Breguet 14 A, to use the Liberty engine.
(Courtesy of Peter Bowers)

Italian Caproni CA-42. This is one of only a hand-full of the model built. This airplane was of similar design to the American-built Ca-5, but a triplane rather than biplane.
(Courtesy of Peter Bowers)

This Italian Caproni CA-43 was a single example of a CA-42 fitted with floats.
(Courtesy of Peter Bowers)

Norway - Russia

The first of four Douglas DT-2b airplanes built in Norway under license. Built in 1927, this airplane remained in service with Liberty power until the mid-1930s. It was then re-engined with an A.S. Panther IIA engine. Note the similar appearence to the Douglas World Cruisers. (Courtesy of Knut Erik Hagen)

Russian R-1 on skis. The R-1 was one of a number of Russian models powered by the Liberty/M-5 engine. With a total of 2,641 produced, it was by far the most used. Considering the long winters in much of Russia, this must have been an often-seen conversion. (Author's collection)

Russian R-1 in standard land configuration. (Author's collection)

U. S. Civil Aircraft with Liberty Power

Aerial Service Mercury. A single airplane built in 1925 by the Aerial Service Corp. Also known as the Aerial Mercury Senior, this airplane was used in mail service by the U.S. Post Office until 1928 with a cargo capacity of 1,000 pounds. Over the years it carried the civil registrations of C40 and N1040. (Courtesy of Peter Bowers)

Aeromarine Model 75 during construction at the factory. Aeromarine photo No. 359 dated June 11, 1920. (Courtesy of Peter Bowers)

Closeup of Aeromarine Model 75 or 80 (probably 80). Factory photo 563 taken February 16, 1922. In this posed photo four factory mechanics are being given instructions by what is probably an Aeromarine executive who knows nothing about the Liberty motor he is pointing at. The more interesting feature of this photo is its 1922 date. Most historians date the construction of the Models 75 and 80 as 1920. It seems odd that one of the planes would be sitting in the Aeromarine factory still looking new and perhaps under construction two years later. (Courtesy of Peter Bowers)

Aeromarine AM-1 Mailplane. A single example was built in 1923 and featured a narrow fuselage and above-wing radiator. Test flights were disapointing as it exhibited stability problems. (Courtesy of Peter Bowers)

Aeromarine AM-2 Mailplane. Built in 1924 as an improvement of the AM-1. The radiator was moved to below the fuselage. Only a single example was built. (Courtesy of Peter Bowers)

Aeromarine AM-3 Mail plane. A further improvement of the series in which a radiator much like the DH-4 was adopted. The radiator and engine installation can be seen in the inset at upper-left. (Courtesy of Peter Bowers)

Atlantic CO-4 Fokker Mailplane. Only this single example was built. (Courtesy of Peter Bowers)

Boeing 40 Mailplane. This single example was built in 1925 for the U.S. Post Office mail plane competition. (Courtesy of Peter Bowers)

Burnelli RB-1 Lifting Fuselage twin-engine airliner. Burnelli built this single Liberty-powered airplane in 1923. (Courtesy of Peter Bowers)

Curtiss HA Mail. One of three built for the U.S. Post Office in 1919 as a variation of the Navy HA-1, HA-2 seaplane. (Courtesy of Peter Bowers)

CD Air Express. This single example was built in 1925 to operate in passenger and freight service. (Courtesy of Peter Bowers)

Dare Variable Camber Monoplane. A single example was built by Melvin E. Dare and a Mr. Carns in 1921. As originally powered by a 140 HP four-cylinder Benze the plane would barely fly. Dare had it repowered with a Liberty which was reportedly supplied by Gar Wood and rebuilt at the Packard factory. It was supposed to be entered in the 1922 Pulitzer but was rejected by officials. The plane flew, but not well. (Courtesy of Thomas Foxworth)

Curtiss Falcon Mailplane. A series of 28 were built in 1928. (Courtesy of Peter Bowers)

Curtiss R-4LM Mailplane 39366 about to leave on a New York-to-Boston flight on June 6, 1918. This was the first flight on the new run and thus the ample number of officials seeing pilot Torrey H. Webb off. (Courtesy of Peter Bowers)

Curtiss R-4LM 39366 after a rough landing at a very rain-soaked Franklin Park Aviation Field near New York. After dropping his left wheel in a soft rut the plane nosed over. No one was hurt and five days later the plane had been repaired and was flown out on its next mail run. (Courtesy of the National Archives)

Dayton-Wright Fire Patrol II. A single example was built in 1922 for fire patrol in Canada and registered there as N-CAED. (Courtesy of Peter Bowers)

Douglas DAM-1 (Douglas Air Mail-One) later changed to Douglas AM-1. This was a conversion of the production Douglas O-2 Army observation plane for air mail service. The conversion consisted of converting the forward cockpit into a mail compartment and moving all pilot's controls to the rear cockpit. Mail access was provided through two hatches. A single unit was built and tested in 1925 but resulted in no immediate U.S. Post Office orders. (Courtesy of Peter Bowers)

Douglas AM-2. Six airplanes of this model were built in 1926. The below engine tunnel radiator of the AM-1 was replaced by a more conventional form radiator and the mail compartment was made convertible to a passenger compartment. Six were ordered by Western Air Express. (Courtesy of Peter Bowers)

Douglas AM-3. The U.S. Post Office initially ordered 50 of this model. Before delivery was complete, Douglas redesigned the AM-3 to increase its payload. The U.S. Post Office liked the change and had the final 40 of the order delivered as AM-4 models. The principal change from the AM-3 was increased wing area which included the elimination of the upper wing cutout above the pilot. (Courtesy of Peter Bowers)

Elias M-1 Mailplane. A single example was built in 1925. (Courtesy of Peter Bowers)

Elias Airmobile. Three of this model were built between 1925 and 1927. (Courtesy of Peter Bowers)

This twin Liberty Friesley Falcon was built by the Friesley Aircraft Corp. of Gridley, California in 1921 and was the only example produced. (Courtesy of Peter Bowers)

Gallaudet Tourist. This 5-passenger airplane was built utilizing primarily DH-4 parts. Marketed by the Johnson Airplane Co. of Dayton, Ohio, apparently only a single airplane of the model was built. It was used by Johnson and eventually carried civil registration N1502. (Courtesy of Peter Bowers)

Huff Daland Duster Model One. This was the first of two built in 1922. It later carried civil registration N2964. (Courtesy of Peter Bowers)

Huff Daland Duster Model Two. Two of this model were built in 1922. Eventually they were registered as N2964 and N2965. (Courtesy of Peter Bowers)

Junkers-Larsen JL-12. A single example was built in 1921 using the Liberty 12 for power. (Author's collection)

Laird Limousine. This single example was built in 1922 for passenger service. (Author's collection)

Lawson 2-engine Model C-2 Airliner. Built by Alfred Lawson in his Green Bay, Wisconsin Lawson Aircraft Company in 1919. Lawson was thinking ahead to what he expected to be the new era of passenger transportation by air. This 26-passenger plane had a range of 400 miles. (Courtesy of Peter Bowers)

Lawson 3 engine Model L-4 which was built in 1920. Lawson, instead of proceeding with production of the C-2, decided to use his remaining funds to build an even larger model, the 32-passenger L-4. Unfortunately he and pilot Charles Wilcox crashed this plane on May 8, 1921 and that ended the Lawson Aircraft Company and what could have been the first series of commercial airliners in America. (Courtesy of Peter Bowers)

Loening Air Yacht M-23. One of seven built in 1922 as the new play toy of the rich and famous. The idea here seemed to be that this was a fast mode of transportation to your favorite fishing spot. (Courtesy of the Smithsonian)

Dayton-Wright KT Cabin Cruiser. This one-of-a-kind airplane was a conversion of a DH-4 built by Dayton-Wright in 1920. The Rinehart-Whelan Co. of Dayton purchased it from General Motors when it discontinued the D-W operation in 1923. R-W converted it to the configuration shown below. (Courtesy of the National Archives)

Rinehart-Whelan Air Cruiser. This is a 1923 conversion of the original Dayton-Wright KT shown above. The airplane remained in active service in the Ohio area until early 1928. (Courtesy of John Underwood)

Sattco DH-4. A single modification of a DH-4 done in 1922. As in the above examples, this was done to enlarge the carrying capacity of the original airplane. The conversion was done by the Service Aviation Training & Transport Co, Wabash, IN. (Courtesy of Peter Bowers)

Sikorsky S-29A twin-engine transport. Built in 1924 and the first American-built Sikorsky and originally powered by 220 HP Hispano engines. It was rebuilt after a forced landing and repowered with Liberties. It was owned by Rosco Turner from 1925-27 and then sold to Howard Hughe's film company where it was rebuilt to look like a German airplane and crashed purposly in the film "Hells Angels." (Courtesy of Peter Bowers)

Sikorsky S-32. Built in 1926 this three-cockpit biplane was sold to a Canadian mining company for use in South American exploration. It was mounted on floats during that use and is pictured in the inset in Colombia, South America. (Courtesy of Peter Bowers)

Stout Pullman 2-AT "Maiden Dearborn IV" at Ford Airport. This was one of 11 built between 1924 and 1925. The first went to the U.S. Post Office and the next 5 were for Ford Air Transport Service. 4 went to Florida Airways and one to retail magnate John Wanamaker. (Courtesy of Peter Bowers.

Waterman 3L-400. A single example of a modification of a LePere LUSAC-11 as an executive transport. The conversion was done for L. C. Brand. (Courtesy of Peter Bowers)

Yackey Transport. Shown is the plane which accompanied the Sterling expedition to New Guinea. At least four were built as conversions of Breguet 14A2, one in 1925 and 3 in 1926. The airplane was still being actively advertised in 1927. (Author's collection)

Pre-Designation U. S. Navy Planes With Liberty Power

Curtiss H-12L A767 during take-off. Official Navy photograph taken on March 24, 1918. (Courtesy of Peter Bowers)

Curtiss H-16A showing armament. Taken at Hampton Roads on 9/7/18. (Courtesy of Peter Bowers)

Curtiss HA-1 A-4110 on floats. (Courtesy of Peter Bowers)

Curtiss HS-2L.
(Courtesy of Peter Bowers)

Curtiss HS-2L modified by Douglas for post-war civil use by Pacific Marine Airways for transportation from the California mainland to Catalina Island. Photographed during use to transport actors to and from the island during the filming of "Old Ironsides" in 1926.
(Paramont Pictures photo courtesy of Peter Bowers)

Curtiss HS-3L.
(Courtesy of Peter Bowers)

Curtiss R6-L. (Courtesy of Peter Bowers)

Curtiss R9. (Courtesy of Peter Bowers)

DH-4 Marine in France in 1918. (Navy photo courtesy of Peter Bowers)

DH-4B Navy observation plane attached to NAS Pensacola. (Courtesy of Peter Bowers)

Gallaudet D-4 during tests. Its Navy number A2653 has yet to be applied. This was a very unique aircraft design in which a midship geared-down propeller is mounted on a ring gear encircling the fuselage. (Courtesy of Peter Bowers)

Martin MT, the Navy version of the Army's MB-2 bomber. (Courtesy of Peter Bowers)

Martin MBT, the Navy version of the Army's MB-1 bomber. (Courtesy of Peter Bowers)

NAF F-5L. One of 225 of this model acquired by the Navy. (Courtesy of Peter Bowers)

NAF N-1. This two-place plane was designed to carry a 2" recoiless Davis gun in the front and pilot in rear. A pusher engine arrangement was used to allow a clear field-of-fire for the gunner. (Courtesy of Peter Bowers)

NAF PT-1. One of 15 built at the National Aircraft Factory. (Courtesy of Peter Bowers)

NAF PT-2 A6046. One of 18 built at the National Aircraft Factory. (Courtesy of Peter Bowers)

NC-1. (Courtesy of Peter Bowers)

NC-2. (Courtesy of Peter Bowers)

NC-3. (Courtesy of Peter Bowers)

NC-4. (Courtesy of Peter Bowers)

NC-6.
(Courtesy of Robert L. Lawson via AAHS. Damaged in an open sea landing off the Nicaragua coast and abandoned on a beach north of San Juan Del Sur in December of 1920)

NC-9.
(Courtesy of Peter Bowers)

NC-10.
(Courtesy of Peter Bowers)

NAF-TG-1.
(Courtesy of Peter Bowers)

NAF-TG-2. (Courtesy of Peter Bowers)

NAF-C1. (Courtesy of Peter Bowers)

Paul Schmitt French seaplane. Purchased from the French in 1918 and repowered with a Liberty engine for tests. The plane was destroyed in a fire in 1920. (Courtesy of Alan Toelle)

Sperry Triplane Amphibian. (Courtesy of Peter Bowers)

Post-Designation Navy Planes With Liberty Power

Boeing O2-B1 (DH-4M-1). One of 30 delivered to the Navy for Marine Corps use in 1925-26. (Courtesy of Peter Bowers)

Douglas DT-1. (Courtesy of Peter Bowers)

Douglas DT-2. One of 57 production versions of the DT built by NAF and L.W.F. (Courtesy of Peter Bowers)

Elias EM-2. One of 6 purchased in 1922. (Courtesy of Peter Bowers)

Fokker FT-1. Three were purchased in 1923. (Courtesy of Peter Bowers)

Loening OL-2. Five were purchased by the Navy for use on the 1925 Arctic Expedition. These were the Navy version of the Army COA-1. (Courtesy of Peter Bowers)

Loening OL-4. Six of this slightly changed version of the OL-2 were purchased in 1926. (Courtesy of Peter Bowers)

Douglas OD-1. Two of this model were purchased for Marine Corps use. They were the Navy versions of the current model Army Douglas O-2 series. (Courtesy of Peter Bowers)

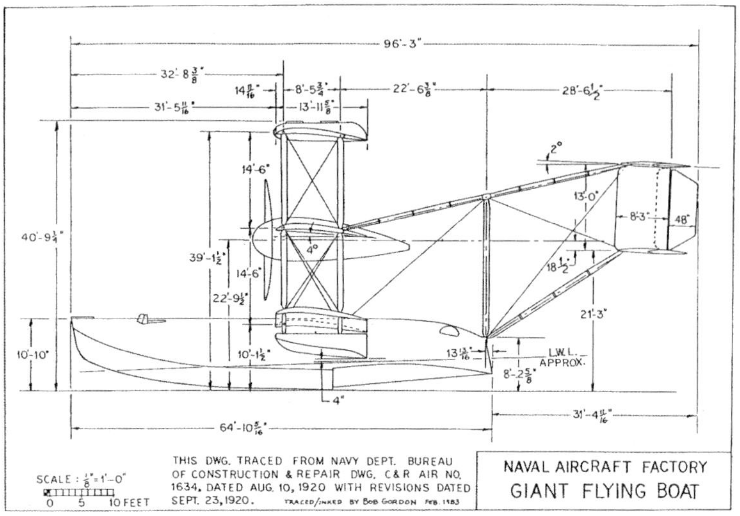

Above drawing was traced from the original Navy drawing. Above-right is a wind tunnel model circa 1920 in the Smithsonian collection. At right are two of the nacelle at the Gallaudet plant ready for shipment. Bottom-left is the first nacelle during test. Bottom-right is a drawing of the nacelle by Bob Gordon from original Navy material. (All photos and drawings courtesy of Bob Gordon)

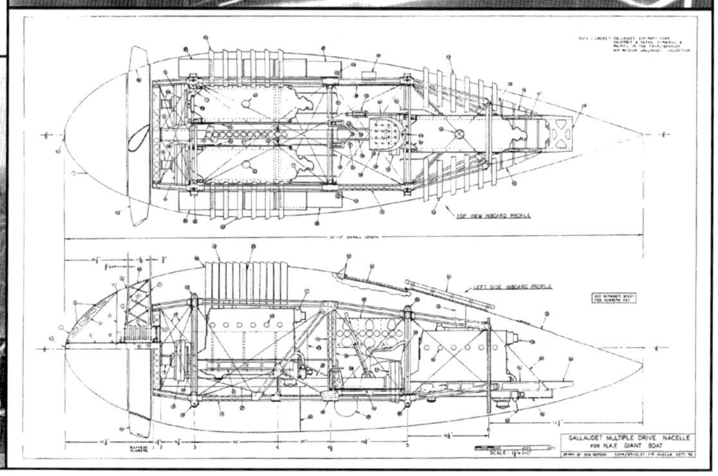

Navy Giant Boat

Commander J. C. Hunstaker and Commander L. W. Atkin were in charge of the design, which started before the Armistice in late 1918. The object was a flying boat capable of carrying a heavy bomb-load over very long distances. By late spring of 1919 it had been decided that 3,600 HP would be required to accomplish this purpose and that 9 Liberty engines grouped in 3 nacelles, each driving a single 18' three blade propeller, would be used. The air frame was to be a scaled up version of the NC boat design but a triplane instead of biplane. Its wing span was 150' (NC-4 126'), empty weight 38,000 lbs. (NC-4 15,874 lbs.), Gross weight 65,000 lbs. (NC-4 28,000 lbs.) and power 3,600 HP from 9 Liberties (NC-4) 1,600 HP from 4 Liberties. The Navy referred to it as the "Giant Boat" and assigned it the designation NAF G-1 BuNo A6059. Because of their prior experience with the type of gear-drive to be used, Gallaudet was given the contract to design and build the three-engine nacelle. The NAF was to design and build the airplane. Actual design work was begun in early 1920 and construction in early 1921. The first nacelle was tested in August 1921 and the last completed nacelle was shipped to the Navy on February 4, 1922. Work on the hull was being carried on at the NAF but was not completed. By this time it had already become evident that Navy funding in the foreseeable future would not support the completion of the Giant Boat and other currently planned high-cost projects, the building of the rigid airship *Shenandoah* in particular. The Giant Boat project was terminated with the nacelle completed but the air frame not. Historian Bob Gordon researched and told the story in detail in Volume No. 98 of *WW-1 Aero* magazine. The 8-engine Caproni CA-60 may have exceeded the Giant Boat in complexity but not in power and the Giant Boat would certainly have been a more successful airplane.

Pre-Designation U. S. Army Planes With Liberty Power

Curtiss R-4L U.S. number 39954. This plane became McCook Field P-15. This is McCook Field photo 79. (Courtesy of Peter Bowers)

Curtiss Bristol USAO-1. 27 of this model were built by Curtiss. Beset by some initial problems, the plane gained what was perhaps an undeserved stigma as prone to failure in flight because of several fatal crashes. Ultimately most could not be blamed on the aircraft but the damage had been done and a large contract was canceled with only 27 delivered. (Courtesy of Peter Bowers)

Curtiss CB "Battler." U.S. 34632. Four were initially ordered but the last three were canceled. (Courtesy of Peter Bowers)

Dayton-Wright-built DH-4 U.S. 32098. This is a very early D-W example. The first of their manufacture was 32059 (the prototype). (Courtesy of Peter Bowers)

USD-9A U.S. 40044 P-64. McCook Field photo 2504. (Courtesy of Peter Bowers)

Engineering Division Pomilio BVL-12. One of six completed in 1919. This example is U.S. 40087 McCook Field P-68. McCook photo 2691. (Courtesy of Peter Bowers)

Captured Fokker D VII put in service for testing engines by the Engineering Division. This example retained its German number of 8530 with AS as a prefix. This one was fitted with a Hall-Scott L-6. (Courtesy of Peter Bowers)

Captured Fokker D VII owned by the Air Service and loaned to Hall-Scott for testing their L-6 engine. Markings on the side of the fuselage read "German Fokker D-VII Type Combat Plane Hall-Scott L-6 200 HP Power Plant Experimental & Research" The SJ-1 in the background appears to also have a Hall-Scott L-6 installed. (Courtesy of Peter Bowers)

Martin MB-1 bomber. (Courtesy of Peter Bowers)

Martin MB-1 converted to three engines. The outboard Liberties have been replaced by Wright Hispano 180 HP engines and a new Liberty installation in the nose. AS 39059 McCook Field P-104. McCook Field photo 10042. (Courtesy of Peter Bowers)

Martin MB-1 Mailplane. One of six built in 1924. (Courtesy of Peter Bowers)

Curtiss-built NBS-1 (Martin MB-2), probably AS 68508, at Langley Field equipped with Form D superchargers during the initial tests of a proposed 20 Curtiss NBS-1 supercharged bombers. In front of the airplane is Major Renolds, Group Commander at Langley. (Courtesy of J. Sloan)

Handley-Page O/400 built by Standard. (Courtesy of Peter Bowers)

Johns Multiplane. This single example was contracted and delivered in 1919. It is shown here either during assembly or after its initial test flight when it failed to get airborne and settled back to the ground tail-down. There appears to be some tail damage and there is a crumpled wing section on the ground at lower-left. (Courtesy of Peter Bowers)

Front view of the Johns Multiplane. As one might suspect, this design was not successful. It even appears so bizarre that one wonders why the Air Service decided to pay for the building of an example. The plane had three Liberty engines, one in the nose of the fuselage, and one in each of two necelles mounted interwing as pushers. It failed to fly upon initial testing in 1919. Several additional attempts were made in 1920 and the plane was then scrapped. (Courtesy of Peter Bowers)

J. V. Martin bomber. This two-engine plane had its power plants mounted in the fuselage and accessible for maintenance, with between-wing-mounted four-blade propellers driven by a shaft with two right angle drives. McCook Field photo 2245. (Courtesy of Peter Bowers)

A closer view of the J. V. Martin bomber which better shows its unique engine/ propeller arrangement. Note that gearing was provided to make the propellers counter-rotating. This equalizes the torque from the two engines and allows better in-flight airplane control. Since all Liberty engines were built with right-hand rotation, this ability did not exist for other multi-engined Liberty-powered airplanes. (Courtesy of Peter Bowers)

LePere LUSAC-11 in standard configuration. Most photos of this model are of the famous supercharged McCook Field P-53. Thus we seldom see a standard model. This particular airplane is somewhat unique in that it is equipped with Packard's fuelizer feature, the air intakes of which can be seen in the bottom of the engine cowling. (Courtesy of Peter Bowers)

Twin engine LUSAO-11 observation plane designed by Captain LePere and built by Packard. This is P-65 AS 40012 and the first of two built. (Author's collection)

Armored ground harassment LUSAGH-11 designed by Captain LePere of the French aircraft mission and built by Packard. This is P-89 AS 40021 and the second of two ordered. (Author's collection)

Lanzius L-II. This rather unusual pursuit design was proposed to the Engineering Division in early 1918 and they evidently felt it had promise as they ordered four from the builder, the Lanzius Airplane Company of New York, NY. The wings had variable camber and incidence controlled by a system of cables and pulleys. This was the first delivered to McCook Field. Taken up for its first test at that location by the company test pilot Lester E. Holt on June 29, 1918, the plane gained some altitude and then shed its wings and crashed, killing the pilot instantly. The remaining three were canceled. (Courtesy of Peter Bowers)

L.W.F.-G2. Used in very early Liberty 12 testing. Only a single example was built. (Courtesy of Peter Bowers)

L.W.F. Owl. The three-engine 105" wing span LWF Owl is posed with a 20' wing span Sperry Messenger for comparison. Only a single "Owl" was built. (Courtesy of Peter Bowers)

Thomas-Morse MB-1. One of two examples purchased by the Air Service. (Courtesy of Peter Bowers)

Post-Designation Army Planes With Liberty Power

Fokker-Atlantic AO-1 seen here in its later life with civil registration 3526. It was originally one of the Fokker CO-4 airplanes (AS 68566) converted to artillery observation configuration for testing. (Courtesy of Peter Bowers)

Cox-Klemin XA-1 ambulance plane AS 23-1248 McCook Field P-421. (Courtesy of Peter Bowers)

Douglas XA-2 AS 25-380 McCook Field P-472 powered by an inverted Liberty V-1410 air-cooled engine. Note the hand crank attached to the right-side of the fuselage. The engine was provided with an Eclipse-Bijur electric starter with backup hand-starting ability. (see page 128) McCook Field photo 33108. (Courtesy of Peter Bowers)

Fokker A-2 AS 64234 McCook Field P-243. (Courtesy of Peter Bowers)

Douglas BT-1. The BT-1 airplanes were "Basic Trainers" converted from various Douglas O-2 models as they became obsolete. Note the rear cockpit hood which could be closed for training pilots in night instrument flying. (Courtesy of Peter Bowers)

Douglas C-1. The first of the new (1924) cargo (C) class airplanes which replaced the old transport (T) class. They were designed to carry either cargo or 8 passengers. Two pilots sat side-by-side in the forward open cockpit. Air Service photo 2265. (Courtesy of Peter Bowers)

Douglas C-1A AS 25-426 McCook Field XP-424. This airplane was used for testing gear reduction Liberty engines of both epicyclic (L-12B) and Allison single reduction spur-gear-type. The L-12B version is seen here. McCook Field photo 30657. (Courtesy of Peter Bowers)

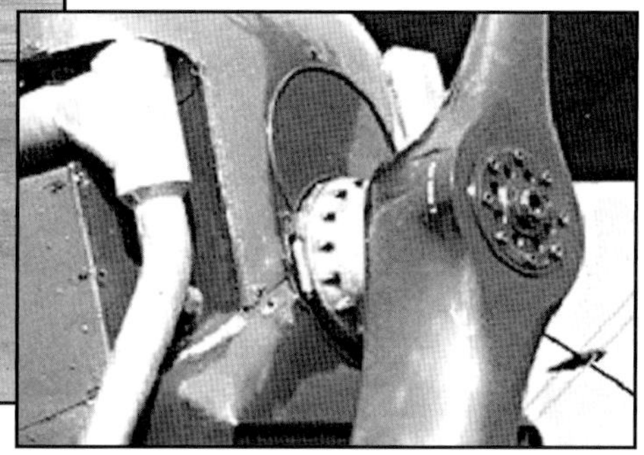

Douglas C-1C, one of 17 production 9-seat versions of the original C-1. (Courtesy of Peter Bowers)

Engineering Division CO-1 AS 64321 McCook Field P-234. McCook Field photo 12334. (Courtesy of Peter Bowers)

Gallaudet CO-1 AS 68587 McCook Field P-281. McCook Field photo 17590. (Courtesy of Peter Bowers)

Engineering Division CO-2 AS 68577 McCook Field P-255. McCook Field photo 14400. (Courtesy of Peter Bowers)

Cox-Klemin CK-CO-2 McCook Field P-379. The plane was delivered to McCook Field for testing but was not accepted and returned to the manufacturer. (Courtesy of Peter Bowers)

Fokker XCO-4 with conventional radiator. (Courtesy of Peter Bowers)

Fokker XCO-4 with tunnel radiator. (Courtesy of Peter Bowers)

Fokker CO-4A AS 23-1209 McCook Field P-325. (Courtesy of Peter Bowers)

Engineering Division XCO-5 AS 23-1204 McCook Field P-305. This plane was actually the second of two planes originally ordered as TP-1 types. As modified to XCO-5 configuration it had larger wings. This was its delivery configuration for tests as a possible new Corps Observation-type. McCook Field photo 18227. (Courtesy of Peter Bowers)

Engineering Division CO-5 AS 23-1204 P-305 with modified wings and new cowling and radiator as finalized as a Corps Observation type. McCook Field photo 21469. (It was never ordered into production.) (Courtesy of Peter Bowers)

Engineering Division XCO-5A AS 23-1204 McCook Field P-305. The plane has now been modified for use in high altitude testing with superchargers. The TP-1, from which it was derived, had proven well-adapted to this type of testing so after completion of Corps Observation design testing the plane was religated to high-altitude work. McCook Field photo 28878. (Courtesy of Peter Bowers)

Engineering Division XCO-5B AS 23-1204 P-305. Shown here during its final months of high-altitude work warming up for a winter test flight. It continued in active use through 1929 and was surveyed in January of 1930. In addition to its enlarged wings, the A and B version of the CO-5 had an enlarged tail surface for better high-altitude control. McCook Field photo 35876. (Courtesy of Peter Bowers)

Engineering Division XCO-6B McCook Field P-360 fitted with a Liberty 12A and tunnel radiator. McCook Field photo 24361. (Courtesy of Peter Bowers)

Engineering Division XCO-6C AS 23-1236 fitted with the first Allison air-cooled V-1410 just before being sent to the 1925 air races at Mitchel Field. McCook Field photo 28388. (Courtesy of Peter Bowers)

Boeing XCO-7A P-332 (DH-4M) AS 23109. A DH-4M with Loening COA-1 wings. McCook Field photo 20577. (Courtesy of Peter Bowers)

Boeing XCO-7B AS 31216 P-368 with inverted water-cooled Liberty 12 and Loening COA-1 wings. (Courtesy of Peter Bowers)

Atlantic XCO-8 AS 23-163 McCook Field P-369. McCook Field photo 24257. (Courtesy of Peter Bowers)

Loening COA-1 AS 24-8 McCook Field P-384. This was the second prototype ordered from Loening. (Courtesy of David Ostrowski)

Loening COA-1 AS 25-229, one of the first production run. (Courtesy of Peter Bowers)

Loening OA-1A during tests by Loening pilots. The plane still carries the Loening logo on the tail. (Courtesy of Peter Bowers)

Loening OA-1B AS 27-325. (Courtesy of Peter Bowers)

Loening OA-1C AS 28-81. (Courtesy of Peter Bowers)

Curtiss Eagle AS 64243. Three of this model were purchased for evaluation. The normal configuration was as a 10-seat transport. The one pictured was tested as a four-litter ambulance. (Courtesy of Peter Bowers)

DH-4B AS 21952 McCook Field P-139 with Allison two-speed spur-gear engine and Form F turbosupercharger for high-altitude testing. McCook Field photo 20906. (Courtesy of Peter Bowers)

Engineering Division GAX (Ground Attack Experimental) AS 63278 McCook Field P-129. Power was by two pusher Liberty 12A engines. One of two of this model procured, resulting in the order for additional airplanes of a refined model, the GA-1. McCook Field photo 8818. (Courtesy of Peter Bowers)

Boeing GA-1, the production version of the Engineering Division GAX. 20 were initially ordered but was reduced to 10 when its performance did not live up to expectations. (Courtesy of Peter Bowers)

Orenco IL-1 AS 63274 McCook Field P-168. Two were ordered for evaluation as Infantry Liaison-types. (Courtesy of Peter Bowers)

Huff-Daland XLB-3 AS 27-333. An Experimental Light Bomber powered by two air-cooled V-1410 Liberty engines Only a single example was built. It was later sent back to Huff-Daland, by now reorgainized as Keystone, and re-engined with P&W R-1340 engines and became the Keystone XLB-3A McCook Field P-498. (Courtesy of Peter Bowers)

Huff-Daland XLB-5 AS 26-208 McCook Field P-453. Another prototype Light Bomber ordered from Huff-Daland. This one was more successful and resulted in following orders for LB-5 and LB-5A airplanes. (Courtesy of Peter Bowers)

Huff-Daland LB-5 AS 27-337 in flight. This is a really nice overhead shot in which all the planes occupants were "watching the camera." The rear observer was even waving. (Courtesy of Peter Bowers)

Keystone LB-5A AC 28-1 and the first of 25 ordered. (Courtesy of Peter Bowers)

Elias XNBS-3 AS 68567 McCook Field P-335. A single test example of a new series of short-range Night Bombers. The Martin MB-1 was re-designated as the NBS-1 and the L.W.F. NBS-2 was designed but canceled. (Courtesy of Peter Bowers)

Curtiss XNBS-4. Based closely on the NBS-1, only two prototypes were built. (Courtesy of Peter Bowers)

Engineering Division/Barling Bomber. This huge six-engine bomber was designed by Walter Barling and constructed by the Whitteman-Lewis Company of Teterboro, NJ. It first flew on August 22, 1923 and made a number of record flights before being dismantled in 1928. The plane had four nacelles with the two inner ones carrying two engines, one tractor and one pusher. It had an empty weight of 27,132 lbs. and a gross weight of 42,569 lbs. (Courtesy of Peter Bowers)

Curtiss XO-1 AS 23-1252 McCook Field P-375. McCook Field photo 24218. (Courtesy of Peter Bowers)

Curtiss O-1A AS 25-333 McCook Field P-449. (Courtesy of Peter Bowers)

Douglas XO-2 AS 23-1251 McCook Field P-374. (Courtesy of Peter Bowers)

Douglas O-2. One of 46 production airplanes. (Courtesy of Peter Bowers)

Douglas O-2A AS 25-387, the first delivered of 18 ordered. (Courtesy of Peter Bowers)

Douglas O-2B AS 25-384. (Courtesy of Peter Bowers)

Douglas O-2C NG 26-12. (Courtesy of David Ostrowski)

Douglas O-2D.
(Courtesy of Peter Bowers)

Douglas O-2E.
(Courtesy of Peter Bowers)

Douglas O-2H National Guard NG 29-400.
(Courtesy of Peter Bowers)

Douglas O-2J AC 28-127. General Patrick's personal aircraft. Note the two stars of a Major General on the tail. (Courtesy of Peter Bowers)

Douglas O-2K in civil registration 13949. (Courtesy of Peter Bowers)

Douglas O-5. One of five ordered as near duplicates of the Douglas DWC but with armament and reduced fuel capacity. Originally called the DOS for Douglas Observation Seaplane (all five were ordered with floats as original equipment for overseas service). The designation was changed to O-5 under the new system of 1924. (Courtesy of Peter Bowers)

XO-6 Thomas-Morse AS 25-435 McCook Field P-439. (Courtesy of Peter Bowers)

Curtiss Falcon XO-11. AS 25-328 McCook Field P-476. (Courtesy of Peter Bowers)

Curtiss Falcon O-1 AC 27-98 McCook Field XP-487. This is one of 66 of this model produced. (Courtesy of Peter Bowers)

Curtiss O-11 Falcon fuselage without covering. We do not often see what is under the skin of early airplanes. This gives the reader a good look at the designs of 1925. (Courtesy of Peter Bowers)

Curtiss Falcon O-11A. A single unit was built and 40 additional ordered but later canceled. (Courtesy of David Ostrowski)

Curtiss XP-17 AS 25-410 McCook Field P-400. Shown here powered by a V-1410 air-cooled Allison-built engine. (Courtesy of Peter Bowers)

Curtiss PN-1 AS 63276 McCook Field P-198. This plane was powered by Liberty L-6A U.S. 47841. McCook Field photo 10038. (Courtesy of Peter Bowers)

Loening S-1 Air Yacht AS 24-1, one of nine ordered in 1924. (Courtesy of Peter Bowers)

Martin T-1 AS 62951, the last of the original MB-1 types was completed as a passenger airplane and designated T-1 as the first of the 1920-23 transport series of airplanes. It could carry 10 passengers plus pilot and co-pilot, who were now housed in an enclosed cockpit rather than in the open. (Courtesy of Peter Bowers)

Martin T-1 transport seen close up. A rather nice looking plane for which the Army saw some use but, in 1920 the civilian market was not quite ready for large passenger planes. (Courtesy of Peter Bowers)

L.W.F. T-3 AS 23-1220 McCook Field P-328. This airplane was powered on a test basis with three different engines, a standard L-12A, an epicyclic L-12B and the Allison single-speed spur-gear. It is pictured here with the latter as power. (Courtesy of Peter Bowers)

Engineering Division TP-1 AS 68578 McCook Field P-263. This plane was designed in 1922 and completed in December of that year. It was intended to be supercharged and is shown here with a Form D-type with honeycomb aftercooler (and side radiators), the last of the front-mounted types and the first type used on it. (Courtesy of Peter Bowers)

Engineering Division TP-1 with Form F supercharger. This later type supercharger was installed by mid-1924. The radiator used when this photo (McCook Field 21667) was taken was mounted on the wheel struts. It was also tested with wing radiators. (Courtesy of Peter Bowers)

Engineering Division TP-1 AS 68578 McCook Field P-263 now with no supercharger and high-type front radiator. It was changed to this configuration probably in early 1925. It was transferred to Kelly Field on 7/14/25. This is McCook Field photo 24122. This configuration is seen in other photos as early as 23027. (Courtesy of Peter Bowers)

Engineering Division TW-1 AS 64317 McCook Field P-200 with a Wright-built L-6B for power. McCook Field photo 10620. (Courtesy of Peter Bowers)

Appendix G

The Liberty Engine Illustrated

The Thomas-Morse Liberty 6 MNFG NO. 5 U.S. NO. 47843

Courtesy of William Lewis

Illustrations from the *Standard Test Report of the Engineering Division* at McCook Field dated 5-15-19. This report was generated by test engineer Iskander Hourwich (who continued work at that location for a number of years and who, along with W. J. Foster, compiled the monumental book *Air Service Engine Handbook* published in 1925 by the Engineering Division), approved by engineer in charge George J. Mead (who would go on to work as an engineer for the Wright-Martin Co. and later become chief engineer of Pratt & Whitney) and Major George E. A. Hallett, chief of the Power Plant Dept.

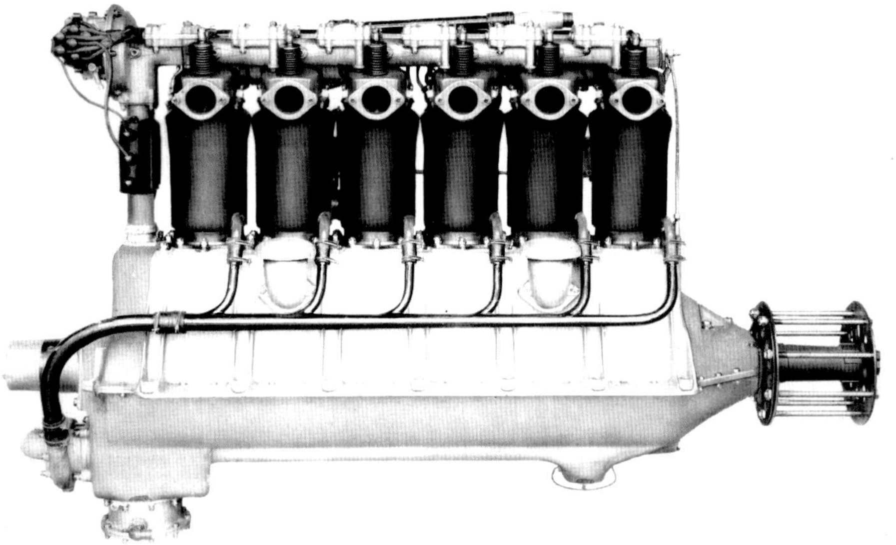

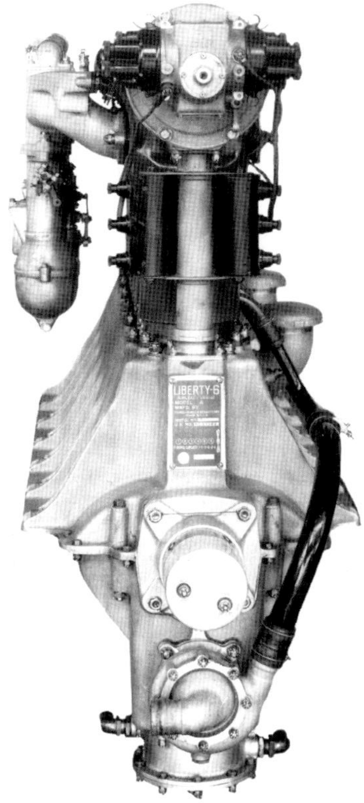

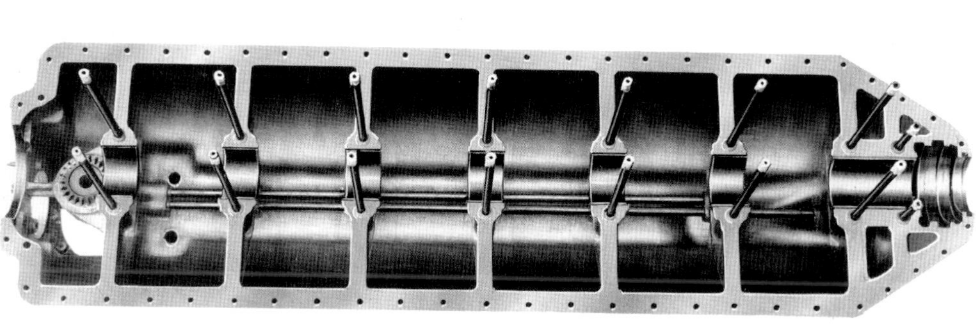

LIBERTY-6
AIRCRAFT ENGINE
MODEL A
MNFD. BY
THOMAS-MORSE AIRCRAFT CORP.
ITHACA, N.Y., U.S.
MNFG. NO. 5
U.S. NO.
1 2 3 4 5 6
FIRING ORDER 1-5-3-6-2-4

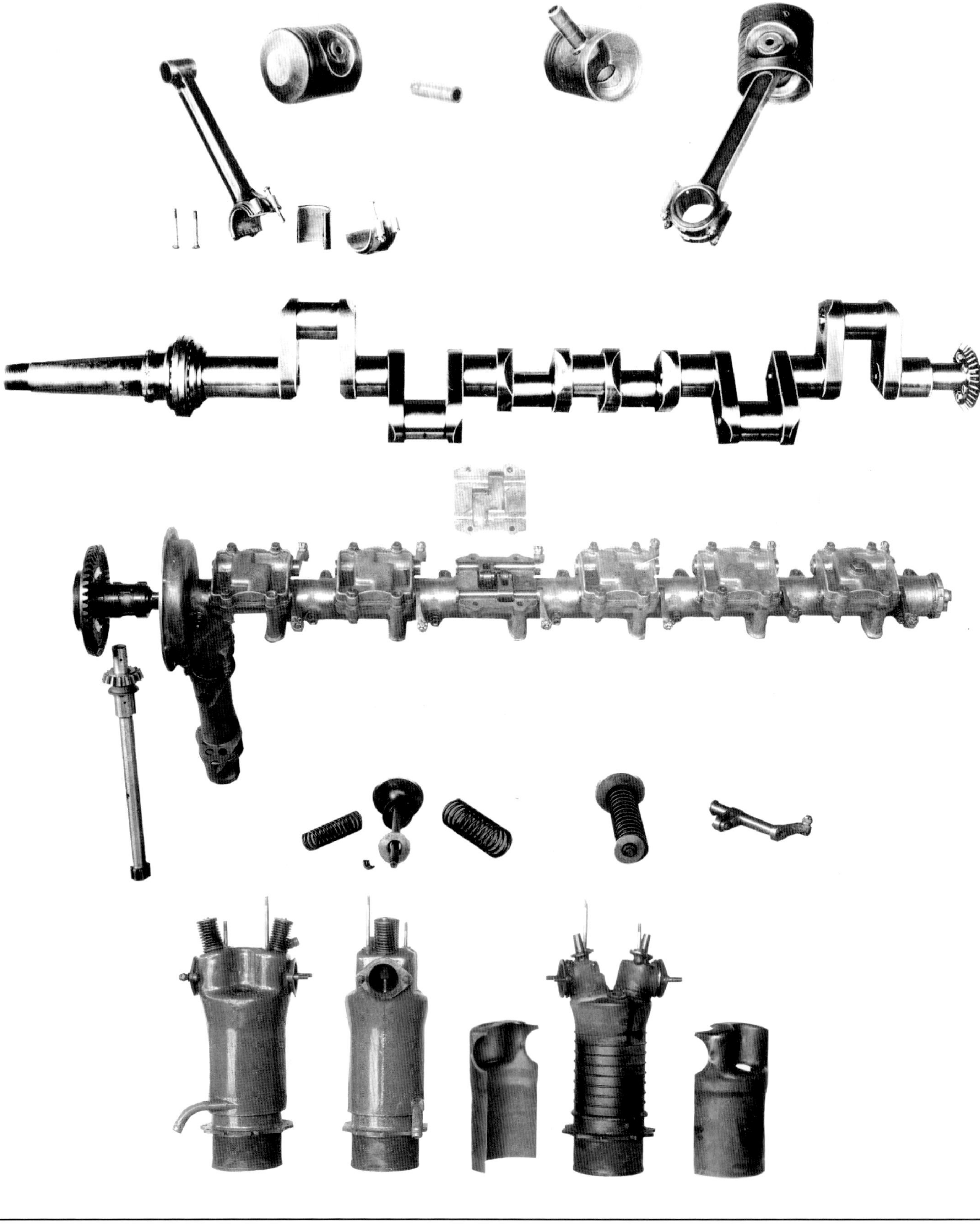

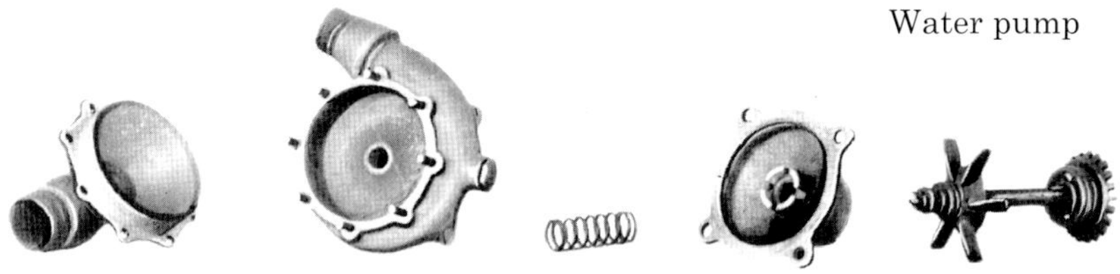

Water pump

Oil pump

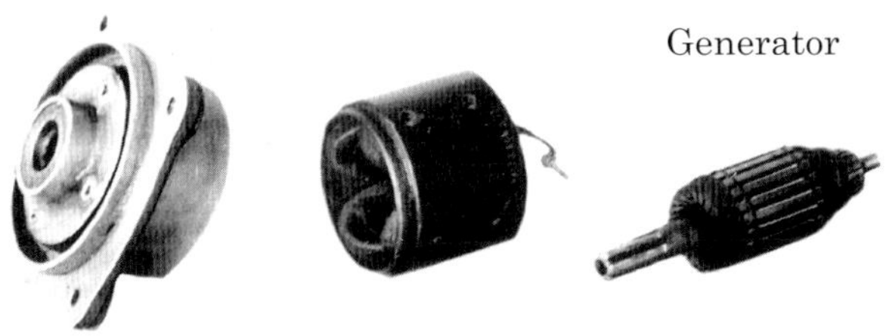

Generator

Distributor

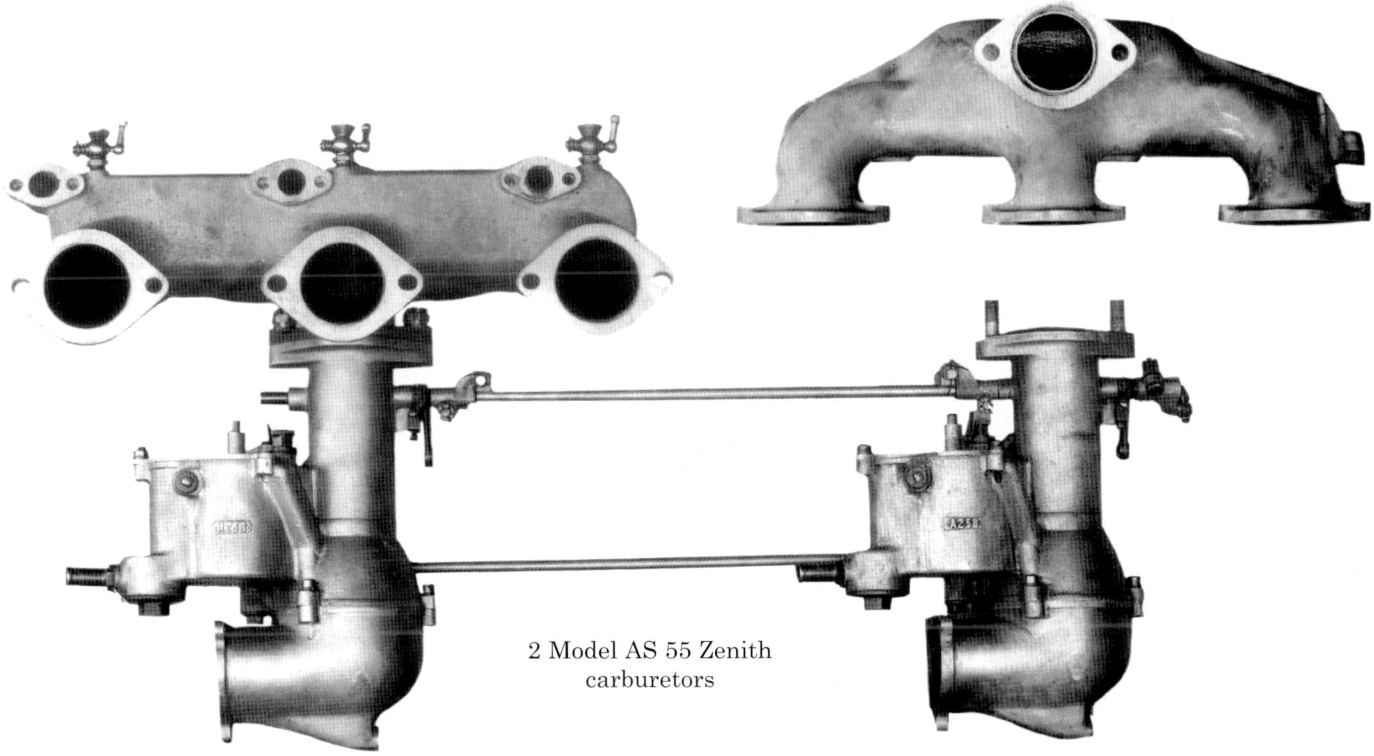

2 Model AS 55 Zenith carburetors

THOMAS MORSE LIBERTY 6 CYLINDER ENGINE
DIAGRAM OF GEAR TRAIN

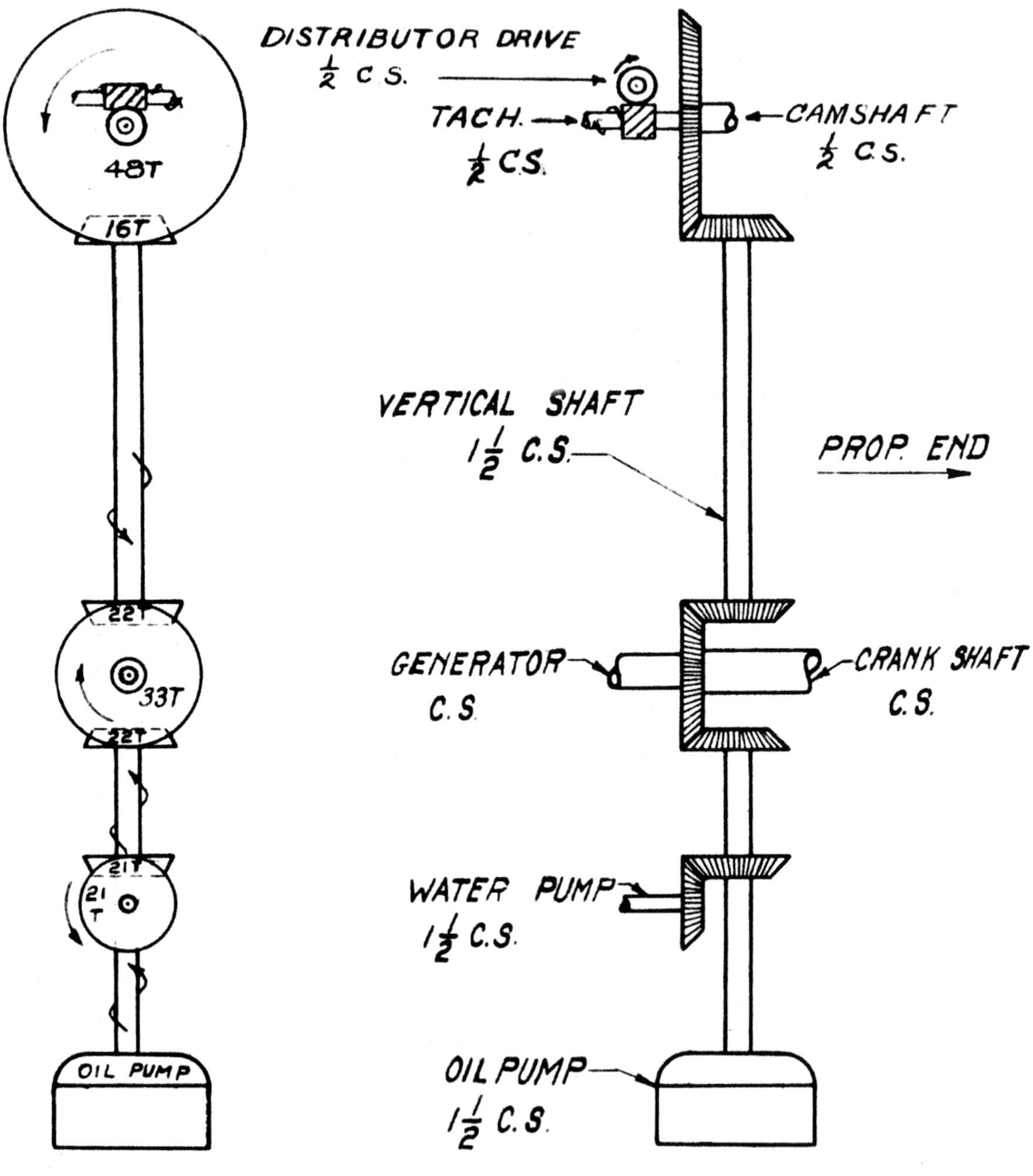

CS = CRANK SHAFT SPEED
T = TEETH

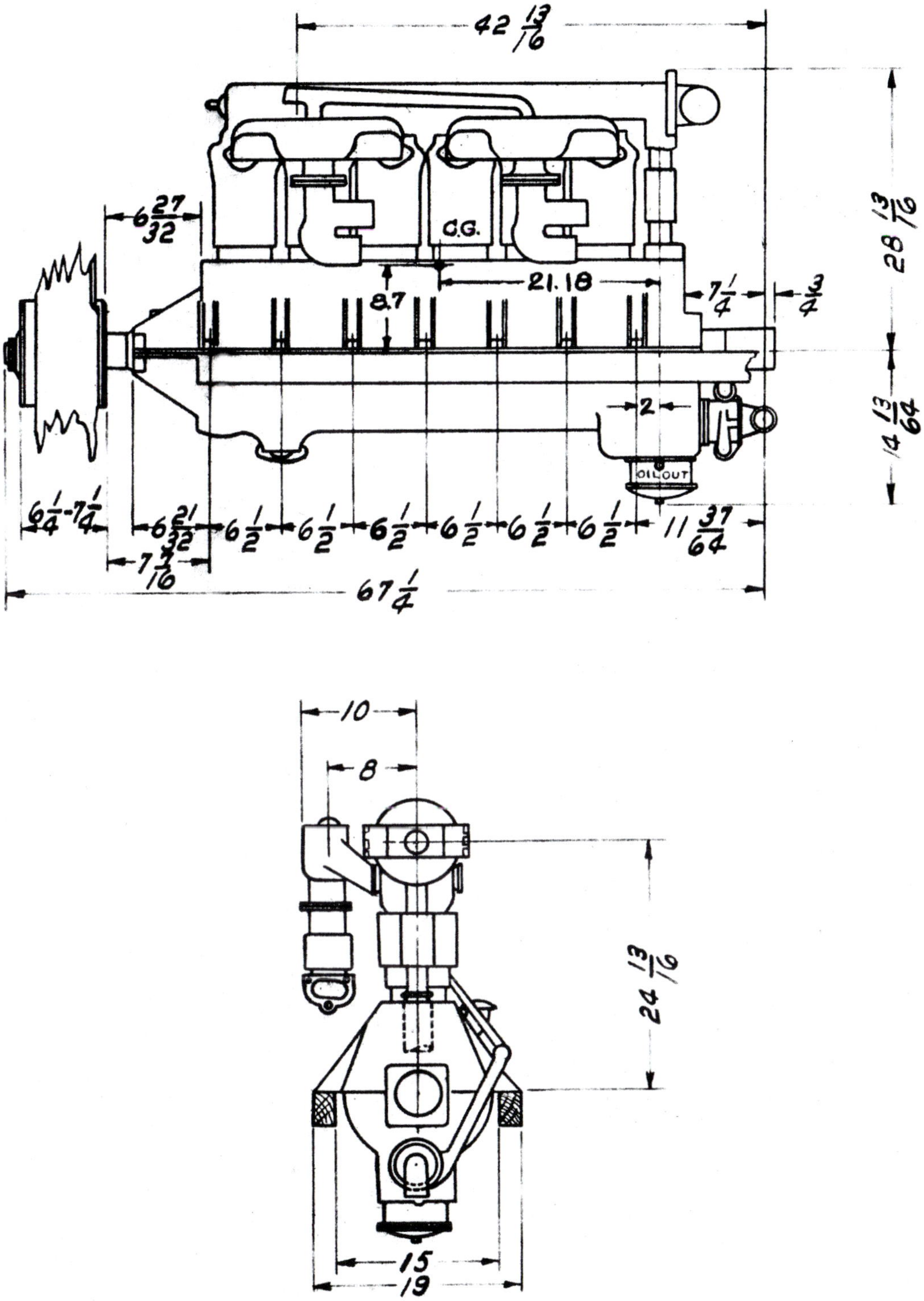

INSTALLATION DRWG.
THOMAS - MORSE LIBERTY SIX CYLINDER AIRCRAFT ENGINE

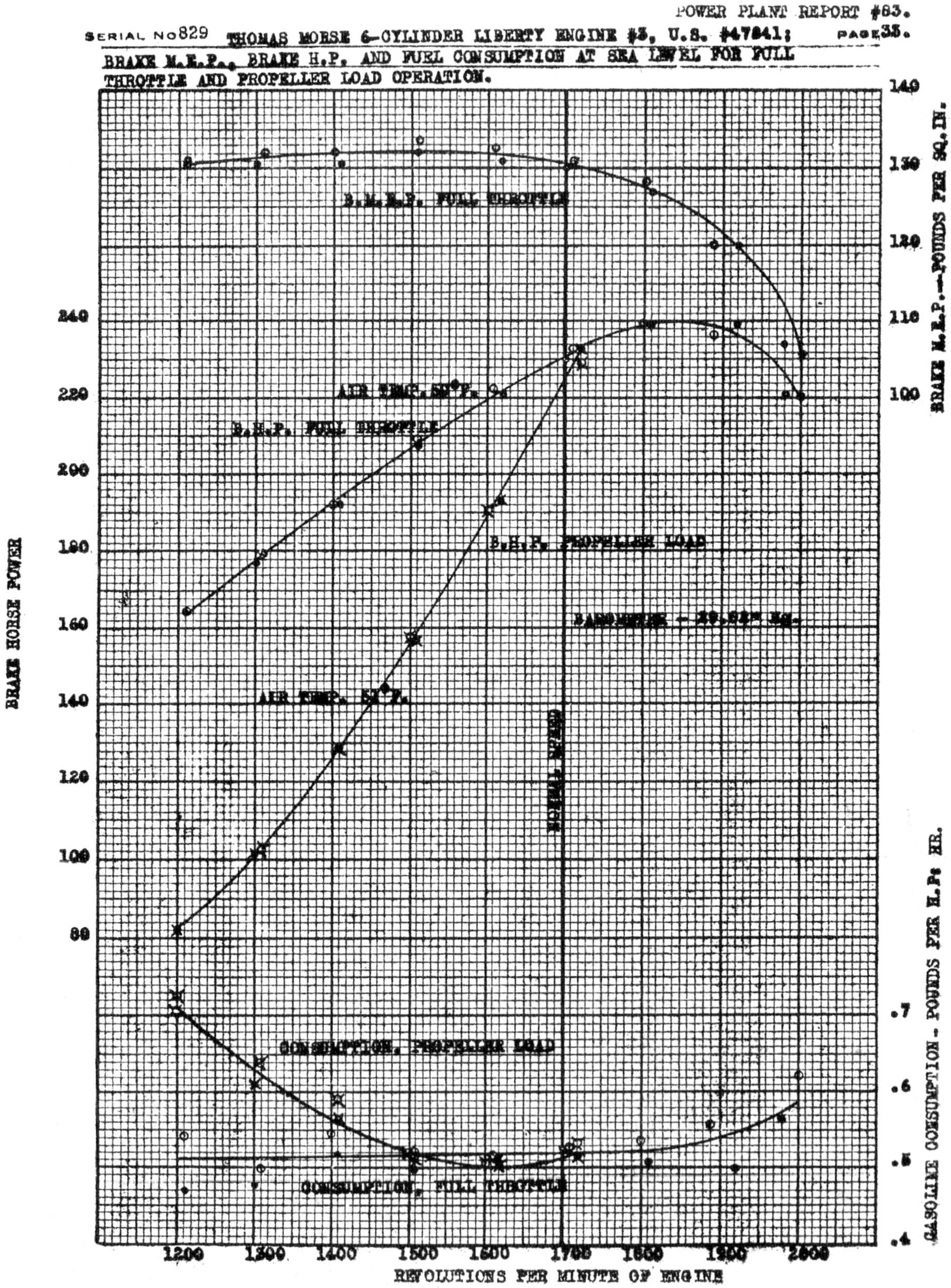

It should be noted that the standard test from which these illustrations came was carried out on Thomas-Morse engine MNFG NO. 3 U.S. NO. 47841 and the power curve from that test is reproduced above. The engine which was disassembled and photographed for illustration in the report was Thomas-Morse MNFG NO. 5 U.S. NO. 47843.

The Buick Liberty 8 MNFG No. BX-5 U.S. NO. -E-
(Later assigned U.S. NO. 94476)

Courtesy of William Lewis
Illustrations from the *Standard Test Report of the Engineering Division* at McCook Field dated 4-18-19.
This report was generated by test engineer Iskander Hourwich, approved by engineer in charge C. F. Taylor, and Major George E. A. Hallett, chief of the Power Plant Dept.

2730

27

LIBERTY-8
AIRCRAFT ENGINE
MODEL A
MNFD. BY
GENERAL MOTORS CORPORATION.
BUICK MOTOR CO. DIV.
MNFG. NO.
U.S. NO.
LEFT 1 7 3 5
FIRING ORDER
RIGHT 6 4 8 2

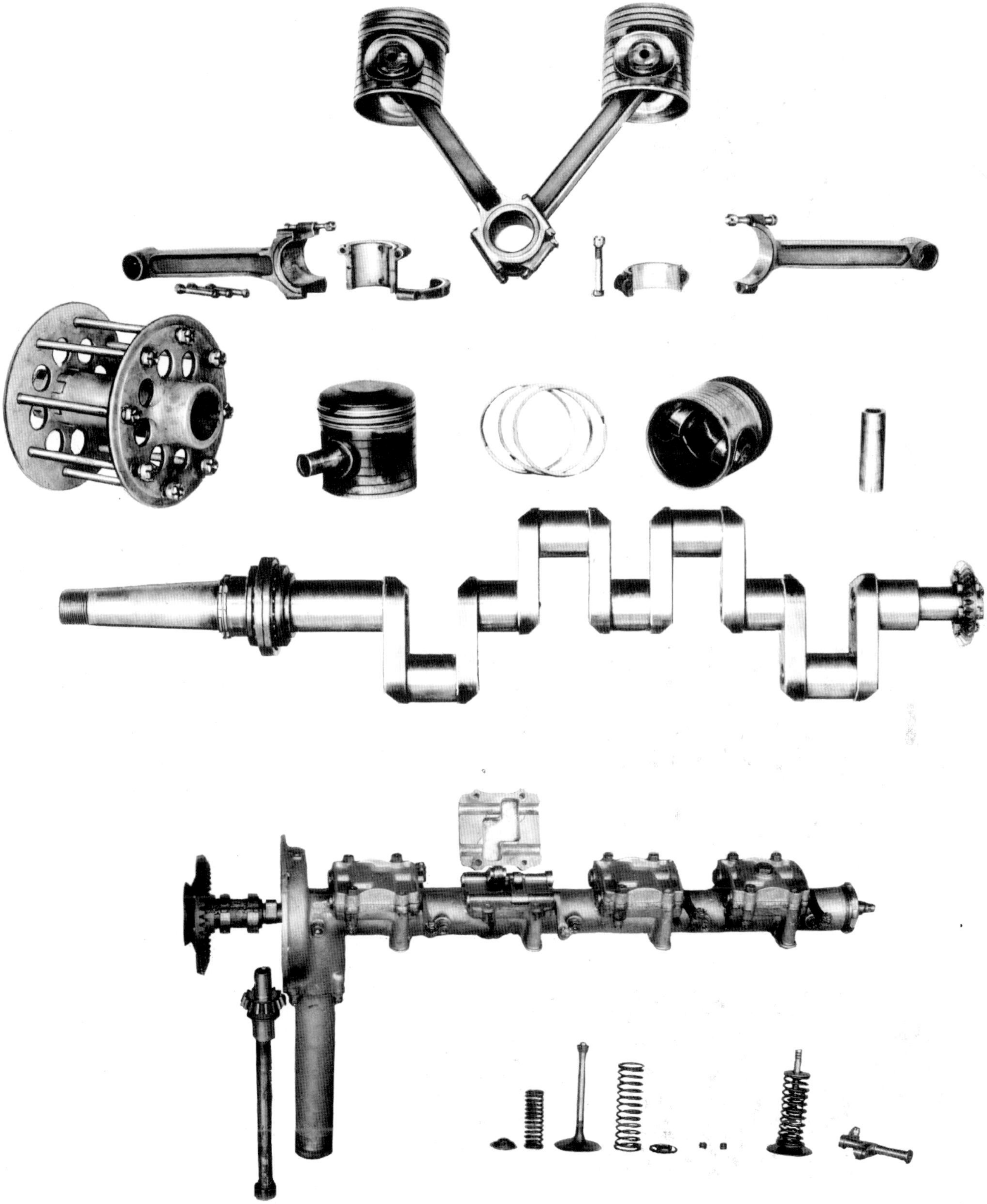

Water pump

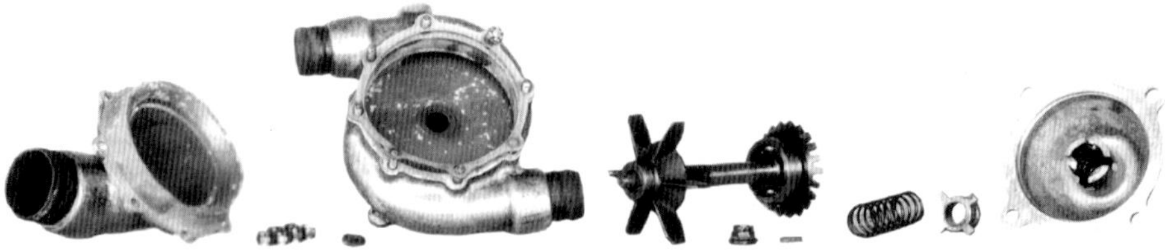

Oil pump

Oil pump

Water pump

Distributor

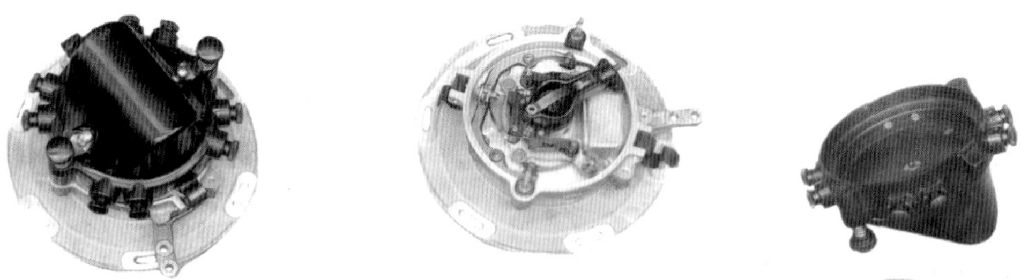

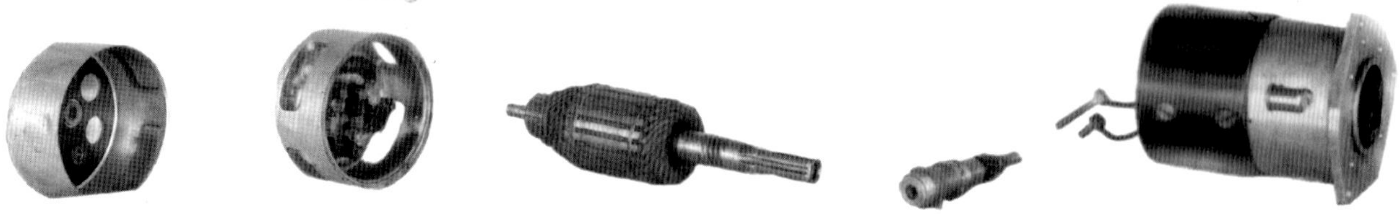

Generator

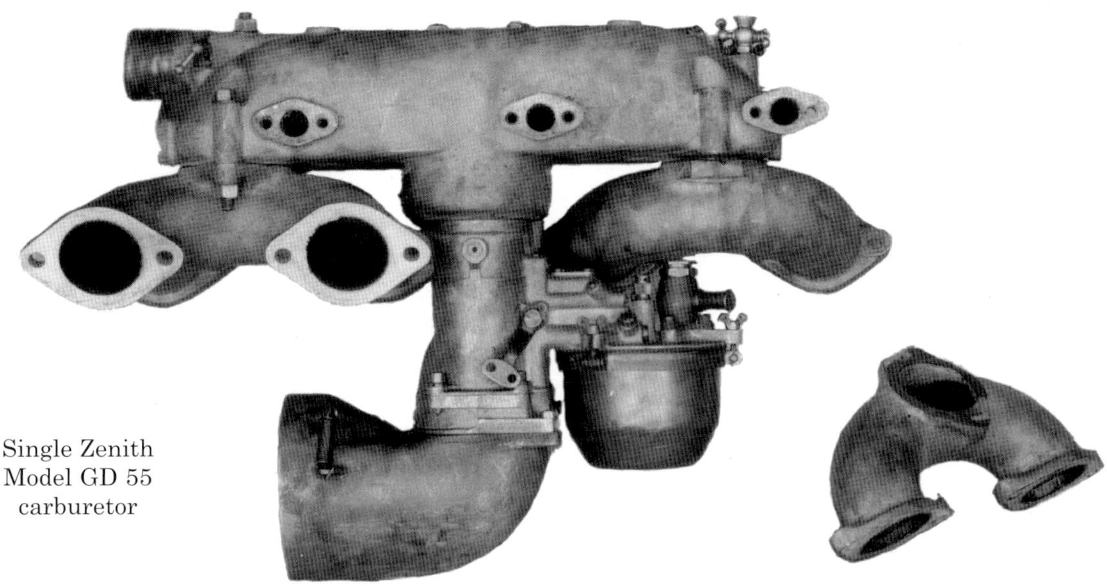

Single Zenith
Model GD 55
carburetor

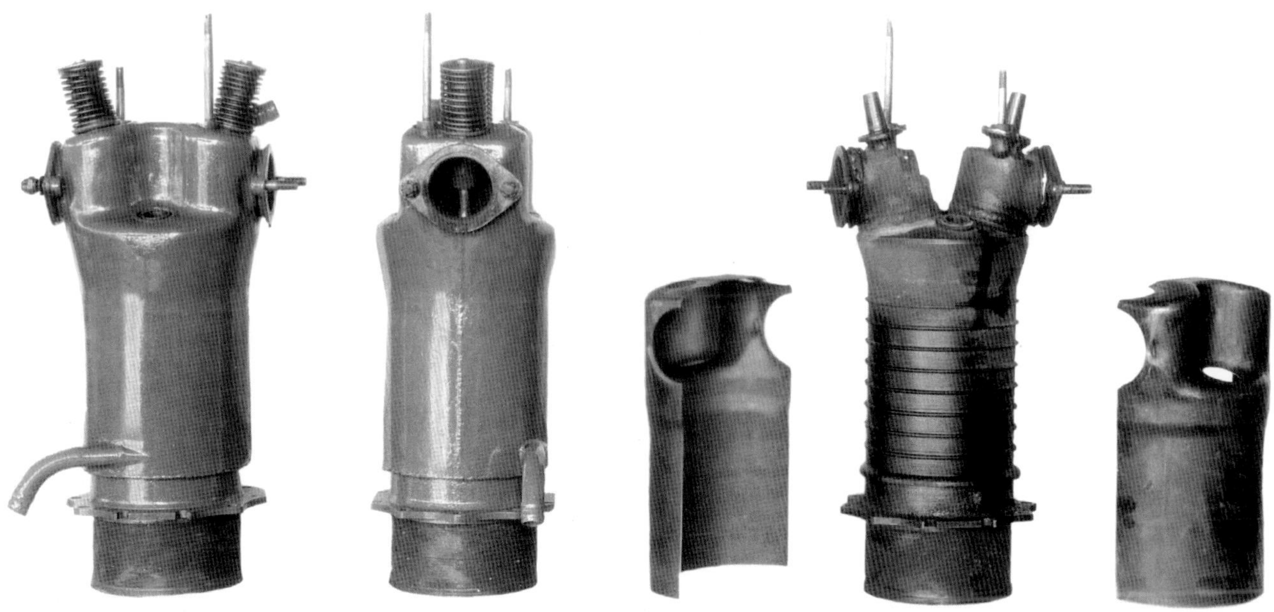

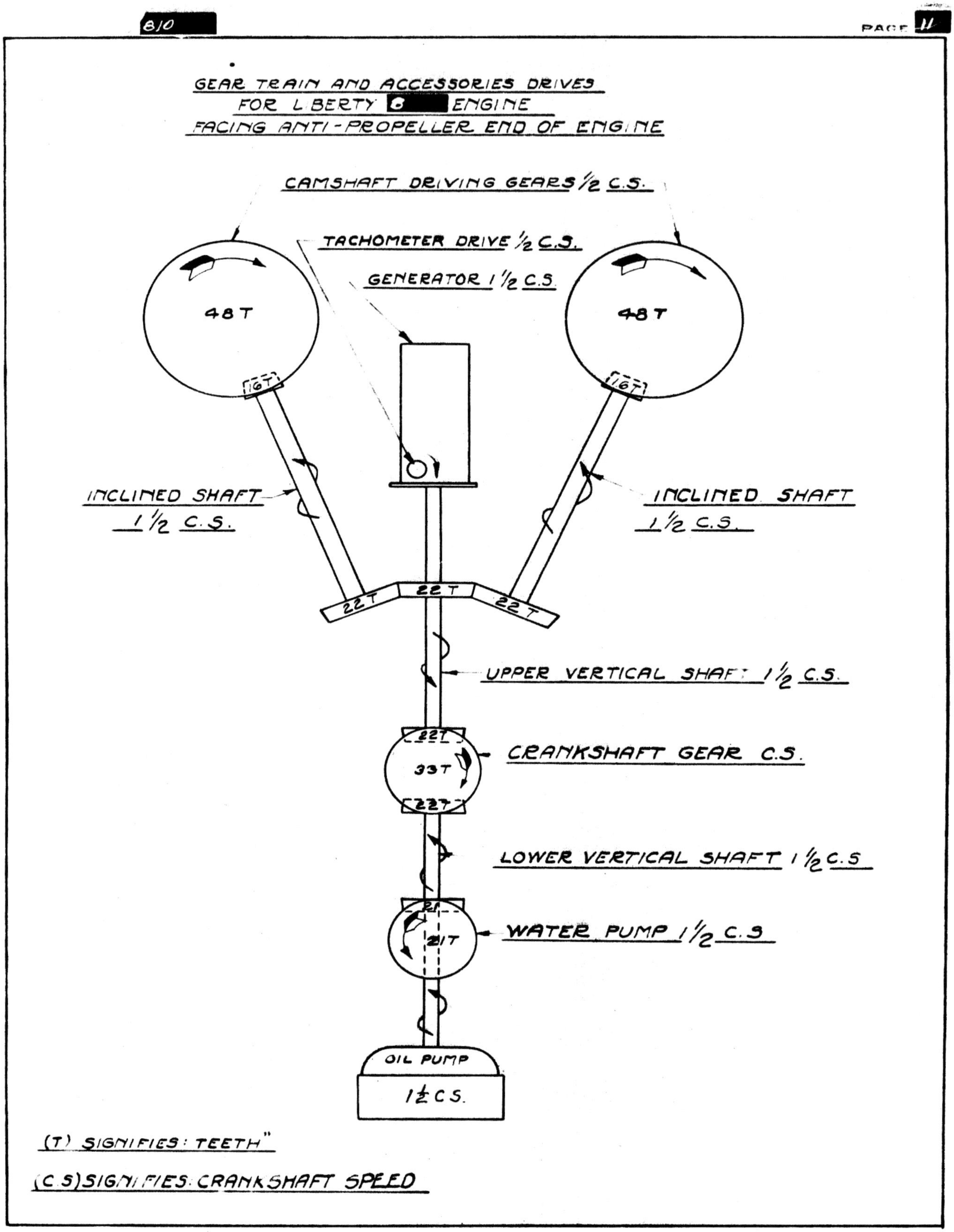
8/0
PAGE 11
GEAR TRAIN AND ACCESSORIES DRIVES
FOR LIBERTY 8 ENGINE
FACING ANTI-PROPELLER END OF ENGINE
CAMSHAFT DRIVING GEARS ½ C.S.
TACHOMETER DRIVE ½ C.S.
GENERATOR 1½ C.S.
48 T
48 T
16 T
16 T
INCLINED SHAFT 1½ C.S.
INCLINED SHAFT 1½ C.S.
22 T
22 T
22 T
UPPER VERTICAL SHAFT 1½ C.S.
22 T
33 T
22 T
CRANKSHAFT GEAR C.S.
LOWER VERTICAL SHAFT 1½ C.S
21
21 T
WATER PUMP 1½ C.S
OIL PUMP
1½ C.S.
(T) SIGNIFIES: TEETH"
(C.S) SIGNIFIES: CRANKSHAFT SPEED

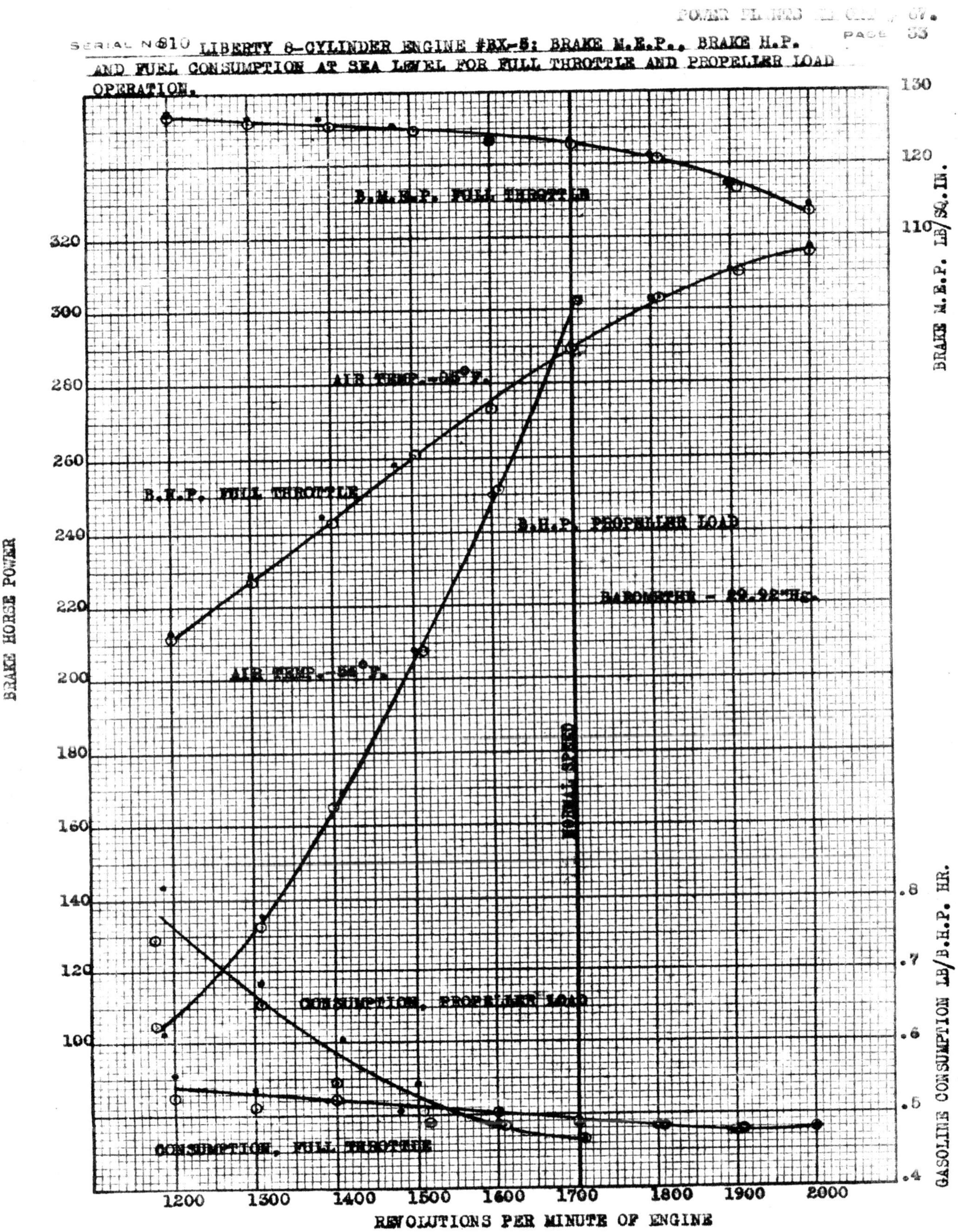
LIBERTY 8-CYLINDER ENGINE #BX-5: BRAKE M.E.P., BRAKE H.P. AND FUEL CONSUMPTION AT SEA LEVEL FOR FULL THROTTLE AND PROPELLER LOAD OPERATION.
B.M.E.P. FULL THROTTLE
B.H.P. FULL THROTTLE
B.H.P. PROPELLER LOAD
CONSUMPTION, PROPELLER LOAD
CONSUMPTION, FULL THROTTLE
NORMAL SPEED
BRAKE HORSE POWER
BRAKE M.E.P. LB/SQ.IN.
GASOLINE CONSUMPTION LB/B.H.P. HR.
REVOLUTIONS PER MINUTE OF ENGINE

The Ford Liberty 12 MNFG NO. 1635 U.S. NO. 31110

Courtesy of William Lewis

Illustrations from the *Standard Test Report of the Engineering Division* at McCook Field dated 3-21-19. This report was generated by engineer in charge George J. Mead and approved by Major George E. A. Hallett, chief of the Power Plant Dept.

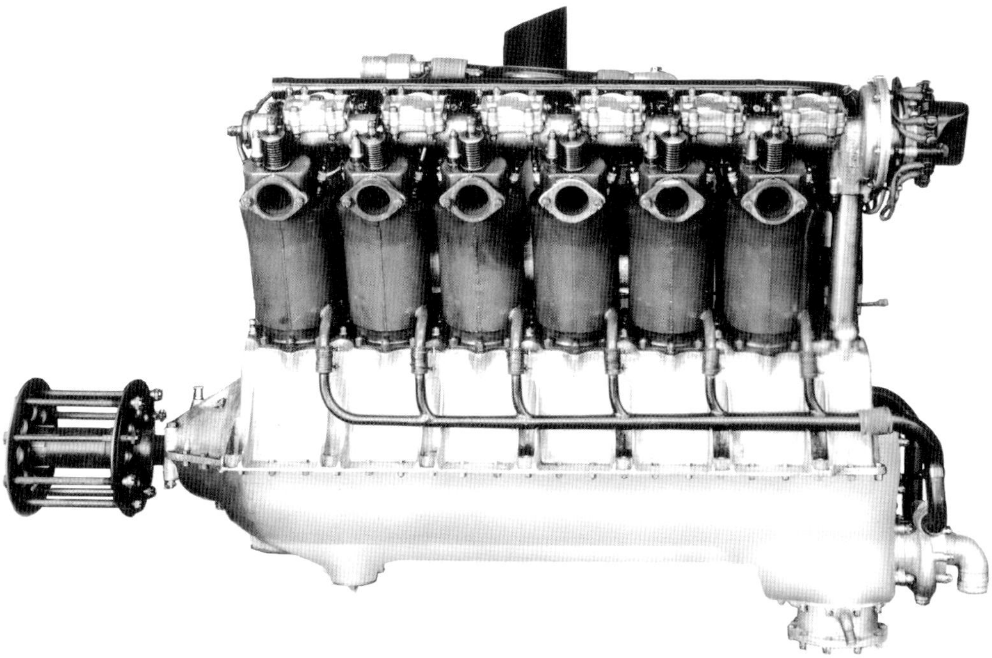

The bottom view above is actually of Packard NO. 1758 Navy NO. 0859. Bottom views are so rare that I thought it should be included in this section. (Courtesy of the Air Force Museum)

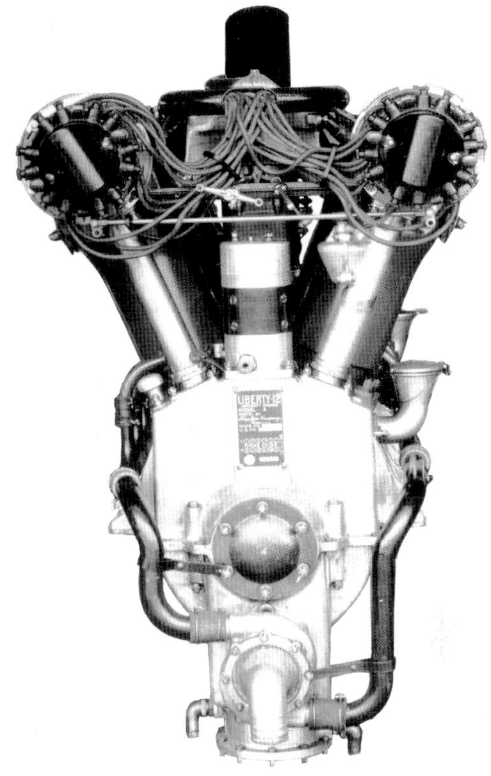

2473

LIBERTY-12
AIRCRAFT ENGINE
MODEL A
MNFD. BY
Ford Motor Company
MNFG. NO.
U. S. NO.
L 1 9 5 11 3 7
FIRING ORDER
R 8 4 12 6 10 2

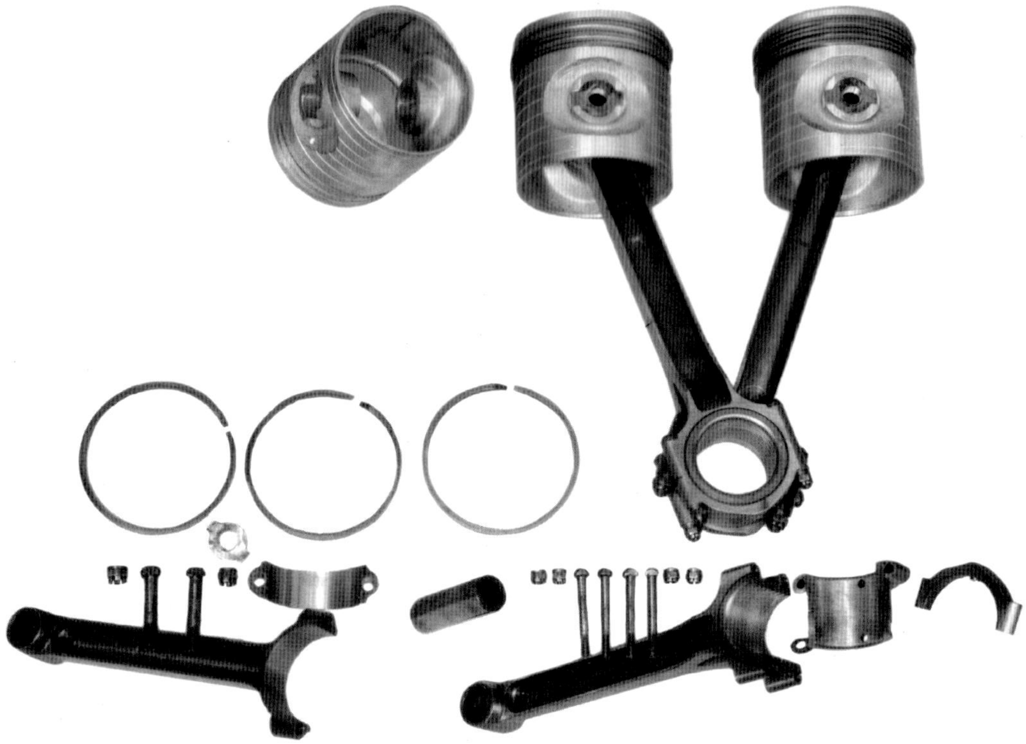

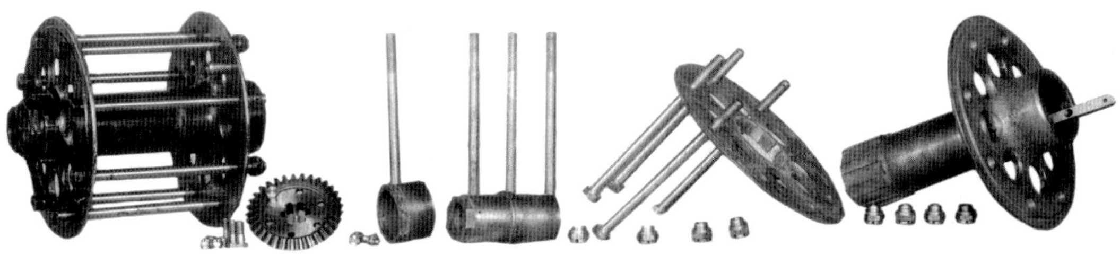

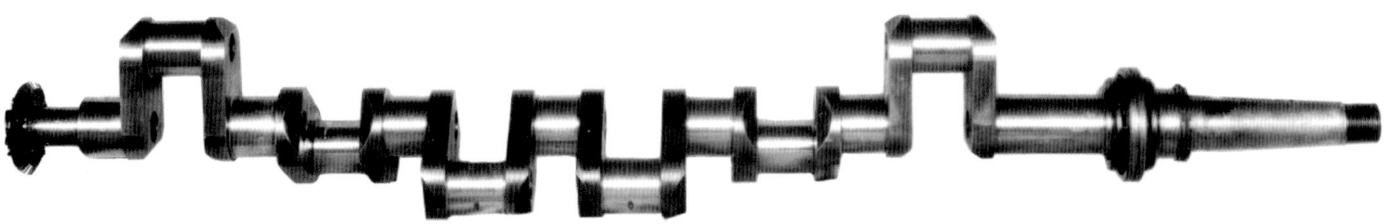

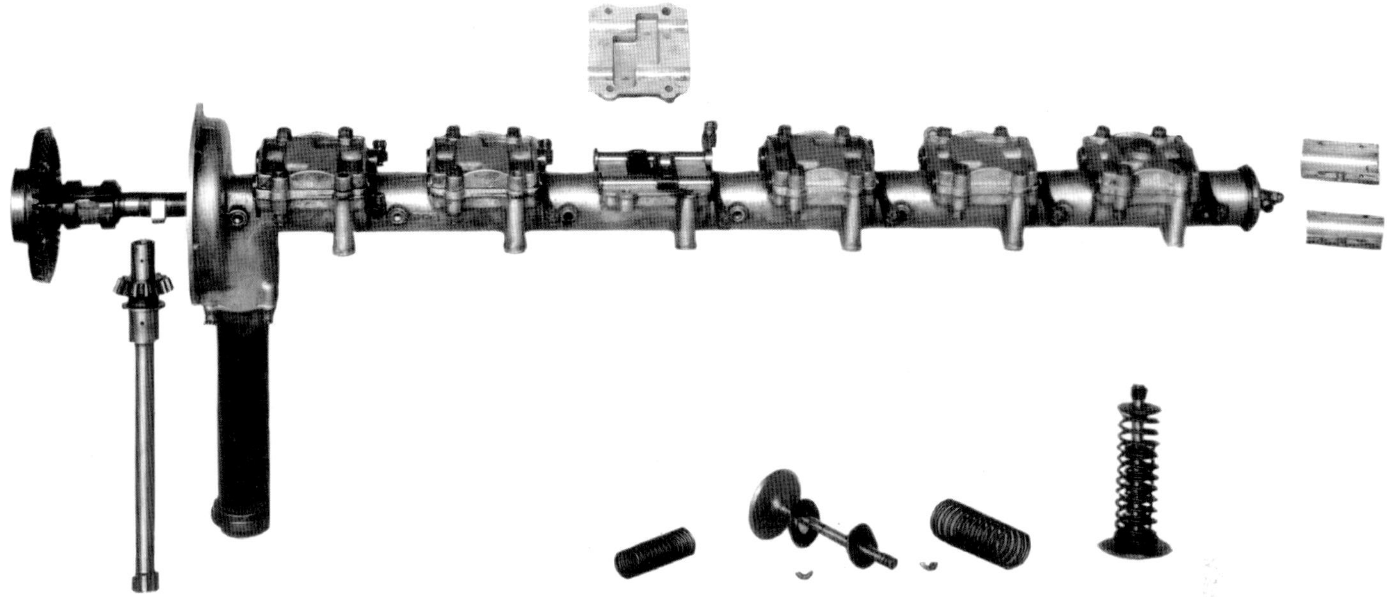

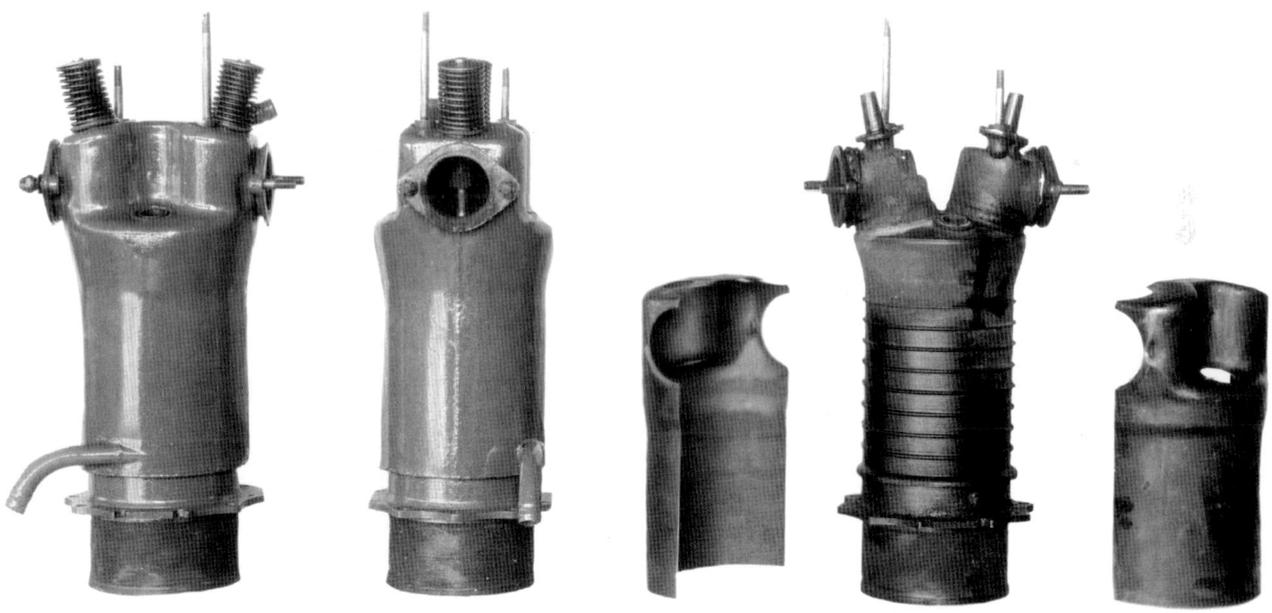

Water pump

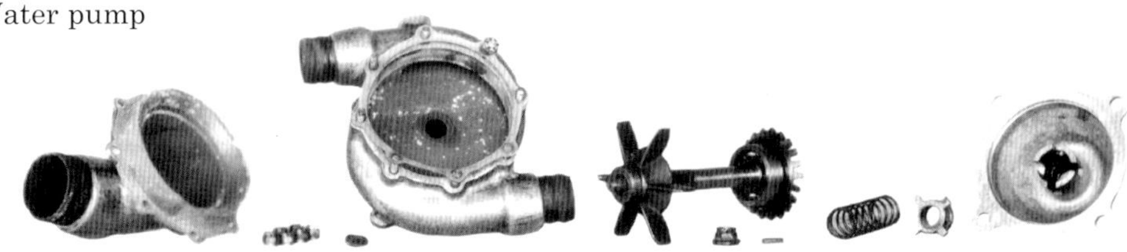

Oil pump

Oil pump

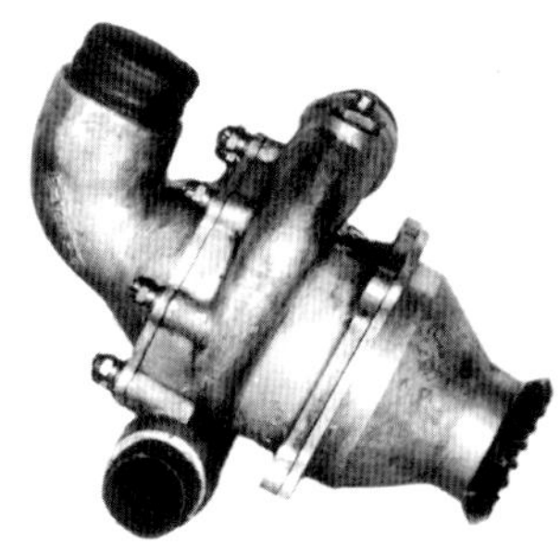

Water pump

Generator

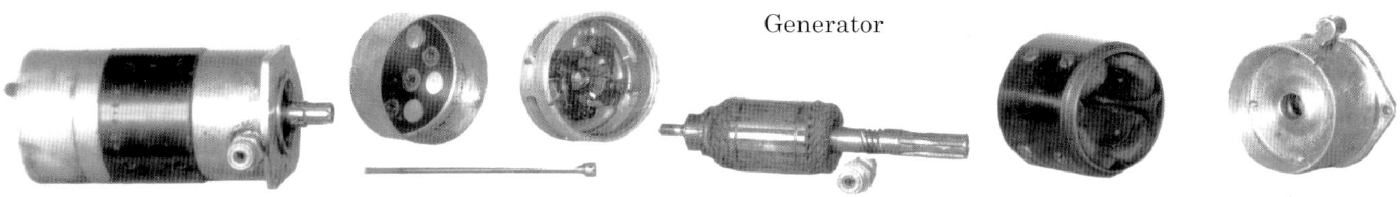

Distributor

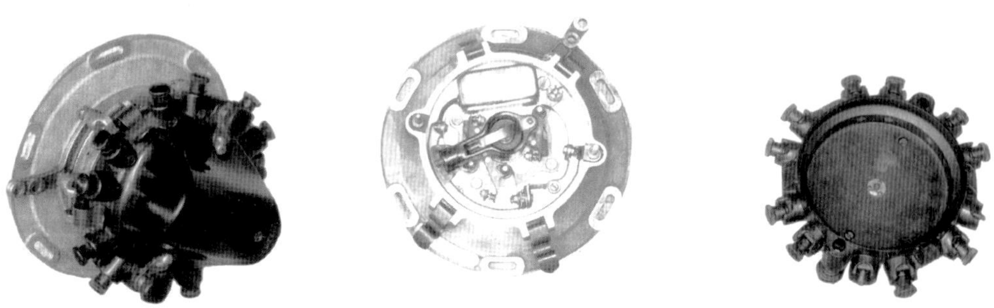

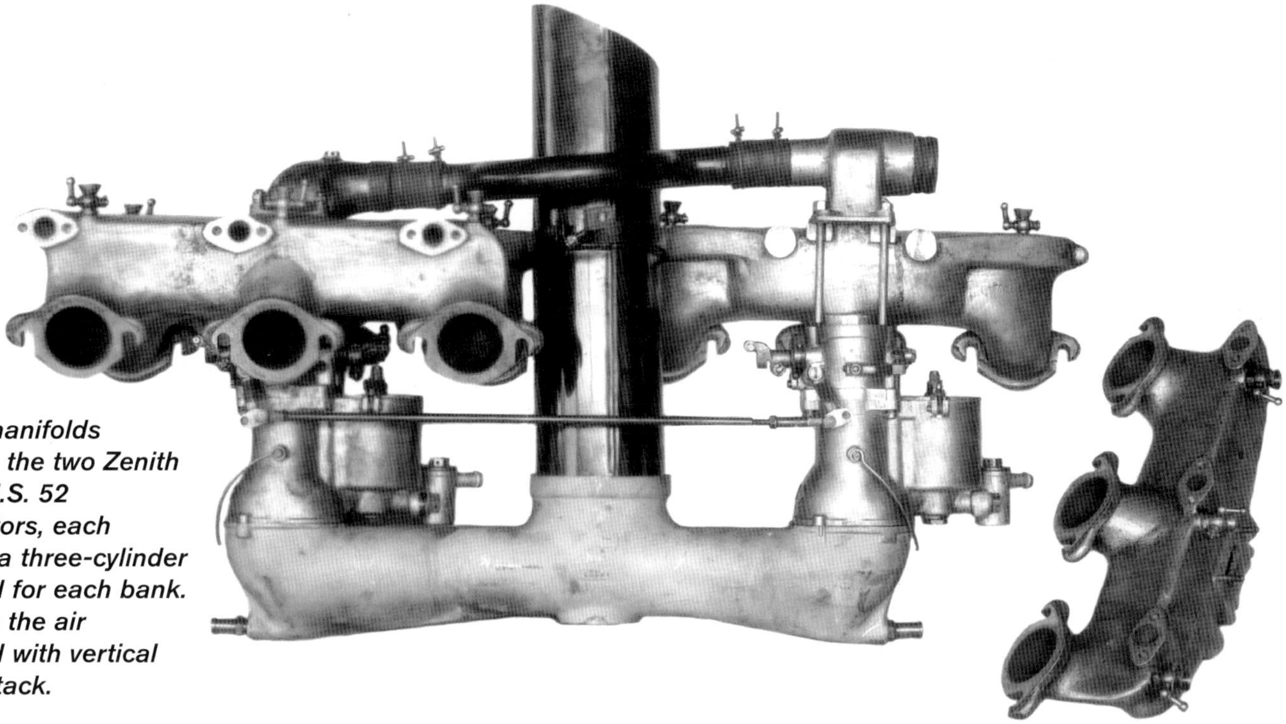

Intake manifolds showing the two Zenith Model U.S. 52 carburetors, each feeding a three-cylinder manifold for each bank. Below is the air manifold with vertical intake stack.

Liberty 12 on dynamometer test at McCook Field.

POWER PLANTS REPORT #52.
PAGE 15.

SERIAL NO 754 GEAR TRAIN AND ACCESSORIES DRIVES
(Facing anti-propeller end of engine)

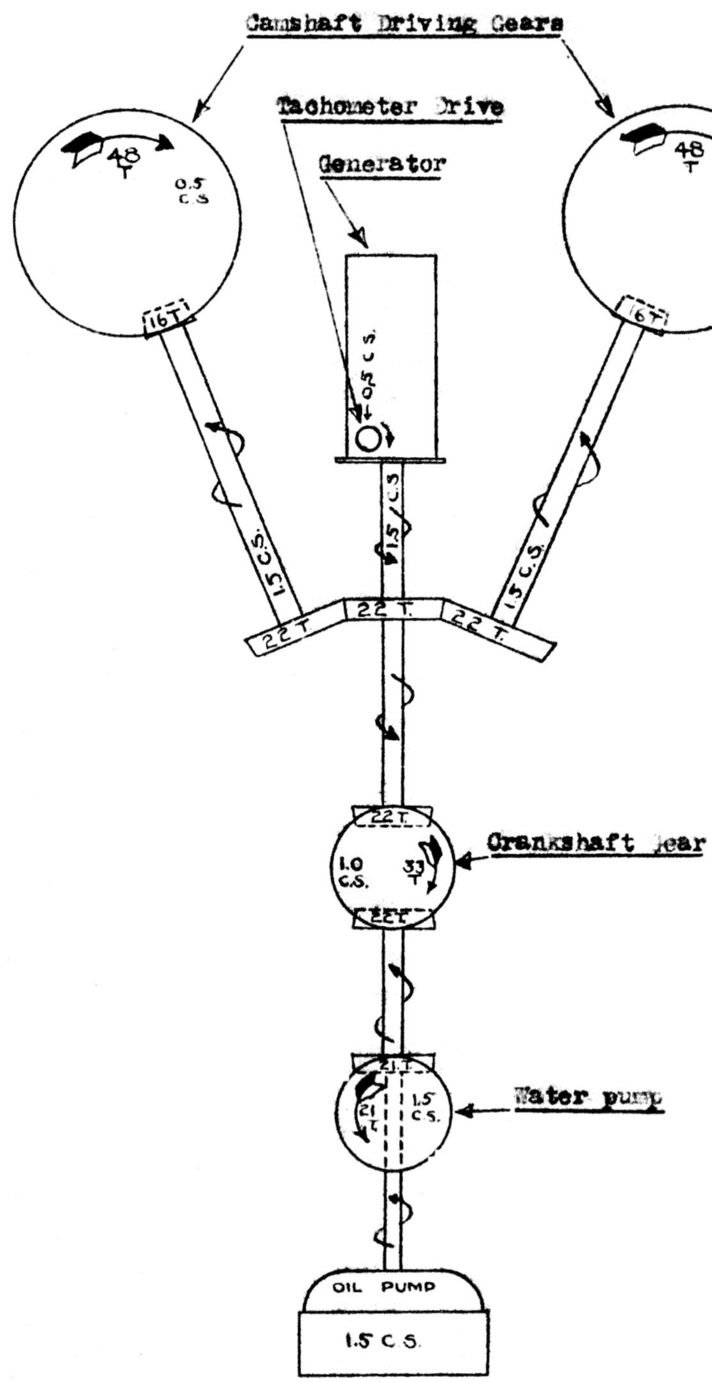

(T) signifies: "Teeth"

(C.S.) signifies: "Crankshaft Speed"

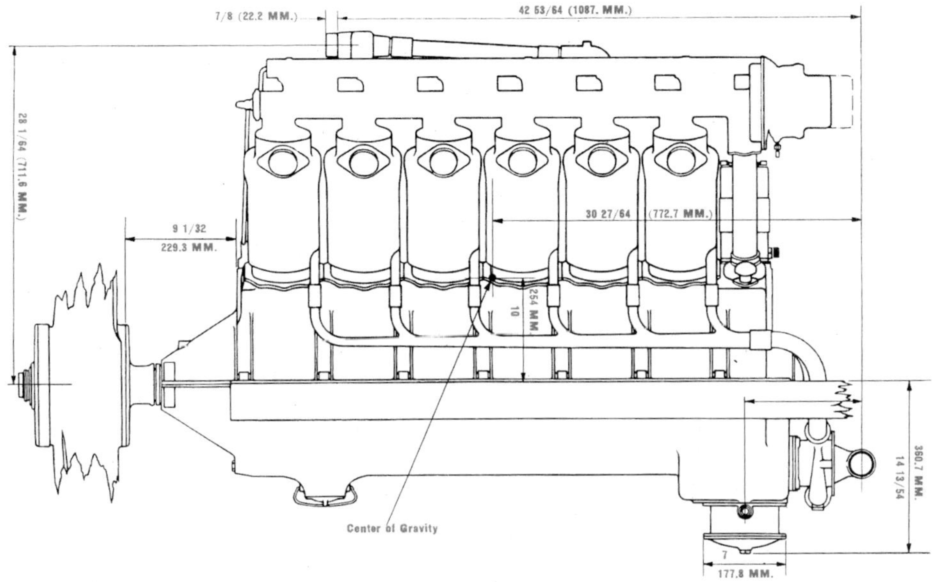
7/8 (22.2 MM.)
42 53/64 (1087. MM.)
28 1/64 (711.6 MM.)
9 1/32
229.3 MM.
30 27/64 (772.7 MM.)
254 MM
10
Center of Gravity
360.7 MM.
7
177.8 MM.

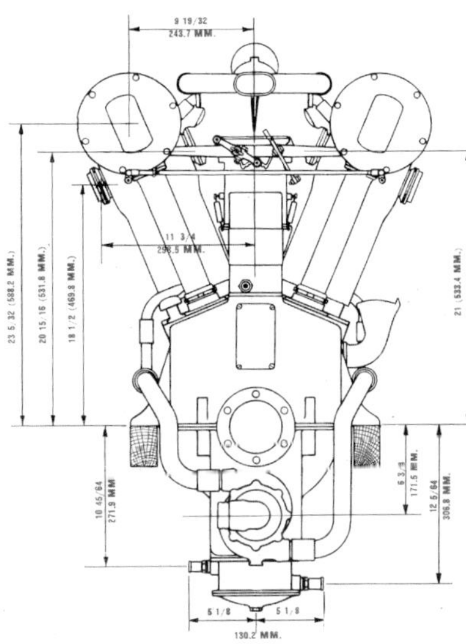
9 19/32
243.7 MM.
11 3/4
130.2 MM.

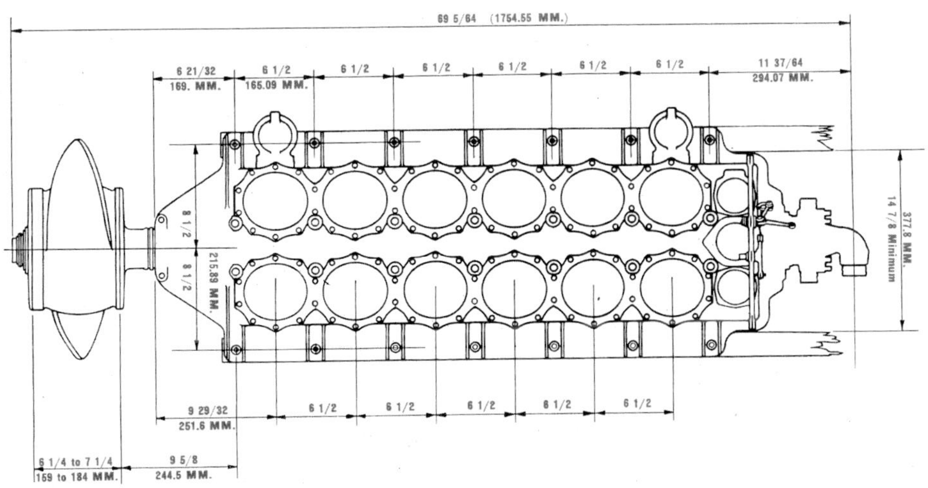
69 5/64 (1754.55 MM.)
6 21/32
169. MM.
6 1/2
165.09 MM.
6 1/2
6 1/2
6 1/2
6 1/2
6 1/2
11 37/64
294.07 MM.
8 1/2
8 1/2
215.89 MM.
377.8 MM.
14 7/8 Minimum
9 29/32
251.6 MM.
6 1/2
6 1/2
6 1/2
6 1/2
6 1/2
6 1/4 to 7 1/4
169 to 184 MM.
9 5/8
244.5 MM.

POWER PLANTS REPORT #52.

SERIAL NO 754 MAXIMUM POWER, PROPELLER LOAD, M. E. P. AND FUEL CONSUMPTION CURVES OF LIBERTY "12" #53, U.S. #23481, AT SEA LEVEL. PAGE 35.

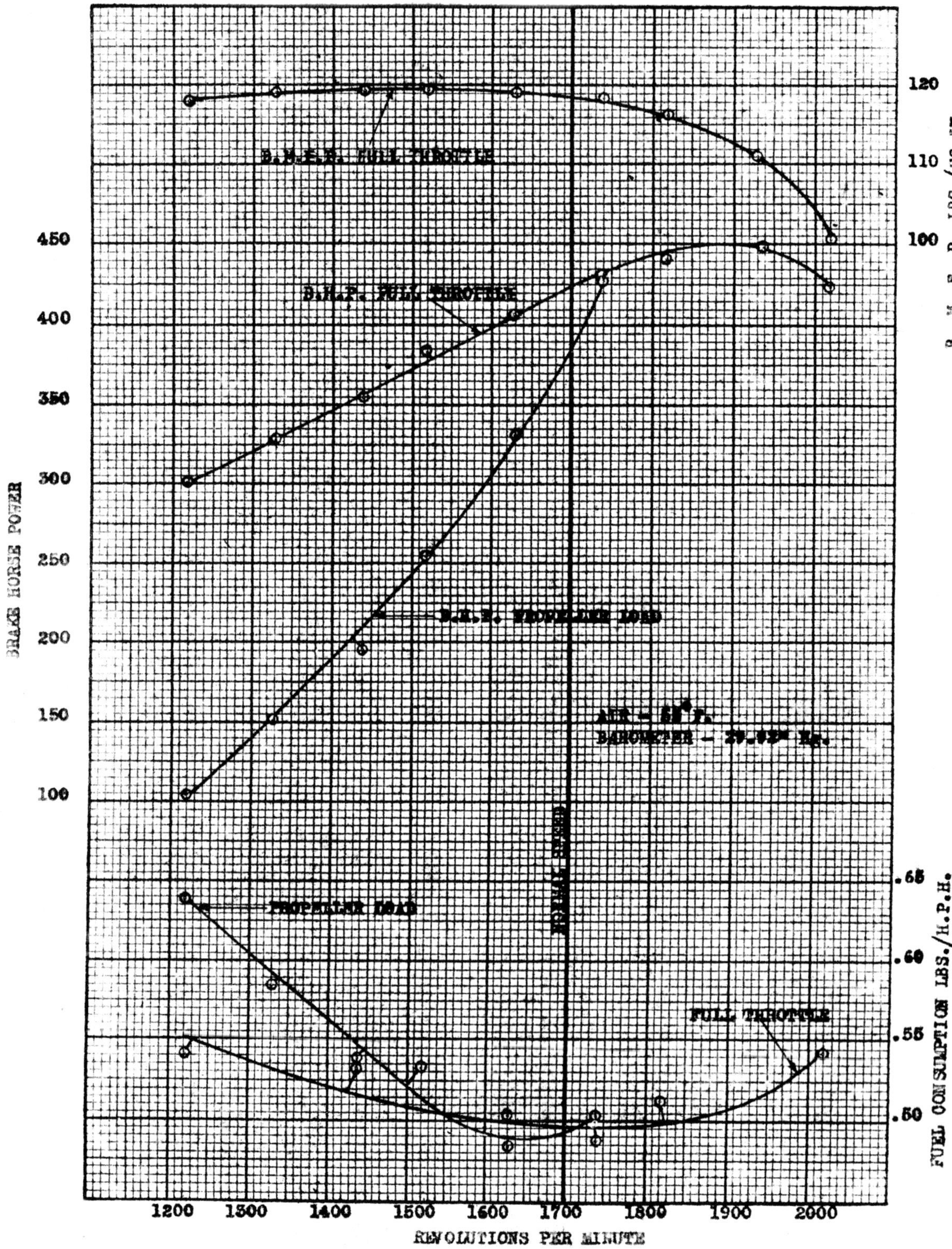

Appendix H

U.S. Military Liberty Engines On-Hand

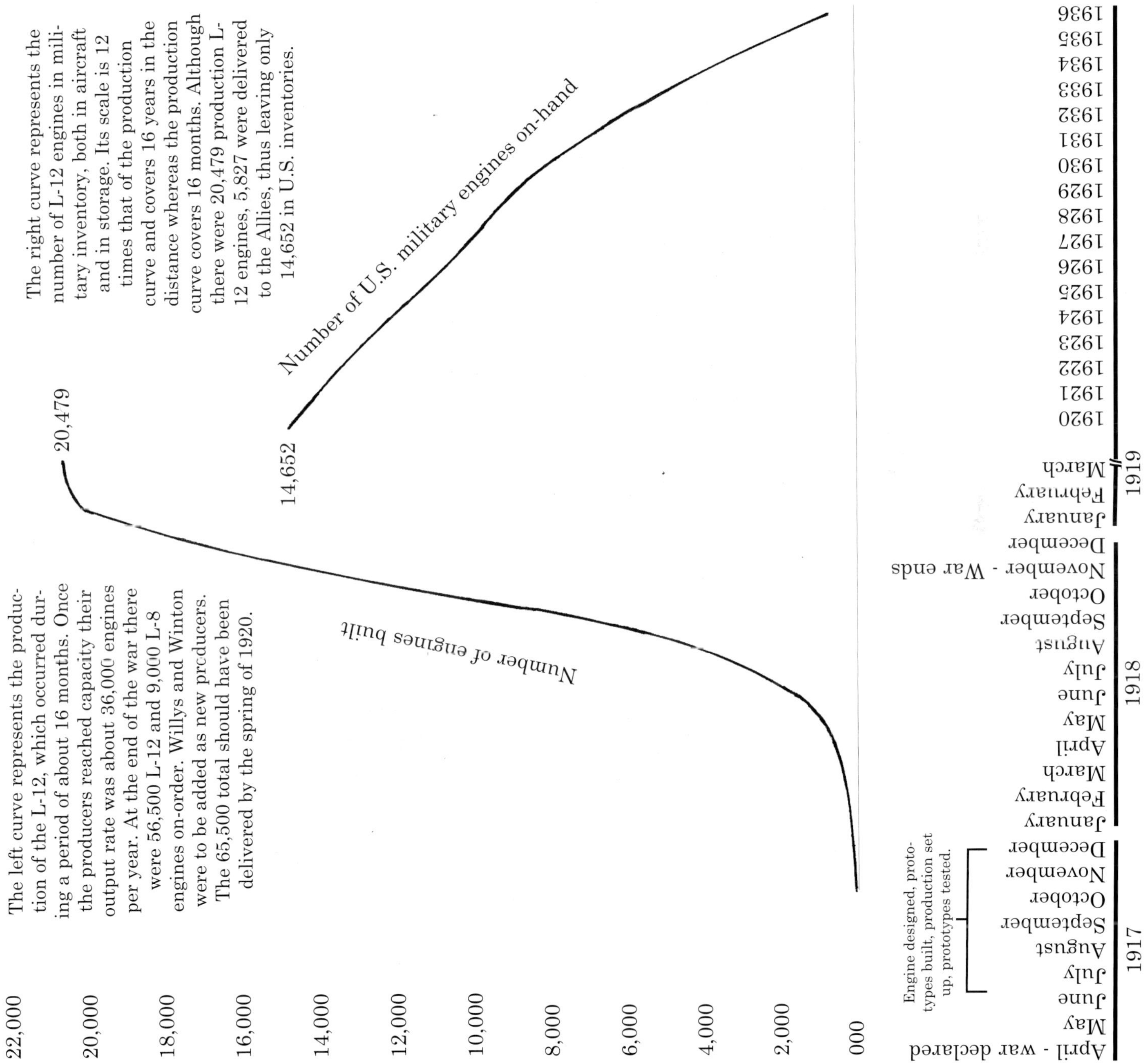

DIARY
LT. COL. J. G. VINCENT.

HISTORY OF THE DEVELOPMENT
OF THE
U.S.A. STANDARDIZED AIRCRAFT
ENGINES.
BY
J. G. VINCENT.

ORIGINAL MINUTES
PACKARD MOTOR CAR COMPANY
FEB. 20, 1917 - SEPT. 19, 1919

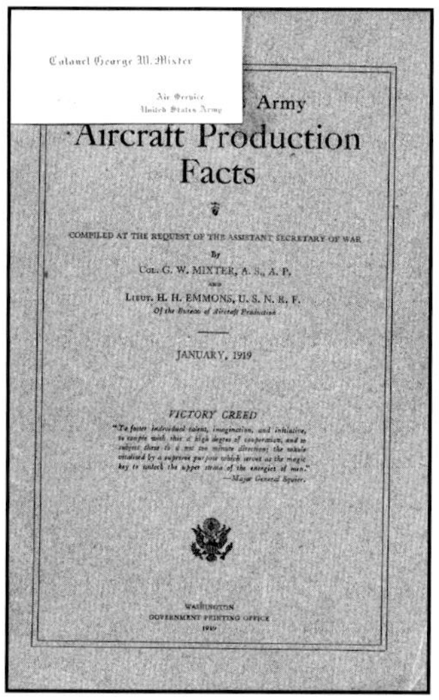
Army
Aircraft Production
Facts
VICTORY CREED

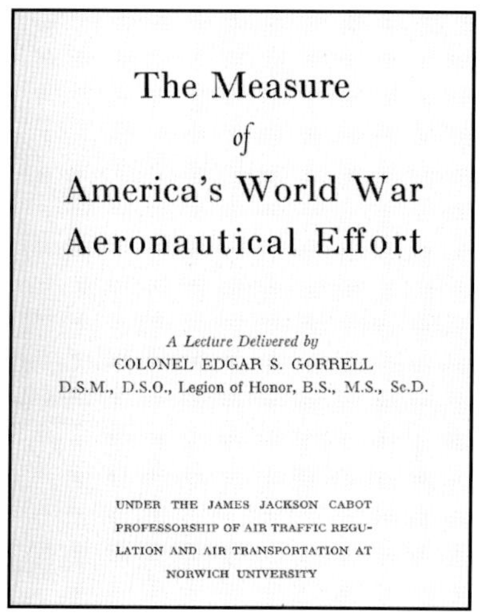
The Measure
of
America's World War
Aeronautical Effort
A Lecture Delivered by
COLONEL EDGAR S. GORRELL
D.S.M., D.S.O., Legion of Honor, B.S., M.S., Sc.D.
UNDER THE JAMES JACKSON CABOT
PROFESSORSHIP OF AIR TRAFFIC REGU-
LATION AND AIR TRANSPORTATION AT
NORWICH UNIVERSITY

THE LIBERTY ENGINE
ORIGIN AND DEVELOPMENT
WORLD WAR I

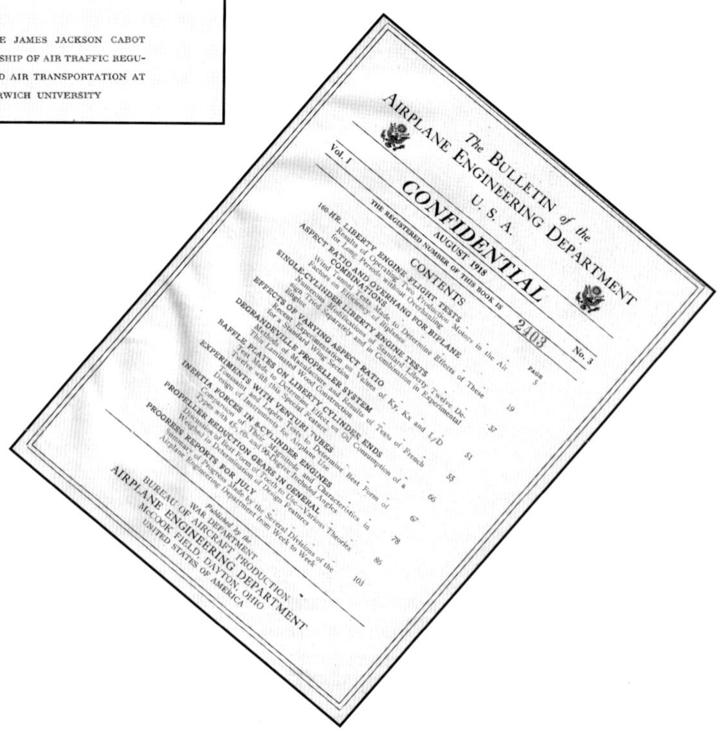
The BULLETIN of the
AIRPLANE ENGINEERING DEPARTMENT
U. S. A.
CONFIDENTIAL
Vol. I
AUGUST 1918
No. 3
CONTENTS
Published by the
WAR DEPARTMENT
BUREAU OF AIRCRAFT PRODUCTION
AIRPLANE ENGINEERING DEPARTMENT
McCOOK FIELD, DAYTON, OHIO
UNITED STATES OF AMERICA

Bibliography

The following material was consulted during the research and writing of this book. Some contributed significantly to the subject under study and some minimally. Comments follow some items in order to advise the reader of the content and possible value of the item.

Andrade, John M. — *U. S. Military Aircraft Designations and Serials Since 1909*
Midland Counties Publications 1979

Andrews, P. W. S. and Brunner, Elizabeth — *The Life of Lord Nuffield A Study in Enterprise and Benevolence*
Basil Blackwell 1955
The "official" biography of Lord Nuffield (William Richard Morris). This is a very informative and, of course, flattering biography of this engineer, industrialist and philanthropist. Lord Nuffield was a man of great intelligence and energy who had enormous business success, built up a group of companies which was eventually brought under the control of what became known as the Nuffield Organization. A great number of British tanks of WW II were built by these companies, as were the Nuffield Liberty engines. As is the case with the biography of E. A. Deeds, a great deal can be learned about the various projects undertaken by the Nuffield Organization, but a secondary source is advised. For example, if one takes the account as presented in this book at face value, he would say Lord Nuffield personally decided the British cruiser tank should be based on the Christie and therefore he personally purchased the sample Christie. Then, as the story was told, after the prototype tanks were put under construction, men from Nuffield looked about for a suitable engine in the 350 HP class to power it. They then remembered the old Liberty on which some had worked during WW I and "they found a works where some half-dozen engines were lying about; some of the old drawings for these engines were located — ." From this set of circumstances Nuffield then developed the Nuffield Liberty. There was some vague truth to the story, but that is not how it happened.

Angle, Glenn D. — *Airplane Engine Encyclopedia*
The Otterbein Press 1921
An invaluable reference which gives brief descriptions of almost all aero engines of which at least one running example had been built from the earliest to the time of publication. Brief design and performance specifications were included as well as illustrations of most engines cited. Its principal short-coming is its failure in many cases to date the various designs. It was, however, the only publication to cover the complete spectrum of aircraft engines and, as such, is a very valuable historical reference.

Angle, Glenn D. — *Aerosphere 1939*
Aircraft Publications 1940
This book is an update and enlargement of Angle's 1921 book. It covers, in the same manner, all the engines found in the 1921 edition plus all designed and built from 1921 through 1939. The book was expanded to also cover aircraft of the period, i.e., late 1930s. As with the 1921 book, this is an invaluable reference to the aero historian.

Air Historical Branch Air Ministry (British) — *H.Q. RFC & RAF Liberty Engines*
Air Ministry Publications
Reports and correspondence related to the Liberty engine in British service from November 2, 1917 to February 4, 1919. This material provided valuable information relative to British procurement and use of the Liberty during WW I.

Beale, Peter — *Death By Design British Tank Development In The Second World War*
Sutton Publishing 1998
The story of British tank development during WW II. As the title indicates, the author does not think the British used the best of methods and logic in this effort. He does do a good job of explaining what was done and how, and thus covers the history well.

Bowers, Peter M. Swanborough, Gordon — *United States Navy Aircraft Since 1911*
Naval Institute Press 1990

Bowers, Peter M. Swanborough, Gordon — *United States Military Aircraft Since 1909*
Smithsonian Institution Press 1989

Boyne, Walter J. — *de Havilland DH-4 – From Flaming Coffin To Living Legend*
Smithsonian Institution Press 1984
The history of the production in this country of the de Havilland DH-4 during World War I. This was Volume 7 of the NASM series on *Famous Aircraft of the National Air and Space Museum* and documents not only the history of the aircraft type as produced in this country but also the history of NASM's DH-4, which was the Dayton-Wright prototype No. 1 aircraft. This 119-page book is probably the most extensive reference so far written on this aircraft.

Bristow, Lt. Col. W. A. — *Report on the First Visit to the United States Technical Departments, Bureau of Aircraft Production.*
British Royal Air Force
Technical Liaison Office 1918
The 40-page report to the RAF of the visit of Col. Bristow, Chief Technical Liaison Officer, British & United States Air Departments, to the the BAP and some of the various plants producing the Liberty. It discusses some of the various difficulties encountered in production and testing of the engine and how they were overcome. It also contains a good deal of information relative to contracts and to experimental engines.

British War Office Reports C.S. (W) 1(c) 1946 — *A13.E1, E2 and E3 Notes on Development*; *A13.E1 Christie Cruiser*; *A13.E2 and E3 Prototype Light Cruiser Tanks*
The above three reports contain a history of the development of the British Christie and the prototypes which led immediately to the British Cruiser tank series. In doing so, they also trace much of the history of the development of the Nuffield Liberty engine.

Bureau of Aircraft Production Liberty Engine Engineering Dept. — *Liberty Engine Test Reports*
Bureau of Aircraft Production 1918
Detroit District Office
On September 15, 1918 a small engineering organization with testing facilities was set up in the old Saxon Automobile plant in Detroit by the Detroit District of the BAP to handle testing of production changes of the Liberty engine. This job was originally handled by BAP and Packard engineers at the Packard plant. It also took on the development of the standardized version of the Liberty 8 engine for eventual production. It was originally referred to as the Saxon Test Plant but was soon designated as the Liberty Engine Experimental Plant. From September 1918 until the end of December 1918 this organization conducted tests, recommended design changes and issued reports separate from those of the Engineering Division based at McCook Field.

Cadillac Motor Car Co. — *Cadillac Participation in the World War*
Cadillac Motor Car Co. 1919
This 70-page booklet was sub-titled "An expression of appreciation to Cadillac men for their loyal services in a great emergency and a historical account of Cadillac accomplishments." This account provides a number of insights to Cadillac and Buick Liberty production and the reasons behind the production division between the two General Motors divisions.

Casari, Robert B. — *Encyclopedia of U. S. Military Aircraft*
Military Aircraft Publications 1970-1975
This outstanding piece of research was aimed at early U. S. Military aircraft of the Army and was split into two subjects, each covered in a series of small books. The first series was subtitled *1908 to April 6, 1917* and the second *The World War I Production Program*. The first series covered Army aircraft procured from 1908 to our entry in World War I and was covered in four volumes

of from 65 to 95 pages. The second subject covered those aircraft procured during the war and was done in three volumes ranging in size from 65 to 150 pages. Unfortunately for WW I aero historians, Casari apparently never finished this second series. Volume 1 begins with an excellent description of the wartime production system. The scheme was to then proceed with detailed histories of the various aircraft and to cover them in alphabetical order by either the name of the designer or the principal producer, whichever seemed most logical. Volume 1 began with the AVRO 504. The last volume published, Vol. 3, ended with the last of the Curtiss Jennies, the JN-5H. A list of the aircraft he intended to eventually cover was listed in Vol. 1 and indicated the first three volumes only covered about 25% of the total named models.

Casari, Robert B. *U. S. Army Aviation Serial Numbers and Orders 1908 – 1923*
Military Aircraft Publications 1995
A compilation of Army aviation orders and serial numbers of aircraft and engines of the period 1908-1923. This invaluable reference lists all orders and deliveries of the period by serial number, even orders which were eventually canceled. It thus gives a reference to all engine and aircraft serials, whether or not the number was actually assigned to a finished article. This is the only accurate compilation of this information as it is not available in either the Army or Air Force archives.

Chamberlain, Peter and Ellis, Chris *British and American Tanks of World War II*
Arco Publishing Co. 1969
An excellent history of the tanks and associated vehicles of these two countries from the late 1930s to the end of World War II.

Christie, J. Edward *Steel Steeds Christie*
Sunflower University Press 1985
A history of J. Walter Christie's lifetime of inventive and production endeavors written by his son. It provides some interesting insights to Christie's prolific output of front-wheel-drive vehicles and tanks, and a few facts that will probably be found nowhere else. It is to some degree influenced by the fact that the author is his son and does contain some errors of fact. It is, however, an invaluable reference if one is doing an in-depth study of Christie's vehicles.

Crowell, Benedict *America's Munitions 1917-1918*
Government Printing Office 1919
A 592-page history of munitions production for the Army for use during World War I. This government history was compiled by Assistant Secretary of War Benedict Crowell at the direction of Secretary of War Newton D. Baker at the end of the war. It includes all types of munitions and thus covers all types of aircraft and their engines and other equipment. Each section was the result of input from those individuals who were either responsible for, or intimately associated with, the material covered. Unfortunately, as was the case with almost all government-sponsored histories related to aircraft, it documents, for the most part, no further than the date of the signing of the Armistice, November 11, 1918. For this reason, little is documented regarding wind-down production of aeronautical material.

Crowell, Benedict *Report of the American Aviation Mission 1919*
Government Printing Office 1919 and
His Majesty's Stationary Office October 1919
This is the report of the American Aviation Mission to France, Italy and England directed by Secretary of War Newton Baker following the end of the war. It was published both here and in England. Its objective was to consult with those men who were in charge of the production and use of aircraft in the Allied governments and draw the best conclusions possible as to how best to preserve the air capabilities America had built up during the war. Most of their recomendations were never carried out following the war and the resulting predicted rapid deterioation of America's air capabilities occurred.

Dickey, Philip S. III *The Liberty Engine 1918 – 1942*
Smithsonian Institution Press 1968
This brief history of the Liberty engine was the only extensive (110 pages) material published on the overall subject until this book was put in print. It was a well-researched monograph, which contains a great deal of material on the subject. One of its few shortcomings is that it has no index, an almost unforgivable omission in a historical reference.

Emmons, Lt. H. H. *History of the Liberty Engine*
Unpublished manuscript 1919
A 23-page report written at the request of Col. S. D. Waldon and completed on January 9, 1919. Although not an exact duplicate, it formed the basis of much of the material published later in the month in *United States Aircraft Production Facts* by Col. G. W. Mixter and Lieut. H. H. Emmons.

Fahey, James C. *U. S. Army Aircraft 1908 – 1946*
Ships and Aircraft 1946
Though one of the earliest significant references published on army aircraft procurements, it remains a valuable fast reference on the subject.

Foulois, Maj. Gen. Benjamin D. with Col. C. V. Glines *From the Wright Brothers to the Astronauts The Memoirs of Major General Benjamin D. Foulois*
McGraw-Hill 1968
The autobiography of Maj. Gen. Benjamin D. Foulois. Foulois was intimately involved in the development of aeronautics in this country from the first Army-purchased Wright aircraft, which he flew and modified, up to 1935 when he retired after his second service as Chief of the Air Service. He served as chairman of the original Army and Navy Joint Technical Board of 1917, first as a Major and then as a Brig. General. He then went to Europe and served as Chief of Air Service A. E. F. After numerous other assignments he ended his Army career with a four-year service as Chief of Air Corps from 1931-1935. He tells the fascinating story of the behind-the-scenes evolution of the Army Air Service from 1908 to 1935. A must-read for anyone interested in early U. S. military aviation.

Gardner, Lester D. *Who's Who in American Aeronautics Second Edition 1925*
Gardner Publications 1925
A valuable reference of biographies of important personalities in the field of aeronautics from its beginning to 1925.

Glines, Carroll V. and Cohen, Stan *The First Flight Around The World*
Pictorial Histories Publishing Co., Inc. 2000
An excellent well-documented and well-illustrated history of this great accomplishment. It contains hundreds of photographs of every aspect of the venture, from the construction of the airplanes to their final disposition. Every significant happening is at least briefly mentioned and almost all documented with photographs. A must for any student of this great feat.

Goddard, Brig. Gen. George W. with Copp, DeWitt S. *Overview – A Lifelong Adventure with Aerial Photography*
Doubleday & Co. 1969
The autobiography of the man who probably did the most to further military aerial photography in this country from 1919 to 1953. He was not only a leader in his field, but a pilot who flew military airplanes from 1918 until his retirement and spent hundreds of hours and flew thousands of miles in the Liberty-powered DH-4. Another "must read" for historians interested in early U. S. military aviation.

Gorrell, Col. Edgar S. *The Measure of America's World War Aeronautical Effort*
Norwich University 1940
This was one of a number of studies on aeronautical subjects sponsored by Norwich University beginning in 1935. Its aim, according to its sponsors, was to document the successes and failures of America's preparedness for war in the air prior to and during World War I with the hope that the many mistakes of unpreparedness made at that time could be avoided in the event of another war. They were not. It was written by Colonel Gorrell, who served in a number of positions connected with the aviation effort during WW I and was eventually appointed official historian of the A.E.F. Air Service.

Grossnick, Roy A. *United States Naval Aviation 1910 – 1995*
Naval Historical Center
Department of the Navy 1997
This is an excellent thumbnail-type history of Naval aviation. One of its best features is its very extensive statistical documentation of Navy aircraft.

Hall-Scott Motor Car Co. *Pertinent Facts About the Liberty Motor*
Hall-Scott Motor Car Co. 1919
This 24-page company-published booklet was put out in an attempt to capitalize on Hall-Scott's involvement with the original design of the Liberty engine and the building of the first prototype engines. E. J. Hall, the company's chief engineer, was its co-designer and a number of design features of prior Hall-Scott engines were incorporated in it. This was brought out, as it

should have been, but the booklet spent far more space and energy in belittling the contributions of "automotive engineers" and "automotive designed aircraft engines" than it did on anything else connected with the Liberty engine. It was a very thinly veiled attack on Vincent and Packard that was entirely uncalled for. In spite of this, the booklet is quite informative and adds to the published history of the Liberty.

Heron, S. D. *History of the Aircraft Piston Engine - A Brief Outline*
Ethyl Corp. 1961
A history of the development of the aircraft piston engine. Considering the rather large subject covered in this book's 130 pages, it is rightly titled "A Brief Outline." It does, however, provide much historical information on the subject.

Hofmann, George F., Ph. D. *Doctrine, Tank Technology, and Execution: I. A. KhalcpsKii and the Red Army's Fulfillment of Deep Offensive Operations*
Journal of Slavic Military Studies Vol.9, No.2 1996
This in-depth study of Russian tank technology is probably the best single source of information on the origin of the Russian BT tank and in the process provides extensive information on the history of the American Christie tank. Hofmann is probably the foremost authority on the history of the American, British and Russian Christie and the BT tank.

Hourwich, Iskander and Foster, W. J. *Air Service Engine Handbook*
Engineering Division McCook Field 1925
This book was compiled as a handbook for aircraft engine engineers and was intended as a reference for design and performance information relative to all current aero engines as well as most significant obsolete types dating as early as 1915. It contains a large amount of specification and performance data relative to a number of variations of the Liberty as well its contemporaries.

Hunnicutt, R. P. *Sherman A History of the American Medium Tank*
Presidio Press
An excellent history of American armour.

Hunnicutt, R. P. *Stuart A History of the American Light Tank*
Presidio Press
An excellent history of American armor.

Jane, Fred T. and Various editors *Jane's All the World's Aircraft 1918 - 1927*
Jane's Publishing Co. Ltd 1918 - 1927
An invaluable reference for aircraft and their engines, worldwide.

Johnson, Mary Ann *McCook Field 1917 - 1927*
Landfall Press 2002
A well-researched and written history of McCook Field from its inception to its closing. The history of this installation is covered in detail and in so doing is the engineering history of the Army Air Service during this period. As a result, there is much material of interest related to the development and use of the Liberty engine.

Kotelnikov, Vladimir *Russian Piston Aero Engines*
The Crowood Press 2005
This is a history of the evolution of the Russian piston aero engine from its beginning to the present. It is quite comprehensive and is the best available on the subject. The development of the Russian M5, their copy of the Liberty, is covered in some detail, as are all the modifications and designs based upon it.

Leland, Mrs. Wilfred C. With Minnie Dubbs Millbrook *Master of Precision – Henry M. Leland*
Wayne State University Press 1966
A biography of Henry M. Leland, one of the men who formed the Cadillac Motor Car Co. in 1902 and who formed the Lincoln Company in 1917 for the purpose of building Liberty aircraft engines. This company produced the Lincoln automobile after the end of the war and was purchased by Ford in 1922.

Lloyd, Ian *Rolls-Royce: The Growth of a Firm*
MacMillan 1978
Lloyd has documented many of the particulars regarding what went on within the company during its early years which may be found nowhere else. This book contains a number of interesting facts regarding the attempted production of Rolls-Royce aero engines in America during the WW I era.

Lumsden, Alex S. C. *British Piston Aero-Engines and Their Aircraft*
Airlift Publishing Co. 1994
Much of the British use of the Liberty is documented in this book.

McFarland, R. M. *History of the Bureau of Aircraft Production.*
Bureau of Aircraft Production 1920
This eight-volume history was compiled at the request of the BAP under the direction of R. M. McFarland. It was begun shortly after the Armistice and completed probably in early 1920. It was compiled from a combination of official government records and reports submitted by many of the various companies who produced military goods under the direction of the BAP. It covers all aspects of aircraft production and thus the report goes far beyond aircraft and their engines and covers everything from guns, to accessories, to wood for building airplanes, to castor oil production, fabric production, dope production, oxygen equipment design and production and balloon production. In addition, the story of the evolution of the organizations put in place to control aircraft engineering and production from 1914 to the end of the war is covered in great detail. Although the *History of the Liberty Engine* completed by Potter, Benet and Kent in June of 1918 contains more detail of the very early period, this must certainly be the most comprehensive single document in existence relative to its subject. This document was apparently completed in manuscript form and was never published as bound volumes. It resided in Air Force archives in that form until 1951, at which time the Historical Office of the Air Material Command evidently decided the original document was deteriorating to such an extent that it should be retyped in order to preserve it. That task was completed in September of 1951 but was again kept in manuscript form. It included only the printed portion of the original and none of the hundreds of photographs which accompanied the original. The 1951 copy contains 2,530 typewritten pages and is available on microfilm from the Air Force History Office at Maxwell Air Force Base.

Marcosson, Isaac F. *Colonel Deeds Industrial Builder*
Dodd, Mead & Co. 1947
A biography of Edward A. Deeds. A very interesting and flamboyant biography of this rather brilliant engineer and industrialist. Marcosson was given to considerable eloquence and seemed too inclined to write in much supposed actual conversation between Deeds and his various associates dating back many years, which would have been difficult, if not impossible, to have been absolutely accurate. He was also inclined to credit Deeds with responsibility for absolutely every decision with which he was in any way connected. It provides much valuable information when used in conjunction with other references but might provide a rather biased outlook if used as a sole source.

Milson, John *Russian Tanks 1900 - 1970*
Galahad Press 1970
One of the better early Russian tank histories.

Morrow, John H. Jr. *The Great War in the Air*
Smithsonian Press
A history of United States military aviation from 1909 to 1921.

Mixter, Col. George W. Emmons, Lt. Harold H. *United States Aircraft Production Facts*
Government Printing Office 1919
This brief history of the production of aircraft and their components was compiled at the direction of the Assistant Secretary of War after the Armistice was signed and was published in January of 1919. Unfortunately, for historians, its primary aim was to document the situation as it stood on the day the Armistice was signed, November 11, 1918. As such, it does not carry the story to its end as neither engine nor aircraft production had ceased abruptly at that date.

Neal, Robert J. *Master Motor Builders*
Aero-Marine History Publishing 2000
The story of the inception, design, production and uses of the non-automotive engines of the Packard Motor Car Co. This history is the only book ever published covering this subject. Since Packard was so closely connected with the design and production of the Liberty, the subject is discussed extensively.

Neal, Robert J. *Packards At Speed*
Aero-Marine History Publishing Co. 1995
The story of the use of Packard aero, marine and automotive engines in performances of all kinds. This history is the only book ever published covering this subject. It discuses the use of the Liberty engine in land, sea and air vehicles for racing and record setting and other performances.

Neal, Robert J. Neal *The Unknown Liberty Engines*
WW I Aero August 2001
A brief history of the various versions of the Liberty other than the standard Liberty Twelve. In brief, this includes the 1, 2, 4, 6, 8 and 24-cylinder versions as well as the experiments undertaken to raise the output of the standard Liberty Twelve from about 440 HP @ 1,800 RPM to 495 HP @ 1,800 RPM and 527 HP @ 2,200 RPM.

Packard Motor Car Co. *The Real Story of the Liberty Motor*
Packard Motor Car Co. 1919
A company-published booklet of 31 pages giving a brief history of Packard's aero engine development from its beginning in 1915 and leading up to the design and production of the Liberty engine. The fact that Packard built the first Liberty as well as the most Liberties is brought out, as are the design similarities between Packard's "905" engine and the Liberty. E. J. Hall is noted as the co-designer but otherwise only Packard and Vincent are spoken of. The booklet was intended to capitalize on Packard's involvement with the Liberty engine and did so. It did not, however, belittle any other company or individual in the process. The same cannot be said of the similar Hall-Scott booklet published shortly thereafter.

Pershing, General John J. *My Experiences In The World War*
Frederick A. Stokes Co. 1931
This book gives a detailed description of the participation of the United States in World War I as seen by the Supreme Commander of the American Expeditionary Forces in Europe. Although the amount of information regarding the use of aircraft in the war is relatively small, this is a book that should be read by any serious student of World War I. The General makes the point that one of the prime purposes of his book is to note the mistakes we made as a country in our failures to be prepared for a war we could see was on the horizon so we might not repeat them in the future. We repeated most of them in the years immediately preceding World War II.

Potter, Lt. Harold E. *History of the Liberty Engine*
Benet, Lt. William R.
Kent, Lt. Warner W.
Bureau of Aircraft Production June 1918
This early history of the Liberty engine was written at the direction of the BAP by the three Lieutenants noted above. It was initiated at the time the engine had been finally put into series production and it was evidently an attempt to record a brief history of the period of time between the initiation of the design and the accomplishment of actually seeing the engine settled into successful production. This history was completed on June 6, 1918 by which time deliveries had totaled 10,030. This is an extremely important reference regarding the early history of government knowledge and decisions relative to the designing and putting into production the Liberty engine.

Royal Air Force *Liberty Engine Reports and Correspondence*
Air Ministry *January 2, 1917 - February 2, 1919*
Great Britain
Air Historical Branch RAF 1919
File of reports and correspondence reference Liberty engines preserved by the Air Historical Branch Air Ministry and now residing in the British National Archives. This file of historical documents contains much information regarding the performance of the engine in British use as well as its contracting.

Royal Air Force *Various Tests and Reports Relative to the*
Air Ministry *Liberty from 1920 to 1932*
Great Britain
Air Historical Branch RAF 1920 - 1932
Contains reports regarding items ranging from the question of royalty payments on purchases of the engine and spare parts to tests of a supercharged Liberty six engine.

Shirley, Noel C. *United States Naval Aviation 1910 – 1918*
Schiffer Military History 2000
A valuable reference of Naval aviation history of the period which contains much information to be found in no other publication, particularly regarding the WW I period. This is one of those rare historical reference works committing the unforgivable sin of having no index.

Shock, James R. *U. S. Army Airships 1908-1942*
Atlantis Productions 2002
Probably the most comprehensive work so far published on Army airships.

Sonnenburg, Paul and *Allison Power of Excellence*
Schoneberger, William A.
Coastline Publishers 1990
A corporate-published history of the Allison Company which contains a brief history of the company's association with the Liberty engine.

Sherbondy, E. H. and *Textbook of Aero Engines*
Wardrop, G. Douglas
Frederick A. Stokes Co. 1920
Although this book was designed to be what the title indicates, a textbook of contemporary aero engines, it also contains extensive information on Sherbondy's supercharger designs, both gear-driven and turbo types. As such, it adds considerably to the information contained in Vol. 4 No. 1 of the *Bulletin of the Experimental Department of the Airplane Engineering Division*, that being the only other source of information on the Sherbondy designs the author has so far found. It contains extensive illustrative and technical information of most American, British, French, German and Italian aero engines of the period, much of it very difficult to find elsewhere.

Squier, Major Gen. *Aeronautics In the United States 1918*
George O.
American Institute of
Electrical Engineers 1919
An address given by Major General George O. Squier before the American Institute of Electrical Engineers on January 10, 1919 and later published by A.I.E.E. The paper documents briefly the history of the Air Service from 1908 to the declaration of war in April of 1917. It then details the history of the U. S. aeronautical efforts from that time until the signing of the Armistice on November 11, 1918. It is accompanied by detailed statistics and charts as well as illustrations of equipment on all phases of the air war, from pilot training to combat use of aircraft and from design of the equipment used to its production. Its only drawback is that it stops on November 11, 1918, the date of the Armistice, and thus does not go to the completion of all aeronautical production.

Sweetser, Arthur *The American Air Service - A Record of Its Difficulties,*
D. Appleton & Co. *Its Failures, and Its Final Achievements*
New York 1919
The author was a Captain in the Air Service and the forward was written by Newton D. Baker, Secretary of War. In this book, the author traces the accomplishments of the Air Service from its beginnings to the end of WW I. In doing so, he covers all facets of the production of aircraft and engines as well as the increase in Air Service personnel and their training. It is a fairly large book (383 5 x 8 pages) and tells the complete story well but necessarily contains a relatively small amount of detail regarding the design and production of the Liberty engine. It does the job intended well and certainly is a necessary reference for any serious student of the Air Service during this period. It does not, however, add any information relative to the Liberty that is not to be found elsewhere.

Taylor, C. Fayette *Aircraft Propulsion: A Review of the Evolution of*
Aircraft Piston Engines
Annals of Flight Volume 1 No. 4
Smithsonian Institute 1971
An excellent overall review of the development of the piston aero engine. Its very extensive bibliography (34 pages listing several hundred items) of related reference material is probably more valuable for historical research relative to the subject than is the text.

Thomas, Senator *Aircraft Production in the United States*
Government Printing Office August 22, 1918
Report of Senator Thomas to the Committee on Military Affairs on the outcome of an investigation of aircraft production in the United States as directed by a Senate resolution adopted on April 30, 1918. This report very definately made every effort possible to seem to prove what the Senator thought about the decisions made regarding aircraft production, that most were bad and resulted in the lack of progress made. It was actually more of a "witch hunt" than a fact-finding mission and it laid a lot of blame but failed to really come up with the correct actual causes.

US Air Service *McCook/Wright Field List of P-numbered Airplanes*
Skyways Magazine 1987-1989 numbers 3, 5 and 9
The editor of *Skyways* magazine compiled a list of the P-numbered airplanes tested at McCook Field and then at its successor, Wright Field, from the beginning of the assignment of such numbers in early 1918 through number 599, which was assigned in March of 1930. Information included was the name and model of the airplane, its assigned Air Service number, its engine

type (and serial from P-1 through P-400) and its assigned P-number. This is a valuable reference for a number of reasons, not the least of which are its usefulness in identifying photos of airplanes when the P-number is visible and telling what engine it probably had installed.

US Air Service *Air Corps Information Circulars*
United States Government Printing Office 1918 – 1932
Published reports of tests and research conducted by the Engineering Division of the Air Service/Air Corps at McCook Field and Wright Field, Ohio from 1918 to 1932. These reports contain much material relative to the design and improvements and testing of the Liberty engine in most variations. 677 circulars were published during this period.

US Air Service *Technical Reports of the Engineering Division – Air Service*
United States Government Printing Office 1918 – 1922
Test and research reports of Air Service Engineering Division issued in printed form but not necessarily published elsewhere. Between January 1918 and June 1920 reports were issued for a total of 1,300 such projects. By January 1922 the number had reached almost 1,900. Some of these reports were then published as Air Service Circulars, but by no means all. These reports, from January 1918 until the end of the war, were issued by the War Department Bureau of Aircraft Production Airplane Engineering Division, Experimental Department (after April 1918, Experimental Engineering). In early 1919, after the BAP was disbanded, they were continued in the same series as Air Service Engineering Division Test Reports.

US Army *Ordnance Committee Meeting Notes 1919-1933*
US Army Historical Archives
Although the *Ordnance Committee Meeting Notes* do not contain a great deal of technical detail, they do provide a fairly clear history of the development of tanks in the United States.

Vincent, Jesse G. *History of the Development of the U. S. A. Standardized Aircraft Engines*
Unpublished MSS 1918
United States Air Force Museum
This history was written by Vincent as the events occurred and was typed by his secretary on a daily or weekly basis. He retained this manuscript and used much of it as a basis for his 1945 history. It was presented to the Air Force Museum along with other material in 1960. It was also reproduced as a part of the official eight-volume *History of the Bureau of Aircraft Production* which was completed probably in early 1920.

Vincent, Jesse G. *The Liberty Engine Origin and Development World War I*
Packard Motor Car Co. 1945
This is a history of Packard and Vincent's involvement in the design and production of the Liberty engine in this country during the First World War. It was written by Vincent, at the end of World War II, from notes he had made and retained from 1914 to 1920. Vincent was a prolific writer and kept a diary from at least 1917 until the end of his business career (most of them have survived) and probably from as early as 1914 or before. The material in this publication was drawn directly from *History of the Development of the U. S. A. Standardized Aircraft Engines*, Vincent's early diaries, early notes and test material.

Vincent, Jesse G. *Reasons Behind the Liberty Aviation Engine*
Society of Automotive Engineers 1918
This paper describes briefly the reasons for the design of the Liberty engine.

Vincent, Jesse G. *The Liberty Aircraft Engine*
Society of Automotive Engineers 1919
This paper goes extensively into both the history leading up to the design of the Liberty engine and many of the technical details of the engine. A number of the various design changes found necessary during the early history of the engine are also discussed.

Vincent, Jesse G. *Diaries 1917 – 1918*
Personal diaries covering the period of time Vincent served in the Air Service. They are now in the Air Force Museum collection.

Whitney, Daniel D. *Vee's For Victory!*
Schiffer Publishing Ltd. 1998
This book tells the story of the inception, design, production and uses of the WW II Allison V-1710 aircraft engine in great detail. The Allison company started as the Indianapolis Speedway Team Company, became the Allison Speedway Team Company and then the Allison Experimental Company in early 1917. The latter company did design, engineering, contract experimental engine construction and engine rebuild for the Air Service related to the Liberty engine from 1917 until the Liberty was removed from use by the services in 1934. This book contains much Liberty-related material from the archives of Allison. Some historians insist that it was Allison's early Liberty work that led the company into the development of the soon-to-be world-famous Allison V-1710.

Zigmunt, Joan E. *Allison The People and the Power*
Turner Publishing Co. 1997
This is a pictorial history of the Allison Company from its beginnings to 1996. As such, it also covers some of the history of Allison's association with the Liberty engine.

Zaloga, Steven J. and Grandsen, James *Soviet Tanks and Combat Vehicles of World War Two*
Arms and Armour Press 1995
This is one of the better histories published on the Soviet development of tanks.

PERIODICALS:

Aerial Age Weekly, 1917 - 1923

Aeronautical Digest, 1921 - 1924

Aero Digest, 1924 - 1940

Automotive Industries, 1917 - 1922

Aviation and Aeronautical Engineering, 1916 - 1920

Aviation and Aircraft Journal, 1920 - 1921

Aviation Magazine, 1922 - 1947

Motor Age, 1917 - 1920

Motor Boat Magazine, 1918 - 1934

Motor Boating Magazine, 1918 - 1946

Packard Magazine, 1916 - 1931
(Packard company publication)

Power Boating Magazine, 1925 - 1928

Rudder Magazine, 1917 - 1940

Society of Automotive Engineers, 1916 - 1940
(Published papers of various members)

Yachting Magazine, 1917 - 1946

OTHER:

Packard Motor Car Company:
Minutes of the Board of Director's meetings, 1916 - 1926
Stockholder's Reports, 1916 - 1926
Packard test reports for their aero engines, 1916 - 1926.
Diaries of Jesse G. Vincent 1917 - 1918 housed in the Air Force Museum; 1921 - 1940 housed in the National Automotive History Collection of the Detroit Public Library.
Charles H. Vincent papers. Roberta (Vincent) Mocabee collection.

Government Sources:

Air Service, Army Air Corps and Navy test reports. 1917-1935

Air Service Information Circulars 1 - 677. 1919 - 1932

Technical Reports of the Eng. Div. 1 - 1300. 1918 - 1920

Air Service Technical Orders - Technical Bulletins. 1920 - 1925

Index

Indexing a book of this type requires some decisions as to the practicality or desirability of listing the key words or names contained in the extensive appendices. My decision is not to do so. I have made a single exception. I do list all personal names contained in the various appendices. As an example, Appendix F lists all the aircraft that flew with Liberty power and pictures most of them. If a picture caption gives the name of an individual in the picture I list the name in the index.

The main text is completely indexed. Any person, airplane, boat, manufacturing company, subject, etc. mentioned in the main body of the book can be found in the index.

Illustration entries for the main text have page numbers in **Bold Face**. Illustrations reproduced in the appendices are not noted in the index.

APPENDIX A:
LIBERTY CONTRACTS

This appendix lists all U.S. production contracts for Liberty engines of any type or model either proposed or actually issued to any potential builder, whether a private company or a government entity. It notes the engine type, number ordered or proposed and number actually delivered. Also given, if known, are the contract number and the serial numbers to be assigned to the engines.

All known Russian contractual information is listed. Unfortunately, such information is still sparse.

Contracts for the British Nuffield Liberty are listed with all known relative data. Also listed are all contracts for production of the tanks which were powered by the Nuffield Liberty.

APPENDIX B:
LIBERTY PRODUCTION

This appendix lists Liberty order/production and is directed to those contracts actually placed. The layout is somewhat different and some additional information is included which is not found in Appendix A. As an example, this appendix provides a good overview of the number of engines of each type produced by every producer, government or private. It also includes a production chart by date and Air Service serial of all standard L-12A/L-12N engines built during WW I.

APPENDIX C:
DOCUMENTED TESTS JULY 1917 THROUGH JUNE 1925

One of the most interesting facets of studying the Liberty engine is its relatively short period of engineering redesign before it was actually put in series production. Another is its long active military aircraft use. Any historian attempting a detailed study of either subject will find very useful the extensive listing of U.S. government tests of this engine or aircraft powered by it.

APPENDIX D: SPECIFICATIONS OF THE LIBERTY 12, 8 AND 6

A complete listing of the mechanical specifications of all major components of the three sizes of engine as well as the performance specifications of the complete engine.

Plotted power graphs of the engines will be found within the main body of the book and are indexed.

APPENDIX E:
THE MILITARY LIFE OF THE LIBERTY WORLDWIDE

Although this subject is covered in great detail in the main text of the book, it is so extensive that it is difficult to comprehend the overall scope of it without seeing it depicted in some kind of chart form. It is shown in this appendix in both flow and bar chart form.

APPENDIX F:
AIRCRAFT THAT USED LIBERTY POWER

The overall number of aircraft and the number of different types of aircraft which were Liberty-powered is astounding, to say the least. This appendix documents them in list-form and provides dates, models, numbers built and users. Users were worldwide. The list contains many one-of-a-kind aircraft of which photographs are seldom seen. I made a point of trying to reproduce an image of every type of Liberty-powered aircraft if at all possible.

The list of all types and variations with foreign applications counting individually comprises 265 entries. 61 are foreign, 51 are U.S. civil aircraft, 42 are U.S. Navy and 111 are U.S. Army.

Included in the list is the page number within the book on which will be found an illustration of the aircraft. Of the total of 204 U.S. entries all but 24 are illustrated. Of the 61 foreign entries 30 are illustrated. Most of the missing 31 are of foreign uses of U.S.-produced airplanes.

If you think a particular aircraft may have been at one time Liberty powered look at this list first. If there is any discussion of the aircraft in the main text it will also be listed in the main index.

APPENDIX G:
THE LIBERTY ENGINE ILLUSTRATED

A keen interest in a particular engine generally leads to a desire to see it pictured from all angles. If that interest is technical as well as historical then one also has a desire to see what all of its components look like. In this appendix the Liberty 6, 8 and 12 are all pictured from all sides and disassembled into their major components.

Also shown is a plotted power graph, installation drawing and a gear train and accessories drive layout for each engine.

APPENDIX H;
U.S. MILITARY LIBERTY ENGINES ON-HAND

This appendix presents graphically a picture of the U.S. military stocks of Liberty engines from the beginning of production in 1917 until the last Liberty was removed from Government storage in 1936.

— A —

HALL-SCOTT A-8
12-cylinder, 450-500 h.p. Airplane Engine. View looking down on cylinders from Magneto end. Shows typical Direct Drive feature. Adopted from proved Hall-Scott practice for Liberty Engine without modification.

HALL-SCOTT A-8
12 cylinder, 450-500 h.p. Airplane Engine completed prior to May 1, 1917. Ready for final tests when E. J. Hall went to Washington to assist in the Aircraft Production Board Program.

BU

Side view of Hal
bearing crankshaf
May 1, 1917.

HALL-SCOTT A-8

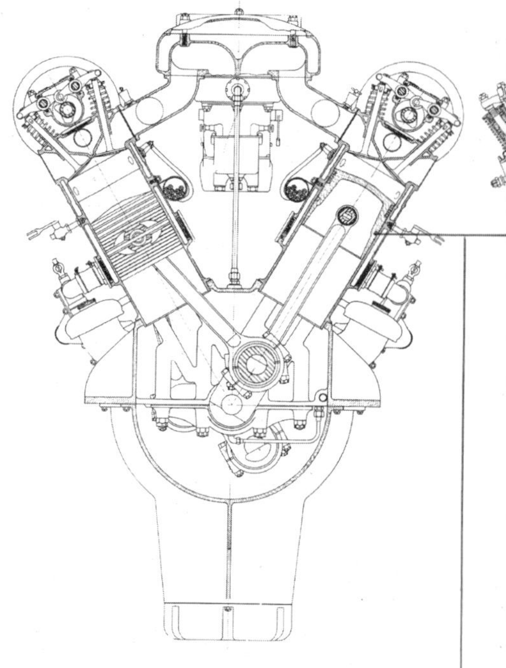

SECTIONAL VIEW O

Two types of die cast aluminum pistons were
low compression engines for naval use at low
pression type for army air service at high altit
compression piston with Hall-Scott piston shown
and note similarity. Design is carefully worked
down the tapering skirts of piston. Pistons for
from Hall-Scott die blocks in the hands of the A
land, Ohio.

U. S. GOVT. EXPERIMENTAL GEARED-DOWN ENGINE

Spur gear reduction in cumbersome housing at the propeller end, of the type discarded by E. J. Hall and Vincent May 31, 1917. Proved unsatisfactory in developing power and speed, over the direct drive practice, perfected by Hall-Scott in ten years of airplane engine building. Never used in a battle plane in France. Built after the Liberty was in production, proved unsatisfactory, and was discarded. Compare with direct drive Hall-Scott A-8 opposite.

No Automotive designed Airc
in a battle plane in France.
designed Aircraft engine is
fundamental part of the Gov